Advances in Antifungal Drug Development

Nikhat Manzoor
Editor

Advances in Antifungal Drug Development

Natural Products with Antifungal Potential

Springer

Editor
Nikhat Manzoor
Department of Biosciences, Faculty of Life Sciences
Jamia Millia Islamia
New Delhi, Delhi, India

ISBN 978-981-97-5164-8 ISBN 978-981-97-5165-5 (eBook)
https://doi.org/10.1007/978-981-97-5165-5

This Springer imprint is published by the registered company Springer Nature Singapore Pte Ltd.
The registered company address is: 152 Beach Road, #21-01/04 Gateway East, Singapore 189721, Singapore

This book is dedicated to my grandfather Maulvi Abdul Latif, who passed away before I was born, yet his lasting influence motivates me to journey on the path of knowledge.

Preface

Fungal infections are common, especially in immunocompromised individuals. The treatment is difficult and often amputation is the only option. The conventional antifungal drugs are limited; fungi being eukaryotic share targets with human host. Moreover, the commonly prescribed azoles are fungistatic, the use of which for extended periods leads to multidrug resistance. Polyenes like Amphotericin B are extremely hepatotoxic and nephrotoxic when used for long durations. Excessive use of antibiotics and other synthetic drugs has caused resistant strains to evolve at an alarming rate, leading to treatment failures. In general, currently available antifungal drugs have several adverse side effects and hence alternative treatment strategies are required to effectively manage mycoses. Plant extracts, essential oils, and their constituent secondary metabolites like terpenoids, phenolic compounds, alkaloids, and sulfur-containing compounds have shown excellent antifungal properties. They have negligible toxicity and can be modified chemically to increase efficacy. Alternative strategies also include chemosensitization of currently available drugs which in turn can reduce dose and toxicity. Therefore, a search for safer phytomedicines is required which are more efficacious, non-toxic, easily available and do not develop resistance in fungal strains.

This book provides a platform to present useful information and new insights for antifungal drug discovery and development. It would be beneficial for faculties, scientists, and students working in the traditional systems of medicine, pharmacognosy, pharmaceutical sciences, and health sciences. There are 27 chapters on the efficacy and antifungal mode of action of plant extracts, essential oils, natural compounds, their derivatives, and plant-based nanoparticles. The book explores the antifungal mode of action and efficacy against the virulence and pathogenicity of *Candida*, *Aspergillus*, *Cryptococcus*, *Histoplasma*, and other pathogenic fungi. The first chapter is an introduction illustrating the various fungal pathogenic species, common fungal diseases, and general mechanisms of action of various antifungal drug classes including the natural products. The subsequent chapters deal with antifungal efficacy of plant extracts, essential oils, and plant-derived natural compounds. The book also discusses the synthesis and characterization of plant-based or green synthesized nanoparticles and their therapeutic application against fungal diseases. Chemical derivatives of natural compounds have been explored further discussing their therapeutic potential. There are two chapters that explore antifungal properties of bioactive compounds derived from microbes and other natural sources.

Majorly non-toxic, the natural compounds also show some toxicity at higher concentrations, but their toxicity is significantly lower than that of conventional drugs. So, two chapters have been included that deal with the issue of natural product toxicity. Plant-based natural antifungals have diverse other applications. One such application is elaborated in the second last chapter which discusses their use in the preservation and conservation of heritage artifacts from fungal infestation. Finally, the bioinformatic tools that can be used to understand the mode of antifungal action of plant products and can be used in the field of antifungal drug discovery are also briefly mentioned. I am grateful to all my contributors, collaborators, and students who have kept my inner fire burning. I must also thank Springer Nature, especially Dr Bhavik Sawhney for his continuous support and patience during manuscript preparation. I am grateful to my parents who have invested their life in me and my husband who always stood by me.

New Delhi, India Nikhat Manzoor

Contents

Editor and Contributors

About the Editor

Nikhat Manzoor is a Professor in the Department of Biosciences, Jamia Millia Islamia, New Delhi, India. She has also served as a Visiting Scientist at the Department of Oral Sciences, School of Dentistry, University of Otago, Dunedin, New Zealand (2009–2010) and Associate Professor in the College of Applied Medical Sciences, Taibah University, Al-Madinah Al-Munawarah, Kingdom of Saudi Arabia (2014–2018). Her research interest includes antifungal drug discovery and development, antifungal mode of action of natural compounds and their derivatives, drug resistance, and antifungal drug targets. She has completed five major research projects funded by the Indian Council of Medical Research (ICMR), University Grants Commission (UGC), Department of Science and Technology (DST) and CCRUM, Ministry of AYUSH, Government of India. She has delivered several invited talks, both in India and abroad and was awarded the prestigious DST-BOYSCAST Fellowship (2009) and DST-Young Scientist (2001–2004, 2004–2007). She has more than 25 years of teaching experience and has published more than 135 research papers in peer-reviewed journals and several book chapters. She has supervised 16 PhD students, 2 post-doctoral fellows, and several post-graduate students. She is life member of the Association of Microbiologists of India and the Society of Biological Chemists.

Contributors

Sageer Abass Centre of Excellence in Unani Medicine (Pharmacognosy and Pharmacology), Bioactive Natural Product Laboratory, School of Pharmaceutical Education and Research, New Delhi, India

Mazen Abdulghani School of Life Sciences, Swami Ramanand Teerth Marathwada University, Nanded, Maharashtra, India

Mohammad Abid Medicinal Chemistry Laboratory, Department of Biosciences, Jamia Millia Islamia, New Delhi, India

Aijaz Ahmad Department of Clinical Microbiology and Infectious Diseases, School of Pathology, Faculty of Health Sciences, University of the Witwatersrand, Johannesburg, South Africa

Division of Infection Control, Charlotte Maxeke Johannesburg Academic Hospital, National Health Laboratory Service, Johannesburg, South Africa

Iqbal Ahmad Department of Agricultural Microbiology, Faculty of Agricultural Sciences, Aligarh Muslim University, Aligarh, India

Sayeed Ahmad Centre of Excellence in Unani Medicine (Pharmacognosy and Pharmacology), Bioactive Natural Product Laboratory, School of Pharmaceutical Education and Research, New Delhi, India

Saiema Ahmedi Medical Mycology Lab, Department of Biosciences, Jamia Millia Islamia, New Delhi, India

Ayesha Aiman Department of Biosciences, Jamia Millia Islamia, New Delhi, India

Danish Alam Enzyme Technology Lab, Department of Biosciences, Jamia Millia Islamia, New Delhi, India

Department of Biosciences, Jamia Millia Islamia, New Delhi, India

Tanveer Alam Sabanci University Nanotechnology Research and Application Center, Istanbul, Turkey

Juned Ali Enzyme Technology Lab, Department of Biosciences, Jamia Millia Islamia, New Delhi, India

Mohd Danish Ansari Laboratory of Green Synthesis, Department of Chemistry, University of Allahabad, Allahabad, India

Mostafa Mohammed Atiyah School of Biosciences, Mar Athanasios College for Advanced Studies, Thiruvalla, Kerala, India

Department of Biology, Thi-Qar Education Directorate, Thi-Qar, Iraq

Kashish Azeem Medicinal Chemistry Laboratory, Department of Biosciences, Jamia Millia Islamia, New Delhi, India

Biji Balan Department of Zoology, Dyal Singh College, University of Delhi, New Delhi, India

Sarika Bano Laboratory for Structural Biology of Membrane Proteins, Dr. B.R. Ambedkar Center for Biomedical Research, University of Delhi, New Delhi, India

Seemi Farhat Basir Department of Biosciences, Jamia Millia Islamia, New Delhi, India

Priyanka Bhardwaj Department of Biosciences, Jamia Millia Islamia, New Delhi, India

José S. Câmara CQM—Centro de Química da Madeira, Universidade da Madeira, Funchal, Portugal

Departamento de Química, Faculdade de Ciências Exatas e Engenharia, Universidade da Madeira, Funchal, Portugal

Sayali A. Chougule Department of Stem Cell and Regenerative Medicine and Medical Biotechnology, Centre For Interdisciplinary Research, D.Y. Patil Education Society (Deemed to be University), Kolhapur, Maharashtra, India

Sandhya Devi Amity Institute of Biotechnology, Amity University Haryana, Gurugram, Haryana, India

Sanjay Kumar Dey Laboratory for Structural Biology of Membrane Proteins, Dr. B.R. Ambedkar Center for Biomedical Research, University of Delhi, New Delhi, India

Amit Singh Dhaulaniya Department of Zoology, Kirori Mal College, University of Delhi, New Delhi, India

Disha Disha Vocational Studies and Applied Sciences, Gautam Buddha University, Noida, Uttar Pradesh, India

Shazia Faridi Enzyme Technology Lab, Department of Biosciences, Jamia Millia Islamia, New Delhi, India

Saif Hameed Amity Institute of Biotechnology, Amity University Haryana, Gurugram, Haryana, India

Harsha Microbial Technology Laboratory, Acharya Narendra Dev College, University of Delhi, New Delhi, India

Syed Farhan Hasany Sabanci University Nanotechnology Research and Application Center, Istanbul, Turkey

Imtaiyaz Hassan Centre for Interdisciplinary Research in Basic Sciences, Jamia Millia Islamia, New Delhi, India

Nazia Hassan Department of Pharmaceutics, School of Pharmaceutical Education and Research, Jamia Hamdard, New Delhi, India

Nasimul Hoda Drug Design and Synthesis Laboratory, Department of Chemistry, Jamia Millia Islamia, New Delhi, India

Nance Hontman CQM—Centro de Química da Madeira, Universidade da Madeira, Funchal, Portugal

Md Waquar Imam CSIR-Central Institute of Medicinal and Aromatic Plants, Lucknow, Uttar Pradesh, India

Academy of Scientific and Innovative Research (AcSIR), Ghaziabad, Uttar Pradesh, India

Zeenat Iqbal Department of Pharmaceutics, School of Pharmaceutical Education and Research, Jamia Hamdard, New Delhi, India

Faiza Iram Centre for Interdisciplinary Research in Basic Sciences, Jamia Millia Islamia, New Delhi, India

Iram Irfan Medicinal Chemistry Laboratory, Department of Biosciences, Jamia Millia Islamia, New Delhi, India

Asimul Islam Centre for Interdisciplinary Research in Basic Sciences, Jamia Millia Islamia, New Delhi, India

Ashwini K. Jadhav Department of Stem Cell and Regenerative Medicine and Medical Biotechnology, Centre For Interdisciplinary Research, DY Patil Education Society (Deemed to be University), Kolhapur, Maharashtra, India

Neha Jaiswal Department of Biotechnology, National Institute of Technology, Raipur, Chhattisgarh, India

M. S. Jisha School of Biosciences, Mahatma Gandhi University, Kottayam, Kerala, India

Kajal School of Biological and Life Sciences, Galgotias University, Greater Noida, Uttar Pradesh, India

Kapinder Department of Zoology, University of Allahabad, Prayagraj, Uttar Pradesh, India

Monisha Khanna Kapur Microbial Technology Laboratory, Acharya Narendra Dev College, University of Delhi, New Delhi, India

S. Mohan Karuppayil Department of Stem Cell and Regenerative Medicine and Medical Biotechnology, Centre For Interdisciplinary Research, DY Patil Education Society (Deemed to be University), Kolhapur, Maharashtra, India

Hardeep Kaur Department of Zoology, Fungal Biology Laboratory, Ramjas College, University of Delhi, New Delhi, India

Luqman Ahmad Khan Medical Mycology Lab, Department of Biosciences, Jamia Millia Islamia, New Delhi, India

Sarah Ahmad Khan Department of Agricultural Microbiology, Faculty of Agricultural Sciences, Aligarh Muslim University, Aligarh, India

Shabana Khatoon Medical Mycology Lab, Department of Biosciences, Jamia Millia Islamia, New Delhi, India

Khyati School of Biological and Life Sciences, Galgotias University, Greater Noida, Uttar Pradesh, India

Awanish Kumar Department of Biotechnology, National Institute of Technology, Raipur, Chhattisgarh, India

K. M. Uma Kumari CSIR-Central Institute of Medicinal and Aromatic Plants, Lucknow, Uttar Pradesh, India

Munendra Kumar Department of Zoology, Rajiv Gandhi University, Doimukh, Arunachal Pradesh, India

Prateek Kumar Department of Zoology, University of Allahabad, Prayagraj, Uttar Pradesh, India

Suaib Luqman CSIR-Central Institute of Medicinal and Aromatic Plants, Lucknow, Uttar Pradesh, India

Academy of Scientific and Innovative Research (AcSIR), Ghaziabad, Uttar Pradesh, India

Hari Madhav Drug Design and Synthesis Laboratory, Department of Chemistry, Jamia Millia Islamia, New Delhi, India

Nikhat Manzoor Department of Biosciences, Faculty of Life Sciences, Jamia Millia Islamia, New Delhi, India

Medical Mycology Lab, Department of Biosciences, Jamia Millia Islamia, New Delhi, India

Rabiya Mehandi Molecular and Biophysical Research Lab (MBRL), Department of Chemistry, Jamia Millia Islamia, New Delhi, India

Kainat Mirza Department of Biosciences, Jamia Millia Islamia, New Delhi, India

Mohd. Aamir Mirza Department of Pharmaceutics, School of Pharmaceutical Education and Research, Jamia Hamdard, New Delhi, India

Sradhanjali Mohapatra Department of Pharmaceutics, School of Pharmaceutical Education and Research, Jamia Hamdard, New Delhi, India

Lubna Najam British International School Kurdistan, Erbil, Iraq

Rubia Noori Enzyme Technology Lab, Department of Biosciences, Jamia Millia Islamia, New Delhi, India

Nouman Molecular and Biophysical Research Lab (MBRL), Department of Chemistry, Jamia Millia Islamia, New Delhi, India

Satish Pandey Department of Conservation, Indian Institute of Heritage (Formerly National Museum Institute), Noida, Uttar Pradesh, India

Parveen Medical Mycology Lab, Department of Biosciences, Jamia Millia Islamia, New Delhi, India

Rabea Parveen Centre of Excellence in Unani Medicine (Pharmacognosy and Pharmacology), Bioactive Natural Product Laboratory, School of Pharmaceutical Education and Research, New Delhi, India

Department of Pharmaceutics, School of Pharmaceutical Education and Research, New Delhi, India

Rosa Perestrelo CQM—Centro de Química da Madeira, Universidade da Madeira, Funchal, Portugal

Aimee Piketh Department of Clinical Microbiology and Infectious Diseases, School of Pathology, Faculty of Health Sciences, University of the Witwatersrand, Johannesburg, South Africa

Rahisuddin Molecular and Biophysical Research Lab (MBRL), Department of Chemistry, Jamia Millia Islamia, New Delhi, India

Nafis Raj Medical Mycology Lab, Department of Biosciences, Jamia Millia Islamia, New Delhi, India

Manish Rana Department of Chemistry, Ramjas College, University of Delhi, New Delhi, India

Diwan S. Rawat Department of Chemistry, University of Delhi, New Delhi, India

Israil Saifi Sharda University, Greater Noida, Uttar Pradesh, India

Meryam Sardar Enzyme Technology Lab, Department of Biosciences, Jamia Millia Islamia, New Delhi, India
Department of Biosciences, Jamia Millia Islamia, New Delhi, India

Jasmine Shakir Department of Conservation, Indian Institute of Heritage (Formerly National Museum Institute), Noida, Uttar Pradesh, India

Mohammad Shakir Medicinal Chemistry Laboratory, Department of Biosciences, Jamia Millia Islamia, New Delhi, India
Department of Chemistry, University of Delhi, New Delhi, India

Manish Sharma Department of Zoology, University of Allahabad, Prayagraj, Uttar Pradesh, India

Shirjeel Ahmad Siddiqi Department of Agricultural Microbiology, Faculty of Agricultural Sciences, Aligarh Muslim University, Aligarh, India

Gajendra Singh Department of Biotechnology, School of Life Sciences, Central University of Rajasthan, Ajmer, Rajasthan, India

Shweta Singh Amity Institute of Biotechnology, Amity University Haryana, Gurugram, Haryana, India

Sreejeeta Sinha Department of Biotechnology, School of Life Sciences, Central University of Rajasthan, Ajmer, Rajasthan, India

Renu Solanki Department of Zoology, Deen Dayal Upadhyaya, University of Delhi, New Delhi, India

Patrícia Sousa CQM—Centro de Química da Madeira, Universidade da Madeira, Funchal, Portugal

Vartika Srivastava Department of Clinical Microbiology and Infectious Diseases, School of Pathology, Faculty of Health Sciences, University of the Witwatersrand, Johannesburg, South Africa

Nargis Taranum School of Biological and Life Sciences, Galgotias University, Greater Noida, Uttar Pradesh, India

Divya Varshney Department of Agricultural Microbiology, Faculty of Agricultural Sciences, Aligarh Muslim University, Aligarh, India

Rashi Verma Medical Mycology Lab, Department of Biosciences, Jamia Millia Islamia, New Delhi, India

Department of Neuroscience, Morehouse School of Medicine, Atlanta, GA, USA

Smitha Vijayan Department of Food Technology and Quality Assurance, Mar Athanasios College for Advanced Studies, Thiruvalla, Kerala, India

Pooja Vijayaraghavan Antimycotic and Drug Susceptibility Laboratory, Amity Institute of Biotechnology, Amity University, Noida, Uttar Pradesh, India

Khushbu Wadhwa Department of Zoology, Fungal Biology Laboratory, Ramjas College, University of Delhi, New Delhi, India

Gajanan Zore Department of Biotechnology, School of Life Sciences, Central University of Rajasthan, Ajmer, Rajasthan, India

Part I

Introduction

An Update on Human Fungal Diseases: A Holistic Overview

1

Shweta Singh, Pooja Vijayaraghavan, Sandhya Devi, and Saif Hameed

1.1 Introduction

1.1.1 Fungal Infections: A Recurring Problem

The fungal kingdom is the largest eukaryotic domain, comprising diverse species of fungi, with an estimated 1.5–5 million species exhibiting various life cycles, morphologies, metabolism and ecologies. Fungi can survive in a vast variety of environmental conditions, depending on the ecological niche within the earth's ecosystem (Choi and Kim 2017). While humans have long used fungi as food and in food processing methods, the weightage of disease-causing fungi is more in comparison with the beneficial fungi. Among the vast fungal group, around 600 species are characterized as human pathogens (Brown et al. 2012a, b). Fungal infections can range from mild skin infections and cutaneous infections to systemic infections and are emerging as a serious threat to a large number of populations, increasing morbidity and mortality rates (Boral et al. 2018). The incidence of fungi causing infections is becoming a serious public health problem due to fungal tropism, which makes single fungi infect various kinds of cells and tissues in a host cell. These infections could be primary or opportunistic. Primary infections affect the human population, which is not exposed to any fungal infections primarily. On the other hand, opportunistic infections affect immunocompromised patients (Dixon et al. 1996).

S. Singh · S. Devi · S. Hameed (✉)
Amity Institute of Biotechnology, Amity University Haryana, Gurugram, Haryana, India
e-mail: shweta20.gbu@gmail.com; yadavsandhya1207@gmail.com; saifhameed@yahoo.co.in; shameed@ggn.amity.edu

P. Vijayaraghavan
Antimycotic and Drug Susceptibility Laboratory, Amity Institute of Biotechnology, Amity University, Noida, Uttar Pradesh, India
e-mail: vrpooja@amity.edu

N. Manzoor (ed.), *Advances in Antifungal Drug Development*, https://doi.org/10.1007/978-981-97-5165-5_1

Every year, more than one billion individuals are affected by fungal infections, with more than 150 million instances resulting in serious and life-threatening infections (Houšť et al. 2020). The ability of many fungi to cause disease and invade the host systems is due to the functioning of various genes and proteins contributing toward pathogenicity named virulence factors (Tomee and Kauffman 2000). Dermatophytoses are the most commonly occurring fungal infections caused by dermatophytes affecting approximately 25% of the world's population (Havlickova et al. 2008; Zhan and Liu 2017). Over more than 600 pathogenic species, 300 million cases have been reported, mentioning the major role of *Cryptococcus*, *Aspergillus*, *Candida* and *Trichophyton* (Boral et al. 2018). The rise in the incidence of invasive fungal infections has been increasing in the immunocompromised population since the 1980s. The mortality rates were estimated up to 67% in patients hospitalized in ICU (Intensive care unit) associated with invasive fungal infections. The incidences of invasive fungal infections could be predisposed by many factors like the use of antibiotics (broad-spectrum), immunosuppressive agents, corticosteroid therapy and prosthetic devices. Patients undergoing organ transplantation, leukopenia patients, HIV (human immunodeficiency viruses) infection and cancer therapy are more prone to develop invasive mycoses. It has been estimated that in developing and developed countries, the major contributors to fungal infections are HIV/AIDS pandemic, chronic obstructive pulmonary disease (COPD), tuberculosis, asthma and cancers (Limper et al. 2017; Brown et al. 2012a, b; Guinea et al. 2010).

The outbreaks of invasive fungal infections (IFIs) could result due to vulnerability to infectious environmental sources or maybe from contaminated medical items. IFIs caused by fungi can thrive in the human body mostly in tropical and subtropical climates (Garcia-Solache and Casadevall 2010). The IFIs outbreaks could be community-acquired, hospital-acquired and medical product contamination. A study shows the surveillance of the percentage of emerging invasive fungal infections found in solid-organ transplant recipients are Candidiasis (53%), invasive Aspergillosis (19%), cryptococcosis and non-Aspergillus molds (8%), endemic fungi (5%) and zygomycosis (2%). The prevalence of candidiasis in invasive fungal infections is shown in Fig. 1.1 (Ravikant et al. 2015). The increase in the number of patients undergoing organ transplantation and low immunity compel the population

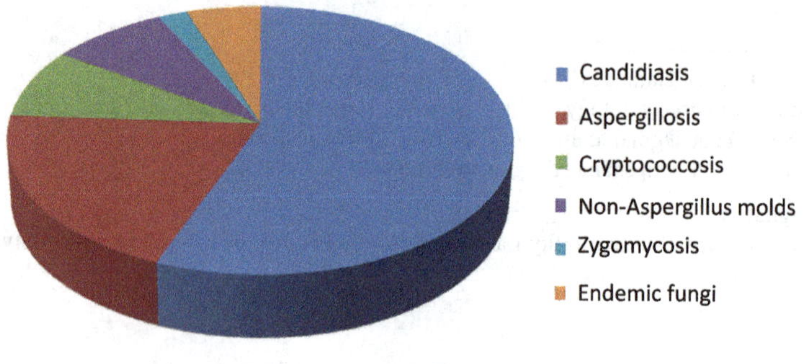

Fig. 1.1 Pie chart showing the distribution of fungal infections

to become susceptible to infections, which in turn puts more pressure on the healthcare sector. The regulation of surveillance and monitoring of the onset of fungal infections also keeps checking on the mortality and morbidity of the cost of cases. An understanding of the procedure of transmission and exposure parameters could help more in identification and awareness of the on-time accurate diagnosis of fungal infections. Also it could lessen the burden of upcoming cases of life-threatening infections. In this book chapter, we have discussed the fungal infections caused by *Candida*, *Cryptococcus* and *Aspergillus* species in subsequent sections.

1.2 *Candida* and Candidiasis

Candidiasis is a fungal infection caused by the opportunistic human fungal pathogen, *Candida*. This commensal generally thrives on the skin and within the body without creating difficulties, such as in the mouth, throat, gut and vagina. On the contrary, it can cause infections if it develops out of control or penetrates deep into the body under immunocompromised conditions (Kim and Sudbery 2011). *Candida* infections are classified into various kinds, including **Vaginal candidiasis**, which is a kind of infection that commonly affects women. It happens when the circumstances inside the vagina alter, allowing *Candida* to flourish. Oral candidiasis (**Thrush**) produces white sores in the mouth, throat, esophagus or tongue. **Cutaneous candidiasis** is a skin infection that causes a raised, red area with tiny, itchy bumps to grow in skin folds such as the underarms, between breasts and around buttocks (diaper rash) or the groin. **Invasive candidiasis (systemic candidiasis)** is when the infection spreads throughout the body, frequently in the bloodstream or on the membrane lining of the heart or skull, because of an immune system that is compromised (Pfaller et al. 2007; Vázquez-González et al. 2013).

1.2.1 Epidemiology

Invasive *Candida* infections vary by geographic location, but the average incidence was roughly 9 per 100,000 persons between 2013 and 2017. According to the Centre for Disease Control and Prevention (CDC), an estimated 34,000 cases of Candidiasis were detected in hospitalized patients in 2017, with around 1700 persons having died. *Candida albicans* is the most commonly recovered *Candida* species (37%) from clinical specimens, followed by *Candida glabrata* (27%), *C. parapsilosis* (14%), *C. krusei* (2%), *C. tropicalis* (8%), *C. dubliniensis* (2%), *C. lusitaniae* (2%) and, most recently, *C. auris*. All these strains were recovered from the bloodstream of patients. *C. auris* is a multidrug-resistant pathogen that is frequently misdiagnosed and is currently a serious problem in healthcare settings. Between 2015 and 2018, the number of reported *C. auris* cases increased by 318% (Bhattacharya et al. 2020; https://www.cdc.gov/drugresistance/pdf/threats-report/candida-508.pdf). Although *C. albicans* is the most common cause of candidiasis, non-*Candida* species have been on the rise in recent years. It is essential to study and understand the non-albicans species also, since therapy is dependent on them, and some drugs, such as the widely used fluconazole, may develop resistance in *Candida* species other than *C. albicans*. According to another study, *C. albicans* was the most common *Candida* species (42/95; 44.21%), followed by *C. lusitaniae* (18/95; 18.95%), *C. parapsilosis* (13/95; 13.69%), *C. glabrata* (8/95; 8.42%), *C. kefyr* (6/95; 6.31%), *C. famata* (5/95; 5.26%), *C. africana* (2/95; 2.11%) and *C. orthopsilosis* (1/95; 1.05%), respectively (Hashemi et al. 2019). *C. albicans* is a commensal and dimorphic fungus that resembles other yeast and resides on gastrointestinal and reproductive mucosal linings in humans showing obligative association. It becomes pathogenic under immunocompromised conditions which makes the host more prone to infections (Yang and Rao 2018). According to a CDC report, 33 cases have been reported in 2018 and 90% of isolates were resistant to one antifungal drug followed by 30% drug resistance to two antifungal drugs (https://www.cdc.gov/drugresistance/pdf/threats-report/candida-auris-508.pdf).

1.2.2 Diagnosis

1.2.2.1 Direct Examination

It is a low-cost, quick approach to diagnosing candidiasis at microscopic level. It entails scraping or swabbing the afflicted region, then putting the swab on a microscope slide. A 10% potassium hydroxide (KOH) solution is used to dissolve the skin cells without disturbing the integrity of the *Candida* cells. The addition of calcofluor white enables the quick identification of fungal components as it adheres to chitin and cellulose in the fungal cell wall and fluoresces when excited by UV light. Periodic acid-Schiff or Gomori methenamine silver stains are also used to stain histological specimens (Gauglitz et al. 2012; Hani et al. 2015).

1.2.2.2 Culture Method

Candida species readily flourish on Sabouraud dextrose agar, used for the isolation of *Candida* species, allowing it to proliferate while suppressing the development of bacteria due to low pH. This procedure involves wiping a sterile swab on the contaminated region, then streaking it on Sabouraud dextrose agar and incubating at 37 °C for a few days to allow the colonies to form. *Candida* is identified based on colony characterization, microscopic appearance and physiological or biochemical properties. Pagano-Levin agar, Nickerson's media, phosphomolybdate agar and chromogenic media are among the differential media available for *Candida* identification. Chromogenic media (*Candida* ID, CHROM agar, Candiselect 4 and *Candida* medium) are also suggested for rapid *Candida* identification (Ilkit and Guzel 2011).

1.2.2.3 (1→3)-β-D-Glucan Detection Method

Candida cell walls include β 1,3-D-glucans as structural components. As this polysaccharide is not found in bacteria, viruses or animals, its presence in patient blood has been utilized to detect invasive illnesses. The activation of a clotting mechanism observed in amebocyte lysates of the Japanese horseshoe crab, *Tachypleus tridentalis*, has been used to build a BDG detection test (Ellepola and Morrison 2005).

1.2.2.4 Serological Methods

Agar gel diffusion (AGD), counter immunoelectrophoresis (CEP), whole cell agglutination (AGGL), latex agglutination (LAT), indirect fluorescent antibody and complement fixation are typical serological procedures. The AGGL test, AGD test and CEP test are the most used. The AGGL test, created by Hasenclever and Mitchell, is one of the most routinely used serological assays for systemic candidiasis. This test identifies antibodies that are largely specific to the mannan component of the *Candida* cell wall (Vicariotto et al. 2012). With high sensitivity (98%) and negative predictive value (95%), β-D-glucan in conjunction with procalcitonin appears promising in the exclusion of invasive candidiasis. Tests based on *Candida* antigens and anti-*Candida* antibody detection in patient serum were the first noncultural diagnostic tools for invasive candidiasis (Barantsevich and Barantsevich 2022).

1.2.2.5 Automated Methods

Recent automated procedures are rapid, dependable and broad spectrum, making it easier to build a new patient care system with novel therapeutic strategies.

Vitek YBC System

The Vitek YBC system (bioMerieux Vitek, Inc., Hazelwood, MO, USA) is a commonly used auto-microbial system used in research centers and labs. This technique can perform 26 biochemical assays with the same inoculum at the same time. Consequently, it is able to distinguish between several *Candida* species, including *C. parapsilosis*, *C. albicans*, *C. glabrata* and *C. tropicalis*. It also includes a computerized evaluation mechanism for more accurate information on the fungal species (Pfaller et al. 1988).

VITEK® 2 IDYST System

VITEK® 2 Yeast Identification (YST) is a fast, basic system that can perform numerous reactions at the same time. This system is capable of performing 47 carbohydrate assimilation-based fluorescent biochemical assays, involving the deamination process and oxidation reactions that interact with several aryl amidases and oxidases of the *Candida* species (Graf et al. 2000).

1.2.2.6 DNA-Based Methods

DNA-based assays can identify and distinguish between the DNA of microbes. These are the oldest microbe identification methods based on DNA components. Hydrolyzer enzymes remove DNA from eukaryotic organisms via cell membrane hydrolysis. The following methods are pulsed-field gel electrophoresis (PFGE), restriction enzyme analysis (REA), random amplified polymorphic DNA (RAPD), amplified fragment length polymorphism (AFLP), polymerase chain reaction (PCR)-based *Candida* detection methods, nucleic acid sequence-based amplification (NASBA), peptide nucleic acid-fluorescent in situ hybridization (PNA-FISH), microsatellite length polymorphism (MLP) typing, multilocus sequence typing (MLST) and DNA microarrays (Arafa et al. 2023).

1.2.2.7 Advanced Methods

Matrix-assisted laser desorption ionization-time of flight mass spectrometry (MALDI-TOF MS) is a cutting-edge technique that is frequently used currently in clinical microbiology laboratories. This technique is a fast, low-cost, dependable and strong identification technology that provides protein fingerprints for each microbe in the sample, which may be easily matched with a reference library (Barantsevich and Barantsevich 2022).

1.2.3 Antifungal Drugs and Mechanisms

Current antifungal drug therapies are confined to limited classes of drugs. They are basically used orally, topically and intravenously for treating fungal infections. They are classified depending on the spectrum of activity and targets as follows:

1.2.3.1 Azoles

They basically target the P450-dependent enzyme 14 α-lanosterol demethylase (CYP51), an enzyme encoded by the *ERG11* gene. Protein Cyp51p (*ERG11*) is most found in all fungal species and plays a crucial role in the synthesis of an important sterol called ergosterol. The inhibition of this enzyme reduces the conversion of lanosterol to ergosterol in the ergosterol biosynthesis pathway. Ergosterol is a key sterol that is present in the fungal membrane and provides many other indirect functions (Shapiro et al. 2011). Cyp51p inhibition could lead to ergosterol synthesis blockage and depletion of sterols from the membranes, resulting in the accumulation of methylated sterols leading to membrane stress. Azoles also inhibit cytochrome P450 which is responsible for sterol Δ^{22}-desaturation encoded by *ERG 5* gene in the ergosterol biosynthesis pathway (Skaggs et al. 1996). Fluconazole is the most prescribed antifungal drug used for treating *Candida* infections.

1.2.3.2 Polyenes

Amphotericin B, nystatin and natamycin are the most commonly used polyenes (Zotchev 2003). Amphotericin B is used for treating infections caused by *Aspergillus*, *Candida*, *Cryptococcus*, *Mucor*, *Fusarium* and *Scedosporium* (Laniado-Laborín and Cabrales-Vargas 2009). Nystatin is mostly used for treating topical infections such as cutaneous, vaginal and esophageal candidiasis. This class of drugs mainly targets the fungal cell membrane by pore formation which leads to cell death. They specifically bind to ergosterol in the fungal cell membrane due to their amphiphilic structure and form a complex. This results in pore formation that disrupts the cell membrane and leads to cellular leakage and loss of ionic balance which ultimately leads to cell death (Sanglard et al. 2009). The antifungals in this class were the first drugs to be used for treating fungal infections. They are broad spectrum with fungicidal nature.

1.2.3.3 Echinocandins

These drugs came into light a decade ago and were approved by the FDA (Food and Drug Administration) and EMA (European Medicines Agency) for clinical use (Vandeputte et al. 2012). They are semisynthetic amphiphilic lipopeptides that are extracted from fungal natural products (Eschenauer et al. 2007). They target the most important component of the cell wall, i.e., 1,3-β-glucans. They are polysaccharides that consist of D-glucose monomers attached to each other by either β-(1,3) or β-(1,6)-glucan linkages (Lorand and Kocsis 2007). They primarily inhibit the synthesis of β-1,3-D-glucan synthase by binding to it in a noncompetitive manner. β-1,3-D-glucan synthase is involved in the synthesis of the 1,3-β-glucan present in the fungal cell wall (Onishi et al. 2000). Currently, three well-known echinocandins are available in the market, namely, Caspofungin, Anidulafungin and Micafungin. They inhibit glucan synthesis in *Candida* spp. and this property makes these drugs fungicidal in nature. On the other hand, they can become fungistatic also, like in the case of *Aspergillus*, where they basically target cell wall synthesis, initially at the apical tip of hypha, causing changes in hyphal morphology, resulting in the modification of the cell wall (Bowman et al. 2002).

1.2.3.4 Pyrimidine Analogues

5-Fluorocytosine, also known as flucytosine, is a fluorinated pyrimidine that inhibits nucleotide synthesis, metabolism and protein synthesis in fungal cells (Waldorf and Polak 1983). Basically, it acts by two mechanisms: In one mode, cytosine deaminase of fungal cells, which can convert 5-fluorocytosine into 5-fluorouracil, is taken by cytosine permease and then incorporated into DNA and RNA, this results in inhibition of DNA synthesis which then ultimately disrupts protein synthesis (Polak and Scholer 1975). In another mode, flucytosine also converts into 5-fluoro deoxyuridine monophosphate, resulting in fungal DNA synthesis (Vermes et al. 2000). Flucytosine is found to be effective against *C. albicans*, *C. tropicalis*, *C. glabrata*, *C. krusei*, *C. parapsilosis* and *C. lusitaniae*. The most used drugs are 5-Fluorocytosine and 5-Fluorouracil. It is currently used along with amphotericin B for treating candidiasis and cryptococcal meningitis (Tassel and Madoff 1968).

1.2.3.5 Allylamines

Naftifine and terbinafine are two well-known drugs of this class. They act by inhibiting ergosterol biosynthesis in *Candida* cells (Birnbaum 1990). Ergosterol biosynthesis is a series of reactions which is governed by many enzymes. Allylamines primarily target the squalene epoxidase (Erg1) which is involved in the conversion of squalene into squalene-2,3-epoxide and then into lanosterol in early steps of ergosterol biosynthesis. Any disturbances in this pathway will result in the accumulation of toxic sterols (Perea et al. 2002). Terbinafine is widely used for treating skin and nail infections like tinea corporis, tinea pedis (Athlete's foot), tinea cruris (Jock itch) and onychomycosis (Chen and Sorrell 2007).

1.2.4 Drug Resistance and Mechanisms

MDR is a multifactorial phenomenon. Microbes can reduce impact of the drug by reducing the drug concentration, modifying metabolic reactions to divert the toxicity exerted by antifungal agents, or by making drug target alterations. The major mechanisms for MDR found in *Candida* are as follows and are depicted in Fig. 1.2.

1.2.4.1 Overexpression of Efflux Pumps

This mechanism is mediated by overexpression of drug transporters like ATP-binding cassettes (ABC), and major facilitator superfamily (MFS), resulting in increased drug efflux that removes toxic drugs out of fungal cells (Cannon et al. 2009). The upregulation of these transporters increase drug efflux and reduce intracellular drug accumulation. CaCdr1p and CaCdr2p are major transporters belonging to the ABC transporter family whereas CaMdr1p transporter belongs to the MFS family. The ABC transporter has membrane domains and regulates ATP hydrolysis for the transport of drugs (Prasad and Goffeau 2012). MFS transporters are transmembrane proteins that use electrochemical protein gradient to efflux the drug out of the cell. The upregulation of genes expressing the efflux pump transporters was reported in azole-resistant strains of *C. albicans* (Perea et al. 2001). *CDR1* and *CDR2* upregulation arbitrates azole resistance via amplified drug efflux which reduces the accumulation of azoles in clinical strains of *C. albicans* (Sanglard et al.

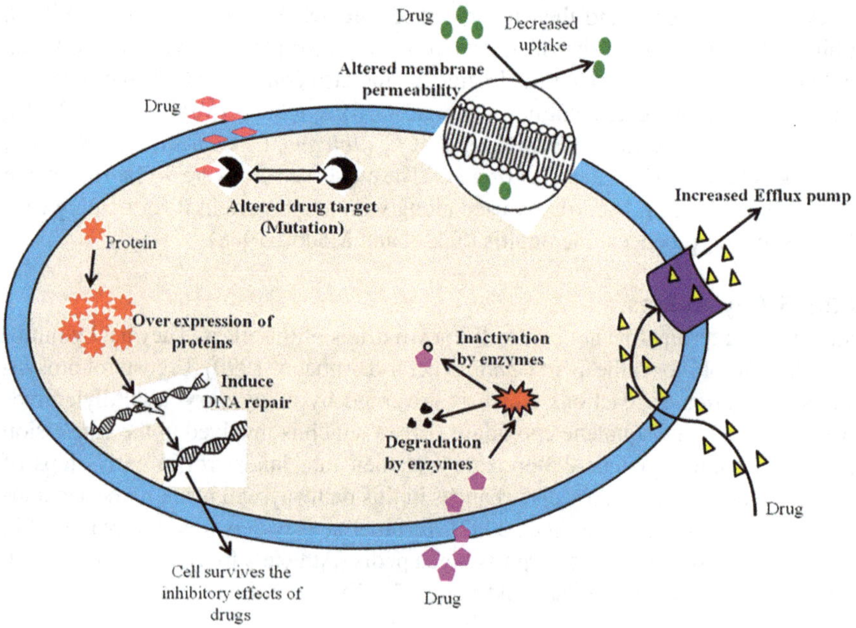

Fig. 1.2 Model showing mechanisms of multidrug resistance in *C. albicans*

2009). Increased azole efflux is also found in *C. albicans* strains due to the upregulation of *MDR1* (Lamping et al. 2007).

1.2.4.2 Alteration in Membrane Permeability

Any changes or alterations in the cell membrane may result in permeability changes which can throw the drug out of the fungal cell. Alteration in membrane permeability can be caused by changes in ergosterol structure which make prevents the binding of drug to the drug target. The drug entry is made easy by passive diffusion as drugs are hydrophobic in nature. Diffusion of drugs via lipid bilayer is an important MDR determinant (Ferté 2000). The alteration in membrane permeability in *C. albicans* was achieved by a mutation in the biosynthesis pathway of sterols which results in decreased drug import. However, variations in membrane fluidity and passive diffusion are the main contributors to this kind of drug resistance. It has been observed that any mutations in the *ERG2* and *ERG6* genes can result in changes in membrane fluidity which alters drug uptake (Mukhopadhyay et al. 2002).

1.2.4.3 Alteration/Mutation in Drug Target

This type of resistance mechanism is found in the azoles and echinocandin-resistant clinical strains (Arendrup 2014). They basically target the *ERG11* gene which is responsible for ergosterol biosynthesis. The target of the azole class is lanosterol 14α-demethylase, an enzyme encoded by the *ERG11* gene and catalyzes the synthesis of ergosterol from lanosterol. Any mutations that occur in this gene can alter the drug target which decreases its affinity for the drug. In the case of echinocandin drugs, a decreased affinity was also found as they basically target the β-1-3-glucan synthase which is encoded by the *FKS* gene. Mutations bring conformational changes at the sites of drug targets of antifungals (Perlin 2015).

1.2.4.4 Overexpression of Drug Target

In this drug resistance mechanism, the target is overexpressed to nullify the effect of the drug. An adequate concentration of the drug is required to act upon the target. Generally, a balance of the drug target and drug concentration is sufficient to inhibit microbial growth. An imbalance could result in higher amounts of drug targets in comparison with the drug concentration which leads to drug resistance (Sanglard et al. 2009). For instance, in azole-resistant strains, the *ERG11* gene was found to be upregulated which develops resistance. This is regulated by a transcription factor called Upc2p (zinc cluster finger transcription factor) (Dunkel et al. 2008).

1.2.5 Virulence and Pathogenicity

The capacity of *Candida* species to survive in extreme environments renders them extremely hazardous that can endanger the lives of immunocompromised persons. Various aggressive mechanisms of these fungal pathogens contribute to disease pathogenesis. *Candida* pathogenicity is enhanced by several virulence mechanisms, including invasion and attachment to inanimate surfaces and bodily tissues, metabolic adaptability, dimorphism, phenotypic switching, hydrolytic enzyme release and biofilm formation depicted in Fig. 1.3.

1.2.5.1 Adhesion and Invasion

Candida spp. adheres to host cells via adhesion proteins on the fungal cell surface (pga1, als1–7, hwp1, als9 and eap1) and immobilized ligands (cadherins, integrins or other microbes). Following attachment, fungal cells enter the tissue. Pathogenicity is defined as epithelial infiltration and injury. Depending on the kind of host cell, it might happen via one of two processes (active penetration or endocytosis). *C. albicans*, for example, invade oral cells by active penetration and endocytosis, whereas intestinal invasion is only conceivable via active penetration (Mayer et al. 2013; Moyes et al. 2015).

1.2.5.2 Dimorphic Switching

Fungal dimorphism is a unique feature of yeast pathogenicity. It has been found that the hyphal form is more invasive than the yeast form. *C. albicans* dimorphism is significant in the pathogenesis of both systemic and superficial infections. It should be emphasized that in *C. albicans* both yeast and filamentous forms were found in the infected tissues (Brown et al. 2012a, b). The *C. albicans* capacity to convert from the yeast to filamentous form incorporates the diverse character of its infection stages, which includes adhesion to epithelial and endothelial cells, invasion, iron acquisition from host sources, biofilm production, phagocyte escape and immunological evasion. Nutrients, pH, temperature (37–40 °C), concentration of CO_2 (5.5%) and amino acids all aid in their morphological transformation, which is critical for pathogenicity. Yeast form may easily spread throughout the host tissues, whereas the filamentous cells with stronger adhesion capabilities aid in host tissue penetration (Nobile et al. 2012).

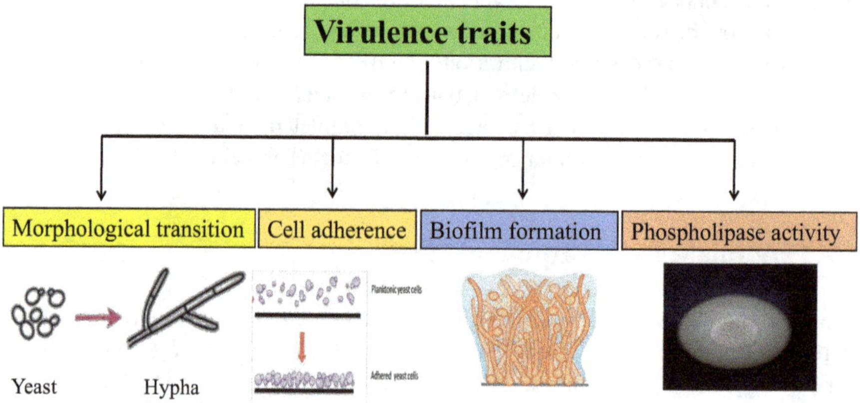

Fig. 1.3 Model showing various virulence traits in *C. albicans*

1.2.5.3 Biofilm Formation

A biofilm is made up of adherent, attached and accumulated microorganisms that produce extracellular polymeric substances (EPS) that serve as a structural matrix. *C. albicans* cell attachment to surfaces is the first stage in the formation of an organized robust extracellular matrix (ECM) structure. The adhesion phase, initiation phase (early phase), maturation phase (middle phase) and dispersal phase (dispersion phase) are the four sequential stages in *C. albicans* biofilm formation (Nobile and Johnson 2015; Aliyu 2022). The regulation of biofilm formation is governed by many transcription factors involving their role in adhesion, hyphal formation and EPS production. Preliminary identification of the main transcription factors regulating biofilm formation has been done by conducting mutant studies (Nobile et al. 2012).

1.2.5.4 Enzymes

The pathogenicity-causing *Candida* virulence factors require extracellular hydrolytic enzymes (phospholipases, proteases, laccases, keratinases, etc.). These enzymes aid fungal invasion by degrading host proteins like hemoglobin and keratin and altering cell membrane structure. By avoiding antimicrobial chemicals, these processes aid in the targeting and penetration of the host immune cells. This pathway is followed by several *Candida* species, including *C. albicans*, *C. tropicalis*, *C. parapsilosis* and *C. dubliniensis* (Mayer et al. 2013).

1.2.5.5 Environmental Factors

Candida is extremely adaptable to a wide range of environmental circumstances, including low oxygen, restricted nourishment, pH variations, nitrosative, cationic, temperature, osmotic pressure and oxidative stressors. This versatility is critical for *C. albicans* pathogenicity (Thompson et al. 2011).

1.2.6 Novel Approaches to Combat MDR

The following section describes some novel approaches to combat MDR for fungal pathogens and has been depicted in Fig. 1.4.

1.2.6.1 Novel Antifungal Drug Targets

There is ongoing research into developing alternative antifungal agents that target the fungal cell wall. Transcription factors (TFs) are appealing as potential antifungal therapeutic targets because they evolved differently between fungi and humans. Targeting fungal virulence-regulating TFs has been proven in studies to be an effective method for producing novel antifungal drugs (Bahn 2015). There are several novel target proteins and pathways being researched, including glycosylphosphatidylinositol biosynthesis, chitin synthase, inositol phosphoceramide synthase, heat shock protein 90 (Hsp90), histone deacetylase 2 and enolase (Zhen et al. 2022). It has been shown that different structures of fungal cells, such as the cell wall and cell membrane, could be targeted to develop novel antifungals. Similarly, metabolic

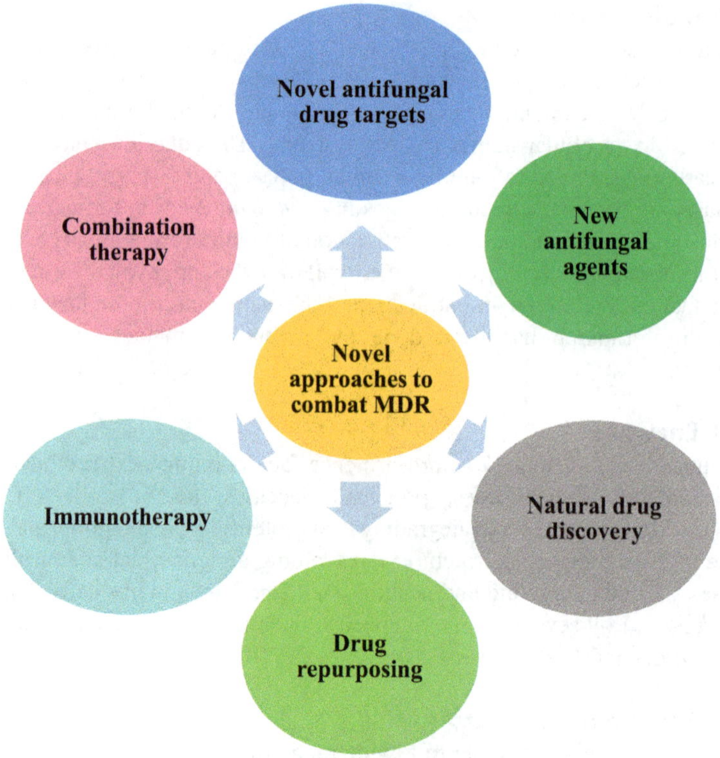

Fig. 1.4 Model showing novel approaches to combat MDR

pathways (glyoxylate cycle, pyrimidine biosynthesis, cytochrome P450 enzymes, iron metabolism, heme biosynthesis and acetate metabolism), signal transduction pathways (MAP kinase, PDK1 and calcium signaling) and gene expression can be explored as antifungal targets. Some novel pathways have recently been presented as potential targets for antifungal research, including enolase, a component of the enolase-plasminogen complex, as well as enzymes involved in mannitol biosynthesis and purine nucleotide biosynthesis (Nguyen et al. 2021). Cell efflux, ergosterol production and Hsp90 have been identified as possible targets for the development of antifungals that might minimize pathogenic fungal mechanisms of resistance (Ivanov et al. 2022).

1.2.6.2 Natural Drug Discovery

Natural compounds are explored for their structures and functional activity so that they could be employed more for treating various infections. Plants can be regarded as a treasure of valuable natural compounds that can be used in the development of drugs serving as anti-inflammatory, antimicrobial and antidiabetic. Microbes are responsible for causing several human diseases (Harvey 2007). Drug discovery has revolutionized the health industry by exploiting plants for extracting natural compounds, belonging to various classes such as terpenoids, alkaloids, phenolic compounds, flavonoids and xanthines (Bakkali et al. 2008; Thomford et al. 2018). It has been also reported that phenolic compounds and essential oils from natural sources have therapeutic potential (Ansari et al. 2014; Saibabu et al. 2015). Natural compounds drug discovery could contribute majorly to solving drug resistance problems (Singh et al. 2020).

1.2.6.3 Drug Repurposing

Drug repurposing or repositioning can be defined as the redirection of a previously employed medicine against another condition to speed up the process and minimize the period required for new drug development. Scientists have been evaluating the current pharmacopeia to lessen the burden of de novo drug manufacture by investigating and using existing medications for battling *Candida* infections (Hua et al. 2022). BQM (bis[1,6-a:5′,6′-g]quinolizinium-8-methyl-salt) was repurposed as an antifungal agent by reversing MDR1-mediated resistance. Calcineurin inhibitors such as cyclosporin and tacrolimus, which are used in dermatology, transplantation and ulcerative colitis, have been repurposed as supplementary antifungal medicines (Katragkou et al. 2016). A study demonstrated the antifungal activity of theophylline against *C. albicans* by inhibiting the glyoxylate cycle. Theophylline is a methylxanthine produced from cocoa beans and tea extracts that is commonly used as a first-line treatment for asthma and other respiratory problems (Singh et al. 2020).

1.2.6.4 Immunotherapy

This approach involves using the body's immune system to fight infections. The in silico work employed an immunotherapeutic peptide vaccine-based strategy to scan the *Candida* proteome and identify the most immunodominant HLA class I, HLA class II and B cell epitopes. To improve immunogenicity, the authors chose the most promising epitopes and produced a multivalent recombinant protein against *C. albicans* (mvPC) with a synthetic adjuvant (RS09) (Tarang et al. 2020).

1.2.6.5 Combination Therapy

This approach involves using multiple drugs to treat infections, which can reduce the evolution of resistance and enhance the longevity of the antimicrobial agents. Another well-known antifungal combination in the treatment of invasive *Candida* infection is echinocandins with azoles. In prospective, multicenter cohort of 40 solid-organ transplant recipients, the combination of caspofungin and voriconazole was utilized as the first treatment for invasive Aspergillosis. A combination of triazoles and polyenes, such as voriconazole and amphotericin B, may be a suitable choice (Johnson and Perfect 2010; Campitelli et al. 2017).

1.2.6.6 New Antifungal Agents

Many new antifungal drugs must pass clinical trials before entering the market. This section mentions few drugs that are under different phases of clinical trials. **APX001 (Fosmanogepix)**, also known as **E1210**, is under phase II of clinical trials and has shown broad-spectrum activity against various fungi such as *Fusarium*, *Aspergillus*,

Scedosporium and *Candida* species. It interferes with the synthesis of glycosylphosphatidylinositol which results in the detachment of mannose proteins in cell wall structure (Watanabe et al. 2012). **AR-12** compound completed phase I of clinical trials and was used to inhibit the acetyl CoA synthetase 1 of *C. albicans* and some other fungal species (Koselny et al. 2016). **CD101** has been derived after modifications in the backbone structure of echinocandins to increase stability under clinical phase II. **SCY078**, which is developed by Scynexis Inc. and is under phase II clinical trial, inhibits 1,3-β glucan synthase (Pfaller et al. 2019). Modifications in pyrimidine analogs have given a new compound **F901318**, which can inhibit dihydroorotate dehydrogenase and interferes with pyrimidine biosynthesis in fungi (Oliver et al. 2016). The other two compounds named **VT-1129** worked against cryptococcosis and were tested in animal models. **VT-1598** has the potential enough to work against cryptococcosis and endemic mycosis (Lockhart and Berkow 2019). Ibrexafungerp belongs to the first of a new class of oral glucan synthase inhibitors that are semi-synthetic derivatives of enfumafungin (McCarty and Pappas 2021).

1.3 *Aspergillus* and Aspergillosis

Aspergillus species encompass a diverse range of characteristics. These opportunistic pathogens are producers of toxins and are of industrial significance also (Barnes and Marr 2006; Mousavi et al. 2016). The pathogenic potential of *Aspergillus* in humans was first recognized by Virchow in 1856. Approximately 20 species of *Aspergillus* have been identified as pathogenic to humans, causing various illnesses from allergies to severe lung infections. Among these, life-threatening infections are primarily caused by the pathogenic species, namely, *Aspergillus fumigatus*, *Aspergillus flavus*, *Aspergillus niger*, *Aspergillus terreus* and *Aspergillus nidulans* (Rudramurthy et al. 2019). Such infections commonly occur in individuals with a weakened immune system, as treatments like chemotherapy can compromise the body's defense mechanisms by targeting rapidly dividing cells like neutrophils (crucial components of the immune system), leading to increased susceptibility to *Aspergillus* infections.

Aspergillus is a mold found widely in nature. These filamentous fungi, known for production of spores, typically thrive on decaying organic matter and obtain nutrients as a saprophyte. Different species of *Aspergillus* can be identified based on the characteristics such as colony color, size, growth rate and tolerance to heat. For instance, *A. flavus* and *A. niger* exhibit yellowish-green and black conidial heads, respectively, while *A. fumigatus* can be easily recognized by its greenish-gray color. *Aspergillus* species are susceptible to changes in air circulation, light exposure and the volume of the growth medium due to their variable morphology (Arastehfar et al. 2021; Viegas et al. 2021). The term "Aspergillus" was introduced by Italian priest and biologist Pier Antonio Micheli, who observed the morphology of conidiophores and likened them to the aspergillum, a sprinkler used in churches for holy water. The first monograph on the *Aspergillus* genus was produced by Thom and Church in 1926, providing detailed instructions for its identification.

1.3.1 Epidemiology

Aspergillosis, a fungal infection caused by *Aspergillus* species, can manifest as either a chronic or acute condition. The most prevalent forms of aspergillosis are chronic pulmonary aspergillosis, invasive aspergillosis and bronchitis (Denning et al. 2016). Invasive aspergillosis accounts for more than 200,000 cases annually. Immunocompromised individuals, including those with severe neutropenia, recipients of bone marrow or solid-organ transplants, advanced AIDS patients and those with chronic granulomatous disease, are particularly susceptible to developing invasive aspergillosis. The invasion of *Aspergillus* into the lungs results in tissue damage, which can lead to sepsis and, in advanced stages, hemoptysis. Classic risk factors for invasive aspergillosis include liver cirrhosis, tuberculosis, diabetes mellitus and persistent lung disease (Agarwal et al. 2014; Komase et al. 2007; Yan et al. 2009). Chronic pulmonary aspergillosis is an infectious disease that progressively damages lung tissue, particularly affecting immunocompromised individuals with pre-existing lung conditions such as tuberculosis or chronic obstructive pulmonary disease (COPD). Awareness of this debilitating and potentially fatal infection is increasing, with an estimated three million cases of chronic pulmonary aspergillosis worldwide (Denning et al. 2016). Among Asian countries, India reported the highest burden of chronic pulmonary aspergillosis with 209,147 cases (Agarwal et al. 2014), followed by Pakistan with 55,509 cases (Jabeen et al. 2017), Bangladesh with 20,720 cases (Gugnani et al. 2017), Nepal with 6611 cases (Khwakhali and Denning 2015) and Sri Lanka with 2886 cases (Jayasekera et al. 2013). Patients with allergic bronchopulmonary aspergillosis often suffer from poorly controlled asthma or recurrent infections associated with bronchiectasis, which can progress to lung damage, respiratory failure and eventual death.

1.3.2 Antifungal Drugs and Mechanisms

Currently, there are three primary categories of antifungal medications employed in the treatment of Aspergillosis: Polyenes, Echinocandins and Azoles. Their mechanism of action as antifungals varies as depicted in Fig. 1.5. However, over time, the occurrence of resistance to these three classes has been progressively growing among various *Aspergillus* species. Furthermore, the conventional antifungal agents have restricted usage due to their associated side effects. The problem of resistance arises from the limited selection of antifungal agents, as certain fungi can exhibit resistance to all three types (Ostrowsky et al. 2020; White et al. 2021). Additionally, amphotericin B, the most effective antifungal agent for systemic infections, is notorious for its nephrotoxicity. The severity of its side effects and toxicity may necessitate discontinuation of therapy, even in cases of life-threatening fungal infections (Laniado-Laborín and Cabrales-Vargas 2009). Given the rise in fungal infections, the emergence of resistance and the limitations of current antifungal agents, there is a growing need for the discovery of novel antifungal agents, preferably with distinct mechanisms of action.

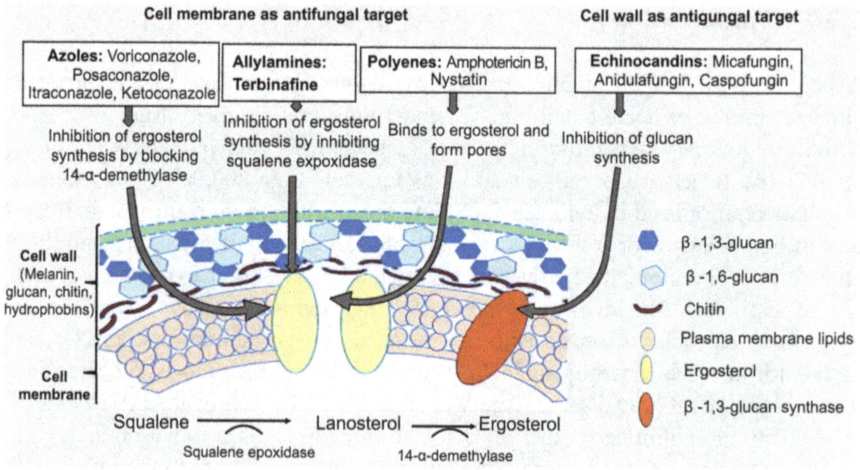

Fig. 1.5 Antifungal drug targets in pathogenic fungi (Scorzoni et al. 2017)

Azoles, particularly triazoles like itraconazole, voriconazole and posaconazole, are commonly used as first-line treatment for invasive aspergillosis. They inhibit the synthesis of ergosterol in the fungal cell membrane, leading to instability and cell death (Kanafani and Perfect 2008). However, azoles can cause hepatotoxicity and have potential drug interactions, especially in immunosuppressed patients. Resistance to azoles has been reported, with varying levels depending on the drug and geographical location. Polyenes, exemplified by amphotericin B, interact with ergosterol, disrupting ionic balance and causing cell death (Laniado-Laborín and Cabrales-Vargas 2009). Recent reports have shown the emergence of widespread amphotericin B resistance among *A. fumigatus* strains, leading to treatment failures (Ashu et al. 2018).

Echinocandins, such as caspofungin, micafungin and anidulafungin, inhibit β-1,3-glucan synthase, an enzyme involved in fungal cell wall synthesis (Aruanno et al. 2019). They have fewer adverse effects and drug interactions compared to other classes. However, echinocandins are less effective against *Aspergillus* spp. and are primarily administered intravenously, limiting their use to hospitalized patients (Birch and Sibley 2017). Resistance to echinocandins has also been reported, causing treatment failures. It is worth noting that many current antifungal agents are derived from natural sources. Amphotericin B is derived from *Streptomyces nodosus* bacteria (Cavassin et al. 2021), while echinocandins are modified compounds derived from fungi. This highlights the potential of natural sources for the discovery of new antifungal agents with better efficacy and low toxicity.

1.3.3 Virulence and Pathogenicity

To evolve from its saprophytic state to a pathogenic state, *A. fumigatus* undergoes a series of adjustments to thrive within the host environment, employing various strategies to enhance its virulence. Virulence refers to the degree of pathogenicity or the ability of a microorganism to cause disease in a host. The virulence of *A. fumigatus* is a complex phenomenon that encompasses a multitude of factors such as genes, molecules and characteristics that contribute to the establishment and progression of infection (Gupta et al. 2021). In the case of *A. fumigatus*, several virulence factors have been identified.

1.3.3.1 Adhesion

Adherence is crucial for the pathogenicity of *A. fumigatus*. Conidia, the airborne spores of *A. fumigatus*, adhere to host tissues through negatively charged carbohydrate motifs and surface proteins (Croft et al. 2016). RodA protein is responsible for assembling the hydrophobic rodlet layer that promotes adherence of conidia to host collagen and albumin (Valsecchi et al. 2017). Adhesins like Asp f 2 and the GPI-anchored CspA protein also play a role in binding to host components. The extracellular protein AfCalAp, facilitates adherence by binding to host laminin (Upadhyay et al. 2009). Additionally, *A. fumigatus* hyphae adhere to host cells through the secretion of the adhesive exopolysaccharide Galactosaminogalactan (GAG). Mutants deficient in GAG exhibit reduced adherence and virulence. GAG is also involved in biofilm formation, antifungal resistance and immune evasion (Gupta et al. 2022; Kowalski et al. 2020).

1.3.3.2 Cell Wall Maintenance

Cell wall maintenance is crucial for fungal pathogenesis as the cell wall serves as the primary barrier against environmental stressors and host immune responses. The cell wall of *A. fumigatus* primarily consists of polysaccharides, including β1,3-glucan, chitin, galactomannan and β1,3-β1,4 glucans, along with the cementing α1,3-glucan and mannans, and various proteins (Fontaine et al. 2010). Deletions in genes responsible for components of cell wall synthesis, such as *chsG*, *glfA* and *afmnt1*, result in reduced virulence in *A. fumigatus*. These findings highlight the importance of maintaining cell wall integrity for fungal pathogenicity and host adaptation (Wagener et al. 2008).

1.3.3.3 Stress Response

A. fumigatus employs various stress response pathways to enhance its survival and virulence. These include the cAMP-PKA, CWI, HOG, calcium-calcineurin, UPR and hypoxia adaptation pathways (Liebmann et al. 2004). Mutations in key genes within these pathways, such as *PkaC1*, *PkaR*, *chsG*, *glfA*, *afmnt1*, *cnaA*, *crzA*, *IreA*, *HacA*, *hrmA* and *PacC*, have been linked to reduced virulence. *A. fumigatus* demonstrates remarkable adaptability to environmental stressors like hypoxia and alkaline pH, influencing its morphology, immune response and antifungal resistance (Dirr et al. 2010). These findings contribute to our understanding of *A. fumigatus* complex stress response mechanisms during infection.

1.3.3.4 Interaction with the Host Immune System

A. fumigatus employs various strategies to manipulate and evade the host immune system. Pigmentation of conidia protects the fungus from reactive oxygen species (ROS) and impairs immune recognition. Deletion of the pksP gene reduces melanin production and attenuates virulence (Chamilos and Carvalho 2020). The antiapoptotic protein AfBIR1 inhibits fungal programmed cell death, enhancing fungal survival and virulence (Shlezinger et al. 2017). Catalase and superoxide dismutase detoxify ROS, impacting virulence in immunosuppressed animals. Gliotoxin, a secreted molecule, suppresses immune functions and contributes to virulence. Deletion of gliP and rglT genes decreases virulence, highlighting the importance of gliotoxin regulation and self-protection. Additionally, *A. fumigatus* secretes GAG, which masks β-glucan, induces neutrophil apoptosis and modulates immune response. These mechanisms enable *A. fumigatus* to subvert the host immune system during infection (Speth et al. 2019).

1.3.4 Drug Resistance Mechanisms in *Aspergillus*

Azole resistance in *Aspergillus* species is a growing concern in clinical practice, as it limits the effectiveness of azole antifungal drugs, which are commonly used to treat *Aspergillus* infections. Understanding the mechanisms underlying azole resistance is crucial for developing strategies to overcome this challenge. The major categories of the mechanism of drug resistance are discussed in the following section.

1.3.4.1 Target-Site Alterations Due to Mutations in the *cyp51A* Gene

One of the primary mechanisms of azole resistance involves mutations or substitutions in the target enzyme lanosterol 14α-demethylase (CYP51A) (Price et al. 2015). CYP51A is responsible for catalyzing the demethylation of lanosterol, a key step in ergosterol biosynthesis. Azole drugs bind to CYP51A, inhibiting its activity and disrupting ergosterol production. However, mutations in the *cyp51A* gene can lead to structural changes in the enzyme, reducing its affinity for azoles. This reduces the effectiveness of the drugs, allowing the fungi to survive and grow despite treatment. Various changes in *cyp51A*, resulting in a pan-azole-resistant phenotype, have been reported in *A. fumigatus* isolates from the environment and clinical sources worldwide. The most frequent mechanisms of resistance reported in environmental and clinical strains of *A. fumigatus* are mutations TR34/L98H and TR46/Y121F/T289A in the *cyp51A* gene. Several point mutations, including P216L, F219C, F219I, A284T, Y431C, G432S and G434C, G54E/R/V and M220I/V/T/K, appear in patients who received long-term azole treatment (Tashiro et al. 2012).

1.3.4.2 Overexpression of *cyp51A*

Furthermore, alterations in the promoter region of the *cyp51A* gene, such as the insertion of tandem repeats or transposable elements, result in the overproduction of *cyp51A*. Notably, the presence of a 34 base-pair tandem repeat (TR34) in conjunction with lysine to histidine substitution at codon 98 (TR34/L98H) leads to a significant up to eightfold elevation in the normal expression of *cyp51A*, as demonstrated by Dhanasobhon et al. (2013). This increase in mRNA expression is believed to directly correlate with an increase in cellular CYP51A levels, ultimately resulting in reduced sensitivity to azole treatment, according to the findings by Price et al. (2015) and Sen et al. (2023).

1.3.4.3 Efflux Transporters

While CYP51A serves as the primary target for azole compounds, there are additional enzymes capable of engendering resistance. Eukaryotic organisms require efflux pumps, including ATP-binding cassette (ABC) transporters and major facilitator superfamily (MFS) transporters, to eliminate toxins from within the cell (Jasinski et al. 2003). Consequently, the overexpression of these genes leads to azole resistance as the fungicide intracellular concentration reduces. The fungus *A. fumigatus* possesses at least 49 genes encoding ABC transporters, with 12 of them demonstrating a significant similarity to the azole-resistant *Saccharomyces cerevisiae* PDR5 and PDR15 proteins (Hagiwara et al. 2016). The Cdr1B efflux transporter, belonging to the PDR subfamily, exhibits overexpression in azole-resistant isolates. Notably, the deletion of the *cdr1B* gene in a resistant strain enhances sensitivity to itraconazole (Hagiwara et al. 2016), highlighting the pivotal role of Cdr1B in azole resistance in *A. fumigatus* (Hagiwara et al. 2016). Moreover, studies have demonstrated that deletion mutants of two distinct ABC transporters (AtrF, AtrI) and a major facilitator superfamily transporter (MdrA) also exhibit susceptibility to azole drugs (Meneau et al. 2016). It is important to note that the mechanisms of azole resistance can vary among different *Aspergillus* species and even within the same species. Additionally, resistance mechanisms can be acquired through the selection pressure of long-term azole therapy or use of azole-based fungicides in agricultural fields.

1.3.5 Antifungals in Pipeline

Aspergillus species are significant fungal pathogens associated with a range of infections, including invasive aspergillosis. The emergence of azole resistance in *Aspergillus* strains has highlighted the need for new antifungal drugs with enhanced efficacy (Scorzoni et al. 2017). Several antifungal drugs are currently in the pipeline and undergoing investigation for their activity against *Aspergillus* species. An overview of some of these promising agents is discussed here, and their chemical structures are shown in Fig. 1.6.

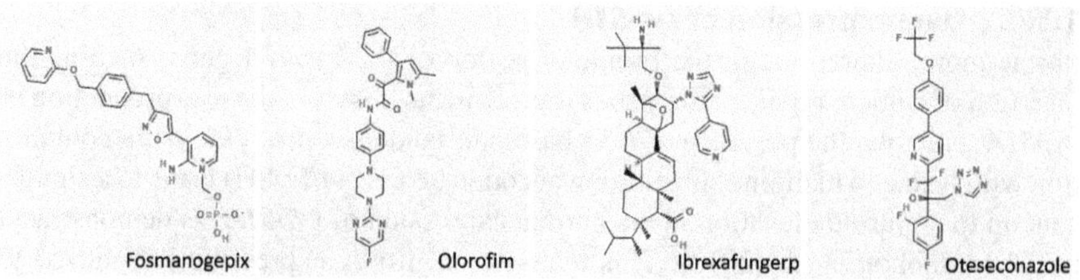

Fig. 1.6 Chemical structures of antifungals in the pipeline

Fosmanogepix (Amplyx) was initially discovered as E1210 by Eisai Co. and exhibits broad clinical activity against yeast and molds (Miyazaki et al. 2011; Nakamoto et al. 2010). It serves as the precursor to the active compound manogepix. This first-in-class antifungal works by inhibiting Gwt1, which blocks the production of glycosylphosphatidylinositol (GPI) compounds. GPI compounds play a crucial role in cell wall construction and the maintenance of homeostasis through a highly conserved pathway.

Olorofim (F2G, LTD) is an investigational antifungal agent from the class of orotomides (Oliver et al. 2016). It has demonstrated potent in vitro and in vivo activity against *Aspergillus* species, including azole-resistant strains. Olorofim works by inhibiting the enzyme dihydroorotate dehydrogenase (DHODH), a key enzyme involved in pyrimidine biosynthesis. It is currently being evaluated in clinical trials for the treatment of invasive fungal infections, including aspergillosis. **Ibrexafungerp (Scynexis, Formerly SCY-078)** is a new class of oral glucan synthase inhibitors, derived from enfumafungin. It belongs to the triterpenoid antifungals and shares structural similarities with echinocandins. It inhibits β-1,3-glucan synthase, like echinocandins, but with a slightly different binding site that overlaps with the echinocandins (Apgar et al. 2021). This structural change allows ibrexafungerp to retain activity against both echinocandin-susceptible and certain echinocandin-resistant *Candida* spp. The drug is available for oral and IV administration, with the oral formulation having been studied in humans. Ibrexafungerp is well-absorbed orally and is considered fungicidal. It has recently received FDA approval for the treatment of vulvovaginal candidiasis (VVC).

Oteseconazole (Mycovia, VT-1161) is a tetrazole agent that demonstrates a significantly higher affinity for fungal CYP51 compared to human cytochrome enzymes. This characteristic aims to enhance selectivity, minimize side effects and improve efficacy compared to currently available azoles. Studies indicate that the affinity for fungal CYP51 is over 2000 times greater than that of the human enzyme counterpart, suggesting a reduced likelihood of drug-drug interactions and direct toxicity (Hoekstra et al. 2014). Although oteseconazole targets the same enzyme as other azoles (14-alpha demethylase), its affinity for the enzyme is superior to triazoles like fluconazole.

Encochleated Amphotericin B (Matinas, MAT2203) is an orally administered lipid nanocrystal designed for the treatment of serious fungal infections. It avoids the significant toxicities associated with intravenously administered amphotericin B deoxycholate (AmB-D) and lipid formulations of AmB, by using a solid lipid bilayer and calcium. AmB molecules are entrapped within the spiral structure of the cochlea (Segarra et al. 2002). When taken up by phagocytic cells, the cochlea unfolds and releases the amphotericin B molecules due to the calcium gradient. The cochlea structure provides protection against degradation in unfavorable environments and enables targeted intracellular delivery.

ATI-2307 (Appili Therapeutics) is a novel antifungal agent with a unique mechanism of action. It is an arylamidine compound that selectively disrupts mitochondrial function in fungal cells by inhibiting respiratory chain complexes. ATI-2307 causes a collapse of the mitochondrial membrane potential and leads to energy depletion and growth inhibition (Shibata et al. 2012; Yamashita et al. 2019). The drug has little metabolic degradation, binds to mitochondrial proteins in tissues for days to weeks and has a relatively short serum half-life. It is excreted slowly in urine and feces and has low potential for drug interactions. Currently, ATI-2307 is available only as a parenteral (intravenous) formulation.

1.4 *Cryptococcus* and Cryptococcosis

There are several fungi that can cause infections, but three are the most prominent cause of concern in developing disease in humans. These include *Candida*, *Aspergillus* and *Cryptococcus* and are mainly pathogenic in patients who already have comorbidities (Brown et al. 2012a, b). Pathogenic fungi cause billions of deaths worldwide and millions of serious deadly infections in immunocompromised patients (Bongomin et al. 2017). It has also become the concern for the health organizations due to its high mortality rates and the treatment costs (Drgona et al. 2014). *Cryptococcus* is a fungus that is omnipresent in the environment, and generally, there are two species that causes infections—*C. gattii* and *C. neoformans*. The last few decades have seen more *Cryptococcus* infections as people are becoming more prone to chronic diseases like diabetes, cancer and increasing number of diseases like HIV/AIDS, which is helping this fungus to become more deadly for humans (Maziarz and Perfect 2016). In the upcoming paragraphs, we will focus on its epidemiology, treatment options and mechanism of novel antifungal drugs.

1.4.1 Epidemiology

Most of the fungi which cause illness are Ascomycetes, whereas *Cryptococcus* belongs to the class Basidiomycetes. This fungus has different morphology and growth pattern as its growth increases in the presence of oxygen (Casadevall and Perfect 1998). *Cryptococcus* is ubiquitously found in the environment all over the world but there are two main species that cause majority of the infections in the human and are more pathogenic: *C. gattii* and *C. neoformans*. Moreover, these two are similar but differences can be found in their epidemiology and pathogenesis. Many reports claim that *C. neoformans* is not contagious in many individuals when they have been exposed to this fungus in childhood (Goldman et al. 2001). *C. neoformans* is the prominent species in case of infections in immunocompromised patients, whereas *C. gattii* plays lesser role. *C. neoformans* affects the lungs and the central nervous system (brain and spinal cord) and can also be adverse for other parts of the body (Chang et al. 2006). If the infection spreads from the lungs to the brain, then symptoms that can be observed are fever, neck pain, nausea, vomiting and headache. However, reports of lethality have highlighted this disease-causing fungus since the nineteenth century (McDonald et al. 2012). Consequently, in the twentieth century, an increasing number of HIV cases added more fuel to the fire of these fungal infections. Fortunately, antiretroviral therapy provides help to most developed countries whereas the underdeveloped and developing countries suffer more as they are unable to control the disease (Deok-jong Yoo et al. 2010). Worldwide its incidence has major concerns in Asia and America. Unfortunately, in sub-Saharan Africa, more than 160,000 deaths have been reported due to these infections (Rajasingham et al. 2017). *C. neoformans* remains in the body immunocompromised patients, and later when the immunity becomes very low, they start multiplying causing problems (Saha et al. 2007). Inhalation of fungal spores from the surroundings causes lung inflammation as the spores settle down. The disease can be treated initially, but becomes a menace when it reaches the brain contributing to high mortality rate. Figure 1.7 shows how the fungal infection disseminates from the environment to humans.

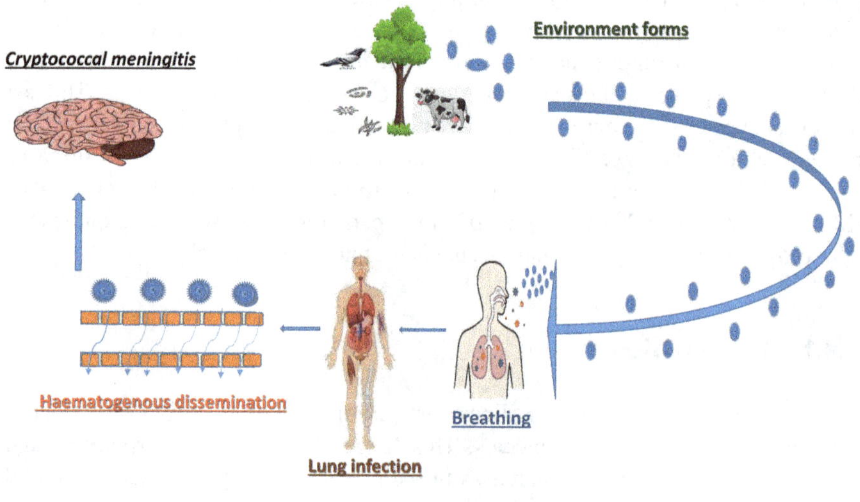

Fig. 1.7 Model showing the dissemination of *Cryptococcus* infections (Bermas and Geddes-McAlister 2020)

1.4.2 Diagnosis

Sometimes *Cryptococcus* strains can be isolated directly from the body fluids of the patient. There are numerous methods that can be employed for the diagnosis of *Cryptococcosis*. Direct detection can be achieved using Indian ink in addition to serology, culture and histopathology.

1.4.2.1 Serology

This method uses blood specimen to detect the existence of antigens and antibodies for diagnosis, so that treatment therapies can be developed to combat the disease. Diagnosis of *Cryptococcus* becomes easier with serology band that help to detect cryptococcal polysaccharide capsular antigens (CrAg) shed during the infection. There are some techniques that have been more accessible like the Latex agglutination and enzyme immunoassay. These techniques that use both serum and cerebrospinal fluid (CSF), have been used frequently till now (Wu and Koo 1983). However, the lateral fluid assay has also been acceptable for both serum and CSF (Lindsley et al. 2011).

1.4.2.2 Histopathology

This helps the clinicians to detect the disease by examining the patient tissue under a microscope. The technique is more sensitive in comparison to the Indian ink method, where this fungus can be recognized by staining the tissue, CSF and other body fluids (Shibuya et al. 2002). The fungus can be recognized by using many kinds of stains like mucicarmine, periodic acid-Schiff, Alcian blue stains, Fontana–Masson stain, Calcofluor and Gomori methenamine silver (Maziarz and Perfect 2016).

1.4.2.3 Indian Ink and Culture

The Indian ink dye is the easiest and quickest method to detect *Cryptococcus* using the CSF samples by direct microscopic study (Diamond and Bennett 1974). In practice, *Cryptococcus* can be easily cultured in media, and isolated from the clinical specimens such as sputum and CSF, after 48–72 h, following which, the colonies can be visualized on the agar plates (Maziarz and Perfect 2016).

1.4.3 Antifungal Drugs and Mechanisms

The *Cryptococcal* infection can be cured according to the intensity of the inflammation. Azoles, polyenes and flucytosine are the antifungal drugs that are used against *Cryptococcus*. Many times these drugs are used in combination for the healing process and to cope with the spreading inflammation (Bermas and Geddes-McAlister 2020). HIV-positive victims with Cryptococcal meningitis are suggested to take two antidotes as primary therapy together with amphotericin B and flucytosine. Sometimes, fluconazole is given for around 8 weeks or for a long period of around 12 months (Srichatrapimuk and Sungkanuparph 2016). Nowadays, echinocandins are used as an antifungal drug that directly targets the cell wall of the fungus but it has no effects on the *Cryptococcus* species. Here, we will discuss the mechanism of these three drugs azoles, polyenes and flucytosine against the *Cryptococcus* and showed diagrammatically in Fig. 1.8.

Polyenes are the first fungicidal agents used against fungal infections. Amphotericin B has a significantly potent effect against *C. neoformans*. Amphotericin B binds to the fungal cell wall and makes the porin passage and helps in the drainage of cytoplasmic elements (Gray et al. 2012). **Flucytosine** is a synthetic compound that has antifungal activity. A combination of 5FC (flucytosine) with amphotericin B has shown more efficacy against *Cryptococcus* as compared to the administration of 5FC (Bermas and Geddes-McAlister 2020). **Azoles** are the antifungal drugs that can be classified based on the presence of 2 and 3 nitrogen in the azole ring (imidazoles or triazoles). Azoles simply affect the ergosterol biosynthesis in fungal cells by inhibiting cytochrome P450-dependent enzyme, lanosterol 14α-sterol demethylase (Sheehan et al. 1999).

1.4.4 Drug Resistance and Mechanisms

Unlike other fungal species, *Cryptococcus* has some special traits that make its treatment more demanding with current antifungal drugs as it shows resistance. When the treatment with conventional drugs is not successful, then *Cryptococcus* fungal pathogen becomes more deadly. A short course of amphotericin B in combination with flucytosine is given to patients for its treatment. Sometimes fluconazole is given as monotherapy, wherein recurrence has been observed in up to 60% of the

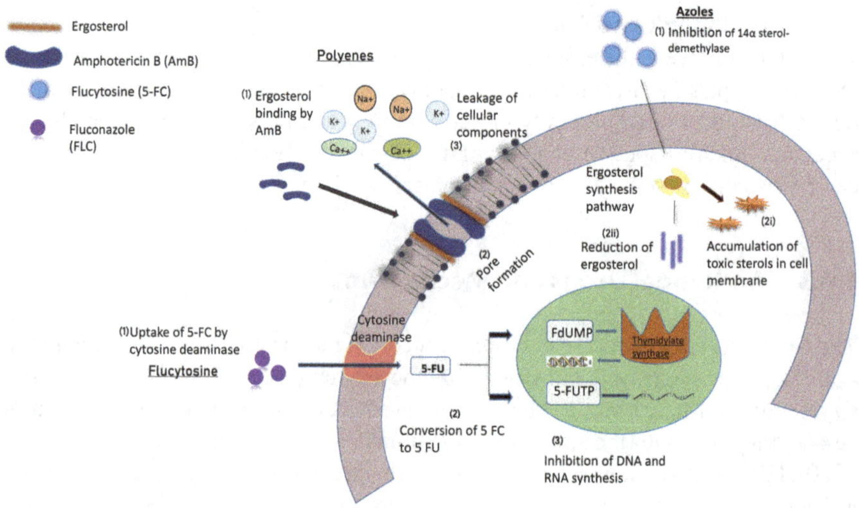

Fig. 1.8 Model showing mechanisms of action for antifungals against *Cryptococcus* (Bermas and Geddes-McAlister 2020)

cases and most of the fluconazole resistance has been reported in such cases (Bicanic et al. 2006). Antifungal drug resistance mechanisms can be of different types. They can be genetic in some fungi or transitory in others. Familial resistance in other fungal species like *Candida* and *Aspergillus* are associated with point mutations, usually in drug transporters and ERG11/CYP51 genes (Flowers et al. 2015). ERG11 single nucleotide polymorphism modifies drug-target interactions and enhances the erg11 protein levels. Moreover, point mutations have also been recognized in other strains but precise drug resistance is still not clearly understood (Stone et al. 2019). On the other hand, transient resistance is recorded in the case of *C. neoformans*, when the drug concentration is reduced, the fluconazole resistance becomes low (Xu et al. 2001).

1.4.5 Virulence and Pathogenicity

In simple words, virulence is the capacity of a pathogen to cause damage and disease in the host (Pirofski and Casadevall 2015). Like other pathogenic fungi and bacteria, *C. neoformans* also possess several virulence factors including secretory enzymes like proteases and lipases (Schaller et al. 2005). Urease is another enzyme secreted by *C. neoformans* that aids in host deterioration and is also considered a virulence factor (Zimmer and Roberts 1979). This particular fungus has two specific virulence factors (capsule and melanin) which play a significant role in protection against human immune responses (Zaragoza 2019).

1.4.5.1 Polysaccharide Capsules

The fungus is enclosed in capsules made up of two types of polysaccharides-glucuronoxylomannogalactan and glucuronoxylomannan (Cherniak et al. 1988; Heiss et al. 2009). Surprisingly capsule formation is very composite, and it depends on the geographical environment too. It develops in a splitting format and attaches to the cell wall of the fungus with the help of covalent bonds (Pierini and Doering 2001). The polysaccharides are synthesized and stored in certain small vesicles that thus play an important role in virulence and disease. They may be involved in affecting *Cryptococcus* virulence during their interaction with the macrophages (Nimrichter et al. 2016).

1.4.5.2 Melanins

These are dark colored polymeric pigments found in almost all types of organisms including animals, plants, bacteria and fungi. In the case of *C. neoformans*, melanin is formed when the conditions are exogenic, precisely in diphenolic amino acid synthesis like L-DOPA (Chaskes and Tyndall 1978). Although, the expression of two genes (LAC1 and LAC2) is responsible for the production of melanin, lac1 is the main protein that synthesizes this pigment (Pukkila-Worley et al. 2005). Melanin synthesis is considered a verifying trait for virulence in this fungus. It is also considered as responsible for the spread of infection from the lung to the brain (Noverr et al. 2004). Melanin plays an essential role in host inflammation during *Cryptococcosis*.

1.4.6 Novel Approaches to Combat MDR

Novel treatment techniques are required to address the difficulties of antifungal resistance and the global incidence of fungal infections. *C. neoformans* is a fungal pathogen that can cause severe infections in immunocompromised individuals. Antifungal resistance in this pathogen is a growing concern, as it can limit treatment options and lead to poor clinical outcomes (Bermas and Geddes-McAlister 2020). However, there are opportunities to overcome antifungal resistance in *C. neoformans* as depicted in Fig. 1.9. One promising approach is to target a nonessential kinase to promote susceptibility to echinocandins, a class of antifungal drugs that are often ineffective against this pathogen due to intrinsic resistance. By inhibiting this kinase, it may be possible to enhance the efficacy of echinocandins and improve treatment outcomes for patients with *C. neoformans* infections. Further research is

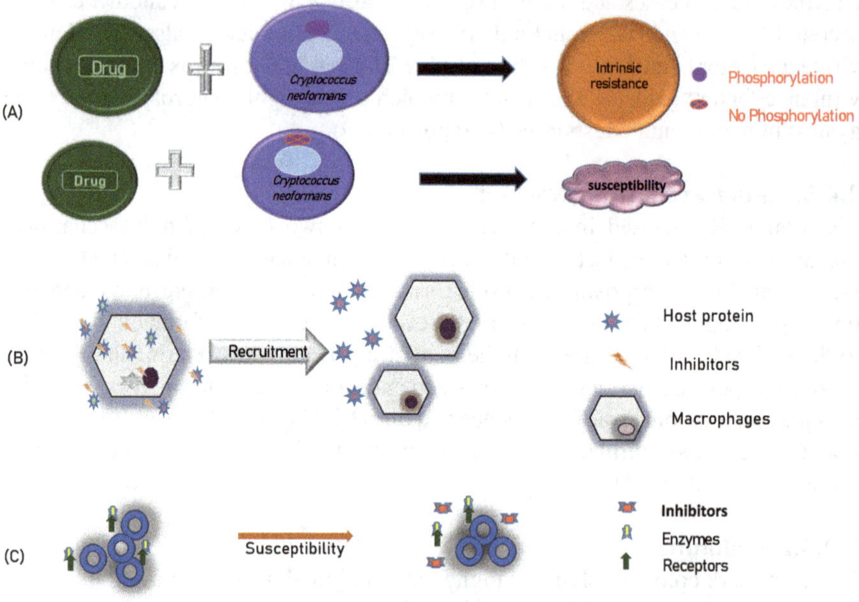

Fig. 1.9 Strategies to overcome antifungal resistance in *C. neoformans* (Bermas and Geddes-McAlister 2020). (**a**) Enhancing the susceptibility of fungal pathogens to echinocandins by inhibiting a nonessential kinase to overcome intrinsic resistance. (**b**) Antivirulence strategy combats fungal infection by disrupting the interaction between virulence factors and host immune response proteins to enhance fungal clearance. (**c**) Enzymatic treatment of biofilms can weaken their protective structure, release internal cells and improve accessibility and susceptibility to current antifungals

needed to fully explore the potential of this approach and identify other strategies for overcoming antifungal resistance in this pathogen (Azevedo et al. 2016).

Antivirulence strategies are emerging as a promising approach to combat fungal infections, including *Cryptococcus*. One such strategy is interfering with the interactions between fungal virulence factors and host immune response proteins. By disrupting these interactions, immune response proteins can be released to recruit additional cells to promote fungal clearance and combat infection. The virulence factors of *Cryptococcus* play a critical role in the pathogenesis of the infection by suppressing the host immune responses. Therefore, targeting these factors can enhance the host immune response and improve the outcome of the infection. Antivirulence strategies can be used to block the interaction between virulence factors and host immune response proteins, allowing the immune response to function normally. This approach has been shown to be effective in preclinical studies, and it holds great promise for the development of new therapies for *Cryptococcus* infections (Vu et al. 2019).

Enzymatic treatment has emerged as a promising strategy to weaken the protective structure of biofilms, enhance the accessibility of internal cells and increase their susceptibility to current antifungals. Enzymes, such as proteases and glycosidases, can be utilized to target the extracellular matrix of biofilms. Proteases break down proteins, while glycosidases degrade complex carbohydrates. By applying these enzymes to the biofilm, they can degrade the matrix, disrupting the structural integrity and weakening the protective barrier. This enzymatic treatment promotes the release of internal cells from the biofilm, making them more accessible to antifungal agents.

Once the biofilm structure is weakened, the internal cells become more susceptible to the action of current antifungals. The increased accessibility allows the antifungal agents to penetrate the biofilm more effectively and reach the fungal cells within. This enhanced susceptibility can lead to improved efficacy of antifungal treatments against *Cryptococcus* biofilms (Upadhya et al. 2016).

1.5 Conclusions

Human fungal diseases are common occurrence and need due attention particularly in a developing country in those people who belong to poor background. Each year, millions of cases of fungal diseases are diagnosed which may become fatal for diseased person. To raise scientific interest and increase global investments into antifungal research, it is pertinent to understand fungal etiology, pathology, epidemiology, disease biology and their economic impact for the betterment of the life of infected people. The development of vaccines for fungal diseases is still an interesting area remained open for all the researchers. The development of vaccines can play a great significance in the prevention of fungal diseases worldwide.

References

Agarwal R, Denning DW, Chakrabarti A (2014) Estimation of the burden of chronic and allergic pulmonary aspergillosis in India. PLoS One 9(12):e114745

Aliyu A (2022) Effects of biofilm formation and plethora of Candida species causing ailments: a mini review. Gadau J Pure Allied Sci 1(2):200–210

Ansari MA, Fatima Z, Hameed S (2014) Sesamol: a natural phenolic compound with promising anticandidal potential. J Pathog 2014:895193

Apgar JM, Wilkening RR, Parker DL, Meng D, Wildonger KJ, Sperbeck D, Greenlee ML, Balkovec JM, Flattery AM, Abruzzo GK, Galgoci AM, Giacobbe RA, Gill CJ, Hsu M-J, Liberator P, Misura AS, Motyl M, Nielsen Kahn J, Powles M et al (2021) Ibrexafungerp: an orally active β-1,3-glucan synthesis inhibitor. Bioorg Med Chem Lett 32:127661

Arafa SH, Elbanna K, Osman GE, Abulreesh HH (2023) Candida diagnostic techniques: a review. J Umm Al-Qura Univ Appl Sci 9:1–18

Arastehfar A, Carvalho A, Houbraken J, Lombardi L, Garcia-Rubio R, Jenks JD, Rivero-Menendez O, Aljohani R, Jacobsen ID, Berman J, Osherov N, Hedayati MT, Ilkit M, Armstrong-James D, Gabaldón T, Meletiadis J, Kostrzewa M, Pan W, Lass-Flörl C et al (2021) Aspergillus fumigatus and aspergillosis: from basics to clinics. Stud Mycol 100:100115

Arendrup MC (2014) Update on antifungal resistance in Aspergillus and Candida. Clin Microbiol Infect 20:42–48

Aruanno M, Glampedakis E, Lamoth F (2019) Echinocandins for the treatment of invasive aspergillosis: from laboratory to bedside. Antimicrob Agents Chemother 63(8):e00399–e00319

Ashu EE, Korfanty GA, Samarasinghe H, Pum N, You M, Yamamura D, Xu J (2018) Widespread amphotericin B-resistant strains of Aspergillus fumigatus in Hamilton, Canada. Infect Drug Resist 11:1549–1555

Azevedo R, Rizzo J, Rodrigues M (2016) Virulence factors as targets for anticryptococcal therapy. J Fungi 2(4):29

Bahn YS (2015) Exploiting fungal virulence-regulating transcription factors as novel antifungal drug targets. PLoS Pathog 11(7):e1004936

Bakkali F, Averbeck S, Averbeck D, Idaomar M (2008) Biological effects of essential oils—a review. Food Chem Toxicol 46(2):446–475

Barantsevich N, Barantsevich E (2022) Diagnosis and treatment of invasive candidiasis. Antibiotics (Basel, Switzerland) 11(6):718

Barnes PD, Marr KA (2006) Aspergillosis: spectrum of disease, diagnosis, and treatment. Infect Dis Clin N Am 20(3):545–561

Bermas A, Geddes-McAlister J (2020) Combatting the evolution of antifungal resistance in Cryptococcus neoformans. Mol Microbiol 114(5):721–734

Bhattacharya S, Sae-Tia S, Fries BC (2020) Candidiasis and mechanisms of antifungal resistance. Antibiotics 9(6):312

Bicanic T, Harrison T, Niepieklo A, Dyakopu N, Meintjes G (2006) Symptomatic relapse of HIV-associated cryptococcal meningitis after initial fluconazole monotherapy: the role of fluconazole resistance and immune reconstitution. Clin Infect Dis 43(8):1069–1073

Birch M, Sibley G (2017) 5.22—Antifungal chemistry review. In: Chackalamannil S, Rotella D, Ward SE (eds) Comprehensive medicinal chemistry III. Elsevier, pp 703–716

Birnbaum JE (1990) Pharmacology of the allylamines. J Am Acad Dermatol 23(4):782–785

Bongomin F, Gago S, Oladele RO, Denning DW (2017) Global and multi-national prevalence of fungal diseases-estimate precision. J Fungi (Basel) 3(4):57

Boral H, Metin B, Döğen A, Seyedmousavi S, Ilkit M (2018) Overview of selected virulence attributes in *Aspergillus fumigatus*, *Candida albicans*, *Cryptococcus neoformans*, *Trichophyton rubrum*, and *Exophiala dermatitidis*. Fungal Genet Biol 111:92–107

Bowman JC, Hicks PS, Kurtz MB, Rosen H, Schmatz DM, Liberator PA, Douglas CM (2002) The antifungal echinocandin caspofungin acetate kills growing cells of Aspergillus fumigatus in vitro. Antimicrob Agents Chemother 46(9):3001–3012

Brown GD, Denning DW, Gow NA, Levitz SM, Netea MG, White TC (2012a) Hidden killers: human fungal infections. Sci Transl Med 4(165):165rv13

Brown GD, Denning DW, Levitz SM (2012b) Tackling human fungal infections. Science 336(6082):647–647

Campitelli M, Zeineddine N, Samaha G, Maslak S (2017) Combination antifungal therapy: a review of current data. J Clin Med Res 9(6):451

Cannon RD, Lamping E, Holmes AR, Niimi K, Baret PV, Keniya MV, Tanabe K, Niimi M, Goffeau A, Monk BC (2009) Efflux-mediated antifungal drug resistance. Clin Microbiol Rev 22(2):291–321

Casadevall A, Perfect JR (1998) Cryptococcus neoformans. ASM Press, Washington, DC

Cavassin FB, Baú-Carneiro JL, Vilas-Boas RR, Queiroz-Telles F (2021) Sixty years of amphotericin B: an overview of the main antifungal agent used to treat invasive fungal infections. Infect Dis Ther 10(1):115–147

Chamilos G, Carvalho A (2020) Aspergillus fumigatus DHN-Melanin. Curr Top Microbiol Immunol 425:17–28

Chang WC, Tzao C, Hsu HH, Lee SC, Huang KL, Tung HJ, Chen CY (2006) Pulmonary cryptococcosis: comparison of clinical and radiographic characteristics in immunocompetent and immunocompromised patients. Chest 129(2):333–340

Chaskes S, Tyndall RL (1978) Pigment production by Cryptococcus neoformans and other Cryptococcus species from aminophenols and diaminobenzenes. J Clin Microbiol 7(2):146–152

Chen SC, Sorrell TC (2007) Antifungal agents. Med J Aust 187(7):404

Cherniak R, Jones RG, Reiss E (1988) Structure determination of Cryptococcus neoformans serotype A-variant glucuronoxylomannan by 13C-n.m.r. spectroscopy. Carbohydr Res 172(1):113–138

Choi J, Kim SH (2017) A genome Tree of Life for the Fungi kingdom. Proc Natl Acad Sci USA 114(35):9391–9396

Croft CA, Culibrk L, Moore MM, Tebbutt SJ (2016) Interactions of Aspergillus fumigatus conidia with airway epithelial cells: a critical review. Front Microbiol 7:472

Denning DW, Cadranel J, Beigelman-Aubry C, Ader F, Chakrabarti A, Blot S, Ullmann AJ, Dimopoulos G, Lange C (2016) Chronic pulmonary aspergillosis: rationale and clinical guidelines for diagnosis and management. Eur Respir J 47(1):45–68

Deok-jong Yoo S, Worodria W, Davis JL, Cattamanchi A, den Boon S, Kyeyune R, Kisembo H, Huang L (2010) The prevalence and clinical course of HIV-associated pulmonary cryptococcosis in Uganda. J Acquir Immune Defic Syndr 54(3):269–274

Dhanasobhon D, Savier E, Lelievre V (2014) To phosphorylate or not to phosphorylate: Selective alterations in tyrosine kinase-inhibited EphB mutant mice. Cell Adhes Migr 8(1):1–4. https://doi.org/10.4161/cam.27478

Diamond RD, Bennett JE (1974) Prognostic factors in cryptococcal meningitis. A study in 111 cases. Ann Intern Med 80(2):176–181

Dirr F, Echtenacher B, Heesemann J, Hoffmann P, Ebel F, Wagener J (2010) AfMkk2 is required for cell wall integrity signaling, adhesion, and full virulence of the human pathogen Aspergillus fumigatus. Int J Med Microbiol 300(7):496–502

Dixon DM, McNeil MM, Cohen ML, Gellin BG, La Montagne JR (1996) Fungal infections: a growing threat. Public Health Rep 111(3):226

Drgona L, Khachatryan A, Stephens J, Charbonneau C, Kantecki M, Haider S, Barnes R (2014) Clinical and economic burden of invasive fungal diseases in Europe: focus on pre-emptive and empirical treatment of Aspergillus and Candida species. Eur J Clin Microbiol Infect Dis 33(1):7–21

Dunkel N, Liu TT, Barker KS, Homayouni R, Morschhäuser J, Rogers PD (2008) A gain-of-function mutation in the transcription factor Upc2p causes upregulation of ergosterol biosynthesis genes and increased fluconazole resistance in a clinical Candida albicans isolate. Eukaryot Cell 7(7):1180–1190

Ellepola AN, Morrison CJ (2005) Laboratory diagnosis of invasive candidiasis. J Microbiol 43(1):65–84

Eschenauer G, DePestel DD, Carver PL (2007) Comparison of echinocandin antifungals. Ther Clin Risk Manag 3(1):71–97

Ferté J (2000) Analysis of the tangled relationships between P-glycoprotein-mediated multidrug resistance and the lipid phase of the cell membrane. Eur J Biochem 267(2):277–294

Flowers SA, Colón B, Whaley SG, Schuler MA, Rogers PD (2015) Contribution of clinically derived mutations in ERG11 to azole resistance in Candida albicans. Antimicrob Agents Chemother 59(1):450–460

Fontaine T, Beauvais A, Loussert C, Thevenard B, Fulgsang CC, Ohno N, Clavaud C, Prevost M-C, Latgé J-P (2010) Cell wall α1–3glucans induce the aggregation of germinating conidia of Aspergillus fumigatus. Fungal Genet Biol 47(8):707–712

Garcia-Solache MA, Casadevall A (2010) Global warming will bring new fungal diseases for mammals. MBio 1(1):10–1128

Gauglitz GG, Callenberg H, Weindl G, Korting HC (2012) Host defence against Candida albicans and the role of pattern-recognition receptors. Acta Derm Venereol 92(3):291–298

Goldman DL, Khine H, Abadi J, Lindenberg DJ, Pirofski LA, Niang R, Casadevall A (2001) Serologic evidence for Cryptococcus neoformans infection in early childhood. Pediatrics 107(5):E66

Graf B, Adam T, Zill E, Göbel UB (2000) Evaluation of the VITEK 2 system for rapid identification of yeasts and yeast-like organisms. J Clin Microbiol 38(5):1782–1785

Gray KC, Palacios DS, Dailey I, Endo MM, Uno BE, Wilcock BC, Burke MD (2012) Amphotericin primarily kills yeast by simply binding ergosterol. Proc Natl Acad Sci USA 109(7):2234–2239

Gugnani HC, Denning DW, Rahim R, Sadat A, Belal M, Mahbub MS (2017) Burden of serious fungal infections in Bangladesh. Eur J Clin Microbiol Infect Dis 36(6):993–997

Guinea J, Torres-Narbona M, Gijón P, Muñoz P, Pozo F, Peláez T et al (2010) Pulmonary aspergillosis in patients with chronic obstructive pulmonary disease: incidence, risk factors, and outcome. Clin Microbiol Infect 16(7):870–877

Gupta L, Hoda S, Vermani M, Vijayaraghavan P (2021) Understanding the fundamental role of virulence determinants to combat Aspergillus fumigatus infections: exploring beyond cell wall. Mycol Prog 20(4):365–380

Gupta L, Sen P, Bhattacharya AK, Vijayaraghavan P (2022) Isoeugenol affects expression pattern of conidial hydrophobin gene RodA and transcriptional regulators MedA and SomA responsible for adherence and biofilm formation in Aspergillus fumigatus. Arch Microbiol 204(4):1–14

Hagiwara D, Watanabe A, Kamei K, Goldman GH (2016) Epidemiological and genomic landscape of azole resistance mechanisms in Aspergillus fungi. Front Microbiol 7:1382

Hani U, Shivakumar HG, Vaghela R, Osmani AM, Shrivastava A (2015) Candidiasis: a fungal infection—current challenges and progress in prevention and treatment. Infect Disord Drug Targets 15(1):42–52

Harvey AL (2007) Natural products as a screening resource. Curr Opin Chem Biol 11(5):480–484

Hashemi SE, Shokohi T, Abastabar M, Aslani N, Ghadamzadeh M, Haghani I (2019) Species distribution and susceptibility profiles of Candida species isolated from vulvovaginal candidiasis, emergence of C. lusitaniae. Curr Med Mycol 5(4):26

Havlickova B, Czaika VA, Friedrich M (2008) Epidemiological trends in skin mycoses worldwide. Mycoses 51:2–15

Heiss C, Klutts JS, Wang Z, Doering TL, Azadi P (2009) The structure of Cryptococcus neoformans galactoxylomannan contains beta-D-glucuronic acid. Carbohydr Res 344(7):915–920. https://doi.org/10.1016/j.carres.2009.03.003

Hoekstra WJ, Garvey EP, Moore WR, Rafferty SW, Yates CM, Schotzinger RJ (2014) Design and optimization of highly-selective fungal CYP51 inhibitors. Bioorg Med Chem Lett 24(15):3455–3458

Houšť J, Spížek J, Havlíček V (2020) Antifungal drugs. Metabolites 10(3):106

Hua Y, Dai X, Xu Y, Xing G, Liu H, Lu T et al (2022) Drug repositioning: progress and challenges in drug discovery for various diseases. Eur J Med Chem 234:114239

Ilkit M, Guzel AB (2011) The epidemiology, pathogenesis, and diagnosis of vulvovaginal candidosis: a mycological perspective. Crit Rev Microbiol 37(3):250–261

Ivanov M, Ćirić A, Stojković D (2022) Emerging antifungal targets and strategies. Int J Mol Sci 23(5):2756

Jabeen K, Farooqi J, Mirza S, Denning D, Zafar A (2017) Serious fungal infections in Pakistan. Eur J Clin Microbiol Infect Dis 36(6):949–956. https://doi.org/10.1007/s10096-017-2919-6

Jasinski M, Ducos E, Martinoia E, Boutry M (2003) The ATP-binding cassette transporters: structure, function, and gene family comparison between rice and Arabidopsis. Plant Physiol 131(3):1169–1177

Jayasekera P, Denning D, Perera P, Fernando A, Kudavidanage S (2013) The burden of serious fungal infections in Sri Lanka. Mycoses 56:103–103

Johnson MD, Perfect JR (2010) Use of antifungal combination therapy: agents, order, and timing. Curr Fungal Infect Rep 4(2):87–95

Kanafani ZA, Perfect JR (2008) Resistance to antifungal agents: mechanisms and clinical impact. Clin Infect Dis 46(1):120–128

Katragkou A, Roilides E, Walsh TJ (2016) Can repurposing of existing drugs provide more effective therapies for invasive fungal infections? Expert Opin Pharmacother 17(9):1179–1182

Khwakhali US, Denning DW (2015) Burden of serious fungal infections in Nepal. Mycoses 58(Suppl 5):45–50

Kim J, Sudbery P (2011) Candida albicans, a major human fungal pathogen. J Microbiol 49:171–177

Komase Y, Kunishima H, Yamaguchi H, Ikehara M, Yamamoto T, Shinagawa T (2007) Rapidly progressive invasive pulmonary aspergillosis in a diabetic man. J Infect Chemother 13(1):46–50

Koselny K, Green J, DiDone L, Halterman JP, Fothergill AW, Wiederhold NP, Patterson TF, Cushion MT, Rappelye C, Wellington M, Krysan DJ (2016) The celecoxib derivative AR-12 has broad-spectrum antifungal activity in vitro and improves the activity of fluconazole in a murine model of cryptococcosis. Antimicrob Agents Chemother 60(12):7115–7127

Kowalski CH, Morelli KA, Schultz D, Nadell CD, Cramer RA (2020) Fungal biofilm architecture produces hypoxic microenvironments that drive antifungal resistance. Proc Natl Acad Sci USA 117(36):22473–22483

Lamping E, Monk BC, Niimi K, Holmes AR, Tsao S, Tanabe K et al (2007) Characterization of three classes of membrane proteins involved in fungal azole resistance by functional hyperexpression in Saccharomyces cerevisiae. Eukaryot Cell 6(7):1150–1165

Laniado-Laborín R, Cabrales-Vargas MN (2009) Amphotericin B: side effects and toxicity. Rev Iberoam Micol 26(4):223–227

Liebmann B, Müller M, Braun A, Brakhage AA (2004) The cyclic AMP-dependent protein kinase A network regulates development and virulence in Aspergillus fumigatus. Infect Immun 72(9):5193–5203

Limper AH, Adenis A, Le T, Harrison TS (2017) Fungal infections in HIV/AIDS. Lancet Infect Dis 17(11):e334–e343

Lindsley MD, Mekha N, Baggett HC, Surinthong Y, Autthateinchai R, Sawatwong P, Harris JR, Park BJ, Chiller T, Balajee SA, Poonwan N (2011) Evaluation of a newly developed lateral flow immunoassay for the diagnosis of cryptococcosis. Clin Infect Dis 53(4):321–325

Lockhart SR, Berkow EL (2019) Antifungal susceptibility testing: the times they are a-changing. Clin Microbiol Newsl 41(10):85–90

Lorand T, Kocsis B (2007) Recent advances in antifungal agents. Mini Rev Med Chem 7(9):900–911

Mayer FL, Wilson D, Hube B (2013) Candida albicans pathogenicity mechanisms. Virulence 4(2):119–128

Maziarz EK, Perfect JR (2016) Cryptococcosis. Infect Dis Clin N Am 30(1):179–206. https://doi.org/10.1016/j.idc.2015.10.006

McCarty TP, Pappas PG (2021) Antifungal pipeline. Front Cell Infect Microbiol 11:732223

McDonald T, Wiesner DL, Nielsen K (2012) Cryptococcus. Curr Biol 22(14):R554–R555. https://doi.org/10.1016/j.cub.2012.05.040

Meneau I, Coste AT, Sanglard D (2016) Identification of Aspergillus fumigatus multidrug transporter genes and their potential involvement in antifungal resistance. Med Mycol 54(6):616–627

Miyazaki M, Horii T, Hata K, Watanabe N, Nakamoto K, Tanaka K, Shirotori S, Murai N, Inoue S, Matsukura M, Abe S, Yoshimatsu K, Asada M (2011) In vitro activity of E1210, a novel antifungal, against clinically important yeasts and molds. Antimicrob Agents Chemother 55(10):4652–4658

Mousavi B, Hedayati M, Hedayati N, Ilkit M, Syedmousavi S (2016) Aspergillus species in indoor environments and their possible occupational and public health hazards. Curr Med Mycol 2(1):36–42

Moyes DL, Richardson JP, Naglik JR (2015) Candida albicans-epithelial interactions and pathogenicity mechanisms: scratching the surface. Virulence 6(4):338–346

Mukhopadhyay R, Rosen BP, Phung LT, Silver S (2002) Microbial arsenic: from geocycles to genes and enzymes. FEMS Microbiol Rev 26(3):311–325

Nakamoto K, Tsukada I, Tanaka K, Matsukura M, Haneda T, Inoue S, Murai N, Abe S, Ueda N, Miyazaki M, Watanabe N, Asada M, Yoshimatsu K, Hata K (2010) Synthesis and evaluation of novel antifungal agents—quinoline and pyridine amide derivatives. Bioorg Med Chem Lett 20(15):4624–4626

Nguyen S, Truong JQ, Bruning JB (2021) Targeting unconventional pathways in pursuit of novel antifungals. Front Mol Biosci 7:621366

Nimrichter L, de Souza MM, Del Poeta M, Nosanchuk JD, Joffe L, Tavares Pde M, Rodrigues ML (2016) Extracellular vesicle-associated transitory cell wall components and their impact on the interaction of fungi with host cells. Front Microbiol 7:1034

Nobile CJ, Johnson AD (2015) Candida albicans biofilms and human disease. Annu Rev Microbiol 69:71–92

Nobile CJ, Fox EP, Nett JE, Sorrells TR, Mitrovich QM, Hernday AD, Tuch BB, Andes DR, Johnson AD (2012) A recently evolved transcriptional network controls biofilm development in Candida albicans. Cell 148(1–2):126–138

Noverr MC, Williamson PR, Fajardo RS, Huffnagle GB (2004) CNLAC1 is required for extrapulmonary dissemination of Cryptococcus neoformans but not pulmonary persistence. Infect Immun 72(3):1693–1699

Oliver JD, Sibley GEM, Beckmann N, Dobb KS, Slater MJ, McEntee L, du Pré S, Livermore J, Bromley MJ, Wiederhold NP, Hope WW, Kennedy AJ, Law D, Birch M (2016) F901318 represents a novel class of antifungal drug that inhibits dihydroorotate dehydrogenase. Proc Natl Acad Sci USA 113(45):12809–12814

Onishi J, Meinz M, Thompson J, Curotto J, Dreikorn S, Rosenbach M et al (2000) Discovery of novel antifungal (1,3)-β-D-glucan synthase inhibitors. Antimicrob Agents Chemother 44(2):368–377

Ostrowsky B, Greenko J, Adams E, Quinn M, O'Brien B, Chaturvedi V, Berkow E, Vallabhaneni S, Forsberg K, Chaturvedi S, Lutterloh E, Blog D, C. auris Investigation Work Group (2020) Candida auris isolates resistant to three classes of antifungal medications—New York, 2019. MMWR Morb Mortal Wkly Rep 69(1):6–9

Perea S, López-Ribot JL, Kirkpatrick WR, McAtee RK, Santillán RA, Martínez M et al (2001) Prevalence of molecular mechanisms of resistance to azole antifungal agents in Candida albicans strains displaying high-level fluconazole resistance isolated from human immunodeficiency virus-infected patients. Antimicrob Agents Chemother 45(10):2676–2684

Perea S, Gonzalez G, Fothergill AW, Sutton DA, Rinaldi MG (2002) In vitro activities of terbinafine in combination with fluconazole, itraconazole, voriconazole, and posaconazole against clinical isolates of Candida glabrata with decreased susceptibility to azoles. J Clin Microbiol 40(5):1831–1833

Perlin DS (2015) Mechanisms of echinocandin antifungal drug resistance. Ann N Y Acad Sci 1354(1):1–11. https://doi.org/10.1111/nyas.12831

Pfaller MA, Preston T, Bale M, Koontz FP, Body BA (1988) Comparison of the Quantum II, API Yeast Ident, and AutoMicrobic systems for identification of clinical yeast isolates. J Clin Microbiol 26(10):2054–2058

Pfaller MA, Diekema DJ, Gibbs DL, Newell VA, Meis JF, Gould IM et al (2007) Results from the ARTEMIS DISK Global Antifungal Surveillance study, 1997 to 2005: an 8.5-year analysis

of susceptibilities of Candida species and other yeast species to fluconazole and voriconazole determined by CLSI standardized disk diffusion testing. J Clin Microbiol 45(6):1735–1745
Pfaller MA, Diekema DJ, Turnidge JD, Castanheira M, Jones RN (2019) Twenty years of the SENTRY antifungal surveillance program: results for Candida species from 1997–2016. In: Open forum infectious diseases, vol 6(Suppl_1). Oxford University Press, pp S79–S94
Pierini LM, Doering TL (2001) Spatial and temporal sequence of capsule construction in Cryptococcus neoformans. Mol Microbiol 41(1):105–115
Pirofski LA, Casadevall A (2015) What is infectiveness and how is it involved in infection and immunity? BMC Immunol 16:13
Polak A, Scholer HJ (1975) Mode of action of 5-fluorocytosine and mechanisms of resistance. Chemotherapy 21(3–4):113–130
Prasad R, Goffeau A (2012) Yeast ATP-binding cassette transporters conferring multidrug resistance. Ann Rev Microbiol 66:39–63
Price CL, Parker JE, Warrilow AG, Kelly DE, Kelly SL (2015) Azole fungicides—understanding resistance mechanisms in agricultural fungal pathogens. Pest Manag Sci 71(8):1054–1058
Pukkila-Worley R, Gerrald QD, Kraus PR, Boily MJ, Davis MJ, Giles SS, Cox GM, Heitman J, Alspaugh JA (2005) Transcriptional network of multiple capsule and melanin genes governed by the Cryptococcus neoformans cyclic AMP cascade. Eukaryot Cell 4(1):190–201
Rajasingham R, Smith RM, Park BJ, Jarvis JN, Govender NP, Chiller TM, Denning DW, Loyse A, Boulware DR (2017) Global burden of disease of HIV-associated cryptococcal meningitis: an updated analysis. Lancet Infect Dis 17(8):873–881
Ravikant KT, Gupte S, Kaur M (2015) A review on emerging fungal infections and their significance. J Bacteriol Mycol Open Access 1(2):00009
Rudramurthy SM, Paul RA, Chakrabarti A, Mouton JW, Meis JF (2019) Invasive aspergillosis by Aspergillus flavus: epidemiology, diagnosis, antifungal resistance, and management. J Fungi 5(3):Article 3
Saha DC, Goldman DL, Shao X, Casadevall A, Husain S, Limaye AP, Lyon M, Somani J, Pursell K, Pruett TL, Singh N (2007) Serologic evidence for reactivation of cryptococcosis in solid-organ transplant recipients. Clin Vaccine Immunol 14(12):1550–1554
Saibabu V, Fatima Z, Khan LA, Hameed S (2015) Therapeutic potential of dietary phenolic acids. Adv Pharmacol Pharm Sci 2015:823539
Sanglard D, Coste A, Ferrari S (2009) Antifungal drug resistance mechanisms in fungal pathogens from the perspective of transcriptional gene regulation. FEMS Yeast Res 9(7):1029–1050
Schaller M, Borelli C, Korting HC, Hube B (2005) Hydrolytic enzymes as virulence factors of Candida albicans. Mycoses 48(6):365–377
Scorzoni L, de Paula e Silva ACA, Marcos CM, Assato PA, de Melo WCMA, de Oliveira HC, Costa-Orlandi CB, Mendes-Giannini MJS, Fusco-Almeida AM (2017) Antifungal therapy: new advances in the understanding and treatment of mycosis. Front Microbiol 8:36
Segarra I, Movshin DA, Zarif L (2002) Pharmacokinetics and tissue distribution after intravenous administration of a single dose of amphotericin B cochleates, a new lipid-based delivery system. J Pharm Sci 91(8):1827–1837
Sen P, Gupta L, Vijay M, Vermani Sarin M, Shankar J, Hameed S, Vijayaraghavan P (2023) 4-Allyl-2-methoxyphenol modulates the expression of genes involved in efflux pump, biofilm formation and sterol biosynthesis in azole resistant Aspergillus fumigatus. Front Cell Infect Microbiol 13:1103957
Shapiro RS, Robbins N, Cowen LE (2011) Regulatory circuitry governing fungal development, drug resistance, and disease. Microbiol Mol Biol Rev 75(2):213–267
Sheehan DJ, Hitchcock CA, Sibley CM (1999) Current and emerging azole antifungal agents. Clin Microbiol Rev 12(1):40–79
Shibata T, Takahashi T, Yamada E, Kimura A, Nishikawa H, Hayakawa H, Nomura N, Mitsuyama J (2012) T-2307 causes collapse of mitochondrial membrane potential in yeast. Antimicrob Agents Chemother 56(11):5892–5897

Shibuya K, Coulson WF, Naoe S (2002) Histopathology of deep-seated fungal infections and detailed examination of granulomatous response against cryptococci in patients with acquired immunodeficiency syndrome. Nippon Ishinkin Gakkai Zasshi 43(3):143–151

Shlezinger N, Irmer H, Dhingra S, Beattie SR, Cramer RA, Braus GH, Sharon A, Hohl TM (2017) Sterilizing immunity in the lung relies on targeting fungal apoptosis-like programmed cell death. Science (New York, N.Y.) 357(6355):1037–1041

Singh S, Fatima Z, Ahmad K, Hameed S (2020) Repurposing of respiratory drug theophylline against Candida albicans: mechanistic insights unveil alterations in membrane properties and metabolic fitness. J Appl Microbiol 129(4):860–875

Skaggs BA, Alexander JF, Pierson CA, Schweitzer KS, Chun KT, Koegel C, Barbuch R, Bard M (1996) Cloning and characterization of the Saccharomyces cerevisiae C-22 sterol desaturase gene, encoding a second cytochrome P-450 involved in ergosterol biosynthesis. Gene 169(1):105–109

Speth C, Rambach G, Lass-Flörl C, Howell PL, Sheppard DC (2019) Galactosaminogalactan (GAG) and its multiple roles in Aspergillus pathogenesis. Virulence 10(1):976–983

Srichatrapimuk S, Sungkanuparph S (2016) Integrated therapy for HIV and cryptococcosis. AIDS Res Ther 13(1):42

Stone NR, Rhodes J, Fisher MC, Mfinanga S, Kivuyo S, Rugemalila J, Segal ES, Needleman L, Molloy SF, Kwon-Chung J, Harrison TS, Hope W, Berman J, Bicanic T (2019) Dynamic ploidy changes drive fluconazole resistance in human cryptococcal meningitis. J Clin Invest 129(3):999–1014

Tarang S, Kesherwani V, LaTendresse B, Lindgren L, Rocha-Sanchez SM, Weston MD (2020) In silico design of a multivalent vaccine against Candida albicans. Sci Rep 10(1):1066

Tashiro M, Izumikawa K, Hirano K, Ide S, Mihara T, Hosogaya N, Takazono T, Morinaga Y, Nakamura S, Kurihara S, Imamura Y, Miyazaki T, Nishino T, Tsukamoto M, Kakeya H, Yamamoto Y, Yanagihara K, Yasuoka A, Tashiro T, Kohno S (2012) Correlation between triazole treatment history and susceptibility in clinically isolated Aspergillus fumigatus. Antimicrob Agents Chemother 56(9):4870–4875

Tassel D, Madoff MA (1968) Treatment of Candida sepsis and Cryptococcus meningitis with 5-fluorocytosine. A new antifungal agent. JAMA 206(4):830–832

Thomford NE, Senthebane DA, Rowe A, Munro D, Seele P, Maroyi A, Dzobo K (2018) Natural products for drug discovery in the 21st century: innovations for novel drug discovery. Int J Mol Sci 19(6):1578

Thompson DS, Carlisle PL, Kadosh D (2011) Coevolution of morphology and virulence in Candida species. Eukaryot Cell 10(9):1173–1182

Tomee JF, Kauffman HF (2000) Putative virulence factors of Aspergillus fumigatus. Clin Exp Allergy 30(4):476–484

Upadhya R, Lam WC, Maybruck B, Specht CA, Levitz SM, Lodge JK (2016) Induction of protective immunity to cryptococcal infection in mice by a heat-killed, chitosan-deficient strain of Cryptococcus neoformans. MBio 7(3):10–1128

Upadhyay SK, Mahajan L, Ramjee S, Singh Y, Basir SF, Madan T (2009) Identification and characterization of a laminin-binding protein of Aspergillus fumigatus: extracellular thaumatin domain protein (AfCalAp). J Med Microbiol 58(Pt 6):714–722

Valsecchi I, Dupres V, Stephen-Victor E, Guijarro JI, Gibbons J, Beau R, Bayry J, Coppee J-Y, Lafont F, Latgé J-P, Beauvais A (2017) Role of hydrophobins in Aspergillus fumigatus. J Fungi 4(1):2

Vandeputte P, Ferrari S, Coste AT (2012) Antifungal resistance and new strategies to control fungal infections. Int J Microbiol 2012:1

Vázquez-González D, Perusquía-Ortiz AM, Hundeiker M, Bonifaz A (2013) Opportunistic yeast infections: candidiasis, cryptococcosis, trichosporonosis and geotrichosis. J Dtsch Dermatol Ges 11(5):381–394

Vermes A, Guchelaar HJ, Dankert J (2000) Flucytosine: a review of its pharmacology, clinical indications, pharmacokinetics, toxicity and drug interactions. J Antimicrob Chemother 46(2):171–179

Vicariotto F, Del Piano M, Mogna L, Mogna G (2012) Effectiveness of the association of 2 probiotic strains formulated in a slow release vaginal product, in women affected by vulvovaginal candidiasis: a pilot study. J Clin Gastroenterol 46:S73–S80

Viegas C, Dias M, Carolino E, Sabino R (2021) Culture media and sampling collection method for Aspergillus spp. assessment: tackling the gap between recommendations and the scientific evidence. Atmosphere 12(1):Article 1

Vu K, Garcia JA, Gelli A (2019) Cryptococcal meningitis and anti-virulence therapeutic strategies. Front Microbiol 10:353

Wagener J, Echtenacher B, Rohde M, Kotz A, Krappmann S, Heesemann J, Ebel F (2008) The putative alpha-1,2-mannosyltransferase AfMnt1 of the opportunistic fungal pathogen Aspergillus fumigatus is required for cell wall stability and full virulence. Eukaryot Cell 7(10):1661–1673

Waldorf AR, Polak A (1983) Mechanisms of action of 5-fluorocytosine. Antimicrob Agents Chemother 23(1):79–85

Watanabe M, Goto K, Sugita-Konishi Y, Kamata Y, Hara-Kudo Y (2012) Sensitive detection of whole-genome differentiation among closely-related species of the genus Fusarium using DNA-DNA hybridization and a microplate technique. J Vet Med Sci 74(10):1333–1336

White JK, Nielsen JL, Poulsen JS, Madsen AM (2021) Antifungal resistance in isolates of Aspergillus from a pig farm. Atmosphere 12(7):Article 7. https://doi.org/10.3390/atmos12070826

Wu TC, Koo SY (1983) Comparison of three commercial cryptococcal latex kits for detection of cryptococcal antigen. J Clin Microbiol 18(5):1127–1130

Xu J, Onyewu C, Yoell HJ, Ali RY, Vilgalys RJ, Mitchell TG (2001) Dynamic and heterogeneous mutations to fluconazole resistance in Cryptococcus neoformans. Antimicrob Agents Chemother 45(2):420–427

Yamashita K, Miyazaki T, Fukuda Y, Mitsuyama J, Saijo T, Shimamura S, Yamamoto K, Imamura Y, Izumikawa K, Yanagihara K, Kohno S, Mukae H (2019) The novel arylamidine T-2307 selectively disrupts yeast mitochondrial function by inhibiting respiratory chain complexes. Antimicrob Agents Chemother 63(8):e00374–e00319

Yan X, Li M, Jiang M, Zou L, Luo F, Jiang Y (2009) Clinical characteristics of 45 patients with invasive pulmonary aspergillosis. Cancer 115(21):5018–5025

Yang B, Rao R (2018) Emerging pathogens of the Candida species. In: *Candida albicans*. IntechOpen

Zaragoza O (2019) Basic principles of the virulence of Cryptococcus. Virulence 10(1):490–501

Zhan P, Liu W (2017) The changing face of dermatophytic infections worldwide. Mycopathologia 182(1–2):77–86

Zhen C, Lu H, Jiang Y (2022) Novel promising antifungal target proteins for conquering invasive fungal infections. Front Microbiol 13:911322

Zimmer BL, Roberts GD (1979) Rapid selective urease test for presumptive identification of Cryptococcus neoformans. J Clin Microbiol 10(3):380–381

Zotchev SB (2003) Polyene macrolide antibiotics and their applications in human therapy. Curr Med Chem 10(3):211–223

Part II

Plant Extracts and Essential Oils as Antifungals

Composition of Different Species of Lamiaceae Plant Family: A Potential Biosource of Terpenoids and Antifungal Agents

2

Rosa Perestrelo, Patrícia Sousa, Nance Hontman, and José S. Câmara

2.1 Introduction

The Lamiaceae family is widely distributed around the world, and many of its species are grown for their aromatic leaves and fascinating flowers. This family is known for having over 200 genera and 6000 species, which are mostly found in Asia and the Mediterranean and have long been used in traditional medicine. *Lavandula angustifolia* Mill. (known as lavender), *Mentha piperita* L. (peppermint), *Origanum vulgare* L. (oregano), *Rosmarinus officinalis* L. (rosemary), and *Thymus vulgaris* L. (thyme) are the most well-known representative of the Lamiaceae family (Rodríguez-Solana et al. 2015; Çelik et al. 2021; Gladikostić et al. 2023). Their essential oils and/or extracts have been employed as anticancer, antioxidant, anti-inflammatory, and antifungal agents; they have also been used as traditional tea and sweeteners (Hassanein et al. 2020; Çelik et al. 2021). Essential oils are described internationally as complex mixtures of chemical compounds produced by plants' secondary metabolism, such as monoterpenoids, sesquiterpenoids, diterpenoids, triterpenoids, phenylpropanoids, and isothiocyanates, among others. Nevertheless, the main composition of essential oils of lavender, peppermint, oregano, rosemary, and thyme are typically monoterpenes, sesquiterpenes, and their oxygenated derivatives (Table 2.1). These terpenoids vary in concentration, and some can account for up to 70% of the total essential oil, making them main components and hence in charge

R. Perestrelo (✉) · P. Sousa · N. Hontman
CQM—Centro de Química da Madeira, Universidade da Madeira, Funchal, Portugal
e-mail: rmp@staff.uma.pt; patricia.sousa@staff.uma.pt; nance.8@hotmail.com

J. S. Câmara
CQM—Centro de Química da Madeira, Universidade da Madeira, Funchal, Portugal

Departamento de Química, Faculdade de Ciências Exatas e Engenharia, Universidade da Madeira, Funchal, Portugal
e-mail: jsc@staff.uma.pt

N. Manzoor (ed.), *Advances in Antifungal Drug Development*,
https://doi.org/10.1007/978-981-97-5165-5_2

Table 2.1 The main terpenoids and minimal inhibitory concentration (MICs) against fungi of Lamiaceae plant essential oils

Plant	Terpenoids		Targeted fungus	MICs (μg/mL)	References
Lavandula angustifolia Mill.	Monoterpenoids	Linalool (7.72–52.6%), lavandulyl acetate (1.32–11.9%), camphor (0.15–16.5%), linalyl acetate (9.27–25.7%), borneol (6.68–10.4%)	*Candida albicans*	10.3–512	Danh et al. (2013), Khoury et al. (2016), Dolatabadi et al. (2019), Wesołowska et al. (2019), Karpiński (2020), Guo and Wang (2020), El-Kharraf et al. (2022), Nedeltcheva-Antonova et al. (2022)
	Sesquiterpenoids	Caryophyllene (0.30–6.67%), β-farnesene (1.37–4.50%), germacrene D (0.13–0.91%), caryophyllene oxide (0.22–1.50%)			
	Diterpenoids	Phytol (0.21%), camphorene (0.26%)			
	Triterpenoids	Lupeol acetate (0.06–0.21%), α-amyrin (0.03–0.14%), β-amyrin (0.03–0.10%)			
Mentha piperita L.	Monoterpenoids	Menthol (3.69–51.0%), menthone (2.65–26.4%), 1,8-cineole (2.27–8.50%), menthyl acetate (0.10–5.41%), pulegone (1.13–5.60%)	*Aspergillus flavus* *Aspergillus niger* *Candida albicans* *Cryptococcus neoformans* *Penicillium minioluteum*	1450–5000 625–10,000 225–1125 313 2050–2200	Hossain et al. (2014, 2016), Camiletti et al. (2014), Rodríguez-Solana et al. (2015), Samber et al. (2015), Gavahian et al. (2015), Powers et al. (2018), Karpiński (2020), Hassanein et al. (2020), Hassan et al. (2023)
	Sesquiterpenoids	Caryophyllene (3.52–4.47%), germacrene D (0–3.42%)			
	Diterpenoids	Phytol (0.48–4.67%)			
	Triterpenoids	α-Amyrin (4.80%), squalene (0.97%)			
Origanum vulgare L.	Monoterpenoids	Carvacrol (3.10–91.2%), sabinene (0.01–7.70%), (*E*)-sabinene hydrate (0.01–1.43%), thymol (0.01–15.9%), γ-terpinene (4.90–10.6%)	*Aspergillus flavus* *Aspergillus niger* *Candida albicans* *Cryptococcus neoformans* *Penicillium ochrochlorom*	0.64–2500 0.32–623 0.32–700 0.16–78 710	Vale-Silva et al. (2012), Rodríguez-Solana et al. (2015), Hossain et al. (2016), Ahamad et al. (2018), Powers et al. (2018), Elansary et al. (2018), Jan et al. (2020), Karpiński (2020), Simirgiotis et al. (2020), Soltani et al. (2021), Ilić et al. (2022), Abdelghffar et al. (2022)
	Sesquiterpenoids	β-Caryophyllene (0.01–1.45%), caryophyllene oxide (0–12.5%), germacrene D (17.4–22.5%)			
	Diterpenoids	Cembrene (0.06%), phytol (2.03%)			
	Triterpenoids	α-Amyrin (4.89%), β-amyrin (5.24%), β-sitosterol (1.30%), lupeol (3.15%)			

Rosmarinus officinalis L.	Monoterpenoids	1,8-Cineole (7.28–55.3%), α-pinene (9.90–39.0%), camphor (2.10–26.8%), α-terpineol (0.47–11.0%), β-pinene (0.38–13.3%)	*Aspergillus flavus* *Aspergillus niger* *Candida albicans* *Cryptococcus neoformans* *Penicillium ochrochlorom*	330 380–10,000 7.5–1000 313 470	Jiang et al. (2011), Rodríguez-Solana et al. (2015), Powers et al. (2018), Elansary et al. (2018), Ali et al. (2019), Sadeh et al. (2019), Karpiński (2020), Ferreira et al. (2020), Binzet et al. (2020), Ibrahim et al. (2022), Sakar et al. (2023), Hashemi et al. (2023)
	Sesquiterpenoids	β-Caryophellene (0.10–5.60%), caryophyllene oxide (0.20–0.30%)			
	Diterpenoids	Carnosol (11.7–17.3%), carnosic acid (1.09–3%), phytol (nd–3.28%)			
	Triterpenoids	α-Amyrin (20.6%)			
Thymus vulgaris L.	Monoterpenoids	Thymol (50.5–55.8%), *p*-cymene (3.21–20.6%), linalool (1.50–48.2%), carvacrol (2.30–21.3%), γ-terpinene (1.37–15.2%)	*Aspergillus flavus* *Aspergillus niger* *Candida albicans* *Cryptococcus neoformans* *Penicillium chrysogenum*	9.35–1500 9.35–1250 0.16–313 78 312.5–1750	Hossain et al. (2016), Sharifzadeh et al. (2016), Powers et al. (2018), Wesołowska and Jadczak (2019), Gedikoğlu et al. (2019), Karpiński (2020), Palmieri et al. (2020), Pandur et al. (2022), Arafa et al. (2022)
	Sesquiterpenoids	Caryophyllene (1.56–4.88%), caryophyllene oxide (0.27–0.97%), germacrene D (0.04–0.11%), β-bisabolene (0.67–2.00%)			
	Diterpenoids	Phytol (0.05–0.08%), geranylgeraniol (3.33%)			
	Triterpenoids	Squalene (3.60%)			

Abbreviations: *MICs* minimum inhibitory concentration values, *nd* not detected of the essential oils claimed antifungal activity (Jiang et al. 2011; Samber et al. 2015; Gavahian et al. 2015; Powers et al. 2018; Lešnik et al. 2021). To overcome the inherent variability of essential oil composition caused by variations in geographical origin, extraction methods, and storage times, as well as to better understand the cell targets of these molecules, the antifungal activity of individual terpenoids of these oils has also been studied (Vale-Silva et al. 2012; Khoury et al. 2016). These investigations suggested that the terpenoids present in several essential oils are responsible for their antifungal properties against *Aspergillus flavus*, *Aspergillus niger*, *Candida albicans*, *Cryptococcus neoformans*, and *Penicillium ochrochlorom* (Table 2.1). In particular, the effectiveness of lavender, peppermint, rosemary, and thyme essential oils was associated with thymol, 1,8-cineole, linalool, camphor, and carvacrol, among other terpenoids (Vale-Silva et al. 2012; Camiletti et al. 2014; Khoury et al. 2016; Powers et al. 2018). In this context, in this chapter, we will explore the information related to the terpenoids identified in Lamiaceae plants, namely lavender, peppermint, oregano, rosemary, and thyme, as well as the most used extraction techniques to obtain the essential oil, in addition to their antifungal properties specifying the minimum inhibitory concentration (MICs). Moreover, the mechanisms of action of antifungal agents are also discussed.

2.2 Terpenoids Identified in Plant Extracts

Terpenoids are a large family of organic compounds characterized by C5 units whose structure is based on the repeating unit of isoprene. The terpenoid building blocks isopentenyl diphosphate (IPP) and dimethylallyl diphosphate (DMAPP) are produced by two metabolic pathways: the mevalonate pathway and the non-mevalonate pathway, also known as the 2-C-methyl-D-erythritol 4-phosphate/1-deoxy-D-xylulose 5-phosphate (MEP/DOXP) pathway. Prenyl transferases then produce higher-order building blocks, such as geranyl diphosphate, farsenyl diphosphate, and geranylgeranyl diphosphate as shown in Fig. 2.1, which are precursors to monoterpenoids (C10), sesquiterpenoids (C15), and diterpenoids (C20), respectively. These building blocks are converted into the precursors of sterols (C30) and carotenoids (C40), respectively (Lešnik et al. 2021; Chandra et al. 2022).

As can be observed in Table 2.1, the chemical composition of essential oils obtained from lavender, peppermint, oregano, rosemary, and thyme is mainly constituted by terpenoids, and several studies have explored these essential oils as potential antifungal agents (Vale-Silva et al. 2012; Samber et al. 2015; Hossain et al. 2016; Sharifzadeh et al. 2016; Powers et al. 2018). Powers and collaborators (2018) screened the antifungal activity of 60 essential oils from medicinal plants against *Aspergillus niger*, *Candida albicans*, and *Cryptococcus neoformans*. The data obtained demonstrated that lavender, peppermint, and rosemary showed MIC values of 625 μg/mL against *A. niger* and *C. albicans*, while thyme showed MIC values of 156 and 313 μg/mL, respectively. On the other hand, Jiang et al. (2011) assessed the antifungal activity of rosemary essential oil and the two main terpenoid compounds, namely α-pinene and 1,8-cineole, against *C. albicans* and *A. niger*. The

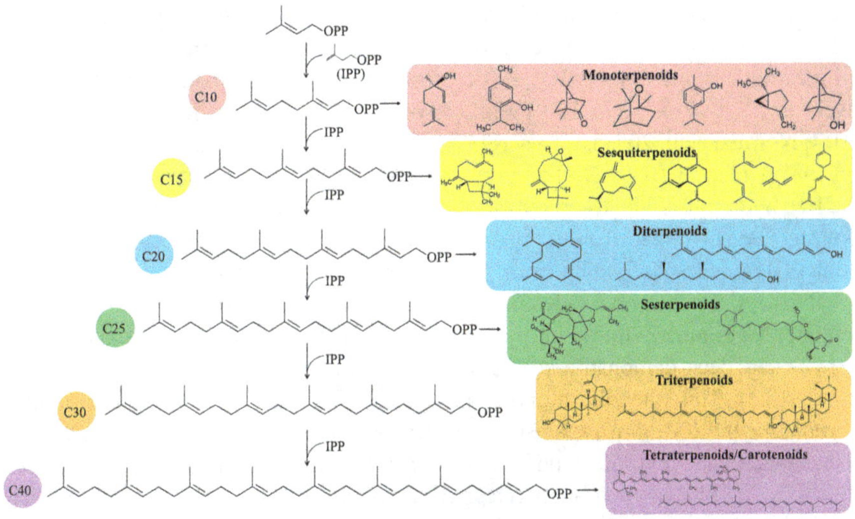

Fig. 2.1 Schematic representation of terpenoid biosynthesis

MICs for the investigated fungi varied from 0.1% (v/v) to 1.0% (v/v) for the rosemary essential oil, and from 0.5% (v/v) to 4.0% (v/v) for α-pinene and 1,8-cineole. It was possible to observe that rosemary essential oils showed greater antifungal activity compared to α-pinene and 1,8-cineole. In the following subsection, the most representative subclasses of terpenoids identified in lavender, peppermint, oregano, rosemary, and thyme essential oils and their respective MIC values will be discussed.

2.2.1 Monoterpenoids

Several hydrophobic or lipophilic monoterpenoids have been found in Lamiaceae essential oils and/or extracts and can be organized into three groups, namely acyclic, monocyclic, and bicyclic. The monoterpenoids within each category might be straightforward unsaturated hydrocarbons or they can include functional groups (e.g., aldehydes, ketones, alcohols). The most common monoterpenoid found in lavender essential oil is linalool, whose contribution to total volatile composition can range from 7.72% to 52.6%, followed by linalyl acetate, borneol, camphor, and lavandulyl acetate. The antifungal activity of linalool was well documented, and some studies report a MIC of 1000 μg/mL against *C. albicans*, and 3.91 μg/mL against *A. niger* (Dias et al. 2018; Bao et al. 2023). A study found out that this monoterpenoid has, theoretically, the capacity to interact with three key enzymes that play an important role in the biosynthesis of the cell wall and plasma membrane of this fungi, resulting in the rupture of the structures (Medeiros et al. 2022).

As can be observed in Table 2.1, menthol (represents 3.69–51.0% of the total volatile essential oil composition) is the main monoterpenoid found in peppermint essential oil, followed by menthone, 1,8-cineole, pulegone, and menthyl acetate.

Menthol has demonstrated antifungal properties against *Candida* spp. According to Waller et al. (2017), menthol represents 51.0% of the total composition of peppermint essential oil and has MIC_{90} (90% of growth inhibition) of 12.5 μg/mL respectively to *C. albicans*, and 150 μg/mL respectively to *A. niger* (Abbaszadeh et al. 2014; Norouzi et al. 2021). Carvacrol is a phenolic monoterpenoid found in essential oils of oregano (3.10–91.2%) and thyme (2.30–21.3%), and due to the presence of the free hydroxyl group, hydrophobicity, and phenol moiety, its antifungal activity is often greater than that of other monoterpenoids found in essential oils (Sharifi-Rad et al. 2018), with MIC value of 0.16 μg/mL relative to *C. albicans*, and 50 μg/mL relative to *A. niger* (Abbaszadeh et al. 2014). Moreover, thymol is a phenolic monoterpene that is mostly found in thyme essential oil (50.5–55.8%), and its antifungal activity has been demonstrated in several studies (Abbaszadeh et al. 2014; Barros et al. 2023). This phenolic monoterpene has a MIC value of 0.16 μg/mL against *C. albicans* and 200 μg/mL against *A. niger* (Abbaszadeh et al. 2014). Monoterpenoids, like carvacrol and thymol, could increase the extracellular pH of the yeast cells which indicates that plasma membrane integrity and permeability are affected given to the ion leakage leading to disruption of cell membrane, and, consequently, cell death (Konuk and Ergüden 2020). On the other hand, 1,8-cineole (7.28–55.3%), followed by α-pinene, camphor, β-pinene, and α-terpineol are the major monoterpenoids identified in rosemary essential oils. The antifungal activity of these monoterpenoids is well documented. According to the reports, the MIC value against *C. albicans* for α-pinene is 3.13 μg/mL (Silva et al. 2012), for camphor is 125 μg/mL, and for 1,8-cineole can range from 5 to 800 μg/mL (Mulyaningsih et al. 2010).

2.2.2 Sesquiterpenoids

Sesquiterpenoids are lipophilic molecules that are colorless. Three isoprene units are the starting point for biosynthesis in plants, which takes place in the endoplasmic reticulum through farnesyl pyrophosphate (FPP). This terpenoid class has a 15-carbon backbone, and although they have a wide range of structural variations, the majority and most useful forms are cyclic (Chadwick et al. 2013). The most abundant sesquiterpenoids found in the five plants chosen for this book chapter are caryophyllene, caryophyllene oxide, germacrene D, β-farnesene, and β-bisabolene. Caryophyllene is present in all the selected plants with its percentage ranging from 0.01% to 6.67%. The study of Barros et al. (2023) reported that caryophyllene, caryophyllene oxide, and germacrene D have antifungal effects, 155 and 500 μg/mL for MIC values of caryophyllene, and 78 μg/mL for caryophyllene oxide (Maggi et al. 2009; Neta et al. 2017). According to EL-Hefny et al. (2019), essential oils containing β-farnesene demonstrate to have great antifungal activity against fungi such as *A. niger* and *C. albicans*.

2.2.3 Diterpenoids

Diterpenoids are a rich and abundant class of uncommon metabolites that may be regarded as the product of natural processes; they can operate as a chemical defense against predators, mechanical damage, phytophages, infections, mechanical damage, and ecological interactions (Barros et al. 2023). Phytol is a diterpene alcohol identified in essential oils from lavender, peppermint, oregano, rosemary, and thyme, and has been explored as a potential antifungal agent. According to Inoue et al. (2005), phytol can cause the fungus' cell membranes to rupture, allowing K^+ ions to flow out and kill the hyphae. The MIC value for phytol was found to be 62.5 μg/mL for *Candida albicans*, *Aspergillus niger*, 0.250–0.040 μg/mL against *Klebsiella pneumoniae*, *Pseudomonas aeruginosa*, *Staphylococcus aureus*, and *Streptococcus pyogenes* (Baloyi et al. 2023). On the other hand, Ljunggren et al. (2020) verified that cembrene, identified in oregano essential oils, did not have a significant antifungal effect against *Coniophora puteana*.

2.3 Extraction and Analytical Approaches to Extract, Isolate, and Identify the Main Terpenoids from Lamiaceae Plant

2.3.1 Extraction Procedure

The most popular techniques for extracting essential oils from Lamiaceae plants are maceration and Soxhlet. Nevertheless, these extraction techniques can be ineffective in recovering the main bioactive compounds (e.g., terpenoids, phenolic compounds, flavonoids) and need extended processing times (Rodríguez-Solana et al. 2015; Jan et al. 2020; Ferreira et al. 2020; Palmieri et al. 2020). The amount and quality of essential oils in plants are still being researched. By utilizing modern technology, efforts are being made to develop and optimize extraction procedures that are more effective than conventional techniques. Hydrodistillation (HD), hydrodistillation with clevenger apparatus (CHD), and steam distillation (SD), have been extensively used to obtain essential oil from the Lamiaceae plants as can be observed in Table 2.2, due to their simplicity and low cost, and wide applicability (Lešnik et al. 2021). It is also well known that the volatile bioactive compounds in essential oils are particularly heat sensitive and susceptible to deterioration, consequently, the bioactivity of the obtained extract was reduced. Researchers have developed alternatives to conventional extraction techniques with the purpose of improving extraction efficiency, reducing terpenoid losses and the presence of hazardous component residues in essential oils (Akdağ and Öztürk 2019; Lešnik et al. 2021). Other distillation technique, known as microwave-assisted hydrodistillation (MAHD) and microwave hydrodiffusion and gravity extraction (MHDG) was developed by integrating microwave technology (Fig. 2.2).

According to Gavahian and collaborators (2015), the extraction procedure using MAHD takes less than 30 min, whereas HD methods take roughly an hour, and the GC-MS analysis did not significantly differ in the constituents of the essential oils produced using the investigated extraction procedure. Drinić et al. (2020) observed that MAHD had a shorter extraction time (24–45 min), better yields of essential oil (2.55–7.10%), higher content of oxygenated compounds (79–85%), and a more environment-friendly approach (0.135–0.240 kWh of electrical consumption) compared to HD (136 min, 5.81%, 76.82%, 1.360 kWh, respectively). Other modern extraction techniques, such as ultrasound-assisted

Table 2.2 Overview of extraction procedures, analytical approaches, and outcomes for Lamiaceae plants

Plants	Extraction procedure/conditions	Analytical approach	Outcomes	References
Lavandula angustifolia Mill. (lavender)	HD (40 g of plant, 500 mL dH_2O, 5 h) Soxhlet (40 g of plants, 300 mL hexane, 3 h) SFE (10 g of plant, 145 bar, 45 °C, 2 mL CO_2/min)	GC-MS	• The yield of 6.7% (dry weight) from SFE was greater than the yield of HD (4.6%), which was equivalent to Soxhlet extraction's yield of 7.6% (dry weight)	Danh et al. (2013)
	SSD (800 g of plant, 1.25 L/h dH_2O flow rate, 4 h)	GC-MS	• With 7.3 mL system productivity, 60% system efficiency, 98% extraction efficiency, and 0.785% (w/w) essential oil yield, SSD has shown to be a successful method for extracting high-quality lavender volatile oil	Radwan et al. (2020)
	HD (50 g of plants, 300 g of ethanol or water, 60 °C, 2 h) SD (100 g of plants, 1500 mL dH_2O, 1.5 h)	GC-MS	• The yield of essential oil extracted by SD (2.21%) is much higher than that obtained by HD	Guo and Wang (2020)
	SHSD (100 g of plant, 100 mL dH_2O, 100 °C) SHSDACD (100 g of plant, 100 mL dH_2O, 100 °C, 224 mL/g CO_2 flow)	GC-MS	• SHSDACD provided higher oil yields than the SHSD at a shorter extraction time: 1.5% at 28 min vs. 1.2% at 100 min, respectively	El-Kharraf et al. (2022)
Mentha piperita L. (peppermint)	CHD (100 g of plant, 100 °C, 3 h) MAH (100 g of plant, 100 °C, 800 W, 60 min)	GC-MS	• The MAH requires lower extraction time compared to CHD • MAH offers better yields than those obtained using CHD extraction	Hassanein et al. (2020)
	HD (30 g of plant, 500 mL dH_2O, 1 h) MAHD (30 g of plant, 1 L salted water (1% NaCl, w/v), 500 W, 27 °C, <30 min) OAHD (30 g of plant, 0.5 L salted water (1% NaCl, w/v), 27 °C, <30 min)	GC-MS	• HD techniques take over an hour for the extraction procedure, whereas OAHD and MAHD methods take less than 30 min • Compared to HD, GC-MS analysis did not reveal any appreciable differences in the compounds of the essential oils derived by the innovative explored techniques	Gavahian et al. (2015)
	Soxhlet (5 g of plant, 150 mL methanol, 4 h) SFE (5 g of plant, 240 bar, 60 °C, 40 g CO_2/min, 3% methanol as cosolvent) ASE (5 g of plant, 15 mL methanol, 30 min)	GC-MS	• The largest quantities of volatile and phenolic compounds were extracted by the ASE and Soxhlet procedures, demonstrating their appropriateness for identifying the chemical composition of aromatic plants	Rodríguez-Solana et al. (2015)
	RFX (2 g of sample, 40 mL of hexane, 10 min) MAE (2 g of sample, 60 mL of ethanol:hexane (7:3 v/v), 150 W, 30 min) UAE (2 g of sample, 80 mL of ethanol:hexane (3:7 v/v), 40 kHz, 60 min)	GC-MS	• Using MAE, extraction time and solvent consumption were reduced, while extraction yields were increased	Dai et al. (2010)

Oreganum vulgare L. (oregano)	HD (1:10 plant:dH_2O ratio, 15–120 min)	GC-MS, GC-FID	• In contrast to nonshaded plants (flowers), which had a low extraction yield (0.21 mL/100 g p.m.), oregano essential oils were most abundantly produced in cultivated shaded plants (flowers) at 0.35 mL/100 g p.m.	Ilić et al. (2022)
	Soxhlet (3 g of plant, 80 mL petroleum ether, 20 mL of ethanol or methanol)	GC-MS	• A yield of 2.31% v/w was attained • Carvacrol, β-caryophyllene, terpinen-4-ol, limonene, thymoquinone, and (*Z*)-β-ocimene were the most abundant components	Jan et al. (2020)
	CHD (30 g of plants, 450 mL dH_2O, 3 h)	GC-MS	• The oregano essential yields ranged from 2.6% to 3.3% (v/w)	Tsitlakidou et al. (2022)
	CHD (50 g of plant, 60 min)	GC-MS	• The oregano essential yield was 5.3%	Simirgiotis et al. (2020)
	MAHD (1:20 plant:dH_2O ratio, 0.135 kWh, 24–45 min) HD (1:20 plant:dH_2O ratio, 1.360 kWh, 136 min)	GC-MS	• MAH had a shorter extraction time, better yields of essential oil, higher content of oxygenated compounds, and is a more environment-friendly approach compared to HD	Drinić et al. (2020)
Rosmarinus officinalis L. (rosemary)	HD (300 g of fresh leaves and stems, 100 °C, 1.5 h) SD (30 kg of fresh leaves and stems, 1.5 h) SE (4 g of dry leaves, 40 mL of solvent, 24 h, 25 °C)	GC-MS	• The interaction of volatile by technique revealed that SE produced more α-pinene, while distillation produced more camphor and 1,8-cineole	Sadeh et al. (2019)
	HD (100 g of leaves, 500 mL dH_2O, 3 h) Maceration (10 g of rosemary, 100 mL dH_2O or 80% (v/v) ethanol solution, 1–4 h, 25 °C) MHG (100 g of plant, 400 W, 20 min, 250 mL of dH_2O)	GC-MS	• The profiles of the extracts produced by HD and MHG differed, with the essential oil obtained by MHG including more oxygenated chemicals	Ferreira et al. (2020)
	CHD (100 g of leaves, 800 mL dH_2O, 3 h, refrigerator) MAHD (100 g of leaves, 200 mL dH_2O, 600 W, 20 min)	GC-MS	• 1,8-Cineole, camphor, and α-pinene were the major components in extracts obtained by HD and MAH • The results indicate that employing the MAH rather than the HC can significantly reduce the time and energy costs associated with essential oil extraction	Elyemni et al. (2019)
	MAHD (150 g of plant, 2.45 GHz, 15 min, heated under atmospheric pressure) CHD (100 g of leaves, 800 mL dH_2O, 3 h, refrigerator)	GC-MS	• The highest yield was achieved with MAH (2.04–2.85%), while the lowest was for CHD (1.10–1.93%)	Sakar et al. (2023)
	UAE (150 μL of extract, 1 mL MeCN with 0.1% FA, 10 min)	UHPLC-ESI-MS^n	• 24 diterpenoids (carnosic acid, carnosol, and rosmanol derivatives), 1 triterpenoid (betulinic acid) were identified	Mena et al. (2016)

(continued)

Table 2.2 (continued)

Plants	Extraction procedure/conditions	Analytical approach	Outcomes	References
Thymus vulgaris L. (thyme)	RSDLE (20 g of leaves and stems, 250 mL ethanol, 2 h, 30 cycles) Soxhlet (20 g of leaves and stems, 250 mL ethanol, 100 °C, 2 h) Maceration (4 g of leaves and stems, 50 mL of ethanol, 30 days, 25 °C) UAE (8 g of leaves and stems, 100 mL of ethanol, 15 min, 40 kHz, 180 W)	GC-MS, HPLC-UV	• The highest yield was achieved with Soxhlet extraction, while the lowest was for maceration and UAE • 21 monoterpenoids and 1 sesquiterpenoid (β-bisabolene) were identified	Palmieri et al. (2020)
	HD (1:10 plant:H_2O ratio, 100 °C, 3 h) UAE-HD (1:10 plant:H_2O ratio, 550 W, 25 °C, 15 min, followed by HD) SFE (1:10 plant:H_2O ratio, 15–30 MPa, 313.15 K, 70 g CO_2/min, with and without cosolvent (10% ethanol))	GC-MS	• At low pressures, SFE yields are significantly greater than HD yields, ranging from 0.66% to 4.75% wt/wt • Compared to the extraction without ethanol, the yield increased by 2–3.3 times with the addition of 10% ethanol cosolvent	Quintana et al. (2021)
	MAH (75 g of plant, 150 mL dH_2O, 550 W, 30 min) CHD (75 g of plant, 750 mL dH_2O, 3 h)	GC-MS	• Compared to HD (1.59%), MAE showed a significantly higher oil yield (2.16%)	Gedikoğlu et al. (2019)
	HD (100 g of plant, 1000 mL dH_2O, 3 h)	GC-MS, GC-FID	• Thymol, *p*-cymene, γ-terpinene, carvacrol, linalool, (*E*)-caryophyllene, myrcene, α-terpinene, and α-thujene were the major compounds identified	Pandur et al. (2022)

Abbreviations: *HD* hydro-distillation, *SD* steam-distillation, *SE* solvent extraction, *dH₂O* distilled water, *MHG* microwave hydro-diffusion and gravity extraction, *MAHD* microwave-assisted hydrodistillation, *CHD* hydro-distillation with clevenger apparatus, *UAE* ultrasound-assisted extraction, *UHPLC-ESI-MSⁿ* ultra-high performance liquid chromatography-electrospray ionization-mass spectrometry, *RSDLE* rapid solid-liquid dynamic extraction, *HPLC-UV* high-performance liquid chromatography-ultraviolet, *NaCl* sodium chloride, *OAHD* Ohmic and microwave-assisted hydrodistillation, *ASE* accelerated solvent extraction, *SFE* supercritical fluid extraction, *RFX* reflux temperature extraction, *MAE* microwave-assisted extraction, *MeCN* acetonitrile, *SSD* solar steam distillation, *SHSD* simultaneous hydro- and steam-distillation, *SHSDACD* simultaneous hydro- and steam-distillation assisted by carbon dioxide

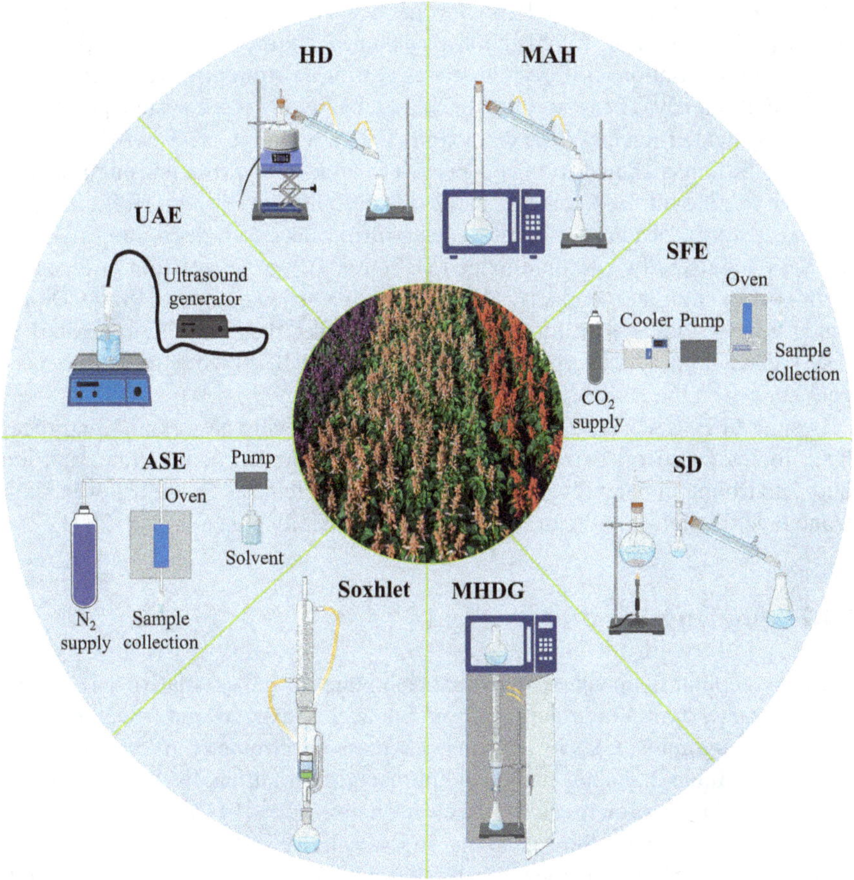

Fig. 2.2 Overview of the extraction techniques to extract essential oils from Lamiaceae plants. *Abbreviations*: *ASE* accelerated solvent extraction, *HD* hydrodistillation, *MAHD* microwave-assisted hydrodistillation, *MHDG* microwave hydrodiffusion and gravity extraction, *SD* steam distillation, *SFE* supercritical fluid extraction, *UAE* ultrasound-assisted extraction

extraction (UAE), supercritical fluid extraction (SFE), and microwave-assisted (MAE) have been used to extract essential oils since they are more efficient, require less chemical and the energy involved is more affordable. For example, Danh and collaborators (2013) verified that the yield of 6.7% (dry weight) from SFE was greater than the yield of HD (4.6%), which was equivalent to Soxhlet extraction's yield of 7.6% (dry weight). Moreover, it was verified that HD extracts exhibited indicators of heat deterioration, while Soxhlet extraction generated oils with the presence of waxes, color pigments, and albuminous components with a semisolid consistency. On the other hand, Rodríguez-Solana et al. (2015) compare the extraction efficiency of accelerated solvent extraction (ASE), Soxhlet, and SFE, and the results showed that the largest quantities of volatile and phenolic compounds were extracted by the ASE and Soxhlet procedures, demonstrating their appropriateness for identifying the chemical composition of aromatic plants. Recently, Radwan et al. (2020) built a system for extracting volatile oils using solar steam distillation (SSD) and it stands out for its straightforward design and excellent efficiency. The results demonstrated that with 7.3 mL system productivity, 60% system efficiency, 98% extraction efficiency, and 0.785% (w/w) essential oil yield, SSD has shown to be a successful method for extracting high-quality lavender volatile oil. El-Kharraf et al. (2022) evaluate the extraction efficiency of two extraction procedures, simultaneous hydro- and steam-distillation (SHSD) and SHSD assisted by carbon dioxide (SHSDACD), on the isolation of essential oil from three Lamiaceae plants. The data obtained showed that SHSDACD presented greater yields with an extraction time lower than 30 min compared to SHSD, consuming less energy (<10 times lower than SHSD) due to the enhancement noticed in the K distillation rate constant.

In sum, SFE, UAE, ASE, and SSD represent promising new extraction procedures for essential oil extraction, since require short extraction time, and are more environment friendly in terms of the contribution to global warming, ozone layer depletion, human toxicity, and ecotoxicity (Lešnik et al. 2021).

2.3.2 Analytical Approach

The most popular hyphenated analytical technique for the qualitative and quantitative characterization of terpenoids from Lamiaceae extracts and essential oils is gas chromatography (GC) in tandem with mass spectrometry (MS), as can be observed in Table 2.2. This analytical approach has traditionally been interfaced using vacuum ionization methods like electron ionization (EI) or chemical ionization (CI). Nevertheless, the selectivity and sensitivity in the detection of terpenoids may be hampered by the typically poor ionization efficiency attained in CI, as well as the considerable fragmentation by EI (Ayala-Cabrera et al. 2023). In addition, the identification of terpenoids sometimes is limited when a single chromatographic peak comprises several chemicals, making it challenging to decipher the recorded mass spectra (Merfort 2002). In this sense, there are several to solve this problem, such as applying a temperature program to improve the separation of the components, consequently allowing the distinction between chemicals that act similarly throughout the GC process, and/or using GC coupled to tandem mass spectrometry (MS/MS) that enables the independent investigation of each component of such complicated peaks (Al-Rubaye et al. 2017). Nan et al. (2021) developed and validated a gas chromatography-triple quadrupole mass spectrometry (GC-MS/MS) method to simultaneously assess the terpene concentrations of *Eupatorium fortunei* Turcz leaves, stems, and roots taken at various growth stages. The findings showed that the GC-MS/MS was precise and reproducible, suggesting that it may be a trustworthy analytical approach for the identification and measurement of terpenoids.

2.4 Techniques and Methods for Assessing the Antifungal Effects

Annually, invasive fungal diseases cause more than 1.5 million fatalities worldwide. Cryptococcosis (*Cryptococcus neoformans*), candidiasis (*Candida albicans*), pneumocystosis (*Pneumocystis jirovecii*), and aspergillosis (*Aspergillus fumigatus*) are life-threatening fungal infections with significant morbidity and mortality rates, with a mortality rate ranging from 20% to 70%, 46% to 75%, 20% to 80%, and 30% to 95%, respectively (Powers et al. 2019; Karpiński 2020). The antifungal effects of Lamiaceae extracts, essential oils, and/or pure terpenoids can be assessed through in vitro or in vivo assays. Regarding in vitro assays several techniques are available, such as checkerboard, agar diffusion, time-kill curves, broth microdilution, bioautography, dilution methods, and flow cytometry (Favre-Godal et al. 2013; Bidaud et al. 2021; Sanchez Armengol et al. 2021). Among these in vitro assays, dilution, and diffusion methods are the most used to assess the antifungal effects. Nevertheless, the nonpolar nature of some terpenoids present in Lamiaceae extracts and essential oils complicates the diffusion across the culture medium, and for this reason, dilution methods appear as a suitable method (Lešnik et al. 2021). The dilution methods were expressed with minimum inhibitory concentration (MIC) values, which correspond to the lowest concentration of the pure compounds, extracts, or essential oils that inhibit fungal growth (Sanchez Armengol et al. 2021). The pure compounds, extracts, or essential oils are dispersed in water, dimethylsulfoxide, or detergents (Tween 80) and mixed in the serial dilution of a liquid or solid inoculated medium, followed by the addition of fungal cells. After incubation, the sample's inhibitory qualities can be calculated using turbidimetric, absorbance, or visual comparison with a control culture. The MIC value is calculated as the lowest active chemical concentration at which microorganism growth is prevented. The approach is adaptable to microplate-based screening procedures. This format, which only needs ng/mg amounts of the test samples, is preferable for analyzing many samples (Favre-Godal et al. 2013; Lešnik et al. 2021).

In the meantime, the antifungal activity of peppermint, oregano, and rosemary is well-documented against aspergillosis, candidiasis, and cryptococcosis as shown in Table 2.1. On the other hand, few studies related to the antifungal activity of lavender were performed. Moreover, analyzing Table 2.1, it was possible to observe significant differences in MIC values determined for the same fungi and Lamiaceae essential oils. For example, the effect of lavender essential oil against *C. albicans*. The research conducted by Khoury and collaborators (2016) determined a MIC value of 512 μg/mL, while in Dolatabadi et al. (2019), the MIC value was 10.3 μg/mL. A similar pattern was observed for the other selected Lamiaceae essential oils. This difference can be explained by chemical composition, extraction procedure, and geographic region, among other factors.

However, it is crucial to validate the in vitro data with in vivo assays. According to Bidaud et al. (2021), no standardized methods were available to assess the antifungal activities in animal models. The mortality rate and the fungal load in the target organs (measured as the number of CFU per gram of tissue by culture) are the two assessment criteria that are most often utilized. In vivo assays showed several disadvantages that limited their application compared to in vitro assays, such as specific infrastructures, time-consuming, and ethical concerns (Bidaud et al. 2021). As far as we know, there are no data available in the literature related to in vivo assays to evaluate the antifungal activity of the Lamiaceae essential oils selected in this book chapter.

2.5 Mechanisms of Terpenoids Action as Antifungal Agents

The oils and plant extracts of the Lamiaceae family have been recognized for many years for their antifungal properties. In addition to the great advantage that is the complex mechanism of action resulting from a rich mixture composition, it is indeed very important to scientifically investigate the plants that have been used in traditional medicine as potential sources of new antifungal compounds (Gucwa et al. 2018). However, the mechanism of antifungal action of active compounds generated from natural products has only been the subject of a small number of studies (Couto et al. 2015). This chapter highlights the antifungal activity of some Lamiaceae plants, including lavender, peppermint, oregano, rosemary, and thyme, against pathogenic fungi, as well as the mechanisms of action of the respective antifungal agents (Fig. 2.3). Terpenoids present different action mechanisms, and can interfere with microbial membrane functions, or suppress virulence factors such as the production of enzymes and toxins, or by acting against fungal biofilm formation (Barros et al. 2023). The hydrophobicity of essential oils and their phytochemicals, which enables them to interact with the fungal cell membrane and compromise its integrity, is an important characteristic. Although the mechanism of action of terpenoids is not fully understood, it is speculated to involve membrane disruption by the lipophilic nature (Arif et al. 2009). As a result, the integrity of the fungal membranes is compromised and they become more permeable, ultimately leading to mycelial death (Lima et al. 2013). Some terpenoids have been identified as major in different extracts and promising for antifungal use as can be observed in Table 2.1, based on the MICs (Karpiński 2020). The terpenoids most found as the main components in essential oils of target Lamiaceae plants, include 1,8-cineole, camphor, carvacrol, *p*-cymene, thymol, α-amyrin, and β-caryophyllene (Table 2.1). Important antifungal chemicals include other monoterpenoids such as linalool, 1,8-cineole, pinene, and terpinene. Among them, 1,8-cineole has been identified in the essential oil of lavender, and rosemary, and has been active against *C. albicans*, *A. niger*, and *P. expansum* by promoting morphological changes through decreased formation of conidia and degradation in its plasmatic membrane, inducing to the disruption of cell membrane and, as a result, fungal cell wall degeneration (Waller et al. 2017).

Several studies (Bona et al. 2016; Skendi et al. 2020) have shown that essential oils containing high concentrations of phenolic monoterpenes (e.g., carvacrol, thymol) have great antifungal activities. Therefore, it is possible that the higher

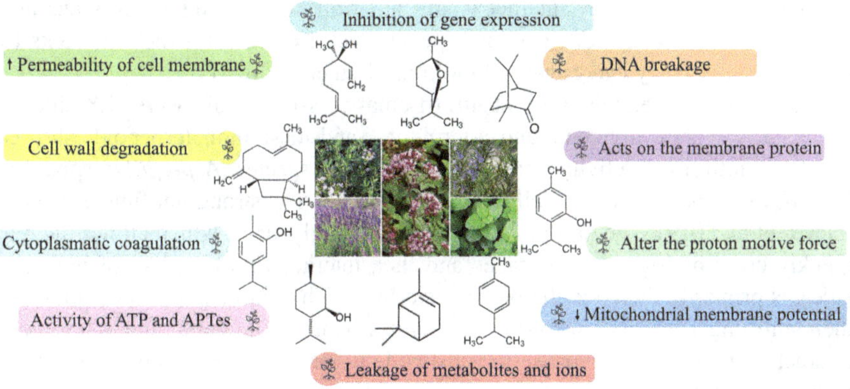

Fig. 2.3 Schematic presentation of the potential mechanism of action of essential oils and plant extracts from different species of Lamiaceae in fungi

inhibition against fungi observed for thyme and oregano when added as essential oils is due to the presence of these terpenoids. As a result of their strong antifungal activity, they interact with sterols, particularly ergosterol (a fungal membrane sterol), and cause damage to cell membranes (Nazzaro et al. 2017). According to Redondo-Blanco and collaborators (2020), carvacrol and thymol are promising compounds with activity against *Candida* spp. species since they are capable of completely blocking ergosterol biosynthesis. This disruption of ergosterol biosynthesis has also been found for other species, such as *Fusarium graminearum*, with thymol (Gao et al. 2016), and the mechanism of action of the compound citral, mentioned previously, is also based on the interruption of the biosynthesis pathway of ergosterol, as Sousa et al. (2016) evidenced testing it against *C. tropicalis*.

In addition, Konuk and Ergüden (2020) showed that the phenolic terpenoids compromise the integrity of cell membranes, which leads to ion leakage and ultimately cell death. This happens not just because of the terpenoids' hydrophobic aromatic portion allowing for their accumulation in the membrane, but also because of the presence of the phenolic –OH group, which is essential for their antifungal activity. These substances might be retained in the yeast plasma membrane and cause conformational modification of the membrane due to the phenolic hydroxyl groups propensity to create hydrogen bonds and exchange protons when coupled to an electron-delocalized system and, consequently, may affect the ion homeostasis in yeast cells.

Lima et al. (2013) also demonstrated the potential antifungal activity of carvacrol against *C. albicans* through morphological alterations of the membrane structure of the fungal cell. Previous studies also revealed that carvacrol, *p*-cymene, and thymol from thyme had fungistatic activity (Ahmad et al. 2011). Since thymol and carvacrol are lipophilic and can act in the fatty acyl chains of membrane lipid bilayers, altering the fluidity and permeability of cell membranes, it is likely that these compounds have antifungal activity (Sharifzadeh et al. 2016). However, the exact mechanism of action by which they work is still not well understood. The specific mechanism by which they interact is still not well known. This process has also been demonstrated with other monoterpenes, such as linalool, which also acts by affecting the stability and growth of biofilms (Hsu et al. 2013).

The major terpenoids present in rosemary essential oil were 1,8-cineole, α-pinene, camphor, myrcene, and γ-terpinene, and these monoterpenoids showed excellent antifungal activity against *Candida* spp. species, *Aspergillus* spp., and *Penicillium* spp., leading to cell death because of their strong antifungal activity (Neves et al. 2019). The lipophilic nature of their aglycones, which allows them to quickly cross biological membranes and then interact with a variety of biomolecules, is primarily linked to the mechanism by which monoterpenoids demonstrate their antifungal effects. According to some research, α-pinene, and γ-terpinene interact with the cell membrane in yeast cells to increase permeability and inhibit cellular respiration and ionic transport processes. This results in changes to the cell membrane's electrical potential, which leads to cell death (Neves et al. 2019).

The hydroalcoholic extract of rosemary was also tested for its antifungal effects on *Trichophyton rubrum* and *Microsporum gypseum*, which revealed significant antifungal activity, according to Sudan and Singh (2019). Except for β-caryophyllene and β-pinene, the terpenic compounds identified in the study by Bellumori et al. (2021) as the main components of rosemary essential oil displayed significant antifungal activity. In the case of camphor and verbenone, these terpenoids suppressed the growth of *Alternaria alternata* at the same rate as copper sulfate, a pesticide that is frequently used to prevent bacterial and fungal pathogens and allowed in organic farming (Van Bruggen and Finckh 2016).

The potential fungicide activity of peppermint essential oils has been evaluated against some common phytopathogens. Menthol and menthone followed by limonene, menthyl acetate, and β-caryophyllene were identified by Camele et al. (2021) as the main terpenoids responsible for antifungal activity. Their potential mechanism of action depends on the disruption of the fungal cell membrane that increases the cell permeability. Additionally, Freire et al. (2017), tested and selected the best product with antifungal activity against *C. albicans* strains isolated from dental prostheses. The study results revealed that citral, out of all essential oils and molecules, had the best antifungal activity against strains nystatin resistant. Furthermore, according to research done by Zheng et al. (2015) on citral's mechanism of action against *P. digitatum*, this compound was found to be involved in inhibiting respiratory metabolism via changing mitochondrial morphology and function.

It is important to consider the possibility that more than one element of the essential oil may be responsible for the antifungal action when identifying the active components. Multiple research projects have demonstrated that some essential oils have a single mechanism of action while others have two or more (Chouhan et al. 2017), indicating that interactions between different constituents can occasionally be crucial for biological activity. As demonstrated for cedrol (Powers et al. 2019), it is feasible that even after identifying the principal component(s) that may be potentially active, full activity may not be found, necessitating the consideration of synergy between several essential oil's components.

2.6 Limitation of the Lamiaceae Plant in Antifungal Activity Research

Most of the studies in the literature are focused on evaluating the antifungal activity of essential oils, neglecting the actual impact of plant material and the potential value of other groups of compounds (Skendi et al. 2020). However, recent studies have attempted to find natural antifungal compounds using natural plants and their herbal extracts for the treatment of a variety of infectious diseases mainly caused by bacteria, fungi, protozoa, and viruses. Among them, Skendi et al. (2020) evaluated the antifungal effect of aromatic plants, such as oregano, thyme, and satureja, in dry form in vitro and in bread against two phytopathogenic fungi found in food (*Aspergillus niger* and *Penicillium*) by calculating the area of mycelium growth. All three of the aromatic plants studied by the researchers exhibited inhibitory effects against both fungi, demonstrating that the addition of aromatic plants in the dry form to the bread recipe is more efficient than using essential oils. Furthermore, the authors observed that aromatic plants were more effective at inhibiting *Penicillium* growth compared to *Aspergillus*. Blank et al. (2020) studied the bioactive compounds and antifungal activity of the aqueous and ethanolic extract of oregano and rosemary against several fungal species and concluded that the ethanolic extract of oregano presents a higher quantity of phenolic compounds and, consequently, an antimicrobial activity potential. In another study, Waller et al. (2018) also used the broth microdilution method, according to the CLSI M38-A2 standard protocol adapted to natural products, to investigate the in vitro antifungal activity of polar extracts of oregano against *Sporothrix brasiliensis*. The results of this study show that all oregano extracts have antifungal activity against all strains tested, with MIC values ranging from ≤0.07 to 40 μg/mL. In a 2021 study, the antifungal assays of five species of the Lamiaceae family were performed on Mueller-Hinton (MH) (Difco) agar through agar well diffusion method, and the findings showed that the methanol extracts exhibit considerable antifungal activity with values in the range of 23–959 μg/mL (Çelik et al. 2021). In another investigation, Cherrate et al. (2022) verified that thyme suppresses the growth of *B. cinerea* with an inhibition rate ranging from 61.87% to 97.94%. Based on the reported studies, it is possible to observe a clear relationship between Lamiaceae essential oils and antifungal activity, but as most of the existing knowledge is based on in vitro research, it is unclear if this relationship holds true in the clinical situation. In vivo effects of substances with in vitro antibacterial activity may be minimal or nonexistent. This could be due to the technology to fully understand this is still being developed (Vaou et al. 2021). Therefore, more study is necessary, particularly in vivo assays and investigations on the toxicity of these chemicals, for them to be recognized as antifungal agents. On the other hand, the lack of treatment standardization is the biggest obstacle to using Lamiaceae plant essential oils as antifungal agents. In this sense, with scientific evolution, it was expected to look deeper into the chemical constitution of Lamiaceae plants and develop more advanced methods for the extraction, fractionation, and identification of bioactive chemicals, which exhibit a variety of chemical structures and modes of action (Vaou et al. 2021).

Acknowledgments This research was supported by FCT-Fundação para a Ciência e a Tecnologia through the CQM Base Fund, UIDB/00674/2020, Programmatic Fund, UIDP/00674/2020, and by ARDITI-Agência Regional para o Desenvolvimento da Investigação Tecnologia e Inovação, through the project M1420-01-0145-FEDER-000005—Centro de Química da Madeira—CQM+ (Madeira 14-20 Program). The authors also acknowledge FCT and Madeira 14-2020 program to the Portuguese Mass Spectrometry Network (RNEM) through the PROEQUIPRAM program, M14-20 M1420-01-0145-FEDER-000008.

References

Abbaszadeh S, Sharifzadeh A, Shokri H et al (2014) Antifungal efficacy of thymol, carvacrol, eugenol and menthol as alternative agents to control the growth of food-relevant fungi. J Mycol Med 24:e51–e56. https://doi.org/10.1016/j.mycmed.2014.01.063

Abdelghffar EAR, El-Nashar HAS, Fayez S et al (2022) Ameliorative effect of oregano (Origanum vulgare) versus silymarin in experimentally induced hepatic encephalopathy. Sci Rep 12:17854–17871. https://doi.org/10.1038/s41598-022-20412-3

Ahamad J, Ali M, Naquvi KJ (2018) Analysis of essential oil of Origanum vulgrae Linn. by GC and GC-MS. J Global Trends Pharm Sci 9:5786–5791

Ahmad A, Khan A, Akhtar F et al (2011) Fungicidal activity of thymol and carvacrol by disrupting ergosterol biosynthesis and membrane integrity against Candida. Eur J Clin Microbiol Infect Dis 30:41–50. https://doi.org/10.1007/s10096-010-1050-8

Akdağ A, Öztürk E (2019) Distillation methods of essential oils. Nisan 45:22–31

Ali J, Tuzen M, Citak D et al (2019) Separation and preconcentration of trivalent chromium in environmental waters by using deep eutectic solvent with ultrasound-assisted based dispersive liquid-liquid microextraction method. J Mol Liq 291:111299. https://doi.org/10.1016/j.molliq.2019.111299

Al-Rubaye AF, Hameed IH, Kadhim MJ (2017) A review: uses of gas chromatography-mass spectrometry (GC-MS) technique for analysis of bioactive natural compounds of some plants. Int J Toxicol Pharmacol Res 9:81–85. https://doi.org/10.25258/ijtpr.v9i01.9042

Arafa NM, Girgis ND, Ibrahim MM et al (2022) Phytochemical profiling by GC-MS analysis and antimicrobial activity potential of in vitro derived shoot cultures of some Egyptian herbal medicinal plants. Egypt J Chem 65:155–169. https://doi.org/10.21608/ejchem.2022.115045.5230

Arif T, Bhosale JD, Kumar N et al (2009) Natural products—antifungal agents derived from plants. J Asian Nat Prod Res 11:621–638

Ayala-Cabrera JF, Montero L, Meckelmann SW et al (2023) Review on atmospheric pressure ionization sources for gas chromatography-mass spectrometry. Part I: Current ion source developments and improvements in ionization strategies. Anal Chim Acta 1238:340353–340365. https://doi.org/10.1016/j.aca.2022.340353

Baloyi IT, Adeosun IJ, Bonvicini F, Cosa S (2023) Biofilm reduction, in-vitro cytotoxicity and computational drug-likeness of selected phytochemicals to combat multidrug-resistant bacteria. Sci Afr 21:e01814–e01826. https://doi.org/10.1016/j.sciaf.2023.e01814

Bao Z, Fan M, Hannachi K et al (2023) Antifungal activity of star anise extract against Penicillium roqueforti and Aspergillus niger for bread shelf life. Food Res Int 172:113225–113233. https://doi.org/10.1016/j.foodres.2023.113225

Barros DB, Nascimento NS, Sousa AP et al (2023) Antifungal activity of terpenes isolated from the Brazilian Caatinga: a review. Braz J Biol 83:e270966. https://doi.org/10.1590/1519-6984.270966

Bellumori M, Innocenti M, Congiu F et al (2021) Within-plant variation in rosmarinus officinalis l. Terpenes and phenols and their antimicrobial activity against the rosemary phytopathogens Alternaria alternata and Pseudomonas viridiflava. Molecules 26:3425–3443. https://doi.org/10.3390/molecules26113425

Bidaud AL, Schwarz P, Herbreteau G, Dannaoui E (2021) Techniques for the assessment of in vitro and in vivo antifungal combinations. J Fungi 7:1–16. https://doi.org/10.3390/jof7020113

Binzet G, Binzet R, Arslan H (2020) The essential oil compositions of Rosmarinus officinalis L. leaves growing in Mersin, Turkey. Eur J Chem 11:370–376. https://doi.org/10.5155/eurjchem.11.4.370-376.2048

Blank DE, Alves GH, Nascente PDS et al (2020) Bioactive compounds and antifungal activities of extracts of Lamiaceae species. J Agric Chem Environ 09:85–96. https://doi.org/10.4236/jacen.2020.93008

Bona E, Cantamessa S, Pavan M et al (2016) Sensitivity of Candida albicans to essential oils: are they an alternative to antifungal agents? J Appl Microbiol 121:1530–1545. https://doi.org/10.1111/jam.13282

Camele I, Gruľová D, Elshafie HS (2021) Chemical composition and antimicrobial properties of mentha × piperita cv. 'kristinka' essential oil. Plants 10:1567–1579. https://doi.org/10.3390/plants10081567

Camiletti BX, Asensio CM, de la Pecci MPG, Lucini EI (2014) Natural control of corn postharvest fungi Aspergillus flavus and Penicillium sp. using essential oils from plants grown in Argentina. J Food Sci 79:M2499–M2506. https://doi.org/10.1111/1750-3841.12700

Çelik G, Kılıç G, Kanbolat Ş et al (2021) Biological activity, and volatile and phenolic compounds from five Lamiaceae species. Flavour Fragr J 36:223–232. https://doi.org/10.1002/ffj.3636

Chadwick M, Trewin H, Gawthrop F, Wagstaff C (2013) Sesquiterpenoids lactones: benefits to plants and people. Int J Mol Sci 14:12780–12805

Chandra M, Kushwaha S, Mishra B, Sangwan N (2022) Molecular and structural insights for the regulation of terpenoids in Ocimum basilicum and Ocimum tenuiflorum. Plant Growth Regul 97:61–75. https://doi.org/10.1007/s10725-022-00796-y

Cherrate M, Echchgadda G, Amiri S et al (2022) Biological control of major postharvest fungal diseases of apple using two Lamiaceae extracts. Arch Phytopathol Plant Protect 55:2356–2381. https://doi.org/10.1080/03235408.2023.2166379

Chouhan S, Sharma K, Guleria S (2017) Antimicrobial activity of some essential oils—present status and future perspectives. Medicines 4:58–78. https://doi.org/10.3390/medicines4030058

Couto CSF, Raposo NRB, Rozental S et al (2015) Chemical composition and antifungal properties of essential oil of Origanum vulgare linnaeus (Lamiaceae) against Sporothrix schenckii and Sporothrix brasiliensis. Trop J Pharm Res 14:1207–1212. https://doi.org/10.4314/tjpr.v14i7.12

Dai J, Orsat V, Vijaya Raghavan GS, Yaylayan V (2010) Investigation of various factors for the extraction of peppermint (Mentha piperita L.) leaves. J Food Eng 96:540–543. https://doi.org/10.1016/j.jfoodeng.2009.08.037

Danh LT, Han LN, Triet NDA et al (2013) Comparison of chemical composition, antioxidant and antimicrobial activity of lavender (Lavandula angustifolia L.) essential oils extracted by supercritical CO_2, hexane and hydrodistillation. Food Bioprocess Technol 6:3481–3489. https://doi.org/10.1007/s11947-012-1026-z

Dias IJ, Trajano ERIS, Castro RD et al (2018) Antifungal activity of linalool in cases of Candida spp. isolated from individuals with oral candidiasis. Braz J Biol 78:368–374. https://doi.org/10.1590/1519-6984.171054

Dolatabadi S, Salari Z, Mahboubi M (2019) Antifungal effects of Ziziphora tenuior, Lavandula angustifolia, Cuminum cyminum essential oils against clinical isolates of Candida albicans from women suffering from vulvovaginal candidiasis. Infectio 23:222–226. https://doi.org/10.22354/in.v23i3.784

Drinić Z, Pljevljakušić D, Živković J et al (2020) Microwave-assisted extraction of O. vulgare L. spp. hirtum essential oil: comparison with conventional hydro-distillation. Food Bioprod Process 120:158–165. https://doi.org/10.1016/j.fbp.2020.01.011

Elansary HO, Abdelgaleil SAM, Mahmoud EA et al (2018) Effective antioxidant, antimicrobial and anticancer activities of essential oils of horticultural aromatic crops in northern Egypt. BMC Complement Altern Med 18:214–224. https://doi.org/10.1186/s12906-018-2262-1

EL-Hefny M, Abo Elgat WAA, Al-Huqail AA, Ali HM (2019) Essential and recovery oils from Matricaria chamomilla flowers as environmentally friendly fungicides against four fungi isolated from cultural heritage objects. Processes 7:809–821. https://doi.org/10.3390/pr7110809

El-Kharraf S, El-Guendouz S, Abdellah F et al (2022) Unassisted and carbon dioxide-assisted hydro- and steam-distillation: modelling kinetics, energy consumption and chemical and biological activities of volatile oils. Pharmaceuticals 15:567–587. https://doi.org/10.3390/ph15050567

Elyemni M, Louaste B, Nechad I et al (2019) Extraction of essential oils of Rosmarinus officinalis L. by two different methods: hydrodistillation and microwave assisted hydrodistillation. Sci World J 2019:1–6. https://doi.org/10.1155/2019/3659432

Favre-Godal Q, Queiroz EF, Wolfender JL (2013) Latest developments in assessing antifungal activity using TLC-bioautography: a review. J AOAC Int 96:1175–1188. https://doi.org/10.5740/jaoacint.SGEFavre-Godal

Ferreira DF, Lucas BN, Voss M et al (2020) Solvent-free simultaneous extraction of volatile and non-volatile antioxidants from rosemary (Rosmarinus officinalis L.) by microwave hydrodiffusion and gravity. Ind Crop Prod 145:112094–112092. https://doi.org/10.1016/j.indcrop.2020.112094

Freire JCP, Júnior JKDO, Silva DDF et al (2017) Antifungal activity of essential oils against Candida albicans strains isolated from users of dental prostheses. Evid Based Complement Alternat Med 2017:1–9. https://doi.org/10.1155/2017/7158756

Gao T, Zhou H, Zhou W et al (2016) The fungicidal activity of thymol against Fusarium graminearum via inducing lipid peroxidation and disrupting ergosterol biosynthesis. Molecules 21:770–782. https://doi.org/10.3390/molecules21060770

Gavahian M, Farahnaky A, Farhoosh R et al (2015) Extraction of essential oils from Mentha piperita using advanced techniques: microwave versus ohmic assisted hydrodistillation. Food Bioprod Process 94:50–58. https://doi.org/10.1016/j.fbp.2015.01.003

Gedikoğlu A, Sökmen M, Çivit A (2019) Evaluation of Thymus vulgaris and Thymbra spicata essential oils and plant extracts for chemical composition, antioxidant, and antimicrobial properties. Food Sci Nutr 7:1704–1714. https://doi.org/10.1002/fsn3.1007

Gladikostić N, Ikonić B, Teslić N et al (2023) Essential Oils from Apiaceae, Asteraceae, Cupressaceae and Lamiaceae Families Grown in Serbia: Comparative Chemical Profiling with In Vitro Antioxidant Activity. Plants 12:745–767. https://doi.org/10.3390/plants12040745

Gucwa K, Milewski S, Dymerski T, Szweda P (2018) Investigation of the antifungal activity and mode of action of Thymus vulgaris, Citrus limonum, Pelargonium graveolens, Cinnamomum cassia, Ocimum basilicum, and Eugenia caryophyllus essential oils. Molecules 23:1116–1134. https://doi.org/10.3390/molecules23051116

Guo X, Wang P (2020) Aroma characteristics of lavender extract and essential oil from Lavandula angustifolia Mill. Molecules 25:5541–5554. https://doi.org/10.3390/molecules25235541

Hashemi SMB, Gholamhosseinpour A, Barba FJ (2023) Rosmarinus officinalis L. essential oils impact on the microbiological and oxidative stability of Sarshir (Kaymak). Molecules 28:4206–4218. https://doi.org/10.3390/molecules28104206

Hassan R, Attia S, Salem S, Elhefny A (2023) Insecticidal effects and chemical composition of lemongrass and peppermint essential oils against the coconut mealybug, Nipaecoccus nipae (Maskell), (Hemiptera: Pseudococcidae). Egypt Acad J Biol Sci F Toxicol Pest Control 15:47–61. https://doi.org/10.21608/eajbsf.2023.287294

Hassanein HD, El-Gendy AENG, Saleh IA et al (2020) Profiling of essential oil chemical composition of some Lamiaceae species extracted using conventional and microwave-assisted hydrodistillation extraction methods via chemometrics tools. Flavour Fragr J 35:329–340. https://doi.org/10.1002/ffj.3566

Hossain MA, Al-Hdhrami SS, Weli AM et al (2014) Isolation, fractionation and identification of chemical constituents from the leaves crude extracts of Mentha piperita L grown in Sultanate of Oman. Asian Pac J Trop Biomed 4:S368–S372. https://doi.org/10.12980/APJTB.4.2014C1051

Hossain F, Follett P, Dang Vu K et al (2016) Evidence for synergistic activity of plant-derived essential oils against fungal pathogens of food. Food Microbiol 53:24–30. https://doi.org/10.1016/j.fm.2015.08.006

Hsu CC, Lai WL, Chuang KC et al (2013) The inhibitory activity of linalool against the filamentous growth and biofilm formation in Candida albicans. Med Mycol 51:473–482. https://doi.org/10.3109/13693786.2012.743051

Ibrahim N, Abbas H, El-Sayed NS, Gad HA (2022) Rosmarinus officinalis L. hexane extract: phytochemical analysis, nanoencapsulation, and in silico, in vitro, and in vivo anti-photoaging potential evaluation. Sci Rep 12:13102–13121. https://doi.org/10.1038/s41598-022-16592-7

Ilić Z, Stanojević L, Milenković L et al (2022) The yield, chemical composition, and antioxidant activities of essential oils from different plant parts of the wild and cultivated oregano (Origanum vulgare L.). Horticulturae 8:1042–1059. https://doi.org/10.3390/horticulturae8111042

Inoue Y, Hada T, Shiraishi A et al (2005) Biphasic effects of geranylgeraniol, teprenone, and phytol on the growth of Staphylococcus aureus. Antimicrob Agents Chemother 49:1770–1774. https://doi.org/10.1128/AAC.49.5.1770-1774.2005

Jan S, Rashid M, Abd-Allah EF, Ahmad P (2020) Biological efficacy of essential oils and plant extracts of cultivated and wild ecotypes of Origanum vulgare L. Biomed Res Int 2020:1–16. https://doi.org/10.1155/2020/8751718

Jiang Y, Wu N, Fu YJ et al (2011) Chemical composition and antimicrobial activity of the essential oil of Rosemary. Environ Toxicol Pharmacol 32:63–68. https://doi.org/10.1016/j.etap.2011.03.011

Karpiński TM (2020) Essential oils of lamiaceae family plants as antifungals. Biomolecules 10:103–137. https://doi.org/10.3390/biom10010103

Khoury M, Stien D, Eparvier V et al (2016) Report on the medicinal use of eleven Lamiaceae species in Lebanon and rationalization of their antimicrobial potential by examination of the chemical composition and antimicrobial activity of their essential oils. Evid Based Complement Alternat Med 2016:1–17. https://doi.org/10.1155/2016/2547169

Konuk HB, Ergüden B (2020) Phenolic –OH group is crucial for the antifungal activity of terpenoids via disruption of cell membrane integrity. Folia Microbiol (Praha) 65:775–783. https://doi.org/10.1007/s12223-020-00787-4

Lešnik S, Furlan V, Bren U (2021) Rosemary (Rosmarinus officinalis L.): extraction techniques, analytical methods and health-promoting biological effects. Phytochem Rev 20:1273–1328. https://doi.org/10.1007/s11101-021-09745-5

Lima IO, De Oliveira Pereira F, De Oliveira WA et al (2013) Antifungal activity and mode of action of carvacrol against Candida albicans strains. J Essent Oil Res 25:138–142. https://doi.org/10.1080/10412905.2012.754728

Ljunggren J, Bylund D, Jonsson BG et al (2020) Antifungal efficiency of individual compounds and evaluation of non-linear effects by recombining fractionated turpentine. Microchem J 153:104325–104324. https://doi.org/10.1016/j.microc.2019.104325

Maggi F, Cecchini C, Cresci A et al (2009) Chemical composition and antimicrobial activity of the essential oil from Ferula glauca L. (F. communis L. subsp. glauca) growing in Marche (Central Italy). Fitoterapia 80:68–72. https://doi.org/10.1016/j.fitote.2008.10.001

Medeiros CIS, de Sousa MNA, Filho GGA et al (2022) Antifungal activity of linalool against fluconazole-resistant clinical strains of vulvovaginal Candida albicans and its predictive mechanism of action. Braz J Med Biol Res 55:e11831–e11841. https://doi.org/10.1590/1414-431X2022e11831

Mena P, Cirlini M, Tassotti M et al (2016) Phytochemical profiling of flavonoids, phenolic acids, terpenoids, and volatile fraction of a rosemary (Rosmarinus officinalis L.) extract. Molecules 21:1576–1590. https://doi.org/10.3390/molecules21111576

Merfort I (2002) Review of the analytical techniques for sesquiterpenes and sesquiterpene lactones. J Chromatogr A 967:115–130

Mulyaningsih S, Sporer F, Zimmermann S et al (2010) Synergistic properties of the terpenoids aromadendrene and 1,8-cineole from the essential oil of Eucalyptus globulus against antibiotic-susceptible and antibiotic-resistant pathogens. Phytomedicine 17:1061–1066. https://doi.org/10.1016/j.phymed.2010.06.018

Nan G, Zhang L, Liu Z et al (2021) Quantitative determination of p-cymene, thymol, neryl acetate, and β-caryophyllene in different growth periods and parts of Eupatorium fortunei Turcz. by GC-MS/MS. J Anal Methods Chem 2021:1–7. https://doi.org/10.1155/2021/2174667
Nazzaro F, Fratianni F, Coppola R, De Feo V (2017) Essential oils and antifungal activity. Pharmaceuticals 10:86–105. https://doi.org/10.3390/ph10040086
Nedeltcheva-Antonova D, Gechovska K, Bozhanov S, Antonov L (2022) Exploring the chemical composition of Bulgarian lavender absolute (Lavandula Angustifolia Mill.) by GC/MS and GC-FID. Plants 11:3150–3162. https://doi.org/10.3390/plants11223150
Neta MCS, Vittorazzi C, Guimarães AC et al (2017) Effects of β-caryophyllene and Murraya paniculata essential oil in the murine hepatoma cells and in the bacteria and fungi 24-h time-kill curve studies. Pharm Biol 55:190–197. https://doi.org/10.1080/13880209.2016.1254251
Neves JS, Lopes-Da-Silva Z, De Sousa Brito Neta M et al (2019) Preparation of terpolymer capsules containing Rosmarinus officinalis essential oil and evaluation of its antifungal activity. RSC Adv 9:22586–22596. https://doi.org/10.1039/c9ra02336d
Norouzi N, Alizadeh F, Khodavandi A, Jahangiri M (2021) Antifungal activity of menthol alone and in combination on growth inhibition and biofilm formation of Candida albicans. J Herb Med 29:100495–100506. https://doi.org/10.1016/j.hermed.2021.100495
Palmieri S, Pellegrini M, Ricci A et al (2020) Chemical composition and antioxidant activity of thyme, hemp and coriander extracts: a comparison study of maceration, soxhlet, UAE and RSLDE techniques. Foods 9:1221–1238. https://doi.org/10.3390/foods9091221
Pandur E, Micalizzi G, Mondello L et al (2022) Antioxidant and anti-inflammatory effects of thyme (Thymus vulgaris L.) essential oils prepared at different plant phenophases on Pseudomonas aeruginosa LPS-activated THP-1 macrophages. Antioxidants 11:1130–1154. https://doi.org/10.3390/antiox11071330
Powers CN, Osier JL, McFeeters RL et al (2018) Antifungal and cytotoxic activities of sixty commercially-available essential oils. Molecules 23:1549–1562. https://doi.org/10.3390/molecules23071549
Powers CN, Satyal P, Mayo JA et al (2019) Bigger data approach to analysis of essential oils and their antifungal activity against Aspergillus niger, Candida albicans, and Cryptococcus neoformans. Molecules 24:2868–2879. https://doi.org/10.3390/molecules24162868
Quintana SE, Llalla O, García-Risco MR, Fornari T (2021) Comparison between essential oils and supercritical extracts into chitosan-based edible coatings on strawberry quality during cold storage. J Supercrit Fluids 171:105198–105209. https://doi.org/10.1016/j.supflu.2021.105198
Radwan MN, Morad MM, Ali MM, Wasfy KI (2020) A solar steam distillation system for extracting lavender volatile oil. Energy Rep 6:3080–3087. https://doi.org/10.1016/j.egyr.2020.11.034
Redondo-Blanco S, Fernández J, López-Ibáñez S et al (2020) Plant phytochemicals in food preservation: antifungal bioactivity: a review. J Food Prot 83:163–171. https://doi.org/10.4315/0362-028X.JFP-19-163
Rodríguez-Solana R, Salgado JM, Domínguez JM, Cortés-Diéguez S (2015) Comparison of soxhlet, accelerated solvent and supercritical fluid extraction techniques for volatile (GC-MS and GC/FID) and phenolic compounds (HPLC-ESI/MS/MS) from lamiaceae species. Phytochem Anal 26:61–71. https://doi.org/10.1002/pca.2537
Sadeh D, Nitzan N, Chaimovitsh D et al (2019) Interactive effects of genotype, seasonality and extraction method on chemical compositions and yield of essential oil from rosemary (Rosmarinus officinalis L.). Ind Crop Prod 138:111419–111425. https://doi.org/10.1016/j.indcrop.2019.05.068
Sakar EH, Zeroual A, Kasrati A, Gharby S (2023) Combined effects of domestication and extraction technique on essential oil yield, chemical profiling, and antioxidant and antimicrobial activities of rosemary (Rosmarinus officinalis L.). J Food Biochem 2023:1–13. https://doi.org/10.1155/2023/6308773
Samber N, Khan A, Varma A, Manzoor N (2015) Synergistic anti-candidal activity and mode of action of Mentha piperita essential oil and its major components. Pharm Biol 53:1496–1504. https://doi.org/10.3109/13880209.2014.989623

Sanchez Armengol E, Harmanci M, Laffleur F (2021) Current strategies to determine antifungal and antimicrobial activity of natural compounds. Microbiol Res 252:126867–126878. https://doi.org/10.1016/j.micres.2021.126867

Sharifi-Rad M, Varoni EM, Iriti M et al (2018) Carvacrol and human health: a comprehensive review. Phytother Res 32:1675–1687. https://doi.org/10.1002/ptr.6103

Sharifzadeh A, Jebeli Javan A, Shokri H et al (2016) Evaluation of antioxidant and antifungal properties of the traditional plants against foodborne fungal pathogens. J Mycol Med 26:e11–e17. https://doi.org/10.1016/j.mycmed.2015.11.002

Silva A, Lopes P, Azevedo M et al (2012) Biological activities of a-pinene and β-pinene enantiomers. Molecules 17:6290–6304

Simirgiotis MJ, Burton D, Parra F et al (2020) Antioxidant and antibacterial capacities of Origanum vulgare L. essential oil from the arid Andean region of Chile and its chemical characterization by GC-MS. Metabolites 10:1–12. https://doi.org/10.3390/metabo10100414

Skendi A, Katsantonis DN, Chatzopoulou P et al (2020) Antifungal activity of aromatic plants of the Lamiaceae family in bread. Foods 9:1642–1655. https://doi.org/10.3390/foods9111642

Soltani S, Shakeri A, Iranshahi M, Boozari M (2021) A review of the phytochemistry and antimicrobial properties of Origanum vulgare L. and subspecies. Iran J Pharm Res 20:268–285. https://doi.org/10.22037/ijpr.2020.113874.14539

Sousa J, Costa A, Leite M et al (2016) Antifungal activity of citral by disruption of ergosterol biosynthesis in fluconazole resistant Candida tropicalis. Int J Trop Dis Health 11:1–11. https://doi.org/10.9734/ijtdh/2016/21423

Sudan P, Singh J (2019) Antifungal potential of Rosmarinus officinalis against Microsporum gypseum and Trichophyton rubrum. Int Res J Pharm 10:205–207. https://doi.org/10.7897/2230-8407.100268

Tsitlakidou P, Papachristoforou A, Tasopoulos N et al (2022) Sensory analysis, volatile profiles and antimicrobial properties of Origanum vulgare L. essential oils. Flavour Fragr J 37:43–51. https://doi.org/10.1002/ffj.3680

Vale-Silva L, Silva MJ, Oliveira D et al (2012) Correlation of the chemical composition of essential oils from Origanum vulgare subsp. virens with their in vitro activity against pathogenic yeasts and filamentous fungi. J Med Microbiol 61:252–260. https://doi.org/10.1099/jmm.0.036988-0

Van Bruggen AHC, Finckh MR (2016) Plant diseases and management approaches in organic farming systems. Annu Rev Phytopathol 54:25–54. https://doi.org/10.1146/annurev-phyto-080615-100123

Vaou N, Stavropoulou E, Voidarou C et al (2021) Towards advances in medicinal plant antimicrobial activity: a review study on challenges and future perspectives. Microorganisms 9:2041–2068. https://doi.org/10.3390/microorganisms9102041

Waller SB, Cleff MB, Serra EF et al (2017) Plants from Lamiaceae family as source of antifungal molecules in humane and veterinary medicine. Microb Pathog 104:232–237. https://doi.org/10.1016/j.micpath.2017.01.050

Waller SB, Hoffmann JF, Madrid IM et al (2018) Polar Origanum vulgare (Lamiaceae) extracts with antifungal potential against Sporothrix brasiliensis. Med Mycol 56:225–233. https://doi.org/10.1093/mmy/myx031

Wesołowska A, Jadczak D (2019) Comparison of the chemical composition of essential oils isolated from two thyme (Thymus vulgaris L.) cultivars. Not Bot Horti Agrobot Cluj Napoca 47:829–835. https://doi.org/10.15835/nbha47311451

Wesołowska A, Jadczak P, Kulpa D, Przewodowski W (2019) Gas chromatography-mass spectrometry (GC-MS) analysis of essential oils from AgNPs and AuNPs elicited Lavandula angustifolia in vitro cultures. Molecules 24:606–618. https://doi.org/10.3390/molecules24030606

Zheng S, Jing G, Wang X et al (2015) Citral exerts its antifungal activity against Penicillium digitatum by affecting the mitochondrial morphology and function. Food Chem 178:76–81. https://doi.org/10.1016/j.foodchem.2015.01.077

Plant Essential Oils and Their Active Ingredients: Antifungal and Therapeutic Potential

3

Sarah Ahmad Khan, Divya Varshney,
Shirjeel Ahmad Siddiqi, and Iqbal Ahmad

3.1 Introduction

The fungi constitute a vast kingdom of more than 3 million species, most of which are non-pathogenic. However, some fungal species are known to be pathogenic to human and plant systems, causing a number of diseases (Sun et al. 2020). Fungal pathogens of the human system, which are constituted of some 200 species of unicellular or filamentous fungi, are known to cause more than a billion infections annually and may result in more than 1.5 million deaths per year (GAFFI 2018). Pathogenic fungal species including *Candida*, *Aspergillus*, *Cryptococcus*, *Pneumocystis*, species of Dermatophytes, etc. may be primary pathogens of the human system or opportunistic pathogens, i.e. they may cause diseases in immunocompromised patients (Geddes-McAlister and Shapiro 2019). According to Walsh and Dixon (1996), fungal infections may be classified as superficial, cutaneous, subcutaneous and deep-seated (aka systemic or invasive) infections, depending upon the site of infection (Walsh and Dixon 1996). While the other three kinds of infections remain within the skin and its underlying mucosal and tissue layers, invasive fungal infections (IFIs) invade the mucosa of the host, and eventually find their way into the respiratory system or the reproductive system, causing serious illnesses in vulnerable patients, resulting in morbidities and mortalities worldwide (Zhang et al. 2023).

Present treatment options against fungal pathogens include five main classes of antifungal drugs, namely, polyenes (includes amphotericin B, nystatin, and natamycin), azoles (fluconazole, ketoconazole, itraconazole, etc.), echinocandins (caspofungin, anidulafungin, and micafungin), allylamines (e.g. terbinafine), and the

S. A. Khan · D. Varshney · S. A. Siddiqi · I. Ahmad (✉)
Department of Agricultural Microbiology, Faculty of Agricultural Sciences, Aligarh Muslim University, Aligarh, India
e-mail: sarahkhan9809@gmail.com; divya9mar@gmail.com; shirjeelahmadsiddiqui786@gmail.com; ahmadiqbal8@yahoo.co.in; iahmad.ag@amu.ac.in

N. Manzoor (ed.), *Advances in Antifungal Drug Development*,
https://doi.org/10.1007/978-981-97-5165-5_3

pyrimidine-derived drugs (e.g. flucytosine) (Jartarkar et al. 2022). Of these classes, polyenes, azoles, and allylamines act mainly on ergosterol biosynthesis, disrupting cell membrane integrity and functions. While echinocandins act on the glucan components of fungal cell wall, the pyrimidines act by inhibiting DNA synthesis (Zhang et al. 2023). These already limited treatment options are further restricted by the development of antifungal resistance by pathogenic fungi and moulds, including emerging pathogens such as *Candida auris* (first identified in 2009), and other species of *Candida*, *Aspergillus*, and some Dermatophytes (CDC 2022c). More recently, pan-drug resistant *Candida auris* (i.e. *C. auris* strains resistant to all available classes of antifungal drugs) was reported as the main cause of several infections, occurring in 30 different countries across six continents (CDC 2023). Such extreme cases of resistance can be linked to the injudicious use of antifungal drugs, not only in the clinical scenario but also in agriculture against phytopathogenic fungi, and the high plasticity and flexibility of the fungal genome, enabling rapid mutations and the development of resistance against stressors (Gow et al. 2022).

Mechanisms of resistance among fungal pathogens include modes similar to that of bacterial pathogens, with increased tendency to form biofilms, expression of efflux pumps, alteration of drug targets such as the cell wall components and ergosterol biosynthesis-related genes, activation of stress signalling pathways, loss of mitochondrial function, etc. (Hokken et al. 2019). Additionally, a major shortcoming of many available drugs is their fungistatic nature, instead of the preferred fungicidal action, which naturally results in increased tolerance and persistence of the pathogenic fungi (Manzoor 2019). Over the years, the issue of antimicrobial resistance in fungal pathogens has emerged as a global public health emergency (Gow et al. 2022). Additionally, resistant fungal infections often call for administration of second-line antifungal drugs, which are limited in their availability, especially in countries of low- and middle income (WHO 2022). The direct cost of treating fungal infections in the USA alone is estimated to be US$11.5 billion, pointing towards the substantiality of the economic burden posed by fungal infections (Benedict et al. 2022). In view of the threats posed by such resistant invasive fungal infections, the WHO launched its first fungal priority pathogens list in 2022 (Table 3.1). The pathogens listed in this report require urgent attention in terms of the distribution of the disease and the mortalities and morbidities posed by such resistant infections (WHO 2022).

Given the steady increase in resistance against frontline antifungal treatments, there is an urgent need to expand the antifungal armamentarium and devise newer, more effective antifungal alternatives (Srinivasan et al. 2014). In the recent times, focus has been shifted to plant-derived natural products and exploring their potential in combating fungal infections (Mishra et al. 2020). Plant-derived natural products including essential oils and other active ingredients offer immense potential to be used as antifungal agents, offering desirable advantages such as low cost, biocompatibility, broad-spectrum bioactivity, low aggression and non-toxicity towards mammalian cells which allows them long-term usage in pharmaceutical and other industries (Khan et al. 2018). Given their significant antifungal activity, this chapter reflects on the role of plant essential oils and their active compounds in combating human fungal pathogens. Additionally, their mechanisms of action and certain

Table 3.1 Fungal priority pathogens (WHO 2022)

Priority group	Fungal pathogens
Critical	*Cryptococcus neoformans* *Candida auris* *Candida albicans* *Aspergillus fumigatus*
High	*Histoplasma* *Candida parapsilosis* *Candida tropicalis* *Candida glabrata*
Medium	*Coccidioides* *Cryptococcus gatti* *Paracoccidioides* spp. *Pneumocystis jirovecii* *Candida krusei* *Talaromyces marnefii* *Lomentospora prolificans* *Scedosporium* spp.

limitations of essential oils that impede their full-fledged applications are also discussed, along with a description of some novel research focusing on improving the properties and antifungal activity of essential oils, in efforts to promote their therapeutic application.

3.2 New Prospects in Phytomedicine: The Antifungal Potential of Plant Essential Oils and Their Active Ingredients

3.2.1 Plant Essential Oils and Their Active Ingredients

Plant-based essential oils are volatile and odorous mixtures of secondary metabolites produced in all kinds of plants and exploited worldwide for different purposes (Hou et al. 2022). In plants, these essential oils are reported to protect the plants against pathogenic microbes, pests, herbivores, and other competing plants as well, i.e. they are allelochemicals (Saroj et al. 2020). Additionally, these compounds, which are insoluble in water and other inorganic solvents, but dissolve well in alcohols, ethers, etc. are also well-recognised for their anti-oxidant, antibacterial, antifungal, anti-inflammatory, and anti-carcinogenic properties, allowing varied applications in pharmaceutical industries (Hanif et al. 2019). More than 3000 plant essential oils are known and at present, and more than 300 commercially recognised are available for applications in perfumery, cosmetics, pharmaceutical, and in the food industry as well (Irshad et al. 2020).

Several parts of the plant can produce essential oils, including the flowers, stems, seeds, barks, leaves, etc., which can be extracted through different processes, including distillation processes such as steam distillation and hydrodistillation, solvent extraction conducted using different solvents such as acetone, methanol, ethanol, petroleum ether, and supercritical CO_2, and solvent-free microwave extraction (Tongnuanchan and Benjakul 2014). Various techniques can be employed to identify the phytocompounds present in essential oils, including Fourier-transform infrared spectroscopy (FTIR), gas chromatography-mass spectrometry (GC-MS), high-performance liquid chromatography (HPLC), liquid chromatography-mass spectrometry (LC-MS), etc. (Aumeeruddy-Elalfi et al. 2016). The constituent compounds present in essential oils may be grouped into two categories, namely, the terpenes and terpenoids, and the phenylpropanoids. The former group of compounds together make up the dominant group of compounds in most essential oils, constituting almost 90% of their chemical makeup (Fokou et al. 2020).

Plant-derived essential oils exhibit desirable antifungal activities, which promote their potential application in formulating alternative treatment options against recalcitrant fungal diseases (Hanif et al. 2019). Essential oils are further preferred because of their general mode of action which offers a broader spectrum of antimicrobial action, are biocompatible with the human system and are not known to cause any adverse reactions in the host, when used in diluted and controlled amounts (de Oliveira Santos et al. 2018). Additionally, given the heterogeneous composition of essential oils, there is very little possibility of resistance development among the target pathogenic yeasts and fungi (Vörös-Horváth et al. 2020). Another advantage of using essential oils as potential antifungal agents in the pharmaceutical and food scenario is that they are Generally Regarded as Safe (GRAS), by the Food and Drug Administration (FDA), facilitating their inconsequential applications in combatting microbial infections, including those caused by fungal pathogens (US FDA 2023). The main plant families whose essential oils are currently under scrutiny for their antifungal action include Lamiacieae, Myrtaceae, Lauraceae, Geraniaceae, and Apiaceae families (D'agostino et al. 2019).

3.2.2 Antifungal Action of Some Essential Oils

Various essential oils have been put to test to assess their antifungal activity. Table 3.2 enlists the recent studies conducted on the antipathogenic activity of certain essential oils against commonly encountered fungal pathogens. As is evident from the table, various essential oils exert considerable antifungal effects against antifungal-resistant pathogens including *Candida* spp., *Aspergillus* spp., *Trichophyton* spp., *Cryptococcus* spp., etc. The most commonly employed preliminary tool for determining the in vitro antifungal potential of any phytocompound, including essential oils is determination of the MIC and MFC values. Further investigations usually involve studies on dimorphic transition, phenotypic switching, altered gene expressions including down/up-regulation of virulence genes, and biofilm formation studies (Manzoor 2019). In the following sections, some exemplar studies on the antifungal effects of some essential oils on common fungal diseases have been discussed. For ease of compilation and accessibility, these studies have been compiled on the basis of the pathogens against which the antifungal activity of essential oils has been studied.

Table 3.2 Some essential oils and their antifungal activity against common human pathogens

Essential oil	Major active ingredient(s) present	Fungal pathogens tested	Observations	Reference
Ferulago capillaris EO	α-Pinene, limonene	*Candida* spp., *Cryptococcus* spp., *aspergillus* spp., *Trichophyton rubrum*, *Microsporum gypseum*, and *Epidermophyton*	FLC-resistant strains of pathogenic fungi inhibited. Reduction in energy production, inhibition of dimorphic transition in a dose-dependent manner	Pinto et al. (2013)
Carum copticum and *Thymus vulgaris* EO	*p*-Cymene, thymol and γ-terpinene in *C. copticum* oil; thymol, γ-terpinene, and α-cymene in *T. vulgaris* oil	*Aspergillus* spp. *Trichophyton* spp.	Thymol exhibited higher antifungal activity than the essential oils and fluconazole (99.75% reduction in growth). Both oils significantly inhibited fungal biomass and mycelial growth	Khan et al. (2014)
Thymus vulgaris EO	Thymol	*Aspergillus*	100 μg/mL of TEO inhibited ergosterol biosynthesis, 150 μg/mL of TEO resulted in total inhibition of aflatoxin biosynthesis, 100 μg/mL of TEO resulted in significant alterations in conidia, conidiophores and hyphae	Kohiyama et al. (2015)
Foeniculum vulgare EO	*Trans*-anethole, pinene and fenchone	*Trichophyton*, *Microsporum gypseum*	FEO demonstrated better antifungal action than amphotericin B and fluconazole, by disrupting cell membrane and intracellular organelles	Zeng et al. (2015)
Eucalyptus citriodora EO	Citronellal, citronellol, and isopulegol	*Microsporum canis*, *M. gypseum*, *Trichophyton rubrum*, and *T. mentagrophytes*	Inhibition of fungal growth at MIC and MFC ranging between 0.6–5 and 1.25–5 μL/mL	Tolba et al. (2015)
Zataria multiflora EO	Carvacrol, thymol cymene	*T. rubrum*, *T. metagrophytes*, *T. schoenleinii*, *M. canis*, and *M. gypseum*	Inhibition of mycelial growth and elastase production	Mahboubi et al. (2017)

(continued)

Table 3.2 (continued)

Essential oil	Major active ingredient(s) present	Fungal pathogens tested	Observations	Reference
Clove and peppermint oils	Eugenol and beta-caryophyllene in clove oil Menthol and menthone in peppermint oil	*Candida*	Reduced or complete loss of enzymatic activity of alkaline phosphatases, alpha-glucosidase, esterase, phospholipase upon treatment with 0.0075% of the oils, suggesting a decrease in the pathogenicity of the tested isolates	Rajkowska et al. (2017)
Myrtus communis EO	α-Pinene, limonene, geraniol, geranyl acetate	*M. furfur* *M. sympodalis*	Highest inhibition at MIC and MFC values of 31.25 and 62.5 μL/mL for *M. furfur* and 62.5 and 125 μL/mL for *M. sympodalis*, respectively	Barac et al. (2018)
Pelargonium graveolens EOs	Citronellol, geraniol, citronellyl formate, linalool	*T. mentagrophytes*, *T. ruburm*, *M. gypseum*, *M. canis*, *T. schoenleinii*	*P. graveolens* essential oils with high amounts of geraniol and citronellol successfully inhibit dermatophytes	Mahboubi and Valian (2019)
Thymus vulgaris oil	Thymol	*C. albicans*, *C. tropicalis*	Significant inhibition of biofilm formation observed at sub-MICs. Synergetic antibiofilm effect between TEO and FLC and amphotericin B was observed	Jafri and Ahmad (2020)
Thymus serpyllum EO	Thymol, phenol, *o*-cymene	*C. albicans*	MIC range between 0.039–0.078%, while thymol yielded an MIC of 2.5–5%. Synergistic effects yielded upon combination of 2 × MIC of EO with ½ MIC of fluconazole, and ½ MIC of EO with 2 × MIC of amphotericin B In silico investigations and docking simulations revealed favourable interaction of thymol with fungal proteins	Salaria et al. (2022)

Table 3.2 (continued)

Essential oil	Major active ingredient(s) present	Fungal pathogens tested	Observations	Reference
Melaleuca alternifolia EO	Terpinenol, and gamma-terpinene	*Trichophyton rubrum*	MIC and MFC ranging between 0.06–0.12% (v/v) for the EO. Terpinenol yielded MIC of 0.06% (v/v) while gamma-terpinene yielded an MIC of 0.5% (v/v). Synergistic combinations of EO with ketoconazole and itraconazole, allowing lower doses of azoles and reduced side-effects	Roana et al. (2021)
M. communis EO	• α-Pinene, 1,8-cineole, myrtenyl acetate limonene and linalool	*A. flavus*	90.5% inhibition of fungal growth at 3.0 μL/mL of EO, spore germination was also inhibited in a dose-dependent manner, with up to 94% inhibition Significant antioxidant potential also observed Couscous samples were fumigated with the vapours of the essential oils, up to 86% protection imparted to the treated samples, over a period of 6 months	Miri et al. (2023)
Seven essential oils from Lamiaceae family	• Citronellol, citronellal, *p*-cymene, thymol, carvacrol, and camphene	*C. albicans*, *C. krusei*, *C. parapsilosis*, *C. tropicalis*, *C. glabrata*, and *C. guilliermondii*	*Oregano vulgare* and *Melissa officinalis* exhibited higher anticandidal activity than others (MIC lower than 3.125 mg/mL) Thyme oil and oregano oil exhibited remarkable antibiofilm efficacy against *Candida* biofilms of more than 90% of the biofilm	Karpiński et al. (2023)

3.2.2.1 Anticandidal Activity of Essential Oils

Although a normal resident of the human system, *Candida* spp. is also known to cause various kinds of superficial and systemic infections, especially in immunocompromised individuals (Ahmedi et al. 2022). Candidiasis is a common term given to any opportunistic infection caused by the species of *Candida*. These include infections of the skin, the oropharynx (aka oral thrush), gastrointestinal tract, and urogenital tract (including vulvovaginal candidiasis), and the bloodstream (i.e. candidemia) and internal organs, including the brain (candida meningitis), heart (endocarditis), lungs (pulmonary candidiasis), kidneys (renal candidiasis), and the liver (hepatic candidiasis) (Kullberg and Arendrup 2015). *C. albicans* is reported as being the causative agent in most candidiasis-related infections (CDC 2022a). Other infectious *Candida* spp. include non-albicans species, such as *C. krusei*, *C. tropicalis*, *C. parapsilosis*, *C. glabrata*, etc. (Arya and Rafiq 2023). These five species of *Candida* are responsible for 90% of all candida infections (Turner and Butler 2014). Their pathogenicity is credited to the expression of adhesins, which cause adhesion, the first step in the establishment of pathogenicity, along with expression of virulence genes, a morphological transition from the yeast to the hyphal state, and secretion of hydrolytic enzymes such as secretory aspartyl proteases, hydrolases, proteases, etc. (Pappas et al. 2018). Severe infections can only be treated by azoles, echinocandins, or amphotericin B (CDC 2023). Presently, the most commonly employed drug against *Candida* infections is fluconazole, although other azoles, which normally target ergosterol biosynthesis, are also used (de Oliveira Santos et al. 2018). However, rapidly increasing cases of fluconazole-resistant strains of both albicans and non-albicans *Candida* are being reported, with mechanisms extending to altered ergosterol biosynthesis and increased expression of efflux pump genes, among others (da Nóbrega Alves et al. 2020). Given such threatening circumstances, it is urgent that alternative methods of treatment against *Candida* spp. be developed.

Various studies have been conducted with the objective of determining the potential of essential oils as effective agents in combating candidal infections and biofilms. For example, in a study conducted by Alshaikh and Perveen (2021) thyme oil, obtained after steam distillation of the plant parts, was tested for its in vitro efficacy against clinical isolates of fluconazole-resistant strains of *C. albicans*. Preliminary analysis via disk diffusion assay revealed favourable anticandidal action of thyme oil against the isolates, with higher antifungal activity than fluconazole. Furthermore, MIC and MFC results showed that thyme oil at concentrations as slow as 0.6 μL/mL was effective in inhibiting candidal growth in vitro, and MFC was recorded at 1.25 μL/mL. Further studies revealed inhibition of budding and germ tube formation in the isolates (Alshaikh and Perveen 2021).

In another recent study conducted by Karpiński et al. (2023), seven essential oils belonging to the family Lamiaceae were assessed for their anticandidal efficacy, against six species of *Candida*—*C. albicans*, *C. krusei*, *C. parapsilosis*, *C. tropicalis*, *C. glabrata*, and *C. guilliermondii*. The seven essential oils tested were *Lavendula stoechas*, *Melissa officinalis*, *Metha × piperita*, *Origanum vulgare*, *Rosmarinus officinalis*, *Salvia officinalis*, and lastly, *Thymus vulgare*. Based on the

MIC and MFC tests, oils of *Oregano vulgare* and *Melissa officinalis* exhibited higher anticandidal activity than the other oils, with an MIC lower than 3.125 mg/mL, along with thyme oil, lavender oil and rosemary oil showing significant antifungal action, inhibiting most of the tested strains. Additionally, the antibiofilm activity of the essential oils against biofilms of *C. albicans*, *C. krusei*, and *C. glabrata* was assessed, revealing high inhibition by thyme oil and oregano oil, with inhibition of more than 90% of the biofilm. It was observed that the essential oil of *M. officinalis* showed the highest activity against the planktonic forms of the fungus, but failed to effectively inhibit the biofilm mode. Possible toxigenic effects of the active compounds present in the Lamiaceae family (including citronellol, citronellal, *p*-cymene, thymol, carvacrol, camphene, etc.) were also evaluated, to validate their potential use in human systems as anti-candidal agents. In silico approaches using ProTox-II and pkCSM softwares were employed to assess the same, revealing higher lethal doses and no significant mutagenic, cytotoxic or carcinogenic effects of the major phytoconstituents present in the essential oils. However, out of the 21 compounds evaluated, 16 showed possible skin sensitisation effects, suggesting possible skin irritation upon their topical usage (Karpiński et al. 2023).

3.2.2.2 Antidermatophytic Activity of Essential Oils

Dermatophytes, comprising fungal pathogens of the skin, hair and nails, are responsible for skin-related mycoses, referred to as dermatophytoses. They include fungi like the *Trichophyton* spp. and others such as *Microsporum* and *Epidermophyton* (Jartarkar et al. 2022). They are collectively associated with a group of skin-related mycoses, clinically known as tenia, and classified according to infection site as tinea corporis (body), tinea manuum (hand), tinea capitis (hair and scalp), tinea barbae (beard), tinea pedis (foot), and tinea unguium (nails), among others (Goldstein and Goldstein 2017). Current treatment options available for dermatophytic diseases include clotrimazole, miconazole, terbinafine, itraconazole, fluconazole and griseofulvin, which are administered orally. Oftentimes, interventions include a combination of oral and topical treatments as well (Brescini et al. 2021; CDC 2021). However, cases of dermatophyte resistance, especially against terbinafine, have increased over the years. Mechanisms of such resistance include expression of ATP-binding cassettes (ABC) transporters in efflux pumps, mutations in squalene epoxidase gene (which is the main target of terbinafine), and reduced biosynthesis of cell membrane components, including ergosterol (Sacheli and Hayette 2021).

In an attempt to study the antidermatophytic action of essential oils, Barac et al. (2018) tested essential oils of *Myrtus communis* were for their potential to treat infections caused by *Malasezzia* spp., a type of yeast, known to cause pityriasis versicolor, a skin infection. Following isolation of yeasts from infected patients and their identification, *M. communis* EO was tested against the yeast cells via the broth microdilution method. Appreciable inhibition of the pathogenic yeast cells was observed, with highest inhibition exhibited against *M. furfur* and *M. sympodalis* (96% and 83%, respectively), with MIC and MFC values of 31.25 and 62.5 μL/mL for *M. furfur* and 62.5 and 125 μL/mL for *M. sympodalis*, respectively (Barac et al.

2018). Such studies suggest the use of essential oils obtained from the family Myrtaceae as a possible treatment option against infections of the skin, hair and mucous membranes (Barac et al. 2018). The authors further warranted the need to conduct appropriate in vivo experiments in order to confirm the safety and inconsequential application of such EOs for topical and systemic applications (Barac et al. 2018).

In another study, members of the family Apiacceae (the Umbellifers) commonly used as spices, namely, *Trachyspermum ammi* (ajowan), *Coriandrum sativum* (coriander), *Carum carvi* (caraway), and *Pimpinella anisum* (anise), also well known for their therapeutic effects, were evaluated for their action against *T. rubrum* and *T. mentagrophytes*. *T. ammi* essential oils demonstrated highest efficacy. Further studies involved analysing possible synergetic interactions between the essential oils and terbinafine, a drug commonly used to treat *Trichophyton* infections. Checkerboard assays revealed synergy between the essential oils and terbinafine, with better synergy obtained for coriander and anise EOs with the antifungal drug. Further studies involved cytotoxicity assessments on human neutrophils, yielding favourable results without any adverse effects on the neutrophil viability. Furthermore, the effects of the resultant synergistic concentrations of EOs on pro-inflammatory cytokine production (IL-1β, IL-8, and TNF-α secretion) were assessed, using an ex-vivo stimulated neutrophil model. Results revealed most potent inhibition of cytokine release by anise EO. Such studies provide us the rationale of developing effective essential oil-based treatment methods which do not result in the development of inflammatory reactions upon their application, promoting their use as safe topical agents (Trifan et al. 2021).

3.2.2.3 Antiaflatoxigenic Activity of Essential Oils

Aflatoxins (AF) are a group of mycotoxins produced by fungi occurring on decaying crops such as cereals such as maize, wheat, sorghum, etc., oilseeds including peanut, soybean, etc., nuts such as almond, pistachio, walnut, etc., among other food crops (Kumar et al. 2021). Although other fungal species also produce these fatal secondary metabolites (including *Penicillum*, *Fusarium*, etc.), the aflatoxins produced by *Aspergillus flavus* and *Aspergillus parasiticus* are mycotoxins that present a great risk to human health, owing to their teratogenic, mutagenic and carcinogenic effects, translated into the human body via dietary exposure (Zhaveh et al. 2015). Aflatoxins are mainly of four kinds, namely, AFB_1, AFB_2, AFG_1, and AFG_2, of which AFB_1 is the most fatal, owing to its hepatocarcinogenic nature (Ramadan and Al-Ameri 2022). Symptoms of aflatoxin poisoning include vomiting, abdominal cramps, and symptoms of acute liver injury, while long-term exposure may result in cirrhosis and hepatocarcinoma (Kumar et al. 2021). Since at present, there is no known antitoxin for aflatoxicosis, and treatment relies on supportive care, the best solution is prevention. Exposure to mycotoxins can be reduced by following appropriate post-harvest treatment strategies, and following some dietary modifications, if there's a risk of an outbreak, while some experts suggest the use of chemopreventive measures upon exposure, including certain phytochemicals (Dhakal et al. 2023).

Essential oils offer considerable potential against such toxigenic fungal strains, as was demonstrated by Kohiyama et al. (2015). Various concentrations of *T. vulgaris* essential oil (TEO), obtained from hydrodistillation of thyme leaves were tested for their action against ergosterol and aflatoxin production. Note that 100 and 150 μg/mL of TEO were found to exhibit antifungal activity in vitro, resulting in a reduction of ergosterol production (49.6% and 98.1%, respectively), due to interference in the growth and proliferation of the fungal cells. Furthermore, the concentrations of TEO also resulted in degenerative alterations in the morphological features of the conidiophores and the hyphae. Additionally, significant activity was demonstrated at low concentrations (50 μg/mL) of TEO against aflatoxin production. Total inhibition of the aflatoxins (AFB_1 and AFB_2) production was observed at a concentration of 150 μg/mL of the TEO, owing to inhibition of toxin production (Kohiyama et al. 2015).

Essential oils have been explored as plant-based preservatives against aflatoxigenic strains of *Aspergillus* spp., as was reported by a recent study conducted by Miri et al. (2023). Three essential oils, namely obtained from *Mentha piperita*, *Mentha pulegium*, and *Myrtus communis* were tested for their antiaflatoxigenic action against strong aflatoxin-producing *Aspergillus flavus* strains, isolated from Couscous. Inhibition of mycelial growth in a dose-dependent manner was found to occur in all cases, with *M. communis* EO exhibiting highest inhibition of fungal growth (90.5%) at 3.0 μL/mL. Spore germination was also inhibited in a dose-dependent manner, with up to 94% inhibition exhibited by *M. communis* oils. Since oxidation of food acts as a trigger stimulating higher toxin production by pathogens, the antioxidant activity of the EOs were also tested using the DPPH radical scavenging assay, revealing significant antioxidant potential in the EOs. Additionally, couscous samples were fumigated with the vapours of the essential oils, to evaluate their applications as food preservatives in the industry. Analysis after 6 months of treatment revealed that *M. communis* has the highest efficacy in preserving the food samples, with up to 86% protection imparted to the treated samples (Miri et al. 2023).

3.2.2.4 Anticryptococcal Activity of Essential Oils

Cryptococcosis is a major systemic fungal infection caused in humans by two species of *Cryptococcus*, viz., *C. neoformans* and *C. gatti*. Of these, *C. neoformans* has been declared a fungal pathogen of critical priority by the WHO FPPL (WHO 2022). Cryptococcosis is known to mainly affect immunocompromised individuals, and manifests itself in the form of a pulmonary infection known as pulmonary cryptococcosis, and may also result in the infection of central nervous system (CNS cryptococcosis), leading to the deadly cryptococcal meningitis. The pathogenicity of this yeast is associated with the production of a virulent antiphagocytic capsule, and melanin production (Maziarz and Perfect 2016). Present treatment options against cryptococcosis include either fluconazole or a combination of flucytosine with amphotericin B, followed by fluconazole, depending on the severity of the infection (CDC 2022b). However, cases of resistance against these drugs are on the rise, calling out for urgent development of novel anti-cryptococcal agents (Bermas and Geddes-McAlister 2020). In an attempt to study the anti-cryptococcal potential of essential oils, Scalas et al. (2018) tested three essential oils, viz., *Pinus sylvestris* EO, *Origanum vulgare* EO, and *Thymus vulgaris* EO against strains of *C. neoformans*. Of the three essential oils, pine oil exhibited the highest inhibitory effect on all the strains tested, suggesting its potential in the treatment of azole-resistant strains of *C. neoformans* (Scalas et al. 2018).

3.2.3 Antifungal Activity of Active Ingredients of Essential Oils

The antifungal action of the commonly reported essential oils is credited to the presence of several bioactive compounds, grouped into the terpenes and terpenoids, and the phenylpropanoids (Abd Rashed et al. 2021). These compounds may be aliphatic (terpenes) or aromatic (terpenoids and phenylpropanoids). Among the aromatic compounds, the commonly occurring groups include aldehydes, phenols, alcohols, ketones, esters, and lactones (Noriega 2020). These aromatic compounds impart a characteristic fruity, floral or herbal smell to the essential oils (Dhifi et al. 2016).

The terpenes and terpenoids constitute the major classes of phytochemicals (may constitute more than 80% of the essential oil), and are known to commonly occur in essential oils of various plants (Baptista-Silva et al. 2020). The terpenes are colourless and highly volatile derivatives of isoprene (C_5H_8) units, and are further grouped on the basis of the number of isoprene units and carbon atoms into the monoterpenes (C_{10}, i.e. 2 isoprene units), sesquiterpenes (C_{15}, i.e. 3 isoprene units), diterpenes (C_{20}, i.e. 4 isoprene units), triterpenes (C30, i.e. 6 isoprenoid units), and the polyterpenes (IUPAC 2006a). Among these, the essential oils are mainly dominated by the monoterpenes and the sesquiterpenes (Noriega 2020). The terpenoids, on the other hand, are modified forms of terpenes, containing functional groups with oxygen (IUPAC 2006b), and are derivatives of aldehydes, phenols, alcohols, esters, ketones and the ethers (Pandey et al. 2017). The phenylpropanoids are the other class of compounds found in various essential oils. These compounds are derived from phenylalanine through a different pathway than that of the terpenes and terpenoids, and comprise of ingredients such as eugenol and cinnamaldehyde, among others (Dhifi et al. 2016).

Some examples of the different compounds found in essential oils have been listed in Table 3.3, along with the chemical structures of some commonly reported active ingredients in Fig. 3.1.

According to most reports, only three to four of these compounds actually make up the majority of the essential oil composition (up to 85%), while others are present only in traces (Chouhan et al. 2017). These compounds are speculated to synergistically contribute to the antifungal and/or therapeutic effects, therefore being referred to as the 'active' components/ingredients of the oil (Cimino et al. 2021).

It must be noted that the chemical composition of an essential oil may vary, depending upon the variety of the plant, the part of the plant used for extraction of oil, and the season of harvest (D'agostino et al. 2019). It should also be acknowledged that the chemical composition of essential oils obtained from different plant sources is different, and as such, different essential oils have different active

Table 3.3 Classification of the active ingredients present in essential oils

Class of compounds	Common compounds present in essential oils
Terpenes	Pinene, myrcene, limonene, cymene, thujene, phellandrene, caryophyllene, curcumene, farnesene, etc.
Terpenoids (oxygenated derivatives of terpenes)	
Aldehydes	Cuminaldehyde, geranial, citral, citronellal, etc.
Phenols	Carvacrol, thymol, cresol, etc.
Alcohols	Geraniol, linalool, terpineol, citronellol, myrtenol, menthol, etc.
Ketones	Menthone, carvone, thujone, camphor, etc.
Lactones	Bergaptrol, umbelliferone, coumarins, etc.
Esters	Benzyl acetate, geranyl acetate, citronellyl butyrate, citronellyl formate, eugenyl acetate, methyl butyrate, etc.
Phenylpropanoids	Eugenol, cinnamaldehyde, chavicol, etc.

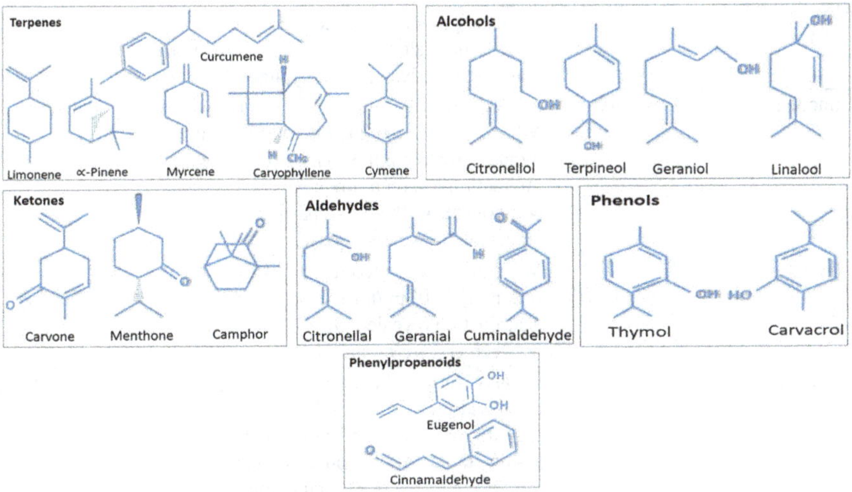

Fig. 3.1 Chemical structures of some active ingredients found in essential oils

ingredients and therefore may demonstrate variable therapeutic effects (NIEHS 2022). Interestingly, 'chemotypes' of plants are also known to exist, which harbour different concentrations of different compounds. For example, different chemotypes of *Thymus vulgaris* are known, according to the major compound (thymol, carvacrol, linalool, etc.) present in their essential oil (Baptista-Silva et al. 2020). Such variation makes it very difficult to standardise the application of a particular oil, necessitating the need to conduct a thorough analysis of the phytochemical composition of the oil, before appropriation of any therapeutic application.

Table 3.4 lists some significant studies on the antifungal activity of some active ingredients found to occur in essential oils. Most of these studies have been conducted in vitro, and are focussed on understanding the antifungal mechanism of action of the active compounds, with *Candida*, *Cryptococcus* and *Trichophyton* spp.

Table 3.4 Antifungal activity of active ingredients of essential oils

Active compound	Organism tested	Observations	References
Thymol	*Aspergillus flavus*	Efficient inhibition of mycelial growth and lysis of fungal spores upon exposure to 80 μg/mL of thymol was observed, along with ROS-mediated antifungal activity, via the production of nitric oxide (NO)	Shen et al. (2016)
Carvacrol	*C. neoformans*	MIC values were recorded at ranges between 25 and 81 μg/mL, while highest MFC was recorded at 102 μg/mL. Ergosterol assays revealed exogenous binding of carvacrol to ergosterol, indicating membrane disruption action	Nóbrega et al. (2016)
Linalool	*T. rubrum*	MIC values recorded between 256 and 512 μg/mL in fluconazole and flucytosine-resistant strains. Significant inhibition of mycelial growth and conidiogenesis and germination, along with leakage of cellular material, and dysmorphic hyphae	de Oliveira Lima et al. (2017)
Limonene	*T. rubrum*	Mycelial growth inhibition was observed at MIC (2 μL/mL), along with destruction of hyphal structures. In silico studies revealed interaction between limonene and cell wall components of the yeast, particularly lanosterol 14α-demethylase	Padhan et al. (2017)
Thymol	*C. neoformans*	MICs and MFCs recorded at 20–51 μg/mL, and 40 and 101 μg/mL, respectively. Further examination revealed probable interaction between ergosterol and thymol	Teixeira et al. (2020)
Geraniol	*C. albicans*	Geraniol targets the drug efflux pumps, specifically CaCdr1p, and also inhibits fungal growth and biofilm formation when administered with fluconazole, increasing the susceptibility of resistant strains as well. Significant and synergistic inhibition of mitochondrial activity and virulence factors also observed. Examinations in C. *elegans* model revealed enhanced survival and	Singh et al. (2018)
Carvacrol	*C. albicans*	MIC value of 247 μg/mL resulted in a 42.6% decrease in the survival of the fungal cells. Further examinations revealed elevated ROS, decreased mitochondrial membrane potential and disruptions in cell membrane integrity and permeability, along with DNA fragmentation and increased metacaspase activity, indicating cellular apoptosis. In vivo investigations in mice models revealed enhanced survival rates in the *Candida*-infected test group upon treatment with carvacrol, compared to the control group, which died within 10 days	Niu et al. (2020)

Table 3.4 (continued)

Active compound	Organism tested	Observations	References
Limonene	*C. albicans*	Inhibition of fungal structures and dimorphic transition observed, along with 73% reduction in the synthesis of proteinases, more than 90% reduction in adhesion and 87% reduction in biofilm formation at an MIC of 300 μg/mL. In silico studies revealed good interaction between limonene and virulence factors Plb1, Tec1, Als3, and Sap2	Ahmedi et al. (2022)
Cinnamaldehyde	*C. glabrata*	MFC was observed at 256 μg/mL, with significant inhibition of biofilm formation on urinary catheters and contact lenses	Gupta et al. (2018)
Cinnamaldehyde	*A. fumigatus*	Immunosuppressed mice models were treated with oral doses of cinnamaldehyde (240 mg/kg day) after infection with spores of *A. fumigatus*. Investigations after 14 days of treatment revealed up to 80% clearance rate in pulmonary fungi, with a possible effect on the 1,3-β-D-glucans component (cell wall integrity) of the fungi	Deng et al. (2018)
Linalool	*C. albicans* and *C. tropicalis*	MIC recorded at 1000 μg/mL and 500 μg/mL, respectively. Linalool found effective against FLC-resistant *C. albicans* as well	Dias et al. (2017)
β-Citronellol	*Candida* spp.	Inhibition of yeast to fungal transition, inhibition of biofilm formation, along with downregulation of virulence genes and ergosterol biosynthesis genes	Sharma et al. (2020)
Cinnamaldehyde	*C. albicans*, *C. tropicalis*	Inhibition of yeast growth and reduction in biofilm formation observed. Molecular docking studies revealed the interaction of cinnamaldehyde with squalene epoxidase, suggesting disruption of ergosterol biosynthesis. Cytotoxicity studies on human keratinocytes and erythrocytes revealed favourable results	da Nóbrega Alves et al. (2020)
Citronellol	*T. rubrum*	MICs recorded at >0.012 μL/mL, indicating excellent antidermatophytic activity	Jain and Sharma (2020)
Eugenol	*C. neoformans*	100% inhibition of fungal growth at 500 μg/mL. RT-PCR revealed significant downregulation (by 54%) *Cxt1-p* gene, indicating inhibition of xylosyltransferase involved in capsule formation and virulence. Synergy-based studies revealed considerable synergetic interaction between eugenol and fluconazole	Hassanpour et al. (2020)

(continued)

Table 3.4 (continued)

Active compound	Organism tested	Observations	References
Limonene	*C. albicans*	A concentration of ≥500 μM yielded inhibition of yeast growth and viability. In vivo evaluation in mice models demonstrated better antifungal activity in the mice models, compared to the fluconazole-treated set. However, upon SEM examination, some minor signs of irritation were observed in the vulvovaginal epithelial tissues	Muñoz et al. (2020)
Thymol	*C. neoformans*	Downregulation of genes involved in ergosterol biosynthesis, disruption in intracellular calcium ion homeostasis and disruption in activity of endoplasmic reticulum	Jung et al. (2021)
Eucalyptol	*C. albicans*, *C. glabrata*	ROS production along with downregulation of genes involved in cell wall biosynthesis, secreted enzymes, and ABC transporters. Eucalyptol further arrested the cell cycle at the G1/G2 phase	Gupta et al. (2021)

included as the test organisms. From these studies, it is quite evident that these compounds act through membrane disruption and ROS production and may demonstrate a better effect when combined with conventional antifungal drugs.

3.2.4 Mechanism of Antifungal Action of Essential Oils and Their Active Ingredients

The antifungal action of essential oils is credited to the variety of bioactive compounds present in them, which possess inherent antifungal activity and are known to have different sites of action within the fungal cell (Nazzaro et al. 2017). Figure 3.2 depicts the common routes of antifungal action of essential oils. As is apparent from the figure, a generally accepted mode of action involves the disruption of the fungal cell membrane due to the irreversible binding of the essential oils with the cell membrane components (Swamy et al. 2016).

The toxicological effects of essential oils or, essentially, their active compounds are credited to the presence of certain functional groups on the compounds, which are usually lipophilic in nature and thus interfere with and cause disruption of cell membrane structure and permeability, and are reported to be the main mode of action of the essential oils (Sharifi-Rad et al. 2017). Permeabilisation of the essential oil or their active compounds by the cell wall and cell membrane leads to further damage, disrupting the mitochondrial activity, and affecting proton pumps and ATP production as well, which further translates to damage of the cellular lipids including ergosterol, proteins and even the nucleic acids (Gupta et al. 2018). Since essential oils and their active compounds mainly exercise their antifungal effects by

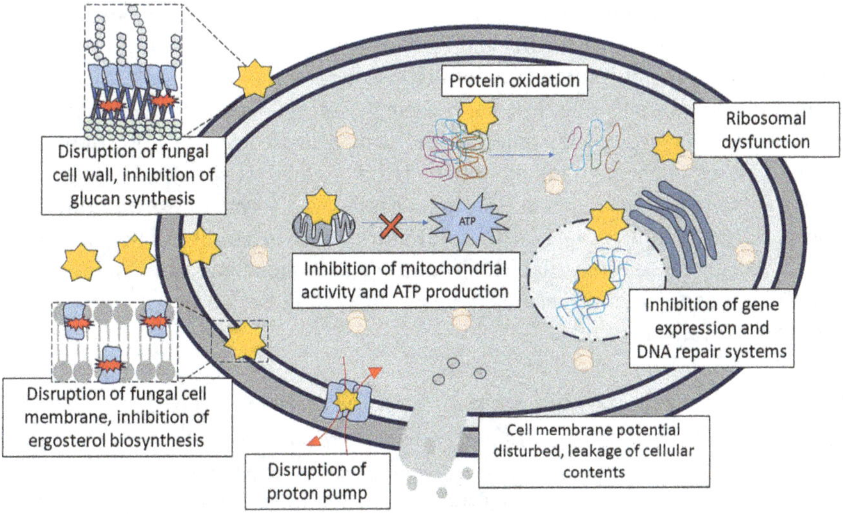

Fig. 3.2 Common routes of antifungal action of essential oils and/or their active ingredients

targeting the cellular membranes, there is ultimately a change in the fluidity of the membranes, leading to leakage of cellular components, alteration in the biochemical metabolic processes, and ultimately, an overall collapse of the cellular system (Mani-López et al. 2021).

Additionally, EOs are also known to inhibit sporulation and germination of the fungal cells, thereby inhibiting fungal growth (Miri et al. 2023). They are also associated with disruption of hyphal growth, inhibition of yeast to hyphal dimorphism, significant alteration of ultrastructural cellular components including mitochondrial membrane and their functions, and an overall pro-apoptotic effect, observed by various researchers in this regard (Niu et al. 2020; Ahmedi et al. 2022). For example, the essential oils of *Antheum graveolens* were tested for their potential against *Candida*, and their mechanism of action was also studied. Experiments revealed a significant decrease in ATPase activity, cytochrome c production, and significant overproduction of ROS, leading to apoptosis (Chen et al. 2014).

3.3 Improved Forms of Essential Oils and Their Active Ingredients

Despite their useful applications in developing alternative medicine and advantages such as ease of availability and broad-spectrum of antimicrobial action, essential oils are faced with certain limitations that restrict their full-fledged applications in the pharmaceutical field (Barros and Casey 2020). These include their volatility, insolubility and chemical instability under normal conditions of humidity, light, heat and oxygen, which significantly impact their antimicrobial properties (Barradas and de Holanda e Silva 2021).

To overcome such challenges, researchers have attempted to develop improved forms of essential oils and/or their active ingredients through various strategies, but are not limited to combination of essential oils with biopolymers/ synthetic polymeric molecules; formulating essential oils into emulsions and liposomes or, manipulating them at the nanoscale to form nanoliposomes or solid lipid nanospheres, among others (Abd Rashed et al. 2021). Table 3.5 enlists some recent attempts at developing improved forms of essential oils for pharmaceutical applications, with regard to fungal infections. Such formulations offer increased bioavailability, stability and reduced volatility and toxicity, and allow a controlled release of the essential oil and/or its active compounds (Cimino et al. 2021).

Table 3.5 Improved forms of some essential oils and their antifungal activity

Essential oil	Improved form	Tested microbes	Observations	References
Orange essential oils	Inclusion complexes with cyclodextrin	*Aspergillus niger* *A. flavus*	EO-cyclodextrin complexes were synthesised. Appreciable inhibition of fungal growth observed. However, overall effect of EO-CD complexes was lower than that of pure EO	Torres-Alvarez et al. (2020)
Bidens tripartita	Topical gel-based formulations	*Candida krusei*, *C. albicans*, *C. tropicalis*, *C. glabrata*, and *C. parapsilosis*	Hydrogel-based formulation exhibited anticandidal activity against all strains tested	Tomczykowa et al. (2018)
Cymbopogon martini	Topical formulations using different bases	*C. albicans*, *Trichophyton rubrum*, *T. verrucosum*, *T. mentagrophytes*, *Microsporum canis*, and *Aspergillus niger*	Hydrophilic ointment demonstrated antifungal effect on all the tested strains. Safety evaluation of the formulation on Guinea pigs implied non-toxicity at 5% concentrations	Gemeda et al. (2018)
Melaleuca alternifolia	Pickering emulsions of EO with tioconazole, with silica nanoparticles as stabiliser	*T. rubrum* and *C. albicans*	Sire-specific and efficient delivery of the antifungal agents, within 10 min of application	Vörös-Horváth et al. (2020)

Table 3.5 (continued)

Essential oil	Improved form	Tested microbes	Observations	References
Satureja montana and *Thymus capitatus*	Complexes with cyclodextrin molecules synthesised into tablets containing 15–25 mg of the CD-EO powders, with polyvinyl pyrollidone (PVP) added in as a binder	*C. albicans* and *C. glabrata*	Significant activity, higher than that observed with clotrimazole Tablet form had a longer dissolving time, indicating the possibility of longer exposure time within patients	Arrais et al. (2021)
Cuminum cyminum	Encapsulation in chitosan-caffeic acid nanogel	*Aspergillus flavus*	Increased efficiency of the essential oils, with controlled release of the active ingredients, and enhanced bioavailability	Zhaveh et al. (2015)
Carvacrol-loaded chitosan NPs		*C. albicans, C. krusei, C. tropicalis, C. glabrata*	Significant inhibition of fungal growth and enhanced inhibition and eradication of biofilms, compared to free carvacrol	Vitali et al. (2021)
Eugenol and linalool loaded onto (poly) acrylic acid based emulgel		*T. rubrum*	Emulgel proved to be more effective than pure eugenol or linalool, producing a higher zone of inhibition. In vivo studies in rabbits support safe topical applications	Akram et al. (2021)

3.3.1 Inclusion Complexes of Essential Oils with Polysaccharides

In a novel attempt, antifungal essential oils of winter savory (*Satureja montana*) and oregano (*Thymus capitatus*), exhibiting good activity against *Candida* strains, were sonicated with lipophilic beta-cyclodextrin polymers, resulting in the formation of 'host–guest complexes', with cyclodextrin molecules hosting the EO 'guest' molecules. The resulting mixture was characterised via FTIR and mass spectrometry. This particular formulation was then synthesised into tablets, containing 15–25 mg of the CD-EO powders, with polyvinyl pyrollidone (PVP) added in as a binder. Antifungal tests using the novel tablets and clotrimazole were carried out via the diffusion assay on two reference strains, *C. albicans* ATCC 14053 and *C. glabrata* ATCC 15126. Interestingly, both EOs and their tablet formulations with cyclodextrin yielded significant activity, higher than that observed with clotrimazole.

However, bare EOs did exhibit higher antifungal effects than the tablet forms. However, the tablet form had a longer dissolving time, indicating the possibility of longer exposure time within patients, thereby potentially having greater efficacy than the bare EOs. This is also listed as an advantage by the authors, since pure EOs may give rise to side effects including inflammatory reactions, and the controlled release of such solid-state EOs could possibly reduce such effects (Arrais et al. 2021).

In another study conducted by Torres-Alvarez et al. (2020), concentrated oils of orange (5×, 10×, and 20×) were employed in the development of inclusion complexes with the oligosaccharide β-cyclodextrin (β-CD). This oligosaccharide is often employed in the pharmaceutical industry to increase the availability of poorly soluble bioactive compounds. This particular study attempted the encapsulation of orange essential oils in β-CD, obtained via a co-precipitation method, and further investigated the antifungal efficiency of the formulation on fungal pathogens, namely, *Aspergillus niger* and *A. flavus*. Three different concentration ratios (EO: CD) were analysed, i.e. 4:96, 12:88, and 16:84. The essential oil–β-CD complexes thus obtained had entrapment efficiencies ranging between 70% and 89.5%, depending upon the concentration ratios of the essential oils to the oligosaccharide, and the highest efficiency and recovery were reported for the 20× concentration of EO, mixed in a 12:88 ratio with the cyclodextrin polymer. Although complete inhibition of fungal growth by the newly formed complexes was not observed in this study, the antifungal effect against *A. niger* was predominantly observed for the encapsulated EO at 10× concentration, in two of three concentration ratios tested. Interestingly, the bare essential oils showed greater inhibition of mycelial growth than the complexed EOs, suggesting that complexation may not always increase bioactivity and bioavailability (Torres-Alvarez et al. 2020).

3.3.2 Gel-Based Formulations of Essential Oils

Topical gel-based formulations in the form of ointments were explored for their efficacy as carriers of antifungal essential oil of *Bidens tripartite.* The EOs obtained after hydrodistillation were further analysed by GC-MS using the pharmacopoeia method. Cymene, linalool, and phellandrene were among the major compounds detected. Different gels, namely, hydrogel (prepared using 5% w/w sodium alginate), oleogel (prepared using Tween80, castor oil and Aerosil® 200), and a bigel formulation (prepared by mixing the hydrogel and the oleogel in a 30:70 ratio by weight) were prepared, supplemented with different concentrations of the extracted essential oils. These formulations were then tested for the antifungal effects against reference and clinical strains of *Candida*, including *C. krusei*, *C. albicans*, *C. tropicalis*, *C. glabrata*, and *C. parapsilosis*, using the agar-well diffusion method. Positive controls used in the study included commercial antifungal products containing ketoconazole and clotrimazole. Among all the tested formulations, the hydrogel-based formulation demonstrated appreciable results, with anticandidal activity against all the tested strains. Through this study, it is evident that gel-based formulations can act as effective carriers for the topical applications of antifungal essential oils (Tomczykowa et al. 2018).

Another approach to developing improved forms of essential oils was adopted by Gemeda et al. (2018). Essential oil of *Cymbopogon martini* was incorporated into five different topical bases, namely, macrogol ointment base, macrogol cream base, hydrophilic ointment base, white petrolatum base, and simple cream base. The essential oil itself showed a good antidermatophytic effect on the fungal strains thus tested. Negative controls comprised simply the topical bases without the essential oils, while commercial topical drugs were used as the positive controls. These formulations were then tested for their antifungal efficacy against skin pathogens including *C. albicans*, *Trichophyton rubrum*, *T. verrucosum*, *T. mentagrophytes*, *Microsporum canis*, and *Aspergillus niger*. Out of the five formulations, only hydrophilic ointment and macrogol base ointment exhibited significant antifungal action, and the hydrophilic ointment demonstrated appreciable antifungal effect on all the tested strains, and at all the concentrations tested. Among the tested strains, *Trichophyton* spp., known to cause dermatophytic infections in humans, was the most susceptible among all the tested strains. Further study involved safety evaluation of the topical formulations of the essential oil, for skin sensitisation and possible irritation in guinea pigs, with favourable results obtained, showing no visible signs of skin damage/irritation upon application of 5% of each of the tested formulations (Gemeda et al. 2018).

3.3.3 Emulsions of Essential Oils

Oil-in-water or water-in-oil emulsions of essential oils using appropriate surfactants can improve the application of essential oils by preserving their desirable properties. Based on their droplet size, emulsions may be classified as macroemulsions and microemulsions, which may be applied to our advantage in improving the therapeutic efficacy of essential oils (Barradas and de Holanda e Silva 2021). In one such study, conducted by Vörös-Horváth et al. (2020), Pickering emulsions of *Melaleuca alternifolia* essential oil and tioconazole were developed as a novel treatment method for onychomycosis caused by *C. albicans* and *T. rubrum*. Pickering emulsions (PE) are different from conventional O/W or W/O emulsions in that these emulsions are particle-stabilised, offering better stability than conventional emulsions. In this study, a PE of tioconazole in *M. alternifolia* EO, stabilised with silica nanoparticles was assessed as a possible topical treatment option against *T. rubrum* and *C. albicans* infections. Diffusion-based assays using artificial agar gel membranes revealed better delivery of the antifungal agents, being site-specific and efficient in delivering the required concentration of the drug (assessed via MIC and MFC) within 10 min (Vörös-Horváth et al. 2020).

3.3.4 Nanoformulations of Essential Oils

Over the past few years, nanotechnological interventions in various fields, including pharmaceuticals, have gained unprecedented impetus—a variety of nanomaterials are actively being researched for their antimicrobial, antibiofilm, antioxidant, anti-inflammatory, and anti-cancer properties (Barros and Casey 2020). Nanoformulation, in the most basic sense, involves manipulating bulk materials, including bioactive phytocompounds at the nanoscale, such that their surface area: volume ratio increases (Makabenta et al. 2021). Such manipulation can be employed to increase the efficacy of several potential antifungal essential oils and their active ingredients as well (Cimino et al. 2021). Considering the volatile nature of the essential oils, such nanoformulation protects the essential oils from degradation and improves their solubility, thereby increasing their bioavailability and the efficiency of the essential oils, and further offering controlled release of the active ingredients (da Silva Gündel et al. 2020).

In one such study, clove oils, obtained after steam distillation using a traditional Clevenger-type apparatus, were assessed for their anti-candidal activity and potential for treatment of vaginal candidiasis. In order to improve their action and stability, the essential oils were encapsulated into O/W (oil-in-water) nanoemulsions. Different carrier vehicles were tested, out of which sodium carboxymethylcellulose and sodium alginate demonstrated better stability and thus were selected for formulation of the essential oil-based nanogel. The nanoformulations were further tested for their emulsion stability, and assessed for their average size and polydispersity index, revealing a particle size ranging between 198 and 321 nm. Further studies into the in vitro efficacy of the formulated nanogel on standard *C. parapsilosis* and *C. krusei*, and fluconazole-resistant clinical strains of *C. albicans* showed appreciable results, revealing significant anti-oxidant and antifungal activity. Such studies elucidate the potential advantages of nanoformulating essential oils in order to improve their efficacy and applications (da Silva Campelo et al. 2021).

Such alternative therapies have gained interest in treatment of resistant diseases such as vulvovaginal candidiasis (VVC), caused by strains of *C. albicans*. In efforts to validate the in vivo effects of nanoformulations of essential oils, murine models of vulvovaginal candidiasis were tested for their susceptibility towards nanoemulsions of eucalyptus and lemongrass. VVC is especially problematic owing to the growing resistance of the candidal pathogens against azole drugs. In this study, the mice models were first infected with the etiologic agent, i.e. *C. albicans* and then treated with the predetermined concentrations of antifungal nanoemulsions of the essential oils for a period of 7 days at every 24 h. On the eighth day, the mice were euthanised, and their vaginal tissue samples were tested for fungal burden and subjected to histopathological examinations. Compared to the control group treated with conventional miconazole, the nanoemulsion-treated group displayed similar or improved antifungal effects, without yeast cells or any signs of inflammation (da Silva Gündel et al. 2020).

3.4 Combinational Therapy: In- vitro Studies of Essential Oils in Combination with Other Antifungal Agents

In view of the on-going issue of antifungal resistance against commercially available antifungal drugs, and the appreciable therapeutic effects of essential oils and/or their active ingredients, combinational approaches have also been undertaken by various researchers in order to study the possible synergetic effects of these oils in combination with each other and with other antifungal agents (Khan et al. 2014). Such combinations exhibit a broad-spectrum and multi-target mode of action along with reduced side effects of the phytoextracts, by minimising the effective dosage (Trifan et al. 2021). This has emerged as an interesting area of research that also helps improve certain limitations in applications of essential oils, primarily their volatility, insolubility, and potential toxicity to mammalian cells at concentrated doses (Tullio et al. 2019). Additionally, such combinations may also improve the activity and reduce the toxic effects of conventional antifungal drugs such as azoles, against which resistance has been encountered (Shaban et al. 2020).

For example, in a study conducted by Cardoso et al. (2016), possible synergistic effects of essential oil obtained from *Ocimum basilicum* and its active ingredients (geraniol and linalool) on fluconazole (FLC) were studied on *Cryptococcus neoformans* and resistant and susceptible strains of *C. albicans*. Individual studies revealed geraniol as the most active antifungal agent against both pathogens. Antifungal action of the EO, geraniol and linalool were linked with inhibition of ergosterol synthesis, disruption of the fungal capsule, reduction in mitochondrial activity, and inhibited biofilm formation. Combinations of FLC and EO (at 1.01 and 156 μg/mL, respectively) were found to exhibit synergistic antifungal action against FLC-resistant *C. albicans*. Similarly, combinations of FLC and linalool (at 2.02 and 197 μg/mL, respectively) were found to be effective against the FLC-resistant strain of *C. albicans*, while combinations of FLC and geraniol successfully inhibited *C. neoformans* (at 4.14 and 17 μg/mL, respectively) and FLC-resistant *C. albicans* (at 1.04 and 38 μg/mL, respectively). Additionally, combinations of geraniol and linalool were able to inhibit all the three strains. Such synergistic combinations significantly reduced the individual MICs of the bioactive phytocompounds as well. For example, the MIC values of geraniol against *C. neoformans* dropped from 76 to 19 μg/mL when combined with FLC. Similarly, the MIC value of the EO dropped from 1250 to 156 μg/mL when combined with FLC, against FLC resistant strain of *C. albicans*. Such studies suggest the effective potential of combinations of essential oils (and/or their active ingredients) with conventional antifungal drugs, against which resistant pathogens have been identified (Cardoso et al. 2016). Such synergistic combinations can also inhibit ergosterol biosynthesis and aid in the downregulation of and inhibition of expression of virulence genes as well, as was observed by Essid and co-workers, when testing out combinations of *Pelargonium graveolens* EO and fluconazole against pathogenic *Candida* spp. (Essid et al. 2017).

Not only essential oils, but their active ingredients too may be combined with antifungal drugs to improve their activity. For example, combinations of carvacrol, a monoterpene phenol, with fluconazole, amphotericin B and nystatin and caspofungin were developed and tested for their efficacy against antifungal-resistant strains of *C. auris*. Carvacrol at its MIC (125 μg/mL) was able to inhibit adherence to epithelial cells, along with an inhibition of proteinase and phospholipase production in majority of the isolates. Combining carvacrol with amphotericin and nystatin further demonstrated synergistic antifungal effects in- vitro, along with reduced doses of both, the phenol and the antifungal drug (Shaban et al. 2020).

Such combinational approaches can also be employed to combat pathogenic fungal biofilms, as was demonstrated by Jafri and Ahmad (2020). Fungal biofilms, especially those of *Candida* spp. can have fatal impacts when formed on medical devices such as catheters, dentures, etc. (Cavalheiro and Texeira 2018). In this study, the synergetic effects of thyme essential oil and its main ingredient, thymol, with conventional antifungal drugs were evaluated against standard and clinical strains of *C. albicans* and *C. tropicalis*, revealing significant synergetic inhibition of the fungal pathogens. Significant synergy was observed between thyme oil with antifungal drugs fluconazole and amphotericin B. Thyme oil showed the highest synergy with fluconazole against the planktonic state of both the test species. Significant reduction in the respective MICs was also observed in the synergy-based studies. Furthermore, the synergetic studies against the biofilm mode revealed the highest synergy between thymol and fluconazole, against clinical strains of *C. albicans* and *C. tropicalis*, with significant reduction observed in the SMIC of thymol (Jafri and Ahmad 2020).

3.5 From the Lab to the Clinic: Limitations in Therapeutic Applications of Essential Oils

The likes of in- vitro studies described in this chapter are limited by certain factors which further restrict their in vivo and clinical application. These factors include the controlled conditions of the in vitro experiments, the culture conditions, the age of the fungal inoculum, the incubation temperature, etc., which are significantly different from the conditions encountered within living systems (Kalemba and Kunicka 2003). Another significant issue faced in developing novel antifungal drugs is the similar cell biology at the structural and biochemical level, of the fungal and human systems, which further limits the fungal drug targets, thereby impeding the development of novel antifungal drugs (Sun et al. 2020). Furthermore, the active ingredients of the essential oils are bound to interact with the biomolecules present within the human/animal systems, which increases the chances of deviation from the in vitro results obtained (Kalemba and Kunicka 2003). Moreover, the inherent properties of essential oils, such as their volatility, instability, hydrophobicity, and potential toxicity in mammalian systems impede in vivo studies and their full-fledged introduction at the clinical scenario (Zuzarte and Salgueiro 2022).

Nevertheless, the fact remains that any novel therapeutic/antimicrobial agent can be validated for clinical trials only after their in vivo efficacy and safety has been confirmed. Although limited, certain studies have been conducted in this regard. In an extensive study conducted by Abd Ellah et al. (2021), essential oils of cumin (*Cuminum cyminum*) were encapsulated within polyethylene glycol (PEG) and formulated into suppositories for the treatment of vulvovaginal candidiasis. The PEG coating acted as a protective layer that prevented evaporation and oxidation of the cumin oil. *C. albicans* strains, isolated from infected patients, were tested for their susceptibility towards the essential oils, with appreciable in vitro results obtained after conducting a standard well-diffusion assay, indicating a controlled release of the potential anti-candidal agent. The next step of the study involved biocompatibility assays, with female rabbits chosen as test models. Histopathological examination of vaginal tissue after treatment showed significant compatibility with no changes in overall tissue structure. This prompted further clinical testing in a pilot-scale study, involving 30 female patients suffering from symptoms of vulvovaginal candidiasis. A pre- and post-treatment microbiological examination revealed negative cultures of *C. albicans* and *C. krusei* after treatment with the essential oil suppositories. Furthermore, there was a significant reduction in general symptoms among 70% of the tested group of patients (Abd Ellah et al. 2021).

3.6 Conclusions and Future Perspectives

Plant products have great potential to be utilised for a variety of purposes, including phytopharmacological applications, owing to their antioxidant, anti-inflammatory, anti-cancer, antibacterial, and antifungal properties. Although in use since ancient times for their various health benefits, the antifungal properties of plant essential oils have only recently been re-visited, owing to the steady increase in the cases of antifungal resistance among various fungal pathogens. A plethora of studies have been conducted, scrutinising the role of essential oils and their active compounds in combatting resistant fungal pathogens including *Candida* spp., *Aspergillus* spp. and Dermatophytes, among others. A vast majority of these studies successfully report appreciable antifungal activity against fungal growth and reproduction, fungal biofilm formation, and anti-mycotoxigenic effects as well.

A major limitation of using essential oils or pure phytocompounds including terpenes and phenylpropanoids for their antifungal effects is the broad spectrum of their mode of action and the similarities in cell structure and biochemistry between human cells and fungal cells. Hence, deeper studies into the mechanisms of action at the molecular and genetic level, ruling out any chances of cross-reactivity and undesired effects on the host systems, are warranted. It must also be taken into consideration that the antifungal action of essential oils is based on their chemical composition, which may vary with the part of the plant used for extraction of the oil, the harvesting season, and the type of extraction method employed. Therefore, it is essential to evaluate the chemical composition of the oils under scrutiny, as their antifungal action is known to vary with composition.

Although a number of significant efforts have been made to improve the physicochemical properties of essential oils to enhance their application and therapeutic administration, a major setback of these studies is the fact that they are, more often than not, claims based strictly on in vitro experiments. Considering a lack of in vivo examinations to validate the therapeutic applications of various essential oils and/or their major compounds, investigation of their pharmacodynamics and pharmacokinetic properties, and their potential side-effects is warranted, which would further help determine their stability and safety within human systems. Although in- silico methods provide sufficient ideas at the preliminary level, elaborate in- vivo studies are required. Subsequently, these lab-based studies must be subjected to clinical trials as well to validate their efficacy and tolerability and therefore require proper scrutiny before being administered to the test subjects.

It is expected that delving deeper into the interactions between the various components of essential oils, and devising new methods to improve their potency may help in increasing their applications as novel antifungal alternatives to treat drug-resistant fungal pathogens.

Acknowledgement The authors are thankful to Department of Science and Technology, Department of Biotechnology, and University Grants Commission, New Delhi, for providing financial assistance to SAK, SAS and IA, and DV respectively.

References

Abd Ellah NH, Shaltout AS, Abd El Aziz SM, Abbas AM, Abd El Moneem HG, Youness EM et al (2021) Vaginal suppositories of cumin seeds essential oil for treatment of vaginal candidiasis: formulation, in vitro, in vivo, and clinical evaluation. Eur J Pharm Sci 157:105602

Abd Rashed A, Rathi DNG, Ahmad Nasir NAH, Abd Rahman AZ (2021) Antifungal properties of essential oils and their compounds for application in skin fungal infections: conventional and nonconventional approaches. Molecules 26(4):1093

Ahmedi S, Pant P, Raj N, Manzoor N (2022) Limonene inhibits virulence associated traits in Candida albicans: in-vitro and in-silico studies. Phytomedicine Plus 2(3):100285

Akram MA, Khan BA, Khan MK, Alqahtani A, Alshahrani SM, Hosny KM (2021) Fabrication and characterization of polymeric pharmaceutical emulgel co-loaded with eugenol and linalool for the treatment of trichophyton rubrum infections. Polymers 13(22):3904

Alshaikh NA, Perveen K (2021) Susceptibility of fluconazole-resistant *Candida albicans* to thyme essential oil. Microorganisms 9(12):2454

Arrais A, Manzoni M, Cattaneo A, Gianotti V, Massa N, Novello G et al (2021) Host–guest inclusion complexes of essential oils with strong antibacterial and antifungal features in beta-cyclodextrin for solid-state pharmaceutical applications. Appl Sci 11(14):6597

Arya NR, Rafiq NB (2023) Candidiasis [updated 2023 May 29]. In: StatPearls. StatPearls Publishing, Treasure Island, FL. Available from: https://www.ncbi.nlm.nih.gov/books/NBK560624/

Aumeeruddy-Elalfi Z, Gurib-Fakim A, Mahomoodally MF (2016) Antimicrobial and antibiotic potentiating activity of essential oils from tropical medicinal herbs and spices. In: Antibiotic resistance. Elsevier, p 271

Baptista-Silva S, Borges S, Ramos OL, Pintado M, Sarmento B (2020) The progress of essential oils as potential therapeutic agents: a review. J Essent Oil Res 32(4):279–295

Barac A, Donadu M, Usai D, Spiric VT, Mazzarello V, Zanetti S et al (2018) Antifungal activity of Myrtus communis against Malassezia sp. isolated from the skin of patients with pityriasis versicolor. Infection 46:253–257

Barradas TN, de Holanda e Silva KG (2021) Nanoemulsions of essential oils to improve solubility, stability and permeability: a review. Environ Chem Lett 19(2):1153–1171

Barros CH, Casey E (2020) A review of nanomaterials and technologies for enhancing the antibiofilm activity of natural products and phytochemicals. ACS Appl Nano Mater 3(9):8537–8556

Benedict K, Whitham HK, Jackson BR (2022, April) Economic burden of fungal diseases in the United States. In: Open forum infectious diseases, vol 9(4). Oxford University Press, p ofac097

Bermas A, Geddes-McAlister J (2020) Combatting the evolution of antifungal resistance in Cryptococcus neoformans. Mol Microbiol 114(5):721–734

Brescini L, Fioriti S, Morroni G, Barchiesi F (2021) Antifungal combinations in dermatophytes. J Fungi 7(9):727

Cardoso NN, Alviano CS, Blank AF, Romanos MTV, Fonseca BB, Rozental S et al (2016) Synergism effect of the essential oil from Ocimum basilicum var. Maria Bonita and its major components with fluconazole and its influence on ergosterol biosynthesis. Evid Based Complement Alternat Med 2016:5647182

Cavalheiro M, Teixeira MC (2018) Candida biofilms: threats, challenges, and promising strategies. Frontiers in medicine 5:28

Centers for Disease Control and Prevention (2021) Treatment for ringworm. CDC. https://www.cdc.gov/fungal/diseases/ringworm/treatment.html

Centers for Disease Control and Prevention (2022a) Candidiasis. CDC. https://www.cdc.gov/fungal/diseases/candidiasis/index.html

Centers for Disease Control and Prevention (2022b) Treatment of Cryptococcosis. CDC. https://www.cdc.gov/fungal/diseases/cryptococcosis-neoformans/treatment.html

Centers for Disease Control and Prevention (2022c) Understanding antifungal resistance. CDC. https://www.cdc.gov/fungal/antifungal-resistance.html

Centers for Disease Control and Prevention (2023) Candida auris. CDC. https://www.cdc.gov/fungal/candida-auris/index.html

Chen Y, Zeng H, Tian J, Ban X, Ma B, Wang Y (2014) Dill (Anethum graveolens L.) seed essential oil induces Candida albicans apoptosis in a metacaspase-dependent manner. Fungal Biol 118(4):394–401

Chouhan S, Sharma K, Guleria S (2017) Antimicrobial activity of some essential oils—present status and future perspectives. Medicines 4(3):58

Cimino C, Maurel OM, Musumeci T, Bonaccorso A, Drago F, Souto EMB et al (2021) Essential oils: pharmaceutical applications and encapsulation strategies into lipid-based delivery systems. Pharmaceutics 13(3):327

D'agostino M, Tesse N, Frippiat JP, Machouart M, Debourgogne A (2019) Essential oils and their natural active compounds presenting antifungal properties. Molecules 24(20):3713

da Nóbrega Alves D, Monteiro AFM, Andrade PN, Lazarini JG, Abílio GMF, Guerra FQS et al (2020) Docking prediction, antifungal activity, anti-biofilm effects on Candida spp., and toxicity against human cells of cinnamaldehyde. Molecules 25(24):5969

da Silva Campelo M, Melo EO, Arrais SP, do Nascimento FBSA, Gramosa NV, de Aguiar Soares S et al (2021) Clove essential oil encapsulated on nanocarrier based on polysaccharide: a strategy for the treatment of vaginal candidiasis. Colloids Surf A Physicochem Eng Asp 610:125732

da Silva Gündel S, de Godoi SN, Santos RCV, da Silva JT, de Menezes Leite LB, Amaral AC, Ourique AF (2020) In vivo antifungal activity of nanoemulsions containing eucalyptus or lemongrass essential oils in murine model of vulvovaginal candidiasis. J Drug Deliv Sci Technol 57:101762

de Oliveira Lima MI, de Medeiros AA, Silva KS, Cardoso GN, de Oliveira Lima E, de Oliveira Pereira F (2017) Investigation of the antifungal potential of linalool against clinical isolates of fluconazole resistant Trichophyton rubrum. J Mycol Med 27(2):195–202

de Oliveira Santos GC, Vasconcelos CC, Lopes AJ, de Sousa Cartágenes MDS, Filho AK, do Nascimento FR et al (2018) Candida infections and therapeutic strategies: mechanisms of action for traditional and alternative agents. Front Microbiol 9:1351

Deng J, Wang G, Li J, Zhao Y, Wang X (2018) Effects of cinnamaldehyde on the cell wall of A. fumigatus and its application in treating mice with invasive pulmonary aspergillosis. Evid Based Complement Alternat Med 2018:5823209

Dhakal A, Hashmi MF, Sbar E (2023) Aflatoxin toxicity. In: StatPearls. StatPearls Publishing

Dhifi W, Bellili S, Jazi S, Bahloul N, Mnif W (2016) Essential oils' chemical characterization and investigation of some biological activities: a critical review. Medicines 3(4):25

Dias IJ, Trajano ERIS, Castro RD, Ferreira GLS, Medeiros HCM, Gomes DQC (2017) Antifungal activity of linalool in cases of Candida spp. isolated from individuals with oral candidiasis. Braz J Biol 78:368–374

Essid R, Hammami M, Gharbi D, Karkouch I, Hamouda TB, Elkahoui S et al (2017) Antifungal mechanism of the combination of Cinnamomum verum and Pelargonium graveolens essential oils with fluconazole against pathogenic Candida strains. Appl Microbiol Biotechnol 101(18):6993–7006

Fokou JBH, Dongmo PMJ, Boyom FF (2020) Essential oil's chemical composition and pharmacological properties. In: Essential oils-oils of nature. IntechOpen

GAFFI (2018) GAFFI annual report 2018 (version 6). Retrieved from: https://gaffidocuments.s3.eu-west-2.amazonaws.com/GAFFI+annual+report+2018+v6.pdf

Geddes-McAlister J, Shapiro RS (2019) New pathogens, new tricks: emerging, drug-resistant fungal pathogens and future prospects for antifungal therapeutics. Ann NY Acad Sci 1435(1):57–78

Gemeda N, Tadele A, Lemma H, Girma B, Addis G, Tesfaye B et al (2018) Development, characterization, and evaluation of novel broad-spectrum antimicrobial topical formulations from Cymbopogon martini (Roxb.) W. Watson essential oil. Evid Based Complement Alternat Med 2018:9812093

Goldstein AO, Goldstein BG (2017) Dermatophyte (tinea) infections. UpToDate, Walthman, MA

Gow NA, Johnson C, Berman J, Coste AT, Cuomo CA, Perlin DS et al (2022) The importance of antimicrobial resistance in medical mycology. Nat Commun 13(1):5352

Gupta P, Gupta S, Sharma M, Kumar N, Pruthi V, Poluri KM (2018) Effectiveness of phytoactive molecules on transcriptional expression, biofilm matrix, and cell wall components of Candida glabrata and its clinical isolates. ACS Omega 3(9):12201–12214

Gupta P, Pruthi V, Poluri KM (2021) Mechanistic insights into Candida biofilm eradication potential of eucalyptol. J Appl Microbiol 131(1):105–123

Hanif MA, Nisar S, Khan GS, Mushtaq Z, Zubair M (2019) Essential oils. In: Essential oil research: trends in biosynthesis, analytics, industrial applications and biotechnological production. Springer, pp 3–17

Hassanpour P, Shams-Ghahfarokhi M, Razzaghi-Abyaneh M (2020) Antifungal activity of eugenol on Cryptococcus neoformans biological activity and Cxt1p gene expression. Curr Med Mycol 6(1):9

Hokken MW, Zwaan BJ, Melchers WJG, Verweij PE (2019) Facilitators of adaptation and antifungal resistance mechanisms in clinically relevant fungi. Fungal Genet Biol 132:103254

Hou T, Sana SS, Li H, Xing Y, Nanda A, Netala VR, Zhang Z (2022) Essential oils and its antibacterial, antifungal and anti-oxidant activity applications: a review. Food Biosci 47:101716

International Union of Pure and Applied Chemistry (2006a) Terpenes. IUPAC Gold Book. https://doi.org/10.1351/goldbook.T06278

International Union of Pure and Applied Chemistry (2006b) Terpenoids. IUPAC Gold Book. https://doi.org/10.1351/goldbook.T06279

Irshad M, Subhani MA, Ali S, Hussain A (2020) Biological importance of essential oils. In: Essential oils-oils of nature, vol 1. IntechOpen, pp 37–40

Jafri H, Ahmad I (2020) Thymus vulgaris essential oil and thymol inhibit biofilms and interact synergistically with antifungal drugs against drug resistant strains of Candida albicans and Candida tropicalis. J Mycol Med 30(1):100911

Jain N, Sharma M (2020) Inhibitory effect of some selected essential oil terpenes on fungi causing superficial infection in human beings. J Essent Oil Bear Plants 23(4):862–869

Jartarkar SR, Patil A, Goldust Y, Cockerell CJ, Schwartz RA, Grabbe S, Goldust M (2022) Pathogenesis, immunology and management of dermatophytosis. J Fungi 8(1):39

Jung KW, Chung MS, Bai HW, Chung BY, Lee S (2021) Investigation of antifungal mechanisms of thymol in the human fungal pathogen, Cryptococcus neoformans. Molecules 26(11):3476

Kalemba DAAK, Kunicka A (2003) Antibacterial and antifungal properties of essential oils. Curr Med Chem 10(10):813–829

Karpiński TM, Ożarowski M, Seremak-Mrozikiewicz A, Wolski H (2023) Anti-Candida and antibiofilm activity of selected lamiaceae essential oils. Front Biosci Landmark 28(2):28

Khan MSA, Ahmad I, Cameotra SS (2014) Carum copticum and Thymus vulgaris oils inhibit virulence in Trichophyton rubrum and Aspergillus spp. Braz J Microbiol 45:523–531

Khan MS, Qais FA, Ahmad I (2018) Quorum sensing interference by natural products from medicinal plants: significance in combating bacterial infection. In: Biotechnological applications of quorum sensing inhibitors. Springer, pp 417–445

Kohiyama CY, Ribeiro MMY, Mossini SAG, Bando E, da Silva Bomfim N, Nerilo SB et al (2015) Antifungal properties and inhibitory effects upon aflatoxin production of Thymus vulgaris L. by Aspergillus flavus Link. Food Chem 173:1006–1010

Kullberg BJ, Arendrup MC (2015) Invasive candidiasis. N Engl J Med 373(15):1445–1456

Kumar A, Pathak H, Bhadauria S, Sudan J (2021) Aflatoxin contamination in food crops: causes, detection, and management: a review. Food Prod Process Nutr 3:1–9

Mahboubi M, Valian M (2019) Anti-dermatophyte activity of Pelargonium graveolens essential oils against dermatophytes. Clin Phytoscience 5(1):1–5

Mahboubi M, HeidaryTabar R, Mahdizadeh E (2017) The anti-dermatophyte activity of Zataria multiflora essential oils. Journal de mycologie medicale 27(2):232–237

Makabenta JMV, Nabawy A, Li CH, Schmidt-Malan S, Patel R, Rotello VM (2021) Nanomaterial-based therapeutics for antibiotic-resistant bacterial infections. Nat Rev Microbiol 19(1):23–36

Mani-López E, Cortés-Zavaleta O, López-Malo A (2021) A review of the methods used to determine the target site or the mechanism of action of essential oils and their components against fungi. SN Appl Sci 3:1–25

Manzoor N (2019) Candida pathogenicity and alternative therapeutic strategies. In: Pathogenicity and drug resistance of human pathogens: mechanisms and novel approaches. Springer, pp 135–146

Maziarz EK, Perfect JR (2016) Cryptococcosis. Infect Dis Clin 30(1):179–206

Miri YB, Benabdallah A, Taoudiat A, Mahdid M, Djenane D, Tacer-Caba Z et al (2023) Potential of essential oils for protection of Couscous against Aspergillus flavus and aflatoxin B1 contamination. Food Control 145:109474

Mishra KK, Kaur CD, Sahu AK, Panik R, Kashyap P, Mishra SP, Dutta S (2020) Medicinal plants having antifungal properties. In: Medicinal plants: use in prevention and treatment of diseases. IntechOpen

Muñoz JE, Rossi DC, Jabes DL, Barbosa DA, Cunha FF, Nunes LR et al (2020) In vitro and in vivo inhibitory activity of limonene against different isolates of Candida spp. J Fungi 6(3):183

National Institute of Environmental Health Sciences (2022) Essential oils. NIEHS. https://www.niehs.nih.gov/health/topics/agents/essential-oils/index.cfm

Nazzaro F, Fratianni F, Coppola R, De Feo V (2017) Essential oils and antifungal activity. Pharmaceuticals 10(4):86

Niu C, Wang C, Yang Y, Chen R, Zhang J, Chen H et al (2020) Carvacrol induces Candida albicans apoptosis associated with Ca2+/calcineurin pathway. Front Cell Infect Microbiol 10:192

Nóbrega RDO, Teixeira APDC, Oliveira WAD, Lima EDO, Lima IO (2016) Investigation of the antifungal activity of carvacrol against strains of Cryptococcus neoformans. Pharm Biol 54(11):2591–2596

Noriega P (2020) Terpenes in essential oils: bioactivity and applications. In: Terpenes and terpenoids: recent advances. IntechOpen

Padhan D, Pattnaik S, Behera AK (2017) Growth-arresting activity of acmella essential oil and its isolated component D-limonene (1, 8 P-Mentha diene) against Trichophyton rubrum (microbial type culture collection 296). Pharmacogn Mag 13(Suppl 3):S555

Pandey AK, Kumar P, Singh P, Tripathi NN, Bajpai VK (2017) Essential oils: sources of antimicrobials and food preservatives. Front Microbiol 7:2161

Pappas PG, Lionakis MS, Arendrup MC, Ostrosky-Zeichner L, Kullberg BJ (2018) Invasive candidiasis. Nat Rev Dis Primers 4(1):1–20

Pinto E, Hrimpeng K, Lopes G, Vaz S, Gonçalves MJ, Cavaleiro C, Salgueiro L (2013) Antifungal activity of Ferulago capillaris essential oil against Candida, Cryptococcus, Aspergillus and dermatophyte species. Eur J Clin Microbiol Infect Dis 32:1311–1320

Rajkowska K, Otlewska A, Kunicka-Styczyńska A, Krajewska A (2017) Candida albicans impairments induced by peppermint and clove oils at sub-inhibitory concentrations. Int J Mol Sci 18(6):1307

Ramadan NA, Al-Ameri HA (2022) Aflatoxins. In Aflatoxins-occurrence, detoxification, determination and health risks. IntechOpen

Roana J, Mandras N, Scalas D, Campagna P, Tullio V (2021) Antifungal activity of Melaleuca alternifolia essential oil (TTO) and its synergy with itraconazole or ketoconazole against Trichophyton rubrum. Molecules 26(2):461

Sacheli R, Hayette MP (2021) Antifungal resistance in dermatophytes: genetic considerations, clinical presentations and alternative therapies. J Fungi 7(11):983

Salaria D, Rolta R, Patel CN, Dev K, Sourirajan A, Kumar V (2022) In vitro and in silico analysis of Thymus serpyllum essential oil as bioactivity enhancer of antibacterial and antifungal agents. J Biomol Struct Dyn 40(20):10383–10402

Saroj A, Oriyomi OV, Nayak AK, Haider SZ (2020) Phytochemicals of plant-derived essential oils: a novel green approach against pests. In: Natural remedies for pest, disease and weed control. Academic Press, pp 65–79

Scalas D, Mandras N, Roana J, Tardugno R, Cuffini AM, Ghisetti V et al (2018) Use of Pinus sylvestris L.(Pinaceae), Origanum vulgare L.(Lamiaceae), and Thymus vulgaris L.(Lamiaceae) essential oils and their main components to enhance itraconazole activity against azole susceptible/not-susceptible Cryptococcus neoformans strains. BMC Complement Alternat Med 18(1):1–13

Shaban S, Patel M, Ahmad A (2020) Improved efficacy of antifungal drugs in combination with monoterpene phenols against Candida auris. Sci Rep 10(1):1162

Sharifi-Rad J, Sureda A, Tenore GC, Daglia M, Sharifi-Rad M, Valussi M et al (2017) Biological activities of essential oils: from plant chemoecology to traditional healing systems. Molecules 22(1):70

Sharma Y, Rastogi SK, Perwez A, Rizvi MA, Manzoor N (2020) β-Citronellol alters cell surface properties of Candida albicans to influence pathogenicity related traits. Med Mycol 58(1):93–106

Shen Q, Zhou W, Li H, Hu L, Mo H (2016) ROS involves the fungicidal actions of thymol against spores of Aspergillus flavus via the induction of nitric oxide. PLoS One 11(5):e0155647

Singh S, Fatima Z, Ahmad K, Hameed S (2018) Fungicidal action of geraniol against Candida albicans is potentiated by abrogated CaCdr1p drug efflux and fluconazole synergism. PLoS One 13(8):e0203079

Srinivasan A, Lopez-Ribot JL, Ramasubramanian AK (2014) Overcoming antifungal resistance. Drug Discov Today Technol 11:65–71

Sun S, Hoy MJ, Heitman J (2020) Fungal pathogens. Curr Biol 30(19):R1163–R1169

Swamy MK, Akhtar MS, Sinniah UR (2016) Antimicrobial properties of plant essential oils against human pathogens and their mode of action: an updated review. Evid Based Complement Alternat Med. https://doi.org/10.1155/2016/3012462

Teixeira APDC, Nóbrega RDO, Lima EDO, Araújo WDO, Lima IDO (2020) Antifungal activity study of the monoterpene thymol against Cryptococcus neoformans. Nat Prod Res 34(18):2630–2633

Tolba H, Moghrani H, Benelmouffok A, Kellou D, Maachi R (2015) Essential oil of Algerian Eucalyptus citriodora: chemical composition, antifungal activity. J Mycol Med 25(4):e128–e133

Tomczykowa M, Wróblewska M, Winnicka K, Wieczorek P, Majewski P, Celińska-Janowicz K et al (2018) Novel gel formulations as topical carriers for the essential oil of Bidens tripartita for the treatment of candidiasis. Molecules 23(10):2517

Tongnuanchan P, Benjakul S (2014) Essential oils: extraction, bioactivities, and their uses for food preservation. J Food Sci 79(7):R1231–R1249

Torres-Alvarez C, Castillo S, Sánchez-García E, Aguilera Gonzalez C, Galindo-Rodríguez SA, Gabaldón-Hernández JA, Báez-González JG (2020) Inclusion complexes of concentrated orange oils and β-cyclodextrin: physicochemical and biological characterizations. Molecules 25(21):5109

Trifan A, Luca SV, Bostănaru AC, Brebu M, Jităreanu A, Cristina RT et al (2021) Apiaceae essential oils: boosters of terbinafine activity against dermatophytes and potent anti-inflammatory effectors. Plants 10(11):2378

Tullio V, Roana J, Scalas D, Mandras N (2019) Evaluation of the antifungal activity of Mentha x piperita (Lamiaceae) of Pancalieri (Turin, Italy) essential oil and its synergistic interaction with azoles. Molecules 24(17):3148

Turner SA, Butler G (2014) The Candida pathogenic species complex. Cold Spring Harb Perspect Med 4(9):a019778

US Food and Drug Administration (2023) Title 21—Food and drugs: Chapter I—Food and Drug Administration, Department of Health and Human Services: Subchapter B—Food for human consumption: Part 182—Substances generally recognized as safe. Accessdata FDA. https://www.accessdata.fda.gov/scripts/cdrh/cfdocs/cfcfr/cfrsearch.cfm?fr=182.20

Vitali A, Stringaro A, Colone M, Muntiu A, Angiolella L (2021) Antifungal carvacrol loaded chitosan nanoparticles. Antibiotics 11(1):11

Vörös-Horváth B, Das S, Salem A, Nagy S, Böszörményi A, Kőszegi T et al (2020) Formulation of tioconazole and melaleuca alternifolia essential oil pickering emulsions for onychomycosis topical treatment. Molecules 25(23):5544

Walsh TJ, Dixon DM (1996) Spectrum of mycoses. In: Baron S (ed) Medical microbiology, 4th edn. University of Texas Medical Branch at Galveston, Galveston, TX. Chapter 75. Available from: https://www.ncbi.nlm.nih.gov/books/NBK7902/

World Health Organization (2022) WHO fungal priority pathogens list to guide research, development and public health action. World Health Organization, Geneva

Zeng H, Chen X, Liang J (2015) In vitro antifungal activity and mechanism of essential oil from fennel (Foeniculum vulgare L.) on dermatophyte species. J Med Microbiol 64(1):93–103

Zhang CW, Zhong XJ, Zhao YS, Rajoka MSR, Hashmi MH, Zhai P, Song X (2023) Antifungal natural products and their derivatives: a review of their activity and mechanism of actions. Pharmacol Res Mod Chin Med 7:100262

Zhaveh S, Mohsenifar A, Beiki M, Khalili ST, Abdollahi A, Rahmani-Cherati T, Tabatabaei M (2015) Encapsulation of Cuminum cyminum essential oils in chitosan-caffeic acid nanogel with enhanced antimicrobial activity against Aspergillus flavus. Ind Crop Prod 69:251–256

Zuzarte M, Salgueiro L (2022) Essential oils in respiratory mycosis: a review. Molecules 27(13):4140

Essential Oils and Their Compounds for Applications in Fungal Diseases: Conventional and Nonconventional Approaches

4

Tanveer Alam, Syed Farhan Hasany, and Lubna Najam

4.1 Introduction

Worldwide, people have been using plants as medicine for thousands of years. There are two major classes that can be used to categorize plant chemical compounds. The main metabolites are those that are present throughout all of the plants and serve as the fundamental building blocks of life. Plants produce a number of secondary metabolites in addition to the primary metabolites that play a crucial role in modulating relationships among plants and their natural environments and are linked to adaption mechanisms. Essential oils are volatile oils derived from plants with potent aromatic components and are composed of various chemical compounds. Alcohols, aldehydes, esters, hydrocarbons, ketones, and phenols for instance, are some of the principal ingredients in essential oils (Younis et al. 2008). Various plant parts can be used to make these volatile oils, e.g., leaves (Eucalyptus, Geranium, Tea Tree, and Tobacco), grass (*Cymbopogon citratus*, *Cymbopogon martinii*, and *Cymbopogon nardus*), fruits (Grapefruit, Black Pepper, and Juniper Berry), bark (Cassia, and cinnamon), root (Angelica, Ginger, Spikenard, and Vetiver), gum (Olibanum, Copaivabalsam, Tolubalsam, and Styrax), berries (Juniper Berry, and May Chang), seed (Anise, Cardamom, and Carrot), flowers (Jasmine, Tuberose, and Lotus), wood (Amyris, and Sandalwood), heartwood (Cedarwood), and Moss/Lichen (Oakmoss) (Burt 2004; Sood et al. 2006; Cava et al. 2007; Dang et al. 2001; Hussain et al. 2008).

Polyphenolic compounds, terpenoids, N-containing alkaloids, and S-containing compounds are the different categories of secondary metabolites found in various

T. Alam (✉) · S. F. Hasany
Sabanci University Nanotechnology Research and Application Center, Istanbul, Turkey
e-mail: tanveer.alam@sabanciuniv.edu; syed.hasany@sabanciuniv.edu

L. Najam
British International School Kurdistan, Erbil, Iraq
e-mail: najam.lubna@gamil.com

N. Manzoor (ed.), *Advances in Antifungal Drug Development*,
https://doi.org/10.1007/978-981-97-5165-5_4

plant species. Terpenoids and shikimates are the two most significant compounds that are connected to essential oils (Baser and Buchbauer 2010). More investigation in novel sources of naturally occurring biologically active compounds with novel and targeted actions has been motivated by the potential use of secondary metabolites for disease management (Lang and Buchbauer 2012; Edris 2007). Around the world, microbial (viral, bacterial, and fungal) infections are the most frequently occurring diseases. The search for new fungicides is a broad endeavor, and the current emphasis is on substances that are naturally occurring. Several experimental studies have been conducted using natural substances that could serve as models for defenses against fungi that attack individuals (Stevic et al. 2014; Tabassum and Vidyasagar 2013).

Pathogenic fungi are eukaryotes that possess cellular and molecular characteristics similar to their hosts. Several immunocompromised people are affected by opportunistic fungal pathogens, *viz.*, the well-known *Aspergillus* species, *Candida* species, and *Cryptococcus* species. The number of medicines available for effective antifungal chemotherapy in the market is limited (Kathiravan et al. 2012).

The prevention and treatment of fungal infections is made more challenging by the emergence of biofilms and device-associated infections, drug-resistant fungi, and the adverse effects of already prescribed medications. As a result, invasive fungal infections are linked to extremely high rates of diseases and death (Sardi et al. 2013). It has been discovered that a variety of yeasts, as well as human and plant pathogenic fungi, are susceptible to essential oils. With respect to the intended microorganisms and the essential oils studied, inhibitory effectiveness varies. For instance, three Apiaceae family members have varying anti-*C. albicans* activities, with coriander > anise > fennel with MICs of 0.25%, 0.5%, and 1%, respectively (Hammer et al. 1999). In general, Cymbopogon species exhibit positive effects on pathogenic fungi (yeast) (Irkin and Korukluoglu 2009). In comparison to other essential oils, cinnamon, clove, geranium, ginger, Japanese mint, and lemongrass were found to be the most effective against *Candida albicans*.

Essential oils, alternatively known as volatile oils, are extensive combinations of odorous components that are kept in special plant cells. Cold pressing and hydro/steam distillation procedures are used to extract these essential oils from aromatic plants. They can be extracted from whole aromatic plants or diverse parts of plants such as roots, wood, bark, leaves, flowers, fruits, and gum resin. It is a strong hydrophobic liquid that contains the volatile aroma components. They have concentrations in the range of about 0.01% to about 0.15% (Hammer and Carson 2011; Devkatte et al. 2005). Eugenol and monocyclic sesquiterpene alcohols likebisabolol, which are rich in phenylpropanoids, readily suppress the growth of dermatophytes and spore development (Maxia et al. 2009; Pragadheesh et al. 2013). Essential oils of plant origin stop the growth and aflatoxin formation of molds like *A. flavus* (Lang and Buchbauer 2012; Kumar et al. 2010). Lemongrass (*C. citratus*) is one of the best oils for combating filamentous fungus, and has active concentrations between 0.006% and 0.03%. At <1% concentration, grapefruit, mandarin, lemon, and orange essential oils inhibit *A. flavus*, *A. niger*, *P. chrysogenum*, and *Penicillium verrucosum* (Viuda-Martos et al. 2008). Additionally, *C. albicans* is inhibited by the effect of essential oils on the cell cycle. For instance, citral, citronellol, geraniol, and geranyl acetate are the primary ingredients of eucalyptus oil, geranium oil, and tea tree oil. These are known to inhibit *C. albicans* in the S phase of the cell cycle (Zore et al. 2011a, b). Likewise, the effects of carvacrol, eugenol, and thymol on Ca^{2+} and pH homeostasis (H^{+}) result in the loss of ion and suppression of *S. cerevisiae* (Rao et al. 2010). Essential oils are important organic substances with various medicinal and biological activities among the secondary metabolites. So, in this chapter, we look into how essential oils affect pathogenic fungi.

4.2 Extraction Methods of Essential Oils

Essential oils/volatile oils are generally utilized in different categories of consumer products namely; alcoholic beverages, cosmetics, confectionery food products, detergents, insecticides, soaps, soft drinks, pharmaceuticals, perfumes, and toilet products. Global production and use of essential oils and fragrances is rapidly increasing. The use of production technology can improve essential oil quality and overall yield. These volatile oils are produced from aromatic plants using a number of extraction techniques (Dick and Starmans 1996; Wang and Weller 2006).

There are various techniques for obtaining essential oils from aromatic plant materials. The basic classical conventional (traditional) and widely used methods include:

(a) Maceration
(b) Cold pressing
(c) Enfleurage
(d) Hydrodistillation
(e) Cohobation
(f) Steam distillation
(g) Solvent extraction

Primary shortcoming of classical and conventional methods is thermolability of the essential oil components, which are subject to chemical alterations (hydrolysis, oxidation, and isomerization) due to the high applied heat. Consequently, the extracted essential oils quality is drastically degraded, when the distillation process takes a long time to complete. It is very important to maintain the composition and naturality at its original state during the distillation process.

The techniques for extracting essential oils have consistently improved as a result of the present industrial production and emphasis on economy, competitiveness, environmental friendliness, sustainability, high efficiency, and high quality requirements. Additionally, new extraction methods need to cut down on

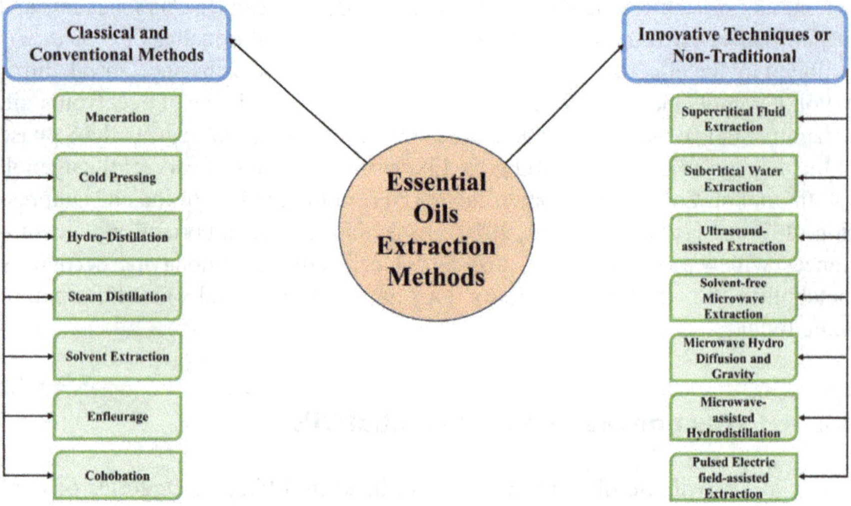

Fig. 4.1 Schematic presentation of extraction methods of essential oils

extraction time, energy utilization, solvent usage, and CO_2 emission (Fig. 4.1). The innovative techniques or nontraditional methods of essential oils extraction include:

(a) Supercritical fluid extraction
(b) Supercritical water extraction
(c) Ultrasound-assisted extraction
(d) Solvent-free microwave extraction
(e) Microwave hydro diffusion and gravity
(f) Microwave-assisted hydrodistillation
(g) Pulsed electric field-assisted extraction

4.3 Chemistry of Essential Oils

Chemically, the components of essential oils fall into two categories: terpenoids and aromatic chemical compounds. This composition is complicated and very changeable. The term "Turpentine" in English is the source of the name "terpene" (Guenther 1952, 1982). Isoprene (C_5H_8) or 2-methyl-1,3-butadiene is the main unit for terpenes. These have the chemical formula $C_{10}H_{16}$ and are made up of two isoprene units that combine through head to tail manner (Fig. 4.2) (Pinder 1960).

Terpenes, terpenoids, phenolic and phenylpropanoids, heterocyclic, and non-terpenoid aliphatic molecules are among some major chemical compounds of essential oils. The essential oil constituents can be divided into two categories: (a) terpene and terpenoids and (b) aliphatic and aromatic compounds (Croteau et al. 2000; Betts 2001; Bowles 2003; Pichersky et al. 2006).

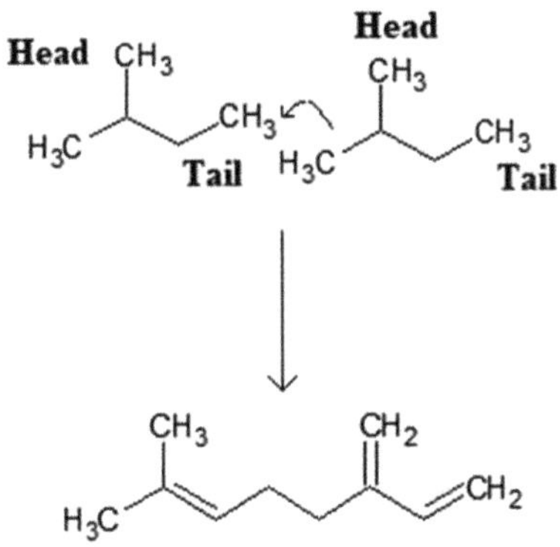

Fig. 4.2 Head-to-tail coupling of two isoprene units

4.3.1 Terpenes and Terpenoids

By linking together through a "head-to-tail" arrangement, terpenes can form monoterpenes, sesquiterpenes, diterpenes, and higher sequences (Pinder 1960). Monoterpenes (C10) and sesquiterpenes (C15) are the essential terpenes; however, hemiterpenes (C5), diterpenes (C20), triterpenes (C30), and tetraterpenes (C40) are also occasionally found in the essential oils. Terpenoid is a type of terpene that contains oxygen atom in its structure.

4.3.1.1 Classification of Terpenes

According to the structural formula $(C_5H_8)_n$, where n is the number of linked isoprene units (Fig. 4.3), terpenes are classified into different groups; for example, monoterpenes (two C5 groups) are made up of two isoprene units and have the chemical formula $C_{10}H_{16}$ (Dorman and Deans 2000). Diterpenes (C20), triterpenes (C30), tetraterpenes (C40), as well as alternate hemiterpenes (C5) and sesquiterpenes (C15) are produced by other conjugations (Table 4.1).

Monoterpenes

The simplest molecule is a monoterpene, which consists of two isoprene units linked together through head-tail manner. In monoterpenoids, there is a carbon chain of ten carbons (C10) and may contain a –OH group at any position along with the chain. α-Terpinene, D-limonene, and β-phellandrene are the common examples of monoterpene (Fig. 4.3) (Grayson 2000).

Sesquiterpenes

Sesquiterpenes are the terpene which contain three isoprene (C_5H_8) units. Zingiberene is the example of a sesquiterpene and is present in ginger essential oil (Fig. 4.3). Sesquiterpenes can change by adding functional groups in their structures. When a functional group is attached to a terpene, it is referred to as a terpenoid. Sesquiterpene with functional group is called Sesquiterpenoid (McGuinness 2007; Bowles 2003).

Diterpenes

Furthermore, occasionally trace levels of four isoprene unit (diterpenes) are included in the essential oils (Fig. 4.4) (Baser and Buchbauer 2010; Bowles 2003).

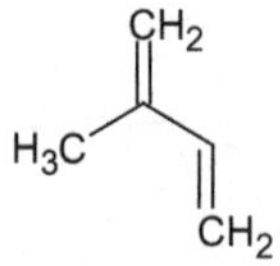

Isoprene Unit

(Hemiterpene)

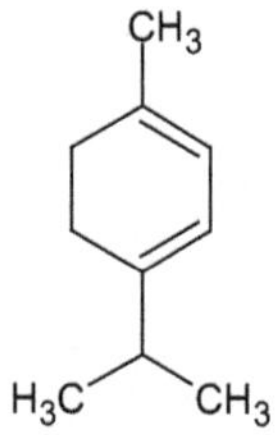

α-terpenene

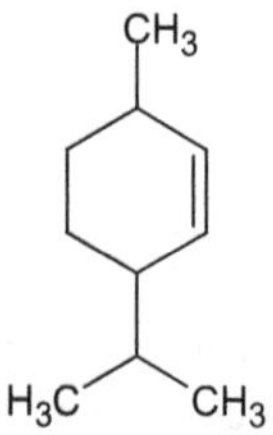

β-phellandrene

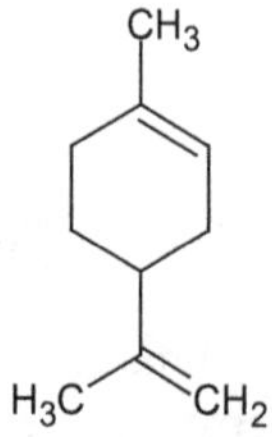

d-limonene

(Monoterpenes)

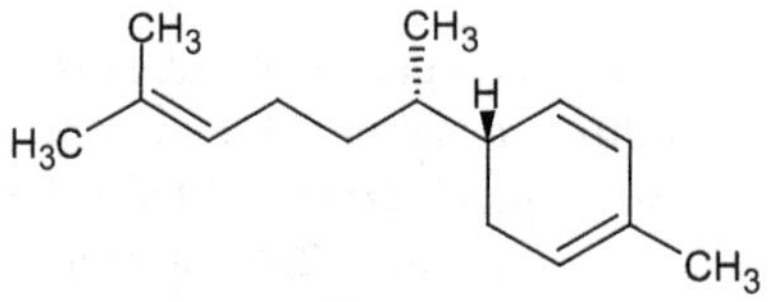

Zingiberene

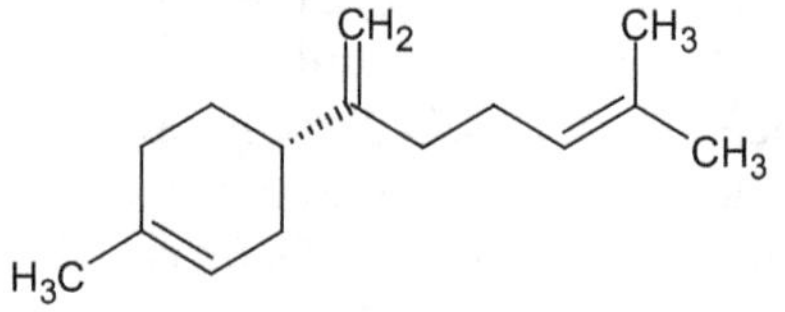

β-bisabolene

(Sesquiterpenes)

Fig. 4.3 Structures of hemiterpene, monoterpenes, and sesquiterpenes

Table 4.1 Classification of terpenes on the basis of the number of isoprene units

S. no.	Number of carbon atoms	Number of isoprene units	Classification
1.	40	8	Tetraterpenes
2.	30	6	Triterpenes
3.	25	5	Sesterterpenes
4.	20	4	Diterpenes
5.	15	3	Sesquiterpenes
6.	10	2	Monoterpenes
7.	5	1	Hemiterpenes

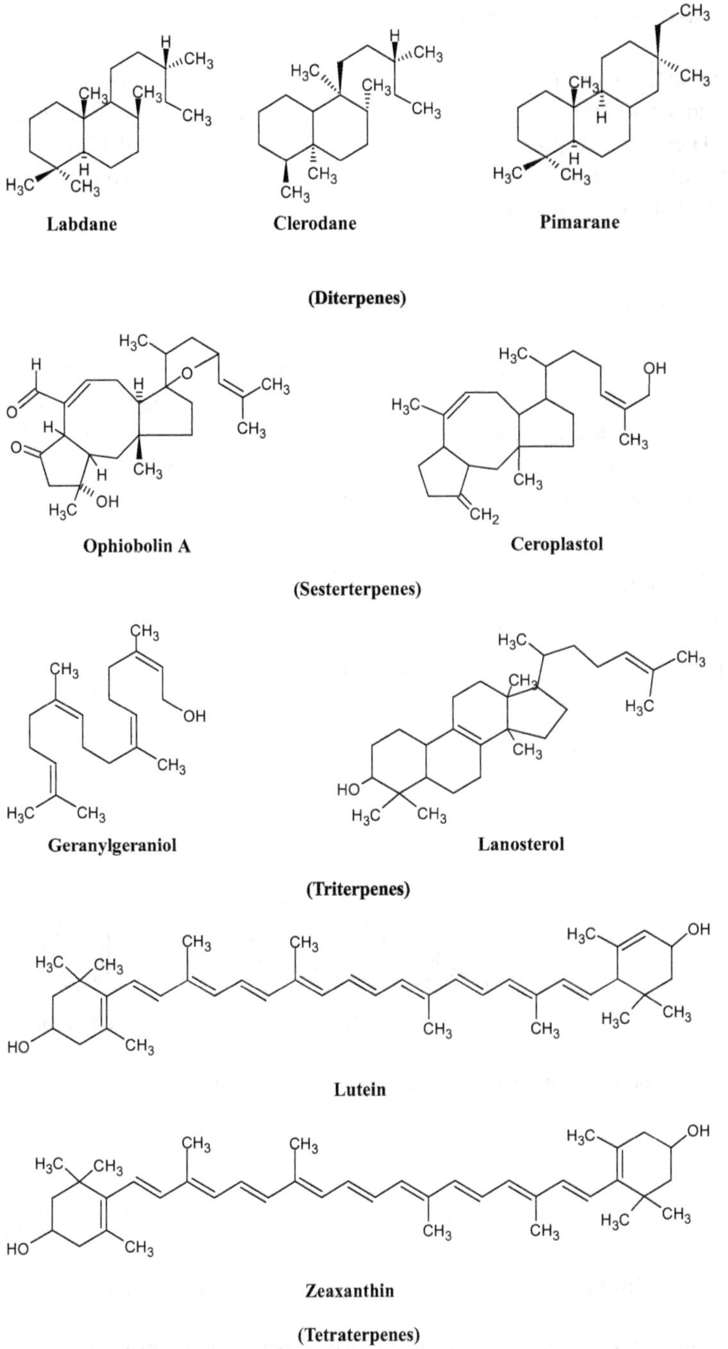

Fig. 4.4 Structures of diterpenes, sesterterpenes, triterpenes, and tetraterpenes

4.3.2 Aromatic Compounds

The applications of aromatic compounds are comparable to those of monoterpenoids and sesquiterpenoids in the essential oils. They are not as common as terpenoids. They include phenols (charvicol), alcohols (cinnamyl alcohol), aldehydes (cynnamaldehyde), methoxy compounds (anethole), and derivatives of methylene dioxy (safrole) (Fig. 4.5) (Grayson 2000).

4.3.3 Nonterpenoid Aliphatic Molecules

The term "aliphatic" refers to carbon-based compounds that are straight-chained and without an aromatic or closed ring. The pungent-smelling C8, C9, and C10 – CHO group that are present in trace quantity in the oil of citrus are examples of aliphatic molecules and the C6 compounds with a green-leafy like fragrance that are present in some floral oils like rose and jasmine (Fig. 4.6).

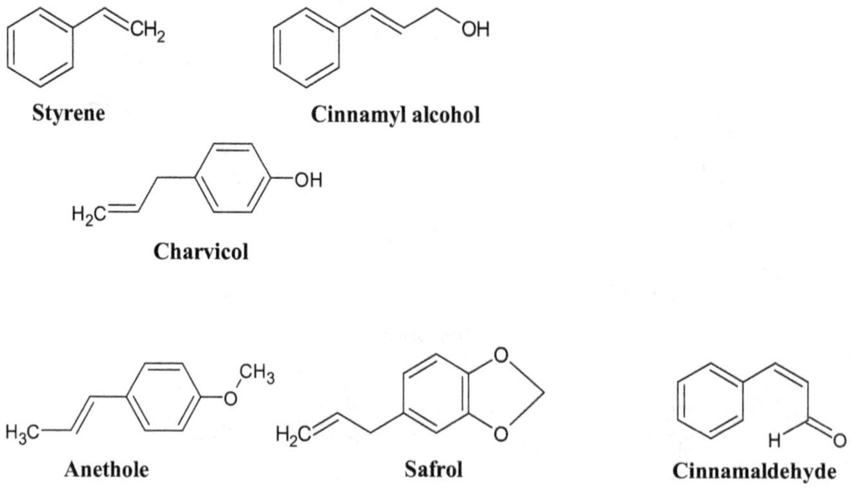

Fig. 4.5 Structures of aromatic compounds present in essential oils

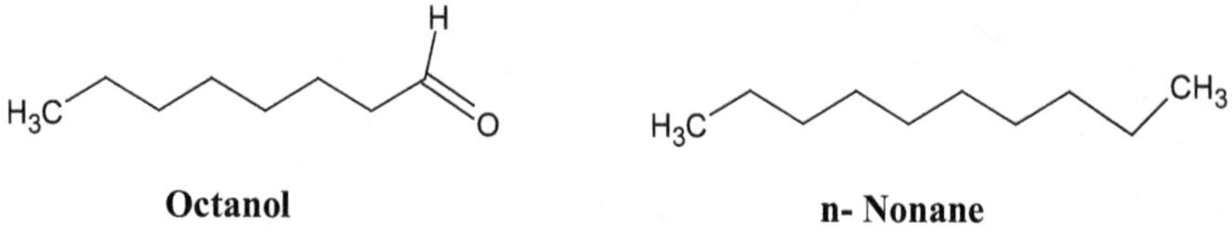

Fig. 4.6 Structures of nonterpenoid aliphatic molecules present in essential oils

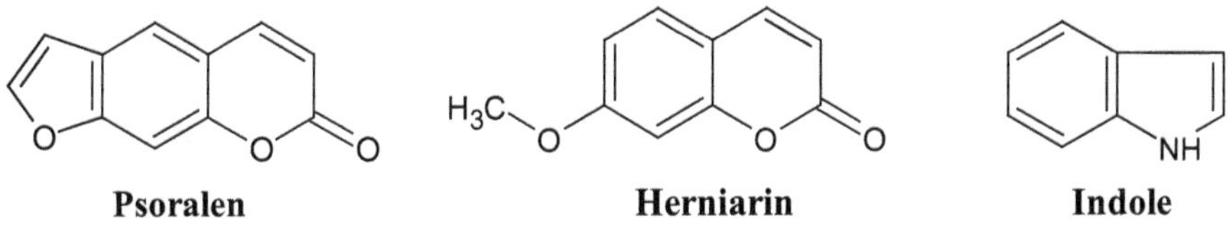

Fig. 4.7 Structures of heterocyclic molecules present in essential oils

4.3.4 Heterocyclic Molecules

In heterocyclic compounds, atoms (N, O, S, etc.) other than carbon are found in closed rings. They include molecules like the N-containing indole and methyl anthranilate as well as the O-containing coumarins, lactones, and furanoid compounds, and they are only present in a small number of oils. The basic components of heterocyclic molecules are rings of C-atoms with either a nitrogen or an oxygen atom attached. Indole is the best example of heterocyclic compounds found in jasmine oil (Fig. 4.7).

4.4 Biosynthesis Pathway of Main Compound of Essential Oils

Essential oils or volatile oils are organic molecules which contain C, H, and O in their structures (McGuinness 2007). Three biosynthetic pathways occur that lead to the formation of primary constituents of essential oils: Mevalonic acid pathway, methylerythritol pathway, and shikimic acid pathway.

4.4.1 Mevalonic Acid Pathway

The cytosolic mevalonic acid pathway produces terpenoid compounds. The isoprenoid pathway and the HMG-CoA reductase pathway are also considered as the mevalonate pathway (Fig. 4.8).

The pathway includes the following steps:

(a) Production of mevalonic acid from a 6-C atoms
(b) Conversion of mevalonic acid through a sequence of enzymatic reactions that produce a branched 5-carbon molecule known as isopentenyl pyrophosphate (C_5H_8 unit)
(c) Linkage of Isoprene to two phosphate groups. The primary element of essential oils that initiate the production of terpenoid molecules is isoprene (Bowles 2003)

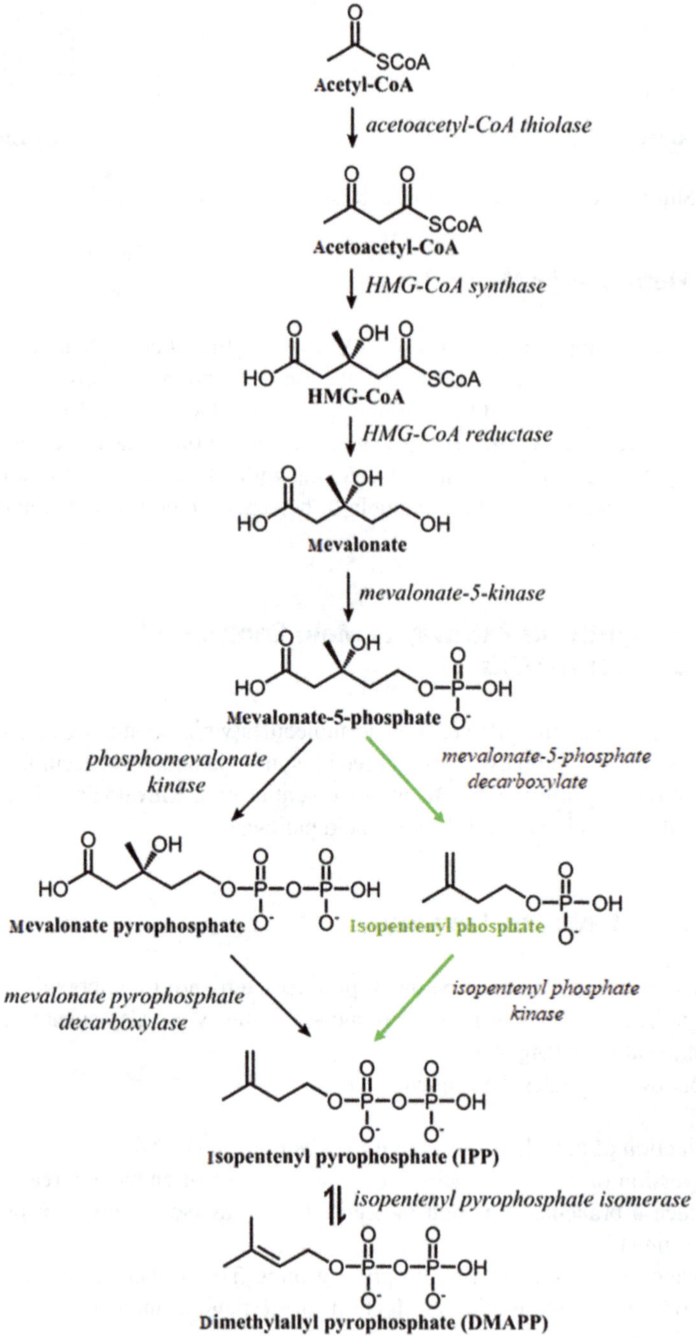

Fig. 4.8 Mevalonic acid pathway reactions in the biosynthesis of isoprenoids

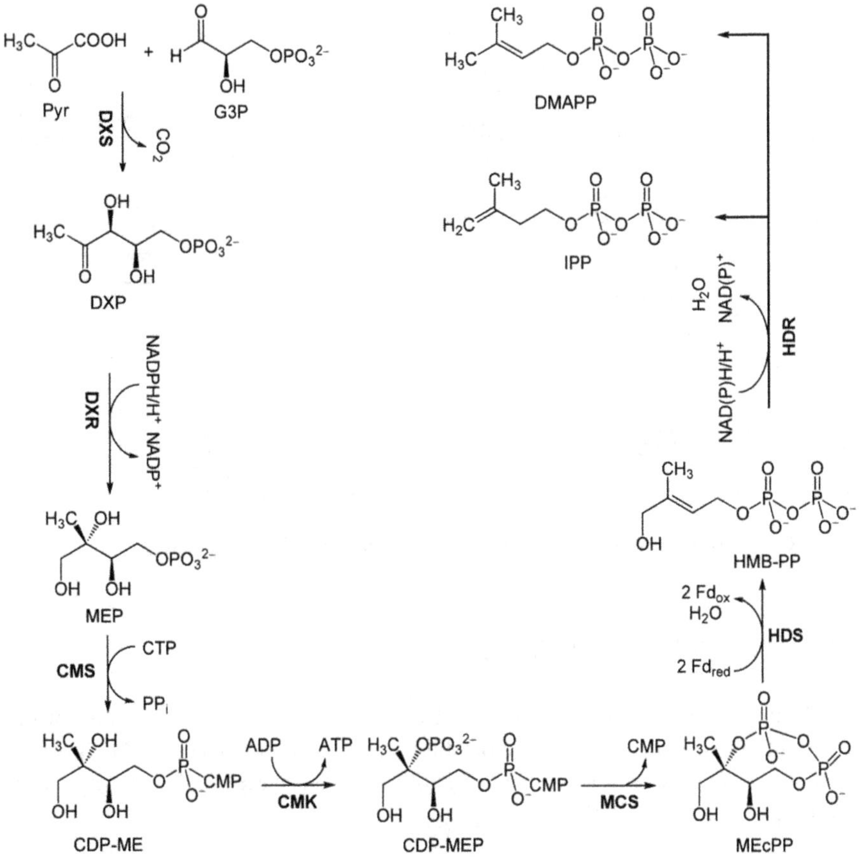

Fig. 4.9 Methylerythritol pathway reactions in the biosynthesis of isoprenoids

4.4.2 Methylerythritol Pathway

The mevalonate-independent pathway, also known as the nonmevalonate pathway and the biosynthesis of the isoprenoid precursors isopentenyl pyrophosphate (IPP) and dimethylallyl pyrophosphate (DMAPP) use an alternate metabolic pathway called the 2-C-methyl-D-erythritol 4-phosphate/1-deoxy-D-xylulose 5-phosphate (MEP/DOXP) pathway. Since MEP is the first committed metabolite formed on the way to IPP, the MEP pathway is a name that is currently most frequently used to refer to this pathway (Fig. 4.9) (Eisenreich et al. 2004; Rohmer et al. 1999; Lichtenthaler 1999).

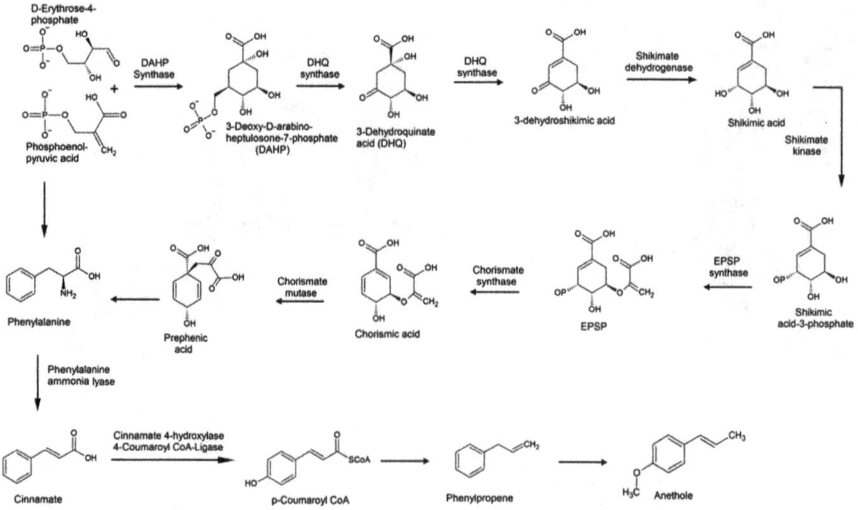

Fig. 4.10 The Shikimic acid biosynthetic pathway in the biosynthesis of phenylpropanoids

4.4.3 The Shikimate Pathway

All phenylpropanoids are biosynthesized *via* the shikimate pathway. Many of the important enzymes originate in the plastids, where it is mostly active. The phenylpropenes are produced through a number of important phases in the synthesis process (Fig. 4.10) (Vogt 2010).

4.5 Pharmacology of Essential Oils

Pharmacology is the scientific study of the effects and actions of essential oils in living things. The two major branches of pharmacology are pharmacodynamics and pharmacokinetics. The actions of the human body toward the essential oils and the effect of the essential oils on the human body, respectively, are referred as pharmacokinetics and pharmacodynamics in the current context. Pharmacodynamics is concerned with how the essential oils attach to drug targets, as well as their biochemical, physiological, and adverse effects. Every essential oil molecule reacts with the body's target molecules to become active. Pharmacokinetics refers to the action of the human body toward the essential oils. It gives an account of how quickly and how much an essential oil is absorbed, distributed, metabolized, and excreted (Gwilt et al. 1994). These processes explain how essential oils circulate throughout the body.

4.5.1 Absorption

The two main methods for applying essential oils topically to the skin or respiratory membranes are inhalation and dermal application. Essential oil constituents are absorbed into the circulatory system using each method at variable rates and amounts (Blaschke and Bjornsson 1995).

4.5.2 Distribution

The movement of essential oil constituents between two locations in the human body is known as distribution (Bryant et al. 2003). The bloodstream maintains the distribution of essential oils to the tissue or body organ, and the ability of the oil to bind to plasma proteins.

4.5.3 Metabolism

Mostly the solubility of essential oils in ethanol is very high. The primary objective of essential oil metabolism is to change the alcohol solubility of the molecules so that they can be eliminated by urination. However, some molecules of essential oils can be eliminated in the urine without going through metabolism. The chemical reactions that transform essential oil molecules into various metabolic activities are often carried out by enzymes that are located in the liver or skin epidermis (Rawlins 1989).

4.5.4 Excretion

Excretion is the process by which an organ or the entire body is able to remove molecules of essential oils and their metabolites from the bloodstream, which is also known as clearance. The body eliminates essential oils and their metabolites through the skin, lungs, kidneys, and feces (Gwilt et al. 1994). The process of an essential oil from absorption to excretion is shown in Fig. 4.11.

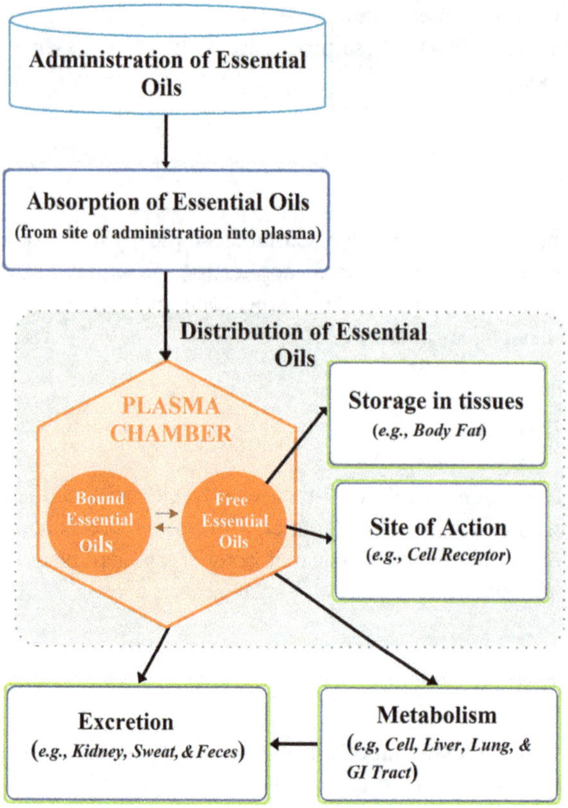

Fig. 4.11 The pathway from absorption to excretion used by essential oils through various body organs

4.6 Mycoses or Fungal Infections

Mycosis is a disease caused by a fungus (mold or yeast). The most common areas for fungus to infect are the skin or nails although they can infect many other areas of the human body like the urinary tract, lungs, mouth, and hair (Parish and Jouni 1991).

4.6.1 Types of Fungal Infections

Fungal infections (Mycoses) are classified into three categories:

1. Superficial or mucocutaneous fungal infections
2. Subcutaneous fungal infections
3. Deep seated or systemic fungal infections

4.6.1.1 Superficial or Mucocutaneous Fungal Infections

In humans, superficial mycosis is quite common. A fungus can infect the skin, nails, or mucous membranes on the outside of the body.

Examples of fungal infections that are superficial include:

Ringworm (Dermatophytosis)

The fungi known as dermatophytes, which feed on skin, hair, and nail cells, cause ringworm. They can affect the groin and inner thighs (tinea cruris), hands (tinea manuum), scalp (tinea capitis), feet (tinea pedis), facial hair and the skin around it (tinea barbae), and other parts of the human body (tinea corporis).

Onychomycosis

Onychomycosis is an infection of the fingernails or toenails caused by a variety of fungus. This may result in brittle and discolored nails.

Candidiasis

Candida (often *Candida albicans*) causes candidiasis, an infection of the skin and mucous membranes (mucocutaneous). These include vaginal yeast infections (vulvovaginitis), candidal intertrigo, esophageal candidiasis, diaper rash, and oral thrush.

Tinea Versicolor/Pityriasis Versicolor

The fungus *Malassezia* is the source of the skin discoloration known as tinea versicolor or pityriasis versicolor.

4.6.1.2 Subcutaneous Fungal Infections

Subcutaneous mycoses are a class of fungi-related illnesses that affect the skin, subcutaneous tissue, and, sometimes underlying tissues and organs. Tropical and subtropical regions of the world have higher prevalence rates of subcutaneous fungal infections.

Examples of fungal infections that are subcutaneous include:

Sporotrichosis

The sporotrichosis is produced by *Sporothrix* fungus. Additionally, sporotrichosis might affect the lungs or other body organs.

Chromoblastomycosis

Chromoblastomycosis (or "chromomycosis") is a persistent fungal infection of the skin and subcutaneous tissue and is brought on by any of several kinds of pigmented fungus. It rarely spreads to other body parts.

Eumycetoma

It is a persistent fungal infection of the epidermis and the tissues just beneath the skin. Eumycetoma, also known as "Madura foot," most frequently affects the feet but can also affect the hands and other body parts.

4.6.1.3 Deep Seated or Systemic Fungal Infections

Other than the skin, systemic fungal infections can affect the lungs, blood, urinary tract, or brain. The invasive or systemic fungal diseases include:

Histoplasmosis

Lungs, brain, or other organs may become infected by the fungus *Histoplasma*, which causes histoplasmosis.

Coccidioidomycosis (Valley Fever)

Coccidioidomycosis, which is brought on by the *Coccidioides* fungus, can also affect lungs and, in rare cases, spread to other areas of your body. Fungal spores are found in the soil at specific areas.

Blastomycosis

The fungus that causes blastomycosis, called *blastomyces*, frequently affects skin, lungs, and bones. Occasionally, it may also infect the brain and spinal cord.

Aspergillosis

A number of lung infections, including chronic pulmonary aspergillosis, and allergic bronchopulmonary aspergillosis, may be brought on by the mold *Aspergillus*. It may also cause an aspergilloma (a fungus ball) or spread to other areas of the body.

Candidal Urinary Tract Infection

The majority of urinary tract infections (UTIs) are caused by bacteria; however, some are brought on by the yeast like *Candida*.

Invasive Candidiasis

Invasive candidiasis is brought on by many different *Candida* species. Blood (candidemia), bones, brain, eyes (endophthalmitis), heart, or other body organs may become infected.

Pneumocystis Pneumonia

This is a lung infection brought on by the fungus *Pneumocystis jirovecii*. People with a weak immune system are usually at risk.

Mucormycosis

Mucormycosis is brought on by a class of molds known as mucormycetes which can infect the skin (cutaneous mucormycosis), sinuses (rhinocerebral mucormycosis), intestines (gastrointestinal mucormycosis), lungs (pulmonary mucormycosis), cutaneous mucormycosis, cutaneous mucormycosis, and many parts of the body simultaneously (disseminated mucormycosis).

Cryptococcosis

Cryptococcosis is brought on by *Cryptococcus neoformans* and *Cryptococcus gattii*. In addition to lungs, they can also infect brain and spinal cord (cryptococcal meningitis).

There are a number of essential oils used against several fungal pathogens (Table 4.2). In the past 10 years, several studies have examined the antifungal activities of essential oils and their constituents against pathogenic fungi (Table 4.3) (Jayant and Sankunny 2014; Filomena et al. 2017).

Table 4.2 List of essential oils and fungal pathogens (Jayant and Sankunny 2014; Filomena et al. 2017)

S. no.	Fungal pathogen	Essential oil producing plants
1.	*Alternaria alternata*	*Allium sativum*; *Carum opticum*; *Carum carvi*; *Cedrus libani*; *C. citratus*; *Cymbopogon martini*; *Echinophora platyloba*; *Eugenia caryophillata*; *Foeniculum vulgare*; *Laurus nobilis*; *Lavandula angustifolia*; *O. Basilicum*; *Origanum vulgare*; *Rosmarinus officinalis*; *Salvia sclarea*; *Tamarix boveana*; *Thuja plicata*; *Thymus vulgaris*; *Thymus zygiis*
2.	*Alternaria humicola*	*Asarum heterotropoides*
3.	*Alternaria solani*	*Genista quadriflora*; *Pinus pinea*; *Echinophora tenuifolia*; *Origanum vulgare*; *Salvia fruticosa*; *Satureja hortensis*; *Zingiber officinale*
4.	*Aspergillus carbonarius*	*Citrus limon*
5.	*Aspergillus flavus*	*Angelica glauca*; *Artemisia nilagirica*; *Carum nigru*; *Cedrus libani*; *Cuminum cyminum*; *Michelia alba*; *Mentha piperita*; *Satureja hortensis*; *Schinus mole*; *Valeriana wallichii*; *Vetiveria zizanioides*; *Zingiber officinale*
6.	*Aspergillus fumigatus*	*Chenopodium ambrosioides*; *Eugenia caryophillata*; *Eugenia caryophyllus*; *Santolina chamaecyparissus*; *Thymus vulgaris*
7.	*Aspergillus niger*	*Allium sativum*; *Artemisia nilagirica*; *Genista quadriflora*; *Juniperi aetheroleum*; *Lallemantia royleana*; *Marrubium vulgare*; *Matricaria chamomilla*; *Ocimum basilicum*; *Solidago canadensis*; *Tamarix boveana*; *Vetiveria zizanioides*; *Zingiber officinale*
8.	*Aspergillus ochraceus*	*Artemisia nilagiric*
9.	*Aspergillus parasiticus*	*Citrus limon*; *Rosmarinus officinalis*; *Satureja hortensis*
10.	*Bipolaris oryzae*	*Piper sarmentosum*
11.	*Bipolaris sorokiniana*	*Eucalyptus erythrocor*; *Pinus pinea*
12.	*Biscogniauxia mediterranea*	*Eucalyptus* spp.
13.	*Botryotinia fuckeliana*	*Thymus zygiis*

(continued)

Table 4.2 (continued)

S. no.	Fungal pathogen	Essential oil producing plants
14.	*Botrytis cinerea*	*Artemisia absinthium*; *Artemisia biennis*; *Angelica archangelica*; *Artemisia judaica*; *Mentha piperita*; *Artemisia judaica*; *A. monosperma*; *Callistemon viminals*; *Citrus aurantifolia*; *Cinnamomum cassia*; *C. Lemon*; *C. Paradisi*; *C. sinensis*; *Cupressus macrocarpa*; *C. Sempervirens*; *Cestrum nocturnum*; *Carum carvi*; *Eugenia caryophyllata*; *Eucalyptus erythrocorys*; *Foeniculum vulgare*; *Marrubium vulgar*; *Melaleuca alternifolia*; *Myrtus communis*; *Melissa officinalis*; *Mentha pulegium*; *Metasequoia glyptostroboides*; *Origanum heracleoticum*; *Origanum majorana*; *Origanum vulgare*; *Pelargonium graveolens*; *Rosmarinus officinalis*; *Solidago canadensis*; *Syzygium cumini*; *Schinus molle*; *S. terebinthifolius*; *Tetraclinis articulata*; *Thymus* spp.; *Thuja occidentalis*; *Tetraclinis articulata*; *Salvia officinalis*; *Thymus zygis*; *T. vulgaris*; *Vitex agnus-castus*
15.	*Botrytis fabae*	
16.	*Candida albicans*	*Croton cajucara*; *Cymbopogon citratus*; *Cymbopogon martini*; *Eugenia caryophyllus*; *Eucalyptus saligna*; *Juniperi aetherole*; *Lavandula* sp.; *Melaleuca alternifolia*; *Mentha piperita*; *M. viridis*; *Melissa officinalis*; *M. longifolia*; *Ocimum* sp.; *Piper nigrum*; *Pimpinella anisum*; *Santolina rosmarinifolia*; *Ziziphora clinopodioides*
17.	*Candida glabrata*	
18.	*Choanephora cucurbitarum*	*Cinnamomum camphora*; *Syzygium cumini*
19.	*Cladosporium cladosporioides*	*Artemisia absinthium*; *Artemisia judaic*; *Artemisia biennis*; *Cedrus libani*; *Citrus limon*; *Lavandula angustifolia*; *Eugenia caryophillata*; *Origanum vulgare*; *Salvia sclarea*; *Thymus vulgaris*; *Thuja plicata*
20.	*Cladosporium herbarum*	*Artemisia biennis*; *Artemisia absinthium*; *Artemisia judaica*; *Cedrus libani*
21.	*Colletotrichum capsici*	*Piper chaba*; *Cestrum nocturnum*; *Metasequoia glyptostroboides*
22.	*Colletotrichum coccoides*	*Echinophora tenuifolia*; *Origanum vulgare*; *Salvia fruticosa*; *Satureja hortensis*
23.	*Colletotrichum gloeosporioides*	*Asarum heterotr*; *Cymbopogon* sp.
24.	*Colletotrichum lindemuthianum*	*Allium sativum*
25.	*Colletotrichum musae*	*Cinnamomum zeylanicum*; *Syzıgium aromaticum*
26.	*Colletotrichum tricbellum*	*Echinophora platyloba*
27.	*Cryptococcus neoformans*	*Lavandula* sp.; *Ziziphora clinopodioides*
28.	*Curvularia fallax*	*Echinophora platyloba*
29.	*Curvularia lunata*	*Curcuma longa*

Table 4.2 (continued)

S. no.	Fungal pathogen	Essential oil producing plants
30.	*Cytospara sacchari*	*Echinophora platyloba*
31.	*Eurotium herbariorum*	*Citrus limon*
32.	*Fonsecaea pedrosoi*	*Artemisia absinthium*; *Artemisia biennis*; *Artemisia judaica*
33.	*Fusarium avenaceum*	*Eucalyptus erythrocorys*
34.	*Fusarium proliferatum*	*Cinnamomum zeylanicum*; *Syzıgium aromaticum*
35.	*Fusarium graminearum*	*Cinnamomum zeylanicum*; *Pimpinella anisum*; *Syzygium aromaticum*
36.	*Fusarium moniliforme*	*Allium sativum*; *Artemisia absinthium*; *Artemisia biennis*; *Artemisia judaica*; *Chenopodium ambrosioides*; *Cymbopogon martini*; *Cymbopogon citratus*; *Salvia officinalis*; *Salvia fruticosa*; *Salvia rosifolia*; *Tamarix boveana*; *Zingiber officinale*
37.	*Fusarium oxysporum*	*Allium sativum*; *Artemisia monosperma*; *Artemisia judaica*; *Artemisia absinthium*; *Artemisia biennis*; *Callistemon viminals*; *Citrus aurantifolia*; *Citrus lemon*; *Citrus paradisi*; *Citrus sinensis*; *Cupressus sempervirens*; *Cupressus macrocarpa*; *Chenopodium ambrosioides*; *Cymbopogon citratus*; *Cymbopogon martini*; *Echinophora platyloba*; *Eucalyptus erythrocorys*; *Eucalyptus globulus*; *Genista quadriflora*; *Metasequoia glyptostroboides*; *Mentha piperita*; *Myrtus communis*; *Mikania scandens*; *Origanum vulgare*; *Pelargonium graveolens*; *Rosmarinus officinalis*; *Piper chaba*; *Salmea scandens*; *Salvia fruticosa*; *Salvia officinalis*; *Salvia sosifolia*; *Syzygium aromaticum*; *Syzygium cumini*; *Tamarix boveana*; *Terebinthifolius*; *Thuja occidentalis*; *Vitex agnus-castus*; *Zingiber officinale*
38.	*Fusarium proliferatum*	*Allium sativum*; *Artemisia absinthium*; *Artemisia biennis*; *Artemisia judaica*; *Chenopodium ambrosioides*; *Cymbopogon citratus*; *Cymbopogon martini*; *Rosmarinus officinalis*; *Salvia fruticosa*; *Salvia sosifolia*; *Salvia officinalis*; *Tamarix boveana*; *Zingiber officinale*
39.	*Fusarium sulphureum*	*Zanthoxylum bungeanum*
40.	*Fusarium verticillioides*	*Curcuma longa*

(continued)

Table 4.2 (continued)

S. no.	Fungal pathogen	Essential oil producing plants
41.	*Fusarium solani*	*Allium sativum*; *Angelica archangelica*; *Artemisia absinthium*; *Artemisia biennis*; *Artemisia judaica*; *Artemisia monosperma*; *Asarum heterotropoides*; *Angelica glauca*; *Callistemon viminals*; *Cestrum nocturnum*; *Citrus lemon*; *Citrus aurantifolia*; *Citrus sinensis*; *Citrus paradisi*; *Cupressus macrocarpa*; *C. Sempervirens*; *Chenopodium ambrosioides*; *Cymbopogon citratus*; *Cymbopogon martini*; *Eucalyptus erythrocorys*; *Marrubium vulgare*; *Metasequoia glyptostroboides*; *Origanum vulgare*; *Myrtus communis*; *Pelargonium graveolens*; *Pinus pinea*; *Piper chaba*; *Plectranthus rugosus*; *Rosmarinus officinalis*; *Salvia fruticosa*; *Salvia rosifolia*; *Salvia officinalis*; *Syzygium cumini*; *Schinus terebinthifolius*; *Thuja occidentalis*; *Tamarix boveana*; *Tetraclinis articulata*; *Valeriana wallichii*; *Vitex agnus-castus*; *Zingiber officinale*
42.	*Geotrichum candidum*	*Artemisia absinthium*; *Artemisia biennis*; *Artemisia judaica*
43.	*Geotrichum citriaurantii*	*Thymus* spp.
44.	*Lasiodiplodia theobromae*	*Myrcia lundiana*
45.	*Macrophomina phaseolina*	*Allium sativum*; *Chenopodium ambrosioides*; *Echinophora platyloba*; *Mentha piperita*; *Ocimum basilicum*; *Z. officinale*
46.	*Microdochium nivale*	*Pinus pinea*
47.	*Microsporum canis*	*Artemisia absinthium*; *Artemisia biennis*; *Artemisia judaica*; *Cinnamomum* sp.; *Croton argyrophylloides*; *Croton cajucara*; *Croton zehntneri*; *Daucus carota*; *Syzigium aromaticum*
48.	*Microsporum gypseum*	
49.	*Monilinia fructicola*	*Mentha pulegium*; *Solidago canadensis*
50.	*Mucor ramannianus*	*Mentha longifolia*; *Mentha piperita*; *Mentha viridis*
51.	*Penicillium cyclopium*	*Allium sativum*; *Artemisia absinthium*; *Artemisia biennis*; *Artemisia judaica*; *Carum nigrum*; *Cymbopogon citratus*; *Cymbopogon martini*; *Tamarix boveana*
52.	*Penicillium digitatum*	*Carum opticum*; *Carum carvi*; *Foeniculum vulgare*; *Marrubium vulgare*; *Origanum dictamnus*; *Origanum vulgare*; *Origanum majorana*; *Ocimum basilicum*; *Satureja hortensis*; *Thymus* spp.
53.	*Penicillium expansum*	*Pulicaria mauritanica*; *Melissa officinalis*; *Warionia saharae*; *Solidago canadensis*
54.	*Penicillium italicum*	*Rosmarinus officinalis*
55.	*Penicillium roquefortii*	*Allium sativum*; *Artemisia absinthium*; *Artemisia biennis*; *Artemisia judaica*; *Carum nigrum*; *Cymbopogon citratus*; *Cymbopogon martini*; *Tamarix boveana*
56.	*Penicillium verrucosum*	*Allium sativum*; *Origanum onites*; *Salvia officinalis*; *Mentha piperita*

Table 4.2 (continued)

S. no.	Fungal pathogen	Essential oil producing plants
57.	*Penicillium viridicatum*	*Allium sativum*; *Artemisia absinthium*; *Artemisia biennis*; *Artemisia judaica*; *Carum nigrum*; *Citrus limon*; *Cymbopogon citratus*; *Cymbopogon martini*; *Tagetes minuta*; *Tamarix boveana*; *Vetiveria zizanioides*
58.	*Phytophthora capsici*	*Allium sativum*; *Thymbra spicata*; *Satureja thymbra*
59.	*Phytophthora infestans*	*Piper longum*
60.	*Phytophthora megakarya*	*Syzygium aromaticum*; *Zanthoxylum xanthoxyloides*
61.	*Puccinia recondite*	*Piper longum*
62.	*Pyrenophora avenae*	*Hyssopus officinalis*
63.	*Pyricularia oryzae*	
64.	*Pythium aphanidermatum*	*Allium sativum*
65.	*Pythium ultimum*	*Allium sativum*; *Mikania scandens*; *Thymus* spp.
66.	*Rhizocotonia solani*	*Artemisia absinthium*; *Artemisia biennis*; *Artemisia judaica*; *Thymus vulgaris*
67.	*Rhizopus stolonifer*	*Thymus vulgaris*; *Foeniculum vulgare*; *Melissa officinalis*; *Ocimum basilicum*; *Satureja hortensis*; *Pulicaria mauritanica*; *Warionia saharae*
68.	*Sclerotinia sclerotiorum*	*Metasequoia glyptostroboides*; *Cestrum nocturnum*; *Zingiber officinale*; *Ziziphora clinopodioides*
69.	*Sclerotium rolfsii*	*Allium cepa*; *Cuminum cyminum*; *Mentha piperita*; *T. vulgaris*
70.	*Trichophyton mentagrophytes*	*Artemisia absinthium*; *Artemisia biennis*; *Artemisia judaica*; *Syzigium aromaticum*; *Daucus carota*; *Cinnamomum* sp.
71.	*Trichophyton roseum*	
72.	*Trichophyton rubrum*	
73.	*Uncinula necator*	*Allium sativum*; *Cuminum cyminum*; *Cymbopogon citratus*
74.	*Uromyces viciae-fabae*	*Hyssopus officinalis*

Table 4.3 Antifungal activity of aromatic plant essential oils

S. no.	Aromatic plant	Plant used	Main active compounds	Pathogenic fungi	Methodology used	100% Inhibitory concentration	MIC	MFC	References
Family Araceae (Acoraceae)									
1.	*Acorus calamus*	Rhizomes	β-Asarone	*Aspergillus niger*	Tube dilution method	–	–	0.117 ± 0.017 mg/mL	Joshi (2016)
				Aspergillus fumigatus		–	–	0.104 ± 0.016 mg/mL	
				Penicillium chrysogenum		–	–	0.208 ± 0.032 mg/mL	
Family Apiaceae									
2.	*Anethum graveolens*	Seeds	Carvone, limonene, apiol	*Candida auris*	Broth microdilution method	–	>4.00 (%v/v)	>4.00 (%v/v)	Di Vito et al. (2023)
3.	*Angelica dahurica*	Roots	α-Pinene, sabinene, myrcene, 1-dodecanol, and terpinen-4-ol	*Rhizopus stolonifer*	Agar diffusion	1000 μL/L	–	–	Yongdong et al. (2021)
4.	*Coriandrum sativum*	Fruits	Linalool	*Candida auris*	Broth microdilution method	–	0.25 (%v/v)	0.50 (%v/v)	Di Vito et al. (2023)
5.	*Cuminum cyminum*	Aerial parts	α-Pinene	*Saprolegnia parasitica*	Disc diffusion method	–	1 μg/mL	2 μg/mL	Adel et al. (2020)
6.	*Cuminum cyminum*	Aerial parts	–	*Aspergillus aculeatus*	Contact phase method	–	67.28 μg/mL	–	Tanapichatsakul et al. (2020)
7.	*Cuminum cyminum*	Aerial parts	–	*Candida auris*	Broth microdilution method	–	>4.00 (%v/v)	>4.00 (%v/v)	Di Vito et al. (2023)
8.	*Eryngium campestre*	Aerial parts	Bornyl acetate	*Saprolegnia parasitica*	Disc diffusion method	–	1 μg/mL	2 μg/mL	Adel et al. (2020)
9.	*Levisticum officinale*	Aerial parts	α-Terpinyl acetate and β-phellandrene	*Rhizopus stolonifer*	Agar diffusion	1000 μL/L	–	–	Yongdong et al. (2021)
10.	*Pimpinella affinis*	Aerial parts	Pregeijene	*Saprolegnia parasitica*	Disc diffusion method	–	2 μg/mL	4 μg/mL	Adel et al. (2020)

Family Lamiaceae									
11.	*Bollota aucheri*	Flowering aerial parts	β-Caryophyllene	*Candida. albicans*	Broth microdilution method	–	0.039–1.25 μL/mL	–	Gharaghani et al. (2023)
				Candida parapsilosis		–	0.039–0.15 μL/mL	–	
				Candida glabrata		–	0.039–0.15 μL/mL	–	
12.	*Ballota hirsute*	Leaves and flowers	Germacrene-D	*Botrytis cinerea*	PF and VA techniques	–	2000 ppm	>2000 ppm	Omar et al. (2023)
				Penicillium expansum		–	2000 ppm	>2000 ppm	
13.	*Citrus clementine*	Leaves	Monoterpene hydrocarbons	*Alternaria alternata*	Spore counting method	–	8000 ± 0.00 μg/mL	–	Affes et al. (2022)
14.	*Citrus grandis*	Fruits	Limonene, myrcene, α-pinene, γ-terpinene	*Alternaria alternata*	Substrate poisoning method	–	13.8 mg/mL	–	Zalewska et al. (2022)
				Botrytis cinerea		–	3.3 mg/mL	–	
				Cladosporium cladosporioides		–	2.63 mg/mL	–	
15.	*Hyssopus officinalis*	Aerial parts	α-Pinene, izopinocampfen, pinocampfen	*Fusarium graminearum*	Broth microdilution method	–	0.4 ± 0 mg/mL	–	Michaela et al. (2021)
16.	*Lavandula angustifolia*	Leaves	Linalool, linalyl acetate	*Alternaria alternata*	Substrate poisoning method	–	19.7 mg/mL	–	Zalewska et al. (2022)
				Botrytis cinerea		–	5.0 mg/mL	–	
				Cladosporium cladosporioides		–	−1.05 mg/mL	–	
17.	*Lavandula angustifolia*	Leaves	–	*Candida auris*	Broth microdilution method	–	1.00	1.00	Di Vito et al. (2023)

(continued)

Table 4.3 (continued)

S. no.	Aromatic plant	Plant used	Main active compounds	Pathogenic fungi	Methodology used	100% Inhibitory concentration	MIC	MFC	References
18.	*Lavandula stoechas*	Leaves and flowers	Fenchone/camphor	*Candida albicans*	Microdilution method	–	1.47 ± 0.9 mg/mL	–	Tomasz et al. (2023)
				Candida glabrata		–	1.08 ± 0.47 mg/mL	–	
				Candida guilliermondii		–	0.78 ± 0.0 mg/mL	–	
				Candida krusei		–	3.12 ± 1.45 mg/mL	–	
				Candida parapsilosis		–	1.37 ± 0.39 mg/mL	–	
				Candida tropicalis		–	1.76 ± 0.98 mg/mL	–	
19.	*Melissa officinalis*	Leaves	–	*Candida albicans*	Microdilution method	–	0.78 ± 0.9 mg/mL	–	Tomasz et al. (2023)
				Candida glabrata		–	0.78 ± 0.34 mg/mL	–	
				Candida guilliermondii		–	0.39 ± 0.0 mg/mL	–	
				Candida krusei		–	1.41 ± 0.83 mg/mL	–	
				Candida parapsilosis		–	0.49 ± 0.2 mg/mL	–	
				Candida tropicalis		–	0.68 ± 0.2 mg/mL	–	

20.	*Mentha piperita*	Aerial parts	Menthol, menton, menthyl acetate, 1.8-cineol, isomenton, isomenthol, trans-caryophyllene, limonene, pulegon, β-pinene, and piperiton	*Alternaria alternata*	Broth microdilution method	–	0.4 μL/mL	1.7 μL/mL	Dragana et al. (2020)
				Aspergillus versicolor		–	0.4 μL/mL	14.2 μL/mL	
				Aspergillus niger		–	1.7 μL/mL	7.1 μL/mL	
				Aspergillus fumigatus		–	0.8 μL/mL	113.6 μL/mL	
				Aspergillus flavus		–	1.7 μL/mL	227.2 μL/mL	
				Cladosporium cladosporioides		–	0.2 μL/mL	1.7 μL/mL	
				Fusarium sporotrichioides		–	0.8 μL/mL	1.7 μL/mL	
				Fusarium proliferatum		–	1.7 μL/mL	3.5 μL/mL	
				Penicillium aurantiogriseum		–	0.8 μL/mL	454.5 μL/mL	
				Penicillium oxalicum		–	0.8 μL/mL	56.8 μL/mL	
				Penicillium expansum		–	0.4 μL/mL	1.7 μL/mL	
21.	*Mentha piperita*	Aerial parts	Menthol	*Saprolegnia parasitica*	Disc diffusion method	–	0.5 μg/mL	0.5 μg/mL	Adel et al. (2020)
22.	*Mentha piperita*	Aerial parts	Menthol	*Trichoderma aggressivum*	Microdilution method	–	12.5 μL/mL	25 mL/mL	Jelena et al. (2020)
23.	*Mentha piperita*	Aerial parts	Menthol	*Botrytis cinerea*	Microdilution method	–	10 μg/mL	20 μg/mL	Josemar et al. (2021)
24.	*Mentha piperita*	Aerial parts	Menthol	*Candida auris*	Broth microdilution method	–	4.00	4.00	Di Vito et al. (2023)

(continued)

Table 4.3 (continued)

S. no.	Aromatic plant	Plant used	Main active compounds	Pathogenic fungi	Methodology used	100% Inhibitory concentration	MIC	MFC	References
25.	*Mentha piperita*	Aerial parts	Menthol	*Cladosporium cladosporoides*	Disc diffusion method	–	1.56 ± 0.00	–	Tomić et al. (2023)
				Aspergillus fumigatus		–	6.75 ± 0.00	–	
				Penicillium chrysogenum		–	1.56 ± 0.00	–	
26.	*Mentha piperita*	Leaves	Menthol	*Candida albicans*	Microdilution method	–	2.27 ± 1.37 mg/mL	–	Tomasz et al. (2023)
				Candida glabrata		–	2.78 ± 0.69 mg/mL	–	
				Candida guilliermondii		–	1.56 ± 0.0 mg/mL	–	
				Candida tropicalis		–	2.73 ± 0.78 mg/mL	–	
				Candida krusei		–	6.64 ± 3.9 mg/mL	–	
				Candida parapsilosis		–	2.34 ± 0.9 mg/mL	–	
27.	*Mentha piperita*	Leaves	Menthol, menthone, and 1,8-cineole	*Alternaria alternata*	Mycelial growth inhibition	1.5 mg/mL	–	–	Affes et al. (2023)
28.	*Mentha pulegium*	Fresh leaves	Pulegone, menthol, menthone, and isopulegol	*Monilinia fructicola*	Mycelial growth inhibition test	–	24.6 ± 1.33 μg/mL	–	Iván et al. (2020)
				Botrytis cinerea		–	301.45 ± 1.49 μg/mL	–	
29.	*Mentha spicata*	Leaves		*Trichoderma aggressivum*	Microdilution method	–	6.25 μL/mL	25 mL/mL	Jelena et al. (2020)
30.	*Mentha spicata*	Leaves		*Botrytis cinerea*	Microdilution method	–	5 μg/mL	10 μg/mL	Josemar et al. (2021)
31.	*Mentha spicata*	Leaves	Carvone, 1,8-cineole, *cis*-dihydrocarvone, limonene, sabinene, and α-thujene	*Penicillium crustosum*	Agar disc diffusion method	–	12.93 ± 0.46 mm (ZOI)	–	Ďúranová et al. (2023)
				Penicillium citrinum		–	9.47 ± 0.59 mm (ZOI)	–	
				Penicillium expansum		–	11.46 ± 0.63 mm (ZOI)	–	

32.	*Mentha spicata*	Leaves	–	*Candida albicans*	Microwell dilution assays	–	0.33 mg/mL	–	Pinar et al. (2023)
				Aspergillus niger		–	0.33 mg/mL	–	
				Penicillium spp.		–	0.09 mg/mL	–	
33.	*Ocimum basilicum*	Leaves and flower	–	*Trichoderma aggressivum*	Microdilution method	–	12.5 μL/mL	25 mL/mL	Jelena et al. (2020)
34.	*Ocimum basilicum*	Leaves and flower	Metylchavicol	*Fusarium graminearum*	Broth microdilution method	–	12.5 ± 0 mg/mL	–	Michaela et al. (2021)
35.	*Origanum vulgare*	Leaves	Carvacrol	*Fusarium graminearum*	Broth microdilution method		0.4 ± 0 mg/mL	–	Michaela et al. (2021)
36.	*Origanum vulgare*	Leaves	Carvacrol, γ-terpinene, and *p*-cymene	*Alternaria alternata*	Poisoned food method	0.1 mg/mL	–	–	Zalewska et al. (2022)
				Botrytis cinerea		0.1 mg/mL	–	–	
				Cladosporium cladosporioides		0.1 mg/mL	–	–	
37.	*Origanum vulgare*	Aerial parts	–	*Cladosporium cladosporoides*	Disc diffusion method	–	0.78 ± 0.00	–	Tomić et al. (2023)
				Aspergillus fumigatus		–	1.56 ± 0.00	–	
				Penicillium chrysogenum		–	1.56 ± 0.00	–	
38.	*Origanum vulgare*	Aerial parts	–	*Candida albicans*	Microdilution method	–	0.83 ± 0.71 mg/mL	–	Tomasz et al. (2023)
				Candida glabrata		–	1.08 ± 0.47 mg/mL	–	
				Candida krusei		–	2.05 ± 0.93 mg/mL	–	
				Candida parapsilosis		–	0.88 ± 0.5 mg/mL	–	
				Candida tropicalis		–	0.83 ± 0.56 mg/mL	–	
				Candida guilliermondii		–	0.59 ± 0.28 mg/mL	–	
39.	*Origanum vulgare*	Aerial parts	–	*Rhizoctonia solani*	Mycelium growth inhibition method	–	22.89 mg/mL	–	Tian-Lin et al. (2023)

(continued)

Table 4.3 (continued)

S. no.	Aromatic plant	Plant used	Main active compounds	Pathogenic fungi	Methodology used	100% Inhibitory concentration	MIC	MFC	References
40.	*Pogostemon cablin*	Aerial parts	–	*Candida albicans*	Microdilution technique	–	4 μg/mL	4 μg/mL	Da Cunha et al. (2023)
				Candida parapsilosis		–	8 μg/mL	8 μg/mL	
41.	*Rosmarins officinalis*	Leaves	–	*Candida albican*	Broth dilution methods	–	0.315 μg/mL	1.25 μg/mL	Majda et al. (2022)
				Aspergillus niger		–	1.25 μg/mL	2.5 μg/mL	
				Alternaria alternata		–	1.25 μg/mL	2.5 μg/mL	
42.	*Rosmarinus officinalis*	Leaves	1,8-Cineole	*Alternaria alternata*	Spore counting method	–	5000 ± 0.00 μg/mL	–	Affes et al. (2022)
43.	*Rosmarinus officinalis*	Leaves	–	*Candida auris*	Broth microdilution method	–	4.00	4.00	Di Vito et al. (2023)
44.	*Rosmarinus officinalis*	Leaves	–	*Candida albicans*	Broth microdilution assay	–	6.25%	25%	Dwi et al. (2023)
				Candida non-albicans		–	12.5%	25%	
45.	*Rosmarinus officinalis*	Leaves	–	*Candida albicans*	Microdilution method	–	1.91 ± 1.38 mg/mL	–	Tomasz et al. (2023)
				Candida glabrata		–	1.56 ± 0.68 mg/mL	–	
				Candida krusei		–	7.81 ± 4.09 mg/mL	–	
				Candida parapsilosis		–	3.52 ± 1.97 mg/mL	–	
				Candida tropicalis		–	2.73 ± 0.78 mg/mL	–	
				Candida guilliermondii		–	2.34 ± 1.11 mg/mL	–	

46.	*Salvia officinalis*	Leaves	–	*Candida albicans*	Microdilution method	–	16.36 ± 11.02 mg/mL	–	Tomasz et al. (2023)
				Candida glabrata		–	20.14 ± 13.18 mg/mL	–	
				Candida krusei		–	29.69 ± 29.08 mg/mL	–	
				Candida parapsilosis		–	50 ± 35.36 mg/mL	–	
				Candida tropicalis		–	14.06 ± 7.86 mg/mL	–	
				Candida guilliermondii		–	18.75 ± 8.84 mg/mL	–	
47.	*Satureja calamintha*	Aerial parts	–	*Aspergillus flavus*	Poisoned food method	–	0.666 μL/mL	0.666 μL/mL	Ammar et al. (2022)
				Aspergillus parasiticus		–	1.333 μL/mL	1.333 μL/mL	
				Aspergillus ochraceus		–	2.666 μL/mL	2.666 μL/mL	
48.	*Thymus serpillum*	Leaves	–	*Trichoderma aggressivum*	Microdilution method	–	6.25 μL/mL	25 mL/mL	Jelena et al. (2020)
49.	*Thymus vulgaris*	Aerial parts	Thymol, ρ-cymene	*Fusarium graminearum*	Broth microdilution method	–	0.4 ± 0 mg/mL	–	Michaela et al. (2021)
50.	*Thymus vulgaris*	Leaves	Carvacrol	*Alternaria alternata*	Spore counting method	–	6000 ± 0.00 μg/mL	–	Affes et al. (2022)
51.	*Thymus vulgaris*	Leaves	–	*Candida albicans*	Microdilution method	–	4.16 ± 3.63 mg/mL	–	Tomasz et al. (2023)
				Candida glabrata		–	1.56 ± 0.68 mg/mL	–	
				Candida krusei		–	6.45 ± 4.13 mg/mL	–	
				Candida parapsilosis		–	0.88 ± 0.49 mg/mL	–	
				Candida tropicalis		–	2.15 ± 1.17 mg/mL	–	
				Candida guilliermondii		–	1.56 ± 0.0 mg/mL	–	
52.	*Thymus vulgaris*	Aerial parts	–	*Candida auris*	Broth microdilution method	–	1.00	2.00	Di Vito et al. (2023)

(continued)

Table 4.3 (continued)

S. no.	Aromatic plant	Plant used	Main active compounds	Pathogenic fungi	Methodology used	100% Inhibitory concentration	MIC	MFC	References
Family Myrtaceae									
53.	*Eucalyptus crebra*	Leaves	–	*Aspergillus flavus*	Agar well diffusion method and broth microdilution method	–	4 μg/mL	8 μg/mL	Siddique et al. (2021)
				Aspergillus niger		–	4 μg/mL	8 μg/mL	
				Fusarium oxysporum		–	4 μg/mL	8 μg/mL	
				Fusarium solani		–	4 μg/mL	8 μg/mL	
				Penicillium digitatum		–	4 μg/mL	8 μg/mL	
54.	*Eucalyptus kitsoniana*	Leaves	–	*Aspergillus flavus*	Agar well diffusion method and broth microdilution method	–	4 μg/mL	8 μg/mL	Siddique et al. (2021)
				Aspergillus niger		–	4 μg/mL	8 μg/mL	
				Fusarium oxysporum		–	4 μg/mL	8 μg/mL	
				Fusarium solani		–	4 μg/mL	8 μg/mL	
				Penicillium digitatum		–	4 μg/mL	8 μg/mL	
55.	*Eucalyptus melanophloia*	Leaves	–	*Aspergillus flavus*	Agar well diffusion method and broth microdilution method	–	4 μg/mL	8 μg/mL	Siddique et al. (2021)
				Aspergillus niger		–	4 μg/mL	8 μg/mL	
				Fusarium oxysporum		–	4 μg/mL	8 μg/mL	
				Fusarium solani		–	4 μg/mL	8 μg/mL	
				Penicillium digitatum		–	4 μg/mL	8 μg/mL	

56.	*Eucalyptus microtheca*	Leaves	–	*Aspergillus flavus*	Agar well diffusion method and broth microdilution method	–	4 μg/mL	8 μg/mL	Siddique et al. (2021)
				Aspergillus niger		–	4 μg/mL	8 μg/mL	
				Fusarium oxysporum		–	4 μg/mL	8 μg/mL	
				Fusarium solani		–	4 μg/mL	8 μg/mL	
				Penicillium digitatum		–	4 μg/mL	8 μg/mL	
57.	*Eucalyptus pruinosa*	Leaves	–	*Aspergillus flavus*	Agar well diffusion method and broth microdilution method	–	4 μg/mL	8 μg/mL	Siddique et al. (2021)
				Aspergillus niger		–	4 μg/mL	8 μg/mL	
				Fusarium oxysporum		–	4 μg/mL	8 μg/mL	
				Fusarium solani		–	4 μg/mL	8 μg/mL	
				Penicillium digitatum		–	4 μg/mL	8 μg/mL	
58.	*Eucalyptus rudis*	Leaves	–	*Aspergillus flavus*	Agar well diffusion method and broth microdilution method	–	4 μg/mL	8 μg/mL	Siddique et al. (2021)
				Aspergillus niger		–	4 μg/mL	8 μg/mL	
				Fusarium oxysporum		–	4 μg/mL	8 μg/mL	
				Fusarium solani		–	4 μg/mL	8 μg/mL	
				Penicillium digitatum		–	4 μg/mL	8 μg/mL	
59.	*Eucalyptus tereticornis*	Leaves	–	*Aspergillus flavus*	Agar well diffusion method and broth microdilution method	–	4 μg/mL	8 μg/mL	Siddique et al. (2021)
				Aspergillus niger		–	4 μg/mL	8 μg/mL	
				Fusarium oxysporum		–	4 μg/mL	8 μg/mL	
				Fusarium solani			4 μg/mL	8 μg/mL	
				Penicillium digitatum		–	4 μg/mL	8 μg/mL	

(continued)

Table 4.3 (continued)

S. no.	Aromatic plant	Plant used	Main active compounds	Pathogenic fungi	Methodology used	100% Inhibitory concentration	MIC	MFC	References
60.	*Melaleuca bracteata*	Leaves	–	*Aspergillus flavus*	Agar well diffusion method and broth microdilution method	–	4 μg/mL	4 μg/mL	Siddique et al. (2021)
				Aspergillus niger		–	4 μg/mL	4 μg/mL	
				Fusarium oxysporum		–	4 μg/mL	4 μg/mL	
				Fusarium solani		–	4 μg/mL	4 μg/mL	
				Penicillium digitatum		–	4 μg/mL	4 μg/mL	
61.	*Melaleuca fulgens*	Leaves	–	*Aspergillus flavus*	Agar well diffusion method and broth microdilution method	–	2 μg/mL	2 μg/mL	Siddique et al. (2021)
				Aspergillus niger		–	2 μg/mL	2 μg/mL	
				Fusarium oxysporum		–	2 μg/mL	2 μg/mL	
				Fusarium solani		–	2 μg/mL	2 μg/mL	
				Penicillium digitatum		–	2 μg/mL	2 μg/mL	
62.	*Melaleuca leucadendron*	Leaves	–	*Aspergillus flavus*	Agar well diffusion method and broth microdilution method	–	4 μg/mL	4 μg/mL	Siddique et al. (2021)
				Aspergillus niger		–	2 μg/mL	2 μg/mL	
				Fusarium oxysporum		–	2 μg/mL	2 μg/mL	
				Fusarium solani		–	2 μg/mL	2 μg/mL	
				Penicillium digitatum		–	2 μg/mL	2 μg/mL	
63.	*Melaleuca alternifolia*	Leaves	Carvacrol, γ-terpinene, and *p*-cymene	*Alternaria alternata*	Poisoned food method	0.5 mg/mL	–	–	Zalewska et al. (2022)
				Botrytis cinerea		0.5 mg/mL	–	–	
				Cladosporium cladosporioides		0.5 mg/mL	–	–	
64.	*Melaleuca alternifolia*	Leaves	–	*Candida auris*	Broth microdilution method	–	4.00	4.00	Di Vito et al. (2023)
65.	*Melaleuca linariifolia*	Leaves	1,8-Cineole	*Alternaria alternata*	Spore counting method	–	5000 ± 0.00 μg/mL	–	Affes et al. (2022)

66.	*Myrtus Communis*	Leaves	1.8 Cineol, D-limonene, myrtenyl acetate, and α-pinene	*Aspergillus niger*	Agar diffusion method	–	0.5 ± 00 1 L/mL	0.0 ± 00 1 L/mL	Fatima et al. (2023)
				Penicillium digitatum		–	0.5 ± 00 1 L/mL	1 ± 00 1 L/mL	
67.	*Myrtus communis*	Leaves	–	*Alternaria alternata*	Mycelial growth inhibition	4 mg/mL	–	–	Affes et al. (2023)
68.	*Syzygium aromaticum*	Flower buds	–	*Rhizopus stolonifer*	Agar diffusion method	1000 µL/L	–	–	Yongdong et al. (2021)
69.	*Syzygium aromaticum*	Flower buds	Eugenol	*Fusarium graminearum*	Broth microdilution method	–	0.4 ± 0 mg/mL	–	Michaela et al. (2021)
Family Poaceae									
70.	*Cymbopogon citratus*	Leaves	–	*Candida auris*	Broth microdilution method	–	>4.00	4.00	Di Vito et al. (2023)
71.	*Cymbopogon citratus*	Leaves	–	*Candida cladosporoides*	Broth microdilution method	–	1.56 ± 0.00	–	Tomić et al. (2023)
				Aspergillus fumigatus		–	6.75 ± 0.00	–	
				Penicillium chrysogenum		–	6.75 ± 0.00	–	
72.	*Cymbopogon flexuosus*	Leaves	–	*Candida parapsilosis MTCC998*	Broth dilution method	–	2.5 µL/mL	5.0 µL/mL	Abdullah (2023)
				Candida tropicalis MTC1000		–	1.25 µL/mL	2.5 µL/mL	
73.	*Cymbopogon martinii*	Leaves	Caryophyllene, geraniol, geranyl acetate, and linalool	*Botrytis cinerea*	Microwell dilution methods	–	5 µg/mL	10 µg/mL	Josemar et al. (2021)
74.	*Cymbopogon martinii*	Bark	–	*Candida auris*	Broth microdilution method	–	1.00	1.00	Di Vito et al. (2023)

(continued)

Table 4.3 (continued)

S. no.	Aromatic plant	Plant used	Main active compounds	Pathogenic fungi	Methodology used	100% Inhibitory concentration	MIC	MFC	References
Family Asteraceae									
75.	*Acritopappus confertus*	Leaves	Myrcene	*Candida albicans*	Broth microdilution method	–	256 μg/mL	–	Rafael et al. (2022)
				Candida tropicalis		–	1024 μg/mL	–	
				Candida krusei		–	1024 μg/mL	–	
76.	*Achillea wilhelmsii*	Aerial parts	1,8-Cineol	*Saprolegnia parasitica*	Disc diffusion method	–	4 μg/mL	8 μg/mL	Adel et al. (2020)
77.	*Chamomilla recutita*	Flower head	α-Bisabolol oxide A, En-yn-dicycloether	*Penicillium chrysogenum*	Agar disc diffusion method	–	1.25 μg/mL	2.5 μg/mL	Ghada and Asmaa (2020)
				Aspergillus flavus		–	1.25 μg/mL	2.5 μg/mL	
				Aspergillus niger		–	2.5 μg/mL	5.0 μg/mL	
78.	*Helichrysum italicum*	Aerial parts	α-Pinene	*Candida albicans*	Broth microdilution method	–	0.6% (v/v)	–	Manoharan et al. (2017)
79.	*Helichrysum italicum*	Aerial parts	α-Pinene	*Candida auris*	Broth microdilution method	–	>4.00	>4.00	Di Vito et al. (2023)
80.	*Santolina pectinata*	Aerial parts	(*Z*)-heptadeca-10,16-dien-7-one	*Botrytis cinerea*	PF and VA methods	–	1.00 μL/mL	1.00 μL/mL	Mounir et al. (2022)
				Penicillium expansum		–	1.00 μL/mL	1.00 μL/mL	
				Rhizopus stolonifer		–	>2.00 μL/mL	–	
81.	*Tagetes erecta*	Shoot	Piperitone, piperitenone, sylvestrene, terpinolene, and (*Z*)-β-ocimene	*Candida albicans*	Broth microdilution method	–	0.08 μL/mL	1.25 μL/mL	Safar et al. (2020)
				Candida glabrata		–	0.32 μL/mL	0.64 μL/mL	
				Candida krusei		–	0.16 μL/mL	0.32 μL/mL	
				Candida tropicalis		–	1.25 μL/mL	2.5 μL/mL	

Family Liliaceae									
82.	*Eucalyptus globulus*	Fresh leaves	*p*-Cimene, α-pinene, α-limonene, γ-terpinene, β-pinene, and β-myrcene	*Candida albicans*	Disk diffusion method	–	77.21 μL/mL	–	Čmiková et al. (2023)
				Candida glabrata		–	245.02 μL/mL	–	
				Candida krusei		–	5.86 μL/mL	–	
				Candida tropicalis		–	2.93 μL/mL	–	
				Aspergillus flavus		–	7.67 ± 0.58 mm	–	
				Botrytis cinerae		–	7.33 ± 2.89 mm	–	
				Penicillium citrinum		–	4.33 ± 1.53 mm	–	
83.	*Eucalyptus globulus*	Leaves	–	*Candida albicans*	Disk diffusion method	–	19.01 ± 1.02	–	Dutta (2023)
Family Rutaceae									
84.	*Citrus bergamia*	Peel	Limonene, linalool, and linalyl acetate	*Aspergillus niger*	Disk evaporation method	–	78.00 ± 0.00	–	Cebi and Erarslan (2023)
				Penicillium expansum		–	32.00 ± 2.82	–	
85.	*Citrus sinensis*	Peel	–	*Candida albicans*	Disk diffusion method	–	9.50 ± 1.09 mm	–	Dutta (2023)
86.	*Zanthoxylum armatum*	Leaves	β-Caryophyllene, ᴅ-limonene, linalool, 4-terpenol, and trans-nerolidol	*Aspergillus flavus*	Plate fumigation method	–	0.8 μL/mL	–	Li et al. (2021)
Family Cupressaceae									
87.	*Cupressus sempervirens*	Leaves	Oxygenated sesquiterpenes	*Alternaria alternata*	Spore counting method	–	8000 ± 0.00 μg/mL	–	Affes et al. (2022)
88.	*Juniperus oxycedrus*	Leaves	–	*Candida krusei*	Disc diffusion method	–	0.10 ± 0.00	60.00 ± 0.01	Mariem et al. (2022)
				Candida albicans		–	0.05 ± 0.00	48.00 ± 0.00	
				Candida glabrata		–	0.05 ± 0.00	62.00 ± 0.00	
				Candida tropicalis		–	0.05 ± 0.00	25.00 ± 0.00	
89.	*Juniperus phoenicea*	Leaves	–	*Candida krusei*	Disc diffusion method	–	0.10 ± 0.00	60.00 ± 0.01	Mariem et al. (2022)
				Candida albicans		–	0.05 ± 0.00	56.00 ± 0.00	
				Candida glabrata		–	0.10 ± 0.00	60.00 ± 0.00	
				Candida tropicalis		–	0.05 ± 0.00	60.00 ± 0.00	

(continued)

Table 4.3 (continued)

S. no.	Aromatic plant	Plant used	Main active compounds	Pathogenic fungi	Methodology used	100% Inhibitory concentration	MIC	MFC	References
90.	*Tetraclinis articulata*	Leaves	–	*Candida krusei*	Disc diffusion method	–	0.05 ± 0.00	25.00 ± 0.03	Mariem et al. (2022)
				Candida albicans		–	0.05 ± 0.00	12.50 ± 0.00	
				Candida glabrata		–	0.05 ± 0.00	18.75 ± 0.14	
				Candida tropicalis		–	0.05 ± 0.00	25.00 ± 0.00	
Family Geraniaceae									
91.	*Pelargonium graveolens*	Leaves and flowers	–	*Rhizopus stolonifer*	Agar diffusion method	1000 μL/L	–	–	Yongdong et al. (2021)
92.	*Pelargonium graveolens*	Leaves and flowers	–	*Candida auris*	Broth microdilution method	–	4.00	4.00	Di Vito et al. (2023)
Family Thymelaeaceae									
93.	*Aquilaria sinensis*	Heartwood	–	*Rhizopus stolonifer*	Agar diffusion method	1000 μL/L	–	–	Yongdong et al. (2021)
Family Aquifoliaceae									
94.	*Ilex aquifolium*	Fruits	–	*Rhizopus stolonifer*	Agar diffusion method	1000 μL/L	–	–	Yongdong et al. (2021)
Family Verbenaceae									
95.	*Lippia citriodora*	Leaves and flowers	–	*Rhizopus stolonifer*	Agar diffusion method	1000 μL/L	–	–	Yongdong et al. (2021)
96.	*Lippia gracilis*	Leaves	Thymol, carvacrol, *p*-cymene and α-pinene	*Lasiodiplodia theobromae*	Agar diffusion method	–	–	1.0 μL/L	Juliana et al. (2022)
97.	*Lippia sidoides*	Leaves	Thymol and carvacrol	*Lasiodiplodia theobromae*	Agar diffusion method	–	–	0.5 μL/L	Juliana et al. (2022)
98.	*Lippia thymoides*	Leaves	–	*Fusarium nucleatum*	Microdilution method	–	1.6 μL/L	–	Carvalho et al. (2021)
				Penicillium intermedia		–	6.5 μL/L	–	
				Candida albicans		–	0.19 μL/L	1.30 μL/L	

Family Pinaceae									
99.	*Pinus halepensis*	Needles	β-Phellandrene, β-caryophyllene, α-pinene and camphor	*Candida krusei*	Disc diffusion method	–	0.10 ± 0.00	50.00 ± 0.02	Mariem et al. (2022)
				Candida albicans		–	0.10 ± 0.00	52.00 ± 0.051	
				Candida glabrata		–	0.10 ± 0.00	58.00 ± 0.01	
				Candida tropicalis		–	0.05 ± 0.00	60.00 ± 0.00	
100.	*Pinus longifolia*	Cones	α-Pinene, β-caryophyllene, δ-3-carene	*Candida Albicans*	Disc diffusion method	–	9.55 ± 1.21 mm (ZOI)	–	Dutta (2023)
101.	*Pinus longifolia*	Cones	α-Pinene, β-caryophyllene, δ-3-carene	*C. albicans*	Microdilution and agar well diffusion methods	–	0.31 ± 0.08	0.65 ± 0.21	Norouzi et al. (2023)
				C. glabrata		–	0.34 ± 0.19	0.74 ± 0.41	
102.	*Pinus eldarica*	Cones	α-Pinene, β-caryophyllene, δ-3-carene	*C. albicans*	Microdilution and agar well diffusion methods	–	0.35 ± 0.12	0.70 ± 0.23	Norouzi et al. (2023)
				C. glabrata		–	0.32 ± 0.10	0.65 ± 0.20	
Family Zingiberaceae									
103.	*Curcuma longa*	Rhizome	α-Phellandrene, terpinolene, 1,8-cineole	*Aspergillus flavus*	Mycelial growth inhibition test	–	4 μL/mL	–	Yichen et al. (2017)
104.	*Elettaria cardamomum*	Pods		*Candida auris*	Broth microdilution method	–	>4.00	>4.00	Di Vito et al. (2023)
105.	*Elettaria cardamomum*	Pods		*Candida albicans*	Disc diffusion method	–	13 mm (ZOI)	–	El-Sayed et al. (2023)
				Aspergillus flavus		–	10 mm (ZOI)	–	
106.	*Zingiber officinale*	Rhizome	Isopulegol acetate	*Penicillium chrysogenum*	Agar disc diffusion method	–	2.5 μg/mL	3.75 μg/mL	Ghada and Asmaa (2020)
				Aspergillus flavus		–	2.5 μg/mL	3.55 μg/mL	
				Aspergillus niger		–	5.0 μg/mL	5.5 μg/mL	
Family Piperaceae									
107.	*Piper auritum*	Leaves and inflorescences	–	*Fusarium oxysporum*	Disk microdiffusion assay	–	6 mg/mL	–	Chacón et al. (2021)
				Fusarium equiseti		–	9 mg/mL	–	

(continued)

Table 4.3 (continued)

S. no.	Aromatic plant	Plant used	Main active compounds	Pathogenic fungi	Methodology used	100% Inhibitory concentration	MIC	MFC	References
108.	*Piper nigrum*	Fruits	3-Carene, D-limonene, copaene, caryophyllene, caryophyllene oxide	*Aspergillus flavus*	Fumigation method	50 and 100 μL/mL	–	–	Zhang et al. (2021)
Family Lauraceae									
109.	*Cinnamomum camphora*	Wood	Linalool	*Botrytis cinerea*	Microwell dilution method	–	20 μg/mL	40 μg/mL	Josemar et al. (2021)
110.	*Cinnamomum camphora*	Leaves	Camphor, camphene, α-pinene	*Aspergillus niger*	Microwell dilution method	–	312.5 μg/mL	–	Poudel et al. (2021)
				Aspergillus fumigatus		–	312.5 μg/mL	–	
				Candida albicans		–	312.5 μg/mL	–	
				Microsporum canis		–	312.5 μg/mL	–	
				Microsporum gypseum		–	312.5 μg/mL	–	
				Trichophyton mentagrophytes		–	156.3 μg/mL	–	
				Trichophyton rubrum		–	78.1 μg/mL	–	
111.	*Cinnamomum camphora*	Wood	Camphor	*Candida auris*	Broth microdilution method	–	>4.00	>4.00	Di Vito et al. (2023)
112.	*Cinnamomum glanduliferum*	Bark	Eucalyptol, terpinen-4-ol, α-terpineol	*Aspergillus fumigatus*	Agar diffusion technique	–	32.5 μg/mL	–	Taha and Eldahshan (2017)
				Geotricum candidum		–	1.95 μg/mL	–	
113.	*Cinnamomum verum*	Leaves	Eugenol, Benzyl benzoate, Trans caryophyllene, Acetyle eugenol, Linalool	*Candida albicans*	Broth microdilution method	–	1.0 mg/mL	–	Wijesinghe et al. (2020)
				Candida tropicalis		–	1.0 mg/mL	–	
				Candida dubliniensis		–	1.0 mg/mL	–	

114.	*Cinnamomum zeylanicum*	Bark	Camphor	*Aspergillus niger*	Broth dilution method	–	25 μg/mL	–	Mingcheng et al. (2023)
115.	*Cinnamomum zeylanicum*	Bark	–	*Candida auris*	Broth microdilution method	–	0.06	0.06	Di Vito et al. (2023)
116.	*Laurus nobilis*	Leaves	1,8-Cineole	*Alternaria alternata*	Mycelial growth inhibition	2 mg/mL	–	–	Affes et al. (2023)
117.	*Litsea cubeba*	Fruits	–	*Aspergillus flavus*	Bacteriostatic circle method and spore counting method	–	26.93 ± 0.55 mm	–	Shiqi et al. (2020)
Family Hypericaceae									
118.	*Hypericum scabrum*	Aerial parts	α-Pinene, limonene, myrcene, β-pinene, and nonane	*Candida albicans*	Broth microdilution method	–	512 μg/mL	–	Fahed et al. (2021)
				Trichophyton rubrum		–	64 μg/mL	–	
				Trichophyton mentagrophytes		–	32 μg/mL	–	
				Trichophyto soudanense		–	32 μg/mL	–	
				Trichophyton. violaceum		–	32 μg/mL	–	
				Trichophyton. tonsurans		–	64 μg/mL	–	
Family Myristicaceae									
119.	*Myristica fragrans*	–	Myristicine, α-pinene, β-pinene, satinene	*Fusarium graminearum*	Broth microdilution method	–	25.0 ± 0 mg/mL	–	Michaela et al. (2021)
Family Atherospermataceae									
120.	*Laurelia sempervirens*	Fresh plant	–	*Candida albicans* ATCC 14053	Agar diffusion method	–	64 μg/mL	–	Olga et al. (2021)

(continued)

Table 4.3 (continued)

S. no.	Aromatic plant	Plant used	Main active compounds	Pathogenic fungi	Methodology used	100% Inhibitory concentration	MIC	MFC	References
Family Phytolaccaceae									
121.	*Gallesia integrifolia*	Fruit	Dimethyl trisulfide, 2,8-dithianonane, lenthionine	*Aspergillus fumigatus*	Microdilution technique	–	0.01 μg/mL	0.01 μg/mL	Raimundo et al. (2018)
				Aspergillus niger		–	0.01 μg/mL	0.01 μg/mL	
				Aspergillus ochraceus		–	0.05 μg/mL	0.05 μg/mL	
				Aspergillus versicolor		–	0.03 μg/mL	0.03 μg/mL	
				Penicillium funiculosum		–	0.09 μg/mL	0.09 μg/mL	
				Penicillium ochrochloron		–	0.01 μg/mL	0.01 μg/mL	
				Penicillium verrucosum		–	0.05 μg/mL	0.05 μg/mL	
				Trichoderma viride		–	0.03 μg/mL	0.03 μg/mL	
Family Anacardiaceae									
122.	*Spondias pinnata*	Fruit peels	Furfural, α-terpineol	*A. fumigatus*	Broth microdilution method	–	16 μg/mL	32 μg/mL	Li et al. (2020)
				C. albicans		–	128 μg/mL	256 μg/mL	
Family Euphorbiaceae									
123.	*Croton eluteria*	Bark	α-Pinene, Camphene, β-pinene	*C. albicans*	Broth microdilution method	–	0.4% (v/v)	–	Manoharan et al. (2017)
Family Menispermaceae									
124.	*Tinomiscium petiolare*	Roots	Geranial, neral	*Aspergillus niger*	Broth microdilution method	–	200 μg/mL	–	Chac et al. (2022)
Family Rubiaceae									
125.	*Morinda citrifolia*	Fruit	Hexanoic acid, Octanoic acid, Methyl octanoate	*C. albicans*	Disc diffusion method	–	39.0 μg/mL	–	Luís et al. (2020)
				C. utilis		–	78.1 μg/mL	–	

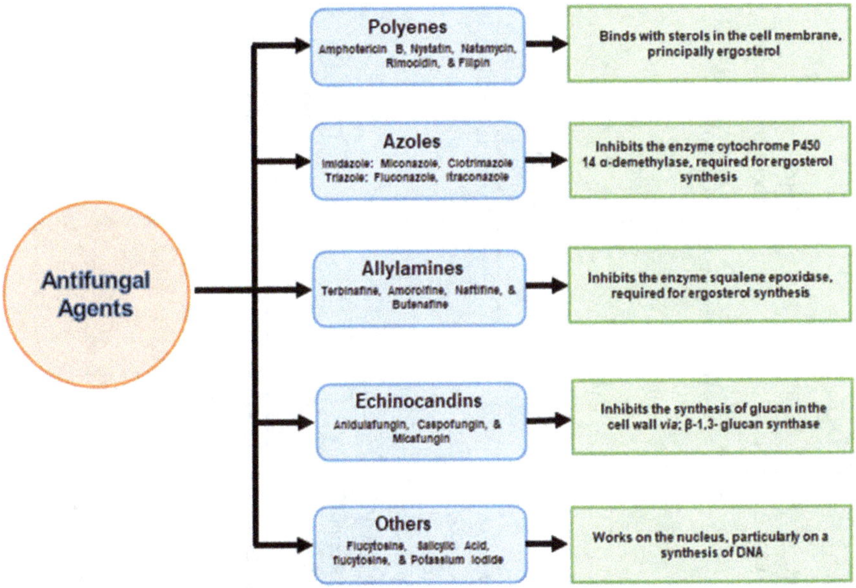

Fig. 4.12 Conventional antifungal agents and their mechanisms of action

4.6.2 Conventional Antifungal Approach

Nowadays, mostly five types of conventional antifungal treatments are commonly used for fungal infections. Figure 4.12 depicts each antifungal drug and its mode of action.

4.6.3 Nonconventional Antifungal Approach

The treatment of fungal infections has become complicated due to growing resistance as a result of the widespread use of antifungal medications. The present situation has led to the realization that nontraditional (nonconventional), alternative approaches are required for effective antifungal treatment. The use of plant essential oils as potential antifungal agents is one of the possible ways. The application of essential oils and their components obtained from aromatic plants as potential antifungal agents are presented in Tables 4.2, 4.3, and 4.4 (Jayant and Sankunny 2014; Filomena et al. 2017).

Table 4.4 Antifungal activity of pure compounds isolated from essential oils

S. no.	Name of active compound	Fungal pathogens	Testing method	MIC	MFC	References
1.	Apiole	*Colletotrichum acutatum*	Poisoned agar method	400 μg/mL	–	Pineda et al. (2018)
2.	β-Asarone	*Aspergillus niger*	Tube dilution method	–	0.416 ± 0.065 mg/mL	Joshi (2016)
		Aspergillus fumigatus		–	0.937 ± 0.139 mg/mL	
		Penicillium chrysogenum		–	1.145 ± 0.104 mg/mL	
3.	β-Pinene	*Aspergillus flavus*	Mycelial growth inhibition test	100 μL	–	Yichen et al. (2015)
4.	β-Pinene	*Candida glabrata*	Broth microdilution method	2.39 mmol/L	–	Iraji et al. (2020)
		Candida tropicalis		5.35 mmol/L	–	
		Candida dubliniensis		10.39 mmol/L	–	
		Candida parapsilosis		0.94 mmol/L	–	
		Candida krusei		1.20 mmol/L	–	
		Candida albicans		30.99 mmol/L	–	
5.	Borneol	*Fusarium oxysporum*	Microdilution method	08.33 μL/mL	26.67 μL/mL	Abderrahmane et al. (2019)
6.	Camphor	*Aspergillus flavus*	Mycelial growth inhibition test	100 μL	–	Yichen et al. (2015)
7.	Camphor	*Candida glabrata*	Broth microdilution method	0.97 mmol/L	–	Iraji et al. (2020)
		Candida tropicalis		1.49 mmol/L	–	
		Candida dubliniensis		1.88 mmol/L	–	
		Candida parapsilosis		2.27 mmol/L	–	
		Candida krusei		2.99 mmol/L	–	
		Candida albicans		7.41 mmol/L	–	

8.	Camphor	*Candida albicans*	Broth microdilution method	0.175 ± 0.02 mg/mL	0.35 ± 0.02 mg/mL	Ivanov et al. (2021)
		Candida tropicalis		0.175 ± 0.02 mg/mL	0.35 ± 0.02 mg/mL	
		Candida parapsilosis		0.125 ± 0.03 mg/mL	0.25 ± 0.008 mg/mL	
		Candida krusei		0.35 ± 0.06 mg/mL	0.7 ± 0.06 mg/mL	
		Candida glabrata		0.175 ± 0.02 mg/mL	0.35 ± 0.04 mg/mL	
9.	(+)-Camphor	*Aspergillus niger*	Microwell dilution method	156.3 μg/mL	–	Poudel et al. (2021)
		Aspergillus fumigatus		312.5 μg/mL	–	
		Candida albicans		156.3 μg/mL	–	
		Microsporum canis		312.5 μg/mL	–	
		Microsporum gypseum		312.5 μg/mL	–	
		Trichophyton mentagrophytes		312.5 μg/mL	–	
		Trichophyton rubrum		312.5 μg/mL	–	
10.	Carvacrol	*Monilinia laxa*	Agar dilution method	0.02 μg/mL	–	Elshafie et al. (2015)
		M. fructigena		0.02 μg/mL	–	
		M. fructicola		0.03 μg/mL	–	
11.	Carvacrol	*Cryptococcus neoformans*	Microdilution method	81 μg/mL	102 μg/mL	de Oliveira Nóbrega et al. (2016)
12.	Carvacrol	*Cryptococcus neoformans*	Broth dilution method	32 μg/mL	–	Kumari et al. (2017)
		Cryptococcus laurentii		16 μg/mL	–	

(continued)

Table 4.4 (continued)

S. no.	Name of active compound	Fungal pathogens	Testing method	MIC	MFC	References
13.	Carvacrol	*Cryptococcus neoformans* (LM-0310)	Microdilution method	64 μg/mL	–	de Oliveira Nóbrega et al. (2016)
		Cryptococcus neoformans (FCF-119)		40 μg/mL	–	
		Cryptococcus neoformans (ICB-59)		40 μg/mL	–	
		Cryptococcus neoformans (FGF-102)		32 μg/mL	–	
		Cryptococcus neoformans (ICB-2601)		51 μg/mL	–	
		Cryptococcus neoformans (JM-10)		51 μg/mL	–	
		Cryptococcus neoformans (LM-5)		25 μg/mL	–	
		Cryptococcus neoformans (LM-120)		64 μg/mL	–	
		Cryptococcus neoformans (LM-39)		51 μg/mL	–	
		Cryptococcus neoformans (LM-22)		81 μg/mL	–	
14.	Carvacrol	*Fusarium oxysporum*	Microdilution method	01.25 μL/mL	02.08 μL/mL	Abderrahmane et al. (2019)
15.	Carvacrol	Lasiodiplodia theobromae	Agar diffusion method	–	0.5 μL/L	Juliana et al. (2022)
16.	Carvone	*C. albicans*	Broth dilution method	248 μg/mL		Neha et al. (2015)
		C. tropicalis		248 μg/mL		
		C. glabrata		497 μg/mL		

17.	(+)-Carvone	*Candida glabrata*	Broth microdilution method	5.54 mmol/L	–	Iraji et al. (2020)
		Candida tropicalis		2.89 mmol/L	–	
		Candida dubliniensis		3.28 mmol/L	–	
		Candida parapsilosis		4.09 mmol/L	–	
		Candida krusei		0.13 mmol/L	–	
		Candida albicans		9.45 mmol/L	–	
18.	(−)-Carvone	*Candida glabrata*	Broth microdilution method	5.54 mmol/L	–	Iraji et al. (2020)
		Candida tropicalis		2.89 mmol/L	–	
		Candida dubliniensis		3.02 mmol/L	–	
		Candida parapsilosis		4.09 mmol/L	–	
		Candida krusei		0.13 mmol/L	–	
		Candida albicans		9.45 mmol/L	–	
19.	1,8-Cinaole	*Aspergillus niger*	Microwell dilution method	156.3 μg/mL	–	Poudel et al. (2021)
		Aspergillus fumigatus		156.3 μg/mL	–	
		Candida albicans		156.3 μg/mL	–	
		Microsporum canis		312.5 μg/mL	–	
		Microsporum gypseum		156.3 μg/mL	–	
		Trichophyton mentagrophytes		156.3 μg/mL	–	
		Trichophyton rubrum		312.5 μg/mL	–	
20.	Cinnamaldehyde	*Cryptococcus neoformans*	Broth dilution method	128 μg/mL	–	Kumari et al. (2017)
		Cryptococcus laurentii		64 μg/mL	–	
21.	Cinnamaldehyde	*C. albicans*	Broth dilution method	125 μg/mL	–	Pootong et al. (2017)
22.	Citral	*Cryptococcus neoformans*	Broth dilution method	64 μg/mL	–	Kumari et al. (2017)
		Cryptococcus laurentii		32 μg/mL	–	
23.	Citral	*Candida albicans*	Broth microdilution method	64 μg/mL	64 μg/mL	Maria et al. (2014)

(continued)

Table 4.4 (continued)

S. no.	Name of active compound	Fungal pathogens	Testing method	MIC	MFC	References
24.	Citral	*Aspergillus flavus*	Vapor phase method	0.5 μL/mL	–	Tang et al. (2018)
		A. ochraceus		0.4 μL/mL	–	
25.	Citral	*Zygosaccharomyces rouxii*	Oxford cup method	0.188 μL/mL	0.375 μL/mL	Rui et al. (2019)
26.	Citral	*Aspergillus flavus*	Spore counting method	30.67 ± 0.42 mm	–	Shiqi et al. (2020)
27.	Citral	*Cladosporium sphaerospermum* (URM 5962)	Broth microdilution method	128 μg/mL	256 μg/mL	Camilla et al. (2022)
		Cladosporium sphaerospermum (URM 5455)		256 μg/mL	512 μg/mL	
		Cladosporium sphaerospermum (URM 5350)		128 μg/mL	256 μg/mL	
		Cladosporium sphaerospermum (URM 6120)		256 μg/mL	1024 μg/mL	
28.	Citral	*Candida albicans*	Broth microdilution method	256 μg/mL	–	Zinnat et al. (2022)
29.	(S)-(−)-Citronellal	*Candida albicans*	Broth microdilution method	256 μg/mL	512 μg/mL	de Oliveira et al. (2017)
30.	Citronellal	*Aspergillus flavus*	Spore counting method	28.83 ± 1.04 mm	–	Shiqi et al. (2020)
		Candida tropicalis		256 μg/mL	512 μg/mL	
31.	Citronellol	*Trichophyton rubrum*	Broth dilution method	1024 μg/mL	–	de Oliveira Pereira et al. (2015)
32.	β-Citronellol (3,7-dimethyl-6-octen-1-ol)	*C. albicans*	Broth dilution method	128 μg/mL	–	Darizy et al. (2018)

33.	(+)-Citronellol	*Candida glabrata*	Broth microdilution method	0.38 mmol/L	–	Iraji et al. (2020)
		Candida tropicalis		0.22 mmol/L	–	
		Candida dubliniensis		0.32 mmol/L	–	
		Candida parapsilosis		0.49 mmol/L	–	
		Candida krusei		1.13 mmol/L	–	
		Candida albicans		0.70 mmol/L	–	
34.	(−)-Citronellol	*Candida glabrata*	Broth microdilution method	0.38 mmol/L	–	Iraji et al. (2020)
		Candida tropicalis		0.22 mmol/L	–	
		Candida dubliniensis		0.28 mmol/L	–	
		Candida parapsilosis		0.49 mmol/L	–	
		Candida krusei		1.99 mmol/L	–	
		Candida albicans		0.70 mmol/L	–	
35.	Cumin aldehyde	*A. aculeatus*	Contact phase method	9.31 μg/mL	–	Tanapichatsakul et al. (2020)
36.	Eucalyptol	*Aspergillus flavus*	Mycelial growth inhibition test	100 μL	–	Yichen et al. (2015)
37.	Eucalyptol	*Fusarium oxysporum*	Microdilution method	40.00 μL/mL	80.00 μL/mL	Abderrahmane et al. (2019)
38.	Eucalyptol	*Candida albicans*	Broth microdilution method	4 ± 0.06 mg/mL	8 ± 0.008 mg/mL	Ivanov et al. (2021)
		Candida tropicalis		4 ± 0.004 mg/mL	8 ± 0.006 mg/mL	
		Candida parapsilosis		2 ± 0.003 mg/mL	4 ± 0.003 mg/mL	
		Candida krusei		4 ± 0.004 mg/mL	8 ± 0.008 mg/mL	
		Candida glabrata		2 ± 0.004 mg/mL	4 ± 0.007 mg/mL	
39.	Eugenol	*Aspergillus niger*	Disc diffusion method	300 μg/mL	300 μg/mL	Mihai and Popa (2015)
		Aspergillus ochraceus		300 μg/mL	300 μg/mL	
		Aspergillus flavus		300 μg/mL	500 μg/mL	
40.	Eugenol	*Cryptococcus neoformans*	Broth dilution method	128 μg/mL	–	Kumari et al. (2017)
		Cryptococcus laurentii		128 μg/mL	–	

(continued)

Table 4.4 (continued)

S. no.	Name of active compound	Fungal pathogens	Testing method	MIC	MFC	References
41.	Eugenol	*Zygosaccharomyces rouxii*	Oxford cup method	0.4 μL/mL	0.8 μL/mL	Rui et al. (2019)
42.	Eugenol	*Candida albicans*	Broth microdilution method	1000 μg/mL	–	Zinnat et al. (2022)
43.	Geranial	*Candida albicans*	Broth microdilution method	225 μg/mL	–	Singh et al. (2016)
44.	Geranial	*Trichophyton rubrum*	Disc diffusion method	55.61 μg/mL	111.23 μg/mL	Zheng (2021)
45.	Geraniol	*Trichophyton rubrum*	Broth dilution method	256 μg/mL	–	de Oliveira Pereira et al. (2015)
46.	Isopulegol	*Monilinia fructicola*	Mycelial growth inhibition test	20.8 ± 1.21 μg/mL	–	Iván et al. (2020)
		Botrytis cinerea		333.84 ± 2.0 μg/mL	–	
47.	Limonene	*Zygosaccharomyces rouxii*	Oxford cup method	0.75 μL/mL	3 μL/mL	Rui et al. (2019)
48.	(+)-Limonene	*Candida glabrata*	Broth microdilution method	4.47 mmol/L	–	Iraji et al. (2020)
		Candida tropicalis		3.72 mmol/L	–	
		Candida dubliniensis		1.20 mmol/L	–	
		Candida parapsilosis		0.69 mmol/L	–	
		Candida krusei		8.69 mmol/L	–	
		Candida albicans		2.83 mmol/L	–	
49.	(−)-Limonene	*Candida glabrata*	Broth microdilution method	3.59 mmol/L	–	Iraji et al. (2020)
		Candida tropicalis		3.02 mmol/L	–	
		Candida dubliniensis		1.26 mmol/L	–	
		Candida parapsilosis		0.50 mmol/L	–	
		Candida krusei		0.82 mmol/L	–	
		Candida albicans		5.29 mmol/L	–	

50.	Limonene	*Trichophyton rubrum*	Broth microdilution method	512 μg/mL	–	Fahed et al. (2021)
51.	D-Limonene	*Candida tropicalis*	Agar dilution method	20 μL/mL	40 μL/mL	Hao et al. (2022)
52.	Linalool	*Candida albicans*	Broth microdilution method	0.5% (v/v)	–	Manoharan et al. (2017)
53.	Linalool	*Candida albicans* 032	Agar dilution method	2000 μg/mL	2000 μg/mL	Dias et al. (2018)
		Candida albicans 051		1000 μg/mL	2000 μg/mL	
		Candida tropicalis 011		500 μg/mL	500 μg/mL	
		Candida krusei 032		2000 μg/mL	2000 μg/mL	
		Candida krusei 031		2000 μg/mL	2000 μg/mL	
54.	Linalool	*Fusarium oxysporum*	Microdilution method	05.00 μL/mL	13.33 μL/mL	Abderrahmane et al. (2019)
55.	Linalool	*Aspergillus flavus*	Spore counting method	17.07 ± 0.31 mm	–	Shiqi et al. (2020)
56.	α-Longipinene	*Candida albicans*	Broth microdilution method	0.3% (v/v)	–	Manoharan et al. (2017)
57.	(+)-Menthone	*Candida glabrata*	Broth microdilution method	7.02 mmol/L	–	Iraji et al. (2020)
		Candida tropicalis		3.31 mmol/L	–	
		Candida dubliniensis		6.79 mmol/L	–	
		Candida parapsilosis		2.15 mmol/L	–	
		Candida krusei		5.34 mmol/L	–	
		Candida albicans		14.21 mmol/L	–	

(continued)

Table 4.4 (continued)

S. no.	Name of active compound	Fungal pathogens	Testing method	MIC	MFC	References
58.	(−)-Menthone	*Candida glabrata*	Broth microdilution method	16.41 mmol/L	–	Iraji et al. (2020)
		Candida tropicalis		16.36 mmol/L	–	
		Candida dubliniensis		17.64 mmol/L	–	
		Candida parapsilosis		20.30 mmol/L	–	
		Candida krusei		36.77 mmol/L	–	
		Candida albicans		26.33 mmol/L	–	
59.	Menthone	*Monilinia fructicola*	Mycelial growth inhibition test	53.4 ± 1.36 μg/mL	–	Iván et al. (2020)
		Botrytis cinerea		444.19 ± 1.57 μg/mL	–	
60.	Menthol	*C. albicans*	Broth dilution method	500 μg/mL	–	Neha et al. (2015)
		C. tropicalis		500 μg/mL	–	
		C. glabrata		500 μg/mL	–	
61.	Menthol	*Cryptococcus neoformans*	Broth dilution method	256 μg/mL	–	Kumari et al. (2017)
62.	(+)-Menthol	*Candida glabrata*	Broth microdilution method	0.57 mmol/L	–	Iraji et al. (2020)
		Candida tropicalis		0.85 mmol/L	–	
		Candida dubliniensis		1.65 mmol/L	–	
		Candida parapsilosis		1.08 mmol/L	–	
		Candida krusei		1.08 mmol/L	–	
		Candida albicans		8.26 mmol/L	–	
63.	(−)-Menthol	*Candida glabrata*	Broth microdilution method	0.57 mmol/L	–	Iraji et al. (2020)
		Candida tropicalis		0.85 mmol/L	–	
		Candida dubliniensis		1.65 mmol/L	–	
		Candida parapsilosis		0.74 mmol/L	–	
		Candida krusei		0.855 mmol/L	–	
		Candida albicans		8.26 mmol/L	–	

64.	Menthol	*Monilinia fructicola*	Mycelial growth inhibition test	33.4 ± 1.23 μg/mL	–	Iván et al. (2020)
		Botrytis cinerea		332.15 ± 2.27 μg/mL	–	
65.	Menthone	*Candida albicans*	Broth dilution method	8400 μg/mL	–	Neha et al. (2015)
		C. tropicalis		8400 μg/mL	–	
		C. glabrata		4200 μg/mL	–	
66.	Myrcene	*Trichophyton rubrum*	Broth microdilution method	512 μg/mL	–	Fahed et al. (2021)
67.	Myristicin	*Aspergillus flavus*	Poison food assay	91 ± 2.9	–	Valente et al. (2015)
		A. ochraceus		82.9 ± 2.6	–	
68.	Myristicin	*Colletotrichum acutatum*	Poisoned agar method	400 μg/mL	–	Pineda et al. (2018)
69.	Myrtenyl acetate	*Fusarium oxysporum*	Microdilution method	80.00 μL/mL	80.00 μL/mL	Abderrahmane et al. (2019)
70.	Neral	*Trichophyton rubrum*	Disc diffusion method	111.23 μg/mL	222.45 μg/mL	Zheng (2021)
71.	Nerol	*Aspergillus niger*	Disc diffusion method	300 μg/mL	300 μg/mL	Mihai and Popa (2015)
		Aspergillus ochraceus		300 μg/mL	500 μg/mL	
		Aspergillus flavus		200 μg/mL	200 μg/mL	
72.	Nonane	*Trichophyton rubrum*	Broth microdilution method	512 μg/mL		Fahed et al. (2021)
73.	α-Pinene	*Fusarium oxysporum*	Microdilution method	20.00 μL/mL	20.00 μL/mL	Abderrahmane et al. (2019)

(continued)

Table 4.4 (continued)

S. no.	Name of active compound	Fungal pathogens	Testing method	MIC	MFC	References
74.	(+)-α-Pinene	*Candida glabrata*	Broth microdilution method	1.86 mmol/L	–	Iraji et al. (2020)
		Candida tropicalis		2.33 mmol/L	–	
		Candida dubliniensis		1.45 mmol/L	–	
		Candida parapsilosis		0.57 mmol/L	–	
		Candida krusei		1.26 mmol/L	–	
		Candida albicans		1.86 mmol/L	–	
75.	(−)-α-Pinene	*Candida glabrata*	Broth microdilution method	49.90 mmol/L	–	Iraji et al. (2020)
		Candida tropicalis		84.73 mmol/L	–	
		Candida dubliniensis		69.43 mmol/L	–	
		Candida parapsilosis		3.46 mmol/L	–	
		Candida krusei		4.22 mmol/L	–	
		Candida albicans		49.90 mmol/L	–	
76.	α-Pinene	*Trichophyton rubrum*	Broth microdilution method	512 μg/mL	–	Fahed et al. (2021)
77.	β-Pinene	*Trichophyton rubrum*	Broth microdilution method	512 μg/mL	–	Fahed et al. (2021)
78.	Piperitone	*Candida glabrata*	Broth microdilution method	6.74 mmol/L	–	Iraji et al. (2020)
		Candida tropicalis		3.62 mmol/L	–	
		Candida dubliniensis		3.78 mmol/L	–	
		Candida parapsilosis		5.67 mmol/L	–	
		Candida krusei		0.88 mmol/L	–	
		Candida albicans		8.38 mmol/L	–	
79.	Pulegone	*Monilinia fructicola*	Mycelial growth inhibition test	69.6 ± 1.36 μg/mL	–	Iván et al. (2020)
		Botrytis cinerea		496.48 ± 1.40 μg/mL	–	

80	Safrole	*Aspergillus niger*	Microwell dilution method	78.1 μg/mL	–	Poudel et al. (2021)
		Aspergillus fumigatus		39.1 μg/mL	–	
		Candida albicans		156.3 μg/mL	–	
		Microsporum canis		312.5 μg/mL	–	
		Microsporum gypseum		312.5 μg/mL	–	
		Trichophyton mentagrophytes		312.5 μg/mL	–	
		Trichophyton rubrum		312.5 μg/mL	–	
81.	α-Terpinen-7-al	*A. aculeatus*	Contact phase method	13.23 μg/mL	–	Tanapichatsakul et al. (2020)
82.	α-Terpineol	*Fusarium oxysporum*	Microdilution method	05.00 μL/mL	10.00 μL/mL	Abderrahmane et al. (2019)
83.	α-Terpineol	*Aspergillus flavus*	Spore counting method	18.63 ± 0.40 mm	–	Shiqi et al. (2020)
84.	Thujone	*Candida glabrata*	Broth microdilution method	18.48 mmol/L	–	Iraji et al. (2020)
		Candida tropicalis		9.91 mmol/L	–	
		Candida dubliniensis		7.43 mmol/L	–	
		Candida parapsilosis		21.80 mmol/L	–	
		Candida krusei		13.11 mmol/L	–	
		Candida albicans		12.74 mmol/L	–	
85.	Thymol	*Monilinia laxa*	Agar dilution method	0.16 μg/L	–	Elshafie et al. (2015)
		M. fructigena		0.16 μg/L	–	
		M. fructicola		0.12 μg/L	–	
86.	Thymol	*C. albicans*	Microdilution method	39.0 μg/mL	39.0 μg/mL	Ricardo et al. (2015)
		C. tropicalis		78.0 μg/mL	78.0 μg/mL	
		C. krusei		39.0 μg/mL	39.0 μg/mL	
87.	Thymol	*Fusarium graminearum*	Mycelial growth inhibition test	26.3 μg/mL	–	Gao et al. (2016)

(continued)

Table 4.4 (continued)

S. no.	Name of active compound	Fungal pathogens	Testing method	MIC	MFC	References
88.	Thymol	*Cryptococcus neoformans*	Broth dilution method	16 μg/mL	–	Kumari et al. (2017)
		Cryptococcus laurentii		8 μg/mL	–	
89.	Thymol	*Fusarium oxysporum*	Microdilution method	00.54 μL/mL	00.94 μL/mL	Abderrahmane et al. (2019)
90.	Thymol	*Cryptococcus neoformans*	Microdilution method	20–51 μg/mL	40–101 μg/mL	Teixeira et al. (2020)
91.	Thymol	*Lasiodiplodia theobromae*	Agar diffusion method	–	1.0 μL/L	Juliana et al. (2022)

4.7 Conclusion

Essential oils are volatile oils derived from plants with potent aromatic components and are composed of various chemical compounds. Alcohols, aldehydes, esters, hydrocarbons, ketones, and phenols, for instance, are some of the principal ingredients in essential oils. Around the world, microbial (viral, bacterial, and fungal) infections are the most frequently occurring diseases. The search for new fungicidals is a broad endeavor, and the current emphasis is on substances that are naturally occurring. Several experimental studies have been conducted using natural substances that could serve as models for defenses against fungi that attack on individuals. The medicines available for effective antifungal chemotherapy in the market are limited. Fungal infection prevention and treatment are made more challenging by the emergence of biofilms and device-associated infections, drug-resistant fungi, and the adverse effects of prescription medications. As a result, invasive fungal infections are linked to extremely high rates of diseases and death. It has been discovered that a variety of yeast, as well as human and plant pathogenic fungi, are susceptible to essential oils. In this regard, further evaluation of essential oils is required in order to investigate their practical therapeutic applications that can be exploited for the benefit of humanity.

References

Abderrahmane R et al (2019) Chemical composition and antifungal activity of five essential oils and their major components against *Fusarium oxysporum* f. sp. albedinis of Moroccan palm tree. Euro Mediterr J Environ Integr 4:27. https://doi.org/10.1007/s41207-019-0117-x

Abdullah AAG (2023) Antifungal activity of *Cymbopogon flexuosus* essential oil and its effect on biofilm formed by *Candida parapsilosis* and *Candida tropicalis* on polystyrene and polyvinyl plastic surfaces. Indian J Pharm Educ Res 57(1):113. https://doi.org/10.5530/001954641705

Adel M et al (2020) Antifungal activity and chemical composition of Iranian medicinal herbs against fish pathogenic fungus *Saprolegnia parasitica*. Iran J Fish Sci 19(6):3239–3254. https://doi.org/10.22092/ijfs.2020.122970

Affes TG et al (2022) A comparative assessment of antifungal activity of essential oils of five medicinal plants from Tunisia. Int J Plant Based Pharm 2(2):220–227. https://doi.org/10.29228/ijpbp.4

Affes TG et al (2023) Biological control of *Citrus* brown spot pathogen, "*Alternaria alternata*" by different essential oils. Int J Environ Health Res 33(8):823–836. https://doi.org/10.1080/09603123.2022.2055748

Ammar RM et al (2022) Chemical composition and antifungal activity of essential oil of *Satureja calamintha* spp. Nepeta (L.) Briq against some toxinogenous mold. Nat Volatiles Essent Oils 9(1):1981–2000

Baser KHC, Buchbauer G (2010) Handbook of Essential Oils: Science, Technology and Applications. CRC Press, Boca Raton

Betts TJ (2001) Chemical characterization of the different types of volatile oil constituents by various solute retention ratios with the use of conventional and novel commercial gas chromatographic stationary phases. J Chromatogr A 936(1–2):33–46

Blaschke T, Bjornsson T (1995) Pharmacokinetics and Pharmacoepidemiology. Sci Am 8:1–14

Bowles EJ (2003) The chemistry of aromatherapeutic oils, 3rd edn. Griffin Press, Salisbury South

Bryant B et al (2003) Pharmacology for health professionals. Mosby, Elsevier (Australia) Pty Ltd., Marrickville, p 2

Burt S (2004) Essential oils: their antimicrobial properties and potential application in foods—a review. Int J Food Microbiol 94:223–253

Camilla PM et al (2022) Investigation on mechanism of antifungal activity of citral against *Cladosporium sphaerospermum* Penz. An Biol 43:43–53. https://doi.org/10.6018/analesbio.44.05

Carvalho TRBC et al (2021) Chemical composition, antimicrobial and antifungal activity of *Lippia thymoide* essential oil in oral pathogens. Brazil J Oral Sci 20:e210219. https://doi.org/10.20396/bjos.v20i00.8660219

Cava R et al (2007) Antimicrobial activity of clove and cinnamon essential oils against *Listeria monocytogenes* in pasteurized milk. J Food Prot 70:2757–2763

Cebi N, Erarslan A (2023) Determination of the antifungal, antibacterial activity and volatile compound composition of *Citrus bergamia* peel essential oil. Food Secur 12:203. https://doi.org/10.3390/foods12010203

Chac LD et al (2022) Chemical composition and antifungal activity of essential oil from the roots of *Tinomiscium petiolare*. Chem Nat Compd 58:760–762. https://doi.org/10.1007/s10600-022-03788-6

Chacón C et al (2021) *In vitro* antifungal activity and chemical composition of *Piper auritum* Kunth essential oil against *Fusarium oxysporum* and *Fusarium equiseti*. Agronomy 11:1098. https://doi.org/10.3390/agronomy11061098

Čmiková N et al (2023) Chemical composition and biological activities of *Eucalyptus globulus* essential oil. Plants 12:1076. https://doi.org/10.3390/plants12051076

Croteau R et al (2000) Natural products (secondary metabolites). Biochem Mol Biol Plants 24:1250–1319

Da Cunha SMD et al (2023) Bioprospecting of the antifungal activity of Patchouli essential oil (*Pogostemon cablin* Benth) against strains of the genus *Candida*. J Med Plants Res 17(1):1–7. https://doi.org/10.5897/JMPR2022.7257

Dang MN et al (2001) Antioxidant activity of essential oils from various spices. Nahrung/Food 45:64–66

Darizy FS et al (2018) Activity anti-*candida albicans* and effects of the association of β-citronellol with three antifungal azolics. Latin Am J Pharm 37(1):182–188

de Oliveira Nóbrega R et al (2016) Investigation of the antifungal activity of carvacrol against strains of *Cryptococcus neoformans*. Pharm Biol 54(11):2591–2596. https://doi.org/10.3109/13880209.2016.1172319

de Oliveira Pereira F et al (2015) Antifungal activity of geraniol and citronellol, two monoterpenes alcohols, against *Trichophyton rubrum* involves inhibition of ergosterol biosynthesis. Pharm Biol 53(2):228–234. https://doi.org/10.3109/13880209.2014.913299

de Oliveira HMBF et al (2017) *In vitro* anti *Candida* effect of (S)-(−)-citronellal. J Appl Pharm Sci 7(11):177–179

Devkatte AN et al (2005) Potential of plant oils as inhibitors of *Candida albicans* growth. FEMS Yeast Res 5(9):867–873

Di Vito M et al (2023) A new potential resource in the fight against *Candida auris*: the *Cinnamomum zeylanicum* essential oil in synergy with antifungal drug. Microbiol Spectr 11(2). https://doi.org/10.1128/spectrum.04385-22

Dias IJ et al (2018) Antifungal activity of linalool in cases of *Candida* spp. isolated from individuals with oral candidiasis. Braz J Biol 78(2). https://doi.org/10.1590/1519-6984.171054

Dick AJ, Starmans HHN (1996) Extraction of secondary metabolites from plant material: a review. Trends Food Sci Technol 7:191–197

Dorman HJD, Deans SG (2000) Antimicrobial agents from plants: antibacterial activity of plant volatile oils. J Appl Microbiol 88:308–316. https://doi.org/10.1046/j.1365-2672.2000.00969.x

Dragana VP et al (2020) Chemical structure and antifungal activity of mint essential oil components. J Serb Chem Soc 85(9):1149–1161

Ďúranová H et al (2023) Antifungal activities of essential oil obtained from *Mentha spicata* var. crispa against selected *Penicillium* species. Bilge Int J Sci Technol Res 7(1):1–8
Dutta SD (2023) Determination of antifungal effect of natural oil and synthetic gutta percha solvents against *Candida albicans*: a disc diffusion assay. J Pharm Bioallied Sci 15:S235–S238
Dwi M et al (2023) Antifungal activity of rosemary essential oil against *Candida* spp. isolates from HIV/AIDS patients with oral candidiasis. J Pak Assoc Dermatol 33(2):437–443
Edris AE (2007) Pharmaceutical and therapeutic potentials of essential oils and their individual volatile constituents: a review. Phytother Res 21(4):308–323. https://doi.org/10.1002/ptr.2072
Eisenreich W et al (2004) Biosynthesis of isoprenoids via the non-mevalonate pathway. Cell Mol Life Sci 61:1401–1426. https://doi.org/10.1007/s00018-004-3381-z
El-Sayed SAH et al (2023) Impact of extraction methods on the chemical composition and biological activity of small cardamom essential oils. Egypt J Chem 65(2):101–109
Elshafie HS et al (2015) Antifungal activity of some constituents of *Origanum vulgare* L. essential oil against postharvest disease of peach fruit. J Med Food 18:1–6
Fahed L et al (2021) Antimicrobial activity and synergy investigation of *Hypericum scabrum* essential oil with antifungal drugs. Molecules 26:6545. https://doi.org/10.3390/molecules26216545
Fatima B et al (2023) Phytochemical composition, antioxidant, and antifungal activity of essential oil from *Myrtus communis*, L. Mater Today Proc 72:3826–3830
Filomena N et al (2017) Essential oils and antifungal activity. Pharmaceuticals 10(86). https://doi.org/10.3390/ph10040086
Gao T et al (2016) The fungicidal activity of thymol against *Fusarium graminearum* via inducing lipid peroxidation and disrupting ergosterol biosynthesis. Molecules 21(6):770. https://doi.org/10.3390/molecules21060770
Ghada AY, Asmaa SM (2020) *In-vitro* antifungal activity of eco-friendly essential oils against pathogenic seed borne fungi. Egypt J Bot 60(2):381–393
Gharaghani M et al (2023) Comparison of the antifungal and antibacterial activity of *Ballota aucheri* essential oil and fluconazole. J Clin Care Skill 4(1):27–31
Grayson DH (2000) Monoterpenoids (mid-1997 to mid-1999). Nat Prod Rep 17:385–419
Guenther E (1952) The essential oils, vol 5. R.E. Krieger Pub. Co., pp 3–38
Guenther E (1982) The production of essential oils: methods of distillation, effleurage, maceration and extraction with volatile solvent. In: Guenther E (ed) The essential oils. History-origin in plants production analysis, vol 1. Krieger Publ. Co., Malabar, pp 85–188
Gwilt PR et al (1994) The effect of garlic extract on human metabolism of acetaminophen. Cancer Epidemiol Biomarkers Prev 3(2):155–160
Hammer K, Carson C (2011) Antibacterial and antifungal activities of essential oils. In: Thormar H (ed) Lipids and essential oils as antimicrobial agents. John Wiley & Sons, pp 255–306
Hammer KA et al (1999) Antimicrobial activity of essential oils and other plant extracts. J Appl Microbiol 86(6):985–990. https://doi.org/10.1046/j.1365-2672.1999.00780.x
Hao Y et al (2022) Antifungal activity and mechanism of D-limonene against foodborne opportunistic pathogen *Candida tropicalis*. LWT 159:113144
Hussain AIF et al (2008) Chemical composition. Antioxidant and antimicrobial activities of basil (*Ocimum basilicum*) essential oils depends on seasonal variations. Food Chem 108:986–995
Iraji A et al (2020) Screening the antifungal activities of monoterpenes and their isomers against *Candida* species. J Appl Microbiol 129(6):1541–1551. https://doi.org/10.1111/jam.14740
Irkin R, Korukluoglu M (2009) Effectiveness of *Cymbopogo citratus* L. essential oil to inhibit the growth of some filamentous fungi and yeasts. J Med Food 12:193–197. https://doi.org/10.1089/jmf.2008.0108
Iván M et al (2020) Antifungal activity of essential oil and main components from *Mentha pulegium* growing wild on the Chilean Central Coast. Agronomy 10:254. https://doi.org/10.3390/agronomy10020254
Ivanov M et al (2021) Camphor and eucalyptol-anticandidal spectrum, antivirulence effect, efflux pumps interference and cytotoxicity. Int J Mol Sci 22:483
Jayant SR, Sankunny MK (2014) A status review on the medicinal properties of essential oils. Ind Crop Prod 62:250–264

Jelena L et al (2020) Antifungal and synergistic activity of five plant essential oils from Serbia against *Trichoderma aggressivum* f. europaeum Samuels and W. Gams. Pest Phytomed (Belgrade) 35(3):173–181. https://doi.org/10.2298/PIF2003173L

Josemar GOF et al (2021) Chemical composition and antifungal activity of essential oils and their combinations against *Botrytis cinerea* in strawberries. J Food Meas Charact 15:1815–1825. https://doi.org/10.1007/s11694-020-00765-x

Joshi RK (2016) Acorus calamus Linn.: phytoconstituents and bactericidal property. World J Microbiol Biotechnol 32:164. https://doi.org/10.1007/s11274-016-2124-2

Juliana OM et al (2022) Essential oils of Lippia gracilis and Lippia sidoides chemotypes and their major compounds carvacrol and thymol: nanoemulsions and antifungal activity against *Lasiodiplodia theobromae*. Res Soc Dev 11(3):e36511326715. https://doi.org/10.33448/rsd-v11i3.26715

Kathiravan MK et al (2012) The biology and chemistry of antifungal agents: a review. Bioorg Med Chem 20(19):5678–5698. https://doi.org/10.1016/j.bmc.2012.04.045

Kumar P et al (2010) Insecticidal properties of *Mentha* species: a review. Ind Crop Res 34:802–817. https://doi.org/10.3390/molecules20058605

Kumari P et al (2017) Antifungal and anti-biofilm activity of essential oil active components against *Cryptococcus neoformans* and *Cryptococcus laurentii*. Front Microbiol 8:2161. https://doi.org/10.3389/fmicb.2017.02161

Lang G, Buchbauer G (2012) A review on recent research results (2008–2010) on essential oils as antimicrobials and antifungals. Flavour Fragr J 27:13–39

Li R et al (2020) Chemical composition and the cytotoxic, antimicrobial, and anti-inflammatory activities of the fruit peel essential oil from *Spondias pinnata* (Anacardiaceae) in Xishuangbanna, Southwest China. Molecules 25:343. https://doi.org/10.3390/molecules2502034ka

Li T et al (2021) Antifungal activity of essential oil from *Zanthoxylum armatum* DC. on *Aspergillus flavus* and aflatoxins in stored platycladi semen. Front Microbiol 12:633714. https://doi.org/10.3389/fmicb.2021.633714

Lichtenthaler HK (1999) The 1-deoxy-D-xylulose5-phosphate pathway of isoprenoid synthesis in plants. Annu Rev Plant Physiol Plant Mol Biol 50:47–65. https://doi.org/10.1146/annurev.arplant.50.1.47

Luís H et al (2020) Potent antifungal activity of essential oil from *Morinda citrifolia* fruits rich in short-chain fatty acids. Int J Fruit Sci 20(2):S448–S454. https://doi.org/10.1080/15538362.2020.1738975

Majda E et al (2022) Chemical composition and antimicrobial activity of essential oil of wild and cultivated *Rosmarinus officinalis* from two Moroccan localities. J Ecol Eng 23(3):214–222. https://doi.org/10.12911/22998993/145458

Manoharan RK et al (2017) Inhibitory effects of the essential oils α-longipinene and linalool on biofilm formation and hyphal growth of *Candida albicans*. Biofouling 33(2):143–155. https://doi.org/10.1080/08927014.2017.1280731

Maria CAL et al (2014) Evaluation of antifungal activity and mechanism of action of citral against *Candida albicans*. Evid Based Complement Alternat Med 378280. https://doi.org/10.1155/2014/378280

Mariem B et al (2022) Chemical composition, antibacterial and antifungal activities of four essential oils collected in the North-East of Tunisia. J Essent Oil Bearing Plants. https://doi.org/10.1080/0972060X.2022.2068971

Maxia A et al (2009) Chemical characterization and biological activity of essential oils from *Daucus carota* L. subsp. *carota* growing wild on the Mediterranean coast and on the Atlantic coast. Fitoterapia 80(1):57–61

McGuinness H (2007) Aromatherapy therapy basics, 2nd edn. Hodder Arnold, New Delhi

Michaela H et al (2021) Comparison of antifungal activity of selected essential oils against *Fusarium graminearum* in vitro. Ann Agric Environ Med 28(3):414–418. https://doi.org/10.26444/aaem/137653

Mihai AL, Popa ME (2015) *In vitro* activity of natural antimicrobial compounds against *Aspergillus* strains. Agric Agric Sci Proc 6:585–592

Mingcheng W et al (2023) Antifungal mechanism of cinnamon essential oil against Chinese Yam-derived *Aspergillus niger*. Hindawi J Food Proc Preserv 2023:5777460. https://doi.org/10.1155/2023/5777460

Mounir M et al (2022) Antifungal activity and chemical composition of essential oil from *Santolina pectinata* against postharvest phytopathogenic fungi in apples. Arab J Med Aromat Plants 8:41

Neha S et al (2015) Synergistic anti-candidal activity and mode of action of *Mentha piperita* essential oil and its major components. Pharm Biol 53(10):1496–1504. https://doi.org/10.3109/13880209.2014.989623

Norouzi Z et al (2023) Antifungal properties of *Pinus eldarica* and *Pinus longifolia* fruit extracts against *Candida* species isolated from vulvovaginal candidiasis patients. Current Drug Therapy 18:333–341

Olga L et al (2021) Antibiofilm and antifungal activities of *Laurelia sempervirens* (Chilean laurel) essential oil. Jundishapur J Nat Pharm Prod 16(4):e113611. https://doi.org/10.5812/jjnpp.113611

Omar OA et al (2023) Chemical composition, in vitro antifungal activity, molecular docking and molecular dynamics simulation studies of the essential oil of *Ballota hirsuta*. J Biol Active Products Nat 13(1):27–48. https://doi.org/10.1080/22311866.2023.2194862

Parish LC, Jouni JU (1991) The sixth Zagazig international conference of dermatology and venereology. Int J Dermatol 30(1):73–76

Pichersky E et al (2006) Biosynthesis of plant volatiles: nature's diversity and ingenuity. Science 311(5762):808–811. https://doi.org/10.1126/science.1118510

Pinar S et al (2023) Promising antimicrobial and antifungal activities of free peppermint (*Mentha piperita* L.) essential oil and its conjugated form with chitosan. Turk J Anal Chem 5(1):77–82. https://doi.org/10.51435/turkjac.1311200

Pinder AR (1960) The chemistry of the terpenes. Wiley, New York

Pineda R et al (2018) Antifungal activity of extracts, essential oil and constituents from *Petroselinum crispum* against *Colletotrichum acutatum*. Rev Fac Nac Agron Medellín 71(3):8563–8572. https://doi.org/10.15446/rfnam.v71n3.68284

Pootong A et al (2017) Antıfungal actıvıty of cınnamaldehyde agaınst *Candıda albıcans*. Southeast Asian J Trop Med Public Health 48(1):150–158

Poudel DK et al (2021) The chemical profiling of essential oils from different tissues of *Cinnamomum camphora* L. and their antimicrobial activities. Molecules 26(17):5132. https://doi.org/10.3390/molecules26175132

Pragadheesh VS et al (2013) Chemical characterization and antifungal activity of *Cinnamomum camphora* essential oil. Ind Crop Prod 49:628–633

Rafael PC et al (2022) Chemical composition and antimicrobial potential of essential oil of *Acritopappus confertus* (Gardner) R.M.King and H.Rob (Asteraceae). Pharmaceuticals 15:1275. https://doi.org/10.3390/ph15101275

Raimundo KF et al (2018) Antifungal activity of *Gallesia integrifolia* fruit essential oil. Braz J Microbiol 49(1):229–235. https://doi.org/10.1016/j.bjm.2018.03.006

Rao A et al (2010) Mechanism of antifungal activity of terpenoid phenols resembles calcium stress and inhibition of the TOR pathway. Antimicrob Agents Chemother 54:5062–5069

Rawlins MD (1989) Clinical pharmacology of the skin. In: Turner P (ed) Recent advances in clinical and pharmacology and toxicology. Churchill Livingstone, Edinburgh, pp 121–135

Ricardo DC et al (2015) Antifungal activity and mode of action of thymol and its synergism with nystatin against *Candida* species involved with infections in the oral cavity: an in vitro study. BMC Complement Alternat Med 15:417

Rohmer M et al (1999) Glyceraldehydes 3-phosphate and pyruvate as precursors of isoprene units in an alternative non-mevalonate pathway for terpenoid biosynthesis. J Am Chem Soc 118:2564–2566

Rui C et al (2019) Antifungal activity and mechanism of citral, limonene and eugenol against *Zygosaccharomyces rouxii*. LWT Food Sci Technol 106:50–56

Safar AA et al (2020) Screening of chemical characterization, antifungal and cytotoxic activities of essential oil constituents of *Tagetes erecta* L. from Erbil, Kurdistan Region-Iraq. Pol J Environ Stud 29(3):2317–2326. https://doi.org/10.15244/pjoes/110612

Sardi JCO et al (2013) *Candida* species: current epidemiology, pathogenicity, biofilm formation, natural antifungal products and new therapeutic options. J Med Microbiol 62(1):10–24. https://doi.org/10.1099/jmm.0.045054-0

Shiqi X et al (2020) Inhibitory effects of *Litsea cubeba* oil and its active components on *Aspergillus flavus*. J Food Qual:8843251. https://doi.org/10.1155/2020/8843251

Siddique S et al (2021) *In vitro* antifungal activities of essential oils from selected species of family myrtaceae. Pharmacology Online Arch 3:1612–1625

Singh S et al (2016) Insights into the mode of action of anticandidal herbal monoterpenoid geraniol reveal disruption of multiple MDR mechanisms and virulence attributes in *Candida albicans*. Arch Microbiol:459–472. https://doi.org/10.1007/s00203-016-1205-9

Sood S et al (2006) Physiological and biochemical studies during flower development in two rose species. Sci Hort 108:390–396

Stevic T et al (2014) Antifungal activity of selected essential oils against fungi isolated from medicinal plant. Ind Crop Prod 55:116–122

Tabassum N, Vidyasagar GM (2013) Antıfungal investigations on plant essential oils. A review. Int J Pharm Pharm Sci 5(2):19–28

Taha AM, Eldahshan OA (2017) Chemical characteristics, antimicrobial, and cytotoxic activities of the essential oil of Egyptian *Cinnamomum glanduliferum* Bark. Chem Biodivers 14:e1600443. https://doi.org/10.1002/cbdv.201600443

Tanapichatsakul C et al (2020) *In vitro* and in vivo antifungal activity of Cuminum cyminum essential oil against Aspergillus aculeatus causing bunch rot of postharvest grapes. PLoS One 15(11):e0242862. https://doi.org/10.1371/journal.pone.0242862

Tang X et al (2018) Antifungal activity of essential oil compounds (geraniol and citral) and inhibitory mechanisms on grain pathogens (*Aspergillus flavus* and *Aspergillus ochraceus*). Molecules 23(9):2108. https://doi.org/10.3390/molecules23092108

Teixeira APC et al (2020) Antifungal activity study of the monoterpene thymol against *Cryptococcus neoformans*. Nat Prod Res 34(18):2630–2633. https://doi.org/10.1080/14786419.2018.1547296

Tian-Lin W et al (2023) Antifungal efficacy of sixty essential oils and mechanism of oregano essential oil against *Rhizoctonia solani*. Ind Crop Prod 191:115975. https://doi.org/10.1016/j.indcrop.2022.115975

Tomasz MK et al (2023) Anti- *Candida* and antibiofilm activity of selected Lamiaceae essential oils. Front Biosci (Landmark Ed) 28(2):28. https://doi.org/10.31083/j.fbl2802028

Tomić A et al (2023) Screening of antifungal activity of essential oils in controlling biocontamination of historical papers in archives. Antibiotics 12:103. https://doi.org/10.3390/antibiotics12010103

Valente VMM et al (2015) Major antifungals in nutmeg essential oil against *Aspergillus flavus* and *A ochraceus*. J Food Res 4(1). https://doi.org/10.5539/jfr.v4n1p51

Viuda-Martos M et al (2008) Antifungal activity of lemon (*Citrus lemon* L.), mandarin (*Citrus reticulata* L.), grapefruit (*Citrus paradisi* L.) and orange (*Citrus sinensis* L.) essential oils. Food Control 19:1130–1138

Vogt T (2010) Phenylpropanoid biosynthesis. Mol Plant 3:2–20

Wang L, Weller CL (2006) Recent advances in extraction of nutraceuticals from plants. Trends Food Sci Technol 17:300–312

Wijesinghe GK et al (2020) Effect of *Cinnamomum verum* leaf essential oil on virulence factors of *Candida* species and determination of the *in-vivo* toxicity with *Galleria mellonella* model. Mem Inst Oswaldo Cruz 115:e200349. https://doi.org/10.1590/0074-02760200349

Yichen H et al (2015) Uncovering the antifungal components from turmeric (*Curcuma longa* L.) essential oil as *Aspergillus flavus* fumigants by partial least squares. RSC Adv 5:41967–41976

Yichen H et al (2017) Mechanisms of antifungal and anti-aflatoxigenic properties of essential oil derived from turmeric (*Curcuma longa* L.) on *Aspergillus flavus*. Food Chem 220(2017):1–8. ISSN: 0308-8146. https://doi.org/10.1016/j.foodchem.2016.09.179

Yongdong X et al (2021) Antifungal activity screening for 32 essential oils against *Rhizopus stolonifer*. IOP Conf Ser Earth Environ Sci 792:012014. https://doi.org/10.1088/1755-1315/792/1/012014

Younis A et al (2008) Extraction and identification of chemical constituents of the essential oil of *Rosa* species. Acta Hort 766:485–492

Zalewska ED et al (2022) Antifungal effects of some essential oils on selected allergenic fungi in vitro. Acta Sci Pol Hortorum Cultus 21(6):115–127. https://doi.org/10.24326/asphc.2022.6.10

Zhang C et al (2021) Antioxidant, hepatoprotective and antifungal activities of black pepper (*Piper nigrum* L.) essential oil. Food Chem 346:128845. https://doi.org/10.1016/j.foodchem.2020.128845

Zheng Y (2021) Antifungal activities of *cis-trans* citral isomers against *Trichophyton rubrum* with ERG6 as a potential target. Molecules 26:4263. https://doi.org/10.3390/molecules26144263

Zinnat S et al (2022) *Candida albicans* reactive oxygen species (ROS)-dependent lethality and ROS-independent hyphal and biofilm inhibition by eugenol and citral. Microbiol Spectr 10(6). https://doi.org/10.1128/spectrum.03183-22

Zore GB et al (2011a) Evaluation of anti-*Candida* potential of geranium oil constituents against clinical isolates of *Candida albicans* differentially sensitive to fluconazole: inhibition of growth, dimorphism and sensitization. Mycoses 54:e99–e109

Zore GB et al (2011b) Terpenoids inhibit *Candida albicans* growth by affecting membrane integrity and arrest of cell cycle. Phytomedicine 18:1181–1190

Antifungal Efficacy of Plant Essential Oils Against *Candida, Aspergillus* and *Cryptococcus* Species

5

K. M. Uma Kumari, Md Waquar Imam, and Suaib Luqman

5.1 Introduction

More than a billion people are infected by fungal diseases globally, with invasive fungal infections having a higher fatality rate and causing more yearly fatalities than tuberculosis or malaria. Mild fungal infection of the skin looks like a rash and is very common and easily treatable but fungal infections in the lungs are often similar to other lung infections such as bacterial and viral lung infections. However, some fungal illnesses such as fungal meningitis and other bloodstream infections are less common than skin and lung infections but they cause high death rates and impact the economy. Despite the alarming impact of these infectious agents on the human healthcare system, current antifungal treatments for invasive fungal infections are restricted to polyenes, azoles and echinocandins. Azoles target and block the ergosterol biosynthesis by inhibiting the function of lanosterol 14α-demethylase whereas polyenes deplete the membrane lipid ergosterol from the cell membrane. The echinocandins inhibit the 1,3-β-D-glucan, a key component of the cell wall, which leads to disruption of the cell wall integrity (Son et al. 2020). Woefully, extensive use of antifungal drugs has fuelled the rapid emergence of resistance to almost all classes of drugs (Bosetti and Neofytos 2023). Additionally, many side effects and toxicity, mild to severe are also associated with antifungal agents. Even though a variety of antifungal drugs are available but the problem persists and affects human physical health as well as mental health leading to a large number of deaths globally. Researchers around the world searching for a novel class of drugs to overcome

KM Uma Kumari and Md Waquar Imam contributed equally with all other contributors.

K. M. U. Kumari
CSIR-Central Institute of Medicinal and Aromatic Plants, Lucknow, Uttar Pradesh, India
e-mail: uma.banku@gmail.com; waquarimam10@gmail.com

M. W. Imam · S. Luqman (✉)
CSIR-Central Institute of Medicinal and Aromatic Plants, Lucknow, Uttar Pradesh, India

Academy of Scientific and Innovative Research (AcSIR), Ghaziabad, Uttar Pradesh, India
e-mail: s.luqman@cimap.res.in

N. Manzoor (ed.), *Advances in Antifungal Drug Development*,
https://doi.org/10.1007/978-981-97-5165-5_5

fungal infection, which have high efficacy and least or no toxicity. In this regard, previous reports suggested that plants have demonstrated enormous potential for antimicrobial action, encouraging researchers to employ natural compounds to combat microbial infections and resistance, and to have little or no toxicity. Since ancient times, plant-derived compounds and essential oils have been used to cure microbial infection. The essential oil shows impressive antifungal activity, due to their high lipophilic/hydrophobic nature it easily disrupts the cell wall and crosses the fungal cell barrier (Hu et al. 2017; Tian et al. 2011a).

5.2 Fungal Infections

Fungus is a tiny but significant component of the human microbiome. While many fungi are friendly to our immune system, others, known as opportunistic pathogens, can give rise to diseases in immunocompromised hosts. Nearly 500 fungal species have been classified as infectious to humans out of approximately 1.5–5 million species (Brown et al. 2012; Naveen et al. 2022). Fungal infections affect more than 25% and kill more than 1.5 million people per year globally (Armstrong-James et al. 2017; Bongomin et al. 2017). A large number of these infections are superficial, and majority of them are possible to cure. On the other hand, it creates invasive illnesses with death rates above 50%. Globally, fungal infections are increasing; however, the level of severity of disease varies from asymptomatic-mild mucocutaneous infections to potentially fatal systemic infections. Asthma, HIV, COVID-19, diabetes, chronic obstructive pulmonary disease (COPD), cancer, tuberculosis, lung disease, organ transplant, chemotherapy and many other diseases or immune-suppressed conditions are behind the increase in fungal infections all over the world (Armstrong-James et al. 2017; Brown et al. 2012; Denning et al. 2016; Guinea et al. 2010; Limper et al. 2017; Marr et al. 2002; Bongomin et al. 2017). Furthermore, the use of wide-ranging antifungal drugs, corticosteroid therapy and immunosuppressive medicines has been linked to an increase in patients with reduced immunity, leading to an increased threat of fungal infections. Even in the current COVID-19 pandemic, opportunistic fungal infections were detected in COVID patients. Likewise, more severe and fatal infections are been observed in COVID-19-positive patients with predisposing conditions including acute respiratory distress syndrome (ARDS), mechanical ventilation, immunosuppressive therapies, organ transplant recipients, diabetes and antibiotics (Fig. 5.1). As a result, the prevalence of opportunistic fungal infection has attracted the interest of scholars all over the world (Raut and Huy 2021; Ventoulis et al. 2020; Naveen et al. 2022).

Above 75% of all invasive fungal fatalities globally are caused by the genera *Aspergillus*, *Candida*, *Pneumocystis*, *Saccharomyces*, *Mucor*, and *Cryptococcus* (Brown et al. 2012; Naveen et al. 2022). Moreover, approximately 30 fungal species are responsible for 99% of the human fungal disease load. The most frequent fungal infections in humans are superficial skin disease and nails and affect approximately 1.7 billion people worldwide. These infections are mainly caused by dermatophytes, that cause most common conditions such as athlete's foot (affects 1 in every 5 adults), ringworm of the scalp (common in young children and is thought to affect

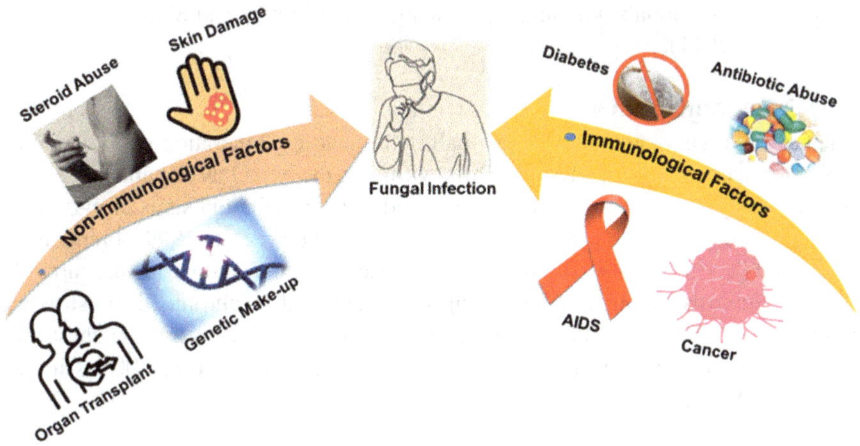

Fig. 5.1 Factors responsible for fungal infection

200 million people around the world) and nail infections (affects 10% of the general population globally, though this incidence increases with age to 50% in individuals 70 years and older) (Havlickova et al. 2008; Thomas et al. 2010; Brown et al. 2012).

5.2.1 *Aspergillus* Infection

The disease, which is caused by *Aspergillus* species known as 'Aspergillosis' (Barnes and Marr 2006). Historically, the first Aspergillus infection was reported almost 180 years ago but the significance and diversity of clinical entities induced by Aspergillosis were not recognised until 1952 (Rinaldi 1983; Rippon 1982). The genus *Aspergillus* possesses nearly 180 species, 34 of which have been linked to human illness. Out of 34, *A. fumigatus* causes 90% *of Aspergillus* infections in humankind. There are several forms of Aspergillosis, some of which are moderate while others are severe, as mentioned below:

5.2.1.1 Allergic Bronchopulmonary Aspergillosis (ABPA)

It is a lung infection which is caused by *A. fumigatus*, and mostly, it affects patients with bronchial asthma, those having cystic fibrosis and immunocompromised (Agarwal et al. 2013; CDC). The colonisation of the airways by *A. fumigatus* can result in type I and III hypersensitivity responses, eventually leading to allergic bronchopulmonary aspergillosis (ABPA). The global burden of ABPA is projected to be 1.4–6.8 million patients (Oguma et al. 2018; Denning et al. 2013).

5.2.1.2 Allergic *Aspergillus* Sinusitis (AAS)

AAS occurs when *Aspergillus* produces sinus inflammation and symptoms of a sinus infection (stuffiness, drainage and headache). It is indicated by features such as radiographic evidence of pansinusitis, nasal plug passage and recurrent nasal polyposis in individuals with an atopic background (Panjabi and Shah 2011; Glass and Amedee 2011).

5.2.1.3 Aspergilloma

Aspergilloma, also known as fungus ball, is a thick agglomeration of *Aspergillus* hyphae, mucus, cellular debris and fibrin within an the bronchus or pulmonary cavity in the lung or may also develop another side such as ethmoid sinus or maxillary or even in the jaw, which is caused by *A. fumigatus* (Fraser et al. 1999). Frequently, it occurs as co-infection with other lung diseases such as histoplasmosis, sarcoidosis, tuberculosis or other bullous lung disorders, and in chronically obstructed paranasal sinuses. Aspergilloma can be diagnosed through X-rays of the chest and serum precipitins (Ferguson 2000; Lee et al. 2004; Stevens et al. 2000; Garcia-Rubio and Alcazar-Fuoli 2018).

5.2.1.4 Azole-Resistant *Aspergillus fumigatus*

Azoles are used as antifungal drugs against a wide range of fungal infections due to their broad-spectrum activity. It was introduced in the 1970s, but after almost two decades in the late 1990s, a resistant strain of *A. fumigatus* was reported in the USA due to long-term use of Itraconazole. Azole-resistant *A. fumigatus* are now identified in Australia, Belgium, Brazil, China, Denmark, France, Germany, Norway, India, Iran, Japan, Kuwait, Portugal, Republic, Spain, Taiwan, the Czech and Turkey (Price et al. 2015; Lelièvre et al. 2013; Rivero-Menendez et al., 2016; Garcia-Rubio et al. 2017; Berger et al. 2017). Surprisingly, azole resistance has also arisen in azole-naive patients. It may occur due to the high use of azoles as a fungicide in agriculture practices (Berger et al. 2017) as shown in Fig. 5.2.

In the biosynthesis of ergosterol, cytochrome P450 14α-sterol demethylase enzyme plays a key role and is regulated by the *cyp51* gene, the product of which is CYP51A. Demethylation of C-14 of lanosterol by cytochrome P450 14α-sterol demethylase leads to the formation of ergosterol. Ergosterol is important for maintaining the permeability and fluidity of the plasma membrane. Azole family antibiotics inhibit the 14α demethylase enzymes by binding with CYP51A enzymes, leading to the accumulation of ergosterol precursors such as 4,14-dimethyl zymosterol and 24-methylene dihydro lanosterol, resulting in the depletion of ergosterol as shown in Fig. 5.3. Ultimately, it affects the membrane integrity and/or membrane structure, and leads to inhibition of fungus growth (Ghannoum and Rice 1999; Chen et al. 2020; Jeanvoine et al. 2020; Berger et al. 2017). However, fungi adopt many mechanisms to skip azole inhibition effects on ergosterol biosynthesis such as: reduced affinity between azole and CYP51A, overexpression of efflux and overexpression of CYP51A.

Point mutation in *cyp51* gene: As we describe above, the *cyp51* gene of *A. fumigatus* plays an important role in the biosynthesis of ergosterol. So, targeting the *cyp51* gene inhibits the fungal infection; however, point mutation in the *cyp51* gene protects the fungus from azole family antibiotics. These mutations are responsible for substitutions that lead to modifications in the structure of the

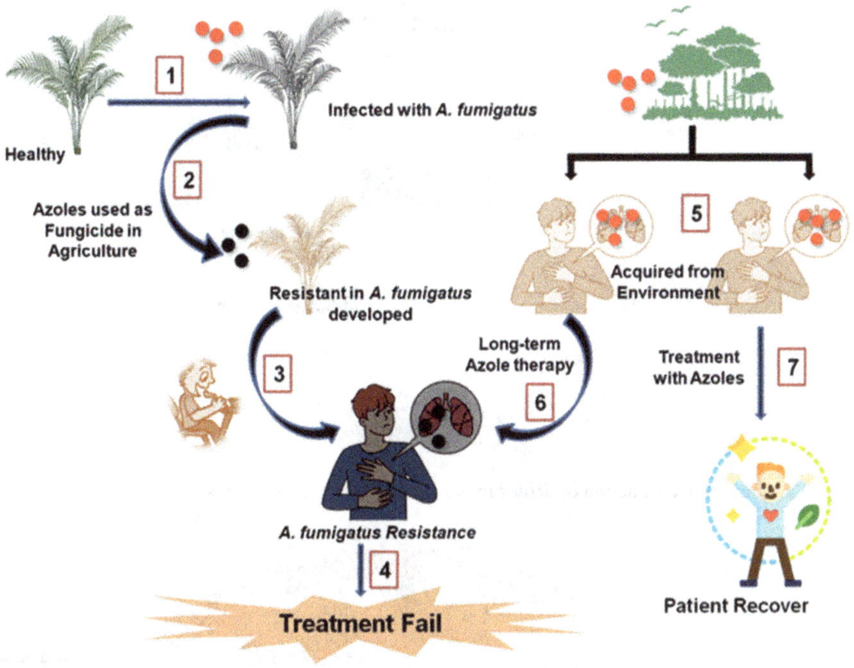

Fig. 5.2 Azole-resistant *Aspergillus fumigatus* infection. (**1**) Healthy plant gets infected with *A. fumigatus*. (**2**) Farmer uses the Azole family antibiotics as fungicides to cure *A. fumigatus* infection but with long-term use of antibiotics, plants become resistant. (**3**) Eating the plant or part of a plant-resistant strain of *A. fumigates* infects the person. (**4**) Treatment will fail. (**5**) *A. fumigatus* infection acquired from the environment. (**6**) Due to long-term use of Azole antibiotics, the person becomes Azole resistant. (**7**) The patient recovered with Azole family antibiotics

CYP51A enzyme. Only one amino acid substitution like G54, F219, G448P216 and M220, in the CYP51A protein leads to an alteration in azole binding affinity for enzyme causing azole tolerance as shown in Fig. 5.4 (Bowyer et al. 2011; Arendrup et al. 2014; Cowen et al. 2015; Sharma et al. 2015; Price et al. 2015; Hagiwara et al. 2016).

Overexpression of efflux: Efflux pumps are responsible for transporting out the toxic substances from the intracellular milieu. So, the overexpression of efflux pumps leads to reduce the concentration of azoles within the cell and ultimately the fungus becomes azole-resistant as shown in Fig. 5.4 (Fraczek et al. 2013; Chen et al. 2020).

Overexpression of CYP51A: Overexpression of CYP51A is caused by alterations in the cyp51A gene at the promoter region caused by the insertion of tandem repeats or transposable elements. The 34 base-pair tandem repeat (TR34), which is invariably present together with a lysine-to-histidine substitution at codon 98 (TR34/L98H), increases the normal production of CYP51A by up to eightfold. It has been proposed that an increase in mRNA levels correlates with the increase in cellular CYP51A levels, resulting in a decrease in azole sensitivity as shown in Fig. 5.4 (Resendiz Sharpe et al. 2018; Ren et al. 2017).

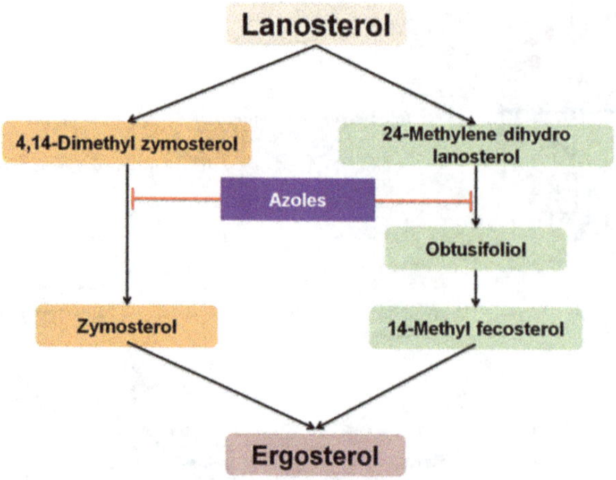

Fig. 5.3 Mechanism of action of azoles in the sterol biosynthesis pathway

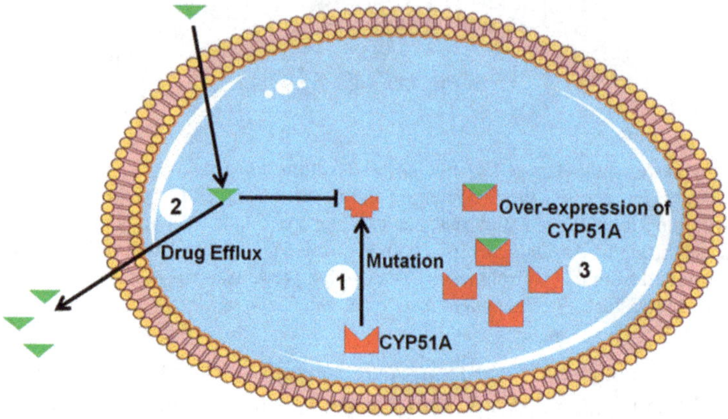

Fig. 5.4 Mechanisms for microbial cells to develop resistance. (**1**) The drug target (CYP51A) is changed such that the medication can no longer attach to it. (**2**) The drug is pumped out by an efflux pump. (**3**) Because the target enzyme is overproduced, the medication does not completely stop the biochemical reaction

5.2.1.5 Chronic Pulmonary Aspergillosis (CPA)

The first time CPA was considered a fatal condition was in 1842 in Edinburgh, UK, but after more than a century, the first successful treatment was reported with Amphotericin B in 1957 (Bennett 1844; Kelmenson 1959; Denning 2001). *A. fumigatus* is the most prevalent *Aspergillus* species, associated with CPA; however, *A. flavus*, *A. nidulans*, *A. terreus*, and *A. niger* also have been linked to this disease (Garcia-Rubio and Alcazar-Fuoli 2018; Maghrabi and Denning 2017). *A. fumigatus* conidia are small in size and optimum growth at 37 °C improves its capacity to germinate and reach the whole airway. Conidia germinate and establish a network of hyphae on the inner surface of a cavity after interacting with and evading host immune responses, causing damage to the surrounding parenchyma. CPA differs from invasive aspergillosis (IA), which occurs when the immune system is compromised, and allergic bronchopulmonary aspergillosis (ABPA) that arises in the setting of atopy (asthma) with immunological hyperactivity (Denning 2001; Latgé 2001, 2003). It affects reportedly immune-competent people, mainly those with preexisting lung disease. People with mycobacterium infection are most often affected with CPA too (Smith and Denning 2011; Maghrabi and Denning 2017). CPA is frequently diagnosed in patients with questionable radiography wherein *A. fumigatus* is cultured from a respiratory sample. The appearance of an aspergilloma is virtually always indicative of CPA. The detected *Aspergillus* IgG diagnostic marker is significant indication of the infection (Denning et al. 2016; Patterson et al. 2016; Maghrabi and Denning 2017).

5.2.1.6 Cutaneous (Skin) Aspergillosis

The occurrence of cutaneous aspergillosis is quite rare. Previous studies classified cutaneous aspergillosis as either primary or secondary. Primary cutaneous aspergillosis typically occurs at or near intravenous access catheter sites, traumatic inoculation sites and areas associated with occlusive dressings, burns or surgery. Secondary cutaneous lesions are caused by either continuous expansion of diseased underlying tissues to the skin or extensive blood-borne seeding of the skin and usually cause infection in immunocompromised patients. Cutaneous aspergillosis can also develop when invasive aspergillosis spreads to the skin from another part of the body, such as the lungs (Van Burik et al. 1998a, b; Allo et al. 1987; Estes et al. 1980). Diagnosis can be done through skin biopsy. Most primary and secondary *Aspergillus* infections need a skin lesion sample for both culture and histology. Because *Aspergillus* tends to penetrate the dermis and subcutis blood vessels, resulting in an ischemic cone above it, for a suspected fungal lesion a skin biopsy specimen should be collected from the lesion centre and should reach the subcutaneous fat (Gupta et al. 1996; Van Burik et al. 1998a).

5.2.1.7 Invasive Aspergillosis

When *Aspergillus* leads to a fatal infection, it mainly affects persons with weaker immune systems, such as those who have undergone stem cell transplant or an organ transplant. Invasive aspergillosis is usually found in the lungs, but it can extend to other regions of the body as well (Barnes and Marr 2006; Agarwal et al. 2013; Glass and Amedee 2011; Denning et al. 2003).

5.2.2 *Candida* Infection

Candida spp. are part of our normal microflora, but they can cause infections in both immunocompetent as well as immunocompromised persons; however, an immunocompromised host is more susceptible to 'candidiasis' (Deorukhkar et al. 2014). In the body, including the mouth, throat, gut and vagina, *Candida* often exists without creating any issues. If circumstances change to favour *Candida* growth, an infection may result. The likelihood of infection might increase due to factors such as hormones, medications, or immune system changes. *Candida* species are the most frequently occurring fungi that cause infections. *Candida albicans*, *Candida glabrata*, *Candida tropicalis*, *Candida parapsilosis*, and *Candida krusei* cause around 90% of Candidiasis. Over 10 years, one of the most extensive investigations (ARTEMIS) of *Candida* utilising data from 142 institutions in 41 countries found 31 species in the clinical samples. According to Pfaller and Diekema (2007), 8% of nosocomial bloodstream infections are brought on by *Candida* species (Pfaller and Diekema 2007). There are several types of Candidiasis reported as follows:

5.2.2.1 Vulvovaginal Candidiasis

Vaginal infection by *Candida* species, termed acute *Candida vaginitis*, has transformed into the concept of vaginal candidiasis in the UK, due to the recognition of a broad range of symptomatic and asymptomatic diseases. Later, the most common site of inflammation and symptom source is the vulva leading to the term vulvovaginal candidiasis (Sobel 1985, 2016; Hong et al. 2014). Vulvovaginal candidiasis (VVC) affects one out of every two women at some point in their lives, although it is seen as a frequent or bothersome ailment since it is easily treated, typically with over-the-counter medications. However, on a population scale, its impact is quite large because of its cost (Mitchell 2004; Dovnik et al. 2015; Foxman et al. 2013; Blostein et al. 2017). Just like the *Lactobacillus* infection, *Candida* also takes the same way to colonise the vagina. This yeast starts moving from the lower GI tract to the surrounding vestibule and vagina. Following menarche, an estrogen-influenced environment promotes colonisation, which reduces throughout the post-menopause period. In healthy women, colonisation may persist asymptomatic for a month and year because *Candida* can live with the vaginal microbiota in symbiotic relations. After the breakdown in this relationship involves either the *Candida* overgrowth or the host protective defence mechanisms alteration that acts to maintain low numbers of *Candida* while also consciously downregulating the mucosal immune inflammatory response to tolerate the low numbers of yeast, these conditions may result in acute symptomatic vulvovaginal infection. Vaginal candidiasis may be characterised by pruritus, excoriation, vulvar erythema, change in odour and an abnormal 'cheese-like' or watery discharge (Ilkit and Guzel 2011; Sobel 1985, 2007, 2016; Goldman et al. 2007; Crum et al. 2015; DeCherney 2019; Blostein et al. 2017). Vaginal candidiasis is uncommon before menarche, rises during the reproductive years and subsequently reduces following menopause. Sexual activity, contraception usage, antibiotic use, carbohydrate intake and diabetes are all risk factors for it (Jack and Sobel 1997; Sobel 1985, 2016; Horowitz et al. 1987; Geiger and Foxman 1996). Several antifungal drugs are commercially available to cure vaginal candidiasis with mild to severe side effects such as topical polyenes (e.g. nystatin), oral azole agents (e.g. fluconazole and itraconazole), etc. (Jack and Sobel 1997; Watson et al. 2002; Gonçalves et al. 2016).

5.2.2.2 Invasive Candidiasis

Unlike candida infections in human body parts such as the mouth, throat (thrush), skin or vagina, invasive candidiasis is a severe infection, that can affect the bones, brain, heart, blood, eyes and other parts of the body. The most common form of Invasive *Candidiasis* infection is bloodstream infection, i.e. Candidemia, and it is the fourth most common infection of blood in the USA which causes long hospital stays and death, so it is responsible for the high medical cost (Kullberg et al. 2005; Nucci et al. 2010; Wisplinghoff et al. 2004; Magill et al. 2014; Morgan et al. 2005). As per the previous reports, invasive candidiasis causes death of more than 50,000 and affects 250,000 people worldwide per year (Cleveland et al. 2015; Kullberg et al. 2014).

In the present day, invasive Candidiasis is an increasing concern in the health care system, particularly in the intensive care unit (ICU). Invasive Candidiasis is primarily a disease of medical progress, owing to great developments in healthcare technology during the last several decades as shown in Fig. 5.5. The use of a wide range of antibiotics indwelling, central venous catheter use and chronic; use of acute haemodialysis in the intensive care setting; widespread use of chronic immunosuppressive agents, including glucocorticosteroids and immunomodulators; use of other internal prosthetic devices; and aggressive chemotherapy for a variety of neoplastic conditions are among the most significant of these advances as shown in Fig. 5.5 (Weinstein and Fridkin 2005; Kao et al. 1999; Diekema et al. 2002;

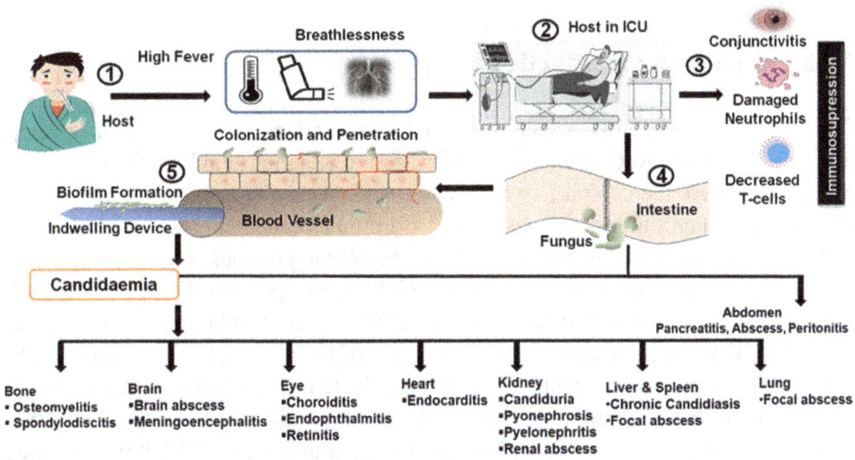

Fig. 5.5 Mechanism of invasive *Candidiasis*. (**1**) The host got infected with *candida* or other pathogens such as SARS-CoV-2, HIV, etc. or the host may cancer patient. (**2**) The host was admitted to the ICU, where a fungal spore inhaled by mechanical equipment or fungal spores goes inside through indwelling device. (**3**) After rapid fungal growth or after massive steroid use, symptoms arise as conjunctivitis and it leads to suppressing the host immune system by damaging the neutrophils and reducing the T-cell production. (**4**) *Candida* spp. can spread directly to the abdominal cavity and enter the bloodstream after gastrointestinal surgery. (**5**) Immunosuppression and other factors lead rapid growth of fungi in the gut and bloodstream, which can promote deep-seated opportunistic infections in various organs

Abi-Said et al. 1997; Morgan et al. 2005; Asmundsdóttir et al. 2002; Pappas et al. 2003; Tortorano et al. 2004; Trick et al. 2002; Wisplinghoff et al. 2004; Malani et al. 2001; Martin et al. 2003; Marr et al. 2000; Wenzel and Edmond 2001; Antoniadou et al. 2003; Pappas 2006).

Several previous reports suggested that *Candida albicans* were dominant species for invasive candidiasis but the distribution of species shifted during the last few decades. This species is found in only half of the isolates today (Arendrup 2010; Cleveland et al. 2015; Guinea et al. 2014; Kullberg et al. 2014). Species distribution depends on region, such as *C. glabrata* is a major pathogen in the northern region of Europe, the USA and Canada, but *C. parapsilosis* is more common in the southern part of Europe, Asia and South America. Given the variations in sensitivity to azoles and echinocandins across various species, changes in the distribution of species may influence treatment recommendations. The pathogenicity of *Candida* species varies greatly. *C. parapsilosis* and *C. krusei* are less dangerous than *Candida albicans*, *Candida glabrata*, and *Candida tropicalis*. This diversity is reflected in the low death rate among patients with *C. parapsilosis* candidiasis, as well as the fact that infection with *C. krusei* is extremely rare, except in individuals with severe immunodeficiency and prior azole exposure (Arendrup et al. 2011; Arendrup et al. 2002; Kullberg et al. 2014). There are varieties of anti-invasive candidiasis drugs commercially available in the market, and they may be prescribed based on the dominant species of *Candida* and the severity of illness such as Fluconazole, Itraconazole, Amphotericin-B, etc. (McCarty and Pappas 2016).

5.2.3 *Cryptococcus* Infection

Cryptococcus spp. are polysaccharide-coated yeasts and include two primary species, i.e. *C. neoformans* and *C. gattii*, known for life-threatening fungal meningoencephalitis. It can invade the central nervous system and cause *Cryptococcal* meningitis in both immunocompromised persons as well as immunocompetent persons but is more vulnerable in immunocompromised persons. As per the previous report, an estimated incidence of cryptococcal meningitis was 223,100 cases per year leading to 181,100 deaths globally each year (Rajasingham et al. 2017; Mourad and Perfect 2018; Perfect et al. 2010; Kwon-Chung et al. 2014; Iyer et al. 2021). In 2014, about 15% of AIDS-related deaths were due to cryptococcal meningitis. In this modern science era, where a variety of antifungal therapies are available, but death due to cryptococcal infection is still high in both developed as well as developing countries. In 2014, it was reported that 70% of death out of total infections in low-income countries were due to cryptococcal meningitis, whereas 40% in middle-income countries and 20–30% in developed countries. Majority of deaths are recorded in resource-limited nations due to a lack of access to medications and the high cost of effective treatments (Taylor-Smith and May 2016; Loyse et al. 2019; Rajasingham et al. 2017; George et al. 2018). The growing number of transplant receivers and patients taking immunosuppressive treatments such as corticosteroids, new biological agents and novel anti-neoplastic therapy has resulted in a high-risk population for cryptococcal infection.

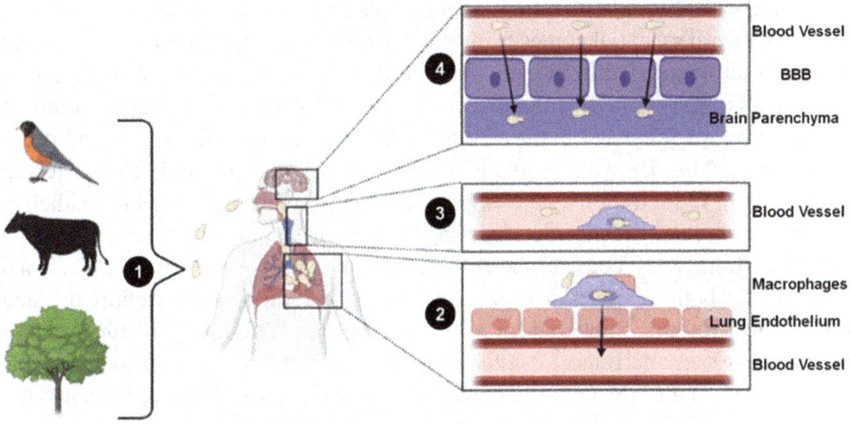

Fig. 5.6 Mechanism of *Cryptococcal* infection. (**1**) Carrier of fungal spore or yeast. (**2**) *Cryptococci* spread through lungs with the help of macrophages. (**3**) *Cryptococci* spp. circulate in the bloodstream as either extra-cellular yeast or within phagocytic cells. (**4**) Cryptococcal invasion of the central nervous system through BBB (blood–brain barrier) crossing

5.2.3.1 Pathogenesis

In humans, macrophages and dendritic cells petrol the lungs and if any invaders are found, the phagocytes engulf the pathogens and destroy them through a series of steps. Interestingly, *C. neoformans* hijack the phagosome and modulate the phagosome maturation process which neutralises it and enables the proliferation. Then, *C. neoformans* circulates in the bloodstream as either extra-cellular yeast or within phagocytic cells such as macrophages. There is invasion of the central nervous system via BBB (blood-brain barrier) crossing. This might be accomplished in different ways including paracellularly crossing the BBB, transcellularly entering brain endothelial cells and leaving cells into the parenchyma, or transcytosis using a Trojan horse phagocyte cell, which leads to *Cryptococcal* meningitis as shown in Fig. 5.6 (Smith and May 2013; Fairn and Grinstein 2012; Smith et al. 2015). Many drugs are available, i.e. Amphotericin-B (Lee et al. 2004), Fluconazole (Zhai et al. 2015), etc. for the treatment of *Cryptococcal* infection, but still problem persists because of high toxicity and drug resistance.

5.3 Essential Oils

5.3.1 Definition of Essential Oils

Essential oils, also known as ethereal oils, are volatile secondary metabolites derived from aromatic plants. The term ‘essential oil’ is believed to have originated from the concept of ‘quintaessentia’ introduced by Paracelsus von Hohenheim, a Swiss medical reformer in the sixteenth century. This term refers to the potent and effective components found in medicinal substances. In essence, essential oils are powerful and natural essences extracted from aromatic plants (Guenther 1952; Macwan et al. 2016). The word essential comes from essence (heavy smell). They are fragrant and volatile liquids extracted from plant material such as wood, bark, roots, peel, seeds, leaves, fruits, flowers and whole plants. Essential oils are products or mixtures of products that are generated in the cytoplasm and are generally found as small droplets between cells. They are aromatic and volatile (Hyldgaard et al. 2012; Mekem Sonwa 2000). Essential oils were employed by the ancient Egyptians in medicine, perfumes and the technique of embalming and preparing remains for burial through mummification. The Vedas of ancient Asia encouraged the use of fragrances and aromatics for both religious and medicinal reasons. Additionally, essential oils and fragrances have been employed for a variety of reasons throughout history, including religious rites, perfume manufacture and medicinal agents against infectious illnesses. The fragrance cultures of the Phoenicians, Jews, Greeks, Romans and other cultures in the Mediterranean basin, as well as the Mayas and Aztecs in the Americas, were all highly developed. With the fall of the Roman Empire and the advent of both Christian and Muslim civilisations, the art and science of fragrance were transported to the Arab world, where it acquired a high level of complexity. In the middle ages, crusaders coming from the holy land carried this knowledge of smells back to Europe and further developed by alchemists as well as monks (Mekem Sonwa 2000; Ríos 2016).

Terpenes and terpenoids are the most common types of substances: nitrogen- and sulphur-containing compounds, coumarins and phenylpropanoids homologues can also be isolated. Terpenes are a broad family of natural hydrocarbons that derive from the isoprene unit (C_5H_8) and have a variety of chemical and biological characteristics. They are synthesised in the plant cell's cytoplasm via the mevalonic acid pathway, which begins with acetyl CoA. Monoterpenes ($C_{10}H_{16}$) and sesquiterpenes ($C_{15}H_{24}$) are the most common terpenes, although longer chains such as diterpenes ($C_{20}H_{32}$), triterpenes ($C_{30}H_{40}$) and others occur as well. Terpenoids are terpene-related chemicals that have some oxygen functionality or rearrangement. The most well-known terpenoids include thymol, geraniol, linalyl acetate, linalool, carvacrol, piperitone, citronellal and menthol (Hyldgaard et al. 2012; Lopez-Reyes et al. 2013).

5.3.2 Antifungal Properties of Essential Oils

Many types of antifungal synthetic drugs are commercially available such as Amphotericin B, Itraconazole, Ketoconazole, Fluconazole, Clotrimazole, Nystatin, Terbinfine, Miconazole, which are widely used in both developing as well as the developed nations. However, extensive use of synthetic drugs causes diverse side effects such as itching or burning, feeling sick, redness, diarrhoea, abdominal pain, etc. as well as pathogens becoming resistant (Macwan et al. 2016). There is a need to limit the use of synthetic drugs as antimicrobial agents, to fight different diseases caused by aggressive and endogenous microbes that are resistant to synthetic drugs. In this way, plant-derived compounds, such as hydro-alcoholic extracts or essential oils, can undoubtedly play an important role. Because of their diverse chemical structures, such compounds have considerable adaptability; the same plant may provide a pool of chemicals with a very wide range of activity (Macwan et al. 2016; Nazzaro et al. 2017). Essential oils are often complex combinations of polar and non-polar natural compounds. They are well-known for their antiseptic and therapeutic characteristics (analgesic, anti-carcinogenic, antimicrobial, sedative, spasmolytic, anti-inflammatory, local anaesthetics) (Macwan et al. 2016; Masango 2005; Tongnuanchan and Benjakul 2014; Bakkali et al. 2008; Božović et al. 2017; Burt and Reinders 2003).

Both humans and fungi are eukaryotic cells, so in comparison to bacterial infection, this makes it more challenging to determine their existence and apply the proper treatment. The fungal cell wall contains a chitin structure, absent in humans and may be considered the primary target for antifungal agents. So, due to high lipophilic in nature and low molecular weight, essential oils can easily disrupt the cell wall and/or cell membrane, leading to cell death or blocking germination and sporulation. There are many other mechanisms of action of essential oils such as dysfunction of fungal mitochondria, inhibition of efflux pump and inhibition of biofilm development, etc. (Hu et al. 2017; Tian et al. 2011b; Bajpai et al. 2011; Chen et al. 2013; Hossain et al. 2016; Chavan and Tupe 2014).

Over the previous years, there has been a significant rise in clinical yeast infections. The widespread utilisation of broad-spectrum antifungal medicines has contributed to a noticeable surge in drug-resistant yeast species (He et al. 2017). Throughout the bygone period, traditional medicine employed plants and their derivatives as remedies for various illnesses. These natural substances have been employed to treat a wide range of health conditions (Lang and Buchbauer 2012). Consequently, the scientific community has shown considerable interest in essential oils as a favoured natural antimicrobial solution. Essential oils are intricate and volatile compounds that occur naturally in different parts of plants in the course of their secondary metabolism, known for their characteristic aroma. These oils have been extensively studied and exploited globally for their medicinal properties. They are extracted from a diverse range of plants, primarily for their antimicrobial effects against fungal, bacterial and viral pathogens. Along with their antimicrobial action essential oils have also been reported their anticancer, antioxidant, anti-inflammatory and immune-modulatory activities. Due to their diverse and potent properties, essential oils have gained significant attention for their various applications (Akthar et al. 2014; Bakkali et al. 2008; Bona et al. 2016).

In the existing scientific literature, there is extensive documentation regarding the antifungal potential of essential oils. Researchers have extensively reported on the intricate chemical composition of these oils and their significant ability to combat fungal pathogens. This growing body of knowledge highlights the promising applications of essential oils as natural antifungal agents in various industries, viz., pharmaceuticals and personal care products. Generally, essential oils with high concentrations of phenols are widely recognised for their potent antifungal activities. Among them, are essential oils extracted from *Thymus* spp. and *Origanum* spp. stand out for their remarkable efficacy against fungal pathogens. Extensive evaluations have been conducted on various microorganisms to assess the antifungal properties of these oils, reaffirming their potential as natural and powerful agents against fungal infections (Manohar et al. 2001; Salgueiro et al. 2003; Pina-Vaz et al. 2004; Pinto et al. 2006, 2013). The mentioned essential oils demonstrated notable effectiveness towards *Candida*, with oregano exhibiting the ability to inhibit the germination and filamentous form of *C. albicans* in vitro (Manohar et al. 2001; Giordani et al. 2004).

Similarly, Raut and Karuppayil (2014) reviewed many research articles, in which essential oils have been reported for their inhibitory activity against several fungal pathogens (Raut and Karuppayil 2014). Arora and Kaur (1999) conducted a study to assess the effectiveness of essential oils from garlic and clove towards various fungal pathogens. They found that garlic and clove extracts exhibited potent inhibitory effects on the growth of *C. acutus*, *C. albicans*, *C. catenulate*, *C. tropicalis*, *C. inconspicua*, and *C. apicola*. Additionally, these essential oils also demonstrated significant inhibitory activity towards *Sacharomyces cerevisae*, *Trignopsis variabilis* and *Rhodotorula rubra* (Arora and Kaur 1999). Essential oils obtained from tea tree oil (*Melaleuca alternifolia*) and *Satureja montana* (*winter savory*), were found effective towards the growth of *Candida* (Hammer et al. 2004; Tampieri et al. 2005). Similarly, Carson et al. reported the potent efficacy of tea tree oil against *Candida* (Carson et al. 2006). Several studies have highlighted the anti-*Candida* properties of mint essential oils derived from three species: *Mentha spicata*, *Mentha cervina*, and *Mentha piperita* (Hammer et al. 1999; Tampieri et al. 2005; Agarwal et al. 2008). Like *Mentha* spp. other oils including laurel (*Laurus nobilis*, Carvalhinho et al. 2012), bergamot (*Citrus bergamia*, Sanguinetti et al. 2007), Anise (*Pimpinella anisum*, Kosalec et al. 2005), basil (*Ocimum basilicum*, Pozzatti et al. 2010), oregano (*Thymus capitatus*, Hosni et al. 2013; Chedia et al. 2013), mountain pepper (*Litsea cubeba*) and Cumin (*Cuminum cyminum*, Li et al. 2022), bay laurel (*Laurus nobilis*, Peixoto et al. 2017), ginger (*Zingiber officinale*, Pozzatti et al. 2010; Takahashi et al. 2011), rosemary (*Rosmarinus officinalis*, Giordani et al. 2004; Bozin et al. 2007), *Origanum onites* (Oregano oil, Hacioglu et al. 2021; Manohar et al. 2001) and lavender (*Lavandula* spp., Schwiertz et al. 2006) are also reported against *Candida* spp. Moreover, the essential oils of *Cymbopogon martini* and *Pelargonium capitatum* can inhibit germ tube formation, a major virulence factor of *C. albicans* (Angiolella 2021). The essential oil of dill seed restricts the growth of *Candida* by damaging the cytoplasmic membrane (Chen et al. 2013). Similarly, Elisa et al. (2021) reported that the essential oil of fennel, cumin and manuka has potential antifungal efficacy against *Candida* strain (Elisa et al. 2021). Sweet basil leaf and Cinnamon bark essential oils restrict the growth of *C. albicans* (Hovijitra et al. 2016). The essential oil of cinnamon revealed anti-candidal activity against *C. orthopsilosis* and *C. parapsilosis* at the concentration of 250 and 500 μg/mL (Pires et al. 2011).

A set of 82 essential oils was reported active towards *Candida albicans*, *Cryptococcus neoformans* and *Aspergillus niger* with MIC value ≤160 mg/mL or ppm (Powers et al. 2019). The essential oil of *Cymbopogon citratus*, commonly known as lemongrass depicted inhibitory effects against several troublesome fungal species, including *Candida albicans*, *Candida tropicalis* and *Aspergillus niger*

(Boukhatem et al. 2014). Likewise, a notable study reported a remarkable antifungal prowess of *Thymus viciosoi* essential oil. This oil demonstrated potent inhibitory effects against a range of challenging fungi, including various *Candida* species, *Cryptococcus neoformans*, as well as the *Aspergillus* strains (*A. fumigatus*, *A. niger* and *A. flavus* (Vale-Silva et al. 2010). Furthermore, additional investigations have highlighted the notable antifungal potential of various essential oils including *Thapsia villosa* (Pinto et al. 2017), *Ferulago capillaris* (Pinto et al. 2013), *Angelica major* (Cavaleiro et al. 2015), *Cryptomeria japonica* (Moiteiro et al. 2013), against the *Candida* spp., *Aspergillus* spp. and *Cryptococcus neoformans. Thymus pulegioides* displayed significant inhibitory effects against *Candida* and *Aspergillus* spp. (Pinto et al. 2006). Moreover, *Achillea wilhelmsii* essential oil found promising antifungal properties against *Candida* species, *Aspergillus* strains and *Cryptococcus neoformans* (Kazemi and Rostami 2015). It was also discovered that *Satureja thymbra* oil has antifungal properties that are effective against *Candida* strains, *Aspergillus* strains and *C. neoformans* (Piras et al. 2011). Tea tree essential oil has been found to have antifungal activity towards strains of *C. albicans* and *C. neoformans* (Mondello et al. 2003). With the concern of fungi *Candida*, *Aspergillus* and *Cryptococcus*, some essential oils are listed in Table 5.1.

5.3.3 Limitation of Essential Oils

Essential oils are concentrated natural extracts extracted from plants. They have been used as alternative remedies since the late twelfth century and gained popularity in the latter half of the sixteenth century (Man et al. 2019). Despite their potential benefits, it is important to note that some adverse effects, viz., cytotoxic, poor stability, high volatility, low solubility in water, burning sensation at the site of application (Natrajan et al. 2015; Sarkic and Stappen 2018; Bleasel et al. 2002; Donato et al. 2020), have been associated with the use of essential oils. These features may significantly reduce the efficacy of essential oils against the pathogens. For instance, the essential oils of *Melaleuca alternifolia* and *Coriandrum sativum* have been found to cause cytotoxicity in HeLa cells (Scazzocchio et al. 2016). Essential oils have a cytotoxic/anti-proliferative effect on the human skin keratinocyte cell lines (HaCaT, Oliveira Ribeiro et al. 2020). In an experiment, it was disclosed that the essential oil of palmarosa (at a concentration of 5.0% w/v) had a sensitivity rate of 50% among the population (Bleasel et al. 2002). Likewise, lemongrass essential oil displays certain limitations, encompassing issues of instability, heightened volatilisation and diminished solubility in water, which may affect its overall performance (Natrajan et al. 2015). Moreover, certain essential oils have been found to cause a burning sensation when applied to the skin. In an experiment to examine sensitisation to some essential oils, lemongrass (at a concentration of 2% w/v) can lead to allergic contact dermatitis (Bleasel et al. 2002). The essential oil of citronella has also been reported for its poor stability in the presence of air and high temperature (Sharma et al. 2019), and allergic contact dermatitis from essential oils has been documented (Rudzki and Grzywa 1985). A number of essential oils and their

Table 5.1 Efficacy of some plant essential oils against fungal pathogens

Essential oil (s)				
Common name	Scientific name	Fungus	Efficacy	Reference (s)
Betel leaf	*Piper betle*	*A. flavus*	0.5 μL/mL and 0.7 μg/mL	Dubey and Tripathi (1987), Prakash et al. (2010)
		C. albicans, *C. tropicalis*, *C. glabrata*, *C. parapsilosis*	31.25–125 μL/mL	Rath and Mohapatra (2015)
Lemongrass	*Cymbopogon citratus*	*A. flavus*, *A. fumigatus*	300–800 mg/L	Nguefack et al. (2004)
		A. niger	500 mg/L	Tzortzakis and Economakis (2007)
		C. albicans, *C. tropicalis*, *A. niger*	20 μL/disc	Boukhatem et al. (2014)
		C. parapsilosis, *C. krusei*	0.16–0.8 mg/L	Córdoba et al. (2019)
		Candida spp.	0.275–2.2 mg/mL	Paiva et al. (2022)
		C. albicans	0.12% (v/v)	Hammer et al. (1998)
Thyme	*Thymus vulgaris*	*A. fumigatus*, *A. flavus*	200–500 mg/L	Nguefack et al. (2004)
		A. flavus	250 μg/mL	Kohiyama et al. (2015)
		C. albicans	1.56–25 μg/mL	Jafri and Ahmad (2020)
		C. albicans, *C. glabrata*, *C. tropicalis*	0.03–0.06% (v/v)	Mandras et al. (2016)
		A. alternata	500 ppm	Feng and Zheng (2007)
		A. flavus	350 ppm	Omidbeygi et al. (2007)
African basil	*Ocimum gratissimum*	*A. flavus*, *A. fumigatus*	200–600 mg/L	Nguefack et al. (2004)
		C. neoformans (25 isolates)	125–250 mg/L	de Aquino Lemos et al. (2005)
Holy basil	*Ocimum sanctum*	*A. flavus*	0.3 μg/mL	Kumar et al. (2010)
Sweet basil	*Ocimum basilicum*	*C. albicans* (planktonic) and (sessile)	0.039% and 0.0391% (v/v)	Hovijitra et al. (2016)
		C. albicans	0.5% (v/v)	Hammer et al. (1998)

Table 5.1 (continued)

Essential oil (s)				
Common name	Scientific name	Fungus	Efficacy	Reference (s)
Marjoram	*Origanum majorana*	*A. niger*	0.5% (w/v) and 0.5 µg/mL	Vági et al. (2005), Kaskatepe et al. (2022)
Oregano	*Origanum onites*	*Candida albicans*	0.025–0.05% (v/v)	Hacioglu et al. (2021)
	Origanum vulgare	*C. albicans*	0.12% (v/v)	Hammer et al. (1998)
		C. krusei, C. albicans, C. dubliniensis	0.01–5.33 µg/mL	Cid-Chevecich et al. (2022), Karaman et al. (2017)
		Cryptococcus neoformans (azole-susceptible and resistant clinical isolates)	0.3–0.6 mg/mL	Scalas et al. (2018)
Orange	*Citrus sinensis*	*A. niger*	500 mg/L	Sharma and Tripathi (2006)
		A. niger, A. flavus	0.94%	Viuda-Martos et al. (2008)
		A. flavus	16,000 mg/L	Velázquez-Nuñez et al. (2013)
		C. albicans	0.078% (v/v)	Hovijitra et al. (2016)
Lemon	*Citrus lemon*	*A. niger, A. flavus*	0.94%	Viuda-Martos et al. (2008)
		C. albicans	0.156% (v/v)	Hovijitra et al. (2016)
Mandarin	*Citrus reticulata*	*A. flavus, A. niger*	0.94%	Viuda-Martos et al. (2008)
Grapefruit	*Citrus paradisi*	*A. flavus, A. niger*	0.94%	Viuda-Martos et al. (2008)
Bitter orange	*Citrus aurantium*	*C. albicans*	0.15–0.31% (v/v)	Nidhi et al. (2020)
Petitgrain	*Citrus aurantium*	*C. albicans*	0.25% (v/v)	Hammer et al. (1998)
Kaffir lime	*Citrus hystrix*	*C. albicans*	0.078% (v/v)	Hovijitra et al. (2016)
Shiluozi	*Cicuta virosa*	*A. flavus, A. oryzae, A. niger*	5.0 µg/mL	Tian et al. (2011a)
Gossolhi	*Lippia rugosa*	*A. flavus*	1000 mg/L	Tatsadjieu et al. (2009)

(continued)

Table 5.1 (continued)

Essential oil (s)				
Common name	Scientific name	Fungus	Efficacy	Reference (s)
Moldenke	*Lippia junelliana*	*C. krusei*, *C. albicans*, *C. parapsilosis*	1.6–6.12 mg/L	Córdoba et al. (2019)
Alecrim pimento	*Lippia sidoid*	*Cryptococcus* spp.	31.25–62.5 μg/mL	de Morais et al. (2012, 2016)
Litsea	*Litsea cubeba*	*A. flavus*	0.5 μg/mL	Li et al. (2016)
Cassia	*Cinnamomum cassia*	*A. alternata*	300–500 ppm	Feng and Zheng (2007)
Thymus	*Thymus vulgaris*	*C. albicans*, *C. famata*	0.05–0.1% (mg/mL)	Ebani et al. (2018)
		C. krusei, *C. parapsilosis*, *C. glabrata*, *C. albicans*	0.16–0.8 mg/L	Córdoba et al. (2019)
		Cryptococcus neoformans (azole-susceptible and resistant clinical isolates)	0.56–1.12 mg/mL	Scalas et al. (2018)
	Thymus viciosoi	*Candida* spp., *Cryptococcus neoformans*, *A. fumigatus*, *A. niger*, *A. flavus*	0.08–0.32 μL/mL	Vale-Silva et al. (2010)
	Thymus pulegioides	*Candida* spp., *Aspergillus*	0.16–0.64 μL/mL	Pinto et al. (2006)
Summer savory	*Satureja hortensis*	*A. flavus*	500 ppm	Omidbeygi et al. (2007)
Santoreggia sarda	*Satureja thymbra*	*Candida* strains, *Aspergillus* strains, *C. neoformans*	0.16–0.32 μL/mL	Piras et al. (2011)
Bay laurel	*Laurus nobilis*	*C. albicans*, *C. krusei*, *C. parapsilosis*	0.8 mg/L	Córdoba et al. (2019)
		C. albicans, *C. tropicalis*, *C. krusei*, *C. glabrata*	250–500 mg/mL	Peixoto et al. (2017)
Peppermint	*Mentha piperita*	*C. parapsilosis*, *C. krusei*, *C. albicans*	1.25–6.12 mg/L	Córdoba et al. (2019)
		C. albicans	0.5% (v/v)	Hammer et al. (1998)
Spearmint	*Mentha spicata*	*C. albicans*	0.313% (v/v)	Hovijitra et al. (2016)
		C. albicans	0.12% (v/v)	Hammer et al. (1998)
Rose geranium	*Pelargonium capitatum*	*C. albicans*	780 mg/mL	Angiolella (2021)

Table 5.1 (continued)

Essential oil (s)				
Common name	Scientific name	Fungus	Efficacy	Reference (s)
Lavender	*Lavandula angustifolia*	*C. albicans*	0.5–2% (v/v)	Mijatovic et al. (2022)
	Lavandula vera	*C. albicans, C. glabrata, C. tropicalis*	0.125–1.0% (v/v)	Mandras et al. (2016)
Tasmanian lavender	*Lavandula angustifolia*	*C. albicans*	0.5% (v/v)	Hammer et al. (1998)
Cinnamon	*Cinnamomum zeylanicum*	*C. albicans, C. auris*	<0.03–0.06% (v/v)	Tran et al. (2020)
		C. albicans (planktonic and sessile)	0.005% and 0.0049% (v/v)	Hovijitra et al. (2016)
	Cinnamomum verum	*C. albicans, C. dubliniensis, C. tropicalis*	1.0 mg/mL	Wijesinghe et al. (2020)
	Cinnamon cassia	*C. albicans*	65 μg/mL	de Fátima Dantas de Almeida et al. (2016)
Fennel	*Fennel vulgare*	*C. albicans*	0.625% (v/v)	Hovijitra et al. (2016)
		C. albicans, C. glabrata, C. tropicalis	0.25–1.0% (v/v)	Mandras et al. (2016)
Citronella	*Cymbopogon winterianus*	*C. albicans*	250 μg/mL	de Fátima Dantas de Almeida et al. (2016)
	Cymbopogon nardus	*C. albicans*	0.25% (v/v)	Hammer et al. (1998)
		C. albicans	250–1000 μg/mL	De Toledo et al. (2016)
Clove	*Syzygium aromaticum*	*C. parapsilosis, C. albicans, C. glabrata, C. krusei*	0.5–2 mg/mL	Biernasiuk et al. (2022)
		C. glabrata, C. albicans, C. tropicalis	0.25% (v/v)	Mandras et al. (2016)
		C. albicans	0.12% (v/v)	Hammer et al. (1998)
Villous deadly carrot	*Thapsia villosa*	*C. albicans, C. parapsilosis, C. glabrata, C. krusei, Cryptococcus neoformans, A. niger, A. fumigatus, A. flavus*	0.16–1.25 μL/mL	Pinto et al. (2017)

(continued)

Table 5.1 (continued)

Essential oil (s)				
Common name	Scientific name	Fungus	Efficacy	Reference (s)
Canaheja/ Canahierro	*Ferulago capillaris*	*Candida, Cryptococcus, Aspergillus*	0.16–1.25 μL/mL	Pinto et al. (2013)
Sweet wormwood	*Artemisia annua*	*Candida* spp.	3.13–25 μL/mL	Santomauro et al. (2016)
Pistachio	*Pistacia vera*	*C. albicans, C. parapsilosis, C. glabrata*	2.50 and 5.0 mg/ mL	D'Arrigo et al. (2019)
Spice coriander/ coriander	*Coriandrum sativum*	*Candida* spp.	31.25–250 μg/ mL	Barbosa et al. (2023)
		C. albicans, C. tropicalis	0.05–0.4% (v/v).	Silva et al. (2011)
		C. albicans	0.25% (v/v)	Hammer et al. (1998)
Papaya	*Carica papaya*	*C. glabrata, C. albicans, C. krusei, C. tropical, C. parapsilosis*	4.0–16.0 μg/mL	He et al. (2017)
Copal	*Bursera morelensis*	*C. albicans*	0.125–0.5 mg/ mL	Rivera-Yañez et al. (2017)
Pine	*Pinus sylvestris*	*C. glabrata, C. albicans, C. tropicalis*	0.015–0.06% (v/v)	Mandras et al. (2016)
		Cryptococcus neoformans (azole-susceptible and resistant clinical isolates)	0.7–0.54 mg/mL	Scalas et al. (2018)
Sage	*Salvia officinalis*	*C. albicans, C. tropicalis, C. glabrata*	1.0% (v/v)	Mandras et al. (2016)
		C. albicans	2.78 g/L	Sookto et al. (2013)
		C. glabrata, C. albicans, C. guilliermondii, C. parapsilosis, C. krusei, C. tropicalis	0.5% (v/v) or 3.125–100 mg/ mL	Hammer et al. (1998), Karpiński et al. (2023)
Lemon balm	*Melissa officinalis*	*C. glabrata, C. albicans, C. tropicalis*	0.25–1.0% (v/v)	Mandras et al. (2016)
Wild celery	*Angelica major*	*Cryptococcus neoformans, C. Albicans, aspergillus* spp.	0.16–>10 μL/mL	Cavaleiro et al. (2015)
Sandalwood	*Santalum album*	*C. albicans*	0.06% (v/v)	Hammer et al. (1998)
Bay	*Pimenta recemosa*	*C. albicans*	0.12% (v/v)	Hammer et al. (1998)

Table 5.1 (continued)

Essential oil (s)				
Common name	Scientific name	Fungus	Efficacy	Reference (s)
Common Myrtle	*Myrtus communis*	*C. albicans*	0.125–1.0% (v/v)	Cannas et al. (2014)
Yarrows	*Achillea wilhelmsii*	*C. albicans, C. parapsilosis, A. niger*	1–3.5 μg/mL	Kazemi and Rostami (2015)
Lemon verbena	*Aloysia triphylla*	*C. albicans, C. dubliniensis, C. krusei, C. glabrata, C. guillermondii, C. tropicalis, C. parapsilosis*	35–140 μg/mL	de las Mercedes Oliva et al. (2011)
Piper	*Piper claussenianum*	*C. albicans*	0.04–0.1%	Curvelo et al. (2014)
Black cumin	*Nigella sativa*	*C. tropicalis, C. albicans, C. parapsilosis, C. glabrata*	15.62–31.25 μL/mL	Rath and Mohapatra (2015)
Curry leaf	*Murraya koenigii*	*C. albicans, C. tropicalis, C. glabrata, C. parapsilosis*	125–250 μL/mL	Rath and Mohapatra (2015)
Ajwain	*Trachispirum ammi*	*C. tropicalis, C. albicans, C. parapsilosis, C. glabrata*	31.25–125 μL/mL	Rath and Mohapatra (2015)
Juniper	*Juniperus* spp.	*Candida, Aspergillus*	0.08–0.16 μL/mL	Cavaleiro et al. (2006)
Sugi	*Cryptomeria japonica*	*C. albicans, C. tropicalis, C. neoformans, A. niger, A. fumigatus, A. flavus*	200–250 μg/mL	Moiteiro et al. (2013)
Nees	*Commiphora myrrha*	*C. neoformans*	0.80–1.40 mg/mL	de Rapper et al. (2012)
Ebap/Ebo/Toab/Poba	*Santiria trimera*	*C. neoformans*	>0.71 μL/mL	Martins et al. (2003)
Tea tree	*Melaleuca alternifolia*	*C. albicans*	0.25% (v/v)	Di Vito et al. (2015)
		C. albicans, C. neoformans	0.03–0.25%	Mondello et al. (2003)

primary constituents have been documented as non-mutagenic and non-genotoxic (Bakkali et al. 2008); however, the adverse impact of the essential oils largely manifests when administered in high doses, through the use of undiluted concentrates or with prolonged exposure. From a toxicological perspective, a considerable number of essential oils do not fall into the category of high-risk substances (Pavela and Benelli 2016).

A substantial number of reports focussing on the biological effects of essential oils on pathogenic micro-organisms exist but a notable absence of research papers addressing the toxicological aspects and the impacts of essential oils on non-target organisms are missing. Despite these flaws, most of the essential oils are safe for the environment and human health if used in a defined amount or quantity. Encapsulating the essential oils with natural or synthetic compounds in the form of formulations has been found to increase efficacy, stability and bioavailability with less or decreased side effects.

Acknowledgement The authors thank Director CSIR-CIMAP and CSIR-Aroma Mission (HCP 007 PI-III).

References

Abi-Said D, Anaissie E, Uzun O, Raad I, Pinzcowski H, Vartivarian S (1997) The epidemiology of hematogenous candidiasis caused by different *Candida* species. Clin Infect Dis 24(6):1122–1128

Agarwal V, Lal P, Pruthi V (2008) Prevention of *Candida albicans* biofilm by plant oils. Mycopathologia 165:13–19

Agarwal R, Vishwanath G, Aggarwal AN, Garg M, Gupta D, Chakrabarti A (2013) Itraconazole in chronic cavitary pulmonary aspergillosis: a randomised controlled trial and systematic review of literature. Mycoses 56(5):559–570

Akthar MS, Degaga B, Azam T (2014) Antimicrobial activity of essential oils extracted from medicinal plants against the pathogenic microorganisms: a review. Issue Biol Sci Pharm Res 2:001. ISSN: 2350-1588

Allo MD, Miller J, Townsend T, Tan C (1987) Primary cutaneous aspergillosis associated with Hickman intravenous catheters. N Engl J Med 317(18):1105–1108

Angiolella L (2021) Synergistic activity of *Pelargonium capitatum* and *Cymbopogon martini* essential oils against *C. albicans*. Nat Prod Res 35(24):5997–6001

Antoniadou A, Torres HA, Lewis RE, Thornby J, Bodey GP, Tarrand JP et al (2003) Candidemia in a tertiary care cancer center: in vitro susceptibility and its association with outcome of initial antifungal therapy. Medicine 82(5):309–321

Arendrup MC (2010) Epidemiology of invasive candidiasis. Curr Opin Crit Care 16(5):445–452

Arendrup M, Horn T, Frimodt-Møller N (2002) In vivo pathogenicity of eight medically relevant Candida species in an animal model. Infection 30:286–291

Arendrup MC, Sulim S, Holm A, Nielsen L, Nielsen SD, Knudsen JD et al (2011) Diagnostic issues, clinical characteristics, and outcomes for patients with fungemia. J Clin Microbiol 49(9):3300–3308

Arendrup MC, Boekhout T, Akova M, Meis JF, Cornely OA, Lortholary O, ESCMID EFISG study group and ECMM (2014) ESCMID and ECMM joint clinical guidelines for the diagnosis and management of rare invasive yeast infections. Clin Microbiol Infect 20:76–98

Armstrong-James D, Brown GD, Netea MG, Zelante T, Gresnigt MS, van de Veerdonk FL, Levitz SM (2017) Immunotherapeutic approaches to treatment of fungal diseases. Lancet Infect Dis 17(12):e393–e402

Arora DS, Kaur J (1999) Antimicrobial activity of spices. Int J Antimicrob Agents 12(3):257–262

Asmundsdóttir LR, Erlendsdóttir H, Gottfredsson M (2002) Increasing incidence of candidemia: results from a 20-year nationwide study in Iceland. J Clin Microbiol 40(9):3489–3492

Bajpai VK, Kang S, Xu H, Lee SG, Baek KH, Kang SC (2011) Potential roles of essential oils on controlling plant pathogenic bacteria Xanthomonas species: a review. Plant Pathol J 27(3):207–224

Bakkali F, Averbeck S, Averbeck D, Idaomar M (2008) Biological effects of essential oils—a review. Food Chem Toxicol 46(2):446–475

Barbosa DHX, Gondim CR, Silva-Henriques MQ, Soares CS, Alves DN, Santos SG, Castro RD (2023) Coriandrumsativum L. essential oil obtained from organic culture shows antifungal activity against planktonic and multi-biofilm Candida. Brazil J Biol 83:e264875

Barnes PD, Marr KA (2006) Aspergillosis: spectrum of disease, diagnosis, and treatment. Infect Dis Clin 20(3):545–561

Bennett JH (1844) XVII. On the parasitic vegetable structures found growing in living animals. Earth Environ Sci Trans R Soc Edinb 15(2):277–294

Berger S, El Chazli Y, Babu AF, Coste AT (2017) Azole resistance in aspergillus fumigatus: a consequence of antifungal use in agriculture? Front Microbiol 8:1024

Biernasiuk A, Baj T, Malm A (2022) Clove essential oil and its main constituent, eugenol, as potential natural antifungals against *Candida* spp. alone or in combination with other antimycotics due to synergistic interactions. Molecules (Basel, Switzerland) 28(1):215

Bleasel N, Tate B, Rademaker M (2002) Allergic contact dermatitis following exposure to essential oils. Australas J Dermatol 43(3):211–213

Blostein F, Levin-Sparenberg E, Wagner J, Foxman B (2017) Recurrent vulvovaginal candidiasis. Ann Epidemiol 27(9):575–582

Bona E, Cantamessa S, Pavan M, Novello G, Massa N, Rocchetti A et al (2016) Sensitivity of *Candida albicans* to essential oils: are they an alternative to antifungal agents? J Appl Microbiol 121(6):1530–1545

Bongomin F, Gago S, Oladele RO, Denning DW (2017) Global and multi-national prevalence of fungal diseases—estimate precision. J Fungi 3(4):57

Bosetti D, Neofytos D (2023) Invasive Aspergillosis and the Impact of Azole-resistance. Current Fungal Infection Reports, 17(2):77–86

Boukhatem MN, Ferhat MA, Kameli A, Saidi F, Kebir HT (2014) Lemon grass (Cymbopogon citratus) essential oil as a potent anti-inflammatory and antifungal drugs. Libyan J Med 9(1):25431

Bowyer P, Moore CB, Rautemaa R, Denning DW, Richardson MD (2011) Azole antifungal resistance today: focus on *Aspergillus*. Curr Infect Dis Rep 13:485–491

Bozin B, Mimica-Dukic N, Samojlik I, Jovin E (2007) Antimicrobial and antioxidant properties of rosemary and sage (Rosmarinus officinalis L. and Salvia officinalis L., Lamiaceae) essential oils. J Agric Food Chem 55(19):7879–7885

Božović M, Garzoli S, Sabatino M, Pepi F, Baldisserotto A, Andreotti E et al (2017) Essential oil extraction, chemical analysis and anti-Candida activity of *Calamintha nepeta* (L.) Savi subsp. glandulosa (Req.) Ball—new approaches. Molecules 22(2):203

Brown GD, Denning DW, Gow NA, Levitz SM, Netea MG, White TC (2012) Hidden killers: human fungal infections. Sci Transl Med 4(165):165rv13

Burt SA, Reinders RD (2003) Antibacterial activity of selected plant essential oils against *Escherichia coli* O157: H7. Lett Appl Microbiol 36(3):162–167

Cannas S, Molicotti P, Usai D, Maxia A, Zanetti S (2014) Antifungal, anti-biofilm and adhesion activity of the essential oil of Myrtus communis L. against *Candida* species. Nat Prod Res 28(23):2173–2177

Carson CF, Hammer KA, Riley TV (2006) Melaleuca alternifolia (tea tree) oil: a review of antimicrobial and other medicinal properties. Clin Microbiol Rev 19(1):50–62

Carvalhinho S, Margarida A, Sampaio A (2012) Susceptibilities of *Candida albicans* mouth isolates to antifungal agents, essential oils and mouth rinses. Mycopathologia 174:69–76

Cavaleiro C, Pinto E, Gonçalves MJ, Salgueiro L (2006) Antifungal activity of Juniperus essential oils against dermatophyte, *Aspergillus* and *Candida* strains. J Appl Microbiol 100(6):1333–1338

Cavaleiro C, Salgueiro L, Gonçalves MJ, Hrimpeng K, Pinto J, Pinto E (2015) Antifungal activity of the essential oil of Angelica major against *Candida*, *Cryptococcus*, *Aspergillus* and dermatophyte species. J Nat Med 69(2):241–248

Chavan PS, Tupe SG (2014) Antifungal activity and mechanism of action of carvacrol and thymol against vineyard and wine spoilage yeasts. Food Control 46:115–120

Chedia A, Ghazghazi H, Dallali S, Houssine S, Brahim H, Abderazzak M (2013) Comparison of chemical composition, antioxidant and antimicrobial activities of Thymus capitatus L. essential oils from two Tunisian localities (Sousse and Bizerte). Int J Agron Plant Prod 4(8):1772–1781

Chen Y, Zeng H, Tian J, Ban X, Ma B, Wang Y (2013) Antifungal mechanism of essential oil from *Anethum graveolens* seeds against *Candida albicans*. J Med Microbiol 62(8):1175–1183

Chen P, Liu J, Zeng M, Sang H (2020) Exploring the molecular mechanism of azole resistance in *Aspergillus fumigatus*. J Mycol Med 30(1):100915

Cid-Chevecich C, Müller-Sepúlveda A, Jara JA, López-Muñoz R, Santander R, Budini M, Escobar A, Quijada R, Criollo A, Díaz-Dosque M, Molina-Berríos A (2022) *Origanum vulgare* L. essential oil inhibits virulence patterns of *Candida* spp. and potentiates the effects of fluconazole and nystatin in vitro. BMC Complement Med Ther 22(1):39

Cleveland AA, Harrison LH, Farley MM, Hollick R, Stein B, Chiller TM et al (2015) Declining incidence of candidemia and the shifting epidemiology of *Candida* resistance in two US metropolitan areas, 2008–2013: results from population-based surveillance. PLoS One 10(3):e0120452

Córdoba S, Vivot W, Szusz W, Albo G (2019) Antifungal activity of essential oils against Candida species isolated from clinical samples. Mycopathologia 184:615–623

Cowen LE, Sanglard D, Howard SJ, Rogers PD, Perlin DS (2015) Mechanisms of antifungal drug resistance. Cold Spring Harb Perspect Med 5(7):a019752

Crum CP, Hirsch MS, Peters WA III, Quick CM, Laury AR (2015) Gynecologic and obstetric pathology e-book: a volume in the high yield pathology series. Elsevier Health Sciences

Curvelo JAR, Marques AM, Barreto ALS, Romanos MTV, Portela MB, Kaplan MAC, Soares RMA (2014) A novel nerolidol-rich essential oil from Piper claussenianum modulates *Candida albicans* biofilm. J Med Microbiol 63(Pt 5):697–702

D'Arrigo M, Bisignano C, Irrera P, Smeriglio A, Zagami R, Trombetta D, Romeo O, Mandalari G (2019) In vitro evaluation of the activity of an essential oil from *Pistacia vera* L. variety Bronte hull against Candida sp. BMC Complement Alternat Med 19(1):6

de Aquino Lemos J, Passos XS, de Fátima Lisboa Fernandes O, de Paula JR, Ferri PH, Souza LKHE, de Aquino Lemos A, do Rosário Rodrigues Silva M (2005) Antifungal activity from *Ocimum gratissimum* L. towards *Cryptococcus neoformans*. Memorias do Instituto Oswaldo Cruz 100(1):55–58

de las Mercedes Oliva M, Carezzano ME, Gallucci MN, Demo MS (2011) Antimycotic effect of the essential oil of *Aloysia triphylla* against *Candida* species obtained from human pathologies. Nat Prod Commun 6(7):1039–1043

de Morais SR, Oliveira TL, Bara MT, da Conceição EC, Rezende MH, Ferri PH, de Paula JR (2012) Chemical constituents of essential oil from Lippiasidoides Cham. (Verbenaceae) leaves cultivated in Hidrolândia, Goiás, Brazil. Int J Anal Chem 2012:363919

de Morais SR, Oliveira TL, de Oliveira LP, Tresvenzol LM, da Conceição EC, Rezende MH, Fiuza TS, Costa EA, Ferri PH, de Paula JR (2016) Essential oil composition, antimicrobial and pharmacological activities of *Lippia sidoides* Cham. (Verbenaceae) from São Gonçalo do Abaeté, Minas Gerais, Brazil. Pharmacogn Mag 12(48):262–270

de Fátima Dantas de Almeida L, de Paula JF, de Almeida RVD, Williams DW, Hebling J, Cavalcanti YW (2016) Efficacy of citronella and cinnamon essential oils on *Candida albicans* biofilms. Acta Odontol Scand 74(5):393–398

de Rapper S, Van Vuuren SF, Kamatou GP, Viljoen AM, Dagne E (2012) The additive and synergistic antimicrobial effects of select frankincense and myrrh oils—a combination from the pharaonic pharmacopoeia. Lett Appl Microbiol 54(4):352–358

De Toledo LG, Ramos MA, Spósito L, Castilho EM, Pavan FR, De Oliveira Lopes É, Zocolo GJ, Silva FA, Soares TH, Dos Santos AG, Bauab TM, De Almeida MT (2016) Essential oil of

Cymbopogon nardus (L.) Rendle: a strategy to combat fungal infections caused by *Candida* species. Int J Mol Sci 17(8):1252

DeCherney, A. H. (2019). L. Nathan, N. Laufer, A. S. Roman AH DeCherney (Eds.)Current diagnosis & treatment: obstetrics & gynecology. McGraw-Hill Education

Denning DW (2001) Chronic forms of pulmonary aspergillosis. Clin Microbiol Infect 7:25–31

Denning DW, Riniotis K, Dobrashian R, Sambatakou H (2003) Chronic cavitary and fibrosing pulmonary and pleural aspergillosis: case series, proposed nomenclature change, and review. Clin Infect Dis 37(Suppl_3):S265–S280

Denning DW, Pleuvry A, Cole DC (2013) Global burden of allergic bronchopulmonary aspergillosis with asthma and its complication chronic pulmonary aspergillosis in adults. Med Mycol 51(4):361–370

Denning DW, Cadranel J, Beigelman-Aubry C, Ader F, Chakrabarti A, Blot S et al (2016) Chronic pulmonary aspergillosis: rationale and clinical guidelines for diagnosis and management. Eur Respir J 47(1):45–68

Deorukhkar SC, Saini S, Mathew S (2014) Non-albicans *Candida* infection: an emerging threat. Interdiscip Perspect Infect Dis 2014:615958

Di Vito M, Mattarelli P, Modesto M, Girolamo A, Ballardini M, Tamburro A, Meledandri M, Mondello F (2015) In vitro activity of tea tree oil vaginal suppositories against *Candida* spp. and probiotic vaginal microbiota. Phytother Res 29(10):1628–1633

Diekema DJ, Messer SA, Brueggemann AB, Coffman SL, Doern GV, Herwaldt LA, Pfaller MA (2002) Epidemiology of candidemia: 3-year results from the emerging infections and the epidemiology of Iowa organisms study. J Clin Microbiol 40(4):1298–1302

Donato R, Sacco C, Pini G, Bilia AR (2020) Antifungal activity of different essential oils against Malassezia pathogenic species. J Ethnopharmacol 249:112376

Dovnik A, Golle A, Novak D, Arko D, Takač I (2015) Treatment of vulvovaginal candidiasis: a review of the literature. Acta Dermatovenerol Alp Pannonica Adriat 24(1):5–7

Dubey P, Tripathi SC (1987) Studies on antifungal, physico-chemical and phytotoxic properties of the essential oil of Piper bette/Untersuchungen über die antimykotischen, physikochemischen und phytotoxischen Eigenschaften des ätherischen Öls von Piper betle. J Plant Dis Prot 94:235–241

Ebani VV, Nardoni S, Bertelloni F, Pistelli L, Mancianti F (2018) Antimicrobial activity of five essential oils against bacteria and fungi responsible for urinary tract infections. Molecules (Basel, Switzerland) 23(7):1668

Elisa B, Aldo A, Ludovica G, Viviana P, Debora B, Nadia M, Giorgia N, Elisa G (2021) Chemical composition and antimycotic activity of six essential oils (cumin, fennel, manuka, sweet orange, cedar and juniper) against different *Candida* spp. Nat Prod Res 35(22):4600–4605

Estes SA, Hendricks AA, Merz WG, Prystowsky SD (1980) Primary cutaneous aspergillosis. J Am Acad Dermatol 3(4):397–400

Fairn GD, Grinstein S (2012) How nascent phagosomes mature to become phagolysosomes. Trends Immunol 33(8):397–405

Feng W, Zheng X (2007) Essential oils to control Alternaria alternata in vitro and in vivo. Food Control 18(9):1126–1130

Ferguson BJ (2000) Fungus balls of the paranasal sinuses. Otolaryngol Clin N Am 33(2):389–398

Foxman B, Muraglia R, Dietz JP, Sobel JD, Wagner J (2013) Prevalence of recurrent vulvovaginal candidiasis in 5 European countries and the United States: results from an internet panel survey. J Low Genit Tract Dis 17(3):340–345

Fraczek MG, Bromley M, Buied A, Moore CB, Rajendran R, Rautemaa R et al (2013) The cdr1B efflux transporter is associated with non-cyp51a-mediated itraconazole resistance in *Aspergillus fumigatus*. J Antimicrob Chemother 68(7):1486–1496

Fraser RS, Müller NL, Colman N, Pare PD (1999) Fraser and Paré's diagnosis of diseases of the chest, vol 1–4, 4th edn. WB Saunders

Garcia-Rubio R, Alcazar-Fuoli L (2018) Diseases caused by Aspergillus fumigatus

Garcia-Rubio R, Cuenca-Estrella M, Mellado E (2017) Triazole resistance in Aspergillus species: an emerging problem. Drugs 77(6):599–613

Geiger AM, Foxman B (1996) Risk factors for vulvovaginal candidiasis: a case-control study among university students. Epidemiology 7(2):182–187

George IA, Spec A, Powderly WG, Santos CA (2018) Comparative epidemiology and outcomes of human immunodeficiency virus (HIV), non-HIV non-transplant, and solid organ transplant associated cryptococcosis: a population-based study. Clin Infect Dis 66(4):608–611

Ghannoum MA, Rice LB (1999) Antifungal agents: mode of action, mechanisms of resistance, and correlation of these mechanisms with bacterial resistance. Clin Microbiol Rev 12(4):501–517

Giordani R, Regli P, Kaloustian J, Mikail C, Abou L, Portugal H (2004) Antifungal effect of various essential oils against *Candida albicans*. Potentiation of antifungal action of amphotericin B by essential oil from Thymus vulgaris. Phytother Res 18(12):990–995

Glass D, Amedee RG (2011) Allergic fungal rhinosinusitis: a review. Ochsner J 11(3):271–275

Goldman JM, Murr AS, Cooper RL (2007) The rodent estrous cycle: characterization of vaginal cytology and its utility in toxicological studies. Birth Defects Res B Dev Reprod Toxicol 80(2):84–97

Gonçalves B, Ferreira C, Alves CT, Henriques M, Azeredo J, Silva S (2016) Vulvovaginal candidiasis: epidemiology, microbiology and risk factors. Crit Rev Microbiol 42(6):905–927

Guenther E (1952) The EOs, vol 3 and 4. D. Van Nostrand & Co., Inc., New York

Guinea J, Torres-Narbona M, Gijón P, Muñoz P, Pozo F, Peláez T et al (2010) Pulmonary aspergillosis in patients with chronic obstructive pulmonary disease: incidence, risk factors, and outcome. Clin Microbiol Infect 16(7):870–877

Guinea J, Zaragoza Ó, Escribano P, Martín-Mazuelos E, Pemán J, Sánchez-Reus F, Cuenca-Estrella M (2014) Molecular identification and antifungal susceptibility of yeast isolates causing fungemia collected in a population-based study in Spain in 2010 and 2011. Antimicrob Agents Chemother 58(3):1529–1537

Gupta M, Weinberger B, Whitley-Williams PN (1996) Cutaneous aspergillosis in a neonate. Pediatr Infect Dis J 15(5):464–465

Hacioglu M, Oyardi O, Kirinti A (2021) Oregano essential oil inhibits *Candida* spp. biofilms. Z Nat C 76(11–12):443–450

Hagiwara D, Watanabe A, Kamei K, Goldman GH (2016) Epidemiological and genomic landscape of azole resistance mechanisms in Aspergillus fungi. Front Microbiol 7:1382

Hammer KA, Carson CF, Riley TV (1998) In-vitro activity of essential oils, in particular *Melaleuca alternifolia* (tea tree) oil and tea tree oil products, against *Candida* spp. J Antimicrob Chemother 42(5):591–595

Hammer KA, Carson CF, Riley TV (1999) Antimicrobial activity of essential oils and other plant extracts. J Appl Microbiol 86(6):985–990

Hammer KA, Carson CF, Riley TV (2004) Antifungal effects of *Melaleuca alternifolia* (tea tree) oil and its components on *Candida albicans*, *Candida glabrata* and *Saccharomyces cerevisiae*. J Antimicrob Chemother 53(6):1081–1085

Havlickova B, Czaika VA, Friedrich M (2008) Epidemiological trends in skin mycoses worldwide. Mycoses 51:2–15

He X, Ma Y, Yi G, Wu J, Zhou L, Guo H (2017) Chemical composition and antifungal activity of Carica papaya Linn.Seed essential oil against Candida spp. Lett Appl Microbiol 64(5):350–354

Hong E, Dixit S, Fidel PL, Bradford J, Fischer G (2014) Vulvovaginal candidiasis as a chronic disease: diagnostic criteria and definition. J Low Genit Tract Dis 18(1):31–38

Horowitz BJ, Edelstein SW, Lippman LEONARD (1987) Sexual transmission of *Candida*. Obstet Gynecol 69(6):883–886

Hosni K, Hassen I, Chaâbane H, Jemli M, Dallali S, Sebei H, Casabianca H (2013) Enzyme-assisted extraction of essential oils from thyme (*Thymus capitatus* L.) and rosemary (*Rosmarinus officinalis* L.): impact on yield, chemical composition and antimicrobial activity. Ind Crop Prod 47:291–299

Hossain F, Follett P, Vu KD, Harich M, Salmieri S, Lacroix M (2016) Evidence for synergistic activity of plant-derived essential oils against fungal pathogens of food. Food Microbiol 53:24–30

Hovijitra RS, Choonharuangdej S, Srithavaj T (2016) Effect of essential oils prepared from Thai culinary herbs on sessile *Candida albicans* cultures. J Oral Sci 58(3):365–371

Hu Y, Zhang J, Kong W, Zhao G, Yang M (2017) Mechanisms of antifungal and anti-aflatoxigenic properties of essential oil derived from turmeric (*Curcuma longa* L.) on *Aspergillus flavus*. Food Chem 220:1–8

Hyldgaard M, Mygind T, Meyer RL (2012) Essential oils in food preservation: mode of action, synergies, and interactions with food matrix components. Front Microbiol 3:12

Ilkit M, Guzel AB (2011) The epidemiology, pathogenesis, and diagnosis of vulvovaginal candidosis: a mycological perspective. Crit Rev Microbiol 37(3):250–261

Iyer KR, Revie NM, Fu C, Robbins N, Cowen LE (2021) Treatment strategies for cryptococcal infection: challenges, advances and future outlook. Nat Rev Microbiol 19(7):454–466

Jack D, Sobel MD (1997) Vaginitis. N Engl J Med 337(26):1896–1903

Jafri H, Ahmad I (2020) Thymus vulgaris essential oil and thymol inhibit biofilms and interact synergistically with antifungal drugs against drug resistant strains of *Candida albicans* and Candida tropicalis. J Mycol Med 30(1):100911

Jeanvoine A, Rocchi S, Bellanger AP, Reboux G, Millon L (2020) Azole-resistant Aspergillus fumigatus: A global phenomenon originating in the environment? Med Mal Infect 50(5):389–395

Kao AS, Brandt ME, Pruitt WR, Conn LA, Perkins BA, Stephens DS et al (1999) The epidemiology of candidemia in two United States cities: results of a population-based active surveillance. Clin Infect Dis 29(5):1164–1170

Karaman M, Bogavac M, Radovanović B, Sudji J, Tešanović K, Janjušević L (2017) Origanum vulgare essential oil affects pathogens causing vaginal infections. J Appl Microbiol 122(5):1177–1185

Karpiński TM, Ożarowski M, Seremak-Mrozikiewicz A, Wolski H (2023) Anti-*Candida* and antibiofilm activity of selected *Lamiaceae* essential oils. Front Biosci (Landmark Ed) 28(2):28

Kaskatepe B, Aslan Erdem S, Ozturk S, Safi Oz Z, Subasi E, Koyuncu M, Vlainić J, Kosalec I (2022) Antifungal and anti-virulent activity of *Origanum majorana* L. essential oil on *Candida albicans* and in vivo toxicity in the *galleria mellonella* larval model. Molecules (Basel, Switzerland) 27(3):663

Kazemi M, Rostami H (2015) Chemical composition and biological activities of Iranian *Achillea wilhelmsii* L. essential oil: a high effectiveness against *Candida* spp. and *Escherichia* strains. Nat Prod Res 29(3):286–288

Kelmenson VA (1959) Treatment of pulmonary aspergillosis. Dis Chest 36(4):442–443

Kohiyama CY, Ribeiro MMY, Mossini SAG, Bando E, da Silva Bomfim N, Nerilo SB et al (2015) Antifungal properties and inhibitory effects upon aflatoxin production of *Thymus vulgaris* L. by *Aspergillus flavus* Link. Food Chem 173:1006–1010

Kosalec I, Pepeljnjak S, Kustrk D (2005) Antifungal activity of fluid extract and essential oil from anise fruits (*Pimpinella anisum* L., Apiaceae). Acta Pharma 55:377–385

Kullberg BJ, Sobel JD, Ruhnke M, Pappas PG, Viscoli C, Rex JH et al (2005) Voriconazole versus a regimen of amphotericin B followed by fluconazole for candidaemia in non-neutropenic patients: a randomised non-inferiority trial. Lancet 366(9495):1435–1442

Kullberg BJ, van de Veerdonk F, Netea MG (2014) Immunotherapy: a potential adjunctive treatment for fungal infection. Curr Opin Infect Dis 27(6):511–516

Kumar A, Shukla R, Singh P, Dubey NK (2010) Chemical composition, antifungal and antiaflatoxigenic activities of *Ocimum sanctum* L. essential oil and its safety assessment as plant based antimicrobial. Food Chem Toxicol 48(2):539–543

Kwon-Chung KJ, Fraser JA, Doering TL, Wang ZA, Janbon G, Idnurm A, Bahn YS (2014) *Cryptococcus neoformans* and *Cryptococcus gattii*, the etiologic agents of cryptococcosis. Cold Spring Harb Perspect Med 4(7):a019760

Lang G, Buchbauer G (2012) A review on recent research results (2008–2010) on essential oils as antimicrobials and antifungals. A review. Flavour Fragr J 27(1):13–39

Latgé JP (2001) The pathobiology of *Aspergillus fumigatus*. Trends Microbiol 9(8):382–389

Latgé JP (2003) *Aspergillus fumigatus*, a saprotrophic pathogenic fungus. Mycologist 17(2):56–61

Lee SH, Lee BJ, Kim JH, Sohn DS, Shin JW, Kim JY et al (2004) Clinical manifestations and treatment outcomes of pulmonary aspergilloma. Korean J Intern Med 19(1):38

Lelièvre L, Groh M, Angebault C, Maherault AC, Didier E, Bougnoux ME (2013) Azole resistant *Aspergillus fumigatus*: an emerging problem. Med Mal Infect 43(4):139–145

Li Y, Kong W, Li M, Liu H, Zhao X, Yang S, Yang M (2016) *Litsea cubeba* essential oil as the potential natural fumigant: inhibition of *Aspergillus flavus* and AFB1 production in licorice. Ind Crop Prod 80:186–193

Li H, Kong Y, Hu W, Zhang S, Wang W, Yang M, Luo Y (2022) *Litsea cubeba* essential oil: component analysis, anti-*Candida albicans* activity and mechanism based on molecular docking. J Oleo Sci 71(8):1221–1228

Limper AH, Adenis A, Le T, Harrison TS (2017) Fungal infections in HIV/AIDS. Lancet Infect Dis 17(11):e334–e343

Lopez-Reyes JG, Spadaro D, Prelle A, Garibaldi A, Gullino ML (2013) Efficacy of plant essential oils on postharvest control of rots caused by fungi on different stone fruits in vivo. J Food Prot 76(4):631–639

Loyse A, Burry J, Cohn J, Ford N, Chiller T, Ribeiro I et al (2019) Leave no one behind: response to new evidence and guidelines for the management of cryptococcal meningitis in low-income and middle-income countries. Lancet Infect Dis 19(4):e143–e147

Macwan SR, Dabhi BK, Aparnathi KD, Prajapati JB (2016) Essential oils of herbs and spices: their antimicrobial activity and application in preservation of food. Int J Curr Microbiol App Sci 5(5):885–901

Maghrabi F, Denning DW (2017) The management of chronic pulmonary aspergillosis: the UK national aspergillosis centre approach. Curr Fungal Infect Rep 11:242–251

Magill SS, Edwards JR, Bamberg W, Beldavs ZG, Dumyati G, Kainer MA et al (2014) Multistate point-prevalence survey of health care–associated infections. N Engl J Med 370(13):1198–1208

Malani PN, Bradley SF, Little RS, Kauffman CA (2001) Trends in species causing fungaemia in a tertiary care medical centre over 12 years. Mycoses 44(11–12):446–449

Man A, Santacroce L, Jacob R, Mare A, Man L (2019) Antimicrobial activity of six essential oils against a group of human pathogens: a comparative study. Pathogens (Basel, Switzerland) 8(1):15

Mandras N, Nostro A, Roana J, Scalas D, Banche G, Ghisetti V, Del Re S, Fucale G, Cuffini AM, Tullio V (2016) Liquid and vapour-phase antifungal activities of essential oils against *Candida albicans* and non-albicans *Candida*. BMC Complement Alternat Med 16(1):330

Manohar V, Ingram C, Gray J, Talpur NA, Echard BW, Bagchi D, Preuss HG (2001) Antifungal activities of *origanum* oil against *Candida albicans*. Mol Cell Biochem 228(1–2):111–117

Marr KA, Seidel K, White TC, Bowden RA (2000) Candidemia in allogeneic blood and marrow transplant recipients: evolution of risk factors after the adoption of prophylactic fluconazole. J Infect Dis 181(1):309–316

Marr KA, Carter RA, Boeckh M, Martin P, Corey L (2002) Invasive aspergillosis in allogeneic stem cell transplant recipients: changes in epidemiology and risk factors. Blood 100(13):4358–4366

Martin GS, Mannino DM, Eaton S, Moss M (2003) The epidemiology of sepsis in the United States from 1979 through 2000. N Engl J Med 348(16):1546–1554

Martins AP, Salgueiro LR, Gonçalves MJ, Proença da Cunha A, Vila R, Cañigueral S (2003) Essential oil composition and antimicrobial activity of *Santiria trimera* bark. Planta Med 69(1):77–79

Masango P (2005) Cleaner production of essential oils by steam distillation. J Clean Prod 13(8):833–839

McCarty TP, Pappas PG (2016) Invasive candidiasis. Infect Dis Clin 30(1):103–124

Mekem Sonwa M (2000) Isolation and structure elucidation of essential oil constituents: comparative study of the oils of Cyperus alopecuroides, Cyperus papyrus, and Cyperus rotundus, Doctoral dissertation. Staats-und Universitätsbibliothek Hamburg Carl von Ossietzky

Mijatovic S, Stankovic JA, Calovski IC, Dubljanin E, Pljevljakusic D, Bigovic D, Dzamic A (2022) Antifungal activity of *Lavandula angustifolia* essential oil against *Candida albicans*: time-kill study on pediatric sputum isolates. Molecules (Basel, Switzerland) 27(19):6300

Mitchell H (2004) Vaginal discharge—causes, diagnosis, and treatment. BMJ 328(7451):1306–1308
Moiteiro C, Esteves T, Ramalho L, Rojas R, Alvarez S, Zacchino S, Bragança H (2013) Essential oil characterization of two Azorean Cryptomeria japonica populations and their biological evaluations. Nat Prod Commun 8(12):1785–1790
Mondello F, De Bernardis F, Girolamo A, Salvatore G, Cassone A (2003) In vitro and in vivo activity of tea tree oil against azole-susceptible and -resistant human pathogenic yeasts. J Antimicrob Chemother 51(5):1223–1229
Morgan J, Meltzer MI, Plikaytis BD, Sofair AN, Huie-White S, Wilcox S et al (2005) Excess mortality, hospital stay, and cost due to candidemia: a case-control study using data from population-based candidemia surveillance. Infect Control Hosp Epidemiol 26(6):540–547
Mourad A, Perfect JR (2018) Present and future therapy of *Cryptococcus* infections. J Fungi 4(3):79
Natrajan D, Srinivasan S, Sundar K, Ravindran A (2015) Formulation of essential oil-loaded chitosan–alginate nanocapsules. J Food Drug Anal 23(3):560–568
Naveen KV, Saravanakumar K, Sathiyaseelan A, MubarakAli D, Wang MH (2022) Human fungal infection, immune response, and clinical challenge—a perspective during COVID-19 pandemic. Appl Biochem Biotechnol 194(9):4244–4257
Nazzaro F, Fratianni F, Coppola R, De Feo V (2017) Essential oils and antifungal activity. Pharmaceuticals 10(4):86
Nguefack J, Leth V, Zollo PA, Mathur SB (2004) Evaluation of five essential oils from aromatic plants of Cameroon for controlling food spoilage and mycotoxin producing fungi. Int J Food Microbiol 94(3):329–334
Nidhi P, Rolta R, Kumar V, Dev K, Sourirajan A (2020) Synergistic potential of Citrus aurantium L. essential oil with antibiotics against *Candida albicans*. J Ethnopharmacol 262:113135
Nucci M, Anaissie E, Betts RF, Dupont BF, Wu C, Buell DN et al (2010) Early removal of central venous catheter in patients with candidemia does not improve outcome: analysis of 842 patients from 2 randomized clinical trials. Clin Infect Dis 51(3):295–303
Oguma T, Taniguchi M, Shimoda T, Kamei K, Matsuse H, Hebisawa A et al (2018) Allergic bronchopulmonary aspergillosis in Japan: a nationwide survey. Allergol Int 67(1):79–84
Oliveira Ribeiro S, Fontaine V, Mathieu V, Zhiri A, Baudoux D, Stévigny C, Souard F (2020) Antibacterial and cytotoxic activities of ten commercially available essential oils. Antibiotics 9(10):717
Omidbeygi M, Barzegar M, Hamidi Z, Naghdibadi H (2007) Antifungal activity of thyme, summer savory and clove essential oils against *Aspergillus flavus* in liquid medium and tomato paste. Food Control 18(12):1518–1523
Paiva LF, Teixeira-Loyola ABA, Schnaider TB, Souza AC, Lima LMZ, Dias DR (2022) Association of the essential oil of *Cymbopogon citratus* (DC) Stapf with nystatin against oral cavity yeasts. Anais da Academia Brasileira de Ciencias 94(1):e20200681
Panjabi C, Shah A (2011) Allergic *Aspergillus sinusitis* and its association with allergic bronchopulmonary aspergillosis. Asia Pac Allergy 1(3):130–137
Pappas PG (2006) Invasive candidiasis. Infect Dis Clin 20(3):485–506
Pappas PG, Rex JH, Lee J, Hamill RJ, Larsen RA, Powderly W et al (2003) A prospective observational study of candidemia: epidemiology, therapy, and influences on mortality in hospitalized adult and pediatric patients. Clin Infect Dis 37(5):634–643
Patterson TF, Thompson GR III, Denning DW, Fishman JA, Hadley S, Herbrecht R et al (2016) Practice guidelines for the diagnosis and management of aspergillosis: 2016 update by the Infectious Diseases Society of America. Clin Infect Dis 63(4):e1–e60
Pavela R, Benelli G (2016) Essential oils as ecofriendly biopesticides? Challenges and constraints. Trends Plant Sci 21(12):1000–1007
Peixoto LR, Rosalen PL, Ferreira GL, Freires IA, de Carvalho FG, Castellano LR, de Castro RD (2017) Antifungal activity, mode of action and anti-biofilm effects of *Laurus nobilis Linnaeus* essential oil against *Candida* spp. Arch Oral Biol 73:179–185
Perfect JR, Dismukes WE, Dromer F, Goldman DL, Graybill JR, Hamill RJ et al (2010) Clinical practice guidelines for the management of cryptococcal disease: 2010 update by the Infectious Diseases Society of America. Clin Infect Dis 50(3):291–322

Pfaller MA, Diekema D (2007) Epidemiology of invasive candidiasis: a persistent public health problem. Clin Microbiol Rev 20(1):133–163

Pina-Vaz C, Goncalves Rodrigues A, Pinto E, Costa-de-Oliveira S, Tavares C, Salgueiro L, Cavaleiro C, Goncalves MJ et al (2004) Antifungal activity of thymus oils and their major compounds. J Eur Acad Dermatol Venereol 18:73–78

Pinto E, Pina-Vaz C, Salgueiro L, Gonçalves MJ, Costa-de-Oliveira S, Cavaleiro C, Palmeira A, Rodrigues A, Martinez-de-Oliveira J (2006) Antifungal activity of the essential oil of *Thymus pulegioides* on *Candida*, *Aspergillus* and dermatophyte species. J Med Microbiol 55(Pt 10):1367–1373

Pinto E, Hrimpeng K, Lopes G, Vaz S, Gonçalves MJ, Cavaleiro C, Salgueiro L (2013) Antifungal activity of *Ferulago capillaris* essential oil against *Candida*, *Cryptococcus*, *Aspergillus* and dermatophyte species. Eur J Clin Microbiol Infect Dis 32(10):1311–1320

Pinto E, Gonçalves MJ, Cavaleiro C, Salgueiro L (2017) Antifungal activity of Thapsiavillosa essential oil against *Candida*, *Cryptococcus*, *Malassezia*, *Aspergillus* and dermatophyte species. Molecules (Basel, Switzerland) 22(10):1595

Piras A, Cocco V, Falconieri D, Porcedda S, Marongiu B, Maxia A, Frau MA, Gonçalves MJ, Cavaleiro C, Salgueiro L (2011) Isolation of the volatile oil from *Satureja thymbra* by supercritical carbon dioxide extraction: chemical composition and biological activity. Nat Prod Commun 6(10):1523–1526

Pires RH, Montanari LB, Martins CH, Zaia JE, Almeida AM, Matsumoto MT, Mendes-Giannini MJ (2011) Anticandidal efficacy of cinnamon oil against planktonic and biofilm cultures of *Candida parapsilosis* and *Candida orthopsilosis*. Mycopathologia 172(6):453–464

Powers CN, Satyal P, Mayo JA, McFeeters H, McFeeters RL (2019) Bigger data approach to analysis of essential oils and their antifungal activity against *Aspergillus niger*, *Candida albicans*, and *Cryptococcus neoformans*. Molecules (Basel, Switzerland) 24(16):2868

Pozzatti P, Loreto ES, Mario DN, Rossato L, Santurio JM, Alves SH (2010) Activities of essential oils in the inhibition of *Candida albicans* and *Candida dubliniensis* germ tube formation. J Mycol Med 20(3):185–189

Prakash B, Shukla R, Singh P, Kumar A, Mishra PK, Dubey NK (2010) Efficacy of chemically characterized Piper betle L. essential oil against fungal and aflatoxin contamination of some edible commodities and its antioxidant activity. Int J Food Microbiol 142(1–2):114–119

Price CL, Parker JE, Warrilow AG, Kelly DE, Kelly SL (2015) Azole fungicides—understanding resistance mechanisms in agricultural fungal pathogens: mode of action and resistance mechanisms to azole fungicides. Pest ManagSci 71(8):1054–1058

Rajasingham R, Smith RM, Park BJ, Jarvis JN, Govender NP, Chiller TM et al (2017) Global burden of disease of HIV-associated cryptococcal meningitis: an updated analysis. Lancet Infect Dis 17(8):873–881

Rath CC, Mohapatra S (2015) Susceptibility characterisation of Candida spp. to four essential oils. Indian J Med Microbiol 33(Suppl):93–96

Raut A, Huy NT (2021) Rising incidence of mucormycosis in patients with COVID-19: another challenge for India amidst the second wave? Lancet Respir Med 9(8):e77

Raut JS, Karuppayil SM (2014) A status review on the medicinal properties of essential oils. Ind Crop Prod 62:250–264

Ren J, Jin X, Zhang Q, Zheng Y, Lin D, Yu Y (2017) Fungicides induced triazole-resistance in *Aspergillus fumigatus* associated with mutations of TR46/Y121F/T289A and its appearance in agricultural fields. J Hazard Mater 326:54–60

Resendiz Sharpe A, Lagrou K, Meis JF, Chowdhary A, Lockhart SR, Verweij PE, ISHAM/ECMM Aspergillus Resistance Surveillance Working Group (2018) Triazole resistance surveillance in *Aspergillus fumigatus*. Med Mycol 56(Suppl_1):S83–S92

Rinaldi MG (1983) Invasive aspergillosis. Rev Infect Dis 5(6):1061–1077

Ríos JL (2016) Essential oils: what they are and how the terms are used and defined. In: Essential oils in food preservation, flavor and safety. Academic Press, pp 3–10

Rippon JW (1982) Medical mycology: the pathogenic fungi and the pathogenic actinomycetes. WB Saunders, London, pp 154–248

Rivera-Yañez CR, Terrazas LI, Jimenez-Estrada M, Campos JE, Flores-Ortiz CM, Hernandez LB, Cruz-Sanchez T, Garrido-Fariña GI, Rodriguez-Monroy MA, Canales-Martinez MM (2017) Anti-Candida activity of *Bursera morelensis* Ramirez essential oil and two compounds, α-pinene and γ-terpinene-an in vitro study. Molecules (Basel, Switzerland) 22(12):2095

Rivero-Menendez O, Alastruey-Izquierdo A, Mellado E, Cuenca-Estrella M (2016) Triazole resistance in *Aspergillus* spp.: a worldwide problem? J Fungi 2(3):21

Rudzki E, Grzywa Z (1985) The value of a mixture of cassia and citronella oils for detection of hypersensitivity of essential oils. Derm Beruf Umwelt (Occup Environ) 33(2):59–62

Salgueiro LR, Cavaleiro C, Pinto E, Pina-Vaz C, Rodrigues AG, Palmeira A et al (2003) Chemical composition and antifungal activity of the essential oil of *Origanum virens* on *Candida* species. Planta Med 69(09):871–874

Sanguinetti M, Posteraro B, Romano L, Battaglia F, Lopizzo T, De Carolis E, Fadda G (2007) In vitro activity of *Citrus bergamia* (bergamot) oil against clinical isolates of dermatophytes. J Antimicrob Chemother 59:305–308

Santomauro F, Donato R, Sacco C, Pini G, Flamini G, Bilia AR (2016) Vapour and liquid-phase Artemisia annua essential oil activities against several clinical strains of Candida. Planta Med 82(11–12):1016–1020

Sarkic A, Stappen I (2018) Essential oils and their single compounds in cosmetics—a critical review. Cosmetics 5(1):11

Scalas D, Mandras N, Roana J, Tardugno R, Cuffini AM, Ghisetti V, Benvenuti S, Tullio V (2018) Use of *Pinus sylvestris* L. (Pinaceae), *Origanum vulgare* L. (Lamiaceae), and *Thymus vulgaris* L. (Lamiaceae) essential oils and their main components to enhance itraconazole activity against azole susceptible/not-susceptible *Cryptococcus neoformans* strains. BMC Complement Alternat Med 18(1):143

Scazzocchio F, Garzoli S, Conti C, Leone C, Renaioli C, Pepi F, Angiolella L (2016) Properties and limits of some essential oils: chemical characterisation, antimicrobial activity, interaction with antibiotics and cytotoxicity. Nat Prod Res 30(17):1909–1918

Schwiertz A, Duttke C, Hild J, Müller HJ (2006) In vitro activity of essential oils on microorganisms isolated from vaginal infections. Int J Aromather 16(3–4):169–174

Sharma N, Tripathi A (2006) Fungitoxicity of the essential oil of Citrus sinensis on post-harvest pathogens. World J Microbiol Biotechnol 22:587–593

Sharma C, Hagen F, Moroti R, Meis JF, Chowdhary A (2015) Triazole-resistant Aspergillus fumigatus harbouring G54 mutation: is it de novo or environmentally acquired? J Glob Antimicrob Resist 3(2):69–74

Sharma R et al (2019) Therapeutic potential of citronella essential oil: a review. Curr Drug Discov Technol 16(4):330–339

Silva F, Ferreira S, Duarte A, Mendonça DI, Domingues FC (2011) Antifungal activity of *Coriandrum sativum* essential oil, its mode of action against *Candida* species and potential synergism with amphotericin B. Phytomedicine 19(1):42–47

Smith NL, Denning DW (2011) Underlying conditions in chronic pulmonary aspergillosis including simple aspergilloma. Eur Respir J 37(4):865–872

Smith LM, May RC (2013) Mechanisms of microbial escape from phagocyte killing. Biochem Soc Trans 41(2):475–490

Smith LM, Dixon EF, May RC (2015) The fungal pathogen *Cryptococcus neoformans* manipulates macrophage phagosome maturation. Cell Microbiol 17(5):702–713

Sobel JD (1985) Epidemiology and pathogenesis of recurrent vulvovaginal candidiasis. Am J Obstet Gynecol 152(7):924–935

Sobel JD (2007) Vulvovaginal candidosis. Lancet 369(9577):1961–1971

Sobel JD (2016) Recurrent vulvovaginal candidiasis. Am J Obstet Gynecol 214(1):15–21

Son SH, Kim JE, Oh SS, Lee JY (2020) Engineering cell wall integrity enables enhanced squalene production in yeast. J Agric Food Chem 68(17):4922–4929

Sookto T, Srithavaj T, Thaweboon S, Thaweboon B, Shrestha B (2013) In vitro effects of Salvia officinalis L. essential oil on *Candida albicans*. Asian Pac J Trop Biomed 3(5):376–380

Stevens DA, Kan VL, Judson MA, Morrison VA, Dummer S, Denning DW et al (2000) Practice guidelines for diseases caused by Aspergillus. Clin Infect Dis 30:696–709

Takahashi M, Inouye S, Abe S (2011) Anti-Candida and radical scavenging activities of essential oils and oleoresins of *Zingiber officinale* Roscoe and essential oils of other plants belonging to the family Zingiberaceae. Drug Discov Therap 5(5):238–245

Tampieri MP, Galuppi R, Macchioni F, Carelle MS, Falcioni L, Cioni PL, Morelli I (2005) The inhibition of *Candida albicans* by selected essential oils and their major components. Mycopathologia 159:339–345

Tatsadjieu NL, Dongmo PJ, Ngassoum MB, Etoa FX, Mbofung CMF (2009) Investigations on the essential oil of *Lippia rugosa* from Cameroon for its potential use as antifungal agent against *Aspergillus flavus* Link ex. Fries. Food Control 20(2):161–166

Taylor-Smith LM, May RC (2016) New weapons in the Cryptococcus infection toolkit. Curr Opin Microbiol 34:67–74

Thomas J, Jacobson GA, Narkowicz CK, Peterson GM, Burnet H, Sharpe C (2010) Toenail onychomycosis: an important global disease burden. J Clin Pharm Ther 35(5):497–519

Tian J, Ban X, Zeng H, He J, Huang B, Wang Y (2011a) Chemical composition and antifungal activity of essential oil from *Cicuta virosa* L. var. *latisecta Celak.* Int J Food Microbiol 145(2–3):464–470

Tian J, Ban X, Zeng H, Huang B, He J, Wang Y (2011b) In vitro and in vivo activity of essential oil from dill (*Anethum graveolens* L.) against fungal spoilage of cherry tomatoes. Food Control 22(12):1992–1999

Tongnuanchan P, Benjakul S (2014) Essential oils: extraction, bioactivities, and their uses for food preservation. J Food Sci 79(7):R1231–R1249

Tortorano AM, Peman J, Bernhardt H, Klingspor L, Kibbler CC, Faure O et al (2004) Epidemiology of candidaemia in Europe: results of 28-month European Confederation of Medical Mycology (ECMM) hospital-based surveillance study. Eur J Clin Microbiol Infect Dis 23:317–322

Tran HNH, Graham L, Adukwu EC (2020) In vitro antifungal activity of *Cinnamomum zeylanicum* bark and leaf essential oils against *Candida albicans* and *Candida auris*. Appl Microbiol Biotechnol 104(20):8911–8924

Trick WE, Fridkin SK, Edwards JR, Hajjeh RA, Gaynes RP, National Nosocomial Infections Surveillance System Hospitals (2002) Secular trend of hospital-acquired candidemia among intensive care unit patients in the United States during 1989–1999. Clin Infect Dis 35(5):627–630

Tzortzakis NG, Economakis CD (2007) Antifungal activity of lemongrass (*Cympopogon citratus* L.) essential oil against key postharvest pathogens. Innovative Food Sci Emerg Technol 8(2):253–258

Vági E, Simándi B, Suhajda A, Hethelyi E (2005) Essential oil composition and antimicrobial activity of *Origanum majorana* L. extracts obtained with ethyl alcohol and supercritical carbon dioxide. Food Res Int 38(1):51–57

Vale-Silva LA, Gonçalves MJ, Cavaleiro C, Salgueiro L, Pinto E (2010) Antifungal activity of the essential oil of *Thymus* x *viciosoi* against *Candida*, *Cryptococcus*, *Aspergillus* and dermatophyte species. Plantamedica 76(9):882–888

Van Burik JAH, Colven R, Spach DH (1998a) Cutaneous aspergillosis. J Clin Microbiol 36(11):3115–3121

Van Burik JAH, Colven R, Spach DH (1998b) Itraconazole therapy for primary cutaneous aspergillosis in patients with AIDS. Clin Infect Dis 27(3):643–644

Velázquez-Nuñez MJ, Avila-Sosa R, Palou E, López-Malo A (2013) Antifungal activity of orange (*Citrus sinensis* var. Valencia) peel essential oil applied by direct addition or vapor contact. Food Control 31(1):1–4

Ventoulis I, Sarmourli T, Amoiridou P, Mantzana P, Exindari M, Gioula G, Vyzantiadis TA (2020) Bloodstream infection by *Saccharomyces cerevisiae* in two COVID-19 patients after receiving supplementation of Saccharomyces in the ICU. J Fungi 6(3):98

Viuda-Martos M, Ruiz-Navajas Y, Fernández-López J, Pérez-Álvarez J (2008) Antifungal activity of lemon (*Citrus lemon* L.), mandarin (*Citrus reticulata* L.), grapefruit (*Citrus paradisi* L.) and orange (*Citrus sinensis* L.), essential oils. Food Control 19(12):1130–1138

Watson MC, Grimshaw JM, Bond CM, Mollison J, Ludbrook A (2002) Oral versus intra-vaginal imidazole and triazole anti-fungal agents for the treatment of uncomplicated vulvovaginal candidiasis (thrush): a systematic review. BJOG Int J Obstet Gynaecol 109(1):85–95

Weinstein RA, Fridkin SK (2005) The changing face of fungal infections in health care settings. Clin Infect Dis 41(10):1455–1460

Wenzel RP, Edmond MB (2001) The impact of hospital-acquired bloodstream infections. Emerg Infect Dis 7(2):174

Wijesinghe GK, Maia FC, de Oliveira TR, de Feiria SNB, Joia F, Barbosa JP, Boni GC, Sardi JCO, Rosalen PL, Höfling JF (2020) Effect of *Cinnamomum verum* leaf essential oil on virulence factors of *Candida* species and determination of the in-vivo toxicity with *Galleria mellonella* model. Memorias do Instituto Oswaldo Cruz 115:e200349

Wisplinghoff H, Bischoff T, Tallent SM, Seifert H, Wenzel RP, Edmond MB (2004) Nosocomial bloodstream infections in US hospitals: analysis of 24,179 cases from a prospective nationwide surveillance study. Clin Infect Dis 39(3):309–317

Zhai B, Wozniak KL, Masso-Silva J, Upadhyay S, Hole C, Rivera A et al (2015) Development of protective inflammation and cell-mediated immunity against *Cryptococcus neoformans* after exposure to hyphal mutants. MBio 6(5):10–1128

Unveiling the Potential of Essential Oils as Antifungal Agents Against Non-albicans *Candida* Species: Mechanisms of Action and Therapeutic Implications

6

Aimee Piketh, Vartika Srivastava, and Aijaz Ahmad

6.1 Introduction

Globally, fungal pathogens are responsible for approximately 13 million infections and more than 1.5 million deaths every year (Rayens and Norris 2022). Fungal diseases began to increase in severity and prevalence during the second half of the twentieth century. The rapid increase in the number of infectious cases leads to these microorganisms being classified as an emerging, major public health concerns. Several factors were hypothesised as being responsible for this complex and multifaceted phenomenon, namely (Rayens and Norris 2022):

1. An increase in international human migration has facilitated and encouraged the spread of pathogenic fungi to a variety of new regions.

A. Piketh
Department of Clinical Microbiology and Infectious Diseases, School of Pathology, Faculty of Health Sciences, University of the Witwatersrand, Johannesburg, South Africa
e-mail: aimee.piketh@gmail.com

V. Srivastava
Department of Clinical Microbiology and Infectious Diseases, School of Pathology, Faculty of Health Sciences, University of the Witwatersrand, Johannesburg, South Africa

Department of Inflammation and Immunity, Lerner Research Institute, Cleveland Clinic, Cleveland, OH, USA
e-mail: vartika.srivastava@wits.ac.za

A. Ahmad (✉)
Department of Clinical Microbiology and Infectious Diseases, School of Pathology, Faculty of Health Sciences, University of the Witwatersrand, Johannesburg, South Africa

Division of Infection Control, Charlotte Maxeke Johannesburg Academic Hospital, National Health Laboratory Service, Johannesburg, South Africa
e-mail: Aijaz.Ahmad@wits.ac.za; Aijaz.Ahmad@nhls.ac.za

N. Manzoor (ed.), *Advances in Antifungal Drug Development*,
https://doi.org/10.1007/978-981-97-5165-5_6

2. Factors such as major advancements in medical technology and practices (outbreaks caused by nosocomial infections in large healthcare facilities, chemotherapy, immunosuppressive drugs and organ transplants), the HIV/AIDS epidemic and an ageing global population are thought to all have contributed to the rise in fungal infections as a result of the increased number of immunocompromised patients. Because fungal infections are opportunistic infections, patients with compromised immune systems are more susceptible to acquiring fungal-associated diseases, as well as suffering from severe cases of these infections.
3. The onset of resistance to antifungal agents, along with antibacterial and antiviral resistance, has sky-rocketed since the mid-twentieth century as a result of the overuse and misuse of medications used to treat infections.
4. Climate change has led to an expansion in the prevalence and geographical distribution of fungi, caused by an increase in atmospheric temperature and humidity patterns.

Additionally, it is important to note that the three fungal genera considered to be the most prevalent and pathogenic are *Candida*, *Pneumocystis* and *Aspergillus*. Table 6.1 gives is a breakdown of the global fungal infection incidences, categorised by the causative agents.

Table 6.1 also makes it abundantly clear that *Candida* is the most burdensome of all the above mentioned pathogens. Even recent literature supports this notion with *Candida* species reportedly accounting for approximately 63–70% of all fungal infections, globally (Mandras et al. 2021). However, while *C. albicans* remains the most virulent species with the highest global prevalence and mortality rates, these infections remain largely unreported. This chapter therefore aims to focus on pathogenic, but non-albicans species of *Candida* and their interactions with natural alternatives to conventional, mainstream drugs. Some of these species include *C. auris*, *C. glabrata*, *C. tropicalis*, *C. krusei* and *C. parapsilosis* as well as other rare yeasts that are considered as emerging diseases, on the rise (Deorukhkar and Saini 2015).

Table 6.1 Annual incidence of the most prevalent fungal infections around the globe (Parker et al. 2022)

Infection	Approximate global annual incidence for 2017
1. Invasive candidiasis	750,000
2. *Pneumocystis jirovecii* pneumonia	500,000
3. Invasive Aspergillosis	300,000
4. AIDs-related cryptococcal infections	223,000
5. Progressive disseminated histoplasmosis	100,000
6. Mucormycosis	>10,000

6.2 *Candida* Species

Candida species can be described as a eukaryotic, yeast that belongs to the kingdom Fungi. Of the millions of species that make up this kingdom, a very small percentage meets the basic conditions which make them capable of causing infections in humans (Köhler et al. 2014). Moreover, these organisms are significantly diverse and can be macroscopic (mushrooms) or microscopic (all pathogenic microorganisms, e.g. *C. auris* and *Aspergillus fumigatus*, just to name a few). Figure 6.1 visualises two of the organisms that belong to each of these groups.

Candida species, however, is commonly found to exist as non-pathogenic microorganisms within the human microbiota. What this means, is that one or multiple species of *Candida* can usually be isolated from areas of a healthy human body such as the mouth, throat, gut and vagina. Disease-causing fungal pathogens are also known to be opportunistic organisms, and only cause infection in humans with immune systems that are weakened or compromised due to other underlying practices of health conditions (Deorukhkar and Saini 2015). Everyday practices that can help prevent *Candida* species infections therefore involve maintaining good physical hygiene, proper and consistent management of underlying medical conditions, as well as avoiding the superfluous use of broad-spectrum antibiotics, which are responsible for disrupting the body's natural microbial balance (Deorukhkar and Saini 2015).

When the causative agent for the disease is *Candida* species, these infections are known as candidiasis. Candidiasis can occur in various parts of the body, and the severity of infection can range anywhere from superficial and mild to invasive and

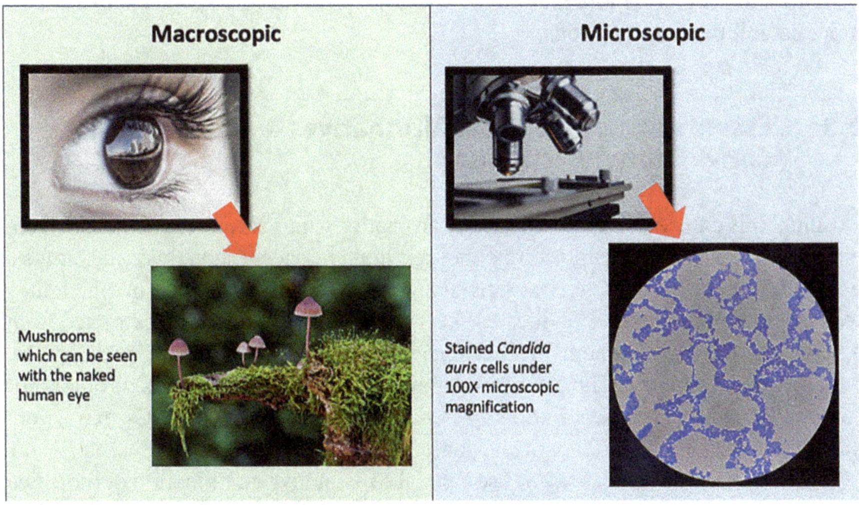

Fig. 6.1 Macroscopic vs. microscopic fungi

Table 6.2 MIC range (μg/mL) of *Candida* species for the mainstream, conventional antifungals (Zapata-Zapata et al. 2022)

	Conventional antifungals (μg/mL)			
Candida species	Amphotericin B	Itraconazole	Fluconazole	Caspofungin
Candida albicans	0.03–0.12	0.03–1.0	4.0–8.0	0.12–0.5
Candida auris	0.06–2.0	0.03–0.5	1.0–32.0	0.5–1.0
Candida krusei	0.12–0.25	0.12–0.5	8.0	1.0
Candida glabrata	0.06–0.12	1.0	2.0–4.0	8.0
Candida tropicalis	0.06–2.0	0.06–16.0	1.0–64.0	0.25–0.5
Candida parapsilosis	0.06	0.25–0.5	0.5–1.0	1.0

potentially fatal, depending on the site and health status of the afflicted individual (Deorukhkar and Saini 2015). There are only four major main classes of antifungal drugs that are used in clinical settings today, namely; polyenes, azoles, allylamines and echinocandins (Chen and Sorrell 2007). Candidiasis can be remedied with topical or oral antifungal medications from any one of these classes with drugs such as fluconazole, amphotericin B, itraconazole and caspofungin just to name a few. However, in some cases, especially with invasive candidiasis, treatment can be challenging due to the development of resistance to these antifungal agents (Deorukhkar and Saini 2015; Srivastava et al. 2018). Table 6.2 gives the reader a sense of the relative levels of antifungal resistance present in the various species of *Candida*, to the currently used, mainstream antifungal drugs.

It is important to remember that as the prevalence of drug resistance started to rise, medical scientists and physicians have been in a race against the clock to discover new and effective treatments for fungal infections. Additionally, as our understanding of the biological processes associated with disease and pharmaceuticals continues to progress, more and more stakeholders are encouraging a shift towards more natural treatment options.

6.3 Essential Oils as Natural Alternatives to Mainstream Conventional Antifungals

Essential oils can be described as natural aromatics which have been extracted from plants. These volatile organic compounds are highly concentrated, and each emits a unique fragrance, that strongly correlates with the species of plant from which they were sourced (Manion and Widder 2017). These precious oils can be extracted from a variety of plant components, namely; leaves, seeds, flowers, fruits, peels, stems, bark or wood and roots. The extraction of essential oils usually occurs through processes like steam or water distillation, cold pressing, solvent or CO_2 extraction, maceration or enfleurage (Fig. 6.2) (Manion and Widder 2017).

These oils have a wide array of uses, and the distinctive combination of bioactive compounds that make up each of these substances is responsible for this. Additionally, humans have capitalised on the multipurpose nature of these

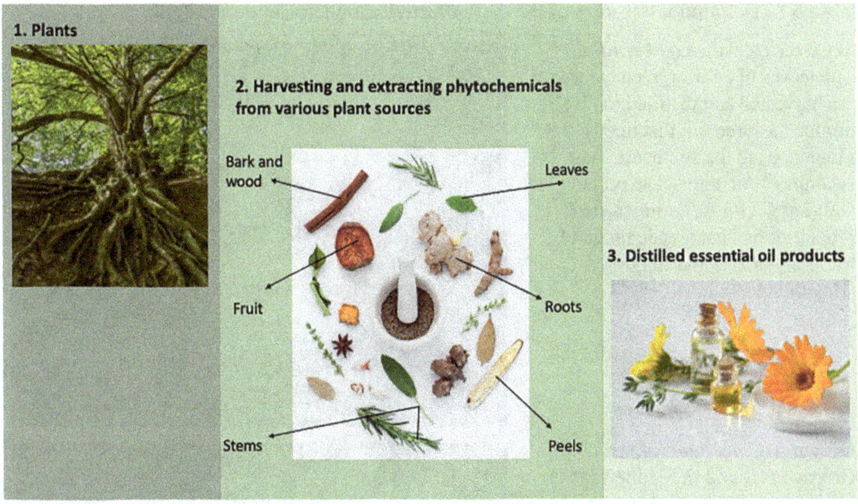

Fig. 6.2 A basic illustration of the major steps involved in the production of essential oils

compounds as they have been used in traditional medicine, food products, cosmetics, perfumes, cleaning products and aromatherapy (Manion and Widder 2017). More specifically, however, their recorded potential health benefits are as follows (Manion and Widder 2017):

1. Mood enhancement
2. Alleviation of stress and anxiety
3. Improved concentration
4. Treatment for insomnia and improved sleep quality
5. Antimicrobial treatments for infection (bacteria, fungi and viruses)
6. Reduced pain and inflammation
7. Relief from nausea and headaches
8. Anti-oxidant effects

Antimicrobial essential oils possess properties that can help inhibit or destroy microorganisms, including bacteria, viruses, fungi and other pathogens. These oils contain natural compounds that exhibit antimicrobial activity and have been traditionally used for their potential therapeutic benefits in various applications (Manion and Widder 2017). There are over 90 types of essential oils known to humans (West 2019), as discussed, all with unique and potentially beneficial properties. These oils are harvested from plants that are indigenous to countries all over the world. Essential oils are, however, highly concentrated and if incorrectly used or diluted, they can cause adverse reactions. From the extensive amount of literature available on these compounds, some essential oils can be shortlisted as having the most well-known and potent antimicrobial activity (Table 6.3).

Table 6.3 Descriptions of some of the most effective antimicrobial essential oils

Tea Tree (*Melaleuca alternifolia*): The leaves of an evergreen tea tree are harvested and distilled to produce tea tree oil; this tree is indigenous to the southeast coast swamps of Australia. The oil is well-known for its antimicrobial properties against a wide range of bacteria, viruses and fungi (Healthline 2023)	
Oregano (*Origanum vulgare*): Oregano is found in Mediterranean countries and western Asia. This herb contains, in large amounts, the potent antimicrobial compounds known as carvacrol and thymol. It has shown efficacy against various bacteria and fungi (Encyclopaedia Britannica 2023a)	
Eucalyptus (*Eucalyptus globulus*): Eucalyptus leaves are harvested from Tasmanian blue gum trees (native to parts of Australia) and are then processed to produce an oil that has strong antiseptic and antiviral properties. It is also often used to alleviate congestion and support respiratory health (Cerasoli et al. 2016)	

(continued)

Table 6.3 (continued)

Peppermint (*Mentha × piperita*): Peppermint oil is extracted from the leaves of the perennial herb commonly known as mint. These plants are indigenous to Europe and the Middle East. The antimicrobial compound in this oil is known as menthol. Additionally, it is used in various applications, including oral care and digestive support (Brazier 2023)	
Lavender (*Lavandula angustifolia*): Lavender oil is extracted and then distilled from the flowers of the shrub-like lavender plant, which are native to the Mediterranean. It is well-known for its calming properties, but it also possesses mild antimicrobial activity. It is often used in skin care and aromatherapy (Missouri Botanical Garden 2023b)	
Thyme (*Thymus vulgaris*): Thyme oil is extracted from the leaves of the herb commonly known as thyme, and contains thymol, a potent antimicrobial compound. It is often found in natural disinfectants and cleaning products. This plant is native to southern Europe (NC State University 2023b)	

(continued)

Table 6.3 (continued)

Lemon (*Citrus limon*): Lemon oil is extracted from the flowers, fruit and twigs of a lemon tree, which are native to north-western India. These plant components are often cold-pressed or steam-distilled to yield lemon essential oil. It has antibacterial and antiviral properties and can be used in cleaning products and to help freshen the air (Missouri Botanical Garden 2023a)	
Lemongrass (*Cymbopogon*): Native to South, and South-east Asia, this tender perennial grass contains the bioactive component known as citral, which has been shown to have strong antimicrobial effects against pathogenic bacteria and fungi (University of Wisconsin 2023)	
Lemon balm (*Melissa officinalis*): Lemon balm is a herb that is native to the Mediterranean region and Central Asia. This bushy perennial plant is the source of lemon balm essential oil. The oil itself is extracted from the flowers, leaves and branches by steam distillation or is chemically extracted (Petruzzello 2022)	
Bay laurel (*Laurus nobilis*): This plant is an evergreen perennial shrub. It is native to the Mediterranean region and is often utilised as a spice (bay leaf) in cooking. The main phytochemical components of this oil are oxygenated monoterpenes (NC State University 2023a)	

(continued)

Table 6.3 (continued)

Pine (*Pinus*): Pine is predominantly native to the northern hemisphere regions, including parts of Europe and the southern parts of the United States of America. These include temperate regions and they are most often found to exist in forests (Encyclopaedia Britannica 2023b)	
Cinnamon (*Cinnamomum verum*): Cinnamon essential oil is extracted and steam distilled from the bark and leaves of a small evergreen, true cinnamon tree that is native to Sri Lanka, the Malabar Coast of India and Myanmar (Petruzzello 2023). The main components of this oil that are thought to be responsible for its antimicrobial activity, are cinnamaldehyde and eugenol	 

(continued)

Table 6.3 (continued)

Clove (*Syzygium aromaticum*): Cloves are the flower buds of a tropical, evergreen tree that are indigenous to the Indonesian Spice Islands (Rodriguez 2023). The main antimicrobial component in this essential oil is eugenol; additionally, studies have reported that this oil is an effective antimicrobial agent against several enteric bacterial pathogens as well as pathogenic fungal species which originated in the environment (Schroder et al. 2017)	

6.4 Essential Oil Components with Antimicrobial Activity

The biological and pharmacological characteristics of each essential oil are determined by (1) the species of plant from which it was extracted and (2) their complex and highly varying phytochemical compositions and concentrations. It is important to note that external factors such as the growing conditions of plants and the methods utilised for essential oil extraction may also play a role in the antimicrobial efficacy of these substances. The general bioactivity can therefore be credited either to the major constituents that make up each individual oil, or to the unique combinations of several bioactive ingredients, which may cause synergistic interactions (Mandras et al. 2021). These phytochemical constituents are largely made up of, terpenes (monoterpenes and sesquiterpenes), oxygenated terpenoids (alcohols and phenols), as well as other aromatic and aliphatic compounds (Mandras et al. 2021). Essential oils also have the potential to exhibit synergistic (antimicrobial, anti-inflammatory or ant-septic) effects in combination with other well-known commercial agents or medications (Serra et al. 2018). As already discussed, we know that essential oils are the natural yields of aromatic plants, and they mainly consist of terpenes and terpenoids. However, of the wide range of phytochemicals that make up essential oils, only a select few are responsible for their antimicrobial properties. Some of the key compounds with notable antimicrobial properties are given in Table 6.4.

Table 6.4 A description of the most significant phytochemicals present in essential oils that are known to exhibit antimicrobial activity (National Centre for Biotechnology Information 2023a, b, c, d, e, f, g, h, i, j, k, l, m, n, o, p, q, r)

Phytochemical	Description	Source
Terpenes		
p-cymene	An organic, aromatic and natural compound, classified as being related to a monoterpene (specifically an alkylbenzene)	Thyme
	Formula: $C_{10}H_{14}$	Eucalyptus
		Tea tree
α-Pinene	An organic, terpene compound which is one of two pinene isomers	Lemon
	Formula: $C_{10}H_{16}$	Eucalyptus
		Tea tree
		Pine
Limonene	An aliphatic hydrocarbon, which is classified as a monoterpene. This compound is a significant component in citrus peel oils	Lemon
	Formula: $C_{10}H_{16}$	Peppermint
		Eucalyptus
Myrcene	Also characterised as β-myrcene, this compound is a monoterpene	Lemon
	Formula: $C_{10}H_{16}$	
γ-terpinene	This compound falls into a group of isomeric hydrocarbons which are characterised as monoterpenes	Thyme
	Formula: $C_{10}H_{16}$	Eucalyptus
		Tea tree
β-ocimene	Ocimenes are also a group of isomeric hydrocarbons which are characterised as monoterpenes	Lavender
	Formula: $C_{10}H_{16}$	Lemon
Linalool	This compound is a natural, terpene alcohol which can be found in many spices and flowers	Lavender
	Formula: $C_{10}H_{18}O$	Thyme
		Lemon
		Eucalyptus
		Bay laurel
Sesquiterpene		
β-Caryophyllene	A natural bicyclic sesquiterpene phytochemical	Clove
	Formula: $C_{15}H_{24}$	Lemon
		Peppermint
α-Caryophyllene (α-humulene)	This compound is known for being a natural, monocyclic sesquiterpene	Clove
	Formula: $C_{15}H_{24}$	
Terpenoids		
Citronellol	A natural, acyclic monoterpenoid	Lemon
	Formula: $C_{10}H_{20}O$	Lemongrass
		Eucalyptus

(continued)

Table 6.4 (continued)

Phytochemical	Description	Source
Menthol	Menthol is an organic compound, more specifically known as a monoterpenoid	Peppermint
	Formula: $C_{10}H_{20}O$	
1,8-Cineole (Eucalyptol)	This compound is a naturally occurring cyclic ether and monoterpenoid	Lavender
	Formula: $C_{10}H_{18}O$	Eucalyptus
		Tea tree
		Rosemary
		Bay laurel
Thymol	This compound is responsible for giving thyme its strong aroma and flavour. It is known for having good antiseptic properties and is classified as a natural monoterpenoid phenol. It is also a derivative of p-Cymene	Oregano
	Formula: $C_{10}H_{14}O$	Thyme
Eugenol	This phytochemical is known for being a naturally occurring, volatile, phenolic monoterpenoid	Clove
	Formula: $C_{10}H_{12}O_2$	Cinnamon
Carvacrol	This phytochemical is a monoterpenoid phenol	Oregano
	Formula: $C_{10}H_{14}O$	
Camphor	A terpenoid compound also known for being a cyclic ketone	Lavender
	Formula: $C_{10}H_{16}O$	
Others		
Cinnamaldehyde	This is a naturally occurring compound that gives cinnamon spice its flavour and aroma	Cinnamon
	Formula: C_9H_8O	
Terpinen-4-ol	Terpinen-4-ol is characterised as an isomer of terpineol	Lavender
	Formula: $C_{10}H_{18}O$	Lemon
		Eucalyptus
		Tea tree

6.5 Mechanisms of Antifungal Action of Essential Oils

Due to the complex nature of essential oils, brought about by the plethora of bioactive constituents present in these phytoextracts, their antimicrobial capabilities are not limited to a single mechanism of action, but rather several mechanisms which can act on internal or external cellular sites (Mandras et al. 2021). Understanding these mechanisms of action is of the utmost importance, particularly when it comes to the development of drugs intended for use in the treatment of human infections. It is for this reason that the diagram below (Fig. 6.3) modestly depicts a number of possible methods for fungal cell function interference or inhibition.

Essential oils are known to be lipophilic (or lipid loving) in nature, it is for this reason that these compounds are characteristically capable of integrating into cell

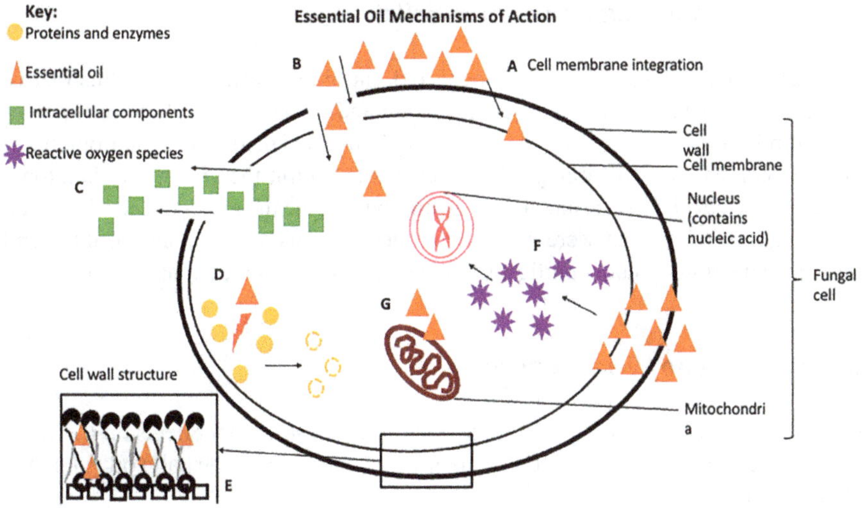

Fig. 6.3 Antifungal mode of action of essential oils

membrane structures (**A**). From here, they can cause disruption and damage to the cell membrane composition which leads to increased cell permeability (**B**); leakage of intracellular components (**C**) and the inactivation or denaturation of enzymes and proteins (**D**). Moreover, studies suggest that when essential oils act against *Candida* species, they are capable of modifying cell wall morphology and inhibiting the biosynthesis processes involved in ergosterol production (**E**) this compound is the homologue to cholesterol in human cell membranes, and actively works to maintain cell membrane integrity (Mandras et al. 2021). Essential oils can also produce oxygen-reactive species (ROS) (**F**), which are substantially reactive elements, formed from diatomic oxygen that can cause damage to fundamental cellular components including DNA, proteins and lipids. Examples include peroxides and superoxide. Furthermore, essential oils can also interact with the membrane surrounding the mitochondria, leading to cell dysfunction and death (**G**) (Mandras et al. 2021).

6.6 Reported Antifungal Efficacy of Various Essential Oils

There has been a myriad of scientific papers published, with data on the antifungal activity of various essential oils, on various strains and species of fungi. These investigations often involve antifungal susceptibility testing techniques such as broth microdilution assays, disk diffusion assays etc. The following section summarises some studies which have been conducted using essential oils as antifungal agents against *Candida* species of interest, namely; *C. auris*, *C. krusei*, *C. glabrata*, *C. tropicalis* and *C. parapsilosis*.

6.6.1 Tea Tree (*Melaleuca alternifolia*)

A recent study looked at the antifungal and antibiofilm effect of four distinctive essential oils, namely tea tree, cajeput, niaouli and white thyme. These four potential antimicrobial agents were tested against *C. auris*. The experimental outcomes clearly demonstrated the ability of all four oils to inhibit the growth of planktonic *C. auris* cells. White thyme and tea tree were found to be the most potent. Moreover, white thyme and cajeput were able to completely eradicated mature biofilms, and tea tree and niaouli could significantly reduce them (Fernandes et al. 2022).

6.6.2 Oregano (*Origanum vulgare*)

Oregano essential oil was also found to have notable antifungal effects on *C. krusei* and *C. parapsilosis*, of a less effective nature than pine and lemon balm oil (Mandras et al. 2021).

6.6.3 Lavender (*Lavandula angustifolia*)

According to an in vitro study, lavender oil exhibited both antifungal and antibiofilm properties against various strains of *C. auris* (de Alteriis et al. 2022). A separate study involving *C. krusei* and *C. parapsilosis* found that lavender also exhibited notable inhibitory effects, these inhibitory and eradication (or total death) concentrations were, however, higher than that of pine oil (Mandras et al. 2021).

6.6.4 Thyme (*Thymus vulgaris*)

Thyme is effective against *C. albicans* and *C. tropicalis*; however, it was found to have more potent antifungal activity against *C. albicans*. Additionally, in a different study, white thyme (*Thymus serpyllum albus*), derived from a similar plant was found to be capable of eradicating not only planktonic, but also sessile or biofilm associated cells of *C. auris* (Mandras et al. 2021).

6.6.5 Lemon Balm (*Melissa officinalis*)

The findings of a study published in 2021 confirmed pre-existing reports of the potent antibacterial and anticandidal effects of lemon balm essential oils. This study evaluated the effect of the essential oil on a number of non-albicans *Candida* species which continuously exhibited peak antifungal activity. The composition of this oil was found to comprise primarily of monoterpene aldehydes, with the major phytochemicals including citronellal and limonene. The main mechanism of action is thought to involve making cell membranes increasingly porous, and in turn increase membrane fluidity (Mandras et al. 2021).

6.6.6 Bay Laurel (*Laurus nobilis*)

The effects of bay laurel essential oil were investigated on four *Candida* species of interest (namely *C. albicans*, *C. tropicalis*, *C. glabrata* and *C. krusei*). The main phytochemical constituent of this oil was reported as isoeugenol. The results showed that this oil had similar effects on all species with lower concentrations of compound needed for inhibition of cellular growth as well as total cell eradication or death in certain strains of *C. albicans* and *C. tropicalis*. Additionally, while the concentration of compound needed for cell inhibition and death was significantly higher than that of the mainstream, conventional antifungal drug (known as Nystatin), the bay laurel essential oil had indifferentiable abilities at disrupting immature and mature biofilms. The main constituents thought to be responsible for this oil's antifungal effect were monoterpenes and sesquiterpenes. The primary mechanisms of action are inhibition or disruption of cell wall synthesis and increased permeability of ions moving in and out of the cell membranes (Peixoto et al. 2017).

6.6.7 Pine (*Pinus*)

Results from a recent study confirmed pine essential oil as an effective antifungal agent against non-albicans *Candida* isolates (namely *C. krusei* and *C. parapsilosis)* as well as rare, non-albicans yeast species (such as *Cryptococcus neoformans* and *Saccharomyces cerevisiae*). Pine oil demonstrated significant microorganism growth inhibition with relatively low concentrations of the compound. The main antimicrobial component found in pine is α-pinene. Previous studies have found that this compound is effective against bacteria and fungi with mechanisms of action on fungal cells (specifically *C. albicans*), decreasing cell integrity, inhibiting cellular respiration, ion transport and hence increasing cell membrane permeability (Mandras et al. 2021).

6.6.8 Cinnamon (*Cinnamomum verum*)

When bark and leaf-derived Cinnamon essential oil was tested against *C. auris* and *C. albicans*, it was found to be an effective antifungal agent. The mechanism of action for these oils, against planktonic *Candida* cells was found to be their ability to damage cell membranes, inhibit hyphae formation and germination as well as the inhibition of *Candida*-associated virulence factors such as haemolysin. This involves the lysing of red blood cells during bloodstream infections, so as to obtain iron from erythrocytes for improved growth and proliferation (Tran et al. 2020).

6.6.9 Clove (*Syzygium aromaticum*)

Clove also demonstrated a notable inhibitory effect against *C. krusei* and *C. parapsilosis;* these inhibitory effects, however, were not as significant as those of pine and lemon balm oil (Mandras et al. 2021).

6.7 Conclusion

From the brief descriptions in this chapter, the following conclusions can be drawn; with the continued and rapid rise of antimicrobial resistance developing in clinical settings around the world, healthcare professionals and researchers are under immense amounts of pressure to discover and successfully develop new, effective and highly biocompatible antifungal drugs. In addition to this, various industries have promoted the development and use of more natural products, this includes the medical industry. For this reason, essential oils have become a popular subject of medical research studies. The plethora of available literature has, to a large extent, offered evidence to support the hypothesis that essential oils show significant potential as natural, effective and biocompatible antimicrobial agents, this includes inhibitory activity against pathogenic microorganisms such as bacteria, viruses and, as discussed above, fungi. More specifically, several essential oils that were extracted from a diverse range of plant species and components have shown notable antifungal activity against pathogenic *Candida* species. Despite these positive results, there remains a significant need for further investments in the research and development of approved, essential oil-based antifungal drugs that are safe for use in clinical settings.

References

Brazier Y (2023) Benefits, uses, forms, and precautions of peppermint. Medical News Today. https://www.medicalnewstoday.com/articles/265214. Accessed 15 Aug 2023

Britannica, The Editors of Encyclopaedia (2023a) Oregano. Encyclopaedia Britannica. https://www.britannica.com/plant/oregano. Accessed 15 Aug 2023

Britannica, The Editors of Encyclopaedia (2023b) Pine. Encyclopaedia Britannica. https://www.britannica.com/plant/pine. Accessed 15 Aug 2023

Cerasoli S, Caldeira M, Pereira J, Caudullo G, de Rigo D (2016) Eucalyptus globulus and other eucalypts in Europe: distribution, habitat, usage and threats. European Atlas of Forest Tree Species. https://forest.jrc.ec.europa.eu/media/atlas/Eucalyptus_globulus.pdf. Accessed 15 Aug 2023

Chen S, Sorrell T (2007) Antifungal agents. Med J Aust 187(7):404–409. https://doi.org/10.5694/j.1326-5377.2007.tb01313.x

de Alteriis E, Maione A, Falanga A, Bellavita R, Galdiero S, Albarano L, Salvatore M, Galdiero E, Guida M (2022) Activity of free and liposome-encapsulated essential oil from *Lavandula angustifolia* against persister-derived biofilm of *Candida auris*. Antibiotics 11:26. https://doi.org/10.3390/antibiotics11010026

Deorukhkar S, Saini S (2015) Non albicans *Candida* species: a review of epidemiology, pathogenicity and antifungal resistance. Pravara Med Rev 7(3):7–15

Fernandes L, Ribeiro R, Costa R, Henriques M, Rodrigues M (2022) Essential oils as a good weapon against drug-resistant *Candida auris*. Antibiotics 11:977. https://doi.org/10.3390/antibiotics11070977

Healthline Editorial Team (2023) 14 everyday uses for tea tree oil. Healthline. https://www.healthline.com/nutrition/tea-tree-oil. Accessed 15 Aug 2023

Köhler J, Casadevall A, Perfect J (2014) The spectrum of fungi that infects humans. Cold Spring Harb Perspect Med 5(1):a019273. https://doi.org/10.1101/cshperspect.a019273

Mandras N, Roana J, Scalas D, Del Re S, Cavallo L, Ghisetti V, Tullio V (2021) The inhibition of non-albicans *Candida* species and uncommon yeast pathogens by selected essential oils and their major compounds. Molecules 26:4937. https://doi.org/10.3390/molecules26164937

Manion C, Widder R (2017) Essentials of essential oils. Am J Health Syst Pharm 74(9):e153–e162. https://doi.org/10.2146/ajhp151043

Missouri Botanical Garden (2023a) *Citrus limon*. https://www.missouribotanicalgarden.org/PlantFinder/PlantFinderDetails.aspx?kempercode=b548. Accessed 15 Aug 2023

Missouri Botanical Garden (2023b) *Lavandula angustifolia*. https://www.missouribotanicalgarden.org/PlantFinder/PlantFinderDetails.aspx?taxonid=281393. Accessed 15 Aug 2023

National Centre for Biotechnology Information (2023a) PubChem compound summary for CID 7463, p-CYMENE. https://pubchem.ncbi.nlm.nih.gov/compound/p-CYMENE. Accessed 15 Aug 2023

National Centre for Biotechnology Information (2023b) PubChem compound summary for CID 6654, alpha-PINENE. https://pubchem.ncbi.nlm.nih.gov/compound/alpha-PINENE. Accessed 15 Aug 2023

National Centre for Biotechnology Information (2023c) PubChem compound summary for CID 22311, limonene. https://pubchem.ncbi.nlm.nih.gov/compound/Limonene. Accessed 15 Aug 2023

National Centre for Biotechnology Information (2023d) PubChem compound summary for CID 31253, myrcene. https://pubchem.ncbi.nlm.nih.gov/compound/Myrcene. Accessed 15 Aug 2023

National Centre for Biotechnology Information (2023e) PubChem compound summary for CID 7461, gamma-terpinene. https://pubchem.ncbi.nlm.nih.gov/compound/gamma-Terpinene. Accessed 15 Aug 2023

National Centre for Biotechnology Information (2023f) PubChem compound summary for CID 18756, beta-ocimene. https://pubchem.ncbi.nlm.nih.gov/compound/beta-Ocimene. Accessed 15 Aug 2023

National Centre for Biotechnology Information (2023g) PubChem compound summary for CID 6549, linalool. https://pubchem.ncbi.nlm.nih.gov/compound/Linalool. Accessed 15 Aug 2023

National Centre for Biotechnology Information (2023h) PubChem compound summary for CID 5281515, caryophyllene. https://pubchem.ncbi.nlm.nih.gov/compound/Caryophyllene. Accessed 15 Aug 2023

National Centre for Biotechnology Information (2023i) PubChem compound summary for CID 5281520, humulene. https://pubchem.ncbi.nlm.nih.gov/compound/Humulene. Accessed 15 Aug 2023

National Centre for Biotechnology Information (2023j) PubChem compound summary for CID 8842, beta-CITRONELLOL. https://pubchem.ncbi.nlm.nih.gov/compound/beta-CITRONELLOL. Accessed 15 Aug 2023

National Centre for Biotechnology Information (2023k) PubChem compound summary for CID 1254, menthol. https://pubchem.ncbi.nlm.nih.gov/compound/Menthol. Accessed 15 Aug 2023

National Centre for Biotechnology Information (2023l) PubChem compound summary for CID 2758, eucalyptol. https://pubchem.ncbi.nlm.nih.gov/compound/Eucalyptol. Accessed 15 Aug 2023

National Centre for Biotechnology Information (2023m) PubChem compound summary for CID 6989, thymol. https://pubchem.ncbi.nlm.nih.gov/compound/Thymol. Accessed 15 Aug 2023

National Centre for Biotechnology Information (2023n) PubChem compound summary for CID 3314, eugenol. https://pubchem.ncbi.nlm.nih.gov/compound/Eugenol. Accessed 15 Aug 2023

National Centre for Biotechnology Information (2023o) PubChem compound summary for CID 10364, carvacrol. https://pubchem.ncbi.nlm.nih.gov/compound/Carvacrol. Accessed 15 Aug 2023

National Centre for Biotechnology Information (2023p) PubChem compound summary for CID 2537, camphor. https://pubchem.ncbi.nlm.nih.gov/compound/Camphor. Accessed 15 Aug 2023

National Centre for Biotechnology Information (2023q) PubChem compound summary for CID 637511, Cinnamaldehyde. https://pubchem.ncbi.nlm.nih.gov/compound/Cinnamaldehyde. Accessed 15 Aug 2023

National Centre for Biotechnology Information (2023r) PubChem compound summary for CID 11230, 4-terpineol. https://pubchem.ncbi.nlm.nih.gov/compound/4-Terpineol. Accessed 15 Aug 2023

NC State University (2023a) *Laurus nobilis*. North Carolina Extension Gardener Plant Toolbox. https://plants.ces.ncsu.edu/plants/laurus-nobilis/. Accessed 15 Aug 2023

NC State University (2023b) *Thymus vulgaris*. North Carolina Extension Gardener Plant Toolbox. https://plants.ces.ncsu.edu/plants/thymus-vulgaris/. Accessed 15 Aug 2023

Parker R, Gabriel K, Graham K, Butts B, Cornelison C (2022) Antifungal activity of select essential oils against *Candida auris* and their interactions with antifungal drugs. Pathogens 11:821. https://doi.org/10.3390/pathogens11080821

Peixoto L, Rosalen P, Ferreira G, Freires I, de Carvalho F, Castellano L, de Castro R (2017) Antifungal activity, mode of action and anti-biofilm effects of *Laurus nobilis Linnaeus* essential oil against *Candida* spp. Arch Oral Biol 73:179–185. https://doi.org/10.1016/j.archoralbio.2016.10.013

Petruzzello M (2022) Lemon balm. Encyclopaedia Britannica. https://www.britannica.com/plant/lemon-balm. Accessed 15 Aug 2023

Petruzzello M (2023) The editors of Encyclopaedia of Britannica. Cinnamon- plant and spice. https://www.britannica.com/plant/cinnamon. Accessed 10 Aug 2023

Rayens E, Norris K (2022) Prevalence and healthcare burden of fungal infections in the United States in 2018. Open Forum Infect Dis 9(1):ofab593. https://doi.org/10.1093/ofid/ofab593

Rodriguez E (2023) The editors of Encyclopaedia of Britannica. Clove-plant and spice. https://www.britannica.com/plant/clove. Accessed 10 Aug 2023

Schroder T, Gaskin S, Ross K, Whiley H (2017) Antifungal activity of essential oils against fungi isolated from air. Int J Occup Environ Health 23(3):181–186. https://doi.org/10.1080/10773525.2018.1447320

Serra E, Hidalgo-Bastida L, Verran J, Williams D, Malic S (2018) Antifungal activity of commercial essential oils and biocides against *Candida albicans*. Pathogens 7:15. https://doi.org/10.3390/pathogens7010015

Srivastava V, Singla R, Dubey A (2018) Emerging virulence, drug resistance and future anti-fungal drugs for *Candida* pathogens. Curr Top Med Chem 18(9):759–778. https://doi.org/10.2174/1568026618666180528121707

Tran H, Graham L, Adukwu E (2020) In vitro antifungal activity of *Cinnamomum zeylanicum* bark and leaf essential oils against *Candida albicans* and *Candida auris*. Appl Microbiol Biotechnol 104(20):8911–8924. https://doi.org/10.1007/s00253-020-10829-z

University of Wisconsin (2023) Lemongrass. https://www.uwsp.edu/sbcb/lemongrass/. Accessed 15 Aug 2023

West H (2019) What are essential oils, and do they work? Healthline. September. https://www.healthline.com/nutrition/what-are-essential-oils. Accessed 15 Aug 2023

Zapata-Zapata C, Loaiza-Oliva M, Martínez-Pabón M, Stashenko E, Mesa-Arango A (2022) In vitro activity of essential oils distilled from Colombian plants against *Candida auris* and other *Candida* species with different antifungal susceptibility profiles. Molecules 27:6837. https://doi.org/10.3390/molecules27206837

Part III

Plant-Derived Natural Compounds as Antifungals

Molecules of Natural Origin as Inhibitors of Signal Transduction Pathway in *Candida albicans*

7

Sayali A. Chougule, S. Mohan Karuppayil, and Ashwini K. Jadhav

7.1 Introduction

About one and a half million people die every year as a result of fungal infections (Bar-Yosef et al. 2017). It is a universal health problem that has been increasing worldwide (Bradshaw and Dennis 2009). Of all the several *Candida* species, *Candida albicans* accounts for 75% of all cases of Candidiasis. Being a commensal, this opportunistic fungal pathogen is found at several places in the human body particularly the skin, oral, genitourinary and gastrointestinal tracts. Its proliferation increases manifold under immunocompromised conditions (Papin et al. 2005). *C. albicans* is polymorphic and can be seen in different forms like yeast, hyphae and pseudohyphae. Under certain conditions, germ tubes form ultimately branching into a hyphal network. Morphological transition can be induced by both intracellular and extracellular signals and plays a significant role in *Candida* pathogenicity. The inhibition of hyphae-specific gene expression can suppress hyphal extension showing that this phenomenon can be modulated by endogenous cellular signals with the help of proteins that function as receptors or sensors (Brown et al. 2014; Villa et al. 2020).

Signaling pathways form coordinated networks as they interact with each other leading to synchronized combinatorial signaling events (Richardson 2022). Such responses include alterations in the transcription, translation, post-translational and conformational changes in the proteins involved in controlling fungal growth, proliferation, metabolism and various other processes. Many transcription factors are responsible for the regulation and coordination of

S. A. Chougule · S. M. Karuppayil · A. K. Jadhav (✉)
Department of Stem Cell and Regenerative Medicine and Medical Biotechnology, Centre For Interdisciplinary Research, D.Y. Patil Education Society (Deemed to be University), Kolhapur, Maharashtra, India
e-mail: bboffice77@gmail.com; prof.karuppayil@gmail.com; ashujadhav09@gmail.com

N. Manzoor (ed.), *Advances in Antifungal Drug Development*,
https://doi.org/10.1007/978-981-97-5165-5_7

transcriptional network with the environment and phenotypic characters of the pathogen (Kornitzer 2019). For example, the Ras gene, a component of Ras signaling pathway, is a principal regulator of cell growth, stress response and mortality in eukaryotic cells like fungi (Pentland et al. 2018). The present chapter reviews recent investigations for the regulation of yeast to hyphal transition in *C. albicans* by natural antifungal molecules with special reference to morphogenetic signals.

7.2 Current Antifungal Therapy and Its Targets

The currently available classes of antifungal drugs, commonly prescribed for the treatment of Candidiasis and other fungal infections, include polyenes, 5-flucytosine, echinocandins and azoles, which is the largest class of antifungals used till date. The major drawback of azoles (e.g., Fluconazole) is their fungistatic nature. Their mode of action is based on their ability to inhibit ergosterol biosynthesis by targeting 14-α-lanosterol demethylase (Pierce et al. 2013). Polyenes bind directly to ergosterol component of cell membranes making it porous and leaky (osmotic lysis). Amphotericin B is a polyene used for severe cases of fungal infection and has been observed to cause cell death by oxidative damage (Kaoud et al. 2012). The Echinocandins, a more recent class with better success rate, target the 1,3-α-D-glucan synthase complex required for the synthesis of glucan polymer in fungal cell walls (Pierce et al. 2013). Lastly, flucytosine or 5-fluorocytosine is an antimetabolite that targets thymidylate synthase and adversely affecting synthesis DNA and RNA molecules (Fig. 7.1).

7.3 Role of Signal Transduction Pathways Involved in the Regulation of Virulence Factors

The RAS-cAMP-PKA pathway has an important role to play during the regulation of *Candida* pathogenicity. RAS proteins behave as master regulators in both the cAMP-PKA signaling and Sre11-Hst7-Cek1/2-mediated pathways in *C. albicans* (Feng et al. 1999). Reports suggest that, deletion of RAS gene in vivo using mouse model with systemic candidiasis, attenuates *Candida* virulence (Rocha et al. 2001). In vivo studies using mouse models (systemic infection and mucosal membranes) have shown that virulence gets fully terminated when the *CYR1* gene was deleted (Park et al. 2021). The catalytic subunit Tpk2 of the enzyme protein kinase A and the transcription factor Efg1 (its target) are crucial for virulence as observed in mouse oropharyngeal candidiasis. On the other hand, *TPK1* is not significant for virulence (Rocha et al. 2001). Interestingly, the virulence of *TPK2/TPK2* mutants was comparable to the wild-type strain in mouse systemic infection model (Park et al. 2005).

Morphological switching from yeast to hyphal form is the most significant to *Candida* virulence (Cao et al. 2005). In *C. albicans*, this transition through

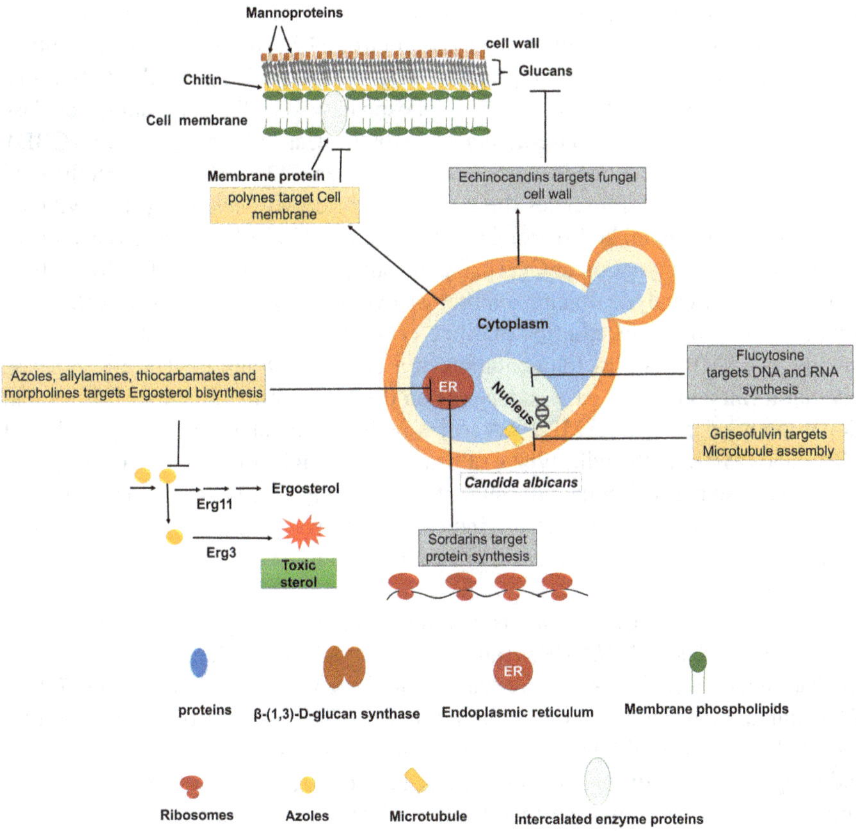

Fig. 7.1 Antifungal drugs and their targets

Ras-cAMP-PKA pathway is regulated by various environmental factors like temperature, serum, CO_2, pH and nutritional conditions (Feng et al. 1999). Deletion of *RAS1* showed defects in hyphal growth, even in the presence of serum, but not in the formation of pseudohyphae (Wang et al. 2018). On the other hand, deletion of *RAS2* had no effect germ tube induction and hyphal growth (Xu et al. 2008). Moreover, deletion of both these genes (*RAS1* and *RAS2*) may have an effect on hyphal growth via both cAMP-dependent and -independent pathways. Filament formation in *C. albicans*, in response to serum, N-acetyl glucosamine and high levels of CO_2, is enhanced on the deletion of *PDE2* or *BCY1* genes and is suppressed after the deletion of *CYR1* or *TPK1* and *TPK2* genes (Cao et al. 2005).

Adenylyl cyclase (Cyr1) is a large membrane bound protein that converts ATP to cAMP. It is considered an epicenter for all environmental cell signal sensing and integration (Rocha et al. 2001). Cyr1 has several functional domains that maybe responsible for these environmental stimuli. Moderate interaction of this protein with *RAS1* increases the level of cAMP and hyphal growth in response to inducers

(Senn et al. 2012). The Leucine-rich repeat domain (LRR) is involved in peptidoglycan sensing and temperature (>37 °C) controlled hyphal growth as a result of physical interaction with HSP90 protein (Davis-Hanna et al. 2008). The C-terminal of *C. albicans* Cyr1 constitutes a protein phosphatase 2C (PP2C) domain, a cyclase catalytic domain and a cyclase-associated protein (Cap1) binding domain (CBD). The cyclase catalytic domain gets stimulated when CO_2 or HCO_3^- directly bind here. This interaction stimulates the production of cAMP enhancing filamentation in *C. albicans* (Li et al. 2020). For CO_2 induced, Lys 1373 of Cyr1 acts receptor for CO_2/bicarbonate regulation of Cyr1p. On the other hand, inhibition of CYR1 activation via cyclase catalytic domain suppresses hyphal growth as observed with farnesol (Lu et al. 2018). The addition of dibutyryl-cAMP, a nonhydrolyzable functional analog of cAMP, restores filamentation in a culture medium containing farnesol. Hyphal formation in this case is depends on the transcription factor *EFG1*, present downstream of the Ras-cAMP-PKA pathway. Farnesol enhances the expression of *CTA1* and *HSP12* in the wild-type strain and interestingly, the *RAS* and *CYR1* genes are also overexpressed in the null mutants (Polke et al. 2017). It was also observed that the deletion of *EED1* causes increased farnesol production in *C. albicans* besides making it more hypersensitive to this quorum sensing molecule (Bockmühl et al. 2001).

Hyphal growth in *C. albicans* is also regulated by the expression of *TPK1* and *TPK2* genes. The *TPK1/TPK1* mutant shows defective filamentation in solid hyphae-inducing media but is normal in liquid media. On the other hand, *TPK2/TPK2* mutant multiplies as yeast in liquid media growth but is partly flawed in solid media (Lowy and Willumsen 1993). This study was done using different hyphae-inducing culture conditions like the Lee's glucose media, Lee's GlcNac media and spider media. It was observed that deletion of *TPK1* or *TPK2* does not affect filamentous growth but no hyphal growth was observed in the *TPK1/TPK1, TPK2/TPK2* double mutants under all inducing conditions. Studies thus show that this catalytic subunit is critical for hyphal growth (Cao et al. 2005).

Morphogenesis in *C. albicans* is regulated by both cAMP–PKA signaling and MAPK pathways collectively. *RAS1* is common to both pathways and gives a signal in response to environmental stimuli encouraging hyphal growth (Park et al. 2021). Similarly, transcription factors Flo8 and Efg1 are critical to dimorphic transition and hence virulence in *C. albicans*. Both these transcription factors are regulated by cAMP pathway and are probable targets of PKA and play a critical role in controlling morphogenesis (Cao et al. 2005). Deletion of EFG1 makes the fungus incapable of filamentous growth. The *EFG1/EFG1* mutant can form elongated cells or pseudohyphae on solid media with serum, but no true hyphae like in the wild-type strains (Stoldt et al. 1997).

Threonine-206 on Efg1 is a potent phosphorylation site for PKA that is desired for germ tube induction in the presence of hyphae-inducing situations (Ding et al. 2017). LisH motif of the transcription factor *FLO8* is involved in microtubule dynamics of hyphal growth induced by serum, Lee's media and CO_2. *FLO8* is supposed to synergistically interact with *EFG1* in both morphological forms (Cao et al. 2005). Increased levels of CO_2 cause the Ras-cAMP-PKA to interact with the tricarboxylic acid (TCA) cycle and transcription factor *SFL2* regulating filamentation in *C. albicans* using ATP and cAMP as molecular linkers (Zhao et al. 2013). Under certain conditions, it has been reported that the Ras–cAMP-PKA pathway may also play a negative regulatory role. In comparison to wild-type strains, the *FLO8/FLO8* and *CYR1/CYR1* mutants showed increased development of mycelia under conditions that require very little free oxygen (Cao et al. 2005). *FLO8* and *EFG1* may also be involved in the repression of hyphal development under certain conditions.

7.4 Natural Products as Inhibitors of Signal Transduction in *C. albicans*

The inhibition of signal transduction pathways by natural molecules is an intriguing area of research, especially in fields like pharmacology and biochemistry. Many natural compounds, derived from plants, microorganisms or marine sources, have been found to modulate signal transduction pathways in various ways. Each of these natural molecules has garnered attention due to its potential health benefits and its ability to modulate specific signaling pathways, which can have implications in treating various diseases and conditions. Some of these molecules are enlisted here (Fig. 7.2; Table 7.1). Also, the targets of these molecules in signal transduction pathway are also discussed (Fig. 7.3).

7.4.1 Tetrandrine (TET)

Tetrandrine, a bis-benzylisoquinoline alkaloid, is a calcium channel blocker. It originates from plant sources, including *Stephania tetrandra.* At low concentrations, it inhibits hyphal growth in both liquid and solid spider media. TET inhibited biofilm formation in *C. albicans* at 16 mg/L and significantly inhibited mature biofilms at 32 mg/L. The inhibition was concentration-dependent, and an exposure of 32 mg/L TET downregulated the expression of hyphae-specific genes *ECE1, HWP1, ALS3, UME6* and *HGC1* by 0.036-, 0.083- and 0.050-fold, respectively. The expression of *ECE1, ALS3* and *HWP1* genes inhibited the induced expression of *EFG1* and *RAS1*, the regulators of hyphal growth. TET suppressed the morphological switching, inhibiting hyphal growth through the Ras1p-cAMP-PKA pathway. It was further confirmed when exogenous cAMP addition caused restoration of the normal phenotype under TET exposure (Zhao et al. 2013).

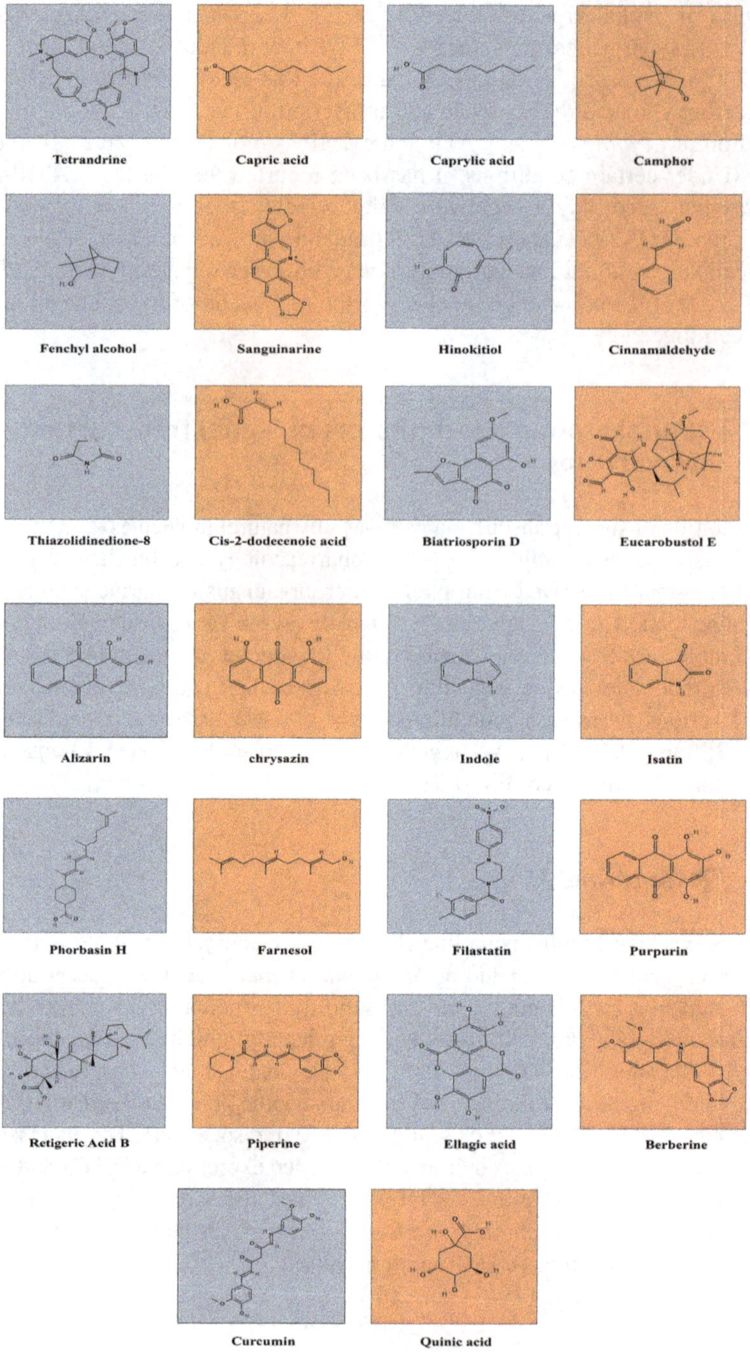

Fig. 7.2 Structure of biomolecules that show antifungal efficacy

Table 7.1 Regulation of genes in the signal transduction pathway of *C. albicans* by natural molecules having antifungal potential

Natural compound	Upregulated genes	Downregulated genes	References
Tetrandrine	–	*HWP1*	Zhao et al. (2013)
		ECE1	
		EFG1	
Capric acid	*EFG1*	*RAS1*	Jadhav et al. (2017a)
	TEC1	*CDC35*	
	ECE1	*PDE2*	
		HWP1	
		BCY1	
Caprylic acid	*RAS1*	*CDC35*	Jadhav et al. (2017a)
	BCY1	*PDE2*	
	EFG1	*TEC1*	
		ECE1	
		HWP1	
Cedar leaf oil			Manoharan et al. (2017b)
1. Camphor	–	*ECE1*	
		ECE2	
2. Fenchyl alcohol	–	*ECE1*	
		ECE2	
Sanguinarine	–	*HWP1*	Zhong et al. (2017)
		ECE1	
		HGC1	
		CYR1	
Hinokitiol	–	*HWP1*	Kim et al. (2017)
		UME6	
		HGC1	
		CYR1	
		RAS1	
		HWP1	
		EFG1	
Cinnamaldehyde	–	*HWP1*	Khan et al. (2017)
Thiazolidinedione-8	*TUP1*	*HWP1*	Feldman et al. (2014)
	NRG1	*UME6*	
		CST20	
		HST7	
		CPH1	
		RAS1	
Cis-2-dodecenoic acid	*EFG1*	–	Tian et al. (2013)

(continued)

Table 7.1 (continued)

Natural compound	Upregulated genes	Downregulated genes	References
Biatriosporin D	*PDE2*	*HWP1*	Zhang et al. (2017)
		ECE1	
		RAS1	
		CDC35	
		EFG1	
		TEC1	
Eucarobustol E	*NRG1*	*ECE1*	Liu et al. (2017)
		UME6	
		HGC1	
		TEC1	
		EFG1	
		CPH1	
Quinones and anthraquinone-related compounds	*EFG1*	*ECE1*	Manoharan et al. (2017a)
		ECE2	
Indole	*NRG1*	*HWP1*	Jadhav et al. (2017b)
	TUP1	*MIG1*	
Isatin	*NRG1*	*HWP1*	
	TUP1		
Phorbasin H	–	*HWP1*	Lee et al. (2013)
		EFG1	
Farnesol	*TUP1*		Lindsay et al. (2012)
Purpurin	–	*HWP1*	Tsang et al. (2012)
		HYR1	
		ECE1	
		RAS1	
Retigeric acid B	*PDE2*	*CDC35*	Chang et al. (2012)
	CST20	*HWP1*	
	CPH1	*ECE1*	
Piperine	–	*HWP1*	Priya and Pandian (2020)
		HST7	
		RAS1	
		ECE1	
		CPH1	
		UME6	
		EFG1	
Ellagic acid	–	*HWP1*	Nejatbakhsh et al. (2020)
Berberine	–	*EFG1*	Huang et al. (2020)
		HWP1	
		ECE1	
Curcumin	*TUP1*	–	Sharma et al. (2010)

Table 7.1 (continued)

Natural compound	Upregulated genes	Downregulated genes	References
Quinic acid	*NRG1*	*HWP1*	Muthamil et al. (2018)
	TUP1	*EFG1*	
		UME6	
		CST20	
		RAS1	
		HST7	
		CPH1	

HWP hyphal wall protein, *ECE* endothelin converting, *EFG* enhanced filamentous growth protein, *RAS* ras-like protein, *CDC* adenylate cyclase, *PDE* 3′,5′-cyclic-nucleotide phosphodiesterase, *TEC1* transcription activator, *BCY1* cAMP-dependent protein kinase regulatory subunit, *HGC1* Hypha-specific G1 cyclin-related protein 1, *CYR1* Adenylate cyclase, *UME6* transcriptional regulatory protein, *CST* serine/threonine-protein kinase, *HST7* Serine/threonine-protein kinase STE7 homolog, *CPH* transcription factor, *NRG1* transcriptional regulator, *MIG1* regulatory protein, *TUP* transcriptional repressor, *HYR1* hyphally regulated cell wall protein

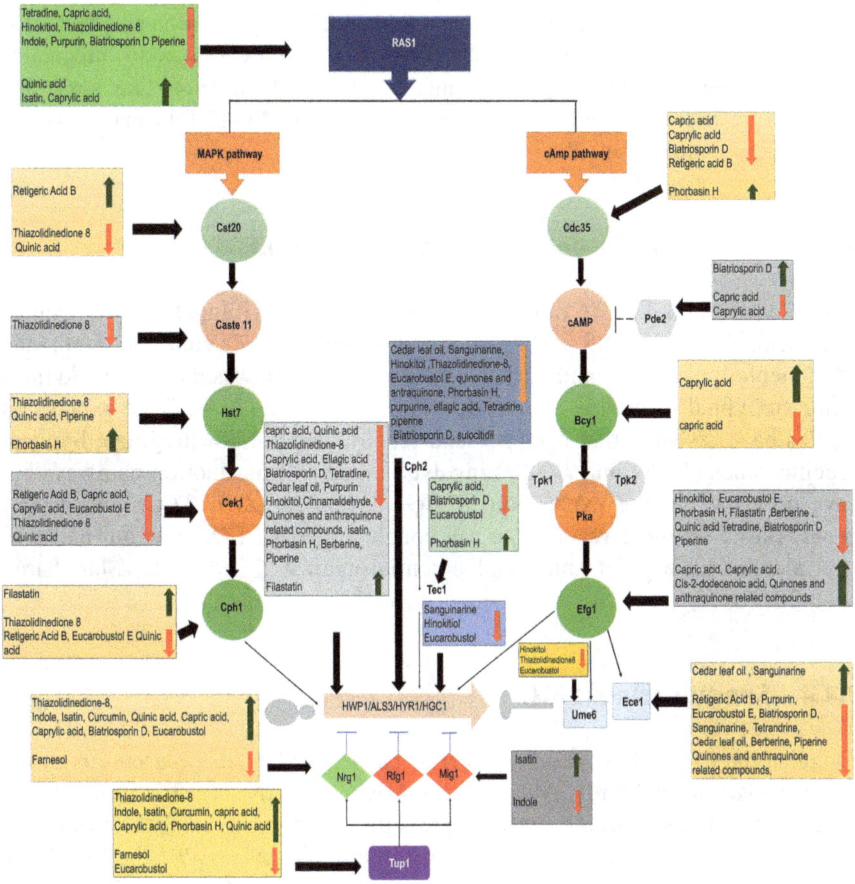

Fig. 7.3 Effect molecules on gene expression in *Candida albicans*

7.4.2 Capric Acid and Caprylic Acid

Capric acid and caprylic acids are constituents of mammalian milk and vegetable oils (e.g., coconut oil and palm oil). Both these compounds were found effective against all the major virulence factors of *C. albicans* like morphogenesis and biofilm formation. They inhibit growth of planktonic cells at low concentrations (0.25–0.5 mg/mL). Moreover, both the natural compounds are fungicidal in nature and affect the Ras-cAMP-Efg1 and MAPK pathways at minimum fungicidal concentrations. Expression of genes involved in serum-induced morphogenesis exhibited reduced expression of Cdc35, which encodes adenylate cyclase and is required for cAMP production. The Ras1-cAMP-PKA and Cek1-MAPK pathways were downregulated after the treatment of both capric acid and caprylic acid by 1.57- and 6.84-fold. After exposure to these acids, Pde2 was reduced by 1.46 and 6.41 times, respectively. *Hwp1* encodes for the hyphal wall proteins, and the genes *HST7* and *CPH1*, that are involved in the Cek1-MAPK pathway, get downregulated after treatment with these two acids. It was observed that capric acid increased the expression of *CEK1* by 2.27 times, whereas caprylic acid decreased it by 1.5-fold. Downregulation of *BCY1*, a gene having a role in cell differentiation and death was also observed. It was also found that cell elongation gene *ECE1* was downregulated by 52-fold. Genes, that are negative regulators of hyphal induction, like *NRG1* and *TUP1,* were overexpressed by the two fatty acids by 11–12-fold and 3–4-fold, respectively (Jadhav et al. 2017a).

7.4.3 Cedar Leaf Essential Oil Components (CLEO)

Cedar leaf essential oil (CLEO) possesses excellent antifungal activity against *C. albicans* biofilms (>85%). Its major components include camphor, fenchone, ethyl alcohol, α-thujone and borneol. Biofilm inhibition was mainly due to inhibition of hyphal formation at concentrations as low as 0.01%. Gene expression studies have revealed that camphor and fenchyl alcohol downregulate hyphae-specific genes (*ECE1* and *HWP1*) involved in signal transduction pathways but have no effect on the expression of *HGC1, HYR1, RAS1, TEC1* and *UME6* genes. Based on these results one can conclude that CLEO, camphor and fenchyl alcohol have great antifungal potential against *C. albicans* (Manoharan et al. 2017b).

7.4.4 Sanguinarine (SAN)

Sanguinarine (SAN) is a quaternary benzo phenanthridine alkaloid present in the *Papaveraceae* plant family. Although the MIC_{50} of SAN was 3.2 g/mL, it was observed that it could significantly suppress the growth of *C. albicans* biofilms at 0.8 g/mL. Treatment with SAN caused suppression of *ALS3, HWP1, ECE1, HGC1* and *CYR1* genes. This compound caused reduction of cAMP levels in *C. albicans,* but the addition of cAMP restores SAN-induced reduction of hyphal growth. It is relatively less cytotoxic as observed with the human umbilical vein endothelial cells (Zhong et al. 2017).

7.4.5 Hinokitiol

Hinokitiol or β-thujaplicin is a monoterpenoid present in the wood of *Chamacyparis taiwanensis* trees (family *Cupressaceae*). With an MIC of 1.6 g/mL and MFC of 100 μg/mL, hinokitiol proved very effective in preventing the development of planktonic *Candida* cells had has fungicidal properties. It could prevent formation of biofilms in both fluconazole-susceptible and fluconazole-resistant *Candida* species. The expression of genes associated with adhesion (*HWP1* and *ALS3*) and long-term hyphal maintenance (*UME6*, *CYR1*, *UME6* and *HGC1*), were downregulated in hinokitiol treated cells. Master regulator *RAS1* was also suppressed by hinokitiol (Kim et al. 2017).

7.4.6 Cinnamaldehyde (CNMA)

Cinnamaldehyde (CNMA), found in *Cinnamomum zeylanicum*, is very effective against planktonic growth with an MFC of 0.62 mM. It was observed that at 0.152 mM, CNMA showed synergy with Fluconazole at 8 μg/mL, inhibiting mature biofilms by more than 90%. Studies have shown that encapsulated preparations of CNMA in multilamellar liposomes (ML) show high effectiveness against *C. albicans*, being more fungicidal than free CNMA. Sustained release of CNMA caused cellular damage by generation of reactive oxygen species (ROS) leading to death of fungal cells. RT-PCR study revealed that ML-CNMA treated cells showed significantly reduced expression of *HWP1* gene, that encodes hyphal wall protein (Khan et al. 2017).

7.4.7 Thiazolidinedione-8

Thiazolidinedione-8 (S-8) is a bacterial quorum sensing quencher molecule found in *Vibrio harveyi* having antibiofilm and antiadhesion properties at concentrations four- to eightfold lower than MIC. Although the MIC of planktonic cell growth was 64 mg/mL, biofilm formation was inhibited by 50% at 8 mg/mL only. This compound, concentration-dependently, downregulated the expression of genes involved in adhesion and hyphae formation, including *HWP1* and *ALS3*. It is interesting to note that S-8 treatment did not affect the expression of *EFG1*, a component of the cAMP-PKA signaling cascade. The *UME6* and *RAS1* genes were also significantly downregulated. S-8 was shown to interfere with the signaling pathways involved in hyphal formation like cAMP-PKA and MAPK pathways. Moreover, the MAPK signal components (*CST20, HST7* and *CPH1*) displayed significant downregulation in treated cells. Significant upregulation of the transcriptional repressors of the filamentation process (*TUP1* and *NRG1*) was also observed in cells treated with this natural compound (Feldman et al. 2014).

7.4.8 Cis-2-Dodecenoic Acid (BDSF)

Cis-2-dodecenoic acid (BDSF) is a quorum-sensing molecule present in *Burkholderi acenocepacia*. It suppresses the hyphal formation and biofilms in clinical isolates of *C. albicans*. The viability of *C. albicans* is not significantly affected by BDSF even at concentrations up to 120 mM, but it reduces the adherence capacity of *C. albicans* by 4–25-fold at 90 mM. The *EFG1* gene was upregulated by roughly 1.8 times, whereas *YWP1* (downstream gene) was overexpressed by more than fourfold (Tian et al. 2013).

7.4.9 Biatriosporin D (BD)

Biatriosporin D (BD) is a small phenolic compound isolated from *Biatriospora* species, a fungus that lives on lichens. It was observed that BD prevents hyphal formation and could restrict the growth of *C. albicans* in the yeast forms after 24 h exposure. Three genes that encode adhesins (*ALS3, HWP1* and *ECE1*), had lower transcriptional levels after 6 h BD treatment. Additionally, genes linked to the Ras1-cAMP-Efg1 pathway (*RAS1, CDC35, EFG1* and *TEC1*), were dramatically downregulated by several orders of magnitude, while *PDE2* was upregulated in the presence of BD. Upregulation of *PDE2* and downregulation of *CDC35* causes decreasing intracellular cAMP concentrations. Reduced intracellular cAMP estimated in treated cells indicates that this natural compound regulates the Ras1-cAMP-Efg1 pathway inhibiting hyphal formation. It also causes downregulation of *ECE1* and stimulates the expression of Dpp3 to produce more farnesol. This straightaway inhibits Cdc35 activity and reduces intracellular cAMP interfering with the morphologic transition of *C. albicans* (Zhang et al. 2017).

7.4.10 Eucarobustol E (EE)

Eucarobustol E (EE) is a formyl phloroglucinol meroterpenoid, an important class of secondary metabolite available from the plants like *Eucalyptus* and *Psidium*. Its MIC50 ranged from 4 to 16 g/mL for fluconazole-susceptible strains and 32–128 g/mL for isolates that were resistant to the drug. EE exerted a strong inhibitory effect against *C. albicans* biofilms and blocked yeast to hyphal transition at 16 μg/mL. Exposure to the same concentration resulted in marked reduction in the expression of genes involved in hyphal growth namely, *EFG1, CPH1, TEC1, UME6, HGC1*, which were downregulated by 5.26-, 6.25-, 10.00- and 12.50-fold, respectively. The expression of genes encoding cell surface proteins (*ALS3* and *HWP*1) was also downregulated. Three key regulators of hyphal initiation, *EFG1, CPH1* and *TEC1*, were each downregulated by 1.58-, 4.35- and 5.00-fold following EE treatment. These findings corroborated well with the results obtained from ergosterol estimation along with farnesol combined with noticeable elevation of negative regulator genes (*TUP1* and *NRG1*) (Liu et al. 2017).

7.4.11 Quinones and Anthraquinone-Related Compounds

Quinones (Fig. 7.2) are found as biological pigments in sea urchins, aphids, lac insects and certain scale insects, while anthraquinones are found in plants of families like *Rubiaceae* and *Leguminoseae*. The presence of a hydroxyl group at the C-1 position is important for antibiofilm and antifilamentation activities. The expression of the hyphae-specific genes *ALS3* (2.4-fold), *ECE1* (3.7-fold) and *ECE2* (6.3-fold) was dramatically downregulated by alizarin (1,2 dihydroxyanthraquinone) at 2 g/mL which further suppressed biofilm formation in *C. albicans*. Furthermore, following 2 g/mL treatment with chrysazin (1,8 dihydroxyanthraquinone), the expression of *ECE1* and *ECE2* was dramatically downregulated by twofold and 2.3-fold, respectively. Alizarin treatment led to an increase in the expression of the hyphae-regulating gene *EFG1* (Manoharan et al. 2017a).

7.4.12 Indole and Isatin

Indole is an aromatic organic compound found in different organism and is a decomposition product of tryptophan. Its derivative, Isatin (1H-indole-2,3-dione), has important physiological functions in humans. Both the molecules are capable of inhibiting yeast to hyphal form switching in *C. albicans* at a concentration of 0.25 mg/mL. Indole downregulated the expression of 11 genes, while isatin upregulated 12 genes involved in signal transduction pathway. Indole only caused an 11.6-fold downregulation of *ECE1*, but isatin caused a 25.5-fold downregulation. In contrast to the 35-fold downregulation caused by isatin, *HWP1* was downregulated 77-folds by Indole. Negative hyphal regulators, such as *NRG1* and *TUP1*, were upregulated by indole and isatin. However, indole reduced *MIG*1 by 1.4 times. Out of the 11 genes downregulated after the indole treatment, four of them were from the Cek1-MAPK pathway. On the other hand, isatin upregulated four genes from this pathway. Both the compounds downregulated the expression of Ece1 and Hwp1 genes. Isatin generated a 78-fold overexpression of *NRG1* but indole only produced a 1.5-fold increase. Isatin and Indole upregulated *TUP1* by 12 and by 5.5 times, respectively. *Nrg1* and *Tup1* are suppressor genes that were significantly upregulated in treated cells. Their enhanced expression affected the expression of multiple other genes in the yeast to hyphal signal transduction pathway in *C. albicans* (Jadhav et al. 2017b).

7.4.13 Phorbasin H

Phorbasin H, a diterpene carboxylic acid isolated from a marine sponge *Phorbas* sp., is an effective inhibitor of yeast-to-hyphal transition in *C. albicans*. Phorbasin H-treated cells showed reduced expression of genes related to the cAMP–Efg1 pathway, but did not inhibit the expressions of *CPH1, EFG1* and *TUP1* genes, as shown by Northern blot analysis. Interestingly, 125 g/mL phorbasin H caused a total decrease in *HWP1* mRNA expression. Addition of a cAMP analog (Dibutyryl-cAMP) did not have any effect on hyphal formation in *C. albicans*. This compound significantly inhibited the expression of hypha-specific genes (*HWP1* and *ALS3*), which are supposed to be positively regulated by *EFG1*, an important regulator of cell wall dynamics. This compound thus inhibited concentration-dependent adherence and morphogenesis in *C. albicans*. (Lee et al. 2013).

7.4.14 Farnesol

C. albicans produces a quorum-sensing molecule extracellularly called farnesol. It also acts as an autoregulatory component that prevents yeast cells from germination. It inhibits the Ras-cAMP pathway through direct inhibition of cyr1. Though it acts as a regulator of yeast to hyphal morphogenesis, the exact mechanism is not clear (Lindsay et al. 2012). Farnesol sensitivity was observed when growth media and conditions were altered. *C. albicans* gives a response to farnesol by altering the gene expression. An increase in *TUP1* expression was observed after treatment with farnesol while the expression of *HWP1* decreased. *NRG1* remain unaffected by the farnesol. The expression of *EFG1* activates *HWP1*. The mRNA levels of *EFG1* were regulated during filamentation, but farnesol had no impact on it (Lindsay et al. 2012).

7.4.15 Purpurin

The food coloring agent purpurin is a natural red anthraquinone pigment usually isolated from madder roots (*Rubia cordifolia*). The sublethal concentration of 3 μg/mL was inhibitory for yeast to hyphal transition and biofilm formation in *C. albicans*. It was observed that purpurin negatively impacted metabolic activity of mature *Candida* biofilms and the reduction was concentration dependent. It was observed that the expression of hyphae-specific genes *ALS3*, *ECE1* and *HYR1* was downregulated by 59%, 61% and 40%, respectively, by purpurin. While the expression of *HWP1* dropped by more than 88%, that of the hyphal growth regulator *RAS1* reduced by only 40% (Tsang et al. 2012).

7.4.16 Retigeric Acid B

Lichens (e.g., Lobaria *kurokawae*) have a pentacyclic triterpene called retigeric acid B (RAB). This natural acid has antifungal potential alone and in combination with the conventional azoles. It is particularly effective against azole-resistant *C. albicans* strains in vitro at 8–16 mg/mL. The gene that encodes phosphodiesterase, *PDE2*, was upregulated after RAB exposure. The results revealed that Ras1 and RAB do not interact directly. RAB did not appear to have any impact on *EFG1*. Phosphodiesterase encoded by PDE2 increased. In the MAPK cascade, *CST20* and *CPH1* were increased as a kind of feedback for the Ras1-cAMPEfg1 pathway that was blocked. RAB also represses *CDC35* activity by stimulating farnesol production which leads to a decrease in cAMP synthesis, leading to defective morphological transition and adhesion (Chang et al. 2012).

7.4.17 Piperine

Piperine is the main bioactive alkaloid present in pepper seeds and is known for giving them their characteristic pungency. Even at the highest tested dose (2–1024 μg/mL), piperine did not significantly affect the development of *C. albicans*. A maximum of 93% of the biofilms were reported to be suppressed by piperine at a dosage of 32 μg/mL. It dramatically reduced the expression of signal transduction genes (*HWP1, HST7, RAS1* and *ECE1*), filamentous growth transcriptional regulators (*CPH1, UME6* and *EFG1*) and biofilm-related genes (ALS3) at biofilm inhibiting concentrations (Priya and Pandian 2020).

7.4.18 Ellagic Acid

Ellagic acid has antioxidant, antimicrobial and anti-inflammatory properties. This phenolic component of several plants and fruits can inhibit *C. albicans* growth at 12.5 μg/mL after 48 h incubation. It inhibits biofilm formation at 25 μg/mL. Gene expression studies revealed that ellagic acid causes downregulation of two hyphal-specific genes that encode for hyphal wall protein (*HWP1*) and adhesin like protein (ALS3) (Nejatbakhsh et al. 2020).

7.4.19 Berberine

Berberine is generally present in the roots, rhizomes, stems and bark of some plants like *Berberis*. It is an isoquinolene alkaloid, a secondary metabolite that belongs to the protoberberine class. Berberine hydrochloride significantly inhibits the biofilms of *C. albicans*. This natural compound can inhibit the expression of four major genes (*EFG1*, *HWP1* and *ECE1*) involved in signal transduction pathways (Huang et al. 2020).

7.4.20 Curcumin

Curcumin is a polyphenol extracted from the rhizome of *Curcuma longa*. It inhibits the growth of *Candida* species and the mode of antifungal action involves increased production of reactive oxygen species (ROS) and induction of early apoptosis. To study the antifungal mechanism of curcumin, various mutants of *C. albicans* were employed. The oxidative stress mutant was more susceptible than the other mutants like that of morphogenesis and ion transport. It was also observed that growth inhibition and elevated ROS levels can be reversed if growth medium is supplemented with natural or synthetic antioxidants. ROS stimulate the proapoptotic controlling apparatus in *Candida* cells and hence increase apoptotic cell death. CUR could significantly block the development of hyphae in both *C. albicans* and other major *Candida* species even at low concentrations. In wild-type strains, CUR increased the transcript levels of *TUP1*, a transcriptional repressor that negatively controls filamentous growth in *C. albicans* (Sharma et al. 2010).

7.4.21 Quinic Acid

Quinic acid is a cyclic polyol obtained from diverse plant sources. This cyclohexanecarboxylic acid can be extracted from Tobacco leaves, *Eucalyptus globules, Hymenocrater calycinus, Tara spinosa, Ageratina adenophora, Urtica dioica,* coffee beans and barks of *Cinchona* trees, and several fruits and vegetables. Quinic acid, when used in combination with an antifungal like undecanoic acid, showed synergy in suppressing the pathogenicity of various *Candida* strains. This synergistic combination significantly inhibited biofilm formation, morphological transition, synthesis of extracellular polymeric substances, secretion of hydrolases and ergosterol biosynthesis. In vivo studies, carried out with *Caenorhabditis elegans*, revealed the nontoxic nature of Quinic acid-undecanoic acid combination and its antivirulence effect against *Candida* spp. The genes crucial to biofilm formation were studied for their expression in the presence of this combination. It was found that *ALS1, HWP1, EFG1* and *UME6* genes were majorly downregulated, while *ALS3, CST20, RAS1, HST7* and *CPH1* genes were moderately downregulated.

NRG1 and *TUP1* were two genes that showed slight upregulation (Muthamil et al. 2018).

7.5 Discussion

Fungal infections create severe problems having adverse effects on the health of a large number of immunocompromised patients that ultimately cause a financial burden on the healthcare system. High mortality rates due to these infections indicate that current antifungal therapy is not so effective due to the toxicity of drugs and increasing resistance to available drugs. These are some major challenges faced by researchers and clinical practitioners. It has now become clear that the signal transduction pathways have an important role to play in many cellular processes involving cell growth, morphogenesis, virulence, cellular mating, stress response and cell death. Phytochemicals have many bioactive properties, and as they come from natural sources may be considered as safe. So, these bioactive molecules may be considered for the development of new therapeutic agents. Many plant molecules have been investigated as inhibitors of *Candida* biofilm. Some of them have been able to inhibit biofilms at very low concentrations and are required to be selected for further evaluation. Since most of the molecules target multiple events, the development of resistance may not happen. The molecules inhibiting the signal transduction pathway should be further studied for in vivo efficiency and toxicity. So, they can be proved to be good candidates for drug development.

7.6 Conclusion

Colonization of prosthetic devices by *Candida albicans* and the formation of biofilms is a significant clinical problem faced by the patients, physicians and pharmaceutical scientists. The biofilms formed are resistant to most of the antifungal antibiotics. As such, there is a necessity to identify novel antibiofilm agents. Natural molecules derived from plant origin have various bioactive properties including antifungal activity. Some of the molecules targeting signal transduction pathways involved on yeast to hyphal morphogenesis are reviewed here. These molecules can be good candidates for developing new antifungal drugs with negligible toxicity.

Acknowledgments Authors are thankful to DY Patil Education Society (Deemed to be University), Kolhapur, 416006, Maharashtra, India, under grant number DYPES/DU/R&D/1154 and DYPES/DU/R&D/2021/273 for providing funding support for research.

References

Bar-Yosef H, Vivanco Gonzalez N, Ben-Aroya S, Kron SJ, Kornitzer D (2017) Chemical inhibitors of *Candida albicans* hyphal morphogenesis target endocytosis. Sci Rep 7(1):5692. https://doi.org/10.1038/s41598-017-05741-y

Bockmühl DP, Krishnamurthy S, Gerads M, Sonneborn A, Ernst JF (2001) Distinct and redundant roles of the two protein kinase A isoforms Tpk1p and Tpk2p in morphogenesis and growth of *Candida albicans*. Mol Microbiol 42(5):1243–1257. https://doi.org/10.1046/j.1365-2958.2001.02688.x

Bradshaw RA, Dennis EA (2009) Cell signaling: yesterday, today, and tomorrow. In: Handbook of cell signaling, 2nd edn. Elsevier, London, pp 1–4

Brown DA, Yang N, Ray SD (2014) Apoptosis. In: Encyclopedia of toxicology, 3rd edn. Elsevier, London, pp 287–294. https://doi.org/10.1016/B978-0-12-386454-3.00242-6

Cao YY, Cao YB, Xu Z, Ying K, Li Y, Xie Y, Zhu ZY, Chen WS, Jiang YY (2005) cDNA microarray analysis of differential gene expression in *Candida albicans* biofilm exposed to farnesol. Antimicrob Agents Chemother 49(2):584–589. https://doi.org/10.1128/AAC.49.2.584-589.2005

Chang W, Li Y, Zhang L, Cheng A, Lou H (2012) Retigeric acid B attenuates the virulence of *Candida albicans* via inhibiting adenylyl cyclase activity targeted by enhanced farnesol production. PLoS One 7(7):41624. https://doi.org/10.1371/journal.pone.0041624

Davis-Hanna A, Piispanen AE, Stateva LI, Hogan DA (2008) Farnesol and dodecanol effects on the *Candida albicans* Ras1-cAMP signalling pathway and the regulation of morphogenesis. Mol Microbiol 67(1):47–62. https://doi.org/10.1111/j.1365-2958.2007.06013.x

Ding X, Cao C, Zheng Q, Huang G (2017) The regulatory subunit of protein kinase A (Bcy1) in candida albicans plays critical roles in filamentation and white-opaque switching but is not essential for cell growth. Front Microbiol 7:2127. https://doi.org/10.3389/fmicb.2016.02127

Feldman M, Al-Quntar A, Polacheck I, Friedman M, Steinberg D (2014) Therapeutic potential of thiazolidinedione-8 as an antibiofilm agent against Candida albicans. PLoS One 9(5):2–9. https://doi.org/10.1371/journal.pone.0093225

Feng Q, Summers E, Guo B, Fink G (1999) Ras signaling is required for serum-induced hyphal differentiation in *Candida albicans*. J Bacteriol 181(20):6339–6346. https://doi.org/10.1128/jb.181.20.6339-6346.1999

Huang X, Zheng M, Yi Y, Patel A, Song Z, Li Y (2020) Inhibition of berberine hydrochloride on *Candida albicans* biofilm formation. Biotechnol Lett 42(11):2263–2269. https://doi.org/10.1007/s10529-020-02938-6

Jadhav A, Mortale S, Halbandge S, Jangid P, Patil R, Gade W, Kharat K, Karuppayil SM (2017a) The dietary food components capric acid and caprylic acid inhibit virulence factors in *Candida albicans* through multitargeting. J Med Food 20(11):1083–1090. https://doi.org/10.1089/jmf.2017.3971

Jadhav AK, Bhan A, Jangid P, Patil R, Gade W, Karuppayil SM (2017b) Modulation of genes involved in yeast to hyphal form signal transduction in *Candida albicans* by Indole and Isatin. Curr Signal Transduct Ther 12(2):116–123. https://doi.org/10.2174/1574888x12666170425100459

Kaoud TS, Park H, Mitra S, Yan C, Tseng CC, Shi Y, Jose J, Taliaferro JM, Lee K, Ren P, Hong J, Dalby KN (2012) Manipulating JNK signaling with (−)-zuonin A. ACS Chem Biol 7(11):1873–1883. https://doi.org/10.1021/cb300261e

Khan SN, Khan S, Iqbal J, Khan R, Khan AU (2017) Enhanced killing and antibiofilm activity of encapsulated cinnamaldehyde against *Candida albicans*. Front Microbiol 8:1641. https://doi.org/10.3389/fmicb.2017.01641

Kim DJ, Lee MW, Choi JS, Lee SG, Park JY, Kim SW (2017) Inhibitory activity of hinokitiol against biofilm formation in fluconazole-resistant *Candida* species. PLoS One 12(2):e0171244. https://doi.org/10.1371/journal.pone.0171244

Kornitzer D (2019) Regulation of *Candida albicans* hyphal morphogenesis by endogenous signals. J Fungi 5(1):21. https://doi.org/10.3390/jof5010021
Lee SH, Jeon JE, Ahn CH, Chung SC, Shin J, Oh KB (2013) Inhibition of yeast-to-hypha transition in *Candida albicans* by phorbasin H isolated from Phorbas sp. Appl Microbiol Biotechnol 97(7):3141–3148. https://doi.org/10.1007/s00253-012-4549-3
Li Y, Shan M, Zhu Y, Yao H, Li H, Gu B, Zhu Z (2020) Kalopanaxsaponin A induces reactive oxygen species mediated mitochondrial dysfunction and cell membrane destruction in *Candida albicans*. PLoS One 15(11):e0243066. https://doi.org/10.1371/journal.pone.0243066
Lindsay AK, Deveau A, Piispanen AE, Hogan DA (2012) Farnesol and cyclic AMP signaling effects on the hypha-to-yeast transition in *Candida albicans*. Eukaryot Cell 11(10):1219–1225. https://doi.org/10.1128/EC.00144-12
Liu RH, Shang ZC, Li TX, Yang MH, Kong LY (2017) In vitro antibiofilm activity of eucarobustol E against *Candida albicans*. Antimicrob Agents Chemother 61(8):e02707-16. https://doi.org/10.1128/AAC.02707-16
Lowy DR, Willumsen BM (1993) Function and regulation of Ras. Annu Rev Biochem 62:851–891
Lu J, Fang K, Wang S, Xiong L, Zhang C, Liu Z, Guan X, Zheng R, Wang G, Zheng J, Wang F (2018) Anti-inflammatory effect of columbianetin on lipopolysaccharide-stimulated human peripheral blood mononuclear cells. Mediat Inflamm 2018:9191743. https://doi.org/10.1155/2018/9191743
Manoharan RK, Lee JH, Kim YG, Lee J (2017a) Alizarin and chrysazin inhibit biofilm and hyphal formation by *Candida albicans*. Front Cell Infect Microbiol 7:447. https://doi.org/10.3389/fcimb.2017.00447
Manoharan RK, Lee JH, Lee J (2017b) Antibiofilm and antihyphal activities of cedar leaf essential oil, camphor, and fenchone derivatives against *Candida albicans*. Front Microbiol 8:1476. https://doi.org/10.3389/fmicb.2017.01476
Muthamil S, Balasubramaniam B, Balamurugan K, Pandian SK (2018) Synergistic effect of quinic acid derived from syzygium cumini and undecanoic acid against candida spp. biofilm and virulence. Front Microbiol 9:1–23. https://doi.org/10.3389/fmicb.2018.02835
Nejatbakhsh S, Ilkhanizadeh-Qomi M, Razzaghi-Abyaneh M, Jahanshiri Z (2020) The effects of ellagic acid on growth and biofilm formation of *Candida albicans*. J Med Microbiol Infect Dis 8(1):14–18. https://doi.org/10.29252/jommid.8.1.14
Papin JA, Hunter T, Palsson BO, Subramaniam S (2005) Reconstruction of cellular signalling networks and analysis of their properties. Nat Rev Mol Cell Biol 6:99–111
Park H, Myers CL, Sheppard DC, Phan QT, Sanchez AA, Edwards JE, Filler SG (2005) Role of the fungal Ras-protein kinase A pathway in governing epithelial cell interactions during oropharyngeal candidiasis. Cell Microbiol 7(4):499–510. https://doi.org/10.1111/j.1462-5822.2004.00476.x
Park YK, Shin J, Lee HY, Kim HD, Kim J (2021) Development of carbazole derivatives compounds against candida albicans: candidates to prevent hyphal formation via the Ras1-MAPK pathway. J Fungi 7(9):688. https://doi.org/10.3390/jof7090688
Pentland DR, Piper-Brown E, Mühlschlegel FA, Gourlay CW (2018) Ras signalling in pathogenic yeasts. Microb Cell 5:63–73
Pierce CG, Srinivasan A, Uppuluri P, Ramasubramanian AK, López-Ribot JL (2013) Antifungal therapy with an emphasis on biofilms. Curr Opin Pharmacol 13:726–730
Polke M, Sprenger M, Scherlach K, Albán-Proaño MC, Martin R, Hertweck C, Hube B, Jacobsen ID (2017) A functional link between hyphal maintenance and quorum sensing in *Candida albicans*. Mol Microbiol 103(4):595–617. https://doi.org/10.1111/mmi.13526
Priya A, Pandian SK (2020) Piperine impedes biofilm formation and hyphal morphogenesis of *Candida albicans*. Front Microbiol 11:756. https://doi.org/10.3389/fmicb.2020.00756
Richardson JP (2022) Candida albicans: a major fungal pathogen of humans. Pathogens 11:459
Rocha CRC, Schröppel K, Harcus D, Marcil A, Dignard D, Taylor BN, Thomas DY, Whiteway M, Leberer E (2001) Signaling through adenylyl cyclase is essential for hyphal growth and viru-

lence in the pathogenic fungus *Candida albicans*. Mol Biol Cell 12(11):3631–3643. https://doi.org/10.1091/mbc.12.11.3631

Senn H, Shapiro RS, Cowen LE (2012) Cdc28 provides a molecular link between Hsp90, morphogenesis, and cell cycle progression in *Candida albicans*. Mol Biol Cell 23(2):268–283. https://doi.org/10.1091/mbc.E11-08-0729

Sharma M, Manoharlal R, Puri N, Prasad R (2010) Antifungal curcumin induces reactive oxygen species and triggers an early apoptosis but prevents hyphae development by targeting the global repressor TUP1 in *Candida albicans*. Biosci Rep 30(6):391–404. https://doi.org/10.1042/BSR20090151

Stoldt VR, Sonneborn A, Leuker CE, Ernst JF (1997) Efg1p, an essential regulator of morphogenesis of the human pathogen *Candida albicans*, is a member of a conserved class of bHLH proteins regulating morphogenetic processes in fungi. EMBO J 16(8):1982–1991. https://doi.org/10.1093/emboj/16.8.1982

Tian J, Weng LX, Zhang YQ, Wang LH (2013) BDSF inhibits *Candida albicans* adherence to urinary catheters. Microb Pathog 64:33–38. https://doi.org/10.1016/j.micpath.2013.07.003

Tsang PWK, Bandara HMHN, Fong WP (2012) Purpurin suppresses *Candida* albicans biofilm formation and hyphal development. PLoS One 7(11):e50866. https://doi.org/10.1371/journal.pone.0050866

Villa S, Hamideh M, Weinstock A, Qasim MN, Hazbun TR, Sellam A, Hernday AD, Thangamani S (2020) Transcriptional control of hyphal morphogenesis in *Candida albicans*. FEMS Yeast Res 20:foaa005

Wang T, Shao J, Da W, Li Q, Shi G, Wu D, Wang C (2018) Strong synergism of palmatine and fluconazole/itraconazole against planktonic and biofilm cells of *Candida* species and efflux-associated antifungal mechanism. Front Microbiol 9:2892. https://doi.org/10.3389/fmicb.2018.02892

Xu XL, Lee RTH, Fang HM, Wang YM, Li R, Zou H, Zhu Y, Wang Y (2008) Bacterial peptidoglycan triggers *Candida albicans* hyphal growth by directly activating the adenylyl cyclase Cyr1p. Cell Host Microbe 4(1):28–39. https://doi.org/10.1016/j.chom.2008.05.014

Zhang M, Chang W, Shi H, Zhou Y, Zheng S, Li Y, Li L, Lou H (2017) Biatriosporin D displays anti-virulence activity through decreasing the intracellular cAMP levels. Toxicol Appl Pharmacol 322:104–112

Zhao LX, Li DD, Hu DD, Hu GH, Yan L, Wang Y, Jiang YY (2013) Effect of tetrandrine against *Candida albicans* biofilms. PLoS One 8(11):e79671. https://doi.org/10.1371/journal.pone.0079671

Zhong H, Hu DD, Hu GH, Su J, Bi S, Zhang ZE, Wang Z, Zhang RL, Xu Z, Jiang YY, Wang Y (2017) Activity of sanguinarine against *Candida albicans* biofilms. Antimicrob Agents Chemother 61(5):e02259-16. https://doi.org/10.1128/AAC.02259-16

8 Harnessing the Antifungal Potential of Natural Products

Neha Jaiswal and Awanish Kumar

8.1 Introduction

In the realm of fungi, there are known to be over 10,000 genera and 100,000 species, the vast majority of which have medicinal properties. The fermentation of food and wine by *Saccharomyces cerevisiae* and the treatment of infections with antibiotics derived from *Penicillium* are two examples of such applications. On the other hand, certain fungi can be lethal to people. These include species of *Aspergillus*, *Candida, Cryptococcus* and *Sporothrix*, which can cause deep systemic infections like aspergillosis, candidiasis, cryptococcosis and sporotrichosis, and *Trichophyton,* which can cause superficial dermatophytosis in humans. Twenty to twenty-five percent people worldwide suffer from dermatophytosis, one of the most serious fungal diseases that can quickly destroy tissue, organ and nerve. Recent years have seen a significant increase in the number of people dying from fungal infections, especially in tropical regions of the world, where between 30% and 50% of the population is affected. Furthermore, pathogenic fungi have become resistant to conventional medications due to the overuse of antifungal medications. Medicinal plants have been used to provide remedies to improve human health since ancient times. They provide an abundance of therapeutic molecules with a broad range of chemical and functional diversity, which makes them perfect for discovering new drugs. Throughout the past few years, there has been a notable increase in research on natural compounds that have antifungal properties. The therapeutic application of natural products is still far off, even though their antifungal effects are often superior to those of commercial medications. The structural modification of naturally occurring compounds that exhibit high fungal selectivity represents a viable resolution to this problem. Thus far, natural product screening has yielded several highly

N. Jaiswal · A. Kumar (✉)
Department of Biotechnology, National Institute of Technology, Raipur, Chhattisgarh, India
e-mail: nehaggv@gmail.com; drawanishkr@gmail.com

N. Manzoor (ed.), *Advances in Antifungal Drug Development*,
https://doi.org/10.1007/978-981-97-5165-5_8

effective antifungal drugs, such as pyrimidine analogues, polyenes and echinocandins. These drugs were created by optimising the molecular scaffolds that were found to have antifungal properties.

Many naturally occurring substances with promiscuous activity and low selectivity have been found in various plant sections. A study documented the mechanisms of action of each type of natural product in terms of their structural categories and gave examples of the most effective natural products in the early stages of research in a report on natural product antifungal agents known from 1990 to 2009 (Di Santo 2010). We collected data on natural products with antifungal effects from the relevant literature, highlighting specific values of their antifungal activity from the perspective of pharmacological antifungal mechanisms, in the light of the development of antifungal natural products over the past 20 years. Using the widely used commercial antifungal medications now in the market as a comparison, we highlight the potential and significance of antifungal natural products and provide an overview of the relevant mechanisms of action, chemical types and antifungal effects. This chapter may provide evidence for the future development of novel antifungal drugs.

8.2 Antifungal Efficacy of Plant Extracts

Primary metabolites, which give rise to secondary plant metabolites, include proteins, lipids, carbohydrates and chlorophyll. It is believed that these substances are biologically active. Terpenes are a class of compounds made up of various isoprene unit (5-carbon base) variations. Spices and herbs are the main sources of monoterpenes (C10) and diterpenes (C20), which are antibacterial and effective against a variety of illnesses. Steroids have a branched carbon chain structure and are derived from tetracyclic triterpenes (C30). Steroids have been reported to have antifungal and antimicrobial qualities (Adetunji et al. 2021). Alkaloids are plant secondary metabolites that have the potential to be toxic as well as beneficial. They are usually found in poisonous plants. These compounds may be effective against bacteria, fungi and parasites in addition to their cytotoxicity towards cancer cells. A phenolic chemical group is made up of any number of compounds with an aromatic ring and at least one hydroxyl substituent. This group of secondary metabolites consists of phenolic acids, tannins and flavonoids. It has been established that phenolic compounds possess antioxidant and antibacterial qualities (Zandavar and Afshari Babazad 2023). Furthermore, it is known that specific isolated groups of phenolic compounds, including flavonoids, phenolic acids and tannins, have antifungal properties (Kumar and Goel 2019). When it comes to antibacterial properties, the types of chemicals present in each extract often matter more than how much of them are flavonoids and phenolic compounds (Álvarez-Martínez et al. 2021). Essential oils have a big impact on extract composition and can change the properties of aromatic plants that are used in food, medicine and aromatherapy. Majority of the complex of essential oils are composed of hydrocarbons (terpenes and sesquiterpenes) and oxidised substances (alcohols, aldehydes, ketones, acids, phenols, oxides, lactones, acetals, ethers and esters (Álvarez-Martínez et al. 2021). The inherent aroma of a plant is derived from the combination of these aromatic components, which fall into these two categories and represent the flavour and aroma of 60 distinct essential oils and extracts. The range of plant extract composition, solubility, pH, volatility, diffusion properties in a growth medium and assessed pathogens all influence the results of the inhibition of microorganisms. For example, hydrolysis of amino acids by *Allium sativum* extract during the hydroalcoholic extraction process yields secondary metabolites with antimicrobial properties; Aloe vera extract has strong antifungal properties due to the presence of anthraquinones (aloin, barbaloin and isobarbaloin); and the antifungal activity of *Glycyrrhiza galbra* extract is determined by the presence of saponins, flavonoids, coumarins and essential oils (Álvarez-Martínez et al. 2021; Baptiste Hzounda Fokou et al. 2020). Numerous studies have examined the antibacterial and antifungal characteristics of plant extracts and essential oils. For example, among the 50 medicinal plants whose antibacterial activity was evaluated by Srinivasan et al. was *Eucalyptus globulus*. When it came to bacteria, *Chromobacterium, Escherichia coli, Klebsiella pneumonia, Enterobacter faecalis, Pseudomonas aeruginosa, Proteus mirabilis, Salmonella partyphi, S. typhi, Bacillus subtilis* and *Staphylococcus aureus* showed no antifungal activity in contrast to the tested fungus (Srinivasan et al. 2001). The antibacterial activity of macrocarpals, phloroglucinol derivatives found in Eucalyptus leaves, was studied and compared in a range of different oral bacteria. Of the microorganisms under study, *P. gingivalis* exhibited the highest sensitivity to macrocarpals. Additionally, macrocarpals decreased the proteinase activity like trypsin and the binding to saliva-coated hydroxyapatite beads (Nagata et al. 2006). The antibacterial effects of *Peganum harmala* leaf extracts was evaluated in ethyl acetate, chloroform, butanol and methanol against a range of pathogens, including 11 Gram-positive and 6 Gram-negative bacteria. Among these, Gram-positive bacteria were more susceptible to the antibacterial effects of methanol and chloroform extracts than the Gram-negative bacteria. The antibacterial qualities of the extracts from *Sophora flavescens* (SE), *Angelica sinensis* (AE) and the herb pair, *A. sinensis* and *S. flavescens* (HPE), were investigated (Han and Guo 2012). The outcomes demonstrated that HPE had strong antibacterial activity against *Shigella Castellani, Staphylococcus aureus, Escherichia coli* and *Chalmers.* Additionally, SE activity against *E. coli* was moderate. Studies also investigated how well *Melia azedarach L.* leaf extracts worked as antimicrobial agents against eight different human pathogens, including *Bacillus cereus, Pseudomonas aeruginosa, Aspergillus flavus, Aspergillus niger, Fusarium oxisporum, Rhizopus stolonifera* and *Staphylococcus aureus.* The alcoholic extract exhibited the highest inhibitory concentration against all microbes, even though all extracts demonstrated significant action against all the infections. While investigating the antibacterial activity of *Alhagi maurorum* leaf extract, it was discovered that the crude extract, chloroform and ethyl acetate fractions showed significant effects against *Bacillus anthrax*, with over 80% inhibition. The crude extract exhibited 80% inhibition when tested against *Shigella dysenteriae* (Hadadi et al. 2020). Similarly, the crude extract and ethyl acetate fraction showed strong suppression of the growth of *Salmonella typhi* by 78.35% and 76.50%, respectively. Antioxidants may also function as free radical scavengers and aid in the prevention of cancer and heart problems because they may neutralise dangerous reactive free radicals in host cells before they can cause protein and lipid oxidation and minimise potential mutation. High concentrations of phytochemical antioxidants found in plants, such as flavonoids, carotenoids, phenolics and tannins, can be utilised by the body to scavenge excess free radicals. Plant extracts and essential oils have been shown in numerous studies to possess antioxidant properties. In a previous study, the antioxidant activity of the leaf extracts from *Peganum harmala* evaluated in ethyl acetate, chloroform, butanol and methanol was found to be highest in the latter. *Hyssopus officinalis* has an antioxidant effect, albeit not as strong as that of ascorbic acid and butylated hydroxytoluene. It was found that the high phenolic plant content was probably the reason why *Alhagi maurorum* leaf extracts and fractions demonstrated potent radical scavenging activity. Analysing the phytochemical composition, antibacterial capabilities and antioxidant effects of several important medicinal plant extracts was the main objective of the current study (Yu et al. 2021; Hadadi et al. 2020).

8.3 Antifungal Efficacy of Plant Essential Oils

Essential oils come from plants through a process called steam distillation or hydrodistillation, as well as by the expression method (citrus peel oils). They are complex, naturally occurring mixtures of volatile secondary metabolites. Aromatic and therapeutic plants derive their biological properties and scent from monoterpenes and sesquiterpenes, which are the main constituents of essential oils, which also include alcohols, ethers, aldehydes and ketones. Because of these characteristics, food has been seasoned with herbs and spices for flavour and preservation since ancient times. Essential oils have long been used for similar purposes after being extracted from other plant sections. The applications for essential oils are very broad (Sharifi-Rad et al. 2017). Numerous essential oils have pharmacological effects and have antioxidant, anti-inflammatory and anticarcinogenic qualities. Others work as biocides against a variety of organisms, including viruses, bacteria, fungi, protozoa, insects and plants. More than 500 publications have examined the antifungal qualities of various essential oils and their compounds against various fungi in recent years. Some of the studies have been summarised in Table 8.1.

Strong antifungal essential oils typically contain significant levels of phenols. The oils of *Thymus* spp. and *Origanum* spp. are more frequently mentioned as being rich in those chemicals. It has been demonstrated that *Origanum* oil and carvacrol, one of its main constituents, can suppress *C. albicans* germination and mycelial development in vitro in a dose-dependent way (Chandrakala et al. 2023). Additional research has demonstrated that *O. virens* samples with high concentrations of carvacrol, which prevent the production of germ tubes, account for a strong antifungal action against *Candida* spp. (Aswandi et al. 2024; Singh et al. 2023). Furthermore, it is well known that *O. vulgare* essential oil, like many other plant extracts, exhibits significant compositional diversity because of the existence of several chemotypes as well as environmental and climatic variations. Compared to the other extracts mentioned above, one made from the leaves of *Origanum vulgare* species in Turkey

Table 8.1 Antifungal activity of some plant extracts and essential oils

Plant	Target microorganisms	Main antifungal effects	References
Baccharis trinervis Pers	*Trichophyton rubum* (4 strains)	MIC: 160 μg mL^{-1}	Sobrinho et al. (2016)
		MFC: 410 μg mL^{-1}	
		Synergistic and additive effects with ketoconazole	
Blepharocalyx salicifolius (Kunth) O. Berg	*T. mentagrophytes* (1 strain) and *Microsporum canis* (1 strain)	MIC: 625 μg mL^{-1} (*T. mentagrophytes*) and 2500 μg mL^{-1} (*M. canis*)	Furtado et al. (2018)
Cymbopogon citratus (DC.) Stapf	*Nannizzia gypsea* (1 strain), *T. mentagrophytes* (1 strain) and *M. canis* (1 strain)	MIC: 16 μg mL^{-1} (*N. gypsea*), 8 μg mL^{-1} (*T. mentagrophytes*), 62 μg mL^{-1} (*M. canis*)	Shah et al. (2011)
Croton nepetaefolius Baill	*M. canis* (10 strains)	Total inhibition of fungal growth from 25,000 μg mL^{-1}	Fontenelle et al. (2008)
Croton argyrophylloides Müll. Arg	*M. canis* (10 strains)	Total inhibition of fungal growth from 25,000 μg mL^{-1}	Neri et al. (2023)
Lippia alba (Mill.) N.E.Br. ex Britton & P. Wilson	*T. mentagrophytes* (1 strain) and *T. rubrum* (1 strain)	MIC: 31.25 μg mL^{-1} (*T. rubrum*) and 125 μg mL^{-1} (*T. mentagrophytes*)	Tangarife-Castaño et al. (2011)
Lippia origanoides Kunth	*T. mentagrophytes* (1 strain) and *T. rubrum* (1 strain)	MIC: > 500 μg mL^{-1} (*T. rubrum*) and 396.85 μg mL^{-1} (*T. mentagrophytes*)	Tangarife-Castaño et al. (2011)
Lippia gracilis Schauer (LGRA 106 genotype)	*T. rubrum* (2 strains)	MIC and MFC: 23.44 μg mL^{-1}	de Melo et al. (2013)
Lippia gracilis Schauer (LGRA 109 genotype)	*T. interdigitale* (2 strains)	MIC: 46.87 μg mL^{-1}	de Melo et al. (2013)
		MFC: 93.75 μg mL^{-1}	
Myrcia guianensis (Aubl.) DC	*M. canis* (1 strain) and *T. rubrum* (1 strain)	Inhibition of fungal growth from dilution 17,375 μg mL^{-1}	Aneke et al. (2018)
Mentha piperita L	*T. rubrum* (2 strains), *T. violaceum* (1 strain) and *T. soudanense* (1 strain)	Total inhibition of fungal growth from 2 μL mL^{-1}	Su et al. (2019)
		MIC: 2 μL mL^{-1}. MFC: not determined	
Mentha piperita L. (chocolate mint)	*T. rubrum* (2 strains) and *M. canis* (2 strains)	MIC: 620 μg mL^{-1} (*T. rubrum*), 2500 μg mL^{-1} (*M. canis*). MFC: 1250 μg mL^{-1} (*T. rubrum*) and >2500 μg mL^{-1} (*M. canis*)	Marwal et al. (2012)

(continued)

Table 8.1 (continued)

Plant	Target microorganisms	Main antifungal effects	References
Mentha piperita L. (grapefruit mint)	*T. rubrum* (2 strains) and *M. canis* (2 strains)	MIC: 620 μg mL^{-1} (*T. rubrum* and *M. canis*). MFC: 1250 μg mL^{-1} (*T. rubrum* and *M. canis*)	Marwal et al. (2012)
Ocimum gratissimum L	*M. canis* (5 strains), *N. gypsea* (5 strains), *T. rubrum* (5 strains) and *T. mentagrophytes* (10 strains)	Total inhibition of fungal growth above 250 μg mL^{-1}	Silva et al. (2005)
Piper aduncum L	*T. mentagrophytes* (1 strain), *T. tonsurans* (1 strain)	Total inhibition of fungal growth above 500 μg mL^{-1}	Taher et al. (2020)
Piper arboreum Aubl	*N. gypsea* (1 strain) and *Epidermophyton floccosum* (1 strain)	Diameter of inhibition zone: 15 mm (*E. floccosum)* and 23 mm (*N. gypsea*)	Taher et al. (2020)
		MIC: 62.5 μg mL^{-1} (*E. floccosum*) and 156.25 μg mL^{-1} (*N. gypsea*)	
		MFC: 125 μg mL^{-1} (*E. floccosum*) and 312.5 μg mL^{-1} (*N. gypsea)*	
Piper obliquum Ruiz & Pav	*T. mentagrophytes* (1 strain) and *T. tonsurans* (1 strain)	No inhibitory effects	Taher et al. (2020)

showed a distinct chemical composition, with caryophyllene and spathulenol being the main constituents (Singh et al. 2023). These results seem to show a novel oil chemotype that is active against *C. albicans*. Due to their extensive research on antifungal properties, essential oils are currently being considered as complementary or alternative treatments for fungal diseases (Baptiste Hzounda Fokou et al. 2020; Chandrakala et al. 2023; Aswandi et al. 2024). The most researched essential oils in terms of their chemical makeup and in vitro/in vivo investigations are included in this section with an eye towards potential clinical applications.

8.4 Antifungal Efficacy of Plant-Derived Natural Compounds

A broad spectrum of fungi, including species of *Candida, Torulopsis, Trichophyton, Aspergillus, Trichosporon* and *Rhodotorula*, were shown to be susceptible to the fungicidal effects of plant extracts. It was recently shown that some plant-extracted compounds inhibited the germination and proliferation of *Rhodotorula mucilaginosa* and *Meyerozyma guilliermondii*. Caprolactam is one of the unique secondary metabolites that actinomycetes make that has antifungal properties (Arif et al. 2009). Plant products have been utilised for food and the treatment of chronic illnesses since ancient times. According to estimates from the World Health Organisation, almost 80% of people on the planet still receive their primary care from plant-based formulations even on this day and age of scientific advancement. Apart from their nutritional role, plants generally contain a complex mixture of bioactive compounds with functional properties claiming beneficial physiological effects (Prakash et al. 2020). Bioactive substances derived from plants, such as isoflavones (found in soy extracts), lutein, zeaxanthin and lycopene (found in tomatoes), β-glucan (found in oats), curcumin (found in haldi), and (−)-epicatechin, (−)-epicatechin-3-gallate, (−)-epigallocatechin and (−)-epigallocatechin-3-gallate (found in green tea), have already been used as dietary supplements to improve health (Prakash et al. 2012). Therefore, plant-derived bioactive compounds could be used to create fortified foods that are effective in treating chronic diseases through diet, as functional foods are foods that offer health benefits beyond those of essential nutrients.

8.5 Antifungal Efficacy of Plant-Based Nanoparticles

The production of nanoparticles from plant extracts is one of the most promising and environment-friendly strategies to expand the applications of antifungals in medicine and agriculture. Like the previously discussed antibacterial and anticancer agents, that also produce free radicals and reactive oxygen species, metallic nanoparticles destroy fungal species by releasing metal ions from their nanostructure (Vanlalveni et al. 2021). As nanoparticles get closer to the surface of microorganisms, they have a number of negative interactions with the cell wall that lead to the leakage of intracellular components (Singh et al. 2019). Specifically, they disrupt redox homeostasis and induce oxidative stress, which weakens fungal cell walls. This breakdown causes a loss of membrane integrity and a homeostatic imbalance, which makes it possible for nanoparticles to enter the cells. By interacting with the phosphorus or sulphur moieties of DNA, nanoparticles can also halt DNA replication, which inhibits the growth of germs and results in cell death (Bakhtiari et al. 2021). They also have a variety of other effects, such as ribosome denature, disruption and inhibition of enzymes, protein denaturation and inhibition of ribosome-synthesised proteins. In particular, glutathione-producing enzymes (GHS), which are antioxidant enzymes that reduce fungal resistance, can be blocked by metallic nanoparticles (Sun et al. 2018). Synthesised Ag and Au nanoparticles from *S. nodiflora* leaf extract perform antifungal actions (Zare et al. 2012; Abbasi et al. 2018). The findings indicated that the inhibition zones of Ag nanoparticles formed against the species *Penicillium* and *Aspergillus* were roughly 10.9 and 12.9 mm, respectively. Under the same conditions, Au nanoparticles show smaller zones inhibition (~9 mm) for both *Penicillium* and *Aspergillus* species. These findings demonstrated that compared to Au nanoparticles, Ag nanoparticles had a higher potential for antifungal activity. The most logical explanation might come from the fact that Ag nanoparticles are smaller than Au nanoparticles (19.4 nm) and can therefore more easily pass through fungal cell membranes and destroy them. Qasim Nasar et al.

(2019) synthesised Ag nanoparticles from *Seriphidium quettense* extract with a much larger particle size between 48.40 and 55.35 nm, in contrast to the findings of Vijayan et al.'s (2018) study (Regidi et al. 2018; Qasim Nasar et al. 2019). However, the antifungal results against *Aspergillus* species seemed more promising because the inhibition zone was larger, measuring 13.2 mm. As a result, Ag nanoparticles biosynthesised from *S. quettense* extract exhibited greater antifungal activity than those biosynthesised from *S. nodiflora*. The significant abundance of 40–56 μg of phenolic and flavonoid components per mg of dried extract in the plant extracts of *S. quettense* is responsible for this outcome. As a result, the bioactive ingredients that were added to the plant extracts during production also affect the ability of metallic nanoparticles to inhibit fungal growth. Other nanoparticles with antifungal qualities against *C. albicans*, *C. tropicalis* and *C. oxysporum* could be used in addition to expensive metallic nanoparticles like Au, Ag and ZnO (Nguyen et al. 2022). A previous study successfully synthesised ZnO nanoparticles from *C. scolymus* (globe artichoke) leaf extract. In this study, the minimum inhibitory concentrations (MIC) of ZnO nanoparticles against *C. albicans* and *C. tropicalis* were roughly 100 and 0.35 μg/mL, respectively. This result indicated that ZnO exhibited about 300 times more antifungal activity against *C. tropicalis* than it did against *C. albicans* (Nguyen et al. 2022). The antifungal efficacy of Ag nanoparticles derived from *M. azedarach* leaf extracts was investigated (Moreno-Vargas et al. 2023). Accordingly, the percentage of Ag nanoparticle inhibition against *V. dahliae* was 18%, 33% and 51% at the concentrations of 20, 40 and 60 ppm, respectively. It can be concluded that greater concentrations of Ag nanoparticles have stronger antifungal effects. In a different study, ZnO nanoparticles were synthesised from *Lactuca aculeata* extract for antifungal effect (Narendhran and Sivaraj 2016). According to the results, the inhibitory zones against *A. flavus* and *F. oxysporum* were found to be at 19 and 21 mm, respectively (Narendhran and Sivaraj 2016). In reference to additional nanoparticles, it has been reported that green CuO nanoparticles prepared from *Moringa oleifera* leaf extract may have antifungal properties (Pagar et al. 2020; Phang et al. 2021). Thus, in general, a range of fungal species can be significantly suppressed by biofabricated metallic nanoparticles (primarily Au, Ag and ZnO) derived from plant extracts.

One of the most promising and eco-friendly ways to increase the antifungal usage in the medical and agricultural domains is the manufacture of nanoparticles from plant extracts. Metallic nanoparticles release metal ions from their nanostructure to destroy fungal species, adopting a similar mechanism to that of the previously discussed antibacterial and anticancer agents that also produces free radicals and reactive oxygen species (Lipovsky et al. 2011). Nanoparticles approach the surface of microorganisms and produce several detrimental interactions with the cell wall, causing intracellular components to seep out (Altammar 2023). In particular, they cause oxidative stress and redox homeostasis to change, which damages cell walls. This breakdown results in a loss of membrane integrity and a homeostatic imbalance, which allows nanoparticles to penetrate the cells. Nanoparticles can also stop DNA replication, which stops the growth of germs and causes cell death through the mechanism of interactions with the phosphorus or sulphur moieties of

DNA. Furthermore, they cause a range of effects, including protein denaturation and damage, disruption and inhibition of enzymes, ribosome denature and inhibition of proteins synthesised by ribosomes. Specifically, metallic nanoparticles have the ability to block glutathione-producing enzymes (GHS), an antioxidant enzyme that lowers fungal resistance (Nethravathy and Dakshayini 2023). The excellent antifungal activity of nanoparticles made from different plant extracts is displayed in Table 8.1. In fact, Ag and Au nanoparticles were created from *S. nodiflora* leaf extract in order to carry out antifungal activities. As a result, the inhibition zones of Ag nanoparticles against *Penicillium* and *Aspergillus species* were found to be approximately 10.9 and 12.9 mm, respectively. Au nanoparticles exhibit the smaller zone inhibition (~9 mm) for both *Penicillium and Aspergillus species* under the same *circumstances* (Vijayan et al. 2018). These results made it clear that Ag nanoparticles had a greater potential for antifungal activity than Au nanoparticles. The plausible explanation could potentially stem from the fact that Ag nanoparticles have a smaller particle size (19.4 nm) than Au nanoparticles (22.01 nm), which enables them to penetrate fungal cell membranes and cause their destruction more easily. Synthesised Ag nanoparticles from *S. quettense* extract with a much bigger particle size between 48.40 and 55.35 nm. Since the inhibition zone was wider—at 13.2 mm—the antifungal results against *Aspergillus species*, however, appeared to be more promising (Qasim Nasar et al. 2019). Therefore, compared to Ag nanoparticles biosynthesised from *S. nodiflora*, those biosynthesised from *S. quettense* extract shown superior antifungal activity. This result can be attributed to the substantial abundance of 40–56 μg of phenolic and flavonoid components per mg of dried extract in the plant extracts of *S. quettense*. Because of this, the antifungal properties of metallic nanoparticles depend not only on their size but also on the bioactive substances that were added to the plant extracts during production. In addition to valuable metallic nanoparticles like Au, Ag and ZnO, other nanoparticles with antifungal properties against *C. albicans, C. tropicalis* and *C. oxysporum* could be employed. In fact, ZnO nanoparticles have been successfully synthesised from *C. scolymus* (globe artichoke) leaf extract (Rather et al. 2022). The MIC of ZnO nanoparticles against *C. albicans* and *C. tropicalis* in this investigation were approximately 100 and 0.35 μg/mL, respectively. This outcome showed that ZnO had antifungal activity against *C. tropicalis* that was approximately 300 times higher than that against *C. albicans*. Ag nanoparticles from *M. azedarach* leaf extracts were investigated for their antifungal efficacy showing that, at the concentrations of 20, 40 and 60 ppm, respectively, the percentage inhibition of Ag nanoparticles against *V. dahliae* was 18%, 33% and 51%, respectively. One could draw the conclusion that, stronger antifungal activity results from higher Ag nanoparticle concentrations. ZnO nanoparticles from *L. aculeata* extract for an antifungal impact in a different investigation. The findings showed that the inhibitory zones formed against *F. oxysporum* and *A. flavus* were of the size 19 and 21 mm, respectively (Narendhran and Sivaraj 2016). Further studies on nanoparticles showed potential antifungal activity of green CuO nanoparticles made with *Moringa oleifera* leaf extract. In general, biofabricated metallic nanoparticles (mostly Au, Ag and ZnO) from plant extracts have significant ability to suppress a variety of fungal species (Pagar et al. 2020).

8.6 Antifungal Mechanism of Action of the Natural Products

Drugs can target the cell wall, cell membrane and nucleus of fungi to decrease their activity or even kill them directly. A number of targets are known, including DNA and protein synthesis (topoisomerases, nucleases, elongation factors), cell wall polymers (glucans, chitin, mannoproteins), signal transduction pathways (protein kinases and protein phosphatases) and myristoylation (addition of myristic acid to the N-terminal glycine of proteins). For instance, azoles can function as sterol synthesis inhibitors, influencing the enzymes responsible for producing it and ultimately damaging the integrity of cell membranes. Echinocandins inhibit (1,3)-β-D-glucan synthetase activity, stop cell wall synthesis and make it harder for the cells to keep their normal structure, which ultimately leads to cell rupture (Szymański et al. 2022). Pyrimidine analogues obstruct the normal DNA or RNA molecular expression. Currently, there are a number of antifungal medications in the market, such as fluconazole, ketoconazole, voriconazole and itraconazole; some of these drugs are prescribed along with amphotericin B to treat skin infections caused by *Candida* and ringworm; however, the majority of these medications are synthetic or semi-synthetic compounds that, although effective against fungi, can also damage the cells of human host (Zhang et al. 2023). Because of insignificant toxicity and their ability to bind specifically to fungi, certain natural compounds have been proven to be preferable to synthetic ones. In addition to acting obstructively and targeting cell membranes, cell walls and different organelles, their underlying molecular targets can result in internal diseases that hinder the ability of fungal cells to proliferate. Furthermore, additional structural optimisation has improved natural product antifungal efficacy, making them comparable to synthetic medications. The natural antifungal agents with therapeutic efficacy and their molecular targets are therefore summarised in this section.

8.6.1 Cell Wall as Antifungal Target

Chitin and glucans make up most of the cell walls found in fungi. Without glucans, a cell finds it difficult to retain its regular shape, which interferes with its regular metabolism and robs it of its protective function. Disrupting the production of its cell wall would therefore be a very simple and effective approach in this situation. Nowadays, echinocandins, which include micafungin and caspofungin, are the primary antifungal medications used to treat this target. Because human cells do not have a cell wall, pharmaceuticals are more selective and have no effect on the human body. Nonetheless, certain fungi exhibit resistance to echinocandins, with the majority of this resistance stemming from modifications in the catalytic subunit of (1,3)-β-D-glucan synthase. This type of fungus can be treated in conjunction with polyenes; pertinent studies have confirmed that the combination of amphotericin B and caspofungin functions as an additive or synergist for over 50% of *Aspergillus* and *Fusarium* species (Healey and Perlin 2018). By blocking the fungal glucan synthase, we compiled the natural products and derivatives with antifungal properties. For instance, 50% of the phospholipase activity and 72% of the esterase activity of *C. neoformans* from interfering cell walls can be inhibited by (+)-α-pinene (Adetunji et al. 2021) at sub-MIC concentrations. Against *Aspergillus* species, anidulafungin functions as a noncompetitive antagonist binding to (1,3)-β-D-glucan synthase, with minimum effective concentration MEC $\leq$ 0.015 μg/mL (Vahedi-Shahandashti and Lass-Flörl 2020). Ibrexafungerp is efficacious against a variety of *Candida* species, including isolates of *C. auris*. The investigated *C. auris* isolates had MIC_{50} and MIC_{90} values of 0.5 and 1.0 μg/mL, respectively. Ibrexafungerp exerts fungicidal action against *Candida* spp. by inhibiting (1,3)-β-D-glucan synthase (Sucher et al. 2022). An acidic terpenoid called arundifungin was initially discovered in *Arthrinium* spp. from tropical and temperate areas of central Spain. Arundifungin preferentially inhibited the (1,3)-β-D-glucan synthase, causing morphological alterations in moulds and yeasts (Cabello et al. 2001).

8.6.2 Cell Membrane as Antifungal Target

Fungal cell membranes are mostly made of proteins, carbohydrates, phospholipids and ergosterols. As the primary structural component, ergosterol adds to the makeup of the cellular membranes. CYP51 (14α-demethylase), an enzyme belonging to the cytochrome P450 family and expressed by the Erg11 gene, is responsible for processing and maturing ergosterol. It plays a mono-oxygenated role in sterol synthesis in vivo, making it a promising target for drugs (Zhang et al. 2019). If inactivated, it directly disrupts the synthesis of ergosterol in fungal cells. The fungus is unable to properly synthesise its membrane, which causes morphological abnormalities in the membrane that impede or kill cell division. For this target, azole medications such as fluconazole, ketoconazole, itraconazole and others have good efficiency and a respectable specificity. These medications work specifically by attaching their nitrogen azole ring to the iron atom in the centre of the iron porphyrin ring. This prevents oxygen from attaching to the iron atom, obstructing oxygen transfer and resulting in cell death. For many years, fluconazole has been the primary medicine of choice for treating *C. albicans* infections and has consistently shown good efficacy (Martin 1999).

However, azole medication resistance has recently developed in *C. albicans*, leading to the emergence of a number of resistant *Candida* species. One of the main causes is that, as a result of long-term drug screening, sterol synthase genes were altered. This, combined with elevated gene levels, cause some sterol synthases, such as CYP51, to become overexpressed, which in turn causes resistance (Becher and Wirsel 2012). Moreover, the medications are continuously expelled from the cells by efflux pumps, which prevents them from forming an efficient concentration within the cells, and the extracellular and intracellular drainage pumps are modified.

The ability of azoles to bind to iron atoms in cytochrome CYP51, which also enables them to bind to iron atoms in heme subunits with similar structures in the human body, has drawn attention to their safety. This binding prevents heme from carrying out its oxygen-carrying function normally, which can have negative effects, including specific hepatic and renal toxicity.

Regarding the natural products, sanguinarine inhibited the formation of ergosterol, making it an antifungal isoquinoline alkaloid with MIC of 32 μg/mL against *Candida* (Hu et al. 2022). When treated with *Candida* spp., berberine was found to be strongly antifungal, with a MIC of around 10 μg/mL. It was also suggested that berberine affects CYP51 (Xie et al. 2020). Tomatidine, which was extracted from *Solanum lycopersicum*, was discovered to have no cytotoxicity towards mammalian cells and to have possible fungistatic action against *Candida* species ($MIC_{50} < 1$ μg/mL) (Bailly 2021). The primary target of tomatidine was found to be Erg6, C-24 sterol methyl-transferase, not found in mammals. Cinnamaldehyde is the primary ingredient and flavouring agent of cinnamon essential oil. This natural compound has been shown by researchers to impede the growth of *Fusarium sambucinum*, maybe by interfering with the process of ergosterol production, which compromises the integrity of cell membranes (Wei et al. 2020). The primary active ingredients in *Melaleuca alternifolia* essential oil are terpinen-4-ol and 1,8-cineole, which have shown to affect *Botrytis cinerea* cell membranes and organelles, respectively. Terpinen-4-ol was found to decrease the ergosterol content of cell membranes, but its exact mechanism of action is unknown; it was be related to CYP51.

8.6.3 Fungal DNA as Antifungal Target

A group of vital enzymes known as topoisomerases are necessary to start DNA replication, and they are also a major target for many antimicrobial substances. Topoisomerase-specific inhibitors have the ability to stabilise the covalent protein–DNA connection, hence demonstrating the reduced catalysis phase and ultimately causing damage to DNA and cellular death (Han et al. 2022). Although the fungal and mammalian cells exhibit significant similarities in DNA replication and RNA translation, DNA/RNA synthesis has historically proven to be a challenging target for the development of selectively toxic antifungal medications. Development in functional genomics and molecular biology are starting to draw attention to significant distinctions between fungal and mammalian cells that may be used to create new antifungal treatments.

8.7 Application of Natural Antifungals

Traditional medicine has been used for a long time and is particularly useful in the conventional treatment of infections, tumours, cardiovascular and cerebrovascular function regulation and bruises. The discovery of some herbal preparations, including six medications like Jinhua Qinggan granules, was found to have a good inhibitory effect on COVID-19. This finding provides traditional Chinese medicine with a good starting point for addressing the challenge of finding mild antifungal drugs and precursor compounds. Numerous identified active ingredients in herbal remedies include polysaccharides, flavonoids, terpenoids and saponins. Since ancient times, people have recognised the antifungal characteristics of plant products and other natural solutions, which are now utilised in antifungal ointments. Essential oils, such as tea tree oil, garlic oil and orange peel liquid, are among the many plant-based goods that are seen to be the most promising natural products due to their mild nature and complex composition of active ingredients. People have applied various mixtures of plant essential oils or broken pieces of garlic to the affected area of fungal infections to considerably limit fungal growth and lessen or even completely eradicate symptoms. It is important to note that certain herbs can be refined into ointments by grinding them and applying them topically to the affected area. These ointments have good antifungal properties and are the main type of herbal preparations available in the market for treating fungal infections. One such product is 'Hua Tuo ointment, the thousand-year wonder herb'. While most herbal prescriptions call for the combination of several herbs, screening and scanning of herbal components—particularly through additional research and development of traditional Chinese medicine—has revealed a number of monomeric compounds with equally potent antifungal activity, like thymol and cinnamaldehyde, which are also specific components of essential oils and were previously mentioned to disrupt fungal cell membranes and affect normal fungal metabolism.

Due to the rise in fungal resistance and the fact that some antifungal medications have major adverse effects in humans, despite the discovery of many commercially viable drugs so far, a growing amount of focus has been directed towards natural products. Animal cells, microbial extracts and plant extracts have all been subjected to high-throughput screening to produce a wide range of exceptional antifungals. Recent years have seen the discovery of novel chemical scaffolds derived from natural products as a result of the quest to identify novel fungal molecular targets. Of natural compounds with strong antifungal activity, some structural modifications have been made to improve antifungal activity. Numerous natural products have been produced by plants, marine life and microbes. Nevertheless, certain monotherapies are becoming outdated due to the rise in drug-resistant fungus. In these situations, medications that have lost their effectiveness may be added to combination treatments in order to re-sensitise the pathogen through the use of suitable collaborating chemicals. Furthermore, compared to pharmaceuticals, natural products typically have milder side effects and will probably cause fewer antagonistic effects. It was observed that *Pellaea calomelanos* extract and benzoyl metronidazole alone did not significantly inhibit *Candida* growth; however, when the two treatments were combined, *P. calomelanos* extract significantly increased the antifungal activity of the drug.

Natural products are a priceless tool for strengthening the antifungal defence. Several novel natural compounds with promising antifungal action against human infections have been reported in literature reviews. The chemicals are derived from a variety of species, including plants, algae, fungus, sponges and bacteria. The distribution of these sources suggests that the most frequent sources of antifungal chemicals were bacteria and fungus. In fact, antifungals that have been licenced, including polyenes and echinocandins, have demonstrated this. A plethora of examples of broad-spectrum antifungal fungal cell secondary metabolites have been reported in recent years. One such example is parnafungin, which was isolated from *Fusarium larvarum* and inhibits polyadenosine polymerase, obstructing the 3′ end of the molecule to prevent fungal mRNA synthesis from occurring. In a similar vein, *Bacillus cereus* produces the new aminopolyol antibiotic Zwittermicin A, which exhibits antifungal action. *Aspergillus nidulans* ferments echinocandin B, which is a derivative of anidulafungin. The most well-known class of natural products with antifungal properties, echinocandins, are used in commercial drugs. Three generations of antifungal drugs—caspofungin, micafungin and anidulafungin—have been created by chemically modifying echinocandins; of these, caspofungin is the most developed and is used in clinical practise the most, while anidulafungin received FDA approval in 2006. Additionally, new targets for antifungal therapy are being identified, which opens the door to the development of antifungal natural products and novel, potentially effective medicines that can potentially overcome drug resistance.

8.8 Conclusion

A comprehensive understanding of the distinctions between a drug's mode of action and target which dictate whether the medicine primarily targets fungal cells or also affects human cells can be gained by summarising the antifungal processes of several natural compounds. We hope to be inspired to perform further structural modifications based on a variety of natural products and believe that in the future, the side effects of drugs alone or even in combination can be reduced to a greater extent by reviewing the antifungal natural products and their underlying targets and finding more specific and compelling targets for fungal pathogens.

References

Abbasi BH, Zaka M, Hashmi SS, Khan Z (2018) Biogenic synthesis of Au, Ag and Au–Ag alloy nanoparticles using *Cannabis sativa* leaf extract. IET Nanobiotechnol 12(3):277–284. https://doi.org/10.1049/iet-nbt.2017.0169

Adetunji CO, Palai S, Ekwuabu CP, Egbuna C, Adetunji JB, Ehis-Eriakha CB et al (2021) General principle of primary and secondary plant metabolites: biogenesis, metabolism, and extraction. In: Preparation of phytopharmaceuticals for the management of disorders. Elsevier, London, pp 3–23. https://doi.org/10.1016/B978-0-12-820284-5.00018-6

Altammar KA (2023) A review on nanoparticles: characteristics, synthesis, applications, and challenges. Front Microbiol 14:1155622. https://doi.org/10.3389/fmicb.2023.1155622

Álvarez-Martínez FJ, Barrajón-Catalán E, Herranz-López M, Micol V (2021) Antibacterial plant compounds, extracts and essential oils: an updated review on their effects and putative mechanisms of action. Phytomedicine 90:153626. https://doi.org/10.1016/j.phymed.2021.153626

Aneke C, Otranto D, Cafarchia C (2018) Therapy and antifungal susceptibility profile of *Microsporum canis*. J Fungi 4(3):107. https://doi.org/10.3390/jof4030107

Arif T, Bhosale JD, Kumar N, Mandal TK, Bendre RS, Lavekar GS et al (2009) Natural products—antifungal agents derived from plants. J Asian Nat Prod Res 11(7):621–638. https://doi.org/10.1080/10286020902942350

Aswandi A, Kholibrina CR, Kuspradini H (2024) Phytochemical, essential oils and product applications from eucalyptus. In: Eucalyptus. Springer Nature, Singapore, pp 163–183. https://doi.org/10.1007/978-981-99-7919-6_11

Bailly C (2021) The steroidal alkaloids α-tomatine and tomatidine: panorama of their mode of action and pharmacological properties. Steroids 176:108933. https://doi.org/10.1016/j.steroids.2021.108933

Bakhtiari MN, Hassan N, Parveen R, Shahnaz G (2021) Nanoparticles induced oxidative stress: a review based approach. Glob Drug Des Dev Rev 6(4):57–66. https://doi.org/10.31703/gdddr.2021(VI-IV).05

Baptiste Hzounda Fokou J, Michel Jazet Dongmo P, Fekam Boyom F (2020) Essential oil's chemical composition and pharmacological properties. In: Essential oils—oils of nature. IntechOpen, London. https://doi.org/10.5772/intechopen.86573

Becher R, Wirsel SGR (2012) Fungal cytochrome P450 sterol 14α-demethylase (CYP51) and azole resistance in plant and human pathogens. Appl Microbiol Biotechnol 95(4):825–840. https://doi.org/10.1007/s00253-012-4195-9

Cabello AM, Platas G, Collado J, Díez TM, Martín I, Vicente F et al (2001) Arundifungin, a novel antifungal compound produced by fungi: biological activity and taxonomy of the producing organisms. Int Microbiol 4(2):93–102. https://doi.org/10.1007/s101230100020

Chandrakala V, Aruna V, Angajala G, Reddy PG (2023) Chemical composition and pharmacological activities of essential oils. In: Essential oils. Wiley, Hoboken, pp 229–268. https://doi.org/10.1002/9781119829614.ch11

de Melo JO, Bitencourt TA, Fachin AL, Cruz EMO, de Jesus HCR, Alves PB et al (2013) Antidermatophytic and antileishmanial activities of essential oils from Lippia gracilis Schauer genotypes. Acta Trop 128(1):110–115. https://doi.org/10.1016/j.actatropica.2013.06.024

Di Santo R (2010) Natural products as antifungal agents against clinically relevant pathogens. Nat Prod Rep 27(7):1084–1098. https://doi.org/10.1039/b914961a

Fontenelle ROS, Morais SM, Brito EHS, Brilhante RSN, Cordeiro RA, Nascimento NRF et al (2008) Antifungal activity of essential oils of Croton species from the Brazilian Caatinga biome. J Appl Microbiol 104(5):1383–1390. https://doi.org/10.1111/j.1365-2672.2007.03707.x

Furtado F, Borges B, Teixeira T, Garces H, Almeida Junior L, Alves F et al (2018) Chemical composition and bioactivity of essential oil from Blepharocalyx salicifolius. Int J Mol Sci 19(1):33. https://doi.org/10.3390/ijms19010033

Hadadi Z, Nematzadeh GA, Ghahari S (2020) A study on the antioxidant and antimicrobial activities in the chloroformic and methanolic extracts of 6 important medicinal plants collected from North of Iran. BMC Chem 14(1):33. https://doi.org/10.1186/s13065-020-00683-5

Han C, Guo J (2012) Antibacterial and anti-inflammatory activity of traditional Chinese herb pairs, Angelica sinensis and Sophora flavescens. Inflammation 35(3):913–919. https://doi.org/10.1007/s10753-011-9393-6

Han S, Lim KS, Blackburn BJ, Yun J, Putnam CW, Bull DA et al (2022) The potential of topoisomerase inhibitor-based antibody–drug conjugates. Pharmaceutics 14(8):1707. https://doi.org/10.3390/pharmaceutics14081707

Healey KR, Perlin DS (2018) Fungal resistance to echinocandins and the MDR phenomenon in *Candida glabrata*. J Fungi 4(3):105. https://doi.org/10.3390/jof4030105

Hu Z, Hu H, Hu Z, Zhong X, Guan Y, Zhao Y et al (2022) Sanguinarine, isolated from macleaya cordata, exhibits potent antifungal efficacy against *Candida albicans* through inhibiting ergosterol synthesis. Front Microbiol 13:908461. https://doi.org/10.3389/fmicb.2022.908461

Kumar N, Goel N (2019) Phenolic acids: natural versatile molecules with promising therapeutic applications. Biotechnol Rep 24:e00370. https://doi.org/10.1016/j.btre.2019.e00370

Lipovsky A, Nitzan Y, Gedanken A, Lubart R (2011) Antifungal activity of ZnO nanoparticles—the role of ROS mediated cell injury. Nanotechnology 22(10):105101. https://doi.org/10.1088/0957-4484/22/10/105101

Martin MV (1999) The use of fluconazole and itraconazole in the treatment of *Candida albicans* infections: a review. J Antimicrob Chemother 44(4):429–437. https://doi.org/10.1093/jac/44.4.429

Marwal A, Meena S, Chandra S, Sharma A (2012) In vitro study of antidermatophytic activity of mint (Mentha Piperita) against trichophyton rubrum and *Microsporum canis*. J Med Sci 12(6):182–187. https://doi.org/10.3923/jms.2012.182.187

Moreno-Vargas JM, Echeverry-Cardona LM, Moreno-Montoya LE, Restrepo-Parra E (2023) Evaluation of antifungal activity of Ag nanoparticles synthetized by green chemistry against *Fusarium solani* and *Rhizopus stolonifera*. Nano 13(3):548. https://doi.org/10.3390/nano13030548

Nagata H, Inagaki Y, Yamamoto Y, Maeda K, Kataoka K, Osawa K et al (2006) Inhibitory effects of macrocarpals on the biological activity of *Porphyromonas gingivalis* and other periodontopathic bacteria. Oral Microbiol Immunol 21(3):159–163. https://doi.org/10.1111/j.1399-302X.2006.00269.x

Narendhran S, Sivaraj R (2016) Biogenic ZnO nanoparticles synthesized using *L. aculeata* leaf extract and their antifungal activity against plant fungal pathogens. Bull Mater Sci 39(1):1–5. https://doi.org/10.1007/s12034-015-1136-0

Neri TS, Silva KWL, Maior LPS, Oliveira-Silva SK, Azevedo PVM, Gomes DCS et al (2023) Phytochemical characterization, antioxidant potential and antibacterial activity of the Croton argyrophylloides Muell. Arg. (Euphorbiaceae). Braz J Biol 83:e236649. https://doi.org/10.1590/1519-6984.236649

Nethravathy V, Dakshayini M (2023) Potential antioxidant enzymes from fungi and their clinical significance. In: Fungal resources for sustainable economy. Springer Nature, Singapore, pp 147–177. https://doi.org/10.1007/978-981-19-9103-5_6

Nguyen NTT, Nguyen LM, Nguyen TTT, Nguyen TT, Nguyen DTC, Van Tran T (2022) Formation, antimicrobial activity, and biomedical performance of plant-based nanoparticles: a review. Environ Chem Lett 20(4):2531–2571. https://doi.org/10.1007/s10311-022-01425-w

Pagar K, Ghotekar S, Pagar T, Nikam A, Pansambal S, Oza R, Sanap D, Dabhane H (2020) Antifungal activity of biosynthesized CuO nanoparticles using leaves extract of moringa oleifera and their structural characterizations. Asian J Nanosci Mater 3(1):15–23. https://doi.org/10.26655/AJNANOMAT.2020.1.2

Phang Y-K, Aminuzzaman M, Akhtaruzzaman M, Muhammad G, Ogawa S, Watanabe A et al (2021) Green synthesis and characterization of CuO nanoparticles derived from papaya peel extract for the photocatalytic degradation of palm oil mill effluent (POME). Sustainability 13(2):796. https://doi.org/10.3390/su13020796

Prakash B, Singh P, Kedia A, Dubey NK (2012) Assessment of some essential oils as food preservatives based on antifungal, antiaflatoxin, antioxidant activities and in vivo efficacy in food system. Food Res Int 49(1):201–208. https://doi.org/10.1016/j.foodres.2012.08.020

Prakash B, Kumar A, Singh PP, Songachan LS (2020) Antimicrobial and antioxidant properties of phytochemicals. In: Functional and preservative properties of phytochemicals. Elsevier, London, pp 1–45. https://doi.org/10.1016/B978-0-12-818593-3.00001-4

Qasim Nasar M, Zohra T, Khalil AT, Saqib S, Ayaz M, Ahmad A et al (2019) *Seripheidium quettense* mediated green synthesis of biogenic silver nanoparticles and their theranostic applications. Green Chem Lett Rev 12(3):310–322. https://doi.org/10.1080/17518253.2019.1643929

Rather GA, Nanda A, Raj E, Mathivanan N, Nayak BK (2022) Green synthesis of ZnO nanoparticles using the leaf extract of lavandula angustifolia and evaluation of their antibacterial activity against human pathogens. Int J Health Sci (Qassim) 6:13478–13485. https://doi.org/10.53730/ijhs.v6nS2.8539

Regidi S, Ravindran S, Vijayan AL, Maya V, Sreedharan L, Varghese J et al (2018) Effect of lyophilization on HRP–antibody conjugation: an enhanced antibody labeling technology. BMC Res Notes 11(1):596. https://doi.org/10.1186/s13104-018-3688-8

Shah G, Shri R, Panchal V, Sharma N, Singh B, Mann A (2011) Scientific basis for the therapeutic use of cymbopogon citratus, stapf (Lemon grass). J Adv Pharm Technol Res 2(1):3. https://doi.org/10.4103/2231-4040.79796

Sharifi-Rad J, Sureda A, Tenore G, Daglia M, Sharifi-Rad M, Valussi M et al (2017) Biological activities of essential oils: from plant chemoecology to traditional healing systems. Molecules 22(1):70. https://doi.org/10.3390/molecules22010070

Silva MRR, Oliveira JG, Fernandes OFL, Passos XS, Costa CR, Souza LKH et al (2005) Antifungal activity of *Ocimum gratissimum* towards dermatophytes. Mycoses 48(3):172–175. https://doi.org/10.1111/j.1439-0507.2005.01100.x

Singh J, Vishwakarma K, Ramawat N, Rai P, Singh VK, Mishra RK et al (2019) Nanomaterials and microbes' interactions: a contemporary overview. 3 Biotech 9(3):68. https://doi.org/10.1007/s13205-019-1576-0

Singh N, Srivastava R, Kanda T, Yadav S, Prajapati R, Yadav S et al (2023) Essential oils: a "potential green" alternative in pharmaceutical, nutritional and agricultural sectors. Bioactivities 92(11):1298–1305. https://doi.org/10.47352/bioactivities.2963-654X.197

Sobrinho ACN, de Souza EB, Rocha MFG, Albuquerque MRJR, Bandeira PN, dos Santos HS et al (2016) Chemical composition, antioxidant, antifungal and hemolytic activities of essential oil from *Baccharis trinervis* (Lam.) Pers. (Asteraceae). Ind Crop Prod 84:108–115. https://doi.org/10.1016/j.indcrop.2016.01.051

Srinivasan D, Nathan S, Suresh T, Lakshmana PP (2001) Antimicrobial activity of certain Indian medicinal plants used in folkloric medicine. J Ethnopharmacol 74(3):217–220. https://doi.org/10.1016/S0378-8741(00)00345-7

Su H, Packeu A, Ahmed SA, Al-Hatmi AMS, Blechert O, İlkit M et al (2019) Species distinction in the trichophyton rubrum complex. J Clin Microbiol 57(9):e00352-19. https://doi.org/10.1128/JCM.00352-19

Sucher AJ, Thai A, Tran C, Mantena N, Noronha A, Chahine EB (2022) Ibrexafungerp: a new triterpenoid antifungal. Am J Health-Syst Pharm 79(24):2208–2221. https://doi.org/10.1093/ajhp/zxac256

Sun G, Zhao Z-J, Mu R, Zha S, Li L, Chen S et al (2018) Breaking the scaling relationship via thermally stable Pt/Cu single atom alloys for catalytic dehydrogenation. Nat Commun 9(1):4454. https://doi.org/10.1038/s41467-018-06967-8

Szymański M, Chmielewska S, Czyżewska U, Malinowska M, Tylicki A (2022) Echinocandins—structure, mechanism of action and use in antifungal therapy. J Enzyme Inhib Med Chem 37(1):876–894. https://doi.org/10.1080/14756366.2022.2050224

Taher M, Amri MS, Susanti D, Abdul Kudos MB, Md Nor NFA, Syukri Y (2020) Medicinal uses, phytochemistry and pharmacological properties of piper aduncum L. Sains Malays 49(8):1829–1851. https://doi.org/10.17576/jsm-2020-4908-07

Tangarife-Castaño V, Correa-Royero J, Zapata-Londoño B, Durán CB, Stanshenko EE, Mesa-Arango AC (2011) Anti-*Candida albicans* activity, cytotoxicity and interaction with antifungal drugs of essential oils and extracts from aromatic and medicinal plants. Infectio 15:160–167

Vahedi-Shahandashti R, Lass-Flörl C (2020) Novel antifungal agents and their activity against Aspergillus species. J Fungi 6(4):213. https://doi.org/10.3390/jof6040213

Vanlalveni C, Lallianrawna S, Biswas A, Selvaraj M, Changmai B, Rokhum SL (2021) Green synthesis of silver nanoparticles using plant extracts and their antimicrobial activities: a review of recent literature. RSC Adv 11(5):2804–2837. https://doi.org/10.1039/D0RA09941D

Vijayan R, Joseph S, Mathew B (2018) Eco-friendly synthesis of silver and gold nanoparticles with enhanced antimicrobial, antioxidant, and catalytic activities. IET Nanobiotechnol 12(6):850–856. https://doi.org/10.1049/iet-nbt.2017.0311

Wei J, Bi Y, Xue H, Wang Y, Zong Y, Prusky D (2020) Antifungal activity of cinnamaldehyde against *Fusarium sambucinum* involves inhibition of ergosterol biosynthesis. J Appl Microbiol 129(2):256–265. https://doi.org/10.1111/jam.14601

Xie Y, Liu X, Zhou P (2020) In vitro antifungal effects of berberine against *Candida* spp. in planktonic and biofilm conditions. Drug Des Devel Ther 14:87–101. https://doi.org/10.2147/DDDT.S230857

Yu M, Gouvinhas I, Rocha J, Barros AIRNA (2021) Phytochemical and antioxidant analysis of medicinal and food plants towards bioactive food and pharmaceutical resources. Sci Rep 11(1):10041. https://doi.org/10.1038/s41598-021-89437-4

Zandavar H, Afshari Babazad M (2023) Secondary metabolites: alkaloids and flavonoids in medicinal plants. In: Herbs and spices—new advances. IntechOpen, London. https://doi.org/10.5772/intechopen.108030

Zare Z, Majd A, Sattari TN, Iranbakhsh A, Mehrabian S (2012) Antimicrobial activity of leaf and flower extracts of Lippia Nodiflora L. (Verbenacea). J Plant Prot Res 52(4):401–403. https://doi.org/10.2478/v10045-012-0065-9

Zhang J, Li L, Lv Q, Yan L, Wang Y, Jiang Y (2019) The fungal CYP51s: their functions, structures, related drug resistance, and inhibitors. Front Microbiol 10:908461. https://doi.org/10.3389/fmicb.2019.00691

Zhang C-W, Zhong X-J, Zhao Y-S, Rajoka MSR, Hashmi MH, Zhai P et al (2023) Antifungal natural products and their derivatives: a review of their activity and mechanism of actions. Pharm Res Modern Chin Med 7:100262. https://doi.org/10.1016/j.prmcm.2023.100262

Clinical Significance, Molecular Formation, and Natural Antibiofilm Agents of *Candida albicans*

9

Mazen Abdulghani and Gajanan Zore

9.1 Introduction

Candida albicans is a common opportunistic fungal pathogen of humans. It is considered a member of the healthy human microbiota, which asymptomatically colonizes the oral cavity, reproductive tract, gastrointestinal (GI) tract, and skin of humans (Lohse et al. 2020). However, *C. albicans* becomes a pathogen when host immunity, physiology, and/or micro-biota changes (Gulati and Nobile 2016). Once it becomes a pathogen, *C. albicans* causes diseases ranging from superficial mucosal and dermal infections, such as thrush, diaper rash, and vaginal yeast infections (75% of women experience a vaginal yeast infection at least once in their lifetime), to more serious hematogenously disseminated infections with mortality rates as high as 47% (Gulati and Nobile 2016; Winter et al. 2016). *Candida* infections are particularly serious in immunocompromised individuals, such as transplant patients receiving immunosuppression therapy, AIDS patients, and patients undergoing chemotherapy treatments, in addition to patients who have implanted medical devices (Perry et al. 2020). Hundreds of millions of symptomatic infections yearly are caused by *C. albicans*, which is also the main cause of nosocomial fungemia (Tong and Tang 2017; Witchley et al. 2019). Almost 80% of total fungal bloodstream infections are associated with *C. albicans* (Ciurea et al. 2020). Life-threatening illnesses can be caused by *C. albicans* in immunocompromised patients and patients

M. Abdulghani
School of Life Sciences, Swami Ramanand Teerth Marathwada University, Nanded, Maharashtra, India
e-mail: mazenmohammed05@gmail.com

G. Zore (✉)
Department of Biotechnology, School of Life Sciences, Central University of Rajasthan, Bandersindri, Kishangarh Dist Ajmer, Rajasthan, India
e-mail: prof.zore_bt@curaj.ac.in

N. Manzoor (ed.), *Advances in Antifungal Drug Development*,
https://doi.org/10.1007/978-981-97-5165-5_9

receiving immunosuppressive therapy with significant mortality rates (Tong and Tang 2017).

The pathogenicity of *C. albicans* is supported by a wide range of virulence factors and fitness attributes (Dadar et al. 2018; Pereira et al. 2021). Several virulence factors include polymorphism, phenotypic switching, adhesins and invasins, biofilm, thigmotropism, and hydrolytic enzymes (Dadar et al. 2018; Pereira et al. 2021). Biofilm is a significant virulence factor contributing mainly to the candidiasis pathogenicity of *C. albicans* (Lohse et al. 2020). In reality, most infections of *C. albicans* are associated with the formation of biofilms on host or abiotic surfaces, such as indwelling medical devices that carry high morbidity and mortality (Tsui et al. 2016). Biofilms formed by *C. albicans* constitute a significant virulence factor and cause clinical problems of concern because of high resistance to antifungal drugs and the evasion of host immune mechanisms, which also protect *C. albicans* cells against other chemical and physical stresses (Gonçalves et al. 2020). Therefore, this chapter focuses on the clinical significance of *C. albicans* biofilm, including immune evasion and drug resistance, and provides an overview of the processes involved in forming a *C. albicans* biofilm. In addition, this chapter discusses the biofilm development regulation of *C. albicans* as well as natural compounds, drug delivery platforms, and synergistic antibiofilm activity with antifungal drugs for clinical use against *C. albicans* biofilm.

9.2 Clinical Significance of Biofilms

Biofilm is a major virulence factor, mainly contributing to candidiasis pathogenicity (Fernandes et al. 2015). The ability of *C. albicans* to form biofilms is an important virulence factor due to significant resistance to antifungal agents and evading host immune mechanisms. Biofilm provides protection against various chemical and physical stresses as well (Gonçalves et al. 2020). Moreover, once the biofilm of *C. albicans* is formed, it can serve as a reservoir to maintain persistent infections and seed new infections in the host (Perry et al. 2020). *C. albicans* can form biofilms on (I) abiotic surfaces (implanted medical devices such as catheters, pacemakers, heart valves, dentures and prosthetic joints) and (II) biotic surfaces (mucosal surfaces and epithelial cell linings) (Dutton et al. 2014; Kojic and Darouiche 2004; Lohse et al. 2020; Mayer et al. 2013). The common *C. albicans* infections associated with biofilm in various sites of the body are summarized in Table 9.1 (Gulati and Nobile 2016).

Table 9.1 Common biofilm-associated infections of *C. albicans*

The medical device directly colonized by a biofilm	Localized or disseminated infection originating from a biofilm
Contact lenses, dentures, central vascular catheter, peripheral vascular catheter, hip prosthetic, orthopedic implant, urinary catheter, stent, heart valve, endotracheal tube, and pacemaker	Wound infection, skin infection, ear infection, dental infection, thrush infection, lung infection, bloodstream infection, and vulvovaginal candidiasis

C. albicans is one of the most fungal species; it can cause hospital-acquired infection in humans, especially on implanted medical devices, including urinary and central venous catheters, heart valves, pacemakers, contact lenses, joint prostheses, and dentures (Fox et al. 2015b). Once *C. albicans* forms a biofilm, it becomes highly drug-resistant, possesses the potential to cause bloodstream infection (candidemia), and leads to invasive infections of many organs (Gulati and Nobile 2016). In the United States of America, approximately five million central venous catheters are inserted annually, and biofilm infection is diagnosed in up to 54% of these venous catheters, with 100,000 deaths and $6.5 billion in costs every year (Fox and Nobile 2012). The growth of biofilm and colonization on medical devices are treated by removing the medical devices and administering high doses of antifungal drugs in some cases (Fox and Nobile 2013; Fox et al. 2015a). Removing medical devices (e.g., artificial heart valves and joints) is not usually safe and costly for patients. In addition, administering antifungal drugs can cause complications and damage some human organs. Hence, searching for better therapeutic is a priority to overcome those complications due to biofilm growth (Fox et al. 2015a; Gulati and Nobile 2016). Besides, mature biofilms of *C. albicans* consist of a mixture of distinct forms of cells, including yeast, pseudohyphal and hyphal, and distinct microenvironments, for example, aerobic niches and hypoxic niches at near-surface and the depths of biofilms, respectively; this heterogeneity also extends to the transcriptomes and proteomes of individual cells (Lohse et al. 2018). This cellular diversity can be crucial in biofilm tolerance to antifungal agents and the host's immune system. Multiple targets must be approached for any threat to the biofilm rather than just a single cell type (Lohse et al. 2018). In addition to the physical architecture of *C. albicans* biofilm, the extracellular matrix of *C. albicans* has several functions, including mechanical stability, adhesive and cohesive interactions. The most critical role of the matrix, perhaps, is related to clinical significance, which acts as a physical barrier that protects biofilm cells from attacks by antifungal agent treatment and the immune system (Pierce et al. 2017). It may also effectively mask cell wall epitopes on the surface of *C. albicans* cells that are necessary for the host to recognize, leading to immune evasion (Pierce et al. 2017). Taken together, these findings give the idea that biofilm development is a state of stressful environments of *C. albicans* in response to host immune system or antifungal drugs. Thus, we will focus on drug resistance and immune system evasion of *C. albicans* biofilm in detail, as the biofilm-associated cells tend to become more resistant to host defenses and antimicrobial agents.

9.2.1 Drug Resistance

The availability of antifungal agents is significantly lower when compared with antibacterial agents. This is mainly because fungi are eukaryotic; thus, identifying drug targets to selectively kill fungal pathogens without toxicity to the host is problematic (Tsui et al. 2016). The five major classes of antifungal agents used to treat *C. albicans* infections are azoles, echinocandins, polyenes, nucleoside analogues, and allylamines (Chong et al. 2018). Azoles (such as fluconazole, clotrimazole, miconazole, ketoconazole, and itraconazole) are fungistatic and inhibit ergosterol biosynthesis by targeting the *ERG11* gene, which encodes 14α-lanosterol demethylase leading to the inhibition of cytochrome P450-dependent conversion of lanosterol to ergosterol. Thus, blocking 14α-demethylase leads to the accumulation of toxic methylated sterols, causing membrane stress (Prasad et al. 2016). The expression of *ERG11* was significantly upregulated after treatment of *C. albicans* biofilm with ketoconazole but not with other classes of drugs so *ERG11* might play a role in the azole resistance of *C. albicans* biofilm (Watamoto et al. 2011). On the other hand, polyenes (such as amphotericin B, nystatin, and natamycin) are unlike azoles, polyenes are fungicidal, and their mechanism of action involves binding to ergosterol of fungal cell membranes leading to the formation of transmembrane channels, resulting in leakage of electrolytes and cytoplasmic contents and leading to fungal cell death (Tsui et al. 2016). However, because ergosterol has structural similarity with cholesterol in the mammalian cell membrane, the major side effects of polyenes include toxicity and dysfunction of the kidney, which limits their use (Jadhav and Karuppayil 2017; Prasad et al. 2016).

Nucleoside analogues, of which 5-flucytosine (5-FC) mimics nucleotides during nucleic acid synthesis, specifically act as a pyrimidine analogue, disrupting fungal RNA, DNA, and protein synthesis, ultimately leading to cell cycle arrest (Fox and Nobile 2013). 5-FC is not intrinsically antifungal by itself, but it becomes antifungal when 5-FC is converted to 5-FU by a fungal cytosine deaminase inside fungal (but not host) cells (Fox and Nobile 2013). The comparatively recent class of antifungal drugs is echinocandins (such as caspofungin, anidulafungin, and micafungin) which have a fungicidal effect in nature, and they are inhibitors of β-1,3-glucan synthase, which requires fungal cell wall synthesis; this inhibition affects the integrity of fungal cells wall resulting in cell wall stress. Generally, echinocandins are nontoxic to mammalian cells because of their specificity toward fungal targets on the cell wall synthesis pathway unique to fungal cells (Jadhav and Karuppayil 2017; Prasad et al. 2016). Finally, allylamines inhibit ergosterol biosynthesis by blocking the enzyme squalene oxidase (Mathé and Van Dijck 2013).

Significantly, biofilms are intrinsically more resistant to triazoles, amphotericin B, and echinocandins than their planktonic counterparts by approximately 1000-fold, 10 to 100-fold, and 2 to 20-fold, respectively (Taff et al. 2013; Dominguez and Andes 2017). The azole, polyenes classic formulations, nucleoside analogues, and allylamines are inactive against biofilm (Mathé and Van Dijck 2013). However, the echinocandins and the polyene amphotericin B lipid formulations for biofilm treatment are effective in vitro and in vivo against biofilms. Therefore, echinocandins and amphotericin B lipid formulations are the only therapeutic options for the treatment of *Candida* biofilm infections (Ghannoum et al. 2015; Mathé and Van Dijck 2013; Tsui et al. 2016; Wall et al. 2019).

Many different mechanisms exhibited by *C. albicans* biofilms contribute to much more resistance to antifungal drugs (Wall et al. 2019). Briefly, the extracellular matrix's presence and persister cells' existence are functional in biofilm-specific. However, cell density, overexpression of drug efflux pumps, modification of drug targets, and altered gene expression are mechanisms functional in both planktonic and biofilm cells. Another mechanism exhibited by *C. albicans* is a low growth rate under only planktonic growth (Taff et al. 2013; Tsui et al. 2016; Pereira et al. 2021)

9.2.1.1 Extracellular Matrix

The biofilm matrix is of great significance due to the importance of clinical repercussions. The *C. albicans* biofilm matrix acts as a physical barrier that defends biofilm cells from immune system attacks and treatment with antifungal drugs (Pierce et al. 2017). β-1,3-Glucan is known as a part of the biofilm matrix that has been related to the protection biofilm from antifungal agents (Nett et al. 2007, 2010c; Taff et al. 2012). The cell walls of biofilm-associated cells contain more total carbohydrates and β-1,3-glucan to planktonic cells; also, the supernatants from the biofilm culture contain at least twofold more β-1,3-glucan (Nett et al. 2007). Thus, the treatment of biofilm by β-1,3-glucanase increases the susceptibility of biofilm to fluconazole, and the addition of exogenous β-1,3-glucans results in increasing the tolerance of planktonic cells to fluconazole (Nett et al. 2007). Moreover, it has also been shown that amphotericin B bind to β-1,3-glucans, thereby preventing the reach of this antifungal drug to fungal cells within the biofilm (Vediyappan et al. 2010). Furthermore, the reduction of *FKS1* expression, encoding a β-1,3-glucan synthase, was found to increase the susceptibility of biofilm to anidulafungin, flucytosine, and amphotericin B (Nett et al. 2010a). In addition, the expression of *FKS1* was significantly upregulated after treatment of *C. albicans* biofilm with caspofungin and amphotericin B (Watamoto et al. 2011). Moreover, mutants of genes are involved in the delivery of β-1,3-glucan to the biofilm matrix, *BGL2, PHR1*, and *XOG1* glucan transferases and exoglucanases, respectively, which encode proteins involved in the modification and transport of β-glucan, which exhibit enhanced susceptibility to fluconazole during biofilm growth only (Taff et al. 2012). Finally, another matrix component is eDNA; Martins et al. (2012) reported that the addition of DNase increases the efficacy of amphotericin B against *C. albicans* biofilm and increases the efficacy of caspofungin in the reduction of mitochondrial activity.

9.2.1.2 Efflux Pumps

High resistance to azoles is due to the overexpression of cell membrane efflux pumps, transporting membrane-associated transport proteins to prevent drug accumulation and prevent toxic levels from killing cells (Pereira et al. 2021). Two main efflux pump classes in *C. albicans* are the major facilitator (MF) transporter (including Mdr1) and the ATP binding cassette (ABC) transporter (including Cdr1 and Cdr2), which play a role in azole resistance (Fox et al. 2015b; Prasad et al. 2016; Taff et al. 2013). These main efflux pumps in the biofilm are upregulated within the first few hours of cells adhesion to the surface and remain upregulated during the development of biofilm, even in the absence of an antifungal agent, whereas, in the presence of antifungal agents, these main efflux pumps are upregulated in planktonic cells (Gulati and Nobile 2016). The transporter genes *CDR1, CDR2,* and

MDR1 are controlled by the transcription factors *TAC1* (control *CDR1, CDR2*) and *MRR1* (control *MDR1*) (Lohberger et al. 2014).

9.2.1.3 Persister Cells

The persister cells are nondividing cells with reduced metabolic activities that make them extremely resistant to antimicrobial agents and, further, likely to reform biofilm infections after drug treatment (Lohse et al. 2018). The persister cells of *C. albicans* were only found in biofilms and discovered by LaFleur et al. (2006). When LaFleur et al. observed a biphasic killing of *C. albicans* biofilm, which was treated with amphotericin B, the majority of the cells were killed at low concentrations of the drug (but above the MIC), whereas about 1% of cells were unaffected even at the high concentration (LaFleur et al. 2006; Mathé and Van Dijck 2013). Recent proteomics research showed that the protein levels of several metabolic enzymes in persister cells were downregulated. In contrast, the protein levels involved in the growth, virulence, and stress response of *Candida* were upregulated compared to the control cells in a biofilm. Therefore, delicate metabolic regulation and coordinated stress adaptation probably cause the antifungal tolerance of persister in *Candida* biofilm (Li et al. 2015).

Recently, it has been found that the expression levels of alkyl hydroperoxide reductase, Ahp1, is closely linked to persister cells that have survived exposure to amphotericin B. Ahp1 is a cell wall peroxidase, which plays a vital role in *C. albicans*, for it plays a key function in defending cells from reactive oxygen species and in responding to stress (Truong et al. 2016). Thus, it is clear that the high resistance by biofilm-associated *C. albicans* cells cannot be attributed to the act of one mechanism. It is a comprehensive mechanism representing the complexity of the lifestyle of the biofilm itself (Tsui et al. 2016).

9.2.1.4 Cell Density

The effect of cell density on *C. albicans* drug resistance was first proposed by Perumal et al. (2007) because biofilm resistance varies with (extreme) inoculum size. At high cell densities, planktonic cells displayed significantly lower susceptibilities to all medications, as demonstrated by testing the effectiveness of several azoles, amphotericin B, and the echinocandin caspofungin on planktonic cells at densities equivalent to those observed in biofilms (up to 1×10^8 cells/mL). A strain lacking mechanisms involved in drug efflux and farnesol quorum sensing showed the same pattern, suggesting that these processes were not involved. The higher resistance was certainly linked to the biofilm architecture since diluted dissociated biofilm cells showed the same susceptibility as planktonic cells at the same cell density (1×10^3 cells/mL) (Perumal et al. 2007; Mathé and Van Dijck 2013). Seneviratne et al. (2008) reported the same findings for the azole ketoconazole and the pyrimidine analog 5-FC. Caspofungin and amphotericin B were not as effective against biofilm-associated and planktonic cells, although they attribute this discrepancy to differences in experimental design. Cell density affects antimicrobial agent efficacy, particularly in biofilms. However, cell density does not seem biofilm-specific, as similar trends were observed in planktonic cells (Perumal et al. 2007; Seneviratne et al. 2008).

9.2.1.5 Altered Gene Expression

Biofilms are characterized by their dynamic response to several stresses (Nobile et al. 2012; Mathé and Van Dijck 2013). When biofilms are exposed to antimicrobials, resistance genes are often induced. This paradigm was investigated in the ergosterol and glucan biosynthesis genes of *C. albicans* biofilms after treatment with azole and echinocandin drugs, respectively. Because of its central role in the membranes of *C. albicans* fungi, the ergosterol molecule is a promising therapeutic target. The ERG-genes play a crucial role in the complicated process of ergosterol production in *C. albicans*, which requires a large number of enzymes (Tsui et al. 2016). To counteract the inhibitory effects of fluconazole, which is in the azole family, *C. albicans* has evolved a number of resistance mechanisms, the most notable of which is the upregulation of genes involved in ergosterol production. Ergosterol biosynthesis (the ERG-genes) and β-1,6-glucan biosynthesis (*SKN1* and the KRE-genes) mRNA levels were evaluated and compared between planktonic and biofilm-associated cells using quantitative RT-PCR in previous work (Khot et al. 2006). A subgroup of amphotericin B-resistant blastospores had a distinct transcript profile, with *ERG1* downregulated and *ERG25, SKN1,* and *KRE1* upregulated. This latter gene's transcription levels likewise correlated positively with increased resistance to amphotericin B. Using reverse transcriptase and real-time polymerase chain reaction, Borecká-Melkusová et al. (2009) studied variations in ERG-gene expression in several *C. albicans* isolates after the addition of the azole fluconazole. They discovered that *ERG9* was upregulated in all strains examined, but *ERG11* was downregulated in fluconazole-susceptible strains. After fluconazole treatment, gene expression data for *ERG1* and *ERG25* indicated modest upregulation in free-floating and biofilm-associated cells (Borecká-Melkusová et al. 2009). Nailis et al. (2010) found that when biofilms were exposed to high concentrations of antifungals, the biofilms responded with a drug-specific transcription response. High fluconazole and amphotericin B concentrations were used to test gene expression patterns in *C. albicans* cells embedded in biofilms. They found that fluconazole significantly upregulated *ERG1, ERG3, ERG11*, and *ERG25* in mature biofilms, whereas amphotericin B significantly upregulated *SKN1, KRE1*, and *ERG1*, suggesting a greater capacity for biofilms to respond to antifungal stress (Nailis et al. 2010). Farnesol is the precursor of ergosterol and a quorum-sensing molecule in *C. albicans*. Yu et al. (2012) discovered that mature biofilms grown in the presence of farnesol were much more susceptible to fluconazole than a farnesol-untreated biofilm. They used real-time PCR to show that the transcription levels of *ERG1, ERG3, ERG6, ERG11*, and *ERG25* were significantly lower in the farnesol-treated group, suggesting that the ergosterol biosynthesis pathway may contribute to the inhibitory effect of farnesol and supporting the idea that increased transcription of the ERG-genes does increase biofilm resistance (Yu et al. 2012).

9.2.1.6 Modification of Drug Targets

Altering the drug target substrate is another strategy used by *C. albicans* to avoid being killed by antifungals. Cross-resistance to azoles and amphotericin B was demonstrated in *C. albicans* by mutations in the *ERG3* and *ERG5* genes, as well as point mutations in the *ERG11* (White 1997; White et al. 1998). Therefore, it is reasonable to assume that changes in gene expression contribute to drug resistance in *C. albicans* cells found in biofilms. However, whether these ERG mutations have ramifications on pathogenesis and resistance in *C. albicans* biofilms remains understudied, even though they are more likely implicated in the establishment of resistance in planktonic populations following prolonged treatment with azoles (Tsui et al. 2016).

9.2.2 Immune Evasion

Cells of *C. albicans* pathogen are recognized by the innate immune system through pathogen-associated molecular patterns (PAMPs) found on the cell wall or secreted by the pathogen (Fox and Nobile 2013; Mathé and Van Dijck 2013). *C. albicans* cell wall structure comprises two layers; the inner layer consists of a network of chitin, β-1,3-glucans, and β-1,6-glucans. This inner network covered by the outer layer mainly consists of O- and N-linked mannoproteins (Gow and Hube 2012; Garcia-Rubio et al. 2020). These PAMPs of *C. albicans* are detected by pattern recognition receptors (PRRs) of the innate immune system. Several host cells recognize *C. albicans*, including epithelial cells, macrophages, neutrophils, and dendritic cells (Fox and Nobile 2013; Mathé and Van Dijck 2013; Cheng et al. 2012). Two member-bound PRRs families are involved in recognizing the cell wall (PAMPs) of *C. albicans*: C type-lectin receptors (CLRs) family and Toll-like receptors (TLRs) family (Cheng et al. 2012; Richardson and Moyes 2015). In addition to membrane-bound receptors PRRs which recognize PAMPs of fungal pathogens, it has been demonstrated that several RPPs recognize the intercellular compounds of *Candida*, such as TLR9 and NLRP3 (Cheng et al. 2012; Richardson and Moyes 2015). Roles of RPPs in recognition PAMPs *C. albicans* are summarized below in Table 9.2. Once binding occurs successfully, between the PAMPs and PRRs initiate several downstream signaling cascades, including increased cytokine and chemokine production and results in phagocytosis of *C. albicans* (Fox and Nobile 2013; Mathé and Van Dijck 2013). Differences in the cell wall polysaccharides chemistry between yeast form and hyphal form may contribute to the evasion of host immune system detection (Lowman et al. 2014).

There are many strategies linked to the biofilm that *C. albicans* use to evade the host immune response (Gulati and Nobile 2016). Since hyphae cells are essential for biofilms, they can invade the epithelial cell layer during invasive growth (Filler and Sheppard 2006). Also, they have the ability to mediate the escape of *C. albicans* cells from inside phagocytic cells by physically piercing the phagocytic cell, hence contributing to evading the host immune response (Ghosh et al. 2009). The interaction of peripheral blood mononuclear cells (PBMCs) with *C. albicans* cells in

Table 9.2 Roles of pattern recognition receptors RPPs and their pathogen-associated molecular patterns PAMPs in *C. albicans*

Family	Receptor	Location	Ligand	References
TLRs	TLR2	Extracellular receptor (on the surface)	Phospholipomannan	Jouault et al. (2003)
	TLR4	Extracellular receptor (on the surface)	O-linked mannan	Tada et al. (2002), Netea (2006)
	TLR9	Intracellular receptor (Endosomal)	*C. albicans* DNA	Miyazato et al. (2009)
CLRs	Dectin-1	Extracellular receptor (on the surface)	β-glucans	Brown et al. (2002)
	Dectin-2	Extracellular receptor (on the surface)	Hyphal highly-mannose α-mannan on both yeast and hyphae	McGreal et al. (2006), Saijo et al. (2010), Sato et al. (2006)
	MR	Extracellular receptor (on the surface)	N-linked mannan	Netea (2006)
	MBL	Extracellular receptor (secreted)	Mannan	Brouwer et al. (2008)
	DC-SIGN	Extracellular receptor (on the surface)	N-linked mannan	Cambi et al. (2008)
	Galectin-3	Extracellular receptor (on the surface)	β-mannosides	Jouault et al. (2006)
	Mincle	Extracellular receptor (on the surface)	Unknown specific ligands	Bugarcic et al. (2008)
NLRs	NLRP3	Intracellular receptor (Cytoplasmic)	β-glucans	Gross et al. (2009), Hise et al. (2009), Joly et al. (2009), Stappers and Brown (2017)

biofilm did not show phagocytose of cells; in contrast, the existence of PBMCs through biofilm growth enhanced *C. albicans* to form thicker biofilm as a consequence of a soluble factor secreted by PBMCs, and these PBMCs localized in the middle and basal layers of biofilm (Chandra et al. 2007). Besides, it has been shown that the biofilm of *C. albicans* alters the profile of cytokines released by PBMCs compared to the cytokines profile of planktonic-PBMCs (Chandra et al. 2007). In addition, the interaction between neutrophils and *C. albicans* biofilm displays higher resistance up to fivefold to killing by neutrophils compared to planktonic cells (Johnson et al. 2016). Although neutrophils in vivo may surround biofilms, mature biofilm is more resistant to killing by neutrophils, which also cannot trigger a reactive oxygen species response (ROS) (Xie et al. 2012). This resistance of

C. albicans biofilm is mediated by the presence of β-glucans in the extracellular matrix, which acts as a decoy mechanism to hinder neutrophil activation and protects *C. albicans* biofilm from killing (Xie et al. 2012). Intriguingly, it has been shown that the biofilm of *C. albicans* fails to release neutrophil extracellular traps (NETs) due to the extracellular matrix inhibiting NET release. This inhibition of NETs has also been related to the suppression of ROS production (Johnson et al. 2016). In support of this, biofilm is formed by a mutant strain (*pmr1Δ/Δ*) defective in extracellular mannan–glucan production, which leads to increased NET release and is more susceptible to killing by neutrophils. Thus, the lack of NETs contributes to immune evasion and gives a survival benefit to cells inside the biofilm as opposed to planktonic cells (Johnson et al. 2016). Also, in the presence of *C. albicans* biofilm, cell culture macrophages tend to have impaired migration compared to their motility in response to planktonic *Candida* (Alonso et al. 2017). Furthermore, members of *C. albicans* secreted aspartyl protease (Sap) family and Pra1 (an antigenic, zinc-binding cell surface protein) have been reported to inhibit the activation of the host complement system (Gropp et al. 2009; Zipfel et al. 2011). The cell wall damage sensor, Msb2, is also involved in protecting the cells of *C. albicans* from host-secreted antimicrobial peptides (AMPs) (Szafranski-Schneider et al. 2012; Swidergall et al. 2013). Both Pra1 and Msb2 proteins are highly expressed in *C. albicans* biofilm, and mutants of these proteins result in defects in biofilm formation (Puri et al. 2012; Kurakado et al. 2018). Additionally, Bcr1 is one of the core transcriptional regulators of biofilm formation. Thus, deletion of Bcr1 has been shown to make *C. albicans* more susceptible to leukocyte damage (Dwivedi et al. 2011).

9.3 Biofilm Formation of *C. albicans*

Biofilms are defined as a community of microorganisms irreversibly attached to a given surface or living tissue and encased in an extracellular polymeric substance, where microbes in biofilm undergo reproduction and self-cloning (Cavalheiro and Teixeira 2018; Chen et al. 2015; Ramage et al. 2009). In *C. albicans*, the biofilm is a highly complicated structure that contains multiple cell types (i.e., yeast, pseudohyphal, and hyphal cells) encased in an extracellular matrix (Fox and Nobile 2012; Lee et al. 2016). Furthermore, *C. albicans* is able to form biofilms on (i) abiotic surfaces (implanted medical devices such as catheters, pacemakers, heart valves, dentures, and prosthetic joints) and (ii) biotic surfaces (mucosal surfaces and epithelial cell linings) (Dutton et al. 2014; Kojic and Darouiche 2004; Lohse et al. 2020; Mayer et al. 2013). Overall, the biofilm formation of *C. albicans* can be divided into four distinct stages: adherence, growth and proliferation, maturation, and dispersion (Tournu and Van Dijck 2012; Mathé and Van Dijck 2013). Several models have been studied in vitro biofilm formation of *Candida*, and they have focused on examining the effect of different types of nutritional media, substrates, and whether flow or static conditions are existent on the production of biofilm (Tournu and Van Dijck 2012). In the laboratory, the biofilm of *C. albicans* can be formed on many different substrates and in several different media types, indicating the potential for biofilm formation in a variety of environmental conditions (Nobile and Johnson 2015).

Generally, biofilm formation in vitro has connected well with biofilm models in vivo and ex vivo. For example, *Candida* biofilm, which was obtained from patients in both denture stomatitis and infected intravascular catheters, confirmed the existence of multiple cells, including yeast and hyphae, and such cells are enclosed in an extracellular matrix (Nobile and Johnson 2015). In addition, in vivo rat central venous catheter model, rat denture stomatitis model, indwelling urinary catheter model, rabbit model, and ex vivo and in vivo vaginal mucosal models have also been observed similar to biofilm architecture in vitro biofilm with many of the same features, including numerous yeast cells and hyphae as well as extracellular matrix throughout the biofilms (Andes et al. 2004; Gulati and Nobile 2016; Harriott et al. 2010; Johnson et al. 2012; Nett et al. 2010b, 2014; Schinabeck et al. 2004).

9.3.1 Adherence

Biofilm formation begins when cells adhere to each other and surfaces; such surfaces may be soft (host tissue) or hard (a medical device). Adhesion is the first and critical phase for all phases of biofilm growth in *C. albicans* (Gulati and Nobile 2016). Therefore, *Candida* cells attach to surfaces by nonspecific interactions, such as electrostatic forces and hydrophobic, and specific adhesin–ligand bonds, mediate the binding of *Candida* species cells to surfaces (Sardi et al. 2013). The cell hydrophobicity and charge depend on the morphology of cell growth and the structure of the cell surface (Li et al. 2007). Adhesins are cell wall proteins that play an important role in biofilm formation and mediate the adhesion of *C. albicans* to each other and various substrates (Garcia et al. 2011). *C. albicans* adhesins can be divided into three gene families (*ALS*, *HWP*, and *IFF/HYR*), which have mainly been addressed in research (de Groot et al. 2013). The members of the *ALS* (agglutinin-like sequence) gene family, as well as Eap1 (enhanced adherence to polystyrene 1) and Hwp1 (hyphal wall protein 1), which follow the *HWP* family, are the best-described cell surface adhesin proteins (McCall et al. 2019). Glycosylphosphatidylinositol-dependent cell wall proteins (GPI-CWPs) are a class of adhesins that have a similar structure characterized by the existence of a C-terminal sequence containing a GPI anchor attachment site and an N-terminal signal peptide (Li et al. 2007). This *ALS* family, Eap1 and Hwp1, follow the class (GPI-CWPs) of adhesins (Desai and Mitchell 2015).

Among adhesins in *C. albicans*, *ALS* family genes have eight distinct loci which are (*ALS*1 to *ALS*7 and *ALS*9); these genes encoded eight large cell surface glycoproteins (Als1-Als7 and Als9) that are suggested to have the same overall structure (Murciano et al. 2012; de Groot et al. 2013; Hoyer and Cota 2016). Als1 and Als3 proteins are specifically involved in the biofilm surface attachment of *C. albicans*, but the expression of these proteins differs depending on the cell morphology (Araújo et al. 2017). Als3, Als1, and Als5 have an important role in adhering to various host constituents due to their broad substrate specificity (Liu and Filler 2011).

Hwp1 is a hypha-associated GPI-linked protein and another important adhesin of *C. albicans* (Mayer et al. 2013). Hwp1 is a substrate for host transglutaminases, and this reaction covalently links *C. albicans* to epithelial cell surfaces (Desai and Mitchell 2015; Mayer et al. 2013). Eap1 is a GPI-linked cell-wall protein that mediates surface binding and involves in cell-cell adhesion in *C. albicans*; its synthesis is regulated by the transcription factor *EFG1* (Araújo et al. 2017; Li et al. 2007; Li and Palecek 2003). In addition to Als1, Als3, Hwp1, and Eap1 (Pga47), several cell surface proteins are required at different stages of biofilm formation, such as Sun41, Pga1, and members of the CFEM (common in several fungal extracellular membranes) family (Pga10/Rbt51, Rbt5, and Wap1/Csa1) (Perez et al. 2006; Norice et al. 2007; Nobile et al. 2008; Klis et al. 2009; Hashash et al. 2011; Cabral et al. 2014).

9.3.2 Proliferation

The next step after the adherence in biofilm development is the growth and proliferation of yeast cells, forming a basal layer of anchoring cells (Nobile et al. 2014). This step is characterized by the initiation of filamentation resulting in the formation of multiple layers of sessile cells of different morphologies, including pseudohyphal and hyphal cells (initiation step) (Tournu and Van Dijck 2012; Tsui et al. 2016; Wall et al. 2019). The ability of *C. albicans* to transition from yeast-to-hyphae cells under several different environmental conditions distinguishes it from many other fungus species; thus, this is the basis of the early classification of *C. albicans* as dimorphic (Nobile and Johnson 2015). *C. albicans* changes its cell morphology and secretes hydrolytic enzymes such as proteinases, hemolysis, and phospholipase during the initiation step to invade either the host mucosal sites or other solid substrates (Chong et al. 2018).

9.3.3 Maturation

In this stage, the growth of hyphae (chains of cylindrical cells) and pseudohyphae (ellipsoid cells joined end to end) takes place with the production of an extracellular matrix (ECM) (Nobile et al. 2012). This ECM is partially self-produced and secreted by *C. albicans* cells grown in the biofilm so that biofilm production can vary depending on the growth conditions; mature *C. albicans* biofilm exhibits complex material with heterogeneous cell types encased in the extracellular matrix (Mathé and Van Dijck 2013; Desai and Mitchell 2015). The extracellular matrix (ECM) is composed of (protein 55%, carbohydrate 25%, lipid 15%, and nucleic acid 5%) in the *C. albicans* biofilm (Zarnowski et al. 2014; Mitchell et al. 2016). In addition, ECM of biofilm in the host may also include environmental aggregates of host cells such as erythrocytes, epithelial cells, urothelial cells and neutrophils, resulting in a diverse biofilm depending on the host location (Pereira et al. 2021). Proteomic analysis of the biofilm matrix in *C. albicans* showed that

565 different proteins were identified with 458 distinct activities (Zarnowski et al. 2014).

The polysaccharides in the biofilm matrix include α-mannan, β-1,6 glucan, and β-1,3 glucan (Mitchell et al. 2015). The most abundant polysaccharides include α-1,2 branched and α-1,6-mannans (87%); these mannans are found to be associated with unbranched β-1,6-glucan (13%) in an apparent mannan–glucan complex (MGCx) (Pierce et al. 2017). The β-1,6-glucan exists in linear conformation in the matrix while highly branched in the cell wall (Kernien et al. 2018). Besides, it was found that polysaccharide mannans in high molecular weight structures of approximately 12,000 residues in the biofilm matrix compared to the outer cell wall layer (~150 residues) (Mitchell et al. 2015). On the other hand, β-1,3-glucan has a minor content in the matrix polysaccharide, while it plays a significant role in biofilm resistance to antifungals by blocking drug diffusion (Taff et al. 2013). In contrast, β-1,6-glucan and β-1,3-glucan in the cell wall of *C. albicans* represent 20% and 40%, respectively, of the dry weight of the cell wall (Reyna-Beltrán et al. 2019). At the same time, chitin is not found in the biofilm matrix, while extant in the cell wall represents 1–2% of the content (Zarnowski et al. 2014). Furthermore, the most abundant monosaccharides in *C. albicans* biofilm are arabinose, mannose, glucose, and xylose, of the total carbohydrate pool (Lopez-Ribot 2014).

The lipids associated with the *C. albicans* biofilm matrix consist of neutral glycerolipids (89.1%), polar glycerolipids (10.4%), and less amounts of sphingolipids (0.5%) (Zarnowski et al. 2014). Eight distinct classes of lipids in the matrix were identified; the most abundant classes in neutral glycerolipids are free fatty acids 75.3% and triacylglycerols 11.1%, while in polar glycerolipids, phosphatidylethanolamine 8.9%. Ergosterol and prostaglandin E_2 were also revealed by mass spectrometry in a small amount (Zarnowski et al. 2014; Alim et al. 2018).

Martins et al. (2010) reported that the addition of exogenous DNA increases biofilm biomass, and conversely, DNase treatment decreases biofilm biomass at later time points of biofilm development; thus, extracellular DNA in biofilm matrix plays a role in the structure and stability of a mature biofilm (Martins et al. 2010). In addition, during *C. albicans* biofilm development, the extracellular DNA (eDNA) induces the morphological transition from yeast to the hyphal growth form (Hirota et al. 2017). The *C. albicans* biofilm matrix DNA is mostly presented as random noncoding sequences (Zarnowski et al. 2014).

9.3.4 Dispersion

The dispersion of yeast cells from the biofilm complex into the surrounding environment and initiate the formation of new biofilms occur at this stage (Nobile et al. 2014). These dispersed cells originate from the topmost hyphal layers of biofilms, and most dispersal cells are yeast cells. The frequency of dispersal directly depends on growing conditions, including carbon source and pH of growth media (Uppuluri et al. 2010a; Wall et al. 2019). *C. albicans* dispersed cells exhibit distinct properties, including enhanced adherence, filamentation, the ability to form biofilm more efficiently, and enhanced virulence-associated characteristics in a murine model and drug resistance as compared to their planktonic counterparts (Uppuluri et al. 2010a, 2018; Wall et al. 2019).

9.4 Regulation of *C. albicans* Biofilm

The whole process of biofilm formation is highly regulated at the molecular level (Wall et al. 2019). Therefore, regulating biofilm formation is governed by several transcription regulators, including those contributing to adhesion, formation of hyphae, and extracellular polymeric substance (EPS) production (Cavalheiro and Teixeira 2018). Over 50 transcriptional regulators have been linked to *C. albicans* biofilm formation (Nobile and Johnson 2015; Lohse et al. 2018). A set of nine regulators (Bcr1, Brg1, Efg1, Flo8, Gal4, Ndt80, Rob1, Rfx2, and Tec1) represent the "core" set of the transcriptional network that is required for biofilm development both in vitro and in vivo (Fox et al. 2015a; Nobile et al. 2012; Pentland et al. 2020; Perry et al. 2020).

In 2012, by screening *C. albicans* library of 165 transcription factor deletion mutants, it has been identified the network of transcriptional regulatory, which consists of six master transcription regulators (Efg1, Tec1, Ndt80, Bcr1, Brg1, and Rob1); they control the formation of *C. albicans* biofilm in vitro and in vivo in two different animal models (Nobile et al. 2012). Forty-four additional transcriptional regulators were identified beside the master six transcriptional regulators, which are essential for *C. albicans* biofilm formation and directly regulated with at least one of the master six transcriptional regulators (Bonhomme and D'Enfert 2013; Fox and Nobile 2012; Kakade et al. 2016; Nobile and Johnson 2015).

In 2015, the six master regulators were expanded with three other transcriptional regulators (Rfx2, Gal4, and Flo8); which have specific roles in biofilm development over time in both in vitro and in vivo (Fox et al. 2015a). Flo8 is required for biofilm formation at all-time points, whereas *flo8Δ/Δ* produces defective biofilm similar to those produced by strain resulting from the deletion of any one of six previously discovered master regulators. At the same time, Rfx2 and Gal4 participated at an intermediate time in biofilm formation. The authors suggested that the Rfx2 and Gal4 are negative regulators for biofilm formation which observed that *rfx2Δ/Δ* and *gal4Δ/Δ* mutant strains enhanced biofilm formation (Fox et al. 2015a).

Further analysis by chromatin immunoprecipitation experiments revealed that these nine regulators form a strongly interconnected transcriptional network in which the individual regulators control each other and also, approximately 1000 target genes (Alim et al. 2018; D'Enfert and Janbon 2015; Fox et al. 2015a; Lohse et al. 2018; Nobile et al. 2012). Based on the literature, Nobile and Johnson (2015) counted 50 transcriptional regulators and 101 nonregulatory genes that functionally validated roles in all stages of biofilm formation (Nobile and Johnson 2015). Also, Lohse et al. made a list of transcriptional regulators with the core regulators binding to the control region of the gene (Lohse et al. 2018).

Based on data from RNA-sequencing and microarray have identified 1599 and 636 genes that were upregulated and downregulated, respectively, at least twofold in biofilm cells compared to their planktonic cell counterparts (Nobile et al. 2012). Moreover, in another study for a specific stage of biofilm formation, 251 genes were upregulated, and 157 genes were downregulated in the cells that have just adhered to the surface compared to cells that do not adhere to the surface in the same medium (Fox et al. 2015a). More than half of these genes were upregulated, corresponding to components involved in DNA synthesis, transcription, and RNA processing or translation. This suggests preparation for a morphological and physiological shift accompanying biofilm formation upon physical adherence (Fox et al. 2015a). Besides, secreted aspartyl protease Sap5 (*Candida* pepsin-5) and Sap-6 (*Candida* pepsin-6) are highly upregulated in biofilm formation (Winter et al. 2016). Finally, dispersed cells (yeast-form) from mature biofilm tend to be more virulent and have an improved capacity to adhere to surfaces to form new biofilm compared to planktonic cells (Uppuluri et al. 2010a). Additionally, two of these differences are previously discussed, viz., drug resistance and immune evasion.

9.4.1 Adherence

The ability of *C. albicans* to adhere to a substrate is a critical step in biofilm development (Lee et al. 2016). Therefore, during biofilm formation, several cell wall proteins have been shown to play essential roles (Cabral et al. 2014). Two classes of adhesion genes have been distinguished (one at the early stages of biofilm formation and the other at a later time of biofilm formation), which are upregulated in biofilm (Table 9.3) (Fox et al. 2015a). Seven of these early genes were upregulated, and their expression was important for the cells to adhere to a solid surface compared to the un-adhered planktonic reference (Table 9.3). While the three remaining genes (*MSB2, ALS3*, and *ORF19.2449*) were upregulated in un-adhered cells, suggesting that their induction did not rely on contact with a solid surface (Fox et al. 2015a).

Further analysis using deletion strains showed that five of these early genes (*ALS1, ALS2, ALS3, EAP1*, and *MSB2*) are directly linked to biofilm formation (Finkel et al. 2012; Li et al. 2007; Lohse et al. 2018; Nobile et al. 2012; Nobile et al. 2006a; Puri et al. 2012; Zhao et al. 2005). Fox et al. (2015a) proposed that early adhesion proteins are mainly involved in binding cells to the substrate, and cell-to-cell contact is mediated by late adhesion proteins (Fox et al. 2015a). Bcr1 is one of the master regulators of biofilm. It controls the expression of some cell wall protein

Table 9.3 Adhesion genes have been distinguished during biofilm formation in early and later times

Ten genes induced early in biofilm formation	Ten genes induced later in biofilm formation
ALS1, ALS2, ALS3, ALS4, EAP1, MSB2, PGA6, SIM1, ORF19.2449, and *ORF19.5126*	*FAV2, HYR1, IFF4, IFF6, PGA32, PGA55, ORF19.3988, ORF19.4906, ORF19.5813*, and *ORF19.7539.1*

genes such as *ALS1, ALS3*, and *HWP1*, which are very important for the adherence process as mediators of cell-cell adherence during biofilm growth (Nobile et al. 2006a, b, 2008; Nobile and Mitchell 2005; Uppuluri and Lopez Ribot 2017; Zhao et al. 2006). The *hwp1/hwp1* mutant and *als1/als1 als3/als3* double mutant are unable to form biofilms individually, but when mixed together, they form a strong biofilm both in vivo and in vitro. This suggests that Hwp1, Als1, and Als3 play distinct and complementary roles in biofilm formation (Nobile et al. 2008). Moreover, the expression of *HWP1* in *S. cerevisiae* enhances its ability to adhere to hyphae of wild-type *C. albicans*. Still, this adherence is reduced when tested with hyphae of a *C. albicans als1/als1 als3/als3* double mutant (Nobile et al. 2008). At this point, a minimal biofilm growth pathway can be proposed. Yeast cells expressing Eap1 and Als1, which facilitate cell-substrate binding, are involved in the first step of the biofilm development pathway. Als3 and Hwp1 mediate cell-cell binding and are expressed and proliferated by surface-bound cells as the second step (Finkel and Mitchell 2011a). Eap1, Sun41, and the CFEM family (Pga10/Rbt51, Rbt5, and Wap1/Csa1) are just some of the cell-surface proteins necessary for biofilm formation alongside Als1, Als3, and Hwp1 (Firon et al. 2007; Hiller et al. 2007; Li et al. 2007; Nobile et al. 2008; Norice et al. 2007; Perez et al. 2006).

Bcr1 is a protein of C_2H_2 zinc finger which has a major role in developing biofilm in *C. albicans* (Nobile et al. 2006a). The Bcr1 transcription factor was shown to control the expression of genes that encode proteins anchored to the cell wall, among which are Als1 and Als3, and hyphal wall protein Hwp1 (Cabral et al. 2014; Nobile et al. 2006a, b; Nobile and Mitchell 2005). The *bcr1/bcr1* mutant shows defective biofilm formation due to altered interactions between cells mediated by cell wall proteins, including Als1, Als3, and Hwp1 in both in vivo and in vitro (Gutiérrez-Escribano et al. 2012; Nobile et al. 2006a). In a study using flow cell assay to screen a library of transcriptional factor mutants in vitro, 30 transcriptional factors were identified as important for the adhesion mechanism of *C. albicans* on a silicone surface (Finkel et al. 2012). Among these transcription factors, only Ace2p, Arg81p, Bcr1p, and Snf5p mutants display in vitro antiadherence properties (Finkel et al. 2012; Chong et al. 2018). Bcr1 acts as a positive regulator of hyphal-specific adhesins but is not required for hyphal morphogenesis (Nobile and Mitchell 2005; Dwivedi et al. 2011). Studies show that biofilm formation in *C. albicans* can happen over a broad range of conditions, and the genetic requirements will probably vary from one condition to another (Gulati and Nobile 2016).

9.4.2 Formation of Hyphal Cells

After the adherence step, the formation of hyphal cells under biofilm conditions takes place. The morphogenetic transition from yeast to hyphae in *C. albicans* is a central characteristic under planktonic and biofilm conditions (Uppuluri and Lopez Ribot 2017). Thus, it is clear that many transcription regulators and structural proteins that contribute to hyphal formation in the suspension culture system are also required for hyphal formation in biofilm growth (Lohse et al. 2018). Its ability to transition between multiple morphogenetic states is tightly coupled to its virulence and ability to cause infection (Sudbery 2011; Shapiro et al. 2011). Hypha morphology and expression of hypha-specific genes are crucial for virulence in various fungal pathogens. *C. albicans* demonstrates a relationship between hyphae and virulence through invasion of epithelial cells, damage to endothelial cells, and lysis of macrophages and neutrophils following phagocytosis (Thompson et al. 2011). In addition, hyphae also use contact sensing, thigmotropism, to identify and penetrate small grooves, crevices, and weak areas during infection in host tissues (Thompson et al. 2011).

Morphogenesis is a prerequisite for forming biofilms, as *C. albicans* biofilms are made up of a heterogeneous population of yeast, pseudohyphae, hyphae, etc. (Douglas 2003; Blankenship and Mitchell 2006). It was reported that biofilm-defective mutants were also defective in hyphal differentiation, citing the significance of morphogenesis in biofilm formation (Richard et al. 2005). The type of cells in biofilm decides the virulence properties; e.g., biofilms with high content of hyphae over 50% exhibit more compressive strength and are more difficult to disrupt. So, the hyphae-yeast ratio of *C. albicans* biofilms modulates its compressive strength (Paramonova et al. 2009). This cellular morphogenetic process is triggered by numerous environmental signals, including temperature, pH level, serum, hypoxia (low oxygen), carbon dioxide, N-acetylglucosamine (GlcNAc), amino acids such as proline and methionine (Chow et al. 2021). Multiple signaling pathways govern morphogenesis in *C. albicans*, such as MAPK (mitogen-activated protein kinase) pathway and the cAMP-PKA (cyclic AMP-protein kinase A) pathway (Sharma et al. 2019). Hsp90 (heat shock protein) uses the cAMP-PKA pathway to regulate temperature-dependent morphogenesis. At the same time, farnesol (a quorum-sensing molecule) can target and inhibit the cAMP-PKA pathway, resulting in decreased cAMP production and, therefore, inhibition of hyphal morphogenesis (Sharma et al. 2019; Chow et al. 2021). Other pathways, including pH, TOR, and cell cycle arrest pathways, all play a role in hyphal gene expression and filamentous growth (Shapiro et al. 2011). Several types of MAPK pathways are critical in morphogenesis, which also play a role in cell wall integrity, such as PKC (protein kinase C) and HOG (high osmolarity glycerol) (Shapiro et al. 2011; Sharma et al. 2019). Please review these references for more details on how these signal transduction pathways activate *C. albicans* hyphal morphogenesis in response to different environmental signals (Chow et al. 2021; Kornitzer 2019).

From the nine core transcriptional regulators, Efg1, Tec1, Ndt80, Flo8, Brg1, and Rob1 are regulated biofilm formation and also regulated hyphal formation (Lohse et al. 2018; Villa et al. 2020). (For more details about transcriptional regulatory factors of hyphal morphogenesis in *C. albicans,* review Villa et al. 2020). The hyphae play an important role in the stability of biofilm architecture and provide a scaffold for yeast cells and the other hyphae. Therefore, the ability of *C. albicans* to form hyphae in biofilm facilitates the adherence of hyphae to one another and to other cells, which are important for the normal growth and maintenance of biofilm (Nobile and Johnson 2015).

9.4.3 Extracellular Matrix of *C. albicans* Biofilm

The accumulation of extracellular matrix is another critical feature in biofilm maturation (Finkel and Mitchell 2011a). The *C. albicans* biofilm matrix under in vitro settings contains a variety of macromolecules, including proteins and glycoproteins (55%), carbohydrates (~25%), lipids (~15%), and nucleic acids (mostly noncoding DNA) (~5%) (Lopez-Ribot 2014; Nett and Andes 2020; Zarnowski et al. 2014). β-1,3-glucan is one specific component of the *C. albicans* biofilm matrix, which is required for biofilm drug resistance (Fox and Nobile 2013; Nett et al. 2010c). Furthermore, biofilm cells exhibit increased susceptibility to antifungal drugs when biofilm is without β-1,3-glucan production (Nett et al. 2011). *FKS1* is the gene responsible for glucan synthase the disruption of which leads to reduced production and deposition of β-1,3-glucan in the biofilm matrix and susceptibility to fluconazole due to reducing β-1,3-glucan in the matrix (Araújo et al. 2017; Nett et al. 2010a). In *C. albicans* matrix, the glucan synthesis by Fks1p is critical for biofilm-specific drug resistance, and Fks1p is responsible for the production of cell wall β-1,3 glucan during biofilm growth as in planktonic cells (Nett et al. 2010c).

In *C. albicans* biofilm, matrix production is regulated through *RLM*1 and *ZAP1* transcriptional regulators (Nett and Andes 2020). The zinc-response transcription factor Zap1 (also known as Csr1) is a negative regulator of extracellular soluble β-1,3 glucan. At the same time, the deletion of *ZAP1* produces a biofilm with 1.5 to 2-fold in vitro and over threefold in vivo more soluble β-1,3 glucan in biofilm than the reference strain (Nobile et al. 2009). *GCA1, GCA2, ADH5, CSH1,* and *IFD6* are target genes of Zap1 regulator, where Zap1 represses the expression of glucoamylase genes (*GCA1, GCA2*) and the alcohol dehydrogenase 5 (*ADH5*), which have positive roles in matrix production. However, Zap1 activates *CSH1* and *IFD6* expression, negatively affecting the matrix production (Finkel and Mitchell 2011a; Nobile et al. 2009).

Rlm1 acts as a positive regulator of the production of β-1-3 glucan, where deletion of Rlm1 reduces matrix glucan, likely through the transcription of Smi1, which regulates Fks1p, glucan production, and drug resistance during biofilm growth (Alim et al. 2018; Nett et al. 2011; Nett and Andes 2020). It has shown that *SMI1, RLM1*, and *FKS1* are downstream genes of *C. albicans* PKC pathway for manufacturing of the cell wall and matrix β-1,3-glucan during biofilm growth where *SMI1* acts as a regulator for glucan production through an expression of *FKS1* glucan synthase (Nett et al. 2011). The extracellular glucoamylases convert long-chain polysaccharides into smaller-chain polysaccharides. Therefore, such glucoamylases might function through the hydrolytic release of soluble β-1,3-glucan fragments (Nobile et al. 2009). It was proposed that alcohol dehydrogenase (Adh5, Csh1, and Ifd6) generate quorum-sensing aryl and acyl alcohols, which control many events of biofilm maturation (Nobile et al. 2009; Bonhomme et al. 2011). Taff et al. (2012) described the role of three genes (*BGL2, PHR1*) to encode two glucosyltransferases and (*XOG1*) to encode the exo-glucanase as glucan modifying genes involved in glucan delivery and matrix incorporation, and these enzymes function independently of Zap1 (Taff et al. 2012). Another gene regulator of biofilm matrix production in *C. albicans* is *CCR4*, where *ccr4Δ/Δ* shows the altered structure of biofilm with morphological changes and hyper-production of extracellular matrix (Verma-Gaur et al. 2015). Thus, the *CCR4* gene is the second negative regulator, and the other is Zap1 (Verma-Gaur et al. 2015). Depleting heat shock protein Hsp90 reduces glucan levels in *C. albicans* biofilm. In addition, Hsp90 is also required for the resistance of *C. albicans* biofilm to fluconazole in a mammalian host (Robbins et al. 2011).

9.4.4 Dispersion

This step is characterized by the dispersal of yeast cells of biofilm, which allows them to go on to colonize new surfaces again; this dispersion occurs throughout biofilm formation, with a more significant number of cells being released when the biofilm proliferates rapidly during the intermediate phase (Uppuluri et al. 2010a). Most dispersed cells are yeast cells; there are three transcriptional regulators of dispersal, namely, Pes1 (also known as Nop7), Ume6, and Nrg1 (Finkel and Mitchell 2011a). *UME6* has been recently identified as a gene that encodes filament-specific transcriptional factors, which is important for extending *C. albicans* hyphae (Banerjee et al. 2008). Thus, while *UME6* levels rise, cells sequentially transition from yeast to pseudohyphae to hyphae, and there is a corresponding increase in biofilm formation (Ume6 serves as a positive regulator for hyphal development) (Banerjee et al. 2013). On the other hand, the transcriptional regulator, Ume6, serves as a negative regulator in the dispersal phase, where overexpression of Ume6 reduces cell dispersal from biofilm. On the contrary, overexpression of Pes1 leads to an increase in the dispersion of yeast cells from the biofilm (Uppuluri et al. 2010a) due to overexpression of *PES1*, which leads to decreased hyphal growth and increased lateral yeast growth (Shen et al. 2008). Besides, the overexpression of Nrg1 leads also to an increase in the dispersal of cells from biofilm (Uppuluri et al. 2010b). *PES1* encodes the *pescadillo* homolog in *C. albicans*, which is critical for the filament-to-yeast switch and lateral yeast growth (Shen et al. 2008). *NRG1* gene encodes Nrg1p, a DNA-binding protein with a zinc finger domain, which plays an important role in biofilm formation as a negative regulator of filamentation (yeast-to-hypha morphogenetic transition) (Braun et al. 2001; Uppuluri et al. 2010b). Nrg1 likely acts through Set3 chromatin-modifying complex (Hnisz et al. 2012). Indeed, testing individual Set 3 complex members mutant strains found that they formed a robust biofilm, which cannot disperse normally (Nobile et al. 2014). Thus, Set 3 shows the central role in biofilm dispersal (Nobile et al. 2014).

The molecular chaperone Hsp90 has also been involved in *C. albicans* biofilm dispersal, as depletion of Hsp90, which dramatically reduces the number of dispersed cells from a biofilm, potentially reducing their ability to serve as reservoirs for persistent infection (Robbins et al. 2011). Depleting Hsp90 also induces filamentous growth by relieving cAMP-PKA signaling repression mediated by Hsp90 (Signaling et al. 2009). Ywp1 is yeast wall protein 1, also called Pga24. It is present in the yeast form and plays a role in biofilm dispersion. The deletion of Ywp1 leads to increased biofilm adhesiveness and decreased biofilm dispersal (Granger et al. 2005; Granger 2012).

9.5 Quorum Sensing

Quorum sensing describes a broad set of phenomena in which microbial behaviors or responses are governed by cell density through signaling by such secreted molecules (Finkel and Mitchell 2011a). Farnesol and tyrosol are the two main quorum-sensing molecules that have been characterized in *C. albicans* and are both found under planktonic and biofilm conditions (Finkel and Mitchell 2011a). *C. albicans* produce them, and they play a role in the morphological transition from yeast to hyphae, where tyrosol and farnesol accelerate and block, respectively, the morphological transition (Alem et al. 2006). Farnesol blocks the transition of yeast-hyphae, which is one of the most important virulence traits of *C. albicans*, at high cell densities, and it does not block the elongation of pre-existing hyphae or impact cell growth rates of *C. albicans* (Sebaa et al. 2019). Furthermore, farnesol acts as a naturally occurring quorum-sensing molecule that inhibits biofilm formation, concomitant with decreased *HWP*1 expression in biofilm; this inhibition is dependent on the initial adherence time and concentration of farnesol (no effect has been noted on biofilm formation after 1–2 h of adherence and after the beginning of hyphae formation at 0.3 mM concentration of farnesol) (Ramage et al. 2002). Microarray analysis of farnesol-treated biofilms shows it controls the expression of genes crucial for hyphal development (*TUP1* and *CRK1*), cell wall integrity (*CHT2* and *CHT3*), and drug resistance (*FCR1* and *PDR16*). By controlling these genes, farnesol may prevent hyphal cell development (Cao et al. 2005; Finkel and Mitchell 2011b). Mutant studies suggest that histidine kinase Chk1 may be part of the farnesol signaling pathway. A *chk1Δ/chk1Δ* mutant strain forms a biofilm in the presence of farnesol, suggesting that this kinase is a key downstream mediator of cellular response to quorum sensing (Kruppa et al. 2004). Farnesol may inhibit hyphae formation and reduce biofilm biomass, suggesting its role in stalling or inhibiting mature biofilm formation (Finkel and Mitchell 2011b).

Tyrosol is a second quorum-sensing molecule and a derivative of the tyrosine molecule in *C. albicans,* which is released into the growth medium continuously during growth (Chen et al. 2004). The biofilm cells produce more tyrosol than planktonic cells calculated as a function of cell dry weight (Alem et al. 2006). Tyrosol accelerates the development of germ tubes under hyphae-inducing conditions (Chen et al. 2004). Exogenous farnesol inhibits biofilm formation up to 33% with 1 mM farnesol when added during the adhesion period, while exogenous tyrosol does not affect biofilm formation (Alem et al. 2006). This study also tested biofilm supernatants for their abilities to block or enhance germ tube formation by planktonic cells. It was suggested that tyrosol stimulates hyphal production during the early and intermediate stages of *C. albicans* biofilm formation in the later stages of biofilm formation. In contrast, farnesol may play a critical role by inducing the release of yeast cells from the biofilm, thus allowing dispersal (Alem et al. 2006). In addition, isoamyl alcohol, E-nerolidol, 1-dodecanol, and 2-phenylethanol small molecules are also detectable in biofilm supernatants; any of these molecules can play a role in morphogenesis by inhibiting the formation of hyphae (Martins et al. 2007). Thus, all the above-mentioned small molecules could help biofilm dispersal by encouraging the formation of yeast cells (Finkel and Mitchell 2011a).

9.6 Natural Compounds as Anti-*Candida albicans* Biofilm

As discussed above, the clinical problems caused by *C. albicans* biofilm are of significant concern due to its high resistance to antifungal drugs and host immune mechanisms. Currently, antifungal drugs have limitations such as narrow spectrum, toxicity, high treatment costs, and increased resistance (Lone and Ahmad 2020). The emergence of multidrug-resistant *Candida* and the need to treat resistant forms like biofilms have led to using natural products as new and effective therapies in clinical treatments (Ahmad et al. 2015). According to recent studies, natural compounds like curcumin, cinnamaldehyde, eugenol, and thymol can prevent the formation of biofilms and also destroy mature biofilms (De Lucena Rangel et al. 2018; Doke et al. 2014). Combining antifungal drugs and natural compounds can effectively treat candidiasis with increased potency, lower drug toxicity and dosages, and reduced risk of resistant strains (Shinde et al. 2013). Therefore, this section will discuss natural compounds, drug delivery platforms, and the synergistic antibiofilm activity of antifungal drugs for clinical use against *C. albicans* biofilm.

9.6.1 Thymol

Thymol is a natural phenol compound (monoterpene) in many plant oils. It has been used for many years in traditional medicine as a result of its different pharmacological potentials, including antifungal, antibacterial, anti-inflammatory, antispasmodic, analgesic, antioxidant, antiseptic, and anticancer (Gholami-Ahangaran et al. 2022). Thymol has been evaluated against a *C. albicans* biofilm, and it was reported that it could inhibit biofilm formation, reduce viable cells, and eliminate mature biofilm (Braga et al. 2008; Dalleau et al. 2008; de Vasconcelos et al. 2014; Shu et al. 2016). In addition, the effect of thymol in combination with other natural compounds such as piperine or with antifungal drugs such as fluconazole against *C. albicans* biofilm has been studied (Jafri and Ahmad 2020; Pemmaraju et al. 2013; Priya et al. 2022). Thymol and piperine have been shown to have synergistic effects on *C. albicans* biofilm; the combination therapy reduced virulence factors related to adhesion, morphological transformation, hyphal extension and controlled hyphal elongation. These findings were confirmed by the gene expression analysis in vitro; results showed that negative transcriptional regulators (*NRG1* and *TUP1*) of filamentous growth are upregulated, while *UME6*, *EFG1*, *CPH1*, and *RAS1*, positive regulators, are downregulated (Priya et al. 2022). In addition, the downregulation of *ALS1* and *EAP1* encodes adhesins that facilitate *C. albicans* adherence to polystyrene and epithelial surfaces. No toxic effects were observed on erythrocytes or human buccal epithelial cells. These results suggest that thymol and piperine can interact synergistically to inhibit *C. albicans* biofilm and hyphal transition (Priya et al. 2022). Thymol and fluconazole, on the other hand, had synergistic effects on both planktonic and biofilm communities (Pemmaraju et al. 2013; Jafri and Ahmad 2020).

9.6.2 Linalool

Linalool was the main ingredient (29%) of Iranian *Satureja macrosiphon* essential oil (EO). This EO exhibited antibiofilm action against *C. albicans* and *C. dubliniensis* at 4–8 μL/mL concentrations (Motamedi et al. 2020). Another investigation showed that purified linalool from *Croton cajucara* plant had inhibited the *C. albicans* biofilm. According to electron microscopy analysis, linalool caused abnormal *C. albicans* germination and decreased cell size. Germ tube is a crucial virulence component for *C. albicans* invasion; thus, *Candida* infections may be treated by inhibiting germ tube development (Alviano et al. 2005). In a recent study, linalool was found to strongly inhibit biofilm development in *C. albicans* without affecting the proliferation of the planktonic cells. It synergistically inhibited biofilm development with α-longipinene (the primary component of cascarilla bark oil) and helichrysum oil (Manoharan et al. 2017a). Linalool also inhibited hypha formation, which is crucial for virulence. *C. albicans* biofilms include yeast, hyphae, and pseudohyphae; the change from the yeast to hyphae stage is an essential virulence component, and abnormal biofilms develop in strains with defects in hyphae formation. Linalool-mediated inhibition could result from interference with filament production and induction. Combining linalool with other natural compounds could be more effective in fungal biofilm elimination (Manoharan et al. 2017a). These findings corroborate those of Hsu et al. (2013), who found that linalool inhibited the development of germ tubes and biofilm of *C. albicans*. Expression analysis revealed that linalool inhibited the expression of multiple *C. albicans* pathogenic genes, including adhesion-related genes *ALS3* and *HWP1*, genes *EED1*, *HGC1*, and *UME6* involved in long-term hyphae maintenance and the genes *CPH1* and *CYR1* involved in the Mitogen-Activated Protein Kinase and cAMP-Protein kinase A hyphal formation regulatory pathways, respectively as well as its upstream regulator gene, *RAS1*, was also down-regulated (Hsu et al. 2013). Thus, linalool suppresses *C. albicans* attachment and biofilm formation while also inhibiting hyphal transformation and maintenance pathways. This highlights its antibiofilm properties for treating *Candida* biofilm-associated infections, including implant medical devices. Further studies are needed to understand better its interactions with *Candida* biofilms (Shariati et al. 2022).

9.6.3 Curcumin

It has been demonstrated that curcumin has greater antiadhesion efficacy against *C. albicans*, decreasing initial adhesion after precoating with the curcumin. However, after 4 h of adhesion, curcumin's inhibitory impact had decreased to 50% of its initial level. It has also been reported that curcumin reduced the expression of

ALS3 and *HWP1* genes (Shahzad et al. 2014). More investigations are required to establish if the compound can be employed to eliminate *Candida* biofilms without harming human cells. In *C. albicans* resistant to fluconazole, curcumin inhibited biofilm formation and yeast-to-hypha morphological transition alone or synergistically with fluconazole. Curcumin also altered membrane permeability and lowered intracellular ATPase synthesis in combination with fluconazole, making it a promising novel treatment for drug-resistant *Candida* strains because of its potential synergistic with fluconazole (Dong et al. 2021). Curcumin's interactions with antifungals are poorly understood; thus, more research is needed. Low oral bioavailability, water insolubility, quick metabolism and degradation, poor absorption from the gut, and low blood plasma levels are reasons curcumin is not more widely used in the therapeutic environment (Shariati et al. 2019). Researchers have been interested in adopting different drug delivery platforms to improve curcumin's antibiofilm effects in an effort to combat these problems. Curcumin-sophorolipid nanocomplex was employed to degrade *C. albicans* biofilm by Rajasekar et al. (2021). Practical candidates for transporting hydrophobic molecules include sophorolipids, which are surfactants produced from biological sources. Curcumin-sophorolipids, which were produced, significantly inhibited adhesion, biofilm formation and filamentation. This chemical also inhibited the expression of several pathogenic genes (*EFG1*, *ALS1*, *SAP8*, and *EAP1*) (Rajasekar et al. 2021). In another study, graphene oxide functionalized with curcumin and polyethylene glycol (PEG) was employed to eliminate *C. albicans* biofilm (Palmieri et al. 2018). This molecule was found to reduce the growth of fungi, their ability to adhere to surfaces, and their capacity to form biofilms. As the authors propose, combining curcumin and PEG molecules on the graphene oxide surface results in a potent antibiofilm action that can also inhibit the local proliferation of *C. albicans* cells that are not adherent to the surface (Palmieri et al. 2018). A recent study showed that curcumin-loaded chitosan nanoparticles (NPs) had good antibiofilm effects against *Staphylococcus aureus* and *C. albicans* mixed-species and mono-species biofilms. Microscopy showed that the produced chemical reduced biofilm thickness and killed microbial cells on silicone surfaces. Thus, curcumin-loaded chitosan NPs could eradicate *C. albicans* biofilm infections, particularly mixed-species biofilms (Ma et al. 2020). However, data in this field are extremely few, and additional in-depth research should be done before the clinical use of curcumin-based drug delivery systems.

9.6.4 Cinnamomum

A recent in vitro study to examine the effect of cinnamon oil on *C. albicans* biofilm was performed on dental devices developed from heat-polymerized polymethyl methacrylate (PMMA) (Choonharuangdej et al. 2021). The cinnamon oil eliminated 99% of the *C. albicans* biofilm after 1 h, while after 24 h, it reduced the biofilm formation by 70%. In addition, cinnamaldehyde was found to have potential activity against the biofilm of *Candida* species isolated from oral candidiasis patients; it reduces the metabolic activity and biomass of mature biofilms (Miranda-Cadena et al. 2021). A recent study has found that cinnamaldehyde from *Cinnamomum verum* EO inhibits *C. albicans* biofilms and reduces the adhesion of clinically isolated samples. It also suppresses the hemolysin and phospholipase activity of *C. albicans*. Molecular docking shows that cinnamaldehyde impacts Als3, making it a promising compound for inhibiting adhesion and biofilm growth (El-Baz et al. 2021). According to another study, cinnamaldehyde inhibits biofilm formation in *Candida* by affecting crucial proteins present in the fungal cell and nucleus (da Nóbrega et al. 2020). Further docking studies are recommended to identify these proteins targeted by cinnamaldehyde accurately. The results of a 2021 study showed that *C. verum* EO at concentrations of 1–2 mg/mL is nontoxic to normal human keratinocytes (Wijesinghe et al. 2021). Cinnamaldehyde's limited effectiveness against active infections is due to its poor solubility, instability, and volatility. Mishra et al. (2021a) addressed this by incorporating it into electrospun gellan (GA)/polyvinyl alcohol (PVA) nanofibers, which eliminated up to 50.45% of *C. albicans* biofilms. The nanofibers had lower activity against *C. albicans* due to its complex biofilm structure with multilayers (Mishra et al. 2021a). Another study found that encapsulated cinnamaldehyde in multilamellar liposomes (CM-ML) showed improved fungicidal effects against *C. albicans* clinical isolates due to the sustained release of cinnamaldehyde is occurring in the encapsulated form (Khan et al. 2017). Recently, researchers loaded cinnamaldehyde onto (PLGA) poly (DL-lactide-co-glycolide) nanoparticles (CA-PLGA NPs) and tested their antibiofilm activity against *C. albicans*. Even at lower doses, CA-PLGA NPs showed a strong antifungal effect, with postbiofilm application proving more effective than prebiofilm application (Gursu et al. 2022). Thus, encapsulating cinnamaldehyde in different carriers may increase its effectiveness and mitigate cytotoxicity, allowing lower doses to eliminate *Candida* biofilms.

9.6.5 Terpinen-4-ol

The terpinen-4-ol, tea tree oil's main component (TTO), suppressed *C. albicans* growth and disrupted biofilm formation. Terpinen-4-ol was effective at a lower concentration (8.86 mg/mL) than TTO (17.92 mg/mL) and penetrated the cytoplasmic membrane, causing cell membrane disruption and interfering with cell integrity and physiology. The authors suggested that the antifungal characteristics of TTO and terpinen-4-ol could be related to their lipophilicity (Francisconi et al. 2020a). The synergistic effects of a combination of terpinen-4-ol and antifungals like nystatin inhibited *C. albicans* biofilm. Nystatin disrupts fungal cell membranes but can cause unpleasant side effects in humans and lead to fungal resistance has been reported (Francisconi et al. 2020a). Hence, combining nystatin with natural compounds may reduce dosage, side effects, prevent resistance, and boost clinical use. In another work, Francisconi et al. (2020b) created a liquid crystalline system (LCS) containing nystatin and terpinen-4-ol to test its antibiofilm and synergistic/modulatory effects against *C. albicans* (Francisconi et al. 2020b). Synergistically, the LCS with nystatin and terpinen-4-ol inhibited *C. albicans* growth and biofilm formation at lower concentrations. Artificial saliva enhanced the LCS's viscosity and mucoadhesive. Thus, LCS with nystatin and terpinen-4-ol might prevent, and therapy *C. albicans* infection, but additional research is needed (Francisconi et al. 2020b). In another investigation, it has been shown that terpinen-4-ol has strong activity on *C. albicans* biofilm. Terpinen-4-ol, the main ingredient of TTO, has advantages over the complete EO regarding safety and product consistency. It is not cytotoxic against OKF6-TERT2 epithelial cells at 0.5 × MIC50, making it useful for preventing and treating various *C. albicans* biofilm-associated infections, particularly established oropharyngeal candidiasis (Ramage et al. 2012). A study found terpinen-4-ol loaded into polyethylene glycol (PEG)-stabilized lipid NP can eliminate *C. albicans* biofilm. This platform effectively suppressed the formation of biofilm by the fungus. Terpinen-4-ol-loaded lipid nanoparticles destroy the cell membrane structure of *C. albicans* and block the respiratory chain by inhibiting succinate dehydrogenase attached to the inner mitochondrial membrane of cells, resulting in antibiofilm activity (Sun et al. 2012). Souza et al. (2017) employed nanostructured lipid carriers to deliver TTO and degrade *Candida* biofilms. The chromatographic profile shows that the oil was consistent with ISO 4730, containing 41.9% Terpinen-4-ol as its major constituent. Various *Candida* species, including *C. albicans* biofilms, were greatly decreased by synthesized TTO NPs. TTO NPs reduced biofilm protein and exopolysaccharides (Souza et al. 2017). Nanobiotechnology could improve the delivery of terpinen-4-ol's effectiveness against *Candida* biofilm infections. More research is needed, but it shows promise for oral hygiene products like mouth rinses and denture cleansers (Shariati et al. 2022).

9.6.6 Geraniol

Singh et al. (2016) investigated the effects of geraniol on the cellular pathways of *C. albicans*. The results showed that geraniol inhibited hyphal growth and biofilm development in *C. albicans*. Iron genotoxicity, disruption of homeostasis, mitochondrial dysfunction, reduction of ergosterol levels, and modification of plasma membrane ATPase activity were also seen as a result of exposure to this compound (Singh et al. 2016). In a 2018 study, geraniol combined with fluconazole showed promise in inhibiting *C. albicans* biofilm. Geraniol reduced biofilm biomass, impaired fungal adherence to the epithelial cells, and synergized with fluconazole (Singh et al. 2018). The authors suggested that geraniol binds to the active site of CaCdr1p and modulates its efflux pump activity. Therefore, combining geraniol and fluconazole has the potential as a promising approach to treatment for candidiasis (Singh et al. 2018). In addition, another study found that geraniol inhibits biofilm formation against both fluconazole-sensitive and resistant *C. albicans* strains at MIC and 2× MIC concentrations (Cardoso et al. 2016). Additionally, Table 9.4 presents further studies that have used natural compounds to inhibit and destroy *C. albicans* biofilm.

Table 9.4 Further findings demonstrate the effectiveness of natural compounds in inhibiting and eradicating *C. albicans* biofilm

Natural compound	Finding	Source of natural compound	References
Piperine	Showed excellent effectiveness against the different stages of biofilm formation (adhesion, development and maturation)	Purchased (Sigma-Aldrich)	Thakre et al. (2020)
	Exhibited synergistic action against *C. albicans* biofilm with fluconazole		
Piperine	Inhibition of hyphal morphogenesis and biofilm growth	Purchased (HiMedia)	Priya and Pandian (2020)
Limonene	Showed excellent effectiveness against the different stages of biofilm formation (adhesion, development and maturation)	Purchased (Sigma-Aldrich)	Thakre et al. (2018)
	Exhibited synergistic action against *C. albicans* biofilm with fluconazole		
Menthol	Showed excellent effectiveness against the different stages of biofilm formation (adhesion, development and maturation)	Purchased (Sigma-Aldrich)	Zore et al. (2022)
	Exhibited synergistic action against *C. albicans* biofilm with fluconazole		
Menthol	Imidazolium ionic liquids based on (−)-menthol inhibited biofilm growth	Purchased (Sigma-Aldrich)	Suchodolski et al. (2021)
Menthol	Inhibited biofilm development	*Mentha piperita* L EO	Saharkhiz et al. (2012)
Eucalyptol	Inhibited adhesion, biofilm formation, morphogenesis, altered microarchitecture, and reduced the viability of the established biofilm	*Lavandula dentata* L EO	Müller-Sepúlveda et al. (2020)
β-Pinene	Inhibited germ tube formation and eradicated mature biofilm	*Santolina impressa*	Alves-Silva et al. (2019)
α-Terpineol	Exhibited rapid antibiofilm properties	Purchased (Sigma-Aldrich)	Ramage et al. (2012)
α-Pinene and β-pinene	These compounds prevented biofilm formation and were highly toxic to *C. albicans*	Purchased (Sigma-Aldrich)	da Silva et al. (2012)
Borneol	Preformed biofilm was disrupted	*Thymus carnosus* EO	Alves et al. (2019)
Borneol	Reduced biofilm growth	Purchased (Sigma-Aldrich)	Manoharan et al. (2017b)
Camphene	Preformed biofilm was disrupted	*Thymus carnosus* EO	Alves et al. (2019)
Camphor	Decreased the established biofilm and hyphal formation	Purchased (Sigma-Aldrich)	Ivanov et al. (2021)

Table 9.4 (continued)

Natural compound	Finding	Source of natural compound	References
Camphor	Biofilm and hyphal formation were significantly reduced. Additionally, some biofilm-related and hypha-specific genes were also downregulated	Purchased (Sigma-Aldrich)	Manoharan et al. (2017b)
Carvacrol	Incorporation of carvacrol into the soft-liner decreased *C. albicans* biofilm growth	Purchased (Sigma-Aldrich)	Baygar et al. (2018)
Citral	The addition of EO to chitosan microparticles has an inhibitory impact on biofilm	Lemongrass EO and geranium EO	Garcia et al. (2021)
Citral	The biofilm's cell viability and biomass decreased across all species. Citral also inhibited virulence factor and hyphal adhesin expression in *C. albicans*	Lemongrass (*Cymbopogon flexuosus)*	Gao et al. (2020)
Citronellol	Showed inhibitory effect on the secretion of extracellular proteinases and phospholipases (planktonic form) and biofilm growth	Purchased (Sigma-Aldrich)	Sharma et al. (2020)
Coumarin	Exhibited inhibition of fungal adhesion and biofilm formation; additionally, eliminated preformed biofilm	Purchased (Sangon Biotech Co., Ltd.)	Xu et al. (2019)
Epigallocatechin gallate	The minimum biofilm inhibitory concentration (MBIC) range of epigallocatechin gallate was lower than fluconazole and ketoconazole	Purchased (Sigma-Aldrich)	Behbehani et al. (2019)
Epigallocatechin gallate	This chemical has been shown to have a synergistic impact with fluconazole, miconazole, and amphotericin B against the biofilm formation	Purchased (Sigma-Aldrich)	Ning et al. (2015)
Ethyl alcohol	Showed inhibition of biofilm development and germ tube formation	Not reported	Chauhan et al. (2011)
Eucalyptol	Eucalyptol/β-cyclodextrin inclusion complex to gellan/polyvinyl alcohol nanofibers inhibited 70% biofilm of *C. albicans*	Purchased (Sigma-Aldrich)	Mishra et al. (2021b)
Eucalyptol	This compound exhibited antibiofilm activity against developing and mature biofilm	Purchased (Sigma-Aldrich)	Gupta et al. (2021)
Saponin	Strong inhibition was seen on biofilm development, existing biofilms, and the yeast-to-hypha transition	*Solidago virgaurea*	Chevalier et al. (2012)
Saponin	Growth of hyphae, adhesion of yeast, development of germ tubes, and formation of biofilm were all inhibited by this compound	*Medicago sativa* and *Saponaria officinali*	Sadowska et al. (2014)

(continued)

Table 9.4 (continued)

Natural compound	Finding	Source of natural compound	References
Saponin	Suppressed adhesion, biofilm growth, transition phase, and phospholipase production. Additionally, saponin led to the generation of endogenous ROS, thus disrupting the cell membrane in planktonic cells	Roots of *Dioscorea panthaica* Prain et Burk	Yang et al. (2018)
Saponin	Disrupted hyphae and biofilm formation	From various natural products	Coleman et al. (2010)
Pyrogallol	Exhibited antibiofilm activity	Not reported	Shahzad et al. (2014)
Fenchone	Reduced biofilm formation	Purchased (Sigma-Aldrich)	Manoharan et al. (2017b)
Fenchyl alcohol	Decreased biofilm and hyphal formation	Purchased (Sigma-Aldrich)	Manoharan et al. (2017b)
Linoleic acid	Formation of biofilms, hyphal development, and cell aggregation of *C. albicans* were all inhibited. Linoleic acid also inhibited biofilms composed of both *S. aureus* and *C. albicans*	Purchased (Sigma-Aldrich)	Kim et al. (2020)
Linoleic acid	Inhibition of *C. albicans* biofilm formation by Zinc oxide NPs coated with Chitosan-linoleic acid was better to that by fluconazole	Purchased (Sigma-Aldrich)	Barad et al. (2017)
Phloretin	Showed that biofilm growth and yeast morphogenesis were suppressed	Purchased (Aladdin)	Liu et al. (2021)
p-coumaric acid	*p*-coumaric acid-loaded liquid crystalline systems showed higher eradication of established biofilms than Amphotericin B and fluconazole	Purchased (Sigma-Aldrich)	Ferreira et al. (2021)

9.6.7 Eugenol

Eugenol inhibits cell adherence and *C. albicans* biofilms by inhibiting morphogenesis according to scanning electron microscopy (SEM) with low cytotoxicity of human erythrocytes (He et al. 2007). According to El-Baz et al. (2021), molecular docking analysis shows that eugenol inhibits adhesion and biofilm formation in *C. albicans* by interacting with the Als3 protein (El-Baz et al. 2021). According to a 2018 study, eugenol found in *Cinnamomum zeylanicum* has promising antibiofilm effects against *Candida*. The study found that *C. zeylanicum* EO (500 μg/mL) significantly inhibited *Candida* biofilm growth without affecting human red blood cell viability. Eugenol comprises 68.96% of the essential oil extracted from *C. zeylanicum* Blume leaves (De Lucena Rangel et al. 2018). A recent study found that eugenol represents 77.22% of *Cinnamomum verum* EO and affects biofilm (Wijesinghe et al. 2020, 2021). In the current study by Jafri et al. (2019), the results show that eugenol inhibits single biofilm of *C. albicans* resistant to itraconazole, AMB and ketoconazole, *S. mutans* and in mixed biofilm as well. Thus, eugenol effectively reduces biofilms formed by drug-resistant strains of two oral pathogens, single and mixed (Jafri et al. 2019). In addition, the synergistic effect of eugenol and fluconazole was studied, and it was found that this combination has a strong effect on the development of *C. albicans* biofilm as a result of exposure to eugenol sensitization of fungal cells inhibited biofilm formation with low fluconazole concentrations (Doke et al. 2014). In line with these observations, another investigation (2012) found the same outcome a synergistic interaction of eugenol with fluconazole against biofilms formed by the test strains (Khan and Ahmad 2012).

References

Ahmad A, Wani MY, Khan A et al (2015) Synergistic interactions of eugenol-tosylate and its congeners with fluconazole against *Candida albicans*. PLoS One 10:e0145053. https://doi.org/10.1371/journal.pone.0145053

Alem MAS, Oteef MDY, Flowers TH, Douglas LJ (2006) Production of tyrosol by *Candida albicans* biofilms and its role in quorum sensing and biofilm development. Eukaryot Cell 5:1770–1779. https://doi.org/10.1128/EC.00219-06

Alim D, Sircaik S, Panwar S (2018) The significance of lipids to biofilm formation in *Candida albicans*: an emerging perspective. J Fungi 4:140. https://doi.org/10.3390/jof4040140

Alonso MF, Gow NAR, Erwig LP, Bain JM (2017) Macrophage migration is impaired within *Candida albicans* biofilms. J Fungi 3:31. https://doi.org/10.3390/jof3030031

Alves M, Gonçalves MJ, Zuzarte M et al (2019) Unveiling the antifungal potential of two Iberian thyme essential oils: effect on *C. albicans* germ tube and preformed biofilms. Front Pharmacol 10:1–11. https://doi.org/10.3389/fphar.2019.00446

Alves-Silva JM, Zuzarte M, Gonçalves MJ et al (2019) Unveiling the bioactive potential of the essential oil of a Portuguese endemism, Santolina impressa. J Ethnopharmacol 244:112120. https://doi.org/10.1016/j.jep.2019.112120

Alviano WS, Mendonca-Filho RR, Alviano DS et al (2005) Antimicrobial activity of Croton cajucara Benth linalool-rich essential oil on artificial biofilms and planktonic microorganisms. Oral Microbiol Immunol 20:101–105. https://doi.org/10.1111/j.1399-302X.2004.00201.x

Andes D, Nett J, Oschel P et al (2004) Development and characterization of an in vivo central venous catheter *Candida albicans* biofilm model. Infect Immun 72:6023–6031. https://doi.org/10.1128/IAI.72.10.6023-6031.2004

Araújo D, Henriques M, Silva S (2017) Portrait of *Candida* species biofilm regulatory network genes. Trends Microbiol 25:62–75. https://doi.org/10.1016/j.tim.2016.09.004

Banerjee M, Thompson DS, Lazzell A et al (2008) UME6, a novel filament-specific regulator of *Candida albicans* hyphal extension and virulence. Mol Biol Cell 19:1354–1365. https://doi.org/10.1091/mbc.E07

Banerjee M, Uppuluri P, Zhao XR et al (2013) Expression of UME6, a key regulator of *Candida albicans* hyphal development, enhances biofilm formation via Hgc1- and Sun41- dependent mechanisms. Eukaryot Cell 12:224–232. https://doi.org/10.1128/EC.00163-12

Barad S, Roudbary M, Omran AN, Daryasari MP (2017) Preparation and characterization of ZnO nanoparticles coated by chitosan-linoleic acid; fungal growth and biofilm assay. Bratislava Med J 118:169–174. https://doi.org/10.4149/BLL_2017_034

Baygar T, Ugur A, Sarac N et al (2018) Functional denture soft liner with antimicrobial and antibiofilm properties. J Dent Sci 13:213–219. https://doi.org/10.1016/j.jds.2017.10.002

Behbehani JM, Irshad M, Shreaz S, Karched M (2019) Synergistic effects of tea polyphenol epigallocatechin 3-O-gallate and azole drugs against oral *Candida* isolates. J Mycol Med 29:158–167. https://doi.org/10.1016/j.mycmed.2019.01.011

Blankenship JR, Mitchell AP (2006) How to build a biofilm: a fungal perspective. Curr Opin Microbiol 9:588–594. https://doi.org/10.1016/j.mib.2006.10.003

Bonhomme J, D'Enfert C (2013) *Candida albicans* biofilms: building a heterogeneous, drug-tolerant environment. Curr Opin Microbiol 16:398–403

Bonhomme J, Chauvel M, Goyard S et al (2011) Contribution of the glycolytic flux and hypoxia adaptation to efficient biofilm formation by *Candida albicans*. Mol Microbiol 80:995–1013. https://doi.org/10.1111/j.1365-2958.2011.07626.x

Borecká-Melkusová S, Moran GP, Sullivan DJ et al (2009) The expression of genes involved in the ergosterol biosynthesis pathway in *Candida albicans* and *Candida* dubliniensis biofilms exposed to fluconazole. Mycoses 52:118–128. https://doi.org/10.1111/j.1439-0507.2008.01550.x

Braga PC, Culici M, Alfieri M, Dal Sasso M (2008) Thymol inhibits *Candida albicans* biofilm formation and mature biofilm. Int J Antimicrob Agents 31:472–477. https://doi.org/10.1016/j.ijantimicag.2007.12.013

Braun BR, Kadosh D, Johnson AD (2001) NRG1, a repressor of flamentous growth in *C. albicans*, is down-regulated during flament induction. EMBO J 20:4753–4761

Brouwer N, Dolman KM, van Houdt M et al (2008) Mannose-binding lectin (MBL) facilitates opsonophagocytosis of yeasts but not of bacteria despite MBL binding. J Immunol 180:4124–4132. https://doi.org/10.4049/jimmunol.180.6.4124

Brown GD, Taylor PR, Reid DM et al (2002) Dectin-1 is a major β-glucan receptor on macrophages. J Exp Med 196:407–412. https://doi.org/10.1084/jem.20020470

Bugarcic A, Hitchens K, Beckhouse AG et al (2008) Human and mouse macrophage-inducible C-type lectin (Mincle) bind *Candida albicans*. Glycobiology 18:679–685. https://doi.org/10.1093/glycob/cwn046

Cabral V, Znaidi S, Walker LA et al (2014) Targeted changes of the cell wall proteome influence *Candida albicans* ability to form single- and multi-strain biofilms. PLoS Pathog 10:e1004542. https://doi.org/10.1371/journal.ppat.1004542

Cambi A, Netea MG, Mora-Montes HM et al (2008) Dendritic cell interaction with *Candida albicans* critically depends on N-linked Mannan. J Biol Chem 283:20590–20599. https://doi.org/10.1074/jbc.M709334200

Cao Y-Y, Cao Y, Xu Z et al (2005) cDNA microarray analysis of differential gene expression in *Candida albicans* biofilm exposed to farnesol. Antimicrob Agents Chemother 49:584–589. https://doi.org/10.1128/AAC.49.2.584-589.2005

Cardoso NNR, Alviano CS, Blank AF et al (2016) Synergism effect of the essential oil from Ocimum basilicum var. Maria Bonita and its major components with fluconazole and its influence on ergosterol biosynthesis. Evid Based Complement Altern Med 2016:1–12. https://doi.org/10.1155/2016/5647182

Cavalheiro M, Teixeira MC (2018) *Candida* biofilms: threats, challenges, and promising strategies. Front Med 5:1–15. https://doi.org/10.3389/fmed.2018.00028

Chandra J, McCormick TS, Imamura Y et al (2007) Interaction of *Candida albicans* with adherent human peripheral blood mononuclear cells increases *C. albicans* biofilm formation and results in differential expression of pro- and anti-inflammatory cytokines. Infect Immun 75:2612–2620. https://doi.org/10.1128/IAI.01841-06

Chauhan NM, Raut JS, Karuppayil SM (2011) A morphogenetic regulatory role for ethyl alcohol in *Candida albicans*. Mycoses 54:e697–e703. https://doi.org/10.1111/j.1439-0507.2010.02002.x

Chen H, Fujita M, Feng Q et al (2004) Tyrosol is a quorum-sensing molecule in *Candida albicans*. Proc Natl Acad Sci 101:5048–5052. https://doi.org/10.1073/pnas.0401416101

Chen Y, Wang XY, Huang YC et al (2015) Study on the structure of *Candida albicans*–Staphylococcus epidermidis mixed species biofilm on polyvinyl chloride biomaterial. Cell Biochem Biophys 73:461–468. https://doi.org/10.1007/s12013-015-0672-y

Cheng S, Joosten LAB, Kullberg B, Netea MG (2012) Interplay between *Candida albicans* and the mammalian innate host defebse. Infect Immun 80:1304–1313. https://doi.org/10.1128/IAI.06146-11

Chevalier M, Medioni E, Prêcheur I (2012) Inhibition of *Candida albicans* yeast–hyphal transition and biofilm formation by Solidago virgaurea water extracts. J Med Microbiol 61:1016–1022. https://doi.org/10.1099/jmm.0.041699-0

Chong PP, Chin VK, Wong WF et al (2018) Transcriptomic and genomic approaches for unravelling *Candida albicans* biofilm formation and drug resistance—an update. Genes (Basel) 9:540

Choonharuangdej S, Srithavaj T, Thummawanit S (2021) Fungicidal and inhibitory efficacy of cinnamon and lemongrass essential oils on *Candida albicans* biofilm established on acrylic resin: an in vitro study. J Prosthet Dent 125:707.e1–707.e6. https://doi.org/10.1016/j.prosdent.2020.12.017

Chow EWL, Pang LM, Wang Y (2021) From jekyll to hyde: the yeast–hyphal transition of *Candida albicans*. Pathogens 10:1–29. https://doi.org/10.3390/pathogens10070859

Ciurea CN, Kosovski IB, Mare AD et al (2020) *Candida* and candidiasis—opportunism versus pathogenicity: a review of the virulence traits. Microorganisms 8:1–17. https://doi.org/10.3390/microorganisms8060857

Coleman JJ, Okoli I, Tegos GP et al (2010) Characterization of plant-derived saponin natural products against *Candida albicans*. ACS Chem Biol 5:321–332. https://doi.org/10.1021/cb900243b

D'Enfert C, Janbon G (2015) Biofilm formation in *Candida* glabrata: what have we learnt from functional genomics approaches? FEMS Yeast Res 16:1–13. https://doi.org/10.1093/femsyr/fov111

da Nóbrega AD, Monteiro AFM, Andrade PN et al (2020) Docking prediction, antifungal activity, anti-biofilm effects on *Candida* spp., and toxicity against human cells of cinnamaldehyde. Molecules 25:5969. https://doi.org/10.3390/molecules25245969

da Silva ACR, Lopes PM, de Azevedo MMB et al (2012) Biological activities of a-Pinene and β-Pinene enantiomers. Molecules 17:6305–6316. https://doi.org/10.3390/molecules17066305

Dadar M, Tiwari R, Karthik K et al (2018) *Candida albicans*—biology, molecular characterization, pathogenicity, and advances in diagnosis and control—an update. Microb Pathog 117:128–138. https://doi.org/10.1016/j.micpath.2018.02.028

Dalleau S, Cateau E, Bergès T et al (2008) In vitro activity of terpenes against *Candida* biofilms. Int J Antimicrob Agents 31:572–576. https://doi.org/10.1016/j.ijantimicag.2008.01.028

de Groot PWJ, Bader O, de Boer AD et al (2013) Adhesins in human fungal pathogens: glue with plenty of stick. Eukaryot Cell 12:470–481

De Lucena Rangel M, De Aquino SG, De Lima JM et al (2018) In vitro effect of cinnamomum zeylanicum blume essential oil on *Candida* spp involved in oral infections. Evid Based Complement Altern Med 2018:4045013. https://doi.org/10.1155/2018/4045013

de Vasconcelos LC, Sampaio FC, Albuquerque AJR, Vasconcelos LCS (2014) Cell viability of *Candida albicans* against the antifungal activity of Thymol. Braz Dent J 25:277–281. https://doi.org/10.1590/0103-6440201300052

Desai JV, Mitchell AP (2015) *Candida albicans* biofilm development and its genetic control. Microbiol Spectr 3:1–16. https://doi.org/10.1128/microbiolspec.mb-0005-2014

Doke SK, Raut JS, Dhawale S, Karuppayil SM (2014) Sensitization of *Candida albicans* biofilms to fluconazole by terpenoids of plant origin. J Gen Appl Microbiol 60:163–168. https://doi.org/10.2323/jgam.60.163

Dominguez EG, Andes DR (2017) *Candida* biofilm tolerance: comparison of planktonic and biofilm resistance mechanisms. In: Prasad R (ed) *Candida albicans:* cellular and molecular biology. Springer International Publishing, Cham, pp 77–92

Dong H, Wang Y, Peng X et al (2021) Synergistic antifungal effects of curcumin derivatives as fungal biofilm inhibitors with fluconazole. Chem Biol Drug Des 97:1079–1088. https://doi.org/10.1111/cbdd.13827

Douglas LJ (2003) *Candida* biofilms and their role in infection. Trends Microbiol 11:30–36

Dutton LC, Nobbs AH, Jepson K et al (2014) O-Mannosylation in *Candida albicans* enables development of interkingdom biofilm communities. MBio 5:1–15. https://doi.org/10.1128/mBio.00911-14

Dwivedi P, Thompson A, Xie Z et al (2011) Role of Bcr1-activated genes Hwp1 and Hyr1 in *Candida albicans* oral mucosal biofilms and neutrophil evasion. PLoS One 6:e16218. https://doi.org/10.1371/journal.pone.0016218

El-Baz AM, Mosbah RA, Goda RM et al (2021) Back to nature: combating *Candida albicans* biofilm, phospholipase and hemolysin using plant essential oils. Antibiotics 10:81. https://doi.org/10.3390/antibiotics10010081

Fernandes T, Silva S, Henriques M (2015) *Candida* tropicalis biofilm's matrix-involvement on its resistance to amphotericin B. Diagn Microbiol Infect Dis 83:165–169. https://doi.org/10.1016/j.diagmicrobio.2015.06.015

Ferreira PS, Victorelli FD, Rodero CF et al (2021) P-Coumaric acid loaded into liquid crystalline systems as a novel strategy to the treatment of vulvovaginal candidiasis. Int J Pharm 603:120658. https://doi.org/10.1016/j.ijpharm.2021.120658

Filler SG, Sheppard DC (2006) Fungal invasion of normally non-phagocytic host cells. PLoS Pathog 2:1099–1105. https://doi.org/10.1371/journal.ppat.0020129

Finkel JS, Mitchell AP (2011a) Genetic control of *Candida albicans* biofilm development. Nat Rev Microbiol 9:109–118. https://doi.org/10.1038/nrmicro2475

Finkel JS, Mitchell AP (2011b) Biofilm formation in *Candida albicans*. In: *Candida* and candidiasis. ASM Press, Washington, DC, pp 299–315

Finkel JS, Xu W, Huang D et al (2012) Portrait of *Candida albicans* adherence regulators. PLoS Pathog 8:e1002525. https://doi.org/10.1371/journal.ppat.1002525

Firon A, Aubert S, Iraqui I et al (2007) The SUN41 and SUN42 genes are essential for cell separation in *Candida albicans*. Mol Microbiol 66:1256–1275. https://doi.org/10.1111/j.1365-2958.2007.06011.x

Fox EP, Nobile CJ (2012) A sticky situation: untangling the transcriptional network controlling biofilm development in *Candida albicans*. Transcription 3:315–322

Fox E, Nobile C (2013) The role of *Candida albicans* biofilms in human disease. In: *Candida albicans* symptoms, causes treat options. Nova Science Publishers, Hauppauge, pp 1–24

Fox EP, Bui CK, Nett JE et al (2015a) An expanded regulatory network temporally controls C andida albicans biofilm formation. Mol Microbiol 96:1226–1239. https://doi.org/10.1111/mmi.13002

Fox EP, Singh-Babak SD, Hartooni N, Nobile CJ (2015b) Biofilms and antifungal resistance. In: Antifungals: from genomics to resistance and the development of novel agents. Caister Academic Press, Poole, pp 71–90

Francisconi RS, Huacho PMM, Tonon CC et al (2020a) Antibiofilm efficacy of tea tree oil and of its main component terpinen-4-ol against *Candida albicans*. Braz Oral Res 34:1–9. https://doi.org/10.1590/1807-3107bor-2020.vol34.0050

Francisconi RS, Maquera-Huacho PM, Tonon CC et al (2020b) Terpinen-4-ol and nystatin co-loaded precursor of liquid crystalline system for topical treatment of oral candidiasis. Sci Rep 10:12984. https://doi.org/10.1038/s41598-020-70085-z

Gao S, Liu G, Li J et al (2020) Antimicrobial activity of lemongrass essential oil (Cymbopogon flexuosus) and its active component citral against dual-species biofilms of staphylococcus aureus and *Candida* species. Front Cell Infect Microbiol 10:1–14. https://doi.org/10.3389/fcimb.2020.603858

Garcia MC, Lee JT, Ramsook CB et al (2011) A role for amyloid in cell aggregation and biofilm formation. PLoS One 6:e17632. https://doi.org/10.1371/journal.pone.0017632

Garcia LGS, da Rocha MG, Lima LR et al (2021) Essential oils encapsulated in chitosan microparticles against *Candida albicans* biofilms. Int J Biol Macromol 166:621–632. https://doi.org/10.1016/j.ijbiomac.2020.10.220

Garcia-Rubio R, de Oliveira HC, Rivera J, Trevijano-Contador N (2020) The fungal cell wall: *Candida*, cryptococcus, and aspergillus species. Front Microbiol 10:1–13. https://doi.org/10.3389/fmicb.2019.02993

Ghannoum M, Roilides E, Katragkou A et al (2015) The role of echinocandins in *Candida* biofilm-related vascular catheter infections: in vitro and in vivo model systems. Clin Infect Dis 61:S618–S621. https://doi.org/10.1093/cid/civ815

Gholami-Ahangaran M, Ahmadi-Dastgerdi A, Azizi S et al (2022) Thymol and carvacrol supplementation in poultry health and performance. Vet Med Sci 8:267–288. https://doi.org/10.1002/vms3.663

Ghosh S, Navarathna DHMLP, Roberts DD et al (2009) Arginine-induced germ tube formation in *Candida albicans* is essential for escape from murine macrophage line RAW 264.7. Infect Immun 77:1596–1605. https://doi.org/10.1128/IAI.01452-08

Gonçalves B, Bernardo R, Wang C et al (2020) Effect of progesterone on *Candida albicans* biofilm formation under acidic conditions: a transcriptomic analysis. Int J Med Microbiol 310:151414. https://doi.org/10.1016/j.ijmm.2020.151414

Gow NAR, Hube B (2012) Importance of the *Candida albicans* cell wall during commensalism and infection. Curr Opin Microbiol 15:406–412. https://doi.org/10.1016/j.mib.2012.04.005

Granger BL (2012) Insight into the antiadhesive effect of yeast wall protein 1 of *Candida albicans*. Eukaryot Cell 11:795–805. https://doi.org/10.1128/EC.00026-12

Granger BL, Flenniken ML, Davis DA et al (2005) Yeast wall protein 1 of *Candida albicans* printed in Great Britain. Microbiology 151:1631–1644. https://doi.org/10.1099/mic.0.27663-0

Gropp K, Schild L, Schindler S et al (2009) The yeast *Candida albicans* evades human complement attack by secretion of aspartic proteases. Mol Immunol 47:465–475. https://doi.org/10.1016/j.molimm.2009.08.019

Gross O, Poeck H, Bscheider M et al (2009) Syk kinase signalling couples to the Nlrp3 inflammasome for anti-fungal host defence. Nature 459:433–436. https://doi.org/10.1038/nature07965

Gulati M, Nobile CJ (2016) *Candida albicans* biofilms: development, regulation, and molecular mechanisms. Microbes Infect 18:310–321. https://doi.org/10.1016/j.micinf.2016.01.002

Gupta P, Pruthi V, Poluri KM (2021) Mechanistic insights into *Candida* biofilm eradication potential of eucalyptol. J Appl Microbiol 131:105–123. https://doi.org/10.1111/jam.14940

Gursu BY, Dag İ, Dikmen G (2022) Antifungal and antibiofilm efficacy of cinnamaldehyde-loaded poly(DL-lactide-co-glycolide) (PLGA) nanoparticles against *Candida albicans*. Int Microbiol 25:245–258. https://doi.org/10.1007/s10123-021-00210-z

Gutiérrez-Escribano P, Zeidler U, Suárez MB et al (2012) The NDR/LATS kinase Cbk1 controls the activity of the transcriptional regulator Bcr1 during biofilm formation in *Candida albicans*. PLoS Pathog 8:e1002683. https://doi.org/10.1371/journal.ppat.1002683

Harriott MM, Lilly EA, Rodriguez TE et al (2010) *Candida albicans* forms biofilms on the vaginal mucosa. Microbiology 156:3635–3644. https://doi.org/10.1099/mic.0.039354-0

Hashash R, Younes S, Bahnan W et al (2011) Characterisation of Pga1, a putative *Candida albicans* cell wall protein necessary for proper adhesion and biofilm formation. Mycoses 54:491–500. https://doi.org/10.1111/j.1439-0507.2010.01883.x

He M, Du M, Fan M, Bian Z (2007) In vitro activity of eugenol against *Candida albicans* biofilms. Mycopathologia 163:137–143. https://doi.org/10.1007/s11046-007-0097-2

Hiller E, Heine S, Brunner H, Rupp S (2007) *Candida albicans* Sun41p, a putative glycosidase, is involved in morphogenesis, cell wall biogenesis, and biofilm formation. Eukaryot Cell 6:2056–2065. https://doi.org/10.1128/EC.00285-07

Hirota K, Yumoto H, Sapaar B et al (2017) Pathogenic factors in *Candida* biofilm-related infectious diseases. J Appl Microbiol 122:321–330. https://doi.org/10.1111/jam.13330

Hise AG, Tomalka J, Ganesan S et al (2009) An essential role for the NLRP3 inflammasome in host defense against the human fungal pathogen *Candida albicans*. Cell Host Microbe 5:487–497. https://doi.org/10.1016/j.chom.2009.05.002

Hnisz D, Bardet AF, Nobile CJ et al (2012) A histone deacetylase adjusts transcription kinetics at coding sequences during *Candida albicans* morphogenesis. PLoS Genet 8:e1003118. https://doi.org/10.1371/journal.pgen.1003118

Hoyer LL, Cota E (2016) *Candida albicans* agglutinin-like sequence (Als) family vignettes: a review of als protein structure and function. Front Microbiol 7:1–16

Hsu C-C, Lai W-L, Chuang K-C et al (2013) The inhibitory activity of linalool against the filamentous growth and biofilm formation in *Candida albicans*. Med Mycol 51:473–482. https://doi.org/10.3109/13693786.2012.743051

Ivanov M, Kannan A, Stojković DS et al (2021) Camphor and eucalyptol—anti *Candidal* spectrum, antivirulence effect, efflux pumps interference and cytotoxicity. Int J Mol Sci 22:483. https://doi.org/10.3390/ijms22020483

Jadhav A, Karuppayil SM (2017) *Candida albicans* biofilm as a clinical challenge. In: Developments in fungal biology and applied mycology. Springer, Singapore, pp 247–264

Jafri H, Ahmad I (2020) Thymus vulgaris essential oil and thymol inhibit biofilms and interact synergistically with antifungal drugs against drug resistant strains of *Candida albicans* and *Candida* tropicalis. J Mycol Med 30:100911. https://doi.org/10.1016/j.mycmed.2019.100911

Jafri H, Khan MSA, Ahmad I (2019) In vitro efficacy of eugenol in inhibiting single and mixed-biofilms of drug-resistant strains of *Candida albicans* and Streptococcus mutans. Phytomedicine 54:206–213. https://doi.org/10.1016/j.phymed.2018.10.005

Johnson CC, Yu A, Lee H et al (2012) Development of a contemporary animal model of *Candida albicans*-associated denture stomatitis using a novel intraoral denture system. Infect Immun 80:1736–1743. https://doi.org/10.1128/IAI.00019-12

Johnson CJ, Cabezas-Olcoz J, Kernien JF et al (2016) The extracellular matrix of *Candida albicans* biofilms impairs formation of neutrophil extracellular traps. PLoS Pathog 12:e1005884. https://doi.org/10.1371/journal.ppat.1005884

Joly S, Ma N, Sadler JJ et al (2009) Cutting edge: *Candida albicans* hyphae formation triggers activation of the Nlrp3 inflammasome. J Immunol 183:3578–3581. https://doi.org/10.4049/jimmunol.0901323

Jouault T, Ibata-Ombetta S, Takeuchi O et al (2003) *Candida albicans* phospholipomannan is sensed through toll-like receptors. J Infect Dis 188:165–172. https://doi.org/10.1086/375784

Jouault T, El Abed-El Behi M, Martínez-Esparza M et al (2006) Specific recognition of *Candida albicans* by macrophages requires galectin-3 to discriminate saccharomyces cerevisiae and needs association with TLR2 for signaling. J Immunol 177:4679–4687. https://doi.org/10.4049/jimmunol.177.7.4679

Kakade P, Sadhale P, Sanyal K, Nagaraja V (2016) ZCF32, a fungus specific Zn(II)2 Cys6 transcription factor, is a repressor of the biofilm development in the human pathogen *Candida albicans*. Sci Rep 6:1–15. https://doi.org/10.1038/srep31124

Kernien JF, Snarr BD, Sheppard DC, Nett JE (2018) The interface between fungal biofilms and innate immunity. Front Immunol 8:1968

Khan MSA, Ahmad I (2012) Antibiofilm activity of certain phytocompounds and their synergy with fluconazole against *Candida albicans* biofilms. J Antimicrob Chemother 67:618–621. https://doi.org/10.1093/jac/dkr512

Khan SN, Khan S, Iqbal J et al (2017) Enhanced killing and antibiofilm activity of encapsulated cinnamaldehyde against *Candida albicans*. Front Microbiol 8:1–15. https://doi.org/10.3389/fmicb.2017.01641

Khot PD, Suci PA, Miller RL et al (2006) A small subpopulation of blastospores in *Candida albicans* biofilms exhibit resistance to amphotericin B associated with differential regulation of ergosterol and β-1,6-glucan pathway genes. Antimicrob Agents Chemother 50:3708–3716. https://doi.org/10.1128/AAC.00997-06

Kim Y-G, Lee J-H, Park JG, Lee J (2020) Inhibition of *Candida albicans* and Staphylococcus aureus biofilms by centipede oil and linoleic acid. Biofouling 36:126–137. https://doi.org/10.1080/08927014.2020.1730333

Klis FM, Sosinska GJ, De Groot PWJ, Brul S (2009) Covalently linked cell wall proteins of *Candida albicans* and their role in fitness and virulence. FEMS Yeast Res 9:1013–1028

Kojic EM, Darouiche RO (2004) *Candida* infections of medical devices. Clin Microbiol Rev 17:255–267. https://doi.org/10.1128/CMR.17.2.255-267.2004

Kornitzer D (2019) Regulation of *Candida albicans* hyphal morphogenesis by endogenous signals. J Fungi 5:21. https://doi.org/10.3390/jof5010021

Kruppa M, Krom BP, Chauhan N et al (2004) The two-component signal transduction protein Chk1p regulates quorum sensing in *Candida albicans*. Am Soc Microbiol 3:1062–1065. https://doi.org/10.1128/EC.3.4.1062-1065.2004

Kurakado S, Arai R, Sugita T (2018) Association of the hypha-related protein Pra1 and zinc transporter Zrt1 with biofilm formation by the pathogenic yeast *Candida albicans*. Microbiol Immunol 62:405–410. https://doi.org/10.1111/1348-0421.12596

LaFleur MD, Kumamoto CA, Lewis K (2006) *Candida albicans* biofilms produce antifungal-tolerant persister cells. Antimicrob Agents Chemother 50:3839–3846. https://doi.org/10.1128/AAC.00684-06

Lee JA, Robbins N, Xie JL et al (2016) Functional genomic analysis of *Candida albicans* adherence reveals a key role for the Arp2/3 complex in cell wall remodelling and biofilm formation. PLoS Genet 12:1–24. https://doi.org/10.1371/journal.pgen.1006452

Li F, Palecek SP (2003) EAP1, a *Candida albicans* gene involved in binding human epithelial cells. Eukaryot Cell 2:1266–1273. https://doi.org/10.1128/EC.2.6.1266-1273.2003

Li F, Svarovsky MJ, Karlsson AJ et al (2007) Eap1p, an adhesin that mediates *Candida albicans* biofilm formation in vitro and in vivo. Eukaryot Cell 6:931–939. https://doi.org/10.1128/EC.00049-07

Li P, Seneviratne CJ, Alpi E et al (2015) Delicate metabolic control and coordinated stress response critically determine antifungal tolerance of *Candida albicans* biofilm persisters. Antimicrob Agents Chemother 59:6101–6112. https://doi.org/10.1128/AAC.00543-15

Liu Y, Filler SG (2011) *Candida albicans* Als3, a multifunctional adhesin and invasin. Eukaryot Cell 10:168–173

Liu N, Zhang N, Zhang S et al (2021) Phloretin inhibited the pathogenicity and virulence factors against *Candida albicans*. Bioengineered 12:2420–2431. https://doi.org/10.1080/21655979.2021.1933824

Lohberger A, Coste AT, Sanglard D (2014) Distinct roles of *Candida albicans* drug resistance transcription factors TAC1, MRR1, and UPC2 in virulence. Eukaryot Cell 13:127–142. https://doi.org/10.1128/EC.00245-13

Lohse MB, Gulati M, Johnson AD, Nobile CJ (2018) Development and regulation of single-and multi-species *Candida albicans* biofilms. Nat Rev Microbiol 16:19–31. https://doi.org/10.1038/nrmicro.2017.107

Lohse MB, Gulati M, Craik CS et al (2020) Combination of antifungal drugs and protease inhibitors prevent *Candida albicans* biofilm formation and disrupt mature biofilms. Front Microbiol 11:1–12. https://doi.org/10.3389/fmicb.2020.01027

Lone SA, Ahmad A (2020) Inhibitory effect of novel Eugenol Tosylate Congeners on pathogenicity of *Candida albicans*. BMC Complement Med Ther 20:131. https://doi.org/10.1186/s12906-020-02929-0

Lopez-Ribot JL (2014) Large-scale biochemical profiling of the *Candida albicans* biofilm matrix: new compositional, structural, and functional insights. MBio 5:1–13. https://doi.org/10.1128/mBio.01781-14

Lowman DW, Greene RR, Bearden DW et al (2014) Novel structural features in *Candida albicans* hyphal glucan provide a basis for differential innate immune recognition of hyphae versus yeast. J Biol Chem 289:3432–3443. https://doi.org/10.1074/jbc.M113.529131

Ma S, Moser D, Han F et al (2020) Preparation and antibiofilm studies of curcumin loaded chitosan nanoparticles against polymicrobial biofilms of *Candida albicans* and Staphylococcus aureus. Carbohydr Polym 241:116254. https://doi.org/10.1016/j.carbpol.2020.116254

Manoharan RK, Lee J-H, Kim Y-G et al (2017a) Inhibitory effects of the essential oils α-longipinene and linalool on biofilm formation and hyphal growth of *Candida albicans*. Biofouling 33:143–155. https://doi.org/10.1080/08927014.2017.1280731

Manoharan RK, Lee J-H, Lee J (2017b) Antibiofilm and antihyphal activities of cedar leaf essential oil, camphor, and fenchone derivatives against *Candida albicans*. Front Microbiol 8:1–12. https://doi.org/10.3389/fmicb.2017.01476

Martins M, Henriques M, Azeredo J et al (2007) Morphogenesis control in *Candida albicans* and *Candida* dubliniensis through signaling molecules produced by planktonic and biofilm cells. Eukaryot Cell 6:2429–2436. https://doi.org/10.1128/EC.00252-07

Martins M, Uppuluri P, Thomas DP et al (2010) Presence of extracellular DNA in the *Candida albicans* biofilm matrix and its contribution to biofilms. Mycopathologia 169:323–331. https://doi.org/10.1007/s11046-009-9264-y

Martins M, Henriques M, Lopez-Ribot JL, Oliveira R (2012) Addition of DNase improves the in vitro activity of antifungal drugs against *Candida albicans* biofilms. Mycoses 55:80–85. https://doi.org/10.1111/j.1439-0507.2011.02047.x

Mathé L, Van Dijck P (2013) Recent insights into *Candida albicans* biofilm resistance mechanisms. Curr Genet 59:251–264. https://doi.org/10.1007/s00294-013-0400-3

Mayer FL, Wilson D, Hube B (2013) *Candida albicans* pathogenicity mechanisms. Virulence 4:119–128

McCall AD, Pathirana RU, Prabhakar A et al (2019) *Candida albicans* biofilm development is governed by cooperative attachment and adhesion maintenance proteins. NPJ Biofilms Microbiomes 5:21. https://doi.org/10.1038/s41522-019-0094-5

McGreal EP, Rosas M, Brown GD et al (2006) The carbohydrate-recognition domain of Dectin-2 is a C-type lectin with specificity for high mannose. Glycobiology 16:422–430. https://doi.org/10.1093/glycob/cwj077

Miranda-Cadena K, Marcos-Arias C, Mateo E et al (2021) In vitro activities of carvacrol, cinnamaldehyde and thymol against *Candida* biofilms. Biomed Pharmacother 143:112218. https://doi.org/10.1016/j.biopha.2021.112218

Mishra P, Gupta P, Pruthi V (2021a) Cinnamaldehyde incorporated gellan/PVA electrospun nanofibers for eradicating *Candida* biofilm. Mater Sci Eng C 119:111450. https://doi.org/10.1016/j.msec.2020.111450

Mishra P, Gupta P, Srivastava AK et al (2021b) Eucalyptol/β-cyclodextrin inclusion complex loaded gellan/PVA nanofibers as antifungal drug delivery system. Int J Pharm 609:121163. https://doi.org/10.1016/j.ijpharm.2021.121163

Mitchell KF, Zarnowski R, Sanchez H et al (2015) Community participation in biofilm matrix assembly and function. Proc Natl Acad Sci U S A 112:4092–4097. https://doi.org/10.1073/pnas.1421437112

Mitchell KF, Zarnowski R, Andes DR (2016) The extracellular matrix of fungal biofilms. Adv Exp Med Biol 931:21–35

Miyazato A, Nakamura K, Yamamoto N et al (2009) Toll-like receptor 9-dependent activation of myeloid dendritic cells by deoxynucleic acids from *Candida albicans*. Infect Immun 77:3056–3064. https://doi.org/10.1128/IAI.00840-08

Motamedi M, Saharkhiz MJ, Pakshir K et al (2020) Chemical compositions and antifungal activities of Satureja macrosiphon against *Candida* and Aspergillus species. Curr Med Mycol 5:20–25. https://doi.org/10.18502/cmm.5.4.2162

Müller-Sepúlveda A, Chevecich CC, Jara JA et al (2020) Chemical characterization of Lavandula dentata essential oil cultivated in Chile and its antibiofilm effect against *Candida albicans*. Planta Med 86:1225–1234. https://doi.org/10.1055/a-1201-3375

Murciano C, Moyes DL, Runglall M et al (2012) Evaluation of the role of *Candida albicans* agglutinin-like sequence (ALS) proteins in human oral epithelial cell interactions. PLoS One 7:1–9. https://doi.org/10.1371/journal.pone.0033362

Nailis H, Vandenbosch D, Deforce D et al (2010) Transcriptional response to fluconazole and amphotericin B in *Candida albicans* biofilms. Res Microbiol 161:284–292. https://doi.org/10.1016/j.resmic.2010.02.004

Netea MG (2006) Immune sensing of *Candida albicans* requires cooperative recognition of mannans and glucans by lectin and toll-like receptors. J Clin Invest 116:1642–1650. https://doi.org/10.1172/JCI27114

Nett JE, Andes DR (2020) Contributions of the biofilm matrix to *Candida* pathogenesis. J Fungi 6:33–38. https://doi.org/10.3390/jof6010021

Nett J, Lincoln L, Marchillo K et al (2007) Putative role of β-1,3 glucans in *Candida albicans* biofilm resistance. Antimicrob Agents Chemother 51:510–520. https://doi.org/10.1128/AAC.01056-06

Nett JE, Crawford K, Marchillo K, Andes DR (2010a) Role of Fks1p and matrix glucan in *Candida albicans* biofilm resistance to an echinocandin, pyrimidine, and polyene. Antimicrob Agents Chemother 54:3505–3508. https://doi.org/10.1128/AAC.00227-10

Nett JE, Marchillo K, Spiegel CA, Andes DR (2010b) Development and validation of an in vivo *Candida albicans* biofilm denture model. Infect Immun 78:3650–3659. https://doi.org/10.1128/IAI.00480-10

Nett JE, Sanchez H, Cain MT, Andes DR (2010c) Genetic basis of *Candida* biofilm resistance due to drug-sequestering matrix glucan. J Infect Dis 202:171–175. https://doi.org/10.1086/651200

Nett JE, Sanchez H, Cain MT et al (2011) Interface of *Candida albicans* biofilm matrix-associated drug resistance and cell wall integrity regulation. Eukaryot Cell 10:1660–1669. https://doi.org/10.1128/EC.05126-11

Nett JE, Brooks EG, Cabezas-Olcoz J et al (2014) Rat indwelling urinary catheter model of *Candida albicans* biofilm infection. Infect Immun 82:4931–4940. https://doi.org/10.1128/IAI.02284-14

Ning Y, Ling J, Wu CD (2015) Synergistic effects of tea catechin epigallocatechin gallate and antimycotics against oral *Candida* species. Arch Oral Biol 60:1565–1570. https://doi.org/10.1016/j.archoralbio.2015.07.001

Nobile CJ, Johnson AD (2015) *Candida albicans* biofilms and human disease. Annu Rev Microbiol 69:71–92. https://doi.org/10.1146/annurev-micro-091014-104330

Nobile CJ, Mitchell AP (2005) Regulation of cell-surface genes and biofilm formation by the *C. albicans* transcription factor Bcr1p. Curr Biol 15:1150–1155. https://doi.org/10.1016/j.cub.2005.05.047

Nobile CJ, Andes DR, Nett JE et al (2006a) Critical role of Bcr1-dependent adhesins in *C. albicans* biofilm formation in vitro and in vivo. PLoS Pathog 2:0636–0649. https://doi.org/10.1371/journal.ppat.0020063

Nobile CJ, Nett JE, Andes DR, Mitchell AP (2006b) Function of *Candida albicans* adhesin hwp1 in biofilm formation. Eukaryot Cell 5:1604–1610. https://doi.org/10.1128/EC.00194-06

Nobile CJ, Schneider HA, Nett JE et al (2008) Complementary adhesin function in *C. albicans* biofilm formation. Curr Biol 18:1017–1024. https://doi.org/10.1016/j.cub.2008.06.034

Nobile CJ, Nett JE, Hernday AD et al (2009) Biofilm matrix regulation by *Candida albicans* Zap1. PLoS Biol 7:e1000133. https://doi.org/10.1371/journal.pbio.1000133

Nobile CJ, Fox EP, Nett JE et al (2012) A recently evolved transcriptional network controls biofilm development in *Candida albicans*. Cell 148:126–138. https://doi.org/10.1016/j.cell.2011.10.048

Nobile CJ, Fox EP, Hartooni N et al (2014) A histone deacetylase complex mediates biofilm dispersal and drug resistance. MBio 5:1–13. https://doi.org/10.1128/mBio.01201-14.Editor

Norice CT, Smith FJ, Solis N et al (2007) Requirement for *Candida albicans* Sun41 in biofilm formation and virulence. Eukaryot Cell 6:2046–2055. https://doi.org/10.1128/EC.00314-07

Palmieri V, Bugli F, Cacaci M et al (2018) Graphene oxide coatings prevent *Candida albicans* biofilm formation with a controlled release of curcumin-loaded nanocomposites. Nanomedicine 13:2867–2879. https://doi.org/10.2217/nnm-2018-0183

Paramonova E, Krom BP, van der Mei HC et al (2009) Hyphal content determines the compression strength of *Candida albicans* biofilms. Microbiology 155:1997–2003. https://doi.org/10.1099/mic.0.021568-0

Pemmaraju SC, Pruthi PA, Prasad R, Pruthi V (2013) *Candida albicans* biofilm inhibition by synergistic action of terpenes and fluconazole. Indian J Exp Biol 51:1032–1037

Pentland DR, Mühlschlegel FA, Gourlay CW (2020) CO_2 enhances the ability of *Candida albicans* to form biofilms, overcome nutritional immunity and resist antifungal treatment. BioRxiv. https://doi.org/10.1101/2020.03.31.018200

Pereira R, Santos Fontenelle RO, Brito EHS, Morais SM (2021) Biofilm of *Candida albicans*: formation, regulation and resistance. J Appl Microbiol 131:11–22. https://doi.org/10.1111/jam.14949

Perez A, Pedros B, Murgui A et al (2006) Biofilm formation by *Candida albicans* mutants for genes coding fungal proteins exhibiting the eight-cysteine-containing CFEM domain. FEMS Yeast Res 6:1074–1084. https://doi.org/10.1111/j.1567-1364.2006.00131.x

Perry AM, Hernday AD, Nobile CJ (2020) Unraveling how *Candida albicans* forms sexual biofilms. J Fungi 6:14. https://doi.org/10.3390/jof6010014

Perumal P, Mekala S, Chaffin WL (2007) Role for cell density in antifungal drug resistance in *Candida albicans* biofilms. Antimicrob Agents Chemother 51:2454–2463. https://doi.org/10.1128/AAC.01237-06

Pierce CG, Vila T, Romo JA et al (2017) The *Candida albicans* biofilm matrix: composition, structure and function. J Fungi 3:14. https://doi.org/10.3390/jof3010014

Prasad R, Shah AH, Rawal MK (2016) Antifungals: mechanism of action and drug resistance. In: Ramos J, Sychrová H, Kschischo M (eds) Yeast membrane transporter. Springer International Publishing, Cham, pp 327–349

Priya A, Pandian SK (2020) Piperine impedes biofilm formation and hyphal morphogenesis of *Candida albicans*. Front Microbiol 11:1–18. https://doi.org/10.3389/fmicb.2020.00756

Priya A, Nivetha S, Pandian SK (2022) Synergistic interaction of piperine and thymol on attenuation of the biofilm formation, hyphal morphogenesis and phenotypic switching in *Candida albicans*. Front Cell Infect Microbiol 11:1–14. https://doi.org/10.3389/fcimb.2021.780545

Puri S, Kumar R, Chadha S et al (2012) Secreted aspartic protease cleavage of *Candida albicans* Msb2 activates Cek1 MAPK signaling affecting biofilm formation and oropharyngeal candidiasis. PLoS One 7:e46020. https://doi.org/10.1371/journal.pone.0046020

Rajasekar V, Darne P, Prabhune A et al (2021) A curcumin-sophorolipid nanocomplex inhibits *Candida albicans* filamentation and biofilm development. Colloids Surf B Biointerfaces 200:111617. https://doi.org/10.1016/j.colsurfb.2021.111617

Ramage G, Saville SP, Wickes BL, López-Ribot JL (2002) Inhibition of *Candida albicans* biofilm formation by farnesol, a quorum-sensing molecule. Appl Environ Microbiol 68:5459–5463. https://doi.org/10.1128/AEM.68.11.5459-5463.2002

Ramage G, Mowat E, Jones B et al (2009) Our current understanding of fungal biofilms. Crit Rev Microbiol 35:340–355. https://doi.org/10.3109/10408410903241436

Ramage G, Milligan S, Lappin DF et al (2012) Antifungal, cytotoxic, and immunomodulatory properties of tea tree oil and its derivative components: potential role in management of oral candidosis in cancer patients. Front Microbiol 3:1–8. https://doi.org/10.3389/fmicb.2012.00220

Reyna-Beltrán E, Isaac Bazán Méndez C, Iranzo M et al (2019) The cell wall of *Candida albicans*: a proteomics view. In: *Candida albicans*. IntechOpen, London

Richard ML, Nobile CJ, Bruno VM et al (2005) *Candida albicans* biofilm-defective mutants. Eukaryot Cell 4:1493–1502. https://doi.org/10.1128/EC.4.8

Richardson JP, Moyes DL (2015) Adaptive immune responses to *Candida albicans* infection. Virulence 6:327–337. https://doi.org/10.1080/21505594.2015.1004977

Robbins N, Uppuluri P, Nett J et al (2011) Hsp90 governs dispersion and drug resistance of fungal biofilms. PLoS Pathog 7:e1002257. https://doi.org/10.1371/journal.ppat.1002257

Sadowska B, Budzyńska A, Więckowska-Szakiel M et al (2014) New pharmacological properties of Medicago sativa and Saponaria officinalis saponin-rich fractions addressed to *Candida albicans*. J Med Microbiol 63:1076–1086. https://doi.org/10.1099/jmm.0.075291-0

Saharkhiz MJ, Motamedi M, Zomorodian K et al (2012) Chemical composition, antifungal and antibiofilm activities of the essential oil of Mentha piperita L. ISRN Pharm 2012:1–6. https://doi.org/10.5402/2012/718645

Saijo S, Ikeda S, Yamabe K et al (2010) Dectin-2 recognition of α-mannans and induction of Th17 cell differentiation is essential for host defense against *Candida albicans*. Immunity 32:681–691. https://doi.org/10.1016/j.immuni.2010.05.001

Sardi JCO, Scorzoni L, Bernardi T et al (2013) *Candida* species: current epidemiology, pathogenicity, biofilm formation, natural antifungal products and new therapeutic options. J Med Microbiol 62:10–24. https://doi.org/10.1099/jmm.0.045054-0

Sato K, Yang X, Yudate T et al (2006) Dectin-2 is a pattern recognition receptor for fungi that couples with the fc receptor γ chain to induce innate immune responses. J Biol Chem 281:38854–38866. https://doi.org/10.1074/jbc.M606542200

Schinabeck MK, Long LA, Hossain MA et al (2004) Rabbit model of *Candida albicans* biofilm infection: liposomal amphotericin B antifungal lock therapy. Antimicrob Agents Chemother 48:1727–1732. https://doi.org/10.1128/AAC.48.5.1727-1732.2004

Sebaa S, Boucherit-Otmani Z, Courtois P (2019) Effects of tyrosol and farnesol on *Candida albicans* biofilm. Mol Med Rep 19:3201–3209. https://doi.org/10.3892/mmr.2019.9981

Seneviratne CJ, Jin LJ, Samaranayake YH, Samaranayake LP (2008) Cell density and cell aging as factors modulating antifungal resistance of *Candida albicans* biofilms. Antimicrob Agents Chemother 52:3259–3266. https://doi.org/10.1128/AAC.00541-08

Shahzad M, Sherry L, Rajendran R et al (2014) Utilising polyphenols for the clinical management of *Candida albicans* biofilms. Int J Antimicrob Agents 44:269–273. https://doi.org/10.1016/j.ijantimicag.2014.05.017

Shapiro RS, Robbins N, Cowen LE (2011) Regulatory circuitry governing fungal development, drug resistance, and disease. Microbiol Mol Biol Rev 75:213–267. https://doi.org/10.1128/MMBR.00045-10

Shariati A, Asadian E, Fallah F et al (2019) Evaluation of Nano-curcumin effects on expression levels of virulence genes and biofilm production of multidrug-resistant Pseudomonas aeruginosa isolated from burn wound infection in Tehran, Iran. Infect Drug Resist 12:2223–2235. https://doi.org/10.2147/IDR.S213200

Shariati A, Didehdar M, Razavi S et al (2022) Natural compounds: a hopeful promise as an antibiofilm agent against *Candida* species. Front Pharmacol 13:1–20. https://doi.org/10.3389/fphar.2022.917787

Sharma J, Rosiana S, Razzaq I, Shapiro R (2019) Linking cellular morphogenesis with antifungal treatment and susceptibility in *Candida* pathogens. J Fungi 5:17. https://doi.org/10.3390/jof5010017

Sharma Y, Rastogi SK, Perwez A et al (2020) β-Citronellol alters cell surface properties of *Candida albicans* to influence pathogenicity related traits. Med Mycol 58:93–106. https://doi.org/10.1093/mmy/myz009

Shen J, Cowen LE, Griffin AM et al (2008) The *Candida albicans* pescadillo homolog is required for normal hypha-to-yeast morphogenesis and yeast proliferation. Proc Natl Acad Sci 105:20918–20923. https://doi.org/10.1073/pnas.0809147105

Shinde RB, Raut JS, Chauhan NM, Karuppayil SM (2013) Chloroquine sensitizes biofilms of *Candida albicans* to antifungal azoles. Braz J Infect Dis 17:395–400. https://doi.org/10.1016/j.bjid.2012.11.002

Shu C, Sun L, Zhang W (2016) Thymol has antifungal activity against *Candida albicans* during infection and maintains the innate immune response required for function of the p38 MAPK signaling pathway in Caenorhabditis elegans. Immunol Res 64:1013–1024. https://doi.org/10.1007/s12026-016-8785-y

Signaling R, Shapiro RS, Uppuluri P et al (2009) Hsp90 orchestrates temperature-dependent *Candida albicans* morphogenesis. Curr Biol 19:621–629. https://doi.org/10.1016/j.cub.2009.03.017

Singh S, Fatima Z, Hameed S (2016) Insights into the mode of action of anti*Candida*l herbal monoterpenoid geraniol reveal disruption of multiple MDR mechanisms and virulence attributes in *Candida albicans*. Arch Microbiol 198:459–472. https://doi.org/10.1007/s00203-016-1205-9

Singh S, Fatima Z, Ahmad K, Hameed S (2018) Fungicidal action of geraniol against *Candida albicans* is potentiated by abrogated CaCdr1p drug efflux and fluconazole synergism. PLoS One 13:e0203079. https://doi.org/10.1371/journal.pone.0203079

Souza ME, Lopes LQS, Bonez PC et al (2017) Melaleuca alternifolia nanoparticles against *Candida* species biofilms. Microb Pathog 104:125–132. https://doi.org/10.1016/j.micpath.2017.01.023

Stappers MHT, Brown GD (2017) Host immune responses during infections with *Candida albicans*. In: Prasad R (ed) *Candida albicans:* cellular and molecular biology. Springer International Publishing, Cham, pp 145–183

Suchodolski J, Feder-Kubis J, Krasowska A (2021) Antiadhesive properties of imidazolium ionic liquids based on (−)-menthol against *Candida* spp. Int J Mol Sci 22:7543. https://doi.org/10.3390/ijms22147543

Sudbery PE (2011) Growth of *Candida albicans* hyphae. Nat Rev Microbiol 9:737–748

Sun L, Zhang C, Li P (2012) Characterization, antibiofilm, and mechanism of action of novel PEG-stabilized lipid nanoparticles loaded with Terpinen-4-ol. J Agric Food Chem 60:6150–6156. https://doi.org/10.1021/jf3010405

Swidergall M, Ernst AM, Ernst JF (2013) *Candida albicans* mucin Msb2 is a broad-range protectant against antimicrobial peptides. Antimicrob Agents Chemother 57:3917–3922. https://doi.org/10.1128/AAC.00862-13

Szafranski-Schneider E, Swidergall M, Cottier F et al (2012) Msb2 shedding protects *Candida albicans* against antimicrobial peptides. PLoS Pathog 8:e1002501. https://doi.org/10.1371/journal.ppat.1002501

Tada H, Nemoto E, Shimauchi H et al (2002) Saccharomyces cerevisiae—and *Candida albicans*—derived mannan induced production of tumor necrosis factor alpha by human monocytes in a CD14- and toll-like receptor 4-dependent manner. Microbiol Immunol 46:503–512. https://doi.org/10.1111/j.1348-0421.2002.tb02727.x

Taff HT, Nett JE, Zarnowski R et al (2012) A *Candida* biofilm-induced pathway for matrix glucan delivery: implications for drug resistance. PLoS Pathog 8:e1002848. https://doi.org/10.1371/journal.ppat.1002848

Taff HT, Mitchell KF, Edward JA, Andes DR (2013) Mechanisms of *Candida* biofilm drug resistance. Future Microbiol 8:1325–1337. https://doi.org/10.2217/fmb.13.101

Thakre A, Zore G, Kodgire S et al (2018) Limonene inhibits *Candida albicans* growth by inducing apoptosis. Med Mycol 56:565–578. https://doi.org/10.1093/mmy/myx074

Thakre A, Jadhav V, Kazi R et al (2020) Oxidative stress induced by piperine leads to apoptosis in *Candida albicans*. Med Mycol 59:366–378. https://doi.org/10.1093/mmy/myaa058

Thompson DS, Carlisle PL, Kadosh D (2011) Coevolution of morphology and virulence in *Candida* species. Eukaryot Cell 10:1173–1182. https://doi.org/10.1128/EC.05085-11

Tong Y, Tang J (2017) *Candida albicans* infection and intestinal immunity. Microbiol Res 198:27–35. https://doi.org/10.1016/j.micres.2017.02.002

Tournu H, Van Dijck P (2012) *Candida* biofilms and the host: models and new concepts for eradication. Int J Microbiol 2012:1–16. https://doi.org/10.1155/2012/845352

Truong T, Zeng G, Lin Q et al (2016) Comparative ploidy proteomics of *Candida albicans* biofilms unraveled the role of the AHP1 gene in the biofilm persistence against amphotericin B. Mol Cell Proteomics 15:3488–3500. https://doi.org/10.1074/mcp.M116.061523

Tsui C, Kong EF, Jabra-Rizk MA (2016) Pathogenesis of *Candida albicans* biofilm. Pathog Dis 74:ftw018. https://doi.org/10.1093/femspd/ftw018

Uppuluri P, Lopez Ribot JL (2017) *Candida albicans* biofilms. In: Prasad R (ed) *Candida albicans:* cellular and molecular biology. Springer International Publishing, Cham, pp 63–75

Uppuluri P, Chaturvedi AK, Srinivasan A et al (2010a) Dispersion as an important step in the *Candida albicans* biofilm developmental cycle. PLoS Pathog 6:e1000828. https://doi.org/10.1371/journal.ppat.1000828

Uppuluri P, Pierce CG, Thomas DP et al (2010b) The transcriptional regulator Nrg1p controls *Candida albicans* biofilm formation and dispersion. Eukaryot Cell 9:1531–1537. https://doi.org/10.1128/EC.00111-10

Uppuluri P, Acosta Zaldívar M, Anderson MZ et al (2018) *Candida albicans* dispersed cells are developmentally distinct from biofilm and planktonic cells. MBio 9:1–16. https://doi.org/10.1128/mBio.01338-18

Vediyappan G, Rossignol T, D'Enfert C (2010) Interaction of *Candida albicans* biofilms with antifungals: transcriptional response and binding of antifungals to beta-glucans. Antimicrob Agents Chemother 54:2096–2111. https://doi.org/10.1128/AAC.01638-09

Verma-Gaur J, Qu Y, Harrison PF et al (2015) Integration of posttranscriptional gene networks into metabolic adaptation and biofilm maturation in *Candida albicans*. PLoS Genet 11:1–28. https://doi.org/10.1371/journal.pgen.1005590

Villa S, Hamideh M, Weinstock A, et al (2020) Transcriptional control of hyphal morphogenesis inC andida albicans. FEMS Yeast Res 20:1–17. https://doi.org/10.1093/femsyr/foaa005

Wall G, Montelongo-Jauregui D, Vidal Bonifacio B et al (2019) *Candida albicans* biofilm growth and dispersal: contributions to pathogenesis. Curr Opin Microbiol 52:1–6. https://doi.org/10.1016/j.mib.2019.04.001

Watamoto T, Samaranayake LP, Egusa H et al (2011) Transcriptional regulation of drug-resistance genes in *Candida albicans* biofilms in response to antifungals. J Med Microbiol 60:1241–1247. https://doi.org/10.1099/jmm.0.030692-0

White TC (1997) Increased mRNA levels of ERG16, CDR, and MDR1 correlate with increases in azole resistance in *Candida albicans* isolates from a patient infected with human immunodeficiency virus. Antimicrob Agents Chemother 41:1482–1487. https://doi.org/10.1128/AAC.41.7.1482

White TC, Marr KA, Bowden RA (1998) Clinical, cellular, and molecular factors that contribute to antifungal drug resistance. Clin Microbiol Rev 11:382–402. https://doi.org/10.1128/CMR.11.2.382

Wijesinghe GK, Maia FC, de Oliveira TR et al (2020) Effect of Cinnamomum verum leaf essential oil on virulence factors of *Candida* species and determination of the in-vivo toxicity with Galleria mellonella model. Mem Inst Oswaldo Cruz 115:1–13. https://doi.org/10.1590/0074-02760200349

Wijesinghe GK, de Oliveira TR, Maia FC et al (2021) Efficacy of true cinnamon (Cinnamomum verum) leaf essential oil as a therapeutic alternative for *Candida* biofilm infections. Iran J Basic Med Sci 24:787–795. https://doi.org/10.22038/ijbms.2021.53981.12138

Winter MB, Salcedo EC, Lohse MB et al (2016) Global identification of biofilm-specific proteolysis in *Candida albicans*. MBio 7:1–13. https://doi.org/10.1128/mBio.01514-16

Witchley JN, Penumetcha P, Abon NV et al (2019) *Candida albicans* morphogenesis programs control the balance between gut commensalism and invasive infection. Cell Host Microbe 25:432–443.e6. https://doi.org/10.1016/j.chom.2019.02.008

Xie Z, Thompson A, Sobue T et al (2012) *Candida albicans* biofilms do not trigger reactive oxygen species and evade neutrophil killing. J Infect Dis 206:1936–1945. https://doi.org/10.1093/infdis/jis607

Xu K, Wang JL, Chu MP, Jia C (2019) Activity of coumarin against *Candida albicans* biofilms. J Mycol Med 29:28–34. https://doi.org/10.1016/j.mycmed.2018.12.003

Yang L, Liu X, Zhuang X et al (2018) Antifungal effects of Saponin extract from rhizomes of Dioscorea panthaica Prain et Burk against *Candida albicans*. Evid Based Complement Altern Med 2018:1–13. https://doi.org/10.1155/2018/6095307

Yu LH, Wei X, Ma M et al (2012) Possible inhibitory molecular mechanism of farnesol on the development of fluconazole resistance in *Candida albicans* biofilm. Antimicrob Agents Chemother 56:770–775. https://doi.org/10.1128/AAC.05290-11

Zarnowski R, Westler WM, Lacmbouh GA et al (2014) Novel entries in a fungal biofilm matrix encyclopedia. MBio 5:1–13. https://doi.org/10.1128/mBio.01333-14

Zhao X, Oh S-H, Yeater KM, Hoyer LL (2005) Analysis of the *Candida albicans* Als2p and Als4p adhesins suggests the potential for compensatory function within the Als family. Microbiology 151:1619–1630. https://doi.org/10.1099/mic.0.27763-0

Zhao X, Daniels KJ, Oh SH et al (2006) *Candida albicans* Als3p is required for wild-type biofilm formation on silicone elastomer surfaces. Microbiology 152:2287–2299. https://doi.org/10.1099/mic.0.28959-0

Zipfel PF, Skerka C, Kupka D, Luo S (2011) Immune escape of the human facultative pathogenic yeast *Candida albicans*: the many faces of the *Candida* Pra1 protein. Int J Med Microbiol 301:423–430. https://doi.org/10.1016/j.ijmm.2011.04.010

Zore G, Thakre A, Abdulghani M et al (2022) Menthol inhibits *Candida albicans* growth by affecting the membrane integrity followed by apoptosis. Evid Based Complement Altern Med 2022:1–10. https://doi.org/10.1155/2022/1297888

10 Futuristic Avenues in *Candida* Treatment: Exploiting Plant-Derived Agents as Potent Inhibitors of Candidiasis

Mazen Abdulghani, Sreejeeta Sinha, Gajendra Singh, and Gajanan Zore

10.1 Introduction

Infections resulting from *Candida* are referred to as candidiasis. *Candida* species are naturally present in the oral cavity, digestive system, and genitourinary tract as part of the human microbiota (Sardi et al. 2013). Under changes in the host's immune condition, microbiome, and similar factors, certain typically harmless *Candida* species can transition into opportunistic infections (Papon et al. 2013). Individuals with compromised immune systems, such as those with diabetes, AIDS, renal failure, extended stays in intensive care units, undergoing chemotherapy, and those with implanted medical devices, face an elevated risk of developing severe *Candida* infections that can be life-threatening (Martins et al. 2014; Perry et al. 2020). People with weakened immune systems are at a heightened vulnerability to developing *Candida* infections of varying severity levels (Pereira et al. 2021). *Candida* infections can

http://www.curaj.ac.in/faculty/gajanan-b-zore

M. Abdulghani
School of Life Sciences, Swami Ramanand Teerth Marathwada University, Nanded, Maharashtra, India
e-mail: mazenmohammed05@gmail.com

S. Sinha · G. Singh · G. Zore (✉)
Department of Biotechnology, School of Life Sciences, Central University of Rajasthan, Ajmer, Rajasthan, India
e-mail: sreejeetasinha123@gmail.com; gajendra.singh@curaj.ac.in; prof.zore_bt@curaj.ac.in

N. Manzoor (ed.), *Advances in Antifungal Drug Development*,
https://doi.org/10.1007/978-981-97-5165-5_10

impact the skin and its associated structures (cutaneous candidiasis), the mucous membranes in the mouth, throat, and genitourinary tract (mucosal candidiasis), or even the circulatory system (systemic candidiasis, encompassing candidemia and invasive candidiasis) (Sadeghi et al. 2018). Depending on the species of *Candida* that causes the illness, candidiasis has a high death rate (15–35%). This makes it the fourth most common cause of nosocomial infections (Sadeghi et al. 2018). Among the *Candida* species, 15 have been identified as potential human pathogens. However, the significance of this group is underscored by the recognition that five predominant species—namely, *C. albicans*, *C. glabrata*, *C. parapsilosis*, *C. krusei*, and *C. tropicalis*—account for over 90% of documented cases of invasive candidiasis (Pappas et al. 2018). Due to the eukaryotic nature, pinpointing pharmacological targets that exclusively impact the fungus without affecting the host is challenging. Recent research has highlighted the existence of only five primary categories of antifungal medications currently in use to address fungal infections. These classes encompass azoles (such as fluconazole), polyenes (like amphotericin B), echinocandins (including caspofungin), nucleoside analogs (such as 5-flucytosine (5-FC)), and allylamines (Vanreppelen et al. 2023). Polyenes bind to sterols, inducing intracellular leakage, while azoles hinder ergosterol formation (Chong et al. 2018). Echinocandins disrupt cell wall integrity by inhibiting the enzyme β-1,3-glucan synthase. Nucleoside analogs interfere with fungal RNA, DNA, and protein synthesis by directly mimicking pyrimidines (Bhattacharya et al. 2020).

Nonetheless, the emergence of resistance to currently employed medications, notably azoles which are extensively utilized, has curtailed their effectiveness. The prevalence of antimicrobial resistance (AMR) within clinical *Candida* isolates and biofilms has reached such levels that the World Health Organization (WHO), in collaboration with other bodies, has designated it as an AMR-threat pathogen. Furthermore, existing antifungal treatments are encumbered by constraints encompassing limited efficacy, toxicity concerns, elevated treatment expenses, and escalating drug resistance (Lone and Ahmad 2020).

The ascent of drug-resistant *Candida* strains and the necessity to combat these strains (Shariati et al. 2022) underscores the potential of antifungal agents derived from natural plant extracts and oils as innovative and effective therapeutic strategies in clinical applications (Yadav et al. 2022). Throughout history, medicinal plants have been relied upon to address an array of medical conditions (Priya and Pandian 2020). Plant-based interventions have demonstrated significant efficacy against resistant and biofilm-forming *Candida* species, employing substances such as essential oils (EOs), extracts, and pure compounds (Shariati et al. 2022). Often, plant compounds exhibit synergistic antifungal effects when combined with traditional antifungal drugs, enhancing potency, reducing drug toxicity and dosages, and diminishing the risk of resistance (Shinde et al. 2013). Hence, the allure of natural compounds that effectively combat infections while exerting minimal harm to host cells is heightened, positioning them as prospective candidates for groundbreaking antimicrobial treatments. This chapter provides an overview of the anti-*Candida* properties, mechanisms of action, and structural attributes of natural compounds like flavonoids, terpenes, terpenoids, and alkaloids—either acting individually or synergistically with conventional antifungal drugs.

10.2 Terpenes and Terpenoids

In contrast to the straightforward hydrocarbon compositions of terpenes—like pinene, myrcene, limonene, terpinene, and p-cymene—terpenoids, which incorporate oxygen into their hydrocarbons, stand as a modified subset. Terpenoids display distinctive functional groups and alterations in the arrangement of oxidized methyl groups at various positions (Masyita et al. 2022). Both terpenes and terpenoids constitute secondary metabolites, synthesized through the mevalonic acid (MVA) and methylerythritol phosphate (MEP) pathways. These pathways culminate in the production of isopentenyl pyrophosphate (IPP) and dimethylallyl pyrophosphate (DMAPP), serving as the foundational building blocks for terpenes and terpenoids (Oldfield and Lin 2012). Figure 10.1 encapsulates the fundamental structures of key compounds exhibiting *Candida*-targeting properties. Within this segment, our focus narrows on the pre-eminent terpenes and terpenoid compounds, such as limonene, menthol, thymol, linalool, eugenol, geraniol, terpinene-4-ol, and cinnamaldehyde, extensively cited in the literature.

Zore et al. (2011) conducted an assessment of six terpenoid compounds—linalool, citral, linalyl acetate (LA), eugenol, citronellal, and benzyl benzoate—against *Candida* species. Linalool and citral exhibited the greatest potency, whereas benzyl

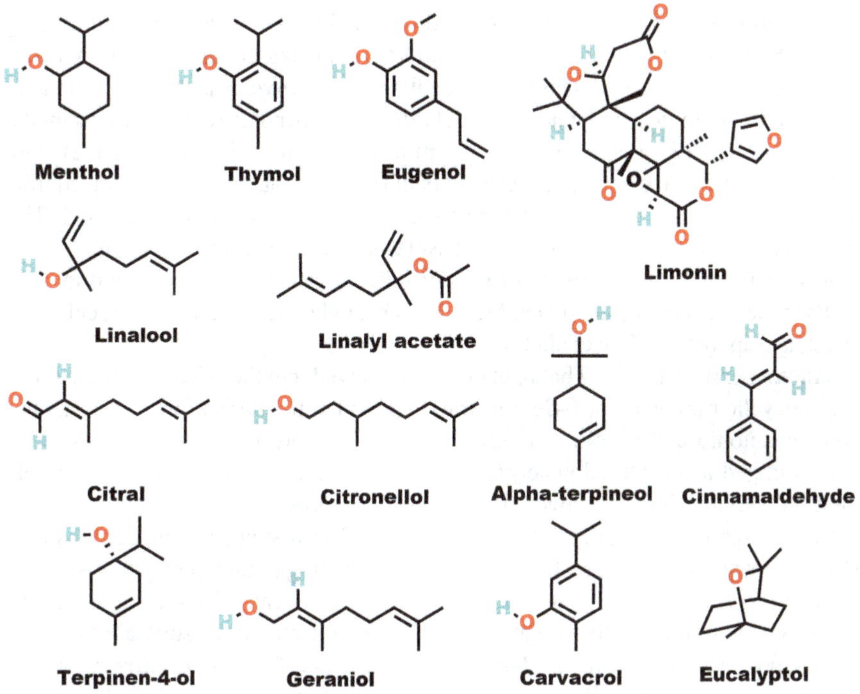

Fig. 10.1 Structures of terpenes and terpenoids bioactive compounds that showed anti-*Candida* activity (structures obtained from PubChem database)

benzoate demonstrated diminished effectiveness against *Candida* isolates (Zore et al. 2011). Linalool and eugenol swiftly achieved a 99.9% eradication rate within 7.5 min, while citral, linalyl acetate, and citronellal required 15 to 120 min for similar outcomes, dependent on the time-kill curve assay of terpenoids. Notably, five out of the six studied terpenoids displayed fungicidal activity (Zore et al. 2011). Through checkerboard assays, all the evaluated terpenoids significantly enhanced the sensitivity of a fluconazole-resistant *C. albicans* strain (ATCC 10231). This enhancement was marked by substantial reductions in fluconazole's MIC, with linalool emerging as the most effective, decreasing the MIC by 64-fold at a concentration of 0.008%. Furthermore, all terpenoids demonstrated strong synergistic activity with fluconazole (Zore et al. 2011).

Flow cytometry analysis divulged that terpenoids elicited cell cycle arrests at varying phases: linalool and linalyl acetate induced G1 arrest, citral and citronellal triggered S phase arrest, and benzyl benzoate caused G2-M arrest. The impact of terpenoids on inhibiting serum-induced hyphae formation in *C. albicans* (ATCC 10231) was explored. Notably, linalool, citral, citronellal, and benzyl benzoate exhibited high inhibitory efficacy, inhibiting over 50% at a concentration of 0.008%. Comparatively, eugenol and LA necessitated concentrations of 0.032% and 0.016%, respectively. Furthermore, all scrutinized terpenoids demonstrated nontoxicity to HeLa cells at MICs against *C. albicans* (Zore et al. 2011).

In a study by Raut et al. (2013), the effects of 28 diverse plant-derived terpenoids on *C. albicans* proliferation, pathogenicity, and biofilm formation were evaluated. Among these, 18 molecules exhibited inhibitory effects on planktonic growth with MICs ≤ 2 mg/mL. Notably, linalool, nerol, isopulegol, menthol, carvone, α-thujone, and farnesol exhibited biofilm-specific actions. Eight distinct terpenoids were shown to inhibit biofilm maturity (Raut et al. 2013). A recent investigation by Zore et al. (2022) highlighted the outstanding anti-*Candida* activity of menthol—an essential oil derived from peppermint (*Mentha canadensis* L), classified as a cyclic monoterpene alcohol. This study included differentially susceptible isolates of *Candida*, encompassing various growth and morphological forms of *C. albicans*. The mechanisms attributed to menthol's mode of action included membrane damage, inducing oxidative stress, cell cycle arrest, and apoptosis (Zore et al. 2022).

Another exploration by Thakre et al. (2018) delved into the effects of limonene—a monocyclic monoterpene—demonstrating potent anti-*Candida* properties against both planktonic and biofilm growth. Limonene notably inhibited morphogenesis and exhibited a substantial synergistic effect with fluconazole against both planktonic and biofilm forms of growth. Proteomic analysis unveiled limonene-responsive proteins, highlighting upregulation in cell wall glucan synthesis, oxidative stress, DNA damage stress, nucleolar stress, while downregulating proteins associated with cytoskeleton organization in response to limonene (Thakre et al. 2018). Additionally, limonene downregulated Tps3 and activated caspase (CaMca1), fostering apoptosis in *C. albicans*. Thus, limonene curbed *C. albicans* growth through cell wall and membrane impairment, DNA damage induction, and modulation of the cell cycle. This culminated in apoptosis via nucleolar stress and a metacaspase-dependent pathway (Thakre et al. 2018). These findings warrant further exploration of terpenoids as a viable treatment avenue against *Candida*-associated infections.

Thymol, a natural monoterpene compound found in plant oils, has numerous pharmacological potentials in traditional medicine (Gholami-Ahangaran et al. 2022). It has been shown to inhibit *C. albicans* biofilm formation, reduce viable cells, and eliminate mature biofilm (Braga et al. 2008; Dalleau et al. 2008; de Vasconcelos et al. 2014; Shu et al. 2016). Thymol and piperine have been shown to have synergistic effects on *C. albicans* biofilm, reducing virulence factors and controlling hyphal elongation (Priya et al. 2022). In vitro, gene expression analysis confirmed these findings, with upregulation of negative transcriptional regulators and downregulation of adhesins (Priya et al. 2022). Thymol and fluconazole also have synergistic effects on planktonic and biofilm growth (Pemmaraju et al. 2013; Jafri and Ahmad 2020).

Terpinen-4-ol, another monoterpene compound, has also been investigated against *Candida*. Terpinen-4-ol has been shown to suppress *C. albicans* growth and disrupt biofilm formation (Francisconi et al. 2020a). Its antifungal properties may be related to its lipophilicity. Combining terpinene-4-ol with antifungals like nystatin can inhibit *C. albicans* biofilm, potentially reducing dosage, side effects, and boosting clinical use. A liquid crystalline system (LCS) containing nystatin and terpinen-4-ol has been tested for its antibiofilm and synergistic effects against *C. albicans*. The LCS with nystatin and terpinen-4-ol may prevent and treat *C. albicans* infection, but further research is needed (Francisconi et al. 2020b). Terpinen-4-ol has advantages over complete essential oils (tea tree oils TTO) in terms of safety and consistency; it is not cytotoxic against OKF6-TERT2 epithelial cells at 0.5 × MIC50, making it useful for preventing and treating various *C. albicans* biofilm-associated infections, particularly established oropharyngeal candidiasis (Ramage et al. 2012). Terpinen-4-ol-loaded lipid nanoparticles can also effectively eliminate *C. albicans* biofilm, reducing the cell membrane structure and blocking the respiratory chain (Sun et al. 2012).

Geraniol, an acyclic monoterpene, has been shown to have antimicrobial activity against the planktonic growth of various *C. albicans* strains (Leite et al. 2015). A study has shown that geraniol suppresses growth and degrades the mature bio-film structure of *C. glabrata* (Gupta et al. 2021). It targets β-glucan and chitin in the cell wall, reduces cell membrane ergosterol content, alters mitochondria activity, increases Ca^{2+} uptake, and blocks ABC drug efflux pumps. Additionally, geraniol downregulates the expression of *ERG4* and *CDR1*, which are involved in the biosynthesis of ergosterol and multidrug efflux transport, respectively (Gupta et al. 2021). Additionally, geraniol has been found to inhibit *C. albicans* biofilm formation and virulence features of hyphal morphogenesis. Further, geraniol decreasing ergosterol levels, changing plasma membrane ATPase activity, inducing iron genotoxicity, disrupting homeostasis, and causing mitochondrial dysfunction (Singh et al. 2016). Biofilm development was also suppressed, and the number of viable biofilm cells was decreased after 48 h of exposure in a clinical strain of *C. tropicalis* (Souza et al. 2016). As a result of its inhibitory effects on *Candida* cellular pathways like ABC transporter-mediated drug transport, ergosterol, chitin, and glucan production, cell cycle progression, and mitochondrial activity, geraniol has been proposed as a novel antifungal agent against infections caused by a wide range of *Candida* species (Shariati et al. 2022). Positive results in inhibiting *C. albicans* biofilm have been seen with the use of geraniol and fluconazole. Geraniol was reportedly synergistic with fluconazole against *C. albicans*, reducing biofilm biomass and impairing fungal adhesion. By binding to CaCdr1p, geraniol could regulate the efflux pump's activity, suggesting that this may be a useful strategy for the combination treatment of candidiasis (Singh et al. 2018). The MIC for geraniol against both fluconazole-sensitive and -resistant *C. albicans* was 152 μg/mL. The antifungal effects of a combination of fluconazole and geraniol were increased, especially against strains that had developed resistance. Geraniol caused cell wall thickening, reduced ergosterol biosynthesis, and inhibited biofilm formation (Cardoso et al. 2016). Combination therapy with geraniol offers new options, but further studies and clinical trials are needed. Additionally, linalool, an unsaturated monoterpene, has antibiofilm activity against *Candida* species and oral *C. albicans* isolate (Alviano et al. 2005; Motamedi et al. 2020). It inhibits *C. albicans* biofilm growth and causes abnormal germination and cell size reduction, potentially offering a treatment option for *Candida* infections (Alviano et al. 2005). Linalool significantly inhibited *C. albicans* biofilm formation without impacting the growth of planktonic cells. The combination of linalool with α-longipinene and helichrysum oil decreases biofilm formation. It also inhibited hypha formation, which is crucial for virulence. Linalool-mediated inhibition could result from interference with filament production and induction. Combining linalool with other natural compounds could be more effective in fungal biofilm elimination (Manoharan et al. 2017a). These findings support the previous results, which found that linalool significantly reduced *C. albicans* biofilm structure by over 80% at 2×MIC, suppressing germ tube and biofilm formation. It downregulated pathogenic genes, including adhesion-related genes (*ALS3* and *HWP1*), *CPH1* and *CYR1* involved in MAPK and cAMP-PKA pathways, and long-term hyphae maintenance-associated genes (*EED1*, *HGC1* and *UME6*) (Hsu et al. 2013).

Another natural compound, eugenol, decreased *C. albicans* cell adhesion capacity and biofilm formation, leading to scant biofilms with suppressed filamentous development. *C. albicans* is a dimorphic fungus, indicating it can shift from its yeast form into its more pathogenic filamentous form. Inhibiting biofilm formation, affecting fungal cell morphogenesis, and decreasing fungal invasiveness were all effects of eugenol (He et al. 2007). Eugenol affects the *C. albicans* Als3 protein, as shown by a molecular docking study (El-Baz et al. 2021). Hoyer and Cota (2016) found that the binding capacity of eugenol to Als3 was greater than that of other natural chemicals like cinnamaldehyde (Hoyer and Cota 2016). This suggests that the mechanism by which eugenol inhibits *C. albicans* adhesion and biofilm formation is through its interaction with Als3 (El-Baz et al. 2021). A major component of *Cinnamomum*, eugenol, has demonstrated effective antibiofilm properties against *Candida*. A 2018 study demonstrated that the use of *Cinnamomum zeylanicum* EO significantly reduced mono- and multispecies biofilm populations and inhibited planktonic growth in *Candida* species (De Lucena Rangel et al. 2018). The primary component of *Cinnamomum verum* EO, eugenol (77.22%), inhibited adhesion, germ tube development, and biofilm growth. However, in the experimental model of *Galleria mellonella*, no fatal effect was seen (Wijesinghe et al. 2020). Further, inhibiting biofilm and hyphal development in *Candida* by the primary component of *Cinnamomum verum* EO, eugenol, results in cell wall damage and shrinking

(Wijesinghe et al. 2021). In several human cell lines, it shows relatively minimal cytotoxicity (Shariati et al. 2022). Jafri et al. (2019) showed that eugenol inhibited single and mixed biofilms of *Streptococcus mutans* and *C. albicans* resistant to AMB, itraconazole, and ketoconazole (Jafri et al. 2019). This means that eugenol can treat oral disorders caused by *Candida* biofilms. The biofilm and planktonic communities of *C. tropicalis* and *Candida dubliniensis* (dose-dependent and fluconazole-resistant strains) isolated from HIV-positive patients were also inhibited. Similarly, eugenol greatly reduced *Candida*'s ability to adhere to polystyrene and HEp-2 cells, which may be useful in treating localized cases of candidiasis (de Paula et al. 2014). Studies have shown that eugenol and fluconazole can be combined to treat *Candida* biofilms. In 2014, a study found a synergistic effect of fluconazole and eugenol against the planktonic and biofilm growth of *C. albicans*. Eugenol destabilized the cytoplasmic membrane, boosting fluconazole penetration and inhibiting biofilm formation (Doke et al. 2014). This combination also enhanced cell membrane integrity, suggesting that eugenol's fungistatic nature can be converted to fungicidal when combined with fluconazole (Khan and Ahmad 2012). In 2020, an Indian investigation found that eugenol suppressed mixed biofilms of *S. mutans* and *C. albicans*. Eugenol was highly synergistic with fluconazole, enhancing treatment efficacy. Microscopic analysis confirmed these findings, suggesting that eugenol may disrupt cell membrane integrity and increase drug entry into microbial cells. Eugenol also inhibited the adhesion capacity of planktonic *Candida* cells, suppressing biofilm formation and destroying established biofilms (Jafri et al. 2020). However, the exact mechanism of antifungal and antibiofilm effects is not fully understood, and further studies are needed to understand its potential. Furthermore, the growth of *C. albicans* biofilm on Polymerized Polymethyl Methacrylate (PMMA) dental devices was prevented by 70.0% using Cinnamomum oil, according to recent research (Choonharuangdej et al. 2021). Additionally, the cinnamaldehyde demonstrated potent antifungal activity against *Candida* species, reducing metabolic activity and biomass in established biofilms (Miranda-Cadena et al. 2021). This reduction may be crucial in handling resistant infections, as biofilms contribute to cell dispersal and virulence (Nobile and Johnson 2015). Cinnamaldehyde has been found to inhibit biofilm formation and affect *Candida* cellular development, with specific features such as the absence of chlamydoconidia and rate of pseudohyphae expression. Its negative ligand-receptor interaction energy targets with the most affinity for squalene epoxidase and thymidylate synthase, suggesting it could restrict biofilm formation in *Candida* (da Nóbrega Alves et al. 2020). Additionally, cinnamaldehyde has been found to destroy biofilms formed by *C. glabrata* on biomaterials, enhance cell lysis, ROS production, plasma membrane ergosterol content, and suppress phospholipase, catalase, and proteinase activity. These findings suggest that interactions between cinnamaldehyde and ergosterol may result in pores affecting cell membrane permeability and integrity, leading to intracellular content leakage and cell lysis (Gupta et al. 2018). Further research is needed to understand the potential of cinnamaldehyde in inhibiting biofilm formation and suppressing *Candida* virulence phenotypes.

Cinnamaldehyde's poor solubility in aqueous solutions makes it unsuitable for active infections. To overcome this, researchers have developed drug platforms like

Table 10.1 Anti-*Candida* activity of terpenes and terpenoid compounds

Terpenes and terpenoids compound	Source	Main finding	References
Eucalyptol	*Lavandula dentata* L EO	Inhibited adhesion, biofilm formation, morphogenesis, altered microarchitecture, and reduced the viability of the established biofilm	Müller-Sepúlveda et al. (2020)
α-Terpineol	Purchased from Sigma-Aldrich	Exhibited rapid antibiofilm properties	Ramage et al. (2012)
Borneol	*Thymus carnosus* EO	The preformed biofilm was disrupted	Alves et al. (2019)
Borneol	Purchased from Sigma-Aldrich	Reduced biofilm growth	Manoharan et al. (2017b)
Camphene	*Thymus carnosus* EO	The preformed biofilm was disrupted	Alves et al. (2019)
Carvacrol	Purchased from Sigma-Aldrich	Incorporation of carvacrol into the soft-liner decreased *C. albicans* biofilm growth	Baygar et al. (2018)
Citral	Lemongrass EO and geranium EO	The addition of essential oil (EO) to chitosan microparticles has an inhibitory impact on biofilm	Garcia et al. (2021)
Citral	Lemongrass (*Cymbopogon flexuosus*)	The biofilm's cell viability and biomass decreased across all species. Citral also inhibited virulence factor and hyphal adhesin expression in *C. albicans*	Gao et al. (2020)
Citronellol	Purchased from Sigma-Aldrich	Showed an inhibitory effect on the secretion of extracellular proteinases and phospholipases (planktonic form) and biofilm growth	Sharma et al. (2020)

electrospun gellan/PVA nanofibers, effectively eliminating *Candida* biofilms (Mishra et al. 2021). Additionally, cinnamaldehyde encapsulated in multilamellar liposomes (CM-ML) has shown improved fungicidal effects (Khan et al. 2017). Finally, cinnamaldehyde loaded onto poly (DL-lactide-co-glycolide) (PLGA) nanoparticles (NPs) has shown strong antifungal effects, with postbiofilm application being more effective than prebiofilm application (Gursu et al. 2022). Thus, drug delivery systems can enhance cinnamaldehyde's effectiveness in eliminating *Candida* biofilms at lower doses, mitigating cytotoxicity, and enhancing biofilm inhibition through various carriers. Table 10.1 lists additional terpenes and terpenoids that exhibit anti-*Candida* properties for those not described in this section.

10.3 Alkaloids

Alkaloids are secondary metabolites with secondary, tertiary, or quaternary nitrogen atoms (Sulaiman et al. 2022). Piperine is a major bioactive alkaloid component found in black pepper seeds, giving pepper its pungent taste. It has been used for centuries in human dietary applications and traditional medicine to treat various disorders such as influenza, cold, fever, rheumatism, etc., and stimulate bodily functions (Gorgani et al. 2017; Priya and Pandian 2020). The investigation of piperine against *Candida albicans* isolates and standard strain has been evaluated via Thakre et al. (2020). Piperine significantly inhibited the growth of *C. albicans* in both planktonic and biofilm forms and showed excellent synergistic effects with fluconazole against both forms, as well as hyphae induction. Proteomic data show that piperine affects membrane integrity, signal transduction, and generating reactive oxygen species (ROS) in *C. albicans* (Thakre et al. 2020). It is suggested that piperine affects membrane integrity, leading to oxidative stress, cell cycle arrest, and apoptosis in *C. albicans* by modulating the involved proteins. Flow cytometry-based assays and RT-qPCR analysis of selected genes confirm piperine-induced oxidative stress, cell cycle arrest, and apoptosis in *C. albicans* (Thakre et al. 2020). Another study by Priya and Pandian (2020) found that piperine exhibits concentration-dependent antibiofilm activity without affecting growth or metabolic activity. It inhibits hyphal development and virulence-associated colony morphologies, as well as regulates morphological transitions between yeast and hyphal forms. Piperine-challenged *C. albicans* show low potential for spontaneous antibiofilm resistance development (Priya and Pandian 2020). In vivo, piperine reduced *C. albicans* colonization and prolonged survival of *Caenorhabditis elegans*, demonstrating antivirulent potential. Gene expression analysis revealed downregulation of biofilm-related and hyphal-specific genes (*ALS3, EFG1, HWP1, CPH1*, etc.). In addition, the tested HBECs and the nematodes showed no signs of acute toxicity when exposed to piperine (Priya and Pandian 2020). Table 10.2 displays additional studies utilizing alkaloid compounds as anti-*Candida* agents, while Fig. 10.2 outlines the basic structures of the main compounds.

Table 10.2 Activity of alkaloid compounds against *Candida* species

Alkaloid Class/Compound	Source	Main finding	References
Phenanthroindolizidine			
Tylophorinine	*Tylophora indica*	Exhibit minimal inhibitory concentrations (MIC) against *Candida* species in the 0.6–2.5 μg/mL range	Dhiman et al. (2012)
Tylophorinidine	*Tylophora indica*	Exhibit minimal inhibitory concentrations (MIC) against *Candida* species in the 2–4 μg/mL range	Dhiman et al. (2012)
Carbazoles			
Koenigine	*Murraya koenigii* (L.) Spreng	Showed activity against a variety of *Candida* species	Joshi et al. (2018)

(continued)

Table 10.2 (continued)

Alkaloid Class/Compound	Source	Main finding	References
Aporphines			
Roemerine	*Fibraurea recisa*	Roemerine significantly inhibited the yeast-to-hyphae transition and C. *albicans* biofilm growth in vitro, although having no fungicidal effect on planktonic growth; the anti-biofilm mechanism may be associated with the cAMP pathway	Ma et al. (2015)
Magnoflorine	Purchased from Sigma-Aldrich	Magnoflorine exhibited potent growth inhibition against *Candida* strains without toxicity to HaCaT cells. It also inhibited alpha-glucosidase activity and reduced *C. albicans* biofilm formation. Cotreatment with magnoflorine and miconazole reduced the miconazole dosage required to kill *C. albicans*	Kim et al. (2018)
Benzophenanthridines			
8-Hydroxydihydrosanguinarine	*Chelidonium majus* Linn	Demonstrate moderate activity against *Candida* species	Meng et al. (2009)
6-Methoxydihydrosanguinarine	*Hylomecon japonica*	Showed inhibitory activity against *Candida albicans*	Xue et al. (2017)
Chelerythrine	*Zanthoxylum rhoifolium*	Showed inhibitory activity against *Candida albicans*	Tavares et al. (2014)
Protoberberines			
Berberine	Purchased from Sigma-Aldrich	Berberine inhibited fluconazole-resistant clinical strains of *Candida* species	da Silva et al. (2016)
Berberine	Purchased from Sigma-Aldrich	The study reveals the molecular mechanism of berberine fungicidal activity and a new role of heat shock factor (*HSF1*) in *Candida* multidrug resistance	Dhamgaye et al. (2014)
Jatrorrhizine	*Mahonia aquifolium*	It has been found to exhibit inhibitory activity against both *C. albicans* and *C. tropicalis*	Volleková et al. (2003)

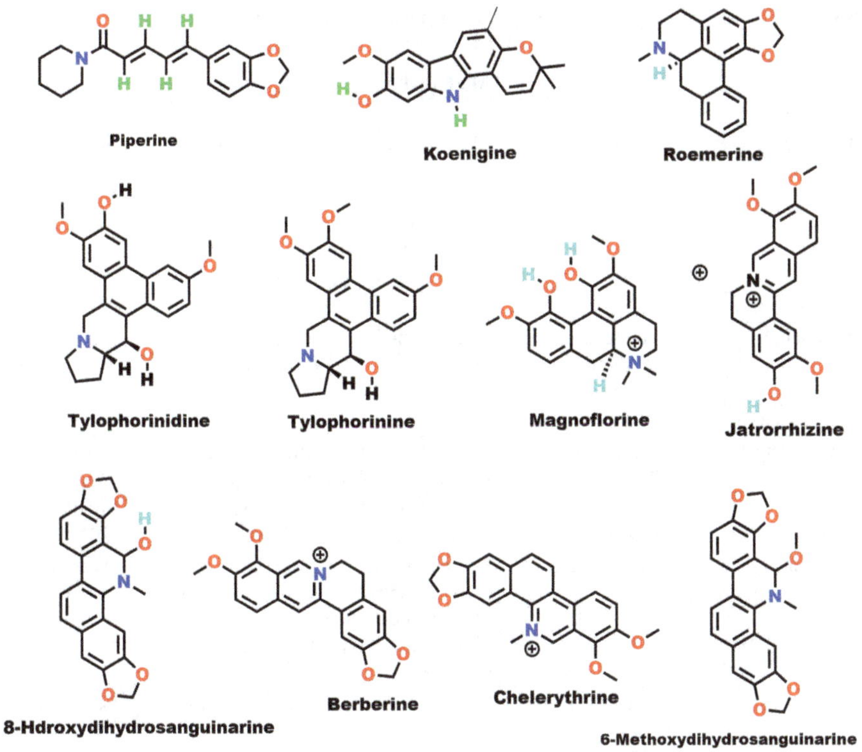

Fig. 10.2 Structures of alkaloid bioactive compounds that showed anti-*Candida* activity (structures obtained from PubChem database)

10.4 Flavonoids

Flavonoids, polyphenol secondary metabolites found in fruits and vegetables, have diverse biological activities, including antifungal, antiviral, and antibacterial effects. Studies have shown synergistic antifungal effects when combined with fluconazole, making them a crucial lead compound in synergistic antifungal drug research (Jin 2019; Nguyen et al. 2021). Flavonoids can be divided into seven subclasses based on specific modifications of the central carbon ring. These subclasses of flavonoids are flavones, isoflavones, chalcones, flavanols, flavonols, flavanones, and anthocyanidins (Jin 2019; Nguyen et al. 2021). In this section, we discuss flavonoid compounds that exhibit anti-*Candida* activity, while those not described are summarized in Table 10.3. Figure 10.3 presents the basic structures of the main compounds.

Table 10.3 Activity of flavonoid compounds against *Candida* species

Flavonoids subclass/ compound	Source	Main finding	References
Chalcones			
Licochalcone-A	Purchased from Sigma-Aldrich	Showed antifungal activity against *C. albicans* isolates and significantly reduced *C. albicans* biofilm. In a mouse model of oral candidiasis, CFU/ mL/mg of tissue was significantly reduced in the LicoA-treatment group	Seleem et al. (2016)
Isoflavanones			
Sedanan A	Methanol extract of *Dalea formosa* roots	Showed inhibitory activity against *C. albicans* and *C. glabrata*	Belofsky et al. (2013)
Flavones and isoflavones			
Licoflavone C	*Retama raetam* flowers	Showed potent activities against *Candida* spp. (MIC = 15.62 μg/mL)	Edziri et al. (2012)
Derrone (*isoflavones*)	*Retama raetam* flowers	Showed potent activities against *Candida* spp. (MIC = 7.81 μg/mL)	Edziri et al. (2012)
Baicalin	Not reported	Baicalin showed weak or indifferent activity against clinical isolates of *C. albicans* but was able to suppress biofilm development by inducing apoptosis.	(Wang et al. 2015)
Apigenin	*Aster yomena*	Showed antifungal activities against *C. albicans* and *C. parapsilosis*. In addition, it inhibits biofilm and induces membrane disturbances, resulting in cell shrinkage and intracellular component leakage of *C. albicans*	Lee et al. (2018)
5-hydroxy-7,4′-dimethoxyflavone	*Combretum zeyheri* leaf extract	Exhibited additive effects on *C. albicans* growth when coupled with miconazole. Time-dependent reduction of ergosterol was observed, and drug efflux pumps (IC_{50} = 51.64 μg/mL) and antioxidant enzymes (5 μM) were both inhibited by this combination	Mangoyi et al. (2015)

10.4.1 Flavanones and Isoflavanones

Recently, 2′, 4′-dihydroxy-5′-(1‴, 1‴-dimethylallyl)-8-prenylpinocembrin (8PP) was reported to have strong inhibitory effects against FLC-sensitive and azole-resistant *C. albicans* strains, with synergic antifungal activity (Barceló et al. 2017). It reduced the MFC of fluconazole (FLC) by fourfold in combination with 125 μM 8PP. 8PP inhibited rhodamine 6-G efflux in azole-resistant *C. albicans*, reversing FLC resistance in cells overexpressing Cdr1, Cdr2 and Mdr1 transporters, but not in azole-sensitive *C. albicans* (Barceló et al. 2017). 8PP has also been shown to strongly inhibit both sensitive and azole-resistant *C. albicans* biofilms at 100 μM. This inhibition was due to accumulating and increasing endogenous reactive oxygen species and reactive nitrogen intermediates (Peralta et al. 2015).

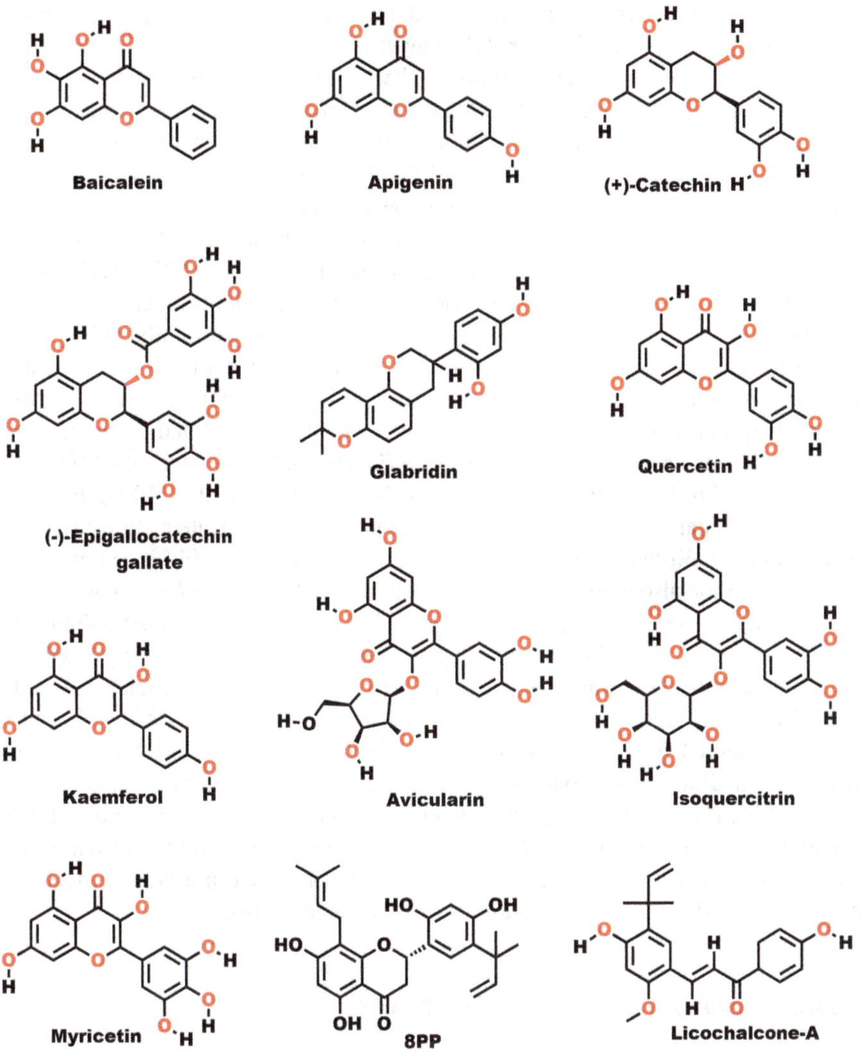

Fig. 10.3 Structures of flavonoid bioactive compounds that showed anti-*Candida* activity (structures obtained from PubChem database)

Biofilms treated with (200–1000 μM) higher concentrations of 8PP showed altered topographic-surface structures, voids, channels, and pores, affecting the flow inside the matrix of *C. albicans* biofilms (Peralta et al. 2018).

10.4.2 Flavones and Isoflavones

The main component of *Scutellaria baicalensis*, baicalein, showed weak inhibition against 30 clinical isolates of *C. albicans* (FLC-resistant) with the MIC50 ≥ 32 μg/mL as an alone agent, but in combination with fluconazole, 90% of isolates exhibit synergism at MIC50. In contrast, at MIC80, all strains showed synergism, with the median of the FIC index being 0.069 (Huang et al. 2008). In a further study by Serpa et al. (2012) against six *Candida* strains, the baicalein showed MIC50 in the range of 13–26 μg/mL, 52–104 μg/mL of *C. albicans*, *C. tropicalis*, respectively, while 13 μg/mL of *C. parapsilosis*. Synergistic inhibition of baicalein and fluconazole observed with *C. parapsilosis* (Serpa et al. 2012). It has been reported that baicalein induces the apoptosis of *C. albicans* cells based on flow cytometry and transmission electron microscopy (TEM) analysis. It also increases intracellular reactive oxygen space (ROS) by upregulation of redox-related genes such as (*CAP1*, *TRR1,* and *SOD2*) (Di Dai et al. 2009). The impact of baicalein on *C. albicans* biofilm was investigated, and the results showed that about 70% of biofilm formation was inhibited by baicalein in the range of 4–32 μg/mL; this inhibition was dependent on dose concentration. In the confocal laser scanning microscopy analysis of *C. albicans* biofilm treated with baicalein, the biofilm was composed only of yeast and pseudohyphae cells. In addition, cell surface hydrophobicity (CSH) and mRNA level of *CSH1* were also decreased in *C. albicans* cells treated with baicalein (Cao et al. 2008). Fu et al. (2011) found that the combination of amphotericin B and baicalein showed synergism against all 30 clinical isolates of *C. albicans*, with a median FIC index of 0.053. The combination also increased ROS production and induced *C. albicans* apoptosis, and baicalein increased caspase activity and *CaMCA1* expression in AmB-induced *C. albicans* (Fu et al. 2011). Moreover, baicalein showed inhibitory effects against non-*Candida albicans* strains (Kang et al. 2010; Tsang et al. 2015; Salazar-Aranda et al. 2015). Isocitrate lyase (ICL), a crucial enzyme for the glyoxylate cycle and a highly conserved protein among fungi, allows *C. albicans* to thrive in nutrient-restricted conditions inside phagocytic cells like neutrophils and macrophages. ICL is a desirable prospective therapeutic target and is necessary for pathogenicity. It has found that apigenin (5,7,4′-trihydroxyflavone) was a potential inhibitor of ICL (with 99.8%) and that it has an effect on *C. albicans* when tested in a glucose-depleted environment (MIC value was 125 μg/mL) (Cheah et al. 2014).

10.4.3 Flavanols (Flavans) and Isoflavans

Combination of fluconazole with (+)-catechin or (−)-epigallocatechin gallate showed synergistic activity against six FLC-resistance *C. tropicalis* strains isolated from biological samples. This combination leads to apoptosis, accumulation of ROS, mitochondrial depolarization, DNA fragmentation and morphological changes (da Silva et al. 2014). A recent study by Hervay et al. (2023) found that catechin lacked antifungal action in the investigated concentration range. Combined with miconazole, it inhibited growth in the sensitive *C. glabrata* isolate and significantly reduced growth in the azole-resistant clinical isolate. Using catechin and miconazole together increases intracellular ROS (Hervay et al. 2023). Glabridin, an active isoflavan from *Glycyrrhiza glabra*, showed strong activity against *Candida* spp. (MIC values: 16–64 μg/mL). The combination with FLC, glabridin had a synergistic impact on drug-resistant *C. albicans* and *C. tropicalis* at lower concentrations (1 to 16 μg/mL), except for *C. glabrata*. The synergy was caused by cell envelope degradation, which reduced cell size and increased membrane permeability (Liu et al. 2014). Another study indicated that apoptosis in *C. glabrata* stains treated with glabridin was linked to increased expression of *MCA1* and *NUC1* genes, DNA damage and chromatin condensation. 8-prenylation may also make glabridin hydrophobic and able to integrate into the cell membrane (Moazeni et al. 2017).

10.4.4 Flavonols

Some flavonol compounds were tested by Salazar-Aranda et al. (2015) against *Candida* sp. Only *C. glabrata* isolates out of five *Candida* species tested showed greater sensitivity to the eight flavonols or glycosides that were evaluated. These compounds myricetin, 3-hydroxyflavone, quercetin, myricitrin, quercitrin, kaempferol, fisetin, and galangin. Furthermore, all of the examined *Candida* isolates were inhibited by myricetin (MIC: 3.9–64 g/mL) (Salazar-Aranda et al. 2015). Myricetin and its derivatives 5-methylmyricetin and myricetin-3-O-β-glucoside also showed good antifungal activity against *C. glabrata*, with MIC50 values of 6.79, 15.37, and 8.53 μg/mL, respectively (Gadetskaya et al. 2015). Although quercetin and kaempferol do not affect *Candida* strains when administered singly (Salazar-Aranda et al. 2015; Oliveira et al. 2016; Shao et al. 2016; da Silva et al. 2017), they exhibit synergistic antifungal ability when combined with fluconazole (da Silva et al. 2014; Gao et al. 2016; Shao et al. 2016; Yang et al. 2017). Gao et al. (2016) found that quercetin (QCT) and fluconazole showed significant synergistic activity against ten planktonic and 17 biofilm growths of *C. albicans* FLC-resistance. This combination at (64 μg/mL QCT and/or 128 μg/mL) inhibits planktonic and biofilm growth, adhesion, morphogenesis, flocculation, thickness, CSH, biofilm dispersion, and metabolism. This result was supported by gene expression analysis, which showed upregulation of *PDE2*, *HSP90*, and *NRG1* and downregulation of *ALS, ALS3, HWP1, UME6*, and *SUN41*. In addition, in a mouse model, the symptoms and the fungus burden were reduced in the vaginal mucosa at this combination of QCT and FLC (Gao et al. 2016). Another investigation of QCT with FLC against another *Candida* species, *C. tropicalis,* showed synergistic activity against six FLC-resistant *C. tropicalis* strains isolated from biological samples. This combination leads to apoptosis, accumulation of ROS, mitochondrial depolarization, DNA fragmentation and morphological changes (da Silva et al. 2014). The derivatives of quercetin demonstrated effective activity against *Candida* strains. Isoquercitrin showed activities against *C. albicans* and *C. parapsilosis* (MIC value of 2.5 μg/mL) through disruption of cell membranes (Yun et al. 2015). In addition, *C. albicans* and *C. parapsilosis* have also been inhibited with avicularin (quercetin-3-*O*-α-arabinofuranoside) at 2–32 μg/mL concentrations (da Silva et al. 2017). *C. albicans* z2003 treated with kaempferol alone showed little rhodamine 6G accumulation and decreased the expression of both *CDR1* and *CDR2*, but FLC and kaempferol concomitantly lowered *CDR1*, *CDR2*, and *MDR1* expression and showed significant fluorescence rhodamine 6G (Shao et al. 2016).

References

Alves M, Gonçalves MJ, Zuzarte M et al (2019) Unveiling the antifungal potential of two iberian thyme essential oils: effect on *C. albicans* germ tube and preformed biofilms. Front Pharmacol 10:1–11. https://doi.org/10.3389/fphar.2019.00446

Alviano WS, Mendonca-Filho RR, Alviano DS et al (2005) Antimicrobial activity of Croton cajucara Benth linalool-rich essential oil on artificial biofilms and planktonic microorganisms. Oral Microbiol Immunol 20:101–105. https://doi.org/10.1111/j.1399-302X.2004.00201.x

Barceló S, Peralta M, Calise M et al (2017) Interactions of a prenylated flavonoid from Dalea elegans with fluconazole against azole- resistant *Candida albicans*. Phytomedicine 32:24–29. https://doi.org/10.1016/j.phymed.2017.05.001

Baygar T, Ugur A, Sarac N et al (2018) Functional denture soft liner with antimicrobial and antibiofilm properties. J Dent Sci 13:213–219. https://doi.org/10.1016/j.jds.2017.10.002

Belofsky G, Kolaczkowski M, Adams E et al (2013) Fungal ABC transporter-associated activity of isoflavonoids from the root extract of Dalea formosa. J Nat Prod 76:915–925. https://doi.org/10.1021/np4000763

Bhattacharya S, Sae-Tia S, Fries BC (2020) Candidiasis and mechanisms of antifungal resistance. Antibiotics 9:1–19

Braga PC, Culici M, Alfieri M, Dal Sasso M (2008) Thymol inhibits *Candida albicans* biofilm formation and mature biofilm. Int J Antimicrob Agents 31:472–477. https://doi.org/10.1016/j.ijantimicag.2007.12.013

Cao Y, Dai B, Wang Y et al (2008) In vitro activity of baicalein against *Candida albicans* biofilms. Int J Antimicrob Agents 32:73–77. https://doi.org/10.1016/j.ijantimicag.2008.01.026

Cardoso NNR, Alviano CS, Blank AF et al (2016) Synergism effect of the essential oil from Ocimum basilicum var. Maria Bonita and its major components with fluconazole and its influence on ergosterol biosynthesis. Evid Based Complement Altern Med 2016:1–12. https://doi.org/10.1155/2016/5647182

Cheah HL, Lim V, Sandai D (2014) Inhibitors of the glyoxylate cycle enzyme ICL1 in *Candida albicans* for potential use as antifungal agents. PLoS One 9:e95951. https://doi.org/10.1371/journal.pone.0095951

Chong PP, Chin VK, Wong WF et al (2018) Transcriptomic and genomic approaches for unravelling *Candida albicans* biofilm formation and drug resistance—an update. Genes (Basel) 9:540

Choonharuangdej S, Srithavaj T, Thummawanit S (2021) Fungicidal and inhibitory efficacy of cinnamon and lemongrass essential oils on *Candida albicans* biofilm established on acrylic resin: an in vitro study. J Prosthet Dent 125:707.e1–707.e6. https://doi.org/10.1016/j.prosdent.2020.12.017

da Nóbrega Alves D, Monteiro AFM, Andrade PN et al (2020) Docking prediction, antifungal activity, anti-biofilm effects on *Candida* spp., and toxicity against human cells of cinnamaldehyde. Molecules 25:5969. https://doi.org/10.3390/molecules25245969

da Silva CR, de Andrade Neto JB, de Sousa CR et al (2014) Synergistic effect of the flavonoid catechin, quercetin, or epigallocatechin gallate with fluconazole induces apoptosis in *Candida tropicalis* resistant to fluconazole. Antimicrob Agents Chemother 58:1468–1478. https://doi.org/10.1128/AAC.00651-13

da Silva AR, de Andrade Neto JB, da Silva CR et al (2016) Berberine antifungal activity in fluconazole-resistant pathogenic yeasts: action mechanism evaluated by flow cytometry and biofilm growth inhibition in *Candida* spp. Antimicrob Agents Chemother 60:3551–3557. https://doi.org/10.1128/AAC.01846-15

da Silva SF, de Paula J, dos Santos P et al (2017) Phytochemical analysis and antimicrobial activity of Myrcia tomentosa (Aubl.) DC. Leaves. Molecules 22:1100. https://doi.org/10.3390/molecules22071100

Dalleau S, Cateau E, Bergès T et al (2008) In vitro activity of terpenes against *Candida* biofilms. Int J Antimicrob Agents 31:572–576. https://doi.org/10.1016/j.ijantimicag.2008.01.028

De Lucena Rangel M, De Aquino SG, De Lima JM et al (2018) In vitro effect of cinnamomum zeylanicum blume essential oil on *Candida* spp. involved in oral infections. Evid Based Complement Altern Med 2018:4045013. https://doi.org/10.1155/2018/4045013

de Paula SB, Bartelli TF, Di Raimo V et al (2014) Effect of eugenol on cell surface hydrophobicity, adhesion, and biofilm of *Candida tropicalis* and *Candida dubliniensis* isolated from oral cavity of HIV-infected patients. Evid Based Complement Altern Med 2014:1–8. https://doi.org/10.1155/2014/505204

de Vasconcelos LC, Sampaio FC, Albuquerque AJR, Vasconcelos LCS (2014) Cell viability of *Candida albicans* against the antifungal activity of thymol. Braz Dent J 25:277–281. https://doi.org/10.1590/0103-6440201300052

Dhamgaye S, Devaux F, Vandeputte P et al (2014) Molecular mechanisms of action of herbal antifungal alkaloid berberine, in *Candida albicans*. PLoS One 9:e104554. https://doi.org/10.1371/journal.pone.0104554

Dhiman M, Parab RR, Manju SL et al (2012) Antifungal activity of hydrochloride salts of tylophorinidine and tylophorinine. Nat Prod Commun 7:1934578X1200700. https://doi.org/10.1177/1934578X1200700916

Di Dai B, Cao YY, Huang S et al (2009) Baicalein induces programmed cell death in *Candida albicans*. J Microbiol Biotechnol 19:803–809. https://doi.org/10.4014/jmb.0812.662

Doke SK, Raut JS, Dhawale S, Karuppayil SM (2014) Sensitization of *Candida albicans* biofilms to fluconazole by terpenoids of plant origin. J Gen Appl Microbiol 60:163–168. https://doi.org/10.2323/jgam.60.163

Edziri H, Mastouri M, Mahjoub MA et al (2012) Antibacterial, antifungal and cytotoxic activities of two flavonoids from retama raetam flowers. Molecules 17:7284–7293. https://doi.org/10.3390/molecules17067284

El-Baz AM, Mosbah RA, Goda RM et al (2021) Back to nature: combating *Candida albicans* biofilm, phospholipase and hemolysin using plant essential oils. Antibiotics 10:81. https://doi.org/10.3390/antibiotics10010081

Francisconi RS, Huacho PMM, Tonon CC et al (2020a) Antibiofilm efficacy of tea tree oil and of its main component terpinen-4-ol against *Candida albicans*. Braz Oral Res 34:1–9. https://doi.org/10.1590/1807-3107bor-2020.vol34.0050

Francisconi RS, Maquera-Huacho PM, Tonon CC et al (2020b) Terpinen-4-ol and nystatin co-loaded precursor of liquid crystalline system for topical treatment of oral candidiasis. Sci Rep 10:12984. https://doi.org/10.1038/s41598-020-70085-z

Fu Z, Lu H, Zhu Z et al (2011) Combination of baicalein and Amphotericin B accelerates *Candida albicans* apoptosis. Biol Pharm Bull 34:214–218. https://doi.org/10.1248/bpb.34.214

Gadetskaya AV, Tarawneh AH, Zhusupova GE et al (2015) Sulfated phenolic compounds from Limonium caspium: isolation, structural elucidation, and biological evaluation. Fitoterapia 104:80–85. https://doi.org/10.1016/j.fitote.2015.05.017

Gao M, Wang H, Zhu L (2016) Quercetin assists fluconazole to inhibit biofilm formations of fluconazole-resistant *Candida* albicans in in vitro and in vivo antifungal managements of vulvovaginal candidiasis. Cell Physiol Biochem 40:727–742. https://doi.org/10.1159/000453134

Gao S, Liu G, Li J et al (2020) Antimicrobial activity of lemongrass essential oil (Cymbopogon flexuosus) and its active component citral against dual-species biofilms of Staphylococcus aureus and *Candida* species. Front Cell Infect Microbiol 10:1–14. https://doi.org/10.3389/fcimb.2020.603858

Garcia LGS, da Rocha MG, Lima LR et al (2021) Essential oils encapsulated in chitosan microparticles against *Candida albicans* biofilms. Int J Biol Macromol 166:621–632. https://doi.org/10.1016/j.ijbiomac.2020.10.220

Gholami-Ahangaran M, Ahmadi-Dastgerdi A, Azizi S et al (2022) Thymol and carvacrol supplementation in poultry health and performance. Vet Med Sci 8:267–288. https://doi.org/10.1002/vms3.663

Gorgani L, Mohammadi M, Najafpour GD, Nikzad M (2017) Piperine-the bioactive compound of black pepper: from isolation to medicinal formulations. Compr Rev Food Sci Food Saf 16:124–140. https://doi.org/10.1111/1541-4337.12246

Gupta P, Gupta S, Sharma M et al (2018) Effectiveness of phytoactive molecules on transcriptional expression, biofilm matrix, and cell wall components of *Candida glabrata* and its clinical isolates. ACS Omega 3:12201–12214. https://doi.org/10.1021/acsomega.8b01856

Gupta P, Gupta H, Poluri KM (2021) Geraniol eradicates *Candida glabrata* biofilm by targeting multiple cellular pathways. Appl Microbiol Biotechnol 105:5589–5605. https://doi.org/10.1007/s00253-021-11397-6

Gursu BY, Dag İ, Dikmen G (2022) Antifungal and antibiofilm efficacy of cinnamaldehyde-loaded poly(DL-lactide-co-glycolide) (PLGA) nanoparticles against *Candida albicans*. Int Microbiol 25:245–258. https://doi.org/10.1007/s10123-021-00210-z

He M, Du M, Fan M, Bian Z (2007) In vitro activity of eugenol against *Candida albicans* biofilms. Mycopathologia 163:137–143. https://doi.org/10.1007/s11046-007-0097-2

Hervay NT, Elias D, Habova M et al (2023) Catechin potentiates the antifungal effect of miconazole in *Candida glabrata*. Folia Microbiol (Praha) 68:835–842. https://doi.org/10.1007/s12223-023-01061-z

Hoyer LL, Cota E (2016) *Candida albicans* agglutinin-like sequence (Als) family vignettes: a review of als protein structure and function. Front Microbiol 7:1–16

Hsu C-C, Lai W-L, Chuang K-C et al (2013) The inhibitory activity of linalool against the filamentous growth and biofilm formation in *Candida albicans*. Med Mycol 51:473–482. https://doi.org/10.3109/13693786.2012.743051

Huang S, Cao YY, Di Dai B et al (2008) In vitro synergism of fluconazole and baicalein against clinical isolates of *Candida albicans* resistant to fluconazole. Biol Pharm Bull 31:2234–2236. https://doi.org/10.1248/bpb.31.2234

Jafri H, Ahmad I (2020) Thymus vulgaris essential oil and thymol inhibit biofilms and interact synergistically with antifungal drugs against drug resistant strains of *Candida albicans* and *Candida tropicalis*. J Mycol Med 30:100911. https://doi.org/10.1016/j.mycmed.2019.100911

Jafri H, Khan MSA, Ahmad I (2019) In vitro efficacy of eugenol in inhibiting single and mixed-biofilms of drug-resistant strains of *Candida albicans* and Streptococcus mutans. Phytomedicine 54:206–213. https://doi.org/10.1016/j.phymed.2018.10.005

Jafri H, Banerjee G, Khan MSA et al (2020) Synergistic interaction of eugenol and antimicrobial drugs in eradication of single and mixed biofilms of *Candida albicans* and Streptococcus mutans. AMB Express 10:185. https://doi.org/10.1186/s13568-020-01123-2

Jin Y-S (2019) Recent advances in natural antifungal flavonoids and their derivatives. Bioorg Med Chem Lett 29:126589. https://doi.org/10.1016/j.bmcl.2019.07.048

Joshi T, Jain T, Mahar R et al (2018) Pyranocarbazoles from Murraya koenigii (L.) Spreng. as antimicrobial agents. Nat Prod Res 32:430–434. https://doi.org/10.1080/14786419.2017.1308363

Kang K, Fong W-P, Tsang PW-K (2010) Antifungal activity of baicalein against *Candida* krusei does not involve apoptosis. Mycopathologia 170:391–396. https://doi.org/10.1007/s11046-010-9341-2

Khan MSA, Ahmad I (2012) Antibiofilm activity of certain phytocompounds and their synergy with fluconazole against *Candida albicans* biofilms. J Antimicrob Chemother 67:618–621. https://doi.org/10.1093/jac/dkr512

Khan SN, Khan S, Iqbal J et al (2017) Enhanced killing and antibiofilm activity of encapsulated cinnamaldehyde against *Candida albicans*. Front Microbiol 8:1–15. https://doi.org/10.3389/fmicb.2017.01641

Kim J, Ha Quang Bao T, Shin Y-K, Kim K-Y (2018) Antifungal activity of magnoflorine against *Candida* strains. World J Microbiol Biotechnol 34:167. https://doi.org/10.1007/s11274-018-2549-x

Lee H, Woo E-R, Lee DG (2018) Apigenin induces cell shrinkage in *Candida albicans* by membrane perturbation. FEMS Yeast Res 18:1–9. https://doi.org/10.1093/femsyr/foy003

Leite MCA, de Brito Bezerra AP, de Sousa JP, de Oliveira LE (2015) Investigating the antifungal activity and mechanism(s) of geraniol against *Candida albicans* strains. Med Mycol 53:275–284. https://doi.org/10.1093/mmy/myu078

Liu W, Li LP, Zhang JD et al (2014) Synergistic antifungal effect of glabridin and fluconazole. PLoS One 9:e103442. https://doi.org/10.1371/journal.pone.0103442

Lone SA, Ahmad A (2020) Inhibitory effect of novel Eugenol Tosylate Congeners on pathogenicity of *Candida albicans*. BMC Complement Med Ther 20:131. https://doi.org/10.1186/s12906-020-02929-0

Ma C, Du F, Yan L et al (2015) Potent activities of Roemerine against *Candida albicans* and the underlying mechanisms. Molecules 20:17913–17928. https://doi.org/10.3390/molecules201017913

Mangoyi R, Midiwo J, Mukanganyama S (2015) Isolation and characterization of an antifungal compound 5-hydroxy-7,4′-dimethoxyflavone from Combretum zeyheri. BMC Complement Altern Med 15:405. https://doi.org/10.1186/s12906-015-0934-7

Manoharan RK, Lee J-H, Kim Y-G et al (2017a) Inhibitory effects of the essential oils α-longipinene and linalool on biofilm formation and hyphal growth of *Candida albicans*. Biofouling 33:143–155. https://doi.org/10.1080/08927014.2017.1280731

Manoharan RK, Lee J-H, Lee J (2017b) Antibiofilm and antihyphal activities of cedar leaf essential oil, camphor, and fenchone derivatives against *Candida albicans*. Front Microbiol 8:1–12. https://doi.org/10.3389/fmicb.2017.01476

Martins N, Ferreira ICFR, Barros L et al (2014) Candidiasis: predisposing factors, prevention, diagnosis and alternative treatment. Mycopathologia 177:223–240. https://doi.org/10.1007/s11046-014-9749-1

Masyita A, Mustika Sari R, Dwi Astuti A et al (2022) Terpenes and terpenoids as main bioactive compounds of essential oils, their roles in human health and potential application as natural food preservatives. Food Chem X 13:100217. https://doi.org/10.1016/j.fochx.2022.100217

Meng F, Zuo G, Hao X et al (2009) Antifungal activity of the benzo[c]phenanthridine alkaloids from Chelidonium majus Linn against resistant clinical yeast isolates. J Ethnopharmacol 125:494–496. https://doi.org/10.1016/j.jep.2009.07.029

Miranda-Cadena K, Marcos-Arias C, Mateo E et al (2021) In vitro activities of carvacrol, cinnamaldehyde and thymol against *Candida* biofilms. Biomed Pharmacother 143:112218. https://doi.org/10.1016/j.biopha.2021.112218

Mishra P, Gupta P, Pruthi V (2021) Cinnamaldehyde incorporated gellan/PVA electrospun nanofibers for eradicating *Candida* biofilm. Mater Sci Eng C 119:111450. https://doi.org/10.1016/j.msec.2020.111450

Moazeni M, Hedayati MT, Nabili M et al (2017) Glabridin triggers over-expression of MCA1 and NUC1 genes in *Candida glabrata*: is it an apoptosis inducer? J Mycol Med 27:369–375. https://doi.org/10.1016/j.mycmed.2017.05.002

Motamedi M, Saharkhiz MJ, Pakshir K et al (2020) Chemical compositions and antifungal activities of Satureja macrosiphon against *Candida* and Aspergillus species. Curr Med Mycol 5:20–25. https://doi.org/10.18502/cmm.5.4.2162

Müller-Sepúlveda A, Chevecich CC, Jara JA et al (2020) Chemical characterization of lavandula dentata essential oil cultivated in Chile and its antibiofilm effect against *Candida albicans*. Planta Med 86:1225–1234. https://doi.org/10.1055/a-1201-3375

Nguyen W, Grigori L, Just E et al (2021) The in vivo anti-*Candida albicans* activity of flavonoids. J Oral Biosci 63:120–128. https://doi.org/10.1016/j.job.2021.03.004

Nobile CJ, Johnson AD (2015) *Candida albicans* biofilms and human disease. Annu Rev Microbiol 69:71–92. https://doi.org/10.1146/annurev-micro-091014-104330

Oldfield E, Lin F-Y (2012) Terpene biosynthesis: modularity rules. Angew Chem Int Ed Engl 51:1124–1137. https://doi.org/10.1002/anie.201103110

Oliveira VM, Carraro E, Auler ME, Khalil NM (2016) Quercetin and rutin as potential agents antifungal against Cryptococcus spp. Braz J Biol 76:1029–1034. https://doi.org/10.1590/1519-6984.07415

Papon N, Courdavault V, Clastre M, Bennett RJ (2013) Emerging and emerged pathogenic *Candida* species: beyond the *Candida albicans* paradigm. PLoS Pathog 9:e1003550. https://doi.org/10.1371/journal.ppat.1003550

Pappas PG, Lionakis MS, Arendrup MC et al (2018) Invasive candidiasis. Nat Rev Dis Prim 4:18026. https://doi.org/10.1038/nrdp.2018.26

Pemmaraju SC, Pruthi PA, Prasad R, Pruthi V (2013) *Candida albicans* biofilm inhibition by synergistic action of terpenes and fluconazole. Indian J Exp Biol 51:1032–1037

Peralta MA, da Silva MA, Ortega MG et al (2015) Antifungal activity of a prenylated flavonoid from Dalea elegans against *Candida albicans* biofilms. Phytomedicine 22:975–980. https://doi.org/10.1016/j.phymed.2015.07.003

Peralta MA, Ortega MG, Cabrera JL, Paraje MG (2018) The antioxidant activity of a prenyl flavonoid alters its antifungal toxicity on *Candida albicans* biofilms. Food Chem Toxicol 114:285–291. https://doi.org/10.1016/j.fct.2018.02.042

Pereira R, Santos Fontenelle RO, Brito EHS, Morais SM (2021) Biofilm of *Candida albicans*: formation, regulation and resistance. J Appl Microbiol 131:11–22. https://doi.org/10.1111/jam.14949

Perry AM, Hernday AD, Nobile CJ (2020) Unraveling how *Candida albicans* forms sexual biofilms. J Fungi 6:14. https://doi.org/10.3390/jof6010014

Priya A, Pandian SK (2020) Piperine impedes biofilm formation and hyphal morphogenesis of *Candida albicans*. Front Microbiol 11:1–18. https://doi.org/10.3389/fmicb.2020.00756

Priya A, Nivetha S, Pandian SK (2022) Synergistic interaction of piperine and thymol on attenuation of the biofilm formation, hyphal morphogenesis and phenotypic switching in *Candida albicans*. Front Cell Infect Microbiol 11:1–14. https://doi.org/10.3389/fcimb.2021.780545

Ramage G, Milligan S, Lappin DF et al (2012) Antifungal, cytotoxic, and immunomodulatory properties of tea tree oil and its derivative components: potential role in management of oral candidosis in cancer patients. Front Microbiol 3:1–8. https://doi.org/10.3389/fmicb.2012.00220

Raut JS, Shinde RB, Chauhan NM, Mohan Karuppayil S (2013) Terpenoids of plant origin inhibit morphogenesis, adhesion, and biofilm formation by *Candida albicans*. Biofouling 29:87–96. https://doi.org/10.1080/08927014.2012.749398

Sadeghi G, Ebrahimi-Rad M, Mousavi SF et al (2018) Emergence of non- *Candida albicans* species: epidemiology, phylogeny and fluconazole susceptibility profile. J Mycol Med 28:51–58. https://doi.org/10.1016/j.mycmed.2017.12.008

Salazar-Aranda R, Granados-Guzmán G, Pérez-Meseguer J et al (2015) Activity of polyphenolic compounds against *Candida glabrata*. Molecules 20:17903–17912. https://doi.org/10.3390/molecules201017903

Sardi JCO, Scorzoni L, Bernardi T et al (2013) *Candida* species: current epidemiology, pathogenicity, biofilm formation, natural antifungal products and new therapeutic options. J Med Microbiol 62:10–24. https://doi.org/10.1099/jmm.0.045054-0

Seleem D, Benso B, Noguti J et al (2016) In vitro and in vivo antifungal activity of lichochalcone-A against *Candida albicans* biofilms. PLoS One 11:e0157188. https://doi.org/10.1371/journal.pone.0157188

Serpa R, França EJG, Furlaneto-Maia L et al (2012) In vitro antifungal activity of the flavonoid baicalein against *Candida* species. J Med Microbiol 61:1704–1708. https://doi.org/10.1099/jmm.0.047852-0

Shao J, Zhang M, Wang T et al (2016) The roles of CDR1, CDR2, and MDR1 in kaempferol-induced suppression with fluconazole-resistant *Candida albicans*. Pharm Biol 54:984–992. https://doi.org/10.3109/13880209.2015.1091483

Shariati A, Didehdar M, Razavi S et al (2022) Natural compounds: a hopeful promise as an antibiofilm agent against *Candida* species. Front Pharmacol 13:1–20. https://doi.org/10.3389/fphar.2022.917787

Sharma Y, Rastogi SK, Perwez A et al (2020) β-citronellol alters cell surface properties of *Candida albicans* to influence pathogenicity related traits. Med Mycol 58:93–106. https://doi.org/10.1093/mmy/myz009

Shinde RB, Raut JS, Chauhan NM, Karuppayil SM (2013) Chloroquine sensitizes biofilms of *Candida albicans* to antifungal azoles. Braz J Infect Dis 17:395–400. https://doi.org/10.1016/j.bjid.2012.11.002

Shu C, Sun L, Zhang W (2016) Thymol has antifungal activity against *Candida albicans* during infection and maintains the innate immune response required for function of the p38 MAPK signaling pathway in Caenorhabditis elegans. Immunol Res 64:1013–1024. https://doi.org/10.1007/s12026-016-8785-y

Singh S, Fatima Z, Hameed S (2016) Insights into the mode of action of anticandidal herbal monoterpenoid geraniol reveal disruption of multiple MDR mechanisms and virulence attributes in *Candida albicans*. Arch Microbiol 198:459–472. https://doi.org/10.1007/s00203-016-1205-9

Singh S, Fatima Z, Ahmad K, Hameed S (2018) Fungicidal action of geraniol against *Candida albicans* is potentiated by abrogated CaCdr1p drug efflux and fluconazole synergism. PLoS One 13:e0203079. https://doi.org/10.1371/journal.pone.0203079

Souza CMC, Pereira Junior SA, Moraes TDS et al (2016) Antifungal activity of plant-derived essential oils on *Candida tropicalis* planktonic and biofilms cells. Med Mycol 54:515–523. https://doi.org/10.1093/mmy/myw003

Sulaiman M, Jannat K, Nissapatorn V et al (2022) Antibacterial and antifungal alkaloids from Asian angiosperms: distribution, mechanisms of action, structure-activity, and clinical potentials. Antibiotics 11:1146. https://doi.org/10.3390/antibiotics11091146

Sun L, Zhang C, Li P (2012) Characterization, antibiofilm, and mechanism of action of novel PEG-stabilized lipid nanoparticles loaded with terpinen-4-ol. J Agric Food Chem 60:6150–6156. https://doi.org/10.1021/jf3010405

Tavares LC, Zanon G, Weber AD et al (2014) Structure-activity relationship of benzophenanthridine alkaloids from zanthoxylum rhoifolium having antimicrobial activity. PLoS One 9:e97000. https://doi.org/10.1371/journal.pone.0097000

Thakre A, Zore G, Kodgire S et al (2018) Limonene inhibits *Candida albicans* growth by inducing apoptosis. Med Mycol 56:565–578. https://doi.org/10.1093/mmy/myx074

Thakre A, Jadhav V, Kazi R et al (2020) Oxidative stress induced by piperine leads to apoptosis in *Candida albicans*. Med Mycol 59:366–378. https://doi.org/10.1093/mmy/myaa058

Tsang PW-K, Chau K-Y, Yang H-P (2015) Baicalein exhibits inhibitory effect on the energy-dependent efflux pump activity in non-albicans *Candida* fungi. J Chemother 27:61–62. https://doi.org/10.1179/1973947814Y.0000000177

Vanreppelen G, Wuyts J, Van Dijck P, Vandecruys P (2023) Sources of antifungal drugs. J Fungi 9:171. https://doi.org/10.3390/jof9020171

Volleková A, Košt'álová D, Kettmann V, Tóth J (2003) Antifungal activity of Mahonia aquifolium extract and its major protoberberine alkaloids. Phyther Res 17:834–837. https://doi.org/10.1002/ptr.1256

Wang T, Shi G, Shao J et al (2015) In vitro antifungal activity of baicalin against *Candida albicans* biofilms via apoptotic induction. Microb Pathog 87:21–29. https://doi.org/10.1016/j.micpath.2015.07.006

Wijesinghe GK, Maia FC, de Oliveira TR et al (2020) Effect of Cinnamomum verum leaf essential oil on virulence factors of *Candida* species and determination of the in-vivo toxicity with Galleria mellonella model. Mem Inst Oswaldo Cruz 115:1–13. https://doi.org/10.1590/0074-02760200349

Wijesinghe GK, de Oliveira TR, Maia FC et al (2021) Efficacy of true cinnamon (Cinnamomum verum) leaf essential oil as a therapeutic alternative for *Candida* biofilm infections. Iran J Basic Med Sci 24:787–795. https://doi.org/10.22038/ijbms.2021.53981.12138

Xue X, Zhang H, Zhang X et al (2017) TLC bioautography-guided isolation and antimicrobial, antifungal effects of 12 alkaloids from Hylomecon japonica roots §. Nat Prod Commun 12:1934578X1701200. https://doi.org/10.1177/1934578X1701200914

Yadav R, Pradhan M, Yadav K et al (2022) Present scenarios and future prospects of herbal nanomedicine for antifungal therapy. J Drug Deliv Sci Technol 74:103430. https://doi.org/10.1016/j.jddst.2022.103430

Yang Y-X, An M-M, Jin Y-S, Chen H-S (2017) Chemical constituents from the rhizome of Polygonum paleaceum and their antifungal activity. J Asian Nat Prod Res 19:47–52. https://doi.org/10.1080/10286020.2016.1196672

Yun J, Lee H, Ko HJ et al (2015) Fungicidal effect of isoquercitrin via inducing membrane disturbance. Biochim Biophys Acta Biomembr 1848:695–701. https://doi.org/10.1016/j.bbamem.2014.11.019

Zore GB, Thakre AD, Jadhav S, Karuppayil SM (2011) Terpenoids inhibit *Candida albicans* growth by affecting membrane integrity and arrest of cell cycle. Phytomedicine 18:1181–1190. https://doi.org/10.1016/j.phymed.2011.03.008

Zore G, Thakre A, Abdulghani M et al (2022) Menthol inhibits *Candida albicans* growth by affecting the membrane integrity followed by apoptosis. Evid Based Complement Altern Med 2022:1–10. https://doi.org/10.1155/2022/1297888

Antifungal Efficacy of Terpenes and Mechanism of Action Against Human Pathogenic Fungi

11

Nafis Raj, Parveen, Shabana Khatoon, and Nikhat Manzoor

11.1 Introduction

Fungi are both ubiquitous and highly diversified. Among the estimated 3.5–5.1 million fungal species only a few hundred are associated with infections (Kim 2016). It can cause diseases from superficial infections of the skin and mucosal surfaces to invasive infections of internal organs. Superficial infections are common and generally not difficult to treat. Twenty to twenty-five percent of the world population has skin infections that are primarily caused by dermatophytes, grouped into three major genera *Microsporum, Trichophyton,* and *Epidermophyton*. In comparison to superficial infections, invasive fungal diseases have a much lower incidence rate but are life-threatening, particularly in immunocompromised patients with HIV/AIDS or autoimmune diseases and in those undergoing anti-cancer chemotherapy or organ transplantation. Invasive infections kill about one and a half million people every year making them a significant global public health problem (Li and Nielsen 2017). The fungal species that often become invasive are primarily *Candida albicans, Cryptococcus neoformans, Aspergillus fumigatus,* and *Histoplasma capsulatum* (Bongomin et al. 2017). *C. albicans* is a commensal that exists as normal microflora in the human body but becomes pathogenic if the host immune defense system is weakened. Mucocutaneous

N. Raj · Parveen · S. Khatoon · N. Manzoor (✉)
Medical Mycology Lab, Department of Biosciences, Jamia Millia Islamia, New Delhi, Delhi, India
e-mail: nafisraj2010@gmail.com; parveenchaprana20@gmail.com; shabana8320@gmail.com; nmanzoor@jmi.ac.in

N. Manzoor (ed.), *Advances in Antifungal Drug Development*,
https://doi.org/10.1007/978-981-97-5165-5_11

infections are frequent in babies, immunocompromised individuals, diabetics, and obese individuals. In their childbearing years, 70–75% of women suffer from vulvovaginal candidiasis (Li and Nielsen 2017). Emerging resistant fungi, such as multidrug-resistant *Candida auris*, presents new threats to public health. *C. neoformans* is a basidiomycete human fungal pathogen that causes life-threatening pneumonia and meningitis in immunocompromised patients. It is unique as it is the only fungus that is encapsulated and can be easily observed using negative staining. Airborne asexual spores of *A. fumigatus*, when inhaled can be effectively cleared by the innate immune system of healthy individuals. However, these spores evade the defense system of an immunocompromised individual and will cause life-threatening invasive aspergillosis. *H. capsulatum* is a thermally induced dimorphic fungus that grows as hyphae in the environment and prefers to grow as budding yeast in the mammalian host. The disease, Histoplasmosis, has clinical and pathogenic features that are similar to tuberculosis. Table 11.1 summarizes the fungal pathogenic species, and represents the fungal pathogenesis in humans and their mortality rate.

Table 11.1 Major fungal species that cause disease in humans and mortality rate

Fungal species	Disease	Mortality rate	References
Candida albicans, Candida dubliniensis, Candida glabrata	Mucocutaneous candidiasis	~138 million	Lionakis et al. (2023)
C. albicans, C. glabrata, Candida tropicalis, Candida auris, Candida parapsilosis	Invasive candidiasis	~750,000	Lionakis et al. (2023)
Aspergillus fumigatus, Aspergillus flavus, Aspergillus terreus, Aspergillus niger	Invasive aspergillosis	~300,000	Lionakis et al. (2023)
Cryptococcus neoformans, C. gattii	Cryptococcosis	~223,000	Lionakis et al. (2023)
Histoplasma capsulatum, Histolasma duboissii	Histoplasmosis	NA	Li and Nielsen (2017)
Coccidiodes posadasii, Coccidiodes immitis	Coccidioidomycosis	NA	Kohler et al. (2015)
Blastomyces dermatitidis, Blastomyces gilcristii, Blastomyces percursus	Blastomycosis	NA	Li and Nielsen (2017)
Pneumocystis jirovecii	Pneumocystosis	~500,000	Lionakis et al. (2023)
Paracoccidioides brasiliensis, Paracoccidioides lutzi	Paracoccidioidomycosis	NA	Kohler et al. (2015)
Talaromyces marneffei	Talaromycoses	NA	Li and Nielsen (2017)

11.2 Plant Essential Oils and Secondary Metabolites

There is growing attention to medicinal and aromatic plants due to their various applications. People are now well aware of medicinal plants and their therapeutic potential (Noriega 2021). Plant essential oils (EO) are complex mixtures of volatile compounds that contribute to the characteristic aroma and flavor of plants. They possess a wide range of chemical constituents, including terpenes, phenolics, and other bioactive compounds. While EOs have long been used in the various cosmetics industries, their potential applications extend much beyond the medicinal arena. A deeper understanding of the chemistry and biological properties of these plant EO and their constituents will have applications in human health, agriculture, and environment (Srivastava and Singh 2019). Concerning human health, EOs have shown potential antimicrobial, antifungal, and antioxidant properties. They are being explored as natural alternatives to conventional drugs or in combination with synthetic compounds or already established drugs offering more sustainable and safer options. EOs are thus being explored for their potential use in the development of new drugs, dietary supplements, and natural remedies for various ailments (Figueiredo et al. 2008). In agriculture, they can be utilized as natural alternatives to synthetic pesticides, reducing the dependency on harmful chemicals and promoting sustainable farming practices (Ootani et al. 2013). To fully understand the potential of EOs, further research is needed to understand their complex composition, synergistic effect, and mechanism of action. By exploring their chemistry and biological properties, EOs can be harnessed as effective and sustainable alternatives or complement synthetic compounds, with the potential to positively impact human health, agriculture, and the environment.

EOs are soluble in alcohol, ether, and fixed oils, but insoluble in water due to their hydrophobic nature. These volatile oils are typically colorless liquids at room temperature and possess a distinct odor. They generally have a density of less than one, except for a few like cinnamon, sassafras, and vetiver oils. They also exhibit a high refractive index and strong optical activity. The EO constituents include different secondary metabolites mainly monoterpenes, sesquiterpenes, and polyterpenes (Elshafie et al. 2023). The pathways involved in the synthesis of terpenes are both the mevalonate pathway and the mevalonate-independent (deoxyxylulose phosphate) pathway. These EOs are synthesized in various specialized structures within plants, and the number and characteristics of their structures can vary significantly. They are produced and stored in specific plant organs, like the secretory hairs or trichomes, epidermal cells, internal secretory cells, and secretory pockets. Within these organs, EO is present in the cytoplasm of certain plant cell secretions. The composition of EO is highly complex and may contain more than 300 different volatile organic compounds with relatively low molecular weight, typically below 300. Due to high vapor pressure at room temperature and atmospheric pressure, these compounds exist in the vapor state partially (Dhifi et al. 2016). They belong to various chemical classes, including alcohols, ethers or oxides, aldehydes, ketones, esters, amines, amides, phenols, heterocycles, and most notably, terpenes.

Terpenes Terpenes are a predominant class of EO constituents that encompass a wide range of chemical structures. These compounds are one of the most diverse families of secondary metabolites, synthesized through the polymerization of isoprene units, specifically 2-methyl-1,3-butadiene. Isoprenes serve as the fundamental building blocks for the synthesis of terpenes, and their polymerization gives rise to a wide array of structures and functions. Terpenoids are another class of organic compounds found in various plants. While terpenes are simple hydrocarbons, terpenoids are terpenes that have undergone chemical modifications, typically through oxidation or the addition of functional groups like oxygen-containing moieties. Terpenes, such as pinene, myrcene, limonene, terpinene, and p-cymene, have been reported to possess antimicrobial properties against both susceptible and drug-resistant organisms (Masyita et al. 2022). These compounds promote cell rupture and inhibit both protein and DNA synthesis in bacteria, contributing to their antimicrobial efficacy (Álvarez-Martínez et al. 2021). Some specific terpenes, including carvacrol, carvone, eugenol, geraniol, and thymol, have also demonstrated antibacterial activity (Masyita et al. 2022). Terpenoids, as secondary metabolites, play a crucial role in the defense of aromatic and medicinal plants against diseases. They are classified as monoterpenoids, sesquiterpenoids, diterpenoids, triterpenoids, tetraterpenoids, and polyterpenoids based on the number of carbon atoms in the skeleton (Yazaki et al. 2017). Monoterpenoids (one isoprene), interfere with the physiological and metabolic processes of microbes and disrupt microbial growth (Elshafie et al. 2023). Several terpenoids, such as azadirachtin, carvone, menthol, ascaridole, methyl eugenol, toosendanin, and volkensin, have been investigated for their antimicrobial, antifungal, and insect/pest repellent properties (Pandey et al. 2017). Monoterpenes are formed by the combination of two isoprene units and contain ten carbon atoms. These compounds are abundant in EOs and contribute to the distinct aromas and flavors of various plants. Monoterpenes like pinene, myrcene, limonene, and thujene contribute to the diverse aromatic profile of plants and are widely utilized in the fragrance, flavor, and pharmaceutical industries (Iraji et al. 2020).

Phenolic Compounds Plants contain around 8000 phenolic compounds with known structures, making them one of the most prevalent families of secondary metabolites in plants. The primary mechanisms for the synthesis of most phenolic compounds are malonic acid and shikimic acid pathways. A combination of one or more aromatic rings and one or more hydroxyl groups forms these molecules. The primary structural components of phenolics are polymeric or monomeric molecules, which can be glycosides, aglycones, substrates, or free-binding substances (Alara et al. 2021). Phenolic compounds contain flavonoids, phenolic acids, polyphenols, and tannins these molecules are mainly found in fruits, tea, vegetables, legumes, and EOs (Alara et al. 2021). Natural plant phenolic compounds have been increasingly identified in recent years, making them a valuable research target for creating innovative antifungal medications (Chtioui et al. 2022).

Flavonoids Flavonoids are hydroxylated phenolic compounds that link two aromatic rings with a heterocyclic ring that has three carbon atoms and connects a C6-C3-C6 carbon skeleton structure. More than 4000 flavonoids have been found in plants since the 1930 discovery of these compounds in orange peel. As a result, flavonoid makes up half of all phenolic compounds and is one of the most significant class within the phenolic family (Tungmunnithum et al. 2018). Furthermore, studies have shown that naturally occurring flavonoids can function as direct antifungal agents and have synergistic action with other antifungals (Jin 2019). Additionally, it has been noted that some flavonoids can significantly lower the spore germination of fungi that are harmful to plants, such *Botryosphaeria* (Ma et al. 2022). It has been demonstrated that several flavonoid compounds have antifungal properties against *Aspergillus*, *Candida*, *Cryptococcus*, *Malassezia* spp., and other human infections (Wang et al. 2021a, b; da Fonseca et al. 2022; Fowler et al. 2011; Alves et al. 2017). Furthermore, curcumin derived from *Curcuma longa* L. exhibits fungistatic properties toward *A. flavus* (Temba et al. 2019). The primary bioactive ingredient of the traditional Chinese medicinal herb *Scutellaria baicalensis*, the flavonoid baicalin, inhibits the production of *C. albicans* biofilms and increases the rate at which this human pathogen undergoes apoptosis, thereby exerting a concentration-dependent antifungal action (Cho et al. 2008). Subsequent research has revealed that the mechanism by which baicalin inhibits biofilms is due to the overexpression of several redox-related genes including *CAP1*, *SOD2*, and *TRR1* which raises the cytosolic Ca^{2+} concentration, damages the ultrastructure of the cell, and accelerates the rate of *C. albicans* apoptosis (Yang et al. 2014). In addition to *C. albicans*, baicalin exhibits antifungal properties against *A. fumigatus*, *T. rubens*, and *T. trichophyton* (Da et al. 2019). These studies indicated that flavonoids will emerge as a key area of study for the development of antifungal medications.

Nitrogenous Secondary Compounds Majority of these plant secondary metabolites are synthesized using amino acids. Their complex metabolism is influenced by stress from the environment and plant hormones (Cho et al. 2008). Secondary metabolites that contain nitrogen include amino acids that are not proteins, cyanogenic glycosides (CNGs), and alkaloids. The α-hydroxynitrile glycosides, or CNGs, are made comprising of an aglycone to which a sugar group is connected (Bolarinwa et al. 2015). Nonprotein amino acids, which constitute significant nitrogen reserves in plants, are another family of nitrogenous molecules found in plants. Apart from their antimicrobial, antifungal, and anticancer properties, nonprotein amino acids mainly aid plants in fending off damaging insects (Huang et al. 2011). More than 20% of plant species contain alkaloids, which are typically found in large amounts in families of plants such as the *Fabaceae*, *Apocynaceae*, *Polygonaceae*, *Papaveraceae*, *Rutaceae*, and *Solanaceae*. These molecules have one or more nitrogen atoms in heterocyclic rings. Alkaloids are classified into several groups based on their chemical structure, including pyridine, isoquinoline, indole, scopolamine, and organic amine alkaloids. Alkaloids, such as acridine, fluoroquinolone, and 4-quinolone have a broad range of biological activities and are isolated from sea buckthorn (*Hippophae rhamnoides*). They limit the growth of *C. albicans* by downregulating the expression of the *ICL1* gene (Adamski et al. 2020; Kamal et al. 2021). Magnoflorine inhibits α-glucosidase activity and reduces biofilm development in *C. albicans*. It is found in the *Acorus calamus*, *Tinospora cordifolia*, and *Celastrus paniculatus* (Kim et al. 2018). One of the furanoquinoline alkaloid, pteleine, has antifungal properties against *C. albicans* (Shang et al. 2018). Additionally, 8-methoxydictamnine and 8-acetylnorchylerythrine from *Zanthoxylum asiatica* (Toddalia) have been reported in recent studies to have antifungal action against *C. tropicalis*, *C. glabrata*, and *C. albicans* (Hu et al. 2014).

11.3 Antifungal Efficacy of Essential Oils and Their Constituent Terpenes (Cyclic & Acyclic)

Infections due to *Candida* spp. are prevalent, especially in the immunocompromised population. Several EOs have been found effective in controlling *Candida* infections. Commercially available lemon EO exhibited antifungal properties against *C. albicans, C. tropicalis,* and *C. glabrata.* The inhibition of *C. albicans* growth was notable across all tested concentrations (Leite et al. 2014). The EO of *P. angulata*, an annual herbaceous plant belonging to the Solanaceae family, displayed remarkable antifungal characteristics against *C. torulopsis, C. stellatoidea,* and *C. albicans* strains, typically resistant to antibiotics. Therefore, further investigation into its anti-infective properties was justified (Cordisco et al. 2019). Several monoterpenes exhibit a wide range of biologically active properties that are useful for treating various diseases and conditions. Terpineol and its isomers (α-terpineol, terpinene-4-ol, and δ-terpineol) exhibited inhibitory effects on gram-negative bacteria, particularly *Shigella flexneri*, by disrupting membrane integrity and leading to the release of nucleic acids and proteins. These compounds have shown potential as natural antibacterial agents, capable of destroying bacterial cell membranes and cell walls (Huang et al. 2021). Borneol and citral have demonstrated synergistic effects as bacteriostatic and antibiofilm activity, particularly against *Listeria monocytogenes* and *Pseudomonas aeruginosa* biofilms. These compounds increase the porosity of bacterial cell membranes (Wang et al. 2021a, b). Various terpenes and terpenoids, including bakuchiol, α-pinene, linalool, champene, 1,8-cineole, α-phellandrene, 3-carene, p-cymene, perillyl alcohol, bornyl acetate, and citral isomers, have also exhibited inhibitory effects on the growth of microorganisms. Studies thus suggest the potential use of these compounds as effective antimicrobial agents used in the food industry (Zahi et al. 2017). Overall, the antimicrobial properties of terpenes and terpenoids have been extensively studied, and these natural compounds show promise as alternatives to combat multidrug-resistant organisms and support various applications in the food, cosmetic, and healthcare industries.

11.4 Antifungal Targets and Mechanism of Inhibition

Over the past few decades, studies have shown that opportunistic fungal infections are a significant cause of human diseases, particularly in the immunocompromised, including cancer patients, HIV/AIDS patients, and organ transplant recipients (Nucci and Marr 2005). Currently, there are only three available antifungal drugs to treat invasive fungal infections (IFIs): triazoles that prevent the synthesis of ergosterol (fluconazole, itraconazole, voriconazole, posaconazole, isavuconazole), Amphotericin B-deoxycholate, a polyene, that binds to ergosterol in fungal cell membranes causing cell lysis and echinocandins that prevent the formation of (1 → 3)-β-D-glucan in fungal cell walls (caspofungin, micafungin, anidulafungin) (Ostrosky-Zeichner et al. 2010). It is difficult to find new medications to treat resistant IFIs since fungi are eukaryotes, which are closely connected to humans more than bacteria or viruses. Recent developments in the fields of fungal life cycle research, functional genomics, proteomics, and gene mapping have made it possible to identify several novel, potential therapeutic targets that may expand the range of treatments for resistant IFIs (Roemer and Boone 2013). Before understanding the mechanism of action of different antimicrobial agents it is necessary to identify potential targets in fact in many cases a clarification of antifungal targets has allowed us to enhance our understanding of the specific mechanism of action. Figure 11.1 and Table 11.2 represent the potential targets of pathogenic fungi.

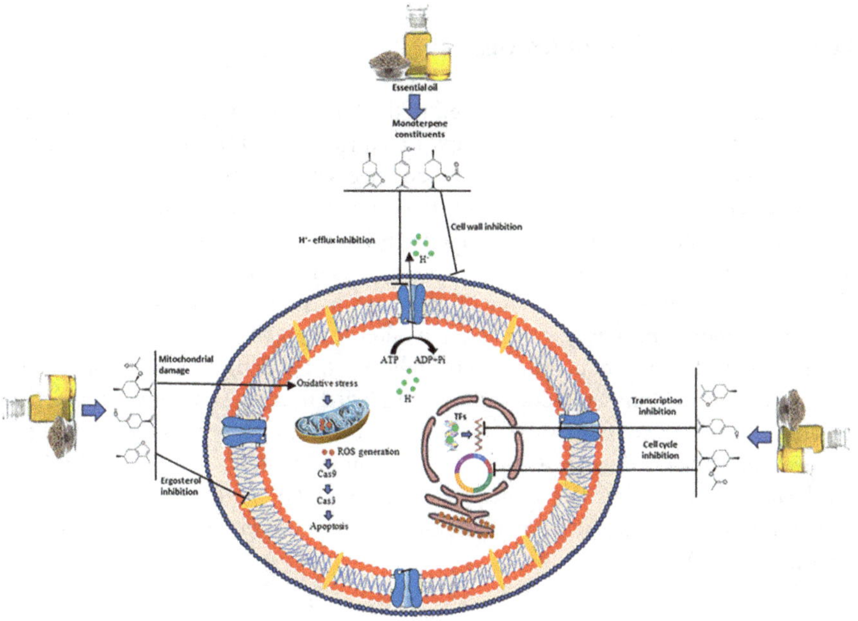

Fig. 11.1 The possible targets and mechanism of action of essential oils and their terpene constituents against the fungal system

Table 11.2 Potential targets present in pathogenic fungi

Molecular targets	Compounds tested	References
C-14 α demethylase	Azoles	Fromtling (1988)
Serine palmitoyl transferase	Sphingofungins, Lipoxamicin, Paramicin, and Viridiofungins	Nimrichter and Rodrigues (2011)
Chitin synthase 1, 2, and 3	HWY-289	Vijayakumar et al. (1996)
DNA replication	5-flucytosine	Vermes et al. (2000)
Topoisomerase I	A-3253 and Camptothecin	Watt and Hickson (1994)
EF3 (elongation factor 3)	Sordarin	Justice et al. (1998)
N-myristoyl transferase (NMT)	SC-59383	Lodge et al. (1997)
Dihydrofolate reductase enzyme	Methotrexate	DeJarnette et al. (2020)
ß-tubulin (microtubule aggreegation)	Griseofulvin	Kilmartin (1981)
Phosphoinositide 3-kinases	Wortmannin	Ui et al. (1995)
Heat-shock protein (HSP) 70 and 90	Herbimycin and Macbecin	Whitesell et al. (1994)
Electron transport chain (ETC)	Antimycin, UK-2A and UK-3A	Ueki et al. (1996)
Calcineurin	CsA and FK-506	Bastidas et al. (2008)

11.4.1 Fungal Cell Membranes

The first azole derivatives were designed in the late 1960s. These were completely synthetic and rapidly expanding groups of antifungal compounds (Lewis 2011; Kathiravan et al. 2012). Azoles have a broad-spectrum but show fungistatic activity against several yeasts and filamentous fungi. They inhibit the ergosterol biosynthesis pathway by inhibiting the enzyme, lanosterol 14α demethylase. Ergosterol levels are drastically reduced making the fungal membranes loose integrity (Fromtling 1988). In both mammalian and fungal cells, sphingolipids are the essential membrane components, primarily located in the outer leaflets of cell membranes (Patton and Lester 1991). In sphingolipid biosynthesis, a condensation reaction occurs between serine and fatty-acyl-coenzyme A catalyzed by serine palmitoyl transferase forming long-chain keto-dihydro sphingosine molecules. This step-in phospholipid synthesis is crucial and hence an important antifungal target. Several natural inhibitors have been identified that target the enzyme serine palmitoyl transferase like sphingofungins, lipoxamicin, paramicin, and viridiofungins (Nimrichter and Rodrigues 2011).

11.4.2 Fungal Cell Walls

The fungal cell is enclosed by a cell wall which provides protection and structural support to the cells. The fact that the mammalian cells lack a cell wall, makes that of the fungal cell a promising target for the development of antifungal agents (Georgopapadakou and Tkacz 1995). β-glucans and chitin are the essential components of the cell wall and provide strength and shape to the cells. The β-(1,3)-glucan synthase has two functional components a catalytic component and a regulatory component. Various known glucan inhibitors specifically inhibit glucan synthesis by directly targeting the enzyme, β-(1,3)-glucan synthase (Cabib and Kang 1987; Hartland et al. 1991). Chitin is a homopolymer of N-acetyl glucosamine (GlcNAc) units linked by β-(1,4) bonds forming the structural component of the cell wall. Chitin synthesis by the enzyme chitin synthase (Chitin synthase 1, 2, and 3) was first identified in *C. albicans* and *Saccharomyces cerevisiae* (Au-Young and Robbins 1990). There are several compounds that act as analogs of the substrate UDP-GlcNAc and bind to the catalytic site of this enzyme (Ruiz-Herrera and San-Blas 2003). HWY-289 binds to chitin synthase 1 and 2 and significantly inhibits chitin synthesis in *C. albicans* (Vijayakumar et al. 1996).

11.4.3 Fungal Nuclear Material

Targeting metabolic processes with compounds that mimic metabolites would be a feasible strategy for drug development. 5-Flucytosine (5-FC) is converted to 5-fluorouracil (5-FU). This phosphorylated form is converted into a deoxynucleoside analog which is incorporated into DNA during replication resulting in the blocking of DNA chain elongation (Vermes et al. 2000). Topoisomerases (TOP) I and II are the most important enzymes which are involved in the replication, transcription, and repair of cellular DNA. TOP I is equivalent to DNA gyrase in bacteria so it would be an excellent target for the identification of inhibitors. It has been reported earlier that A-3253 and camptothecin had great inhibitory potential against fungal TOP I (Watt and Hickson 1994). Elongation factor 3 (EF3) is the only member of the family of elongation factors that is found in fungal cells including the pathogenic *C. albicans* and *P. carinii*. Hence, EF3 also has great potential as an antifungal target for polypeptide synthesis. Besides protein synthesis, posttranslational modification is also an essential process during protein trafficking. N-myristoylation is one of the major posttranslational modifications that occur in most fungi during which myristate from CoA get transferred to the amino-terminal of glycine residues of proteins (Duronio et al. 1989), the reaction being catalyzed by the enzyme N-myristoyl transferase (NMT). The fungal NMT may thus serve as an antifungal target against pathogenic fungi. Similarly, the enzyme dihydrofolate reductase is another significant target for antifungal drug development as it catalyzes the formation of dihydrofolate acid from tetrahydrofolic acid (Chan and Anderson 2006).

11.4.4 Other Possible Targets

α and β tubulins are monomers used for the formation of the microtubule polymer. Microtubule aggregation is a key process that contributes to cell structure, function, and mobility. β tubulin is a highly conserved protein in the eukaryotic system. Griseofulvin is an antifungal drug that binds to β tubulin and inhibits microtubule aggregation (Kilmartin 1981). Wortmannins, a steroid metabolite extracted from the fungi *Talaromyces wortmannii* and *Penicillium funiculosum*, inhibits the phosphoinositide 3 kinase (PI-3-kinase) in *S. cerevisiae*. The binding of this natural product is covalent but nonspecific (Ui et al. 1995). Heat-shock proteins (HSP) 70 and 90 are prominent antifungal targets due to their highly conserved sequences. Studies have showed that geldanamycin and its structural analogs (herbimycin and macbecin) bind to HSP90 and inhibit its function (Whitesell et al. 1994). Antimycins are antibiotics isolated from the *Streptomycetes* spp. their structural analogues, UK-2A and UK-3A, are also isolated from the same species that block the electron transport chain (ETC) and suppress respiratory machinery in yeast (Ueki and Taniguchi 1997). In stress conditions, calcineurin maintains homeostasis by increasing the intracellular calcium that ultimately activates the calmodulin that controls cell survival. Two immunosuppressive compounds CsA and FK-506 have shown inhibitory potential against fungal calcineurin (Bastidas et al. 2008).

11.5 Disruption of Fungal Cell Membranes

Fungal cell membranes play a crucial role in maintaining the intracellular and extracellular environment. The plasma membrane primarily contains a lipid bilayer (amphipathic nature) embedded with proteins (extrinsic and intrinsic) and a small number of oligosaccharides that provide membrane fluidity and integrity to the cell. Ergosterol is the principal sterol present in the lipid bilayer of the fungal cell membranes. It is necessary for normal growth function, membrane fluidity, permeability (exchange of materials), and regulation of other physiological conditions in fungal cells (McGinnis and Tyring 1996). Till date, various antifungals have been developed that bind to ergosterol directly or interfere with the ergosterol biosynthetic pathway. Plant EOs and their bioactive compounds especially monoterpenes and their derivatives have shown efficacy using fungal plasma membranes as primary targets resulting in the alteration of membrane permeability and integrity.

11.5.1 Ergosterol Biosynthesis

Several studies suggest that plant-based EOs and their terpenoid constituents target the ergosterol biosynthetic pathway disrupting the membrane homeostasis. The cell membranes become hypersensitive as the ergosterol levels decrease (Nazzaro et al. 2017). The ergosterol content can be estimated by UV scanning at 230–300 nm and HPLC equipped with a UV detection component (Essid et al. 2017). The effect of selected monoterpenes on the ergosterol content, as quantified by UV scanning spectrum, has been summarized in Table 11.3. All the compounds show dose-dependent decrease in ergosterol levels, and the reduction was in the range of 18–90%. Interestingly for some compounds reduction in ergosterol levels was even greater than FLC.

Table 11.3 Effect of some plant essential oils and their constituent terpenes on the level of ergosterol in fungi

Compound	Microorganism	Ergosterol inhibition in percentage	References
Linalool	*C. neoformans* T-444	57%	Cardoso et al. (2016)
Geraniol	*C. neoformans* T-444	25%	Cardoso et al. (2016)
	T. rubrum ATCC 1683	Similar inhibition comparable to ketoconazole	Pereira et al. (2015)
	C. albicans	14–100%	Sharma et al. (2016)
	C. glabrata	12.5–78.5%	
	C. tropicalis	48–100%	
Carvacrol	*C. albicans* 14,053	57%	Alizadeh et al. (2018)
	C. albicans	24–90%	Ahmad et al. (2011a, b)
β-citronellol	*T. rubrum ATCC 1683*	Similar inhibition comparable to ketoconazole	Pereira et al. (2015)
	C. albicans	17.95–100%	Sharma et al. (2020)
	C. albicans D-27	12–100%	
	C. albicans S-1	2–92.77%	
Eugenol	*T. rubrum* ATCC 1683	More effective than ketoconazole	Pereira et al. (2013)
Thymol	*C. albicans*	19–98%	Ahmad et al. (2011a, b)
Verbenol	*Aspergillus flavus*	37.95–62.77%	Singh et al. (2021)
α-Bisabolol	*A. fumigatus*	26.32–73.77%	Jahanshiri et al. (2017)
Thymus vulgaris EO	*Aspergillus flavus*	49.6–98.1%	Kohiyama et al. (2015)
Curcuma longa EO	*Aspergillus flavus*	14.8–80.27%	Hu et al. (2017)
Anethum graveolens	*Aspergillus flavus*	27.3–79.4%	Tian et al. (2012a, b)
Cinnamomum jensenianum EO	*Aspergillus flavus*	56.9–85.2%	
Ocimum sanctum EO	*Candida* sensitive and resistant isolates	30–46% for sensitive and resistant isolates	Khan et al. (2014a, b)
Coriaria nepalensis leaves EO	*Candida* sensitive and resistant isolates	32–98% (sensitive strain) and 18–97% (resistant strain)	Ahmad et al. (2011a, b)

(continued)

Table 11.3 (continued)

Compound	Microorganism	Ergosterol inhibition in percentage	References
Anethum graveolens EO	*Candida* isolates	70–75%	Chen et al. (2013)
Cinnamomun verum EO	*C. albicans* ATCC 10231	83%	Essid et al. (2017)
Thymus pulegioides EO	*C. albicans* ATCC 10231	80–100%	Pinto et al. (2006)
Thymus pulegioides EO	*Trichophyton rubrum*	70%	
Melaleuca alternifolia EO	*Botrytis Cinerea*	54.5%	Yu et al. (2015)
Litsea cubeba fruits EO	*Botrytis Cinerea*	34.81%	Wang et al. (2019)
Ocimum basilicum EO	*C. neoformans*	79%	Cardoso et al. (2016)
Coreopsis tinctoria Nutt. Flowers EO	*C. neoformans*	73%	Zeng et al. (2016)
Fennel seeds EO	*Trichophyton rubrum*	Dose-dependent reduction	Zeng et al. (2015)

11.5.2 Plasma Membrane H^+-ATPase Activity

In fungal cells, the plasma membrane H^+-ATPase (Pma1p) is a major protein that regulates the cytosolic pH via the active transport of protons from the cytosol to the external environment with the expenditure of ATP. This function maintains the electrochemical gradient between the internal and external environment of the cell which is crucial for ions transport, nutrient uptake, and growth of the cells. Pma1p belongs to a P-type ATPase pump that is similar to mammalian Na^+, K^+, and Ca^+-ATPase (Ambesi et al. 2000). The permeability of the cell membrane is also expressed in terms of extracellular pH. Previous studies have reported a dramatic change in the intracellular pH of the fungal cells. One study measured the extracellular pH drop in *Fusarium sporotrichioides* in response to *Mentha piperita* EO treatment that lower the pH from 0.95 to 1.2 of cell suspension at a contact time of 30 to 90 min (Rachitha et al. 2017). In the presence of geraniol, at its MIC value (125 μg/mL), the glucose-induced acidification decreased to 36.35%, 39.7%, and 38.2% in *C. albicans, C. glabrata,* and *C. tropicalis*, respectively (Sharma et al. 2016). Cinnamaldehyde and its derivatives (sinapaldehyde, coniferyl aldehyde) show great impact on H^+ extrusion by *Candida* isolates. The inhibition was 36–41%, 41–17%, and 43–51% for cinnamaldehyde, coniferyl aldehyde, and sinapaldehyde, respectively (Shreaz et al. 2013). Isoeugenol and o-methoxy cinnamaldehyde showed inhibition of ATPase-mediated proton pumping against different clinical *Candida* isolates. The inhibition was in the range of 70–82% for isoeugenol and 43% for o-methoxy cinnamaldehyde (Bhatia et al. 2012). Thymol and eugenol also inhibit the acidification of the external medium. The pH of the extracellular medium was reduced by 56–61% and 40–54% for thymol and eugenol compared to the control (Ahmad et al. 2010). Reports thus highlight the importance of the membrane protein, Pma1p. This proton pump is vital to fungal cell viability and hence is a potential target for drug development.

11.5.3 Chitin Synthesis

The fungal cell wall is another crucial component that greatly influences the growth and survival of fungi. It consists of three main structural elements: glucans, chitin, and mannans. These elements are widely recognized as potential targets for therapeutic interventions. Chitin, specifically, is a long linear homopolymer composed of β-1,4-linked N-acetylglucosamine (GlcNAc) units, and its synthesis is catalyzed by the enzyme chitin synthase (Wu et al. 2008). It is a critical component in the construction of the fungal cell wall. Inhibiting the chitin polymerization may affect bud ring formation, septum formation, and cell wall maturation. Anise oil and its major constituent trans-anethole inhibits chitin synthesis in filamentous fungi *Mucor mucedo* IFO 7684 (Yutani et al. 2011). Thymoquinone of *Nigella sativa* seed oil damages the cell wall of *C. albicans* irreversibly (İşcan et al. 2016). The leaf extract of *Inula viscosa* inhibits the chitin synthesis in *M. canis, T. rubrum,* and *C. albicans.* On the other hand, miconazole had no effect on chitin synthesis in dermatophytes and *C. albicans* (Maoz and Neeman 2000).

11.6 Modulation of Virulence-Related Proteins/Genes

A virulence factor is a pathogenicity determinant and as per the damage-response framework. It is defined as a microbial component that causes harm to the host. This broad definition allows for many leeways because it includes microbial components that are directly hazardous to the host and antigens that generate detrimental immune responses. Virulence factors play a significant role in microbial pathogenesis (Casadevall and Pirofski 2009). Several defined virulence factors include hydrolytic secretory enzymes, adhesion, and biofilm formation.

11.6.1 Extracellular Secretory Enzymes

Hydrolytic enzymes are the most extensively studied group of enzymes and are major determinants of fungal pathogenesis. Secretory aspartyl proteases (SAP) and phospholipases (PL) are enzymes that degrade host tissues by cleavage proteins and phospholipids, respectively. They support tissue invasion, hyphal formation, and lead to severe *Candida* infection (Schaller et al. 2005). The SAP family consists of ten members, of which SAP 1–8 are secreted extracellularly while sap9 and sap10 are membrane-bound. Among the SAP family, sap2 and sap5 are reported to be the significant virulence factors that cause superficial as well as systemic fungal infections (Borelli et al. 2008). The PL family also contains different classes of phospholipases (pla, plb, plc, and pld). Of these, plb1 has shown active participation during the invasion of animal models of candidiasis. It is also highly expressed in the hyphal tip of cells during the invasion process (Calderone and Fonzi 2001; Ghannoum 2000). These virulence-associated proteins sap2, sap5, and plb1 are thus the prime targets associated with epithelial tissue invasion. As shown in Fig. 11.2, the 3D crystallographic structure of sap2 (PDB ID: 1EAG) and sap5 (PDB ID: 2QZX) are available in the PDB (protein data bank) database with good resolution (2.10 Å and 2.50 Å). The N-terminal and C-terminal of both the proteins have aspartic acid-rich catalytic sites (DTGS 32-35 and DSGT 218-221) with 340 and 342 amino acid residues (Cutfield et al. 1995; Borelli et al. 2008). However, the crystallographic 3D structure of PLB1 is not available. A study reported that the synergistic effect of quinic acid and undecanoic acid downregulates the expression of sap1, sap2, and sap5 genes leading to a decrease in extracellular protease activity in

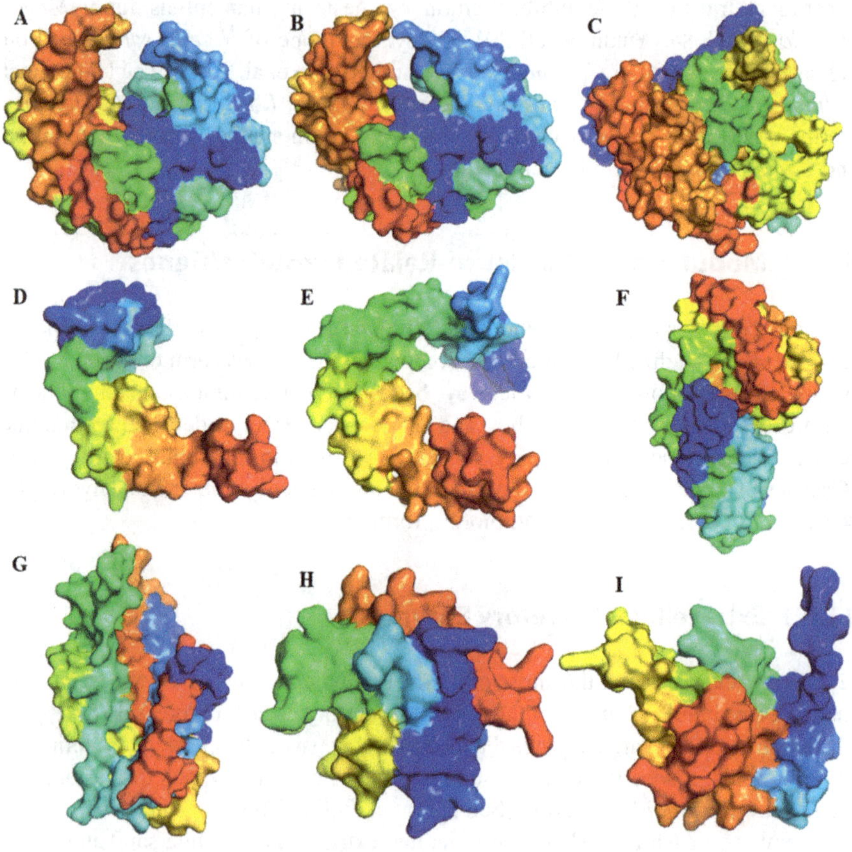

Fig. 11.2 3-Dimensional crystallographic structures of the virulent associated proteins. (**a**) sap2, (**b**) sap5, (**c**) plb1, (**d**) bcr1, (**e**) ace2, (**f**) als3, (**g**) efg1p, (**h**) tec1p, and (**i**) cph1p

C. albicans (Muthamil et al. 2018). At the MIC of geraniol, proteinase secretion dropped in *C. albicans*, *C. tropicalis*, and *C. glabrata* by 32.5–38.7%, 46.51%, and 31.83%, respectively. However, the inhibitory effect on phospholipase secretory activity was dropped only by 10–20% (Sharma et al. 2018). *Origanum vulgare* EO significantly lowers the secretion of phospholipases produced by *C. albicans* strains (Brondani et al. 2018). Sub-MIC concentrations of several EOs were used to inhibit phospholipase activity in six isolates of *C. albicans* that are capable of producing the enzyme. It was found that rosemary had the highest antiphospholipase activity followed by cinnamon oil, jasmine oil, and clove oil (El-Baz et al. 2021).

11.6.2 Biofilm Formation

Fungal cells form biofilms on both animate and inanimate surfaces. This phenomenon usually follows adhesion and leads to severe infections in the human host. Various fungi like *Candida*, *Trichosporon*, *Coccidioides*, *Pneumocystis*, and *Cryptococcus* form biofilms. Fungal cells in these communities are resistant to drugs due to the formation of a protective extracellular matrix around the cells, and the expression of various intrinsic factors. Fungi develop resistance to conventional drugs. *Candida* spp. show resistance to fluconazole, amphotericin B, voriconazole, and nystatin. *A. fumigatus*, however, showed more resistance to itraconazole in comparison to caspofungin (LaFleur et al. 2006). The pathogenicity of fungal strain depends upon the ability to adhere and invade host tissues. Several EOs, including those derived from *Coriandrum sativum* and *Ocimum americanum*, showed intriguing results when they were used to suppress the formation of biofilm in *C. albicans*. Various compounds, including xanthorrizol, linalool, and cinnamon oil have biofilm-inhibiting properties (Thaweboon and Thaweboon 2009; Hsu et al. 2013). Eugenol was found to exhibit antibiofilm action against *C. albicans* cells. The MIC_{50} of eugenol for sessile cells was 500 mg/L, which was two times greater than the MIC_{50} of eugenol for planktonic *C. albicans* cells, after incubation of 48 h (Rukayadi and Hwang 2013). According to a study, EO of eucalyptus, ginger grass, and clove can effectively combat biofilms produced by *C. albicans*. Moreover, eucalyptus oil, in particular, proved to be a promising antifungal agent in comparison to fluconazole (He et al. 2007). Viable cell counts and confocal laser scanning microscopy analysis demonstrated that *Rosmarinus officinalis* EO, when coated with nanoparticles, significantly reduced the capacity of *C. albicans* and *C. tropicalis* to attach to catheter surfaces and form biofilms (Stringaro et al. 2014). In *Alternaria*, *Aureobasidium*, and *Penicillium*, thyme EO altered the "molds" to form vegetative hyphae by again affecting the production of biofilms (Mironescu et al. 2010).

There are a few transcription factors and proteins that play a fundamental role in the regulation of biofilm formation via different surface receptors/proteins required for adherence (Chin et al. 2016). In *C. albicans*, bcr1, and ace2 are C_2H_2 zinc finger transcription factors that play a major role in adherence and biofilm formation (Fig. 11.2) (Ding and Butler 2007; Mulhern et al. 2006). The Agglutinin-like sequences (ALS) are a superfamily of adhesins that have an important role to play during adherence to host tissues. In the adhesin superfamily, als3 is the most suitable target for the discovery of new antifungals from natural sources. The deletion in *ALS3* gene leads to the loss of adhesion to different extracellular matrices (fibronectin, collagen, laminin, fibrinogen, etc.) of the host (Hoyer et al. 2008). The 3D crystallographic structure of als3 is available in PDB databases with a good resolution of 1.4 Å; deposited in 2013 with PDB: ID 4LEB containing 307 amino acid residues (Lin et al. 2014). These properties make als3 an important adhesion factor in comparison to other adhesin proteins (Fig. 11.2).

11.6.3 Morphological Switching

A morphological transition of fungal cells from the yeast form to multicellular filamentous hyphae plays a major role in pathogenesis. This dimorphic switching occurs in response to the host environment. Several signaling pathways regulate the morphological transition. Ras1 mitogen-activated phosphorylation kinase (MAPK) and protein kinase A (PKA) signaling cascade pathways play a vital role in switching. The fungal pathogens that belong to the class ascomycetes, have the potential to switch into the mycelial form. These include *Blastomyces dermatitidis*, *Coccidioides* spp., *Candida* spp. *Histoplasma capsulatum*, *Talaromyces marneffei*, *Paracoccidioides* spp., and *Sporothrix schenckii* and have a serious impact on human health. The class III HHK (hybrid histidine kinases) and class IV HHK proteins are encoded by *DRK1* and *SLNA* virulence genes in *Histoplasma capsulatum* and *Talaromyces marneffei* (Boyce and Andrianopoulos 2015). The HHK family proteins have a crucial role in switching, adaptation to osmotic stress, and pathogenicity via the MAPK pathway (Hagiwara et al. 2016). In *C. albicans*, hyphal morphogenesis is tightly regulated by various transcription factors (TFs) driven by MAPK and PKA signaling pathways. Their activation is completely dependent on a wide range of environmental factors including N-acetylglucosamine (GlcNAc), serum, carbon dioxide, and temperature (Villa et al. 2020). The cAMP and MAPK pathways are highly regulated by efg1p (enhanced filamentous growth factor 1), cph1p, and tec1p (Fig. 11.2) (Desai et al. 2018; Du et al. 2012). Despite the existence of other TFs, these three have an immense role to play in hyphal formation and significantly contribute to the virulence of *C. albicans* (Ponde et al. 2021). A study showed that the EO of *Cymbopogon nardus* effectively suppressed the growth of *C. albicans* strains by preventing its transition from yeast to a hyphal form at concentrations ranging from 15.8 μg/mL to 1000 μg/mL (De Toledo et al. 2016). The action of formyl-phloroglucinol meroterpenoids, specifically eucarobustol E, at 16 μg/mL was able to decrease the hydrophobicity of the *Candida* biofilm cellular surface and prevent the morphological transition (Liu et al. 2017). A EO high in nerolidol was extracted from inflorescences and leaf of *Piper claussenianum*. Its efficacy against *C. albicans* showed that morphological transition was downregulated by 81% (Curvelo et al. 2014).

11.7 Mitochondrial Dysfunction

Mitochondrion in eukaryotic cells is the main site for oxidoreductive reactions coupled with ATP synthesis via the kreb cycle (TCA). The mitochondrial matrix contains all the enzymes and coenzymes required for the synthesis of four to five-carbon intermediates. Mitochondrial dehydrogenase is the crucial enzyme required for the biosynthesis of ATP via oxidative phosphorylation. Various dehydrogenases are active during the oxidative reaction like lactate dehydrogenase (LDH), succinate dehydrogenase (SDH), and malate dehydrogenase (MDH). LDH catalyzes the conversion of pyruvate into lactate anaerobically, MDH converts malate to oxaloacetate, while SDH oxidizes succinate to fumarate in the TCA cycle (Zeng et al. 2015). Plant EOs can deplete ATP levels by inhibiting the activity of mitochondrial dehydrogenase/reductase of the TCA cycle (Nazzaro et al. 2017).

11.7.1 Mitochondrial Dehydrogenase/Reductase Activity

The enzymes, LDH, MDH, and SDH are markers that determine the oxidative activity of mitochondria. The enzyme activity is quantified by MTT assay in which the viable fungal cells reduce the substrate into the water-soluble, purple colored product (formazan) which can be assessed by colorimetric methods (Tian et al. 2012a, b). At 0.0625–4 μL, the EO of *Anethum graveolens* seeds inhibit the activity of MDH in *Aspergillus flavus* by up to 68% (Chen et al. 2013). Similarly, the EO of *C. tinctorial* flower (0.391–1.563 μL/mL) and *Acorus calamus* (0.2–1.0 μg/mL) reduce dehydrogenase activity in *C. neoformans* and *Aspergillus niger* (Zeng et al. 2016; Iswarya et al. 2016). Another study revealed that the EO of *Origanum compactum*, *Artemisia alba,* and *Cinnamomum camphor* caused a cytoplasmic petite mutation that led to mitochondrial damage in *Sacchomycetes cerevesiase* (Bakkali et al. 2006). Similarly, the MIC concentration of triterpenoids (100 μg/mL) and tetrapenoids (120 μg/mL) caused mitochondrial dysfunction that depletes the ATP levels (Haque et al. 2016). Cinnamon EO (0.1 μL/mL) drastically reduced the activity of SDH and MDH within 24 h in *Rhizopus nigaricus* (Li et al. 2014). The EO of Fennel seeds (0.018–0.078 μL/mL) and *C. tinctorial* flowers (0.391–1.563 μL/mL) decreased the SDH and MDH activity in *T. rubrum* and *C. neoformans* (Zeng et al. 2015, 2016). The EO of *F. cappilaris* at 0.08–0.64 μL/mL reduced the activity of mitochondrial reductase in a concentration-dependent manner (Pinto et al. 2013).

11.8 Generation of Reactive Oxygen Species (ROS)

Toxic reactive oxygen species are generated as a by-product of aerobic cellular metabolism. These species may include peroxide ion (O_2^-), hydrogen peroxide (H_2O_2) nitric acid (NO), and hydroxyl radical (OH). Excessive accumulation of ROS causes hyperpolarization of the mitochondrial membrane, nuclear fragmentation, enzyme inactivation, and membrane disruption leading to cell death by apoptosis (Pozniakovsky et al. 2005; Zorova et al. 2018). EO and their constituents can trigger the production of ROS and cause various cellular abnormalities resulting in cell death. ROS production was estimated in *C. albicans* and *A. flavus* treated with dill seed EO (Tian et al. 2012a, b; Chen et al. 2013). Similarly, the ROS production was estimated when EO of *Coreopsis tinctorial* (0.391–1.563 μL/mL) was used against *C. neoformans* (Zeng et al. 2016); thymol (100–200 μL/mL) and geraniol (0.02% v/v) were used against *A. flavus* (Shen et al. 2016); EO of *Cymbopogon martini* (100–600 μL/mL) was used against *Fusarium graminearum* (Kalagatur et al. 2018) and eugenol (256 μg/mL) against *C. glabrata* (Gupta et al. 2018). The ROS production was measured using the luminol-chemiluminescence method with the treatment of trans-cinnamaldehyde (0.626 μg/mL) against *P. italium*. Wang et al. estimated H_2O_2 production in the presence of eugenol against *B. cinerea* (Wang et al. 2010). Shen et al. evaluated the effect of thymol in nitric acid (NO) production using DAF-FMDA (3-amino, 4-aminoethyl-2′,7′-diflurocetein diacetate) method against *A. flavus* (Shen et al. 2016).

11.9 Modulation in Gene Expression

Quantitative real-time Polymerase Chain Reaction (qRT-PCR) analysis can assess the expression level of specific genes. The pattern of expression (downregulation or upregulation) was quantified by qRT-PCR analysis. It has been reported that MIC and MIC/2 concentrations of carvacrol, respectively downregulate the expression of *ERG11* and *ERG3* genes twofold. Similarly, a study showed the downregulation of *AUS1* (sterol importer), *KRE1* (GPI-anchored protein), and *FKS1* (1,3 β-glucan synthase) genes after the treatment of cinnamaldehyde (32 μg/mL) and eugenol (64 μg/mL) against *C. glabrata.* Cinnamaldehyde also had an impact on the expression of *CHS2* and *UAP1* genes of *G. citri-aurantti* having an impact on chitin biosynthesis (Mani-López et al. 2021). Another study has shown that cinnamaldehyde downregulates the levels of *pks*, *nrps*, *laeA*, *veA*, and *velB* by 98%, 96%, 84%, 76%, and 74%, respectively. These genes are associated with the ochratoxin biosynthesis pathway in *Aspergillus ochraceous*. Thymol downregulated the expression of *EST2* gene (telomerase reverse transcriptase) significantly and disrupted the associated biological pathway resulting in disrupted telomerase activity throughout the cell cycle (Darvishi et al. 2013). β-citronellal (200 μg/mL) has been shown to alter the expression analysis of *ERG*-related genes. *ERG1* and *ERG2* were upregulated while *ERG11* was downregulated. *ERG11* gene expression was studied and reported after the treatment with eugenol tosylate congeners (ETC-5, ETC-6, and ETC-7). In comparison to ETC-6 and ETC-7, ETC-5 showed significant downregulation of *ERG11* gene (Lone et al. 2020). The subinhibitory concentrations of *Ocimum sanctum* EO significantly lowered the expression of virulence-associated genes *HWP1*, *SAP2*, and *PLB2* which ultimately mitigated the virulence factors in *C. albicans* (Khan et al. 2014a, b).

11.10 Combinational Therapies and Synergistic Effects

Combinational antifungal therapy is a better alternative to monotherapy for overcoming multidrug resistance and controlling fungal diseases (King et al. 2014). The combinational strategy adopts the administration of synergistic drug combinations to allow effective treatment at low doses, shorter duration and reduced host-related toxicity (Zimmermann et al. 2007). Moreover, compounds given in combinations enhance the antifungal efficacy of currently available drugs resulting in enhanced therapeutic effects in comparison to drugs given alone (Onyewu et al. 2003). The checkerboard microdilution method is commonly used to assess drug interaction (Odds 2003). It was reported that the combined antifungal effect of harmine hydrochloride with azoles, in the early stages of resistant *Candida* biofilm formation, was highly synergistic (Li et al. 2019). Amorolfine has synergistic antifungal efficacy in combination with different conventional drugs such as ketoconazole, terbinafine, itraconazole, and griseofulvin. With the addition of amorolfine, the fungistatic activity of all these antifungals increased against dermatophytes (Polak-Wyss 1995). The combinational study of *Ocimum sanctum* EO, linalool, and geraniol with fluconazole increased the antifungal efficacy, especially against the resistant strain and their MIC reduced from 790 to 105 μg/mL and 152 to 38 μg/mL with linalool and geraniol, respectively (Cardoso et al. 2016). The fractional inhibitory concentration index (FICI) of thymol, eugenol, and methanol in combination with fluconazole was reported to be 0.31, 0.37, and 0.5 FIC index. This study also showed the excellent synergistic antifungal activity in the reduction of biofilms (Pemmaraju et al. 2013). The effect of *A. cryptantha* EO in combination with fluconazole has synergistic effect against *C. albicans* (FICI = 0.31). However, the EO of *A. cryptantha* and *L. integerifolia* showed an additive effect (FICI 0.75) (Lima et al. 2022). Similarly, another study showed the synergistic effect of perillyl alcohol with the antifungal drug miconazole against *C. glabrata* using a checkerboard assay (Gupta and Poluri 2021). The synergistic antifungal potential of turmeric EO with antifungal cream against *C. glabrata* has also been shown (Ogidi et al. 2021).

11.11 Conclusion

Innumerable research reports have shown that plant essential oils and their terpene constituents have fungicidal nature and their mechanism of action involves multiple targets. The advanced knowledge in the area of traditional medicine encourages us to explore natural compounds more for therapeutic effects and to reduce the use of synthetic drugs and their chemically synthesized derivatives. Although a lot of information is currently available, more detailed studies are still required to be done to emphasize the better application of these natural EO constituents to control fungal pathogenesis.

References

Adamski Z, Blythe LL, Milella L, Bufo SA (2020) Biological activities of alkaloids: from toxicology to pharmacology. Toxins 12(4):210

Ahmad A, Khan A, Yousuf S, Khan LA, Manzoor N (2010) Proton translocating ATPase mediated fungicidal activity of eugenol and thymol. Fitoterapia 81(8):1157–1162

Ahmad A, Khan A, Akhtar F, Yousuf S, Xess I, Khan LA, Manzoor N (2011a) Fungicidal activity of thymol and carvacrol by disrupting ergosterol biosynthesis and membrane integrity against Candida. Eur J Clin Microbiol Infect Dis 30:41–50

Ahmad A, Khan A, Kumar P, Bhatt RP, Manzoor N (2011b) Antifungal activity of Coriaria nepalensis essential oil by disrupting ergosterol biosynthesis and membrane integrity against Candida. Yeast 28(8):611–617

Alara OR, Abdurahman NH, Ukaegbu CI (2021) Extraction of phenolic compounds: a review. Curr Res Food Sci 4:200–214

Alizadeh F, Khodavandi A, Esfandyari S, Nouripour-Sisakht S (2018) Analysis of ergosterol and gene expression profiles of sterol Δ5, 6-desaturase (ERG3) and lanosterol 14α-demethylase (ERG11) in *Candida albicans* treated with carvacrol. J Herbmed Pharmacol 7(2):79–87

Álvarez-Martínez FJ, Barrajón-Catalán E, Herranz-López M, Micol V (2021) Antibacterial plant compounds, extracts and essential oils: an updated review on their effects and putative mechanisms of action. Phytomedicine 90:153626. https://doi.org/10.1016/j.phymed.2021.153626

Alves DR, de Morais SM, Tomiotto-Pellissier F, Miranda-Sapla MM, Vasconcelos FR, Silva INGD, Freire FDCO (2017) Flavonoid composition and biological activities of ethanol extracts of Caryocar coriaceum Wittm., a native plant from Caatinga biome. Evid Based Complement Alternat Med 2017:6834218

Ambesi A, Miranda M, Petrov VV, Slayman CW (2000) Biogenesis and function of the yeast plasma-membrane H+-ATPase. J Exp Biol 203(1):155–160

Au-Young J, Robbins PW (1990) Isolation of a chitin synthase gene (CHS1) from *Candida albicans* by expression in *Saccharomyces cerevisiae*. Mol Microbiol 4(2):197–207

Bakkali F, Averbeck S, Averbeck D, Zhiri A, Baudoux D, Idaomar M (2006) Antigenotoxic effects of three essential oils in diploid yeast (Saccharomyces cerevisiae) after treatments with UVC radiation, 8-MOP plus UVA and MMS. Mutat Res 606(1–2):27–38

Bastidas RJ, Reedy JL, Morales-Johansson H, Heitman J, Cardenas ME (2008) Signaling cascades as drug targets in model and pathogenic fungi. Curr Opin Investig Drugs 9(8):856

Bhatia R, Shreaz S, Khan N, Muralidhar S, Basir SF, Manzoor N, Khan LA (2012) Proton pumping ATPase mediated fungicidal activity of two essential oil components. J Basic Microbiol 52(5):504–512

Bolarinwa IF, Orfila C, Morgan MR (2015) Determination of amygdalin in apple seeds, fresh apples and processed apple juices. Food Chem 170:437–442

Bongomin F, Gago S, Oladele RO, Denning DW (2017) Global and multi-national prevalence of fungal diseases—estimate precision. J Fungi 3(4):57

Borelli C, Ruge E, Lee JH, Schaller M, Vogelsang A, Monod M et al (2008) X-ray structures of Sap1 and Sap5: structural comparison of the secreted aspartic proteinases from *Candida albicans*. Proteins 72(4):1308–1319

Boyce KJ, Andrianopoulos A (2015) Fungal dimorphism: the switch from hyphae to yeast is a specialized morphogenetic adaptation allowing colonization of a host. FEMS Microbiol Rev 39(6):797–811

Brondani LP, da Silva Neto TA, Freitag RA, Lund RG (2018) Evaluation of anti-enzyme properties of Origanum vulgare essential oil against oral *Candida albicans*. J Mycol Med 28(1):94–100

Cabib E, Kang MS (1987) [55] Fungal 1, 3-β-glucan synthase. In: Methods in enzymology, vol 138. Academic Press, London, pp 637–642

Calderone RA, Fonzi WA (2001) Virulence factors of *Candida albicans*. Trends Microbiol 9(7):327–335

Cardoso NN, Alviano CS, Blank AF, Romanos MTV, Fonseca BB, Rozental S, Alviano DS (2016) Synergism effect of the essential oil from Ocimum basilicum var. Maria Bonita and its major components with fluconazole and its influence on ergosterol biosynthesis. Evid Based Complement Alternat Med 2016:5647182

Casadevall A, Pirofski LA (2009) Virulence factors and their mechanisms of action: the view from a damage–response framework. J Water Health 7(S1):S2–S18

Chan DCM, Anderson AC (2006) Towards species-specific antifolates. Current medicinal chemistry 13(4):377–398

Chen Y, Zeng H, Tian J, Ban X, Ma B, Wang Y (2013) Antifungal mechanism of essential oil from Anethum graveolens seeds against *Candida albicans*. J Med Microbiol 62(8):1175–1183

Chin VK, Lee TY, Rusliza B, Chong PP (2016) Dissecting Candida albicans infection from the perspective of *C. albicans* virulence and omics approaches on host–pathogen interaction: a review. Int J Mol Sci 17(10):1643

Cho HY, Son SY, Rhee HS, Yoon SYH, Lee-Parsons CW, Park JM (2008) Synergistic effects of sequential treatment with methyl jasmonate, salicylic acid and yeast extract on benzophenanthridine alkaloid accumulation and protein expression in Eschscholtzia californica suspension cultures. J Biotechnol 135(1):117–122

Chtioui W, Balmas V, Delogu G, Migheli Q, Oufensou S (2022) Bioprospecting phenols as inhibitors of trichothecene-producing Fusarium: sustainable approaches to the management of wheat pathogens. Toxins 14(2):72

Cordisco E, Sortino M, Svetaz L (2019) Antifungal activity of traditional medicinal plants from Argentina: effect of their combination with antifungal drugs. Curr Tradit Med 5(1):75–95

Curvelo JAR, Marques AM, Barreto ALS, Romanos MTV, Portela MB, Kaplan MAC, Soares RMA (2014) A novel nerolidol-rich essential oil from Piper claussenianum modulates *Candida albicans* biofilm. J Med Microbiol 63(5):697–702

Cutfield SM, Dodson EJ, Anderson BF, Moody PCE, Marshall CJ, Sullivan PA, Cutfield JF (1995) The crystal structure of a major secreted aspartic proteinase from *Candida albicans* in complexes with two inhibitors. Structure 3(11):1261–1271

da Fonseca STD, Teixeira TR, Ferreira JMS, Lima LARDS, Luyten W, Castro AHF (2022) Flavonoid-rich fractions of *Bauhinia holophylla* leaves inhibit *Candida albicans* biofilm formation and hyphae growth. Plan Theory 11(14):1796

Da X, Nishiyama Y, Tie D, Hein KZ, Yamamoto O, Morita E (2019) Antifungal activity and mechanism of action of Ou-gon (Scutellaria root extract) components against pathogenic fungi. Sci Rep 9(1):1683

Darvishi E, Omidi M, Bushehri AA, Golshani A, Smith ML (2013) Thymol antifungal mode of action involves telomerase inhibition. Med Mycol 51(8):826–834

De Toledo LG, Ramos MADS, Spósito L, Castilho EM, Pavan FR, Lopes ÉDO, De Almeida MTG (2016) Essential oil of Cymbopogon nardus (L.) Rendle: a strategy to combat fungal infections caused by Candida species. Int J Mol Sci 17(8):1252

DeJarnette C, Luna-Tapia A, Estredge LR, Palmer GE (2020) Dihydrofolate reductase is a valid target for antifungal development in the human pathogen *Candida albicans*. Msphere 5(3):10–1128

Desai PR, Lengeler K, Kapitan M, Janßen SM, Alepuz P, Jacobsen ID, Ernst JF (2018) The 5′ untranslated region of the EFG1 transcript promotes its translation to regulate hyphal morphogenesis in *Candida albicans*. MSphere 3(4):10–1128

Dhifi W, Bellili S, Jazi S, Bahloul N, Mnif W (2016) Essential oils' chemical characterization and investigation of some biological activities: a critical review. Medicines 3(4):25

Ding C, Butler G (2007) Development of a gene knockout system in Candida parapsilosis reveals a conserved role for BCR1 in biofilm formation. Eukaryot Cell 6(8):1310–1319

Duronio RJ, Towler DA, Heuckeroth RO, Gordon JI (1989) Disruption of the yeast N-myristoyl transferase gene causes recessive lethality. Science 243(4892):796–800

Du H, Guan G, Xie J, Sun Y, Tong Y, Zhang L, Huang G (2012) Roles of *Candida albicans* Gat2, a GATA-type zinc finger transcription factor, in biofilm formation, filamentous growth and virulence. PLoS One 7(1):e29707

El-Baz AM, Mosbah RA, Goda RM, Mansour B, Sultana T, Dahms TE, El-Ganiny AM (2021) Back to nature: combating *Candida albicans* biofilm, phospholipase and hemolysin using plant essential oils. Antibiotics 10(1):81

Elshafie HS, Camele I, Mohamed AA (2023) A comprehensive review on the biological, agricultural and pharmaceutical properties of secondary metabolites based-plant origin. Int J Mol Sci 24(4):3266

Essid R, Hammami M, Gharbi D, Karkouch I, Hamouda TB, Elkahoui S, Tabbene O (2017) Antifungal mechanism of the combination of Cinnamomum verum and Pelargonium graveolens essential oils with fluconazole against pathogenic Candida strains. Appl Microbiol Biotechnol 101(18):6993–7006

Figueiredo AC, Barroso JG, Pedro LG, Scheffer JJ (2008) Factors affecting secondary metabolite production in plants: volatile components and essential oils. Flavour Fragr J 23(4):213–226

Fowler ZL, Shah K, Panepinto JC, Jacobs A, Koffas MA (2011) Development of non-natural flavanones as antimicrobial agents. PLoS One 6(10):e25681

Fromtling RA (1988) Overview of medically important antifungal azole derivatives. Clin Microbiol Rev 1(2):187–217

Georgopapadakou NH, Tkacz JS (1995) The fungal cell wall as a drug target. Trends Microbiol 3(3):98–104

Ghannoum MA (2000) Potential role of phospholipases in virulence and fungal pathogenesis. Clin Microbiol Rev 13(1):122–143

Gupta P, Poluri KM (2021) Elucidating the eradication mechanism of perillyl alcohol against Candida glabrata biofilms: insights into the synergistic effect with azole drugs. ACS Bio Med Chem Au 2(1):60–72

Gupta P, Gupta S, Sharma M, Kumar N, Pruthi V, Poluri KM (2018) Effectiveness of phytoactive molecules on transcriptional expression, biofilm matrix, and cell wall components of Candida glabrata and its clinical isolates. ACS Omega 3(9):12201–12214

Hagiwara D, Sakamoto K, Abe K, Gomi K (2016) Signaling pathways for stress responses and adaptation in Aspergillus species: stress biology in the post-genomic era. Biosci Biotechnol Biochem 80(9):1667–1680

Haque E, Irfan S, Kamil M, Sheikh S, Hasan A, Ahmad A, Mir SS (2016) Terpenoids with antifungal activity trigger mitochondrial dysfunction in Saccharomyces cerevisiae. Microbiology 85:436–443

Hartland RP, Emerson GW, Sullivan PA (1991) A secreted β-glucan-branching enzyme from *Candida albicans*. Proc R Soc London B Biol Sci 246(1316):155–160

He M, Du M, Fan M, Bian Z (2007) In vitro activity of eugenol against *Candida albicans* biofilms. Mycopathologia 163:137–143

Hoyer LL, Green CB, Oh SH, Zhao X (2008) Discovering the secrets of the *Candida albicans* agglutinin-like sequence (ALS) gene family–a sticky pursuit. Med Mycol 46(1):1–15

Hsu CC, Lai WL, Chuang KC, Lee MH, Tsai YC (2013) The inhibitory activity of linalool against the filamentous growth and biofilm formation in *Candida albicans*. Med Mycol 51(5):473–482

Hu J, Shi X, Chen J, Mao X, Zhu L, Yu L, Shi J (2014) Alkaloids from Toddalia asiatica and their cytotoxic, antimicrobial and antifungal activities. Food Chem 148:437–444

Hu Y, Zhang J, Kong W, Zhao G, Yang M (2017) Mechanisms of antifungal and anti-aflatoxigenic properties of essential oil derived from turmeric (Curcuma longa L.) on Aspergillus flavus. Food Chem 220:1–8

Huang T, Jander G, de Vos M (2011) Non-protein amino acids in plant defense against insect herbivores: representative cases and opportunities for further functional analysis. Phytochemistry 72(13):1531–1537

Huang J, Yang L, Zou Y, Luo S, Wang X, Liang Y et al (2021) Antibacterial activity and mechanism of three isomeric terpineols of Cinnamomum longepaniculatum leaf oil. Folia Microbiol 66:59–67

Iraji A, Yazdanpanah S, Alizadeh F, Mirzamohammadi S, Ghasemi Y, Pakshir K et al (2020) Screening the antifungal activities of monoterpenes and their isomers against Candida species. J Appl Microbiol 129(6):1541–1551

İşcan G, İşcan A, Demirci F (2016) Anticandidal effects of thymoquinone: mode of action determined by transmission electron microscopy (TEM). Nat Prod Commun 11(7):1934578X1601100726

Iswarya S, Subha TS, John Sundar V, Gnanamani A (2016) Microsphere formulations of essential oil of Acorus calamus L. controls Aspergillus Niger growth in finished leather at an extended period of storage: featuring extraction, formulation and evaluation. Int J Pharmacognosy 3(7):295–305

Jahanshiri Z, Shams-Ghahfarokhi M, Asghari-Paskiabi F, Saghiri R, Razzaghi-Abyaneh M (2017) α-Bisabolol inhibits Aspergillus fumigatus Af239 growth via affecting microsomalΔ 24-sterol methyltransferase as a crucial enzyme in ergosterol biosynthesis pathway. World J Microbiol Biotechnol 33:1–8

Jin YS (2019) Recent advances in natural antifungal flavonoids and their derivatives. Bioorg Med Chem Lett 29(19):126589

Justice MC, Hsu MJ, Tse B, Ku T, Balkovec J, Schmatz D, Nielsen J (1998) Elongation factor 2 as a novel target for selective inhibition of fungal protein synthesis. J Biol Chem 273(6):3148–3151

Kalagatur NK, Nirmal Ghosh OS, Sundararaj N, Mudili V (2018) Antifungal activity of chitosan nanoparticles encapsulated with Cymbopogon martinii essential oil on plant pathogenic fungi Fusarium graminearum. Front Pharmacol 9:610

Kamal LZM, Adam MAA, Shahpudin SNM, Shuib AN, Sandai R, Hassan NM, Sandai D (2021) Identification of alkaloid compounds Arborinine and Graveoline from Ruta angustifolia (L.) Pers for their antifungal potential against Isocitrate lyase (ICL 1) gene of *Candida albicans*. Mycopathologia 186:221–236

Kathiravan MK, Salake AB, Chothe AS, Dudhe PB, Watode RP, Mukta MS, Gadhwe S (2012) The biology and chemistry of antifungal agents: a review. Bioorg Med Chem 20(19):5678–5698

Khan A, Ahmad A, Khan LA, Manzoor N (2014a) *Ocimum sanctum* (L.) essential oil and its lead molecules induce apoptosis in *Candida albicans*. Res Microbiol 165(6):411–419

Khan A, Ahmad A, Xess I, Khan LA, Manzoor N (2014b) Ocimum sanctum essential oil inhibits virulence attributes in *Candida albicans*. Phytomedicine 21(4):448–452

Kilmartin JV (1981) Purification of yeast tubulin by self-assembly in vitro. Biochemistry 20(12):3629–3633

Kim JY (2016) Human fungal pathogens: why should we learn? J Microbiol 54:145–148

Kim J, Ha Quang Bao T, Shin YK, Kim KY (2018) Antifungal activity of magnoflorine against Candida strains. World J Microbiol Biotechnol 34:1–7

King AM, Reid-Yu SA, Wang W, King DT, De Pascale G, Strynadka NC, Wright GD (2014) Aspergillomarasmine A overcomes metallo-β-lactamase antibiotic resistance. Nature 510(7506):503–506

Kohiyama CY, Ribeiro MMY, Mossini SAG, Bando E, da Silva Bomfim N, Nerilo SB, Machinski M Jr (2015) Antifungal properties and inhibitory effects upon aflatoxin production of Thymus vulgaris L. by Aspergillus flavus link. Food Chem 173:1006–1010

Kohler JR, Casadevall A, Perfect J (2015) The spectrum of fungi that infects humans. Cold Spring Harb Perspect Med 5(1):a019273

LaFleur MD, Kumamoto CA, Lewis K (2006) *Candida albicans* biofilms produce antifungal-tolerant persister cells. Antimicrob Agents Chemother 50(11):3839–3846

Leite MCA, Bezerra APDB, Sousa JPD, Guerra FQS, Lima EDO (2014) Evaluation of antifungal activity and mechanism of action of citral against *Candida albicans*. Evid Based Complement Alternat Med 2014:378280

Lewis RE (2011) Current concepts in antifungal pharmacology. Mayo Clin Proc 86(8):805–817

Li Z, Nielsen K (2017) Morphology changes in human fungal pathogens upon interaction with the host. J Fungi 3(4):66

Li Y, Nie Y, Zhou L, Li S, Tang X, Ding Y, Li S (2014) The possible mechanism of antifungal activity of cinnamon oil against Rhizopus nigricans. J Chem Pharm Res 6(5):12–20

Li X, Wu X, Gao Y, Hao L (2019) Synergistic effects and mechanisms of combined treatment with harmine hydrochloride and azoles for resistant *Candida albicans*. Front Microbiol 10:2295
Lima B, Sortino M, Tapia A, Feresin GE (2022) Synergistic antifungal effectiveness of essential oils from andean plants combined with commercial drugs. Int J Pharm Sci Dev Res 8(1):23–31
Lin J, Oh SH, Jones R, Garnett JA, Salgado PS, Rusnakova S et al (2014) The peptide-binding cavity is essential for Als3-mediated adhesion of *Candida albicans* to human cells. J Biol Chem 289(26):18401–18412
Lionakis MS, Drummond RA, Hohl TM (2023) Immune responses to human fungal pathogens and therapeutic prospects. Nat Rev Immunol 23:433–452
Liu RH, Shang ZC, Li TX, Yang MH, Kong LY (2017) In vitro antibiofilm activity of eucarobustol E against *Candida albicans*. Antimicrob Agents Chemother 61(8):10–1128
Lodge JK, Jackson-Machelski E, Devadas B, Zupec ME, Getman DP, Kishore N, Gordon JI (1997) N-myristoylation of Arf proteins in *Candida albicans*: an in vivo assay for evaluating antifungal inhibitors of myristoyl-CoA: protein N-myristoyltransferase. Microbiology 143(2):357–366
Lone SA, Khan S, Ahmad A (2020) Inhibition of ergosterol synthesis in *Candida albicans* by novel eugenol tosylate congeners targeting sterol 14α-demethylase (CYP51) enzyme. Arch Microbiol 202(4):711–726
Ma Y, Wang L, Lu A, Xue W (2022) Synthesis and biological activity of novel oxazinyl flavonoids as antiviral and anti-phytopathogenic fungus agents. Molecules 27(20):6875
Mani-López E, Cortés-Zavaleta O, López-Malo A (2021) A review of the methods used to determine the target site or the mechanism of action of essential oils and their components against fungi. SN Appl Sci 3:1–25
Maoz M, Neeman I (2000) Effect of Inula viscosa extract on chitin synthesis in dermatophytes and *Candida albicans*. J Ethnopharmacol 71(3):479–482
Masyita A, Sari RM, Astuti AD, Yasir B, Rumata NR, Emran TB et al (2022) Terpenes and terpenoids as main bioactive compounds of essential oils, their roles in human health and potential application as natural food preservatives. Food Chem X 13:100217
McGinnis MR, Tyring SK (1996) Introduction to mycology. In: Baron S (ed) Medical microbiology. University of Texas Medical Branch at Galveston, Galveston
Mironescu M, Mironescu ID, Georgescu C (2010) Microstructural changes induced by five new biocidal formulations on moulds. Ann Rom Soc Cell Biol 40:162–167
Mulhern SM, Logue ME, Butler G (2006) *Candida albicans* transcription factor Ace2 regulates metabolism and is required for filamentation in hypoxic conditions. Eukaryot Cell 5(12):2001–2013
Muthamil S, Balasubramaniam B, Balamurugan K, Pandian SK (2018) Synergistic effect of quinic acid derived from Syzygium cumini and undecanoic acid against Candida spp. biofilm and virulence. Front Microbiol 9:2835
Nazzaro F, Fratianni F, Coppola R, De Feo V (2017) Essential oils and antifungal activity. Pharmaceuticals 10(4):86
Nimrichter L, Rodrigues ML (2011) Fungal glucosylceramides: from structural components to biologically active targets of new antimicrobials. Front Microbiol 2:212
Noriega P (2021) Terpenes in essential oils: bioactivity and applications. IntechOpen, London. https://doi.org/10.5772/intechopen.93792
Nucci M, Marr KA (2005) Emerging fungal diseases. Clin Infect Dis 41(4):521–526
Odds FC (2003) Synergy, antagonism, and what the chequerboard puts between them. J Antimicrob Chemother 52(1):1–1
Ogidi CO, Ojo AE, Ajayi-Moses OB, Aladejana OM, Thonda OA, Akinyele BJ (2021) Synergistic antifungal evaluation of over-the-counter antifungal creams with turmeric essential oil or Aloe vera gel against pathogenic fungi. BMC Complement Med Ther 21:1–12
Onyewu C, Blankenship JR, Del Poeta M, Heitman J (2003) Ergosterol biosynthesis inhibitors become fungicidal when combined with calcineurin inhibitors against *Candida albicans*, *Candida glabrata*, and *Candida krusei*. Antimicrob Agents Chemother 47(3):956–964
Ootani MA, Aguiar RW, Ramos ACC, Brito DR, Silva JBD, Cajazeira JP (2013) Use of essential oils in agriculture. J Biotechnol Biodivers 4(2):162–174

Ostrosky-Zeichner L, Casadevall A, Galgiani JN, Odds FC, Rex JH (2010) An insight into the antifungal pipeline: selected new molecules and beyond. Nat Rev Drug Discov 9(9):719–727

Pandey AK, Kumar P, Singh P, Tripathi NN, Bajpai VK (2017) Essential oils: sources of antimicrobials and food preservatives. Front Microbiol 7:2161

Patton JL, Lester RL (1991) The phosphoinositol sphingolipids of Saccharomyces cerevisiae are highly localized in the plasma membrane. J Bacteriol 173:3101–3108

Pemmaraju SC, Pruthi PA, Prasad R, Pruthi V (2013) *Candida albicans* biofilm inhibition by synergistic action of terpenes and fluconazole. Indian J Exp Biol 51(11):1032–1037

Pereira FDO, Mendes JM, de Oliveira Lima E (2013) Investigation on mechanism of antifungal activity of eugenol against Trichophyton rubrum. Med Mycol 51(5):507–513

Pereira FDO, Mendes JM, Lima IO, Mota KSDL, Oliveira WAD, Lima EDO (2015) Antifungal activity of geraniol and citronellol, two monoterpenes alcohols, against Trichophyton rubrum involves inhibition of ergosterol biosynthesis. Pharm Biol 53(2):228–234

Pinto E, Pina-Vaz C, Salgueiro L, Gonçalves MJ, Costa-de-Oliveira S, Cavaleiro C, Martinez-de-Oliveira J (2006) Antifungal activity of the essential oil of Thymus pulegioides on Candida, Aspergillus and dermatophyte species. J Med Microbiol 55(10):1367–1373

Pinto E, Hrimpeng K, Lopes G, Vaz S, Gonçalves MJ, Cavaleiro C, Salgueiro L (2013) Antifungal activity of Ferulago capillaris essential oil against Candida, Cryptococcus, Aspergillus and dermatophyte species. Eur J Clin Microbiol Infect Dis 32:1311–1320

Polak-Wyss A (1995) Mechanism of action of antifungals and combination therapy. J Eur Acad Dermatol Venereol 4:S11–S16

Ponde NO, Lortal L, Ramage G, Naglik JR, Richardson JP (2021) *Candida albicans* biofilms and polymicrobial interactions. Crit Rev Microbiol 47(1):91–111

Pozniakovsky AI, Knorre DA, Markova OV, Hyman AA, Skulachev VP, Severin FF (2005) Role of mitochondria in the pheromone-and amiodarone-induced programmed death of yeast. J Cell Biol 168(2):257–269

Rachitha P, Krupashree K, Jayashree GV, Gopalan N, Khanum F (2017) Growth inhibition and morphological alteration of Fusarium sporotrichioides by Mentha piperita essential oil. Pharm Res 9(1):74

Roemer T, Boone C (2013) Systems-level antimicrobial drug and drug synergy discovery. Nat Chem Biol 9(4):222–231

Ruiz-Herrera J, San-Blas G (2003) Chitin synthesis as a target for antifungal drugs. Curr Drug Targets Infect Disord 3(1):77–91

Rukayadi Y, Hwang JK (2013) In vitro activity of xanthorrhizol isolated from the rhizome of Javanese turmeric (Curcuma xanthorrhiza Roxb.) against *Candida albicans* biofilms. Phytother Res 27(7):1061–1066

Schaller M, Borelli C, Korting HC, Hube B (2005) Hydrolytic enzymes as virulence factors of *Candida albicans*. Mycoses 48(6):365–377

Shang XF, Morris-Natschke SL, Liu YQ, Guo X, Xu XS, Goto M, Lee KH (2018) Biologically active quinoline and quinazoline alkaloids part I. Med Res Rev 38(3):775–828

Sharma Y, Khan LA, Manzoor N (2016) Anti-*Candida* activity of geraniol involves disruption of cell membrane integrity and function. Journal de mycologie medicale 26(3):244–254

Sharma Y, Kumar Rastogi S, Saadallah Amin Saadallah S, Al Fadel K, Manzoor N (2018) Anti-Candida activity of geraniol: effect on hydrolytic enzyme secretion and biofilm formation. J Pure Appl Microbiol 12(3):1337–1349

Sharma Y, Rastogi SK, Perwez A, Rizvi MA, Manzoor N (2020) β-citronellol alters cell surface properties of *C andida albicans* to influence pathogenicity related traits. Medical mycology 58(1):93–106

Shen Q, Zhou W, Li H, Hu L, Mo H (2016) ROS involves the fungicidal actions of thymol against spores of Aspergillus flavus via the induction of nitric oxide. PLoS One 11(5):e0155647

Shreaz S, Bhatia R, Khan N, Muralidhar S, Manzoor N, Khan LA (2013) Influences of cinnamic aldehydes on H+ extrusion activity and ultrastructure of Candida. J Med Microbiol 62(2):232–240

Singh PP, Jaiswal AK, Kumar A, Gupta V, Prakash B (2021) Untangling the multi-regime molecular mechanism of verbenol-chemotype Zingiber officinale essential oil against Aspergillus flavus and aflatoxin B1. Sci Rep 11(1):6832

Srivastava AK, Singh VK (2019) Biological action of essential oils (terpenes). Int J Biol Med Res 10(3):6854–6859

Stringaro A, Vavala E, Colone M, Pepi F, Mignogna G, Garzoli S, Angiolella L (2014) Effects of Mentha suaveolens essential oil alone or in combination with other drugs in *Candida albicans*. Evid Based Complement Alternat Med 2014:125904

Temba BA, Fletcher MT, Fox GP, Harvey J, Okoth SA, Sultanbawa Y (2019) Curcumin-based photosensitization inactivates Aspergillus flavus and reduces aflatoxin B1 in maize kernels. Food Microbiol 82:82–88

Thaweboon S, Thaweboon B (2009) In vitro antimicrobial activity of Ocimum americanum L. essential oil against oral microorganisms. Southeast Asian J Trop Med Public Health 40(5):1025–1033

Tian J, Ban X, Zeng H, He J, Chen Y, Wang Y (2012a) The mechanism of antifungal action of essential oil from dill (Anethum graveolens L.) on Aspergillus flavus. PLoS One 7(1):e30147

Tian J, Huang B, Luo X, Zeng H, Ban X, He J, Wang Y (2012b) The control of Aspergillus flavus with Cinnamomum jensenianum Hand.-Mazz essential oil and its potential use as a food preservative. Food Chem 130(3):520–527

Tungmunnithum D, Thongboonyou A, Pholboon A, Yangsabai A (2018) Flavonoids and other phenolic compounds from medicinal plants for pharmaceutical and medical aspects: an overview. Medicines 5(3):93

Ueki M, Abe K, Hanafi M, Shibata K, Tanaka T, Taniguchi M (1996) UK-2A, B, C and D, novel antifungal antibiotics from Streptomyces sp. 517-02 I. Fermentation, isolation, and biological properties. J Antibiot 49(7):639–643

Ueki M, Taniguchi M (1997) The mode of action of UK-2A and UK-3A, novel antifungal antibiotics from *Streptomyces* sp. 517-02. The Journal of Antibiotics 50(12):1052–1057

Ui M, Okada T, Hazeki K, Hazeki O (1995) Wortmannin as a unique probe for an intracellular signalling protein, phosphoinositide 3-kinase. Trends Biochem Sci 20(8):303–307

Vermes A, Guchelaar HJ, Dankert J (2000) Flucytosine: a review of its pharmacology, clinical indications, pharmacokinetics, toxicity and drug interactions. J Antimicrob Chemother 46(2):171–179

Vijayakumar EKS, Roy K, Chatterjee S, Deshmukh SK, Ganguli BN, Fehlhaber HW, Kogler H (1996) Arthrichitin. A new cell wall active metabolite from Arthrinium phaeospermum. J Org Chem 61(19):6591–6593

Villa S, Hamideh M, Weinstock A, Qasim MN, Hazbun TR, Sellam A et al (2020) Transcriptional control of hyphal morphogenesis in *Candida albicans*. FEMS Yeast Res 20(1):foaa005

Wang C, Zhang J, Chen H, Fan Y, Shi Z (2010) Antifungal activity of eugenol against Botrytis cinerea. Trop Plant Pathol 35:137–143

Wang L, Hu W, Deng J, Liu X, Zhou J, Li X (2019) Antibacterial activity of Litsea cubeba essential oil and its mechanism against Botrytis cinerea. RSC Adv 9(50):28987–28995

Wang F, Chen L, Chen S, Chen H, Liu Y (2021a) Microbial biotransformation of Pericarpium Citri Reticulatae (PCR) by Aspergillus Niger and effects on antioxidant activity. Food Sci Nutr 9(2):855–865

Wang Z, Sun Q, Zhang H, Wang J, Fu Q, Qiao H, Wang Q (2021b) Insight into antibacterial mechanism of polysaccharides: a review. LWT 150:111929

Watt PM, Hickson ID (1994) Structure and function of type II DNA topoisomerases. Biochem J 303(Pt 3):681

Whitesell L, Mimnaugh EG, De Costa B, Myers CE, Neckers LM (1994) Inhibition of heat shock protein HSP90-pp60v-src heteroprotein complex formation by benzoquinone ansamycins: essential role for stress proteins in oncogenic transformation. Proc Natl Acad Sci 91(18):8324–8328

Wu XZ, Cheng AX, Sun LM, Lou HX (2008) Effect of plagiochin E, an antifungal macrocyclic bis (bibenzyl), on cell wall chitin synthesis in *Candida albicans*. Acta Pharmacol Sin 29(12):1478–1485

Yang S, Fu Y, Wu X, Zhou Z, Xu J, Zeng X, Zeng Y (2014) Baicalin prevents *Candida albicans* infections via increasing its apoptosis rate. Biochemical and biophysical research communications 451(1):36–41

Yazaki K, Arimura GI, Ohnishi T (2017) 'Hidden' terpenoids in plants: their biosynthesis, localization and ecological roles. Plant Cell Physiol 58(10):1615–1621

Yu D, Wang J, Shao X, Xu F, Wang H (2015) Antifungal modes of action of tea tree oil and its two characteristic components against Botrytis cinerea. J Appl Microbiol 119(5):1253–1262

Yutani M, Hashimoto Y, Ogita A, Kubo I, Tanaka T, Fujita KI (2011) Morphological changes of the filamentous fungus Mucor mucedo and inhibition of chitin synthase activity induced by anethole. Phytother Res 25(11):1707–1713

Zahi MR, El Hattab M, Liang H, Yuan Q (2017) Enhancing the antimicrobial activity of d-limonene nanoemulsion with the inclusion of ε-polylysine. Food Chem 221:18–23

Zeng H, Chen X, Liang J (2015) In vitro antifungal activity and mechanism of essential oil from fennel (Foeniculum vulgare L.) on dermatophyte species. J Med Microbiol 64(1):93–103

Zeng H, Li T, Ding H, Tian J (2016) Anti-cryptococcus activity and mechanism of essential oil from Coreopsis tinctoria flowering on Cryptococcus. Fungal Genom Biol 6(1):132

Zimmermann GR, Lehar J, Keith CT (2007) Multi-target therapeutics: when the whole is greater than the sum of the parts. Drug Discov Today 12(1–2):34–42

Zorova LD, Popkov VA, Plotnikov EJ, Silachev DN, Pevzner IB, Jankauskas SS, Zorov DB (2018) Functional significance of the mitochondrial membrane potential. Biochem Moscow Suppl Ser A 12:20–26

Part IV

Plant-Based Nanoparticles as Antifungals

Exploration of New Plant-Based Nanoparticles with Potential Antifungal Activity and their Mode of Action

12

Hardeep Kaur and Khushbu Wadhwa

12.1 Introduction

Candidiasis is one of the most common fungal infections, generally caused by *C. albicans* and non-*albicans Candida* species. Normally *Candida* is defined as a commensal organism but can turn into an opportunistic pathogen in an immunocompromised patient (Gonçalves et al. 2016). *Candida* infections can be systemic resulting into life threatening candidemia (Bhattacharya et al. 2020). The annual incidence of *Candida* infections is approximately four million cases worldwide (Bongomin et al. 2017). Among all the etiologic agents of candidiasis, *C. albicans* is considered as one of the major pathogens responsible for fungal infections causing sepsis and is also associated with 40% of the mortality cases (Xiao et al. 2019). The colonization and infection by *C. albicans* are determined by many attributes including, their ability to grow at 37 °C, morphological switching from yeast to virulent hyphae form, secretion of hydrolytic enzymes (phospholipases, proteinases, lipases), tissue adhesion, invasion, hemolytic activity, ability to escape from the host immune system and biofilm formation (Mba et al. 2022). For the treatment of fungal infections, several commercially available antifungal drugs are used, including azoles (fluconazole, itraconazole, voriconazole and posaconazole), polyenes (amphotericin B) and echinocandins (caspofungin, micafungin and anidulafungin). However, these antifungal drugs exhibit severe toxicity in host body, have limited spectrum, can cause relapse infections and have high cost (Tragiannidis et al. 2021; Peron et al. 2016). Moreover, emergence of resistant strains can complicate the treatment leading to more frequent hospital visits of patient with

H. Kaur (✉) · K. Wadhwa
Department of Zoology, Fungal Biology Laboratory, Ramjas College, University of Delhi, New Delhi, Delhi, India
e-mail: hardeepkaur@ramjas.du.ac.in; khushbuwadhwa9@gmail.com

N. Manzoor (ed.), *Advances in Antifungal Drug Development*,
https://doi.org/10.1007/978-981-97-5165-5_12

subsequent increase in the cost of treatment (Benedict et al. 2019). Furthermore, antifungal resistance is considered as severe clinical problem that receives major attention due to the emergence of clinically resistant pathogenic strains (Peron et al. 2016). Considering these above points, it is very important to develop effective antifungal agents that should be able to overcome all the challenges of commercially used drugs. Nanoparticles based antifungal drugs have several advantages with better therapeutic value. Among all nanomaterials, metallic nanoparticles are considered to be much superior as they exhibit larger surface area to volume ratio, higher biocompatibility, higher stability, higher surface charge, crystalline and amorphous structures and their size ranges from 10 to 100 nm (Salavati-Niasari et al. 2008). Metallic nanoparticles are widely studied to explore their antifungal mode of action against different fungal pathogens (Bansal et al. 2020). In this chapter, we have discussed different types of metal nanoparticles and their synthesis from plant extracts, their mechanism of action and their ability to treat fungal infections caused by *Candida* species (Table 12.1).

12.2 Green Synthesis of Metal Nanoparticles

Green synthesis is considered as a promising approach in the field of nanobiotechnology. Generally, there are three methods for the synthesis of nanoparticles including physical methods, chemical methods and biological methods. The synthesis of silver nanoparticles can be done by employing physical methods comprising of evaporation-condensation and laser ablation methods; or chemical reduction methods using inorganic and organic reducing agents, namely polyethylene glycol, sodium borohydride, N-dimethylformamide and hydrazine (Iravani et al. 2014). These two methods of synthesis are effective in nature but utilize hazardous chemicals and can release toxic by-products into the environment. On the other hand, biological methods utilize safe and ecofriendly means. Moreover, these methods require low cost of production, less energy and simple routes without much involvement of machines. Further, there are three methods of nanoparticle synthesis from biosynthetic pathways, which include microorganisms, plants and different types of biomolecules (Banasiuk et al. 2020). Biologically synthesized nanoparticles exhibit wide applications in the field of catalysis, biomedical and environmental sciences, cosmetics and space industries, health care, optics, drug delivery and food industries (Vinatier et al. 2009). The use of plant extracts ensures decline in the usage of toxic chemicals (reducing agents and stabilizers) for the synthesis of nanoparticles. Presence of various phytochemicals in the plant extracts help to enhance the stability of nanoparticles by acting as bio-capping and bio-reducing agents in the fabrication process. Various types of plant tissues are used for the synthesis of different types of metallic nanoparticles (Fig. 12.1).

Table 12.1 List of plants used in the preparation of nanoparticles with antifungal activity against *Candida* spp.

S. No.	Name of plant	Plant parts used	Size of nanoparticle (nm)	Shape of nanoparticle	Evaluation method used	Species of fungi tested	References
Silver nanoparticles (Ag-NPs)							
1.	*Annona muricata*	Leaf	22–28	Spherical	Agar well diffusion	*C. albicans*	Akintelu et al. (2019)
2.	*Cassia fistula*	Leaf extract	40–50	Spherical	Agar well diffusion	*C. krusei*	Mohanta et al. (2016)
3.	*Aloe vera*	Aqueous leaf extract	80.31 ± 10.03	Cubic	Agar well diffusion	Clinical strains of *C. albicans* (C1, C2, C3 and C4) and Reference strain *C. albicans* ATCC 10231	Arsène et al. (2023)
4.	*Calotropis giganteam*	Aqueous leaf extract	10–70	Spherical	Disc diffusion	*C. albicans* isolated from sputum samples	Ali and Abdallah (2020)
5.	*Lotus lalambensis*	Aqueous leaf extract	6–26	Spherical	Disc diffusion	*C. albicans*	Abdallah and Ali (2021)
6.	*Encephalartos laurentianus*	Aqueous leaf extract	13.04–23	Spherical	Disc diffusion	*C. albicans* clinical isolates	Alherz et al. (2022)
7.	*Syzygium cumini*	Seed extract	10–100	Hexagonal	Agar well diffusion and MIC by microbroth dilution method	Clinical isolates of *Candida* (*C. albicans*, *C. tropicalis*, *C. dubliniensis*, *C. parapsilosis* and *C. krusei*)	Jalal et al. (2019)
8.	*Erodium glaucophyllum*	Leaf extract	50–100	Irregular spherical forms	Agar well diffusion and MIC by microbroth dilution method	*C. albicans* isolated from oral cavity	Abdallah and Ali (2022)
9.	*Angelica gigas*	Hot melt extrusion-Angelica gigas Nakai (HME-AGN)	<100	Spherical	Colony forming unit assays	*C. albicans* (KCTC 7965/ ATCC 10231)	Ryu et al. (2022)

(continued)

Table 12.1 (continued)

S. No.	Name of plant	Plant parts used	Size of nanoparticle (nm)	Shape of nanoparticle	Evaluation method used	Species of fungi tested	References
10.	*Nerium oleander*	Aqueous leaf extract	37.5–75	Spherical	Agar well diffusion method and MIC by microbroth dilution method	*C. albicans*	Hadi et al. (2020)
11	*Caesalpinia ferrea*	Seed extract	30–50	Spherical	MIC by microbroth dilution method	*C. albicans* ATCC 10231, *C. krusei* (FTI) Collection of Tropical Cultures (CCT) 1517, *C. glabrata* (Taniwaki, M.H.) (CCT) O728, and *C. guilliermondii* (CCT), 1890	Soares et al. (2018)
12	*Enantia chloranta*	Bark extract	–	–	Agar well diffusion and MIC by microbroth dilution method	*C. albicans* clinical isolate and *C. albicans* ATCC 10231	Arsene et al. (2022)
13	*Lamium album*	Aqueous leaf extract	10–40.5	Spherical	MIC by microbroth dilution method	*C. albicans* ATCC 10261, *C. krusei* ATCC 6258, *C. glabrata* ATCC 90030, *C. tropicalis* ATCC 750, *C. parapsilosis* ATCC 4344, *C. dubliniensis* CBS 8501	Zareshahrabadi et al. (2021)
14	Tulsi plant *Ocimum tenuiflorum*	Aqueous leaf extract	2–7	Spherical	MIC by microbroth dilution method	*C. albicans, C. glabrata, C. tropicalis, C. parapsilosis, C. krusei, C. guilliermondii, C. lusitaniae, C. famata, C. kefyr, C. norvegensis, C. utilis*	Khatoon et al. (2015)

15	*Camellia sinensis*	Aqueous tea powder	52	Aggregate of varying shapes	Agar well diffusion and MIC by microbroth dilution method	*C. albicans, C. glabrata, C. parapsilosis, C. tropicalis*	Ali et al. (2022)
16	*Alhagi graecorum*	Aqueous leaf extract	22–36	Spherical	Agar well diffusion	*C. albicans, C. glabrata, C. parapsilosis, C. tropicalis, C. krusei*	Hawar et al. (2022)
17	*Furcraea foetida*	Aqueous extract of roots	15	Spherical	Agar well diffusion	*C. albicans* (183), *C. glabrata* (3019), *C. parapsilosis* (7043), *C. tropicalis* (184), *C. krusei* (9215) acquired from MTCC and Gene Bank	Sitrarasi et al. (2022)
18	*Sesuvium portulacastrum*	Callus and leaf extracts	5–20	Spherical	Disc diffusion	*C. albicans*	Nabikhan et al. (2010)
19	*Punica granatum*	Peel extract	18.425 ± 1.12	Spherical	Disc diffusion and MIC by microbroth dilution method	*C. albicans* ATCC 18804, *C. glabrata* ATCC 15545, and *C. tropicalis* ATCC 13803	Yassin et al. (2022)
20	*Casuarina equisetifolia*	Aqueous extract of leaf and fruit	90.97 and 71.28	Spherical	Agar well diffusion	*C. albicans*	Moustafa et al. (2021)
21	*Hyptis suaveolens*	Aqueous leaf extract	2–85	Spherical	Disc diffusion and MIC by microbroth dilution method	Two clinical strains of *C. albicans* 5314 and 10,261	Malathi et al. (2022)
22	*Azadirachta indica*	Crude latex	17.4–40.9	Triangular, tetragonal, pentagonal and hexagonal	Disc diffusion assay and antibiofilm assay	Fluconazole-resistant clinical isolate of *C. tropicalis*	Al Aboody (2019)

(continued)

Table 12.1 (continued)

S. No.	Name of plant	Plant parts used	Size of nanoparticle (nm)	Shape of nanoparticle	Evaluation method used	Species of fungi tested	References
23	*Terminalia catappa*	Aqueous leaf extract	10.06 ± 0.84	Spherical	Well diffusion and MIC by microbroth dilution method	*C. albicans*	Ansari et al. (2021)
24	*Borojoa patinoi*	*Borojoa patinoi* extract	Size of Ag-NP-43 is 18–45 nm (for spherical shapes) and 18–45 nm for rod shapes, size of Ag-NP-31 is 20.6–330 nm	Spherical and rod shape	Inhibitory effect on filamentation was studied by calcofluor white dye method	*C. albicans* ATCC 10231, *C. albicans* SC5314	Gómez-Garzón et al. (2021)
25	*Tagetes erecta*	Aqueous flower extract	10–90	Spherical and hexagonal	Disc diffusion method	*C. albicans* and *C. glabrata*	Padalia et al. (2015)
Gold nanoparticles (AuNPs)							
26	*Crinum latifolium*	Leaf extract	1–10	Spherical	Well diffusion and MIC by microbroth dilution method	*C. albicans*, *C. glabrata*, *C. tropicalis*, *C. parapsilosis*, *C. krusei*, *C. dubliniensis*	Jalal et al. (2023)
27	*Abelmoschus esculentus*	Seed extract	62	Spherical	Well diffusion	*C. albicans*	Jayaseelan et al. (2013)
28	*Justicia glauca*	Aqueous leaf extract	32.5 ± 0.25	Hexagonal and spherical shaped	Well diffusion	*C. albicans*	Emmanuel et al. (2017)
Iron oxide nanoparticles (IONPs)							
29	*Psidium guajava (guava)*	Aqueous guava extract	12.64	Almost cubic and partly spherical	Agar well diffusion method	*C. tropicalis* and *C. albicans*	Adhikari et al. (2022)
Copper oxide nanoparticles (CuONPs)							
30	*Thespesia populnea*	Aqueous bark extract	61–69	Spherical	Disc diffusion method	*C. albicans* MTCC 183	Narayanan et al. (2022)

31	*Acalypha indica*	Leaf extract	26–30	Spherical	Well diffusion	*C. albicans*	Sivaraj et al. (2014)
32	*Viburnum opulus*	*Viburnum opulus* methanolic extract	–	Spherical	Disc diffusion and MIC by microbroth dilution method	*C. albicans* ATCC 10231 and *C. glabrata* ATCC 90030	Ildiz et al. (2017)
Palladium nanoparticles (PdNPs)							
33	*Bauhinia variegata*	Aqueous bark extract	2–9	Irregular shape (mostly cylindrical)	Agar well diffusion method	*C. albicans* MTCC 183	Vaghela et al. (2018)
34	*Rosmarinus officinalis*	Leaf extract	–	–	Disc diffusion	*C. parapsilosis*, *C. glabrata*, *C. krusei*, *C. albicans*	Rabiee et al. (2020)
Aluminum oxide nanoparticles (Al_2O_3NP)							
35	*Cymbopogon citratus*	Aqueous extract of lemongrass leaves	Average diameter 58.5	Spherical	Well diffusion and MIC by microbroth dilution method	Fluconazole-resistant *Candida* spp.	Jalal et al. (2016)
36	*Citrus aurantium*	Peels of the fruit	28	Spherical	Agar well diffusion	*C. albicans* MTCC 227	Nagarajan et al. (2023)
Selenium oxide nanoparticles (SeNP)							
37	*Spirulina platensis*	*Spirulina platensis* extracts	65.23	Spherical	Disc diffusion and MIC by microbroth dilution method	*C. tropicalis*, *C. albicans*, *C. glabrata*	Abdel-Moneim et al. (2022)
38	*Portulaca oleracea*	Aqueous leaf extract	2–22	Spherical	Agar well diffusion	Clinical *Candida* spp.	Fouda et al. (2022)
39	*Urtica dioica* (Stinging nettle)	Leaf extract	–	Spherical	MIC by microbroth dilution method	*C. albicans*	Hashem and Salem (2022)
Zinc oxide nanoparticles (ZnONP)							

(continued)

Table 12.1 (continued)

S. No.	Name of plant	Plant parts used	Size of nanoparticle (nm)	Shape of nanoparticle	Evaluation method used	Species of fungi tested	References
40	*Pongamia pinnata*	Aqueous seed extracts	30.4–40.8	Spherical	Antibiofilm activity against *C. albicans* MTCC 3017	*C. albicans* MTCC 3017	Malaikozhundan et al. (2017)
41	*Aspalathus linearis* and *Musa paradisiaca*	Banana peel extract and Rooibos and Buchu tea leaves	–	Rod-shaped	Disc diffusion	*C. albicans* ATCC 90028	Lyimo et al. (2022)
42	*Atalantia monophylla*	Aqueous leaf extract	30	Spherical and hexagonal	Agar well diffusion method	*C. albicans* MTCC 227	Vijayakumar et al. (2018a)
43	*Chelidonium majus*	*Chelidonium majus* extract	>10	Spherical	MIC and MFC by microbroth dilution method	*C. albicans* ATCC 10231	Dobrucka et al. (2018)
44	*Glycosmis pentaphylla*	Aqueous leaf extract	32–40	Spherical	Agar well diffusion	*C. albicans* MTCC 227	Vijayakumar et al. (2018b)
45	*Azadirachta indica*	Leaf extract	20	Spherical	Disc diffusion and MIC by microbroth dilution method	*C. albicans* and *C. tropicalis*	Elumalai and Velmurugan (2015)
46	*Cassia auriculata*	Leaf extract	68.64	Spherical	Agar well diffusion, synergistic activity, disc diffusion	*C. albicans* ATCC 2091 and *C. glabrata* NCIM 3438	Padalia et al. (2018)
47	*Vinca rosea*	Leaf extract	21	Hexagonal crystals	Disc diffusion	*C. albicans*	Chandrasekaran and Anbazhagan (2024)
Monometallic nanoparticles							
48	*Moringa oleifera*	Flower extract	10–50	Spherical	Agar well diffusion	*C. albicans*, *C. utilis*	Anand et al. (2016)
Bimetallic nanoparticles							

49	*Beta vulgaris*	*Beta vulgaris* aqueous extract	33	Semispherical morphology	MIC by microbroth dilution method	Four clinical Fluconazole (FLC) susceptible strains of *C. albicans* (4554, 4251, 4175, 4180) and four clinical FLC-resistant strains (4324, 4106, 5112, 4085) and *C. albicans* (SC5314) (Laboratory control strain)	Kamli et al. (2021)
50	*Ephedra aphylla*	Stem extract	13.95–26.26	Spherical and tetragonal shape	Agar well diffusion	*C. albicans* EMCC number-105	El-Zayat et al. (2021)

MIC minimum inhibitory concentration, *MFC* minimal fungicidal concentration, *FLC* fluconazole, *MTCC* microbial type culture collection, *ATCC* American type culture collection

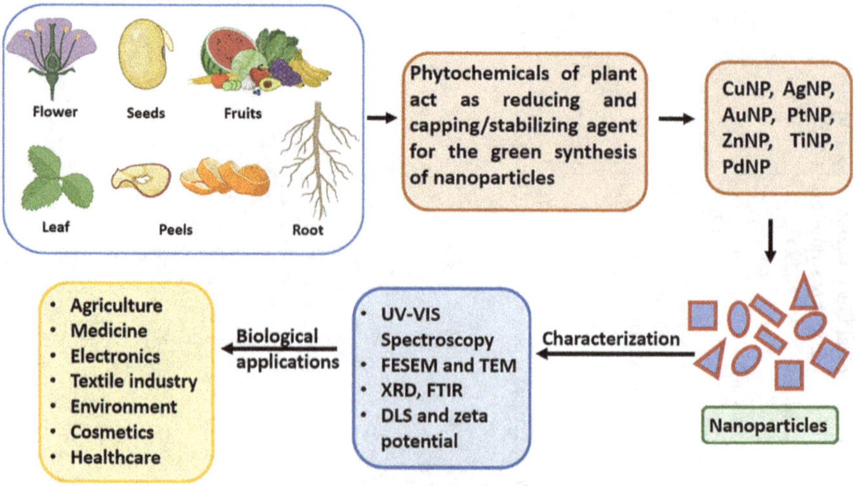

Fig. 12.1 Schematic representation of biological synthesis of nanoparticles and their uses

12.3 Possible Mechanism of Plant-Mediated Synthesis of Nanoparticles

Fabrication of metal nanoparticles requires three major constituents such as reducing agent, stabilizing agent and solvent medium. There are many reports that suggest different types of mechanisms responsible for the synthesis of nanoparticles from plant extracts (Rai et al. 2008; Thakkar et al. 2010). However, due to the presence of various types of phytochemicals in the plant extract, it is very difficult to identify which phytochemical is responsible for the reduction and stabilization process. Phytocompounds such as polyphenols (flavonoids, terpenoids and phenolic acids), organic acids and proteins are defined to be the major constituents that are responsible for the fabrication and reduction of nanoparticles. It has been reported that all the phytocompounds act in a synergistic manner for the bio-reduction of metallic ions into nanoparticles (Shankar et al. 2004). In one such study (Sathishkumar et al. 2009a), it was reported that both polyols and polysaccharides present in the bark extract of *Cinnamom zeylanicum* work together for the reduction of metallic ions. The role of some of the phytochemicals in the bio-reduction and stabilization process of nanoparticle synthesis is described as follows:

12.3.1 Flavonoids

Flavonoids are defined as water-soluble secondary metabolites. They comprise of six major subgroups including: anthoxanthins, flavanones, flavanonols, flavans, anthocyanidins and isoflavonoids. Flavonoids are supposed to play a major role in the biosynthesis of metallic nanoparticles from plant extracts. The oxygen scavenging capacity of flavonoids is dependent upon their ability to donate electron and hydrogen atoms. As per one study (Ahmad et al. 2010), a total of eight hydrogen atoms are released during the keto-enol conversion of flavonoids (luteolin and rosmarinic acid) that are responsible for the synthesis of metal nanoparticles. In addition to this, it was also reported (Ghoreishi et al. 2011) that hydroxyl group of flavonoids (quercetin and myricetin) is oxidized to carbonyl group during reduction of metal ions.

12.3.2 Phenolic Acids

Phenolic acids belong to the family of polyphenols. Antioxidant activities of these compounds are related to the metal chelating ability of highly nucleophilic aromatic rings of phenolic acids. Different types of plant phenolic acids, including gallic acid, caffeic acid, ellagic acid and protocatechuic acid, are identified as bio-reducing agents for the fabrication of metallic nanoparticles. It has been reported (Edison and Sethuraman 2012) that synthesis of silver nanoparticles (Ag-NPs) occurs due to the formation of an intermediate complex of silver ions with the phenolic hydroxyl groups of gallic acid, which undergoes further oxidation to quinone that results in the reduction of silver ions to Ag-NPs. In order to understand the mechanism of green-synthesized nanoparticles, Liu et al. (2020) have observed that polyphenols play major role in the synthesis of gold nanoparticles (AuNPs). The polyphenols when compared to all the components of plant extract, help in both capping and reducing activities. The study also examined the relationship between polyphenols concentration and their capping and reductive ability. Pyrogallic acid was used in this study to understand the functional mechanism of phenolic acids. The concentration of polyphenols was found to be an important factor for the synthesis of AuNPs and positively correlates with the yield of AuNPs, thus indicating their capability to reduce Au^{3+} to Au^0. The authors also examined the role of flavonoids, sugars and proteins in the synthesis of AuNPs. From the statistical data and analysis of the results, it was observed that different components of plant extracts play a synergistic role in the synthesis of AuNPs but polyphenols were found to play lead role in the reduction, and the order of the reductive capability was shown as follows: polyphenols>flavonoids> sugars>proteins. The proteins were found to have very less reductive ability as compared to others because of their large size. As proteins have large volume, they exhibit steric hindrance while trying to contact the active reduction sites. Therefore, proteins in plant extracts do not play a crucial role in the reduction process for the synthesis of AuNPs.

12.3.3 Terpenoids

Terpenoids (isoprenoids) are diverse class of naturally occurring small molecular weight organic compounds synthesized by plants. They are responsible for the aroma, taste and color of various plant species. A study (Shankar et al. 2003) has reported the ability of hydroxyl functional groups, citronellol and geraniol present in *Pelargonium graveolens* leaf extract for the reduction of silver ions. Similarly, terpenoids are also present in the bark extract of *Cinnamon zeylanicum* including linalool, eugenol and methyl chavicol that were found to have ability to reduce metal ions (Sathishkumar et al. 2009b). Another study (Singh et al. 2010) has suggested the role of eugenol present in the extract of *Szyygium aromaticum* that act as bio-reducing agent for fabrication of AuNPs and Ag-NPs. Eugenol changes itself to an anionic form due to the release of proton from the hydroxyl group of eugenols. Furthermore, the reducing power of eugenol is significantly improved due to the inductive effect imposed by the electron withdrawing methoxy and allyl functional groups present at the para- and ortho- positions of the hydroxyl group. The two electrons released during the progression of reaction are found to be responsible for the reduction of metal ions.

12.3.4 Proteins

Bio-reduction mediated by proteins is very complicated in nature due to their complex structure. It is observed that nanoparticles are able to bind proteins through their free amino and carboxylate groups. A study (Tan et al. 2010) has tested the reducing and binding capacity of 20 amino acids with metal ions. The bio-reducing and capping nature of cyclic peptides present in the latex of *Jatropha curcas* have also been tested for the synthesis of Ag-NPs (Bar et al. 2009).

12.3.5 Organic Acids

Various secondary metabolites including organic acids are reported as bio-reducing agents for the fabrication of different metallic nanoparticles. Ascorbic acid present in the *Citrus sinensis* peel extract is found to be an effective reducing agent for the synthesis of Ag-NPs (Konwarh et al. 2011). Similar study (Sood et al. 2016) has also described the use of ascorbic acid for the synthesis of iron oxide and gold core shell nanoparticles.

12.4 Antifungal Properties of Metal Nanoparticles

The most commonly used metals for the synthesis of metallic nanoparticles include silver (Ag), copper (Cu), gold (Au), iron (Fe), cobalt (Co), lead (Pb), zinc (Zn), aluminum (Al) and cadmium (Cd). Among them, silver nanoparticles are defined as

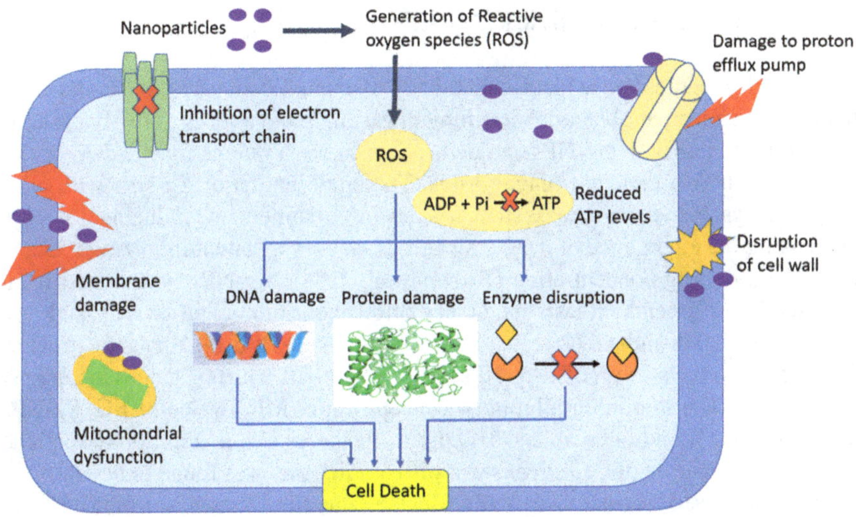

Fig. 12.2 Putative scheme of fungal cell response on exposure to nanoparticles

the most commercialized nanomaterial widely used in health care settings (Wijnhoven et al. 2009). Further, in plant-based nanoparticles, both primary and secondary metabolites are involved in the redox reaction to synthesize eco-friendly metallic nanoparticles.

Metal nanoparticles exhibit unique characteristics that include both physical and chemical properties. The stability of these particles can be altered by adjusting their synthetic pathways due to the change in their size, charge, deformability, hydrophobicity, porosity, biocompatibility, circulation time, biodistribution, cellular localization and internalization. Metallic nanoparticles exhibit antifungal activity through various mechanisms including oxidative stress by inducing the formation of reactive oxygen species (ROS) in cell, nitrosative stress, damage to cell membrane, cell wall and DNA, inhibition of enzymatic activities, reduction in ATP levels, dysfunction of mitochondria, damage to proton efflux pumps and inhibition of electron transport chain (Fig. 12.2) (Abdallah and Ali 2022).

Based on the different mechanisms by which metallic nanoparticles act on fungal cell, they can also be used in the drug delivery of antifungal drugs. The combination of nanoparticles with commercially available antifungal drugs including Amphotericin B (AmB), Echinocandins (caspofungin-CFG) and Fluconazole (FLC) may enhance their activity against *C. albicans* (Salehi et al. 2021; Alshahrani et al. 2022). In this article, recent research focusing on the use of metallic nanoparticles to treat fungal infections caused mainly by *Candida* spp., have been emphasized, with special attention given to silver (Ag), gold (Au), iron (Fe), copper (Cu), palladium (Pd), aluminum (Al), selenium (Se), zinc oxide (ZnO) and bimetallic nanoparticles.

12.5 Silver Nanoparticles (Ag-NPs)

Silver nanoparticles are widely studied for their applications in the biomedical field as anticancer, antibacterial, antifungal, vaccines and as drug delivery systems (Sintubin et al. 2012). Ag-NPs can be used as carrier systems for the delivery of commercially available antifungal drugs. The combination of Ag-NPs with antifungal drugs helps to increase the efficacy of treatment by reducing toxicity, decreasing the emergence of drug-resistant *Candida* spp. and minimizing the cost and effective drug concentration (Brown et al. 2013). Ag-NPs synthesized from ecofriendly or green synthesis methods exhibit more biological activity and less toxicity. Abdallah and Ali (2022) synthesized the Ag-NPs using the leaf extract of *Erodium glaucophyllum* (EG-Ag-NPs) to determine its antifungal activity against *C. albicans*. The minimum inhibitory concentration (MIC) value of EG-Ag-NPs and AmB was found to be 50 and 100 μg/mL respectively. In addition to this, the dimorphic transition of *C. albicans* was also studied and was found to be inhibited by 56.36% at ½ MIC of EG-Ag-NPs. Besides this, the viability of biofilms was decreased by 52% at 50 μg/mL concentration of EG-Ag-NPs (Abdallah and Ali 2022). In addition to this, Miškovská et al. (2022) have synthesized Ag-NPs from the viticulture waste of *Vitis vinifera* canes, by using whole fruits or fruit skin extracts of this plant. The antifungal activity of Ag-NPs was assessed by broth microdilution method by growing yeast cells in the presence of different concentrations of Ag-NPs. The antifungal properties of Ag-NPs are affected by their size, shape, volume of plant extract, concentration of silver nitrate, zeta potential and surface charge. Therefore, in this study, authors have demonstrated the antifungal activity of two different biosynthesized Ag-NP dispersions such as small monodispersed Ag-NPs (mAg-NPs) and large poly-dispersed Ag-NPs (pAg-NPs), both being prepared by using *V. vinifera* cane extract and showing prominent antifungal activity against *C. albicans*. mAg-NPs exhibited 20 mg/L MIC while pAg-NPs displayed 10 mg/L MIC values, significantly inhibiting the growth of *C. albicans*. Interactions between Ag-NPs and cell wall of *C. albicans* were studied by Scanning Electron Microscopy (SEM) by treating *C. albicans* with both types of mAg-NPs and pAg-NPs at MIC concentration. The mechanism of action of Ag-NPs includes release of ROS to create oxidative stress inside the fungal cell. The interaction of Ag-NPs with the cell wall of fungi can damage the cellular organization by interfering with the cell stability and integrity processes that are required to maintain pathogenicity of fungal pathogen. The disruption in the cell wall of *C. albicans* allows the entry of Ag-NPs inside the cell and this further leads to inhibition of electron chain and DNA replication (Miškovská et al. 2022). The extracellular hydrolytic enzymes of *Candida* spp. named as proteinases, phospholipases and hemolysins are known to play a very crucial role in the biofilm formation. These enzymes help the *Candida* spp. in adherence and tissue penetration that further results into invasion of host tissues (Pereira et al. 2015). Biofilm can be defined as three-dimensional (3D) architecture of microbes that are surrounded by an extracellular matrix (ECM) formed on either biotic or abiotic surfaces. The freely floating cells are called as planktonic cells that are mainly

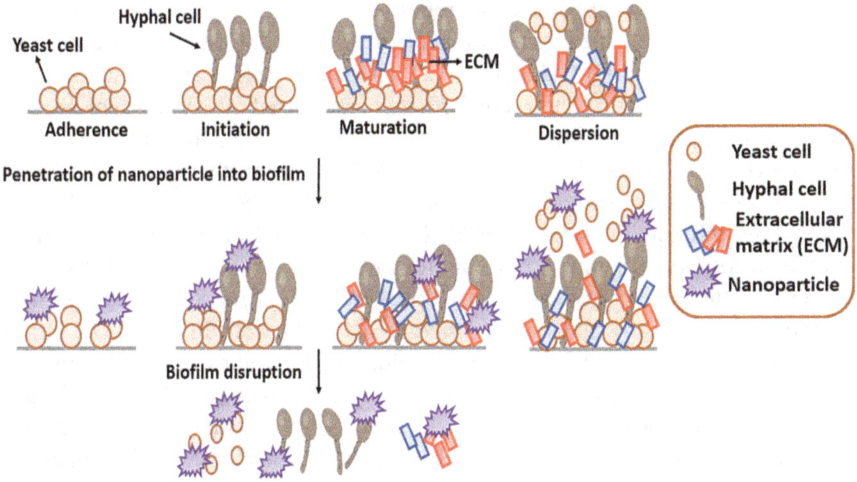

Fig. 12.3 Diagrammatic representation of biofilm disruption of *Candida* spp. by nanoparticles

responsible for the formation of biofilm by an initial attachment. It has been studied that biofilm cells exhibit resistance to antifungal drugs (Marak and Dhanashree 2018) due to increased expression of efflux pump genes that does not allow the drug molecule to act on fungal cells. As there is production of different types of hydrolytic enzymes during the formation of biofilms, hence there is an urgent need to synthesize novel drug molecules that can target the pathogenic pathways of *Candida* spp. including morphogenetic switching. A recent study (Jalal et al. 2019) has reported the antifungal activity of Ag-NPs biosynthesized *from Syzygium cumini* that were able to affect virulence features of *Candida* spp. The effects of Ag-NPs were quantitatively assessed by using Congo Red method and then further visualized by confocal laser microscopy (CFLM). The LIVE/DEAD biofilm viability assay were performed by using florescent stains named as ConA-FITC (Concanavalin A- fluorescein isothiocyanate) and PI (propidium iodide). ConA is defined as carbohydrate binding lectin protein and when combined with FITC, it is used to study its binding with extracellular matrix present on biofilms; while PI is a red fluorescent nucleic acid stain that is able to penetrate inside those cells that have damaged cell membranes. From the study it was found that *S. cumini* Ag-NPs (ScAg-NPs) were able to inhibit biofilm formation at a concentration of 50 μg/mL. The diagrammatic representation of the mechanism of Ag-NPs on biofilm of *Candida* spp. is shown in Fig. 12.3.

12.6 Gold Nanoparticles (AuNPs)

AuNPs synthesized from plant extracts have also been extensively studied for their antifungal activity against *Candida* spp. because they exhibit unique physicochemical properties, chemical resistivity, ease of synthesis and small size (Wani and Ahmad 2013). Recent study has reported the antifungal activity of AuNPs synthesized from the leaf extract of olive plant by using 0.02 M solution of hydrogen tetrachloroaurate (III) ($HAuCL_4 \cdot 4H_2O$) as metallic salt (Kareem and Samaka 2021). The synthesized AuNPs were further characterized by UV-Visible spectrophotometer, SEM, X-ray diffraction (XRD) and Fourier transform infrared spectroscopy (FTIR). The antifungal activity of AuNPs was further assessed by agar well diffusion method against *C. albicans*. A similar study (Jalal et al. 2023) has reported the anti-*Candida* activity of AuNPs synthesized from the aqueous extract of *Crinum latifolium* using 1 mM solution of $AuCl_3$. The antifungal activity of AuNPs was further determined against *C. albicans, C. parapsilosis, C. glabrata, C. tropicalis, C. krusei* and *C. dubliniensis* by using well diffusion and Broth Microdilution Method (BMD). The MIC values of biosynthesized AuNPs ranged between 250 and 500 μg/mL, while MFC values ranged between 500 and 1000 μg/mL. Furthermore, the biosynthesized AuNPs were able to inhibit germ tube development, hinder the secretion of various hydrolytic enzymes and also had an effect on the surface morphology of *C. albicans* by disrupting the cell wall and cell membrane integrity.

12.7 Iron Oxide Nanoparticles (IONPs)

IONPs, prepared by green synthesis, have captured much attention as antifungal agents due to their safety and ease of use. Magnetite (Fe_3O_4) and maghemite (γ-Fe_2O_3) are considered as the two most studied iron oxides. IONPs were synthesized from *Psidium guajava* (guava) and tested for their antifungal activity against *C. tropicalis* and *C. albicans* (Adhikari et al. 2022). The extract of guava contains many types of phytochemicals including polysaccharides, minerals, oleanolic acid, lyxopyranoside, arabopyranoside, guaijavarin, quercetin, essential oils, vitamins, triterpenoid acid containing tannins, flavonoids, alkaloids, steroids, glycosides, saponins, vitamin C and pectin as a dietary fiber (Arima and Danno 2002). The antifungal activity of biosynthesized IONPs was determined by using agar well diffusion method and the diameter of the zone of inhibition was found to be greater in *C. tropicalis* as compared to that in *C. albicans*. The possible mechanism for the interaction of biosynthesized IONPs with the fungi includes generation of ROS, which will further create oxidative stress in fungal cells. IONPs form electron–hole pairs that are activated by UV or visible light. These electron-hole pairs further split the water molecules into OH^- and H^+, and electron addition to dissolved oxygen molecules will further generate superoxide radical anions (O_2^-). These free radicals including OH^- and O_2^- further depolymerize polysaccharides, cause DNA damage, lead to lipid peroxidation, enzyme inhibition and subsequently cell death (Groiss et al. 2017). In addition to this, there is one more plausible mechanism of IONPs that includes their binding to cell membrane proteins through electrostatic and Van-der wall interactions leading to disruption in cellular pathways and disorganization of cell wall and cell membrane. IONPs are defined as paramagnetic as well as di-magnetic substances. Fe_3O_4NP is called as super-paramagnetic IONPs, having very high paramagnetic activity. In addition to the usual mechanism of action with fungal cells, they also cause cell death and biofilm disorganization in the presence of alternating magnetic field, leading to vibrational damage, ROS generation and local hyperthermia. All these pathways lead to cell wall damage of the pathogen, biofilm disruption, membrane disorganization and eventual cell death (Gudkov et al. 2021).

12.8 Copper Oxide Nanoparticles (CuONPs)

"Tamra jal" is well known for its therapeutic use and is prepared by storing drinking water in the vessel made up of copper or brass metal. Storage of water in the copper vessel has been reported since ancient period of time in Ayurveda (Sudha et al. 2012). *Thespesia populnea* commonly called as Portia tree, its leaves, flowers, bark exhibits curative effects against skin infections. CuONPs were biosynthesized by using bark extract of *T. populnea* and were tested for their antibacterial and antifungal activity. *T. populnea* contains variety of phytochemical compounds including gallic acid, catechin, myricetin, protocatechuic acid, epigallocatechin gallate, rosmarinic acid, ellagic acid, rutin and naringenine that are responsible for capping and reduction of nanoparticles (Rangani et al. 2019). Biosynthesized CuONPs were synthesized by mixing aqueous bark extract solution of *T. populnea* with copper acetate solution. Antifungal activity of CuONPs was determined by using disc diffusion method against *C. albicans* MTCC 183 and the diameter of zone of inhibition was found to be 9 mm (Narayanan et al. 2022).

12.9 Palladium Nanoparticles (PdNPs)

PdNPs exhibit significant chemical and thermal stabilities, electronic and optical properties. The biosynthesis of PdNPs from plant received much attention in biomedical field. The initial step for the biosynthesis of PdNPs involves the bioreduction and nucleation of metal ions during which metal ions are reduced into their zero-oxidation states, the second step further leads to growth and agglomeration of the small PdNPs into larger thermodynamically stable PdNPs (Fig. 12.4) (Malik et al. 2014). After this, PdNPs acquire various shapes depending upon the volume of plant extract solution and concentration of metal salt used for the synthesis of PdNPs. The capping and stabilization of biosynthesized PdNPs was mainly done by the plant extract that contains different types of functional groups-alcohols, amines, carboxylic acids, aldehydes, ketones present in the phytochemicals. The phytochemicals mediate capping and stabilization of PdNPs by combination of hydrophobic interactions, hydrogen bonding, covalent bonding and ion dipole interactions (Akhtar et al. 2013).

A study by Vaghela et al. (2018) has demonstrated the antifungal activity of biosynthesized PdNPs from the bark extract of *Bauhinia variegata* and 1 mM solution of $PdCl_2$. Antifungal activity was determined by agar well diffusion method against

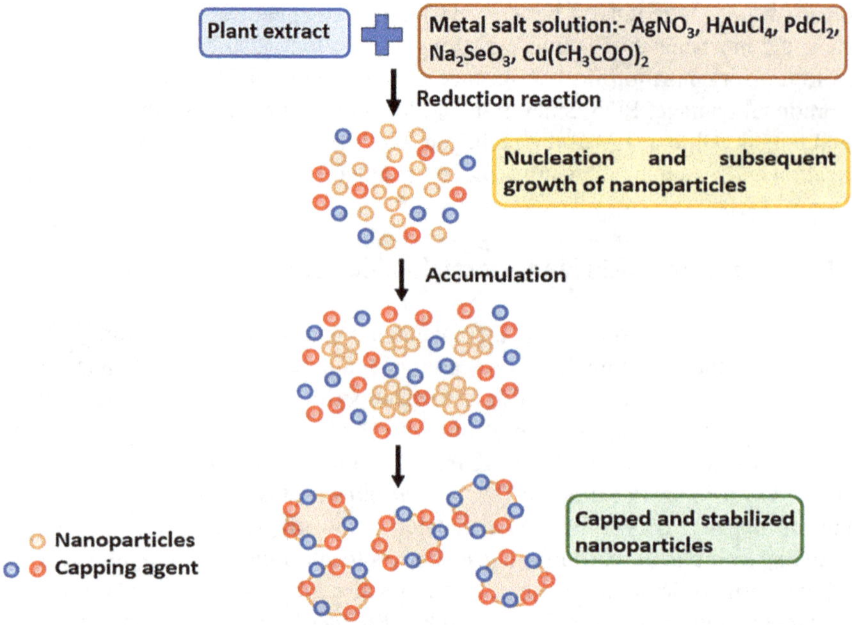

Fig. 12.4 Schematic diagram showing the mechanism of green synthesis of metallic nanoparticles

C. albicans MTCC 183 and the average diameter of zone of inhibition was found to be 7 mm. In another study (Rabiee et al. 2020), PdNPs were synthesized from the leaf extracts of *Rosmarinus officinalis* and 0.1 M of palladium acetate solution. The biosynthesized PdNPs were screened against four major human fungal pathogens named as *C. parapsilosis*, *C. glabrata*, *C. krusei* and *C. albicans* by disc diffusion method. At the concentration of 10 μg/mL, the diameter of zone of inhibition was found to be 19.4 mm for *C. parapsilosis*, 7.1 mm for *C. glabrata*, 21.8 mm for *C. krusei* and 8.5 mm for *C. albicans*.

12.10 Aluminum Oxide Nanoparticles (Al_2O_3NPs)

Al_2O_3NPs are an important component in ceramic industry as they are used as an absorbent in heterogeneous catalysis. Al_2O_3NPs are defined as thermodynamically stable particles, over a wide range of temperatures. They exhibit corundum-like atoms containing oxygen atoms that are closely packed hexagonally while aluminum ions are covering two-thirds of the lattice octahedral spaces. Al_2O_3NPs were synthesized by using leaf extract of lemon grass by reacting with boehmite (γ-AlO(OH) which was prepared from aluminum nitrate nonahydrate (ANN). Antifungal activity of Al_2O_3NPs was determined by agar disc diffusion methods and by MIC, MFC values. The interaction between *Candida* cells and Al_2O_3NPs was examined by SEM and High resolution transmission electron microscopy

(HR-TEM). The MIC and MFC values of Al_2O_3NPs was found to be 283.3 ± 84.9 μg/mL and 766.6 ± 205.4 μg/mL respectively against *C. albicans* MTCC 227 (Jalal et al. 2016). In a similar study (Nagarajan et al. 2023) it had been reported that the biofabricated Al_2O_3NPs synthesized from the peel extracts of *C. aurantium* showed 22 mm of zone of inhibition against *C. albicans* MTCC 227.

12.11 Selenium Nanoparticles (SeNPs)

Selenium is regarded as an essential trace element and has received attention because it exhibits vital functions in biological systems. Selenium is an important component of selenoproteins and plays crucial role in antioxidant defense system (Hosnedlova et al. 2018). According to a study (Abdel-Moneim et al. 2022), biosynthesized SeNP from the extract of *Spirulina platensis* (an edible blue green alga) contains large number of proteins, vitamins, essential fatty acids, pigments, carotenoids and minerals (Kavisri et al. 2023). SeNPs were synthesized from the extract of *S. platensis* and sodium selenite solution. The antifungal activity of SeNPs was determined by using disc diffusion and broth microdilution methods against *C. tropicalis*, *C. albicans* and *C. glabrata*. At 100 μg/mL concentration of SeNPs, the zone of inhibition was found to be 11.8, 13.1 and 9.9 mm for *C. tropicalis*, *C. albicans* and *C. glabrata* respectively. Plant mediated green synthesis of SeNPs was also done (Fouda et al. 2022) by using aqueous leaf extract of *Portulaca oleracea* and 3 mM solution of sodium selenite (Na_2SeO_3). The efficacy of phyto-synthesized SeNPs was determined against clinical *Candida* spp. using the agar well diffusion method. The maximum inhibition zones were reported at 300 μg/mL with the values of 15.6 ± 0.6, 19.3 ± 0.8, 18.3 ± 0.6 and 16.7 ± 0.7 mm for *C. albicans*, *C. glabrata*, *C. tropicalis* and *C. parapsilosis* respectively. For the first time, aqueous leaf extract of stinging nettle (*Urtica dioica*) was used to synthesize SeNPs and the MIC value was found to be 62.5 μg/mL against *C. albicans* (Hashem and Salem 2022).

12.12 Zinc Oxide Nanoparticles (ZnONPs)

The antifungal activity of green-biosynthesized ZnONPs, using seed extract of *Pongamia pinnata,* was determined against *C. albicans* MTCC 3017. The anti-biofilm activity of ZnONPs was found at 50 μg/mL against *C. albicans* MTCC 3017. The morphology and thickness of biofilm formed in treated cells were examined under confocal laser microscopy (CFLM) (Malaikozhundan et al. 2017). Furthermore, Lyimo et al. (2022) have synthesized ZnONPs by using zinc nitrate hexahydrate [$Zn(NO_3)_2{\cdot}6H_2O$] as inorganic metal salt precursor and the extracts of banana (*Musa paradisiaca*) peel and tea leaves of Rooibos (*Aspalathus linearis*), infused with the Buchu (*Agathosma betulina*). To determine antifungal activity of ZnONPs against *C. albicans* ATCC 90028, Kirby-Bauer disc diffusion method was used. The mode of action of ZnONPs was based on their interaction with fungal cells and their physicochemical properties include (i) release of Zn^{2+} ions (ii) ROS generation leading to oxidative stress and damage (iii) inhibition of electron transport chain (iv) lipid peroxidation (v) inhibition of DNA replication (Jin and Jin 2021)

12.13 Bimetallic Nanoparticles

Metallic nanoparticles can be of two different types including monometallic and bimetallic. Monometallic NPs are made up of only one type of metal having unique physical and chemical properties. They can be synthesized by various methods, mostly by chemical ones and their surface can be modified by different types of functional groups. While bimetallic NPs are synthesized by combining two different types of metal NPs and they also exhibit wide variety of morphologies and structures (Argueta-Figueroa et al. 2014). Bimetallic NPs exhibit more intriguing properties as compared to the monometallic NPs, which is mainly due to the synergistic properties between two different metals. Higher performance of bimetallic NPs can be achieved by selecting the proper metal type and their optimization of preparation with each metal type. In a recent study, Padilla-Cruz et al. (2021) have synthesized Ag-Fe bimetallic NPs (Ag-FeNPs) by using extracts of *Gardenia jasminoides* as capping and reducing agent. The synthesized bimetallic NPs have an average diameter of 13 nm with spherical core-shell structures. The antifungal activities of Ag-NPs, FeNPs, Ag-FeNPs were determined against *C. albicans*. The MIC value of Ag-NPs was observed to be 125 ppm and FeNPs do not exhibit any activity at the maximum concentration tested up to 250 ppm, while Ag-FeNPs showed fungicidal activity against *C. albicans* at 62.5 ppm. This study has demonstrated the antifungal property of bimctallic Ag-FeNPs as a potent nanomaterial for application in the medical field. Kamli et al. (2022) have also demonstrated the antifungal activity of silver and nickel bimetallic NPs (Ag-NiNPs), which were synthesized by using leaf extracts of *Salvia officinalis*. Bimetallic NPs were able to inhibit the biofilm formation and hyphae development in *C. albicans*. Additionally, Ag-NiNPs acted in a synergistic manner with fluconazole drug against *C. albicans* thus helping to regain the susceptibility of the fungi to fluconazole. Moreover, Ag-NiNPs was also able to inhibit multidrug-resistant efflux transporters by reducing the expression of ATP-dependent efflux pumps in *C. albicans* at 1.56 μg/mL concentration. Interaction of Ag-NiNPs with fungal cells lead to loss of integrity of fungal cell membrane followed by generation of ROS, as was observed via SEM studies. These findings suggest that bimetallic NPs can be considered as interesting platform for the development of antifungal drugs, as they work against different *Candida* spp. by multiple pathways and help to reduce the emergence of resistance in fungal cells.

12.14 Conclusion

Several reports have demonstrated the role of plant-synthesized nanoparticles against *Candida* spp. by in vitro studies. The biosynthesized silver, gold, copper, iron and other metallic nanoparticles have the ability to inhibit the growth of *Candida* spp. by employing various mechanisms. The electron microscopic study reported by various authors have shown that metallic nanoparticles not only attach to the fungal cell surface but also penetrates inside the *Candida* cell and results in the disruption of morphology and physiology of *Candida* cell. Analyzing various reports and articles, it has been proven that the green synthesis approach for synthesizing different types of metallic nanoparticles is an ecofriendly, cost effective, low maintenance and non-toxic method. These nanoparticles can be widely employed as antimicrobial agents in medicine due to their multiple advantages. Metallic nanoparticles, synthesized from different plant extracts, can also be used in drug delivery systems for the controlled release of drugs. Besides this, they can also be applied on various implanted devices used in orthopedic and dentistry healthcare facilities to combat microbial infections. Application of metallic and bimetallic nanoparticles on the surface of urinary catheters and other medical devices can prevent the growth of biofilm forming *Candida* spp. Thus, the development of metallic nanoparticles can be an interesting and novel approach to overcome emergence of antifungal drug resistance in *Candida* spp. and recurrent fungal infections. Furthermore, in-vivo safety profile studies should be conducted in order to determine the biocompatibility and negative effects of plant derived metallic nanoparticles to enhance their use in healthcare settings. Additionally, there is need to determine the molecular methods and pathways involved in inhibition of virulence factor of *Candida* spp. by metallic nanoparticles.

References

Abdallah BM, Ali EM (2021) Green synthesis of silver nanoparticles using the Lotus lalambensis aqueous leaf extract and their anti-candidal activity against oral candidiasis. ACS Omega 6(12):8151–8162

Abdallah BM, Ali EM (2022) Therapeutic effect of green synthesized silver nanoparticles using erodium glaucophyllum extract against oral candidiasis: in vitro and in vivo study. Molecules 27(13):4221

Abdel-Moneim AME, El-Saadony MT, Shehata AM, Saad AM, Aldhumri SA, Ouda SM, Mesalam NM (2022) Antioxidant and antimicrobial activities of Spirulina platensis extracts and biogenic selenium nanoparticles against selected pathogenic bacteria and fungi. Saudi J Biol Sci 29(2):1197–1209

Adhikari A, Chhetri K, Acharya D, Pant B, Adhikari A (2022) Green synthesis of iron oxide nanoparticles using Psidium guajava L. leaves extract for degradation of organic dyes and antimicrobial applications. Catalysts 12(10):1188

Ahmad N, Sharma S, Alam MK, Singh VN, Shamsi SF, Mehta BR, Fatma A (2010) Rapid synthesis of silver nanoparticles using dried medicinal plant of basil. Colloids Surf B: Biointerfaces 81(1):81–86
Akhtar MS, Panwar J, Yun YS (2013) Biogenic synthesis of metallic nanoparticles by plant extracts. ACS Sustain Chem Eng 1(6):591–602
Akintelu SA, Folorunso AS, Oyebamiji AK, Erazua EA (2019) Antibacterial potency of silver nanoparticles synthesized using Boerhaavia diffusa leaf extract as reductive and stabilizing agent. Int J Pharma Sci Res 10(12):374–380
Al Aboody MS (2019) Silver/silver chloride (Ag/AgCl) nanoparticles synthesized from Azadirachta indica lalex and its antibiofilm activity against fluconazole resistant Candida tropicalis. Artif Cells Nanomed Biotechnol 47(1):2107–2113
Alherz FA, Negm WA, Elekhnawy E, El-Masry TA, Haggag EM, Alqahtani MJ, Hussein IA (2022) Silver nanoparticles prepared using encephalartos laurentianus de wild leaf extract have inhibitory activity against candida albicans clinical isolates. J Fungi 8(10):1005
Ali EM, Abdallah BM (2020) Effective inhibition of candidiasis using an eco-friendly leaf extract of calotropis-gigantean-mediated silver nanoparticles. Nanomaterials (Basel) 10(3):422
Ali SG, Jalal M, Ahmad H, Sharma D, Ahmad A, Umar K, Khan HM (2022) Green synthesis of silver nanoparticles from Camellia sinensis and its antimicrobial and antibiofilm effect against clinical isolates. Materials 15(19):6978
Alshahrani SM, Khafagy ES, Riadi Y, Al Saqr A, Alfadhel MM, Hegazy WA (2022) Amphotericin B-peg conjugates of ZnO nanoparticles: enhancement antifungal activity with minimal toxicity. Pharmaceutics 14(8):1646
Anand K, Tiloke C, Phulukdaree A, Ranjan B, Chuturgoon A, Singh S, Gengan RM (2016) Biosynthesis of palladium nanoparticles by using Moringa oleifera flower extract and their catalytic and biological properties. J Photochem Photobiol B 165:87–95
Ansari MA, Kalam A, Al-Sehemi AG, Alomary MN, AlYahya S, Aziz MK et al (2021) Counteraction of biofilm formation and antimicrobial potential of Terminalia catappa functionalized silver nanoparticles against Candida albicans and multidrug-resistant Gram-negative and Gram-positive bacteria. Antibiotics 10(6):725
Argueta-Figueroa L, Morales-Luckie RA, Scougall-Vilchis RJ, Olea-Mejía OF (2014) Synthesis, characterization and antibacterial activity of copper, nickel and bimetallic Cu–Ni nanoparticles for potential use in dental materials. Prog Nat Sci Mater Int 24(4):321–328
Arima H, Danno GI (2002) Isolation of antimicrobial compounds from guava (Psidium guajava L.) and their structural elucidation. Biosci Biotechnol Biochem 66(8):1727–1730
Arsene MM, Viktorovna PI, Alla MV, Mariya MA, Sergei GV, Cesar E et al (2022) Optimization of ethanolic extraction of Enantia chloranta bark, phytochemical composition, green synthesis of silver nanoparticles, and antimicrobial activity. Fermentation 8(10):530
Arsène MMJ, Viktorovna PI, Alla M, Mariya M, Nikolaevitch SA, Davares AKL et al (2023) Antifungal activity of silver nanoparticles prepared using Aloe vera extract against Candida albicans. Vet World 16(1):18
Banasiuk R, Krychowiak M, Swigon D, Tomaszewicz W, Michalak A, Chylewska A et al (2020) Carnivorous plants used for green synthesis of silver nanoparticles with broad-spectrum antimicrobial activity. Arab J Chem 13(1):1415–1428
Bansal SA, Kumar V, Karimi J, Singh AP, Kumar S (2020) Role of gold nanoparticles in advanced biomedical applications. Nanoscale Adv 2(9):3764–3787
Bar H, Bhui DK, Sahoo GP, Sarkar P, De SP, Misra A (2009) Green synthesis of silver nanoparticles using latex of Jatropha curcas. Colloids Surf A Physicochem Eng Asp 339(1–3):134–139
Benedict K, Jackson BR, Chiller T, Beer KD (2019) Estimation of direct healthcare costs of fungal diseases in the United States. Clin Infect Dis 68(11):1791–1797

Bhattacharya S, Sae-Tia S, Fries BC (2020) Candidiasis and mechanisms of antifungal resistance. Antibiotics 9(6):312

Bongomin F, Gago S, Oladele RO, Denning DW (2017) Global and multi-national prevalence of fungal diseases—estimate precision. J Fungi 3(4):57

Brown PK, Qureshi AT, Moll AN, Hayes DJ, Monroe WT (2013) Silver nanoscale antisense drug delivery system for photoactivated gene silencing. ACS Nano 7(4):2948–2959

Chandrasekaran S, Anbazhagan V (2024) Green synthesis of ZnO and V-doped ZnO nanoparticles using vinca rosea plant leaf for biomedical applications. Appl Biochem Biotechnol 196(1):50–67

Dobrucka R, Dlugaszewska J, Kaczmarek M (2018) Cytotoxic and antimicrobial effects of biosynthesized ZnO nanoparticles using of Chelidonium majus extract. Biomed Microdevices 20:1–13

Edison TJI, Sethuraman MG (2012) Instant green synthesis of silver nanoparticles using Terminalia chebula fruit extract and evaluation of their catalytic activity on reduction of methylene blue. Process Biochem 47(9):1351–1357

Elumalai K, Velmurugan S (2015) Green synthesis, characterization and antimicrobial activities of zinc oxide nanoparticles from the leaf extract of Azadirachta indica (L.). Appl Surf Sci 345:329–336

El-Zayat MM, Eraqi MM, Alrefai H, El-Khateeb AY, Ibrahim MA, Aljohani HM et al (2021) The antimicrobial, antioxidant, and anticancer activity of greenly synthesized selenium and zinc composite nanoparticles using Ephedra aphylla extract. Biomol Ther 11(3):470

Emmanuel R, Saravanan M, Ovais M, Padmavathy S, Shinwari ZK, Prakash P (2017) Antimicrobial efficacy of drug blended biosynthesized colloidal gold nanoparticles from Justicia glauca against oral pathogens: a nanoantibiotic approach. Microb Pathog 113:295–302

Fouda A, Al-Otaibi WA, Saber T, AlMotwaa SM, Alshallash KS, Elhady M et al (2022) Antimicrobial, antiviral, and in-vitro cytotoxicity and mosquitocidal activities of Portulaca oleracea-based green synthesis of selenium nanoparticles. J Funct Biomater 13(3):157

Ghoreishi SM, Behpour M, Khayatkashani M (2011) Green synthesis of silver and gold nanoparticles using Rosa damascena and its primary application in electrochemistry. Physica E 44(1):97–104

Gómez-Garzón M, Gutiérrez-Castañeda LD, Gil C, Escobar CH, Rozo AP, González ME, Sierra EV (2021) Inhibition of the filamentation of Candida albicans by Borojoa patinoi silver nanoparticles. SN Appl Sci 3:1–8

Gonçalves SS, Souza ACR, Chowdhary A, Meis JF, Colombo AL (2016) Epidemiology and molecular mechanisms of antifungal resistance in Candida and Aspergillus. Mycoses 59(4):198–219

Groiss S, Selvaraj R, Varadavenkatesan T, Vinayagam R (2017) Structural characterization, antibacterial and catalytic effect of iron oxide nanoparticles synthesised using the leaf extract of Cynometra ramiflora. J Mol Struct 1128:572–578

Gudkov SV, Burmistrov DE, Serov DA, Rebezov MB, Semenova AA, Lisitsyn AB (2021) Do iron oxide nanoparticles have significant antibacterial properties? Antibiotics 10(7):884

Hadi RG, Samaka HM, Abdulridha WAM (2020) Green synthesis of silver nanoparticles from Nerium oleander and evaluate its effects as an antifungal agent. Biochem Cell Arch 20(2):1–7

Hashem AH, Salem SS (2022) Green and ecofriendly biosynthesis of selenium nanoparticles using Urtica dioica (stinging nettle) leaf extract: antimicrobial and anticancer activity. Biotechnol J 17(2):2100432

Hawar SN, Al-Shmgani HS, Al-Kubaisi ZA, Sulaiman GM, Dewir YH, Rikisahedew JJ (2022) Green synthesis of silver nanoparticles from Alhagi graecorum leaf extract and evaluation of their cytotoxicity and antifungal activity. J Nanomater 2022:1–8

Hosnedlova B, Kepinska M, Skalickova S, Fernandez C, Ruttkay-Nedecky B, Peng Q et al (2018) Nano-selenium and its nanomedicine applications: a critical review. Int J Nanomedicine 13:2107–2128

Ildiz N, Baldemir A, Altinkaynak C, Özdemir N, Yilmaz V, Ocsoy I (2017) Self assembled snowball-like hybrid nanostructures comprising Viburnum opulus L. extract and metal ions for antimicrobial and catalytic applications. Enzym Microb Technol 102:60–66

Iravani S, Korbekandi H, Mirmohammadi SV, Zolfaghari B (2014) Synthesis of silver nanoparticles: chemical, physical and biological methods. Res Pharm Sci 9(6):385

Jalal M, Ansari MA, Shukla AK, Ali SG, Khan HM, Pal R et al (2016) Green synthesis and antifungal activity of Al_2O_3 NPs against fluconazole-resistant Candida spp isolated from a tertiary care hospital. RSC Adv 6(109):107577–107590

Jalal M, Ansari MA, Alzohairy MA, Ali SG, Khan HM, Almatroudi A, Siddiqui MI (2019) Anticandidal activity of biosynthesized silver nanoparticles: effect on growth, cell morphology, and key virulence attributes of Candida species. Int J Nanomedicine 14:4667–4679

Jalal M, Ansari MA, Alshamrani M, Ali SG, Jamous YF, Alyahya SA et al (2023) Crinum latifolium mediated biosynthesis of gold nanoparticles and their anticandidal, antibiofilm and antivirulence activity. J Saudi Chem Soc 27(3):101644

Jayaseelan C, Ramkumar R, Rahuman AA, Perumal P (2013) Green synthesis of gold nanoparticles using seed aqueous extract of Abelmoschus esculentus and its antifungal activity. Ind Crop Prod 45:423–429

Jin SE, Jin HE (2021) Antimicrobial activity of zinc oxide nano/microparticles and their combinations against pathogenic microorganisms for biomedical applications: from physicochemical characteristics to pharmacological aspects. Nanomaterials 11(2):263

Kamli MR, Malik MA, Lone SA, Sabir JS, Mattar EH, Ahmad A (2021) Beta vulgaris assisted fabrication of novel Ag-Cu bimetallic nanoparticles for growth inhibition and virulence in Candida albicans. Pharmaceutics 13(11):1957

Kamli MR, Alzahrani EA, Albukhari SM, Ahmad A, Sabir JS, Malik MA (2022) Combination effect of novel bimetallic Ag-Ni nanoparticles with fluconazole against Candida albicans. J Fungi 8(7):733

Kareem HA, Samaka HM (2021) Evaluation of the effect of the gold nanoparticles prepared by green chemistry on the treatment of cutaneous candidiasis. Curr Med Mycol 7(1):1

Kavisri M, Abraham M, Prabakaran G, Elangovan M, Moovendhan M (2023) Phytochemistry, bioactive potential and chemical characterization of metabolites from marine microalgae (Spirulina platensis) biomass. Biomass Conv Bioref 13:10147–10154

Khatoon N, Mishra A, Alam H, Manzoor N, Sardar M (2015) Biosynthesis, characterization, and antifungal activity of the silver nanoparticles against pathogenic Candida species. BioNanoScience 5:65–74

Konwarh R, Gogoi B, Philip R, Laskar MA, Karak N (2011) Biomimetic preparation of polymer-supported free radical scavenging, cytocompatible and antimicrobial "green" silver nanoparticles using aqueous extract of Citrus sinensis peel. Colloids Surf B: Biointerfaces 84(2):338–345

Liu H, Zhang X, Xu Z, Wang Y, Ke Y, Jiang Z et al (2020) Role of polyphenols in plant-mediated synthesis of gold nanoparticles: identification of active components and their functional mechanism. Nanotechnology 31(41):415601

Lyimo GV, Ajayi RF, Maboza E, Adam RZ (2022) A green synthesis of zinc oxide nanoparticles using Musa paradisiaca and Rooibos extracts. MethodsX 9:101892

Malaikozhundan B, Vaseeharan B, Vijayakumar S, Pandiselvi K, Kalanjiam MAR, Murugan K, Benelli G (2017) Biological therapeutics of Pongamia pinnata coated zinc oxide nanoparticles against clinically important pathogenic bacteria, fungi and MCF-7 breast cancer cells. Microb Pathog 104:268–277

Malathi S, Manikandan D, Nishanthi R, Jagan EG, Riyaz SUM, Palani P, Simal-Gandara J (2022) Silver nanoparticles, synthesized using Hyptis suaveolens (L) poit and their antifungal activity against Candida spp. ChemistrySelect 7(47):e202203050

Malik P, Shankar R, Malik V, Sharma N, Mukherjee TK (2014) Green chemistry based benign routes for nanoparticle synthesis. J Nanoparticles 2014:1–14

Marak MB, Dhanashree B (2018) Antifungal susceptibility and biofilm production of Candida spp. isolated from clinical samples. Int J Microbiol 2018:7495218

Mba IE, Nweze EI, Eze EA, Anyaegbunam ZKG (2022) Genome plasticity in Candida albicans: a cutting-edge strategy for evolution, adaptation, and survival. Infect Genet Evol 99:105256

Miškovská A, Rabochová M, Michailidu J, Masák J, Čejková A, Lorinčík J, Maťátková O (2022) Antibiofilm activity of silver nanoparticles biosynthesized using viticultural waste. PLoS One 17(8):e0272844

Mohanta YK, Panda SK, Biswas K, Tamang A, Bandyopadhyay J, De D et al (2016) Biogenic synthesis of silver nanoparticles from Cassia fistula (Linn.): in vitro assessment of their antioxidant, antimicrobial and cytotoxic activities. IET Nanobiotechnol 10(6):438–444

Moustafa M, Al-Emam A, Sayed M, Alamri S, Alghamdii H, Shati A et al (2021) Green synthesis of Ag nanoparticles from aqueous extracts of leaves and fruit of Casuarina equisetifolia against Candida albicans and other clinical isolates. Int J Agric Biol 25(01):117–122

Nabikhan A, Kandasamy K, Raj A, Alikunhi NM (2010) Synthesis of antimicrobial silver nanoparticles by callus and leaf extracts from saltmarsh plant, Sesuvium portulacastrum L. Colloids Surf B: Biointerfaces 79(2):488–493

Nagarajan P, Subramaniyan V, Elavarasan V, Mohandoss N, Subramaniyan P, Vijayakumar S (2023) Biofabricated aluminium oxide nanoparticles derived from Citrus aurantium L.: antimicrobial, anti-proliferation, and photocatalytic efficiencies. Sustainability 15(2):1743

Narayanan M, Srinivasan B, Sambantham MT, Al-Keridis LA, Al-Mekhlafi FA (2022) Green synthesizes and characterization of copper-oxide nanoparticles by Thespesia populnea against skin-infection causing microbes. J King Saud Univ Sci 34(3):101885

Padalia H, Moteriya P, Chanda S (2015) Green synthesis of silver nanoparticles from marigold flower and its synergistic antimicrobial potential. Arab J Chem 8(5):732–741

Padalia H, Moteriya P, Chanda S (2018) Synergistic antimicrobial and cytotoxic potential of zinc oxide nanoparticles synthesized using Cassia auriculata leaf extract. Bionanoscience 8:196–206

Padilla-Cruz AL, Garza-Cervantes JA, Vasto-Anzaldo XG, García-Rivas G, León-Buitimea A, Morones-Ramírez JR (2021) Synthesis and design of Ag–Fe bimetallic nanoparticles as antimicrobial synergistic combination therapies against clinically relevant pathogens. Sci Rep 11(1):5351

Pereira CA, Costa AC, Silva MP, Back-Brito GN, Jorge AO (2015) Candida albicans and virulence factors that increases its pathogenicity. Battle Again Microb Pathog 2:631–636

Peron IH, Reichert-Lima F, Busso-Lopes AF, Nagasako CK, Lyra L, Moretti ML, Schreiber AZ (2016) Resistance surveillance in Candida albicans: a five-year antifungal susceptibility evaluation in a Brazilian university hospital. PLoS One 11(7):e0158126

Rabiee N, Bagherzadeh M, Kiani M, Ghadiri AM (2020) Rosmarinus officinalis directed palladium nanoparticle synthesis: investigation of potential anti-bacterial, anti-fungal and Mizoroki-Heck catalytic activities. Adv Powder Technol 31(4):1402–1411

Rai M, Yadav A, Gade A (2008) CRC 675—current trends in phytosynthesis of metal nanoparticles. Crit Rev Biotechnol 28(4):277–284

Rangani J, Kumari A, Patel M, Brahmbhatt H, Parida AK (2019) Phytochemical profiling, polyphenol composition, and antioxidant activity of the leaf extract from the medicinal halophyte Thespesia populnea reveal a potential source of bioactive compounds and nutraceuticals. J Food Biochem 43(2):e12731

Ryu S, Nam SH, Baek JS (2022) Green synthesis of silver nanoparticles (Ag-NPs) of Angelica gigas fabricated by hot-melt extrusion technology for enhanced antifungal effects. Materials 15(20):7231

Salavati-Niasari M, Davar F, Mir N (2008) Synthesis and characterization of metallic copper nanoparticles via thermal decomposition. Polyhedron 27(17):3514–3518

Salehi Z, Fattahi A, Lotfali E, Kazemi A, Shakeri-Zadeh A, Nasrollahi SA (2021) Susceptibility pattern of caspofungin-coated gold nanoparticles against clinically important Candida species. Adv Pharm Bull 11(4):693

Sathishkumar M, Sneha K, Kwak IS, Mao J, Tripathy SJ, Yun YS (2009a) Phyto-crystallization of palladium through reduction process using Cinnamom zeylanicum bark extract. J Hazard Mater 171(1–3):400–404

Sathishkumar M, Sneha K, Won SW, Cho CW, Kim S, Yun YS (2009b) Cinnamon zeylanicum bark extract and powder mediated green synthesis of nano-crystalline silver particles and its bactericidal activity. Colloids Surf B: Biointerfaces 73(2):332–338
Shankar SS, Ahmad A, Sastry M (2003) Geranium leaf assisted biosynthesis of silver nanoparticles. Biotechnol Prog 19(6):1627–1631
Shankar SS, Rai A, Ahmad A, Sastry M (2004) Rapid synthesis of Au, Ag, and bimetallic Au core–Ag shell nanoparticles using Neem (Azadirachta indica) leaf broth. J Colloid Interface Sci 275(2):496–502
Singh AK, Talat M, Singh DP, Srivastava ON (2010) Biosynthesis of gold and silver nanoparticles by natural precursor clove and their functionalization with amine group. J Nanopart Res 12:1667–1675
Sintubin L, Verstraete W, Boon N (2012) Biologically produced nanosilver: current state and future perspectives. Biotechnol Bioeng 109(10):2422–2436
Sitrarasi R, Nallal VUM, Razia M, Chung WJ, Shim J, Chandrasekaran M et al (2022) Inhibition of multi-drug resistant microbial pathogens using an eco-friendly root extract of Furcraea foetida mediated silver nanoparticles. J King Saud Univ Sci 34(2):101794
Sivaraj R, Rahman PK, Rajiv P, Narendhran S, Venckatesh R (2014) Biosynthesis and characterization of Acalypha indica mediated copper oxide nanoparticles and evaluation of its antimicrobial and anticancer activity. Spectrochim Acta A Mol Biomol Spectrosc 129:255–258
Soares MR, Corrêa RO, Stroppa PHF, Marques FC, Andrade GF, Correa CC et al (2018) Biosynthesis of silver nanoparticles using Caesalpinia ferrea (Tul.) Martius extract: physicochemical characterization, antifungal activity and cytotoxicity. PeerJ 6:e4361
Sood A, Arora V, Shah J, Kotnala RK, Jain TK (2016) Ascorbic acid-mediated synthesis and characterisation of iron oxide/gold core–shell nanoparticles. J Exp Nanosci 11(5):370–382
Sudha VP, Ganesan S, Pazhani GP, Ramamurthy T, Nair GB, Venkatasubramanian P (2012) Storing drinking-water in copper pots kills contaminating diarrhoeagenic bacteria. J Health Popul Nutr 30(1):17
Tan YN, Lee JY, Wang DI (2010) Uncovering the design rules for peptide synthesis of metal nanoparticles. J Am Chem Soc 132(16):5677–5686
Thakkar KN, Mhatre SS, Parikh RY (2010) Biological synthesis of metallic nanoparticles. Nanomedicine 6(2):257–262
Tragiannidis A, Gkampeta A, Vousvouki M, Vasileiou E, Groll AH (2021) Antifungal agents and the kidney: pharmacokinetics, clinical nephrotoxicity, and interactions. Expert Opin Drug Saf 20(9):1061–1074
Vaghela H, Shah R, Pathan A (2018) Palladium nanoparticles mediated through bauhinia variegata: potent in vitro anticancer activity against mcf-7 cell lines and antimicrobial assay. Curr Nanomater 3(3):168–177
Vijayakumar S, Mahadevan S, Arulmozhi P, Sriram S, Praseetha PK (2018a) Green synthesis of zinc oxide nanoparticles using Atalantia monophylla leaf extracts: characterization and antimicrobial analysis. Mater Sci Semicond Process 82:39–45
Vijayakumar S, Krishnakumar C, Arulmozhi P, Mahadevan S, Parameswari N (2018b) Biosynthesis, characterization and antimicrobial activities of zinc oxide nanoparticles from leaf extract of Glycosmis pentaphylla (Retz.) DC. Microb Pathog 116:44–48
Vinatier C, Mrugala D, Jorgensen C, Guicheux J, Noël D (2009) Cartilage engineering: a crucial combination of cells, biomaterials and biofactors. Trends Biotechnol 27(5):307–314
Wani IA, Ahmad T (2013) Size and shape dependant antifungal activity of gold nanoparticles: a case study of Candida. Colloids Surf B: Biointerfaces 101:162–170
Wijnhoven SW, Peijnenburg WJ, Herberts CA, Hagens WI, Oomen AG, Heugens EH et al (2009) Nano-silver—a review of available data and knowledge gaps in human and environmental risk assessment. Nanotoxicology 3(2):109–138
Xiao Z, Wang Q, Zhu F, An Y (2019) Epidemiology, species distribution, antifungal susceptibility and mortality risk factors of candidemia among critically ill patients: a retrospective study from 2011 to 2017 in a teaching hospital in China. Antimicrob Resist Infect Control 8:1–7

Yassin MT, Mostafa AAF, Al-Askar AA, Al-Otibi FO (2022) Synergistic antifungal efficiency of biogenic silver nanoparticles with itraconazole against multidrug-resistant candidal strains. Crystals 12(6):816

Zareshahrabadi Z, Karami F, Taghizadeh S, Iraji A, Amani AM, Motamedi M, Zomorodian K (2021) Green synthesis of silver nanoparticles using aqueous extract of Lamium album and their antifungal properties. J Nano Res 67:55–67

Green-Synthesized Nanoparticles: Characterization and Antifungal Mechanism of Action

13

Sageer Abass, Rabea Parveen, and Sayeed Ahmad

13.1 Introduction

The realm of infectious diseases has long been a critical concern in global public health. These diseases, which can be triggered by various pathogens such as bacteria, viruses, fungi and parasites, have the potential to wreak havoc on human populations. They can be broadly categorized based on their location within the host's body, their mode of infection and the presence of biofilms or medical device involvement. Infectious diseases, whether they are intracellular or extracellular infections, biofilm-mediated, or associated with medical devices, have consistently posed challenges and threats to public health worldwide. These threats manifest as significant morbidity, mortality and economic burdens, with millions of lives lost annually (Bouz and Doležal 2021). Fungal infections have posed significant challenges in healthcare management for many years. This difficulty arises due to the limited range and potent side effects of commonly employed antifungal medications, prolonged treatment periods and the notable increase in resistance against existing treatments. The severity of fungal infections regained attention amid the unfortunate COVID-19 pandemic, where they manifested as critical secondary infections with

S. Abass · S. Ahmad (✉)
Centre of Excellence in Unani Medicine (Pharmacognosy and Pharmacology), Bioactive Natural Product Laboratory, School of Pharmaceutical Education and Research, New Delhi, India
e-mail: sageer.rather@gmail.com; rabea_nd62@yahoo.co.in

R. Parveen
Centre of Excellence in Unani Medicine (Pharmacognosy and Pharmacology), Bioactive Natural Product Laboratory, School of Pharmaceutical Education and Research, New Delhi, India

Department of Pharmaceutics, School of Pharmaceutical Education and Research, New Delhi, India
e-mail: sahmad_jh@yahoo.co.in

N. Manzoor (ed.), *Advances in Antifungal Drug Development*,
https://doi.org/10.1007/978-981-97-5165-5_13

life-threatening implications, particularly in intensive care settings. *Candida, Cryptococcus* and *Aspergillus* are key culprits behind life-threatening fungal infections in humans (Chowdhary et al. 2020). Candida species cause a series of infections, from mild to invasive *candidemia. Cryptococcus* can lead to dangerous cryptococcosis, particularly in immunocompromised individuals. *Aspergillus* species cause invasive aspergillosis, mainly affecting those with weakened immune systems. These infections, with their severe consequences and resistance to treatments, highlight the urgency for research and innovative approaches to combat them effectively (Boral et al. 2018). Fungal microorganisms give rise to severe invasive illnesses (such as fungaemia, meningitis and pneumonia), as well as challenging long-term health conditions, including complex chronic respiratory issues. These pathogens also contribute to repetitive infections, like oral and vaginal candidiasis. A significant number of these invasive fungal infections (IFIs) result from underlying health conditions connected to weakened immune systems (Brown et al. 2012).

Antifungal agents are crucial in preventing and treating a range of fungal infections, safeguarding both vulnerable patients and public health. By combating drug resistance, preventing outbreaks, enhancing surgical procedures and supporting agriculture, these agents play a vital role in maintaining well-being and safety. Moreover, they enable essential research and discovery, underscoring their significance in modern medicine and beyond (Hossain et al. 2022). Nanotechnology has led to the creation of various nanostructures for use as antifungal agents. Among these, silver nanoparticles (Ag NPs), palladium nanoparticles (Pd NPs), zinc oxide nanoparticles (ZnO NPs) and selenium nanoparticles (Se NPs), etc. have gained significant attention for their antibacterial and antifungal properties. While multiple chemical methods exist for producing Ag NPs, the 'green synthesis' approach is particularly noteworthy due to its avoidance of hazardous waste, generation of purified products, economic advantages and low energy consumption (Ahmed et al. 2016). The quest for innovative antifungal compounds heavily depends on the utilization of ethnobotanical knowledge and exploration within the field of ethnopharmacology. Plants possess remarkable capabilities in synthesizing metallic nanoparticles, and their utilization in lieu of the conventional methods for nanoparticle production has gained prominence. Extracts can be derived from different parts of plants such as stems, leaves, roots and seeds. Moreover, the plant-mediated green chemistry approach, employing phytochemically rich plants, has been shown to offer swifter, more resilient and cost-effective outcomes (Abass et al. 2022).

13.2 Plant-Based Nanoparticles: Green Synthesis Approaches

In recent times, nanoparticles have expanded substantial attention for exploration, owing to their broad utilization across various provinces such as diagnostics, antimicrobial agents, the treatment of cancer, etc. In this context, nanoparticles within the size range of 1 to 100 nm have been identified as the most pragmatic and efficient (Amooaghaie et al. 2015) (Elangovan et al. 2015). These nanoparticles were wholly created using chemical methods. Several of these nanoparticles, including zinc oxide (ZnO), ferrous oxide (FeO), silicon dioxide (SiO_2) and titanium dioxide (TiO_2), are commonly employed because of their desirable photocatalytic characteristics. Nano-sized elemental metals like silver, gold, iron, copper, platinum, nickel, cobalt, etc. are extensively utilized for diverse applications such as antimicrobial activity, targeted drug delivery, sensing, etc. (Goswami et al. 2017). In the process of nanoparticle biosynthesis, the principle of environmentally approved 'green chemistry' has been employed to create nanoparticles in a clean and eco-friendly manner. This approach encompasses various biological agents such as bacteria, fungi and plants, collectively referred to as green synthesis (Pal et al. 2018). The process of producing nanoparticles, particularly AgNPs, using plant extracts or organic materials, has attracted significant consideration due to their plentiful capabilities and diverse collection of bioactive metabolites that aid in reduction. Plants are highly favoured sources for the creation of nanoparticles. In contrast to bacteria and algae, plants exhibit greater resistance to metal toxicity, presenting an environmentally friendly option for the production of AgNPs (Rasheed et al. 2017; Jadoun et al. 2021). Various plant parts have been extensively employed for the fabrication of diverse nanoparticles. Plant extracts create nanoparticles with defined properties and phytochemicals in extracts act as natural stabilizers and reducers during synthesis. Plant-derived nanoparticles have lower human health risks than chemically synthesized ones, offering wide-ranging applications in agriculture, cosmetics and nanomedicine (Hano and Abbasi 2022).

The process of synthesizing nanoparticles using plant extracts is a cost-effective and environmentally conscious method that eliminates the need for intermediary chemical groups. Current literature specifies that certain plants have the ability to accumulate, detoxify and remediate toxic metals through phytoremediation (Carolin et al. 2017). The metallic ion bioreduction is attributed to the phytocompounds found within the plant extract, including alkaloids, terpenoids and polyphenols (Jadoun et al. 2021).

13.3 Nanoparticles as a Delivery System for Antifungal Drugs

Fungal infections are serious threats to human health that impact millions of people worldwide. Antifungal medication must frequently be used in order to treat fungal infections. These medications may, however, have drawbacks such poor solubility, low absorption and probable toxicity. Researchers have used nanotechnology to overcome these obstacles, utilizing the special qualities of nanoparticles to improve the delivery of antifungal medicines (Hossain et al. 2022). The use of nanoparticles as a delivery mechanism for antifungal medications is one of nanotechnology's most fascinating uses, ushering in a new age of medical innovation. The traditional ways of giving antifungal drugs recurrently fall short in terms of effectiveness and safety, and fungal infections can vary from minor annoyances to potentially fatal illnesses. An ground-breaking strategy made possible by nanoparticles has the potential to transform the way we treat fungal infections (Nami et al. 2021). The main goal of developing nanoparticles as a drug delivery system is to control particle size, alter surface properties and ensure precise administration of a specific medication at a specific location and time, with the goal of achieving the greatest impact (Mu and Feng 2003). By using nanoparticles, fungal infections may be treated, and because of their unique properties, these particles have more inhibitory power than pure antibiotics, even at subordinate doses. Antifungal drug delivery systems are being developed in response to prominent and significant factors such as declined drug efficacy, suboptimal drug distribution, incomplete tissue penetration, inadequate water solubility and diminished drug stability, within the body (Hassanpour et al. 2020). They can be made of a variety of materials, including polymers, lipids, metals and ceramics. When employed as a delivery mechanism for antifungal medications, the tiny size of nanoparticles has various advantages, enhanced drug solubility, improve bioavailability, targeted delivery, reduced toxicity (Han et al. 2022).

13.4 Characterization of Nanoparticles

Characterization of nanoparticles is an important step towards understanding and enhancing their characteristics for a diversity of applications, including drug delivery. Here are some distinctive methodologies through which the characterization of nanoparticles can be done.

13.4.1 Dynamic Light Scattering

Dynamic Light Scattering (DLS) is an important method for characterizing plant-based nanoparticles. It is used to investigate the size distribution and hydrodynamic characteristics of suspended nanoparticles. Because of its non-destructive nature, ability to measure in a liquid media and appropriateness for relatively small-sized particles, this approach calculates the Polydispersity Index (PDI) and Z-Average, which show the width of the overall particle distribution and size distribution based on the intensity of individual particles, respectively (Stetefeld et al. 2016). This method helps researchers understand the characteristics of these nanoparticles and optimize their formulation and uses, particularly in medication delivery, agriculture and environmental clean-up (Jia et al. 2023; Fissan et al. 2014). There are different studies which have shown that the characterization of antifungal plant based nanoparticles have been performed through DLS methods (Anandalakshmi et al. 2016). One of the research study showed that the plant extracts from *Amphipterygium glaucum* leaves and *Calendula officinalis* flowers were used to successfully synthesize selenium nanoparticles. The nanoparticles were then analysed using DLS (Lazcano-Ramírez et al. 2023). Another study shows that antifungal activities of grape seed extract and the biosynthesized AgNPs loaded with extracts against different pathogens were analysed through the DLS method (Al-Otibi et al. 2021).

13.4.2 Fourier Transform Infrared Spectroscopy and X-Ray Photoelectron Spectroscopy

Fourier transform infrared spectroscopy (FTIR) is a popular method for detecting functional groups and chemical bonding in nanoparticles. It analyses the sample's absorption of infrared light, which corresponds to the vibrations of various chemical bonds. FTIR plays a great role for analysing the chemical identification, surface modifications, quality control and interaction studies of nanoparticles. XPS is also highly useful for nanoparticle characterization due to its surface-sensitive nature. FTIR and XPS are crucial methods in nanoparticle characterization, offering complementary information about the chemical composition, surface characteristics and interactions of nanoparticles (Alharbi et al. 2022). Different studies have been done through which plant based antibacterial nanoparticles were analysed through the FTIR studies (Krithiga et al. 2015). Another study showed that AgNPs were green synthesized in vitro using *A. graecorum*, and its antifungal activity was performed and the characterization was performed by FTIR analysis (Hawar et al. 2022).

13.4.3 Transmission Electron Microscopy and Scanning Electron Microscopy

Transmission electron microscopy (TEM) and scanning electron microscopy (SEM) are strong tools for analysing nanoparticles in a variety of domains, including plant-based nanoparticle characterization (Malatesta 2021). SEM and TEM both employ electron beams to interact with materials, creating detailed pictures with distinct sorts of information. SEM gives surface morphological features by scanning the sample's surface, whereas TEM provides high-resolution pictures of interior structures by beaming electrons through thin samples. These approaches are critical for visualizing and comprehending the properties of nanoparticles, especially those synthesized using green methods in plant-based systems (Dikshit et al. 2021). The results of TEM and SEM studies demonstrated that the copper nanoparticles synthesized utilizing *Celastrus paniculatus* had a uniform and constant spherical shape. The particle sizes ranged from 2 to 10 nm (Mali et al. 2020). SEM investigation revealed that the diameter of AgNPs produced from *Glycyrrhiza glabra* root extract particles was 20–30 nm (Dinesh et al. 2012).

13.4.4 X-Ray Diffraction

X-ray diffraction (XRD) is an essential technique in nanoparticle characterization, providing information about crystal structure, particle size, lattice strain, crystallinity, phase composition and phase transformations. X-rays impact the surface of the crystal, engaging with the atoms. These atoms organize themselves at an appropriate spacing on the crystal's plane and exhibit a distinct diffraction pattern (Noah

2018). The XRD characterization technique has been applied in various studies to assess the level of crystallinity in NPs synthesized through plant extracts (Naseer et al. 2020).

13.5 Mechanism of Antifungal Activity of Nanoparticles Derived from Plants

Numerous researchers have recently drawn attention to the downsides of antifungal therapies, focusing in particular on their negative effects and the rising resistance of the organisms to these treatments (Arsene et al. 2022). The development and survival of fungal infections are inhibited by a number of complex mechanisms that make up the antifungal action of nanoparticles made from plants. These nanoparticles, also known as phyto-nanoparticles or plant-derived nanoparticles, have drawn a lot of interest since they have the potential to replace traditional antifungal drugs in a sustainable and environmentally benign way (Hiba and Thoppil 2022; Chopra et al. 2022). The antifungal activity of nanoparticles is attributed to a variety of causes and mechanisms of action, including cell wall disruption, cell membrane perturbation, genetic material interference, biofilm disruption, nanoparticle-induced apoptosis (Anand et al. 2022). The detailed mechanism of action of plant-based nanoparticles is mentioned below.

13.5.1 Cell Wall Disruption

Cell walls in fungi are essential for protection of cell shape and integrity. Plant-derived nanoparticles have the aptitude to interrelate with fungal cell wall components like chitin, chitosan, glycosylated proteins and β-glucans. These interfaces can cause structural damage, fading the cell wall and uncovering the fungus to environmental stresses and other antifungal drugs (Garcia-Rubio et al. 2020). A study investigated the effect of AgNPs on the development of *Aspergillus niger* utilizing natural polyphenol catechin (Ct). AgNPs-Ct produced physiological alterations that activated stress responses, resulting in a decrease in fungal biomass. Stress effects were measured using reactive oxygen species produced by fungal hyphae, which confirmed oxidative stress and membrane damage as the primary antifungal effects (Molina-Hernández et al. 2022).

AgNPs made from *Vitis vinifera* cane extract (VV-AgNPs) were discovered to be effective against *C. albicans* biofilms and planktonic cells. The antifungal action of VV-AgNPs was further supported by scanning electron microscopy images, which showed that *C. albicans* cells treated with it visibly adhered to the surface while obviously disrupting other cells (Miškovská et al. 2022). Another study showed that the green synthesis of palladium nanoparticles using quercetin showed antifungal activity against *C. gloeosporioides* and *F. oxysporum* by effecting the cell wall synthesis (Osonga et al. 2020). AgNPs and the *Lotus lalambensis Schweinf* leaf extract nanoparticles were explored for their anti-candidal potency against *C. albicans*. It was shown that the nanoparticles have an effect on the cell wall (Abdallah and Ali 2021).

13.5.2 Cell Membrane Perturbation

The cell membrane is essential for maintaining integrity and controlling a variety of cellular functions in fungal cells. The disruption of cell membrane leads to ion leakage, osmotic imbalances and finally causes cell death (Mani-López et al. 2021). Plant-based nanoparticles with special physicochemical characteristics can interact with fungal cell membranes and the structure and operation of the membrane may become disrupted as a result of these interactions, endangering the survival of the fungus cell. Numerous researchers have recently drawn attention to the downsides of antifungal therapies, focusing in particular on their negative effects and the rising fungal resistance to these treatments.

The leaf extract from *Erodium glaucophyllum* was utilized to synthesize AgNPs (EG-AgNPs) and their potential antifungal impact on *C. albicans* was evaluated. EG-AgNPs hindered the dimorphic transition of *C. albicans* and decreased the viability of biofilms. They also reduced the enzymatic activity of both proteinases and phospholipases. TEM images revealed that EG-AgNPs treatment caused alterations in the morphology of fungal cells, such as cytoplasmic disintegration, vacuolation, perinuclear changes and granular nuclear modifications. Furthermore, the cell wall of *C. albicans* appeared swollen, with the outer layer appearing to detach from the cell (Abdallah and Ali 2022). Another study showed that the antifungal activity test of the synthesized AgNPs using *Citrus limetta* peel extract against Candida species discovered that the nanoparticle have the capacity of cell membrane alteration (Dutta et al. 2020).

13.5.3 Ergosterol Biosynthesis Inhibition

A crucial element of the fungal cell membrane is ergosterol. By blocking the enzyme lanosterol 14-demethylase, which is involved in converting lanosterol into ergosterol, different types of antifungal drugs target this route. The fungal cell membrane becomes more permeable as a result of this disturbance, which also causes cell death (Ghannoum and Rice 1999). A recent study outlines a straightforward process for producing ZnO NPs using a water-based extract from *Salvia officinalis* L. leaves as a highly effective stabilizing and capping agent. The research also investigates the antifungal properties of these bio-fabricated ZnO NPs and suggests that their antifungal effects stem from their ability to hinder the production of ergosterol and disrupt the integrity of the membranes in various clinical *Candida albicans* strains (Abomuti et al. 2021). Another study showed that green synthesis of Se NPs have promising effects on the decreasing expression of ERG3, ERG11 and FKS1 antifungal resistance genes which play a main role in ergosterol biosynthesis pathway in both *C. albicans* and *C. glabrata* (Hosseini Bafghi et al. 2022).

13.5.4 Biofilm Disruption

Biofilms are complex colonies of microorganisms, including fungus that attach to surfaces and are enclosed in a protective extracellular matrix. Biofilm production adds to the virulence and resistance of many fungal diseases. In recent years, researchers have started studying the suppression of fungal biofilms by antifungal medicines (Ramage et al. 2012). One of the studies looked at the antibiofilm and antifungal properties of AgNPs generated by using *Lycopersicon esculentum* extracts against *Candida* species. The capability of AgNPs to inhibit biofilm growth was found through scanning electron microscopy. The researchers hypothesized that the contact of Ag ions with the cell surface caused membrane abnormalities, which inhibited biofilm formation and *Candida* development. Furthermore, it has long been shown that quorum sensing is critical in the establishment of biofilm. Several investigations have found that anti-QS drugs can reduce biofilm growth (Choi et al. 2019). One more study showed that AgNPs using *beech bark* extract (BBE) and acetate and nitrate silver salts. The generation of germ tubes in *C. albicans* was observed to be hindered by AgNP BBE, which resulted in the suppression of SAP2 expression (Mare et al. 2021). Another study evaluated the anti-candidal activity of AgNPs biosynthesized using the aqueous leaf extract of *Calotropis gigantean* against *C. albicans* by inhibition of biofilm (Ali and Abdallah 2020). Research involving the utilization of *Eucalyptus camaldulensis* leaf extract to synthesize AgNPs has demonstrated a reduction in the activity of genes ALS3, HWP1, ECE1, EFG1, TEC1 and ZAP1. These genes are responsible for controlling the growth of hyphae and the development of biofilms in the fungus *C. albicans* (Wunnoo et al. 2021).

13.5.5 Protein Synthesis Inhibition

Some antifungal drugs can impair fungal protein synthesis by binding to the ribosome, interrupting translation and reducing the generation of critical proteins. Protein synthesis is the critical step for the growth of fungal cells (Scorzoni et al. 2017). One of the research studies find that quercetin-assisted AgNPs interacts with intracellular structures and biomolecules (proteins, DNA, ribosomes and enzymes) in the fungal strains and lead to cell death (Chahardoli et al. 2021). Another research work evaluated the antifungal activity of aloe leaf broth extract and zinc nitrate, and it is found that these green-synthesized nanoparticles affect the protein synthesis in different fungal strains (Gunalan et al. 2012). Figure 13.1 describes over all mechanism of action of plant-based nanoparticles. Table 13.1 presents the information about different plant extracts, nanoparticles and their respective antifungal mechanisms.

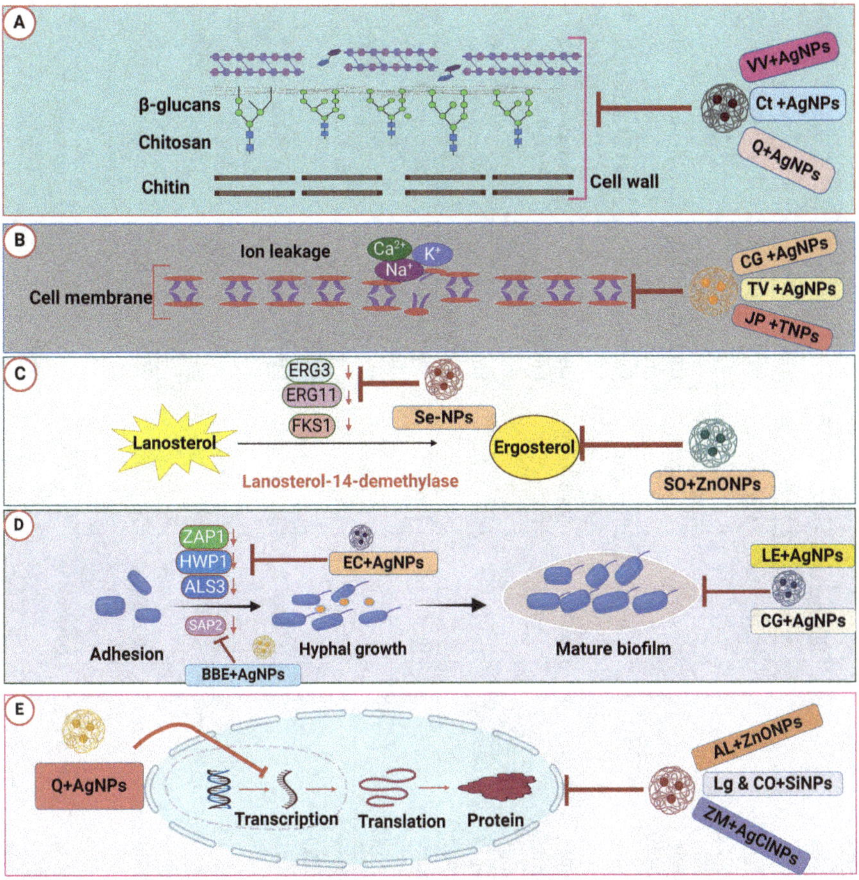

Fig. 13.1 Mechanisms of antifungal activity of plant extract-based nanoparticles. (**a**) Cell wall disruption: *Vitis vinifera* silver nanoparticles (VV+AgNPs), catechin silver nanoparticles (Ct+AgNPs) and quercetin silver nanoparticles (Q+AgNPs) interact with fungal cell wall components, leading to structural damage and disruption of the fungal cell wall. (**b**) Cell membrane perturbation: *Calotropis gigantea* silver nanoparticles (CG+AgNPs), *Juniperus phoenicea* titanium dioxide nanoparticles (JP+TNPs) and *Thymus vulgaris* silver nanoparticles (TV+AgNPs) compromise membrane integrity, causing ion leakage and ultimately resulting in fungal cell death. (**c**) Ergosterol biosynthesis inhibition: Selenium nanoparticles (Se+NPs) and *Salvia officinalis* zinc oxide nanoparticles (SO+ZnONPs) inhibit ergosterol production, a vital component of fungal cell membranes, leading to fungal growth inhibition. (**d**) Biofilm disruption: *Eucalyptus camaldulensis* silver nanoparticles (EC+AgNPs), *beech bark* extract silver nanoparticles (BBE+AgNPs), *Lycopersicon Esculentum* silver nanoparticles (LE+AgNPs) and *Calotropis gigantea* silver nanoparticles (CG+AgNPs) disrupt biofilm formation, a crucial virulence factor of fungi. **(E)** Protein synthesis inhibition: Quercetin silver nanoparticles (Q+AgNPs), *lemongrass* and *clove oil* silica nanoparticles (Lg & CO+SiNPs), *Aloe leaf* zinc nitrate nanoparticles (AL+ZnONPs) and *Ziziphus Mauritiana* silver chloride nanoparticles (ZM+AgCLNPs) inhibit protein synthesis, ultimately leading to fungal cell death

Table 13.1 Antifungal mechanisms of plant-based nanoparticles

S. No.	Plant	Type of nanoparticle	Strain	In vitro/In vivo	MIC/IC_{50}/ Volume	Mechanism	References
1	*Aegle marmelos*	Silver nanoparticles	*C. albicans*	In vitro	100 µL	NA	Patil and Muthusamy (2020)
2	*Calotropis gigantean*	Silver nanoparticles	*C. albicans*	In vitro	200 µg/mL	Blocking of hyphal growth	Ali and Abdallah (2020)
3	*Dodonaea viscosa* and *Hyptis suoveolens*	Silver nanoparticles	*C. albicans*	In vitro/in vivo	10 µg/mL	Inhibit biofilm	Muthamil et al. (2018)
4	*Artemisia annua*	Silver nanoparticles	*C. albicans*	In vitro	80–120 µg/mL	Disfigured morphology	Khatoon et al. (2019)
			C. tropicalis				
			C. glabrata				
5	*Calotropis procera*	Silver nanoparticles	*T. rubrum, C. albicans* and *A. terreus*	In vitro	10 µL	Disturbance of permeability	Mohamed et al. (2014)
6	*Calotropis gigantean*	Silver nanoparticles	*C. albicans*	In vitro	2 mM	Dysfunctional mitochondria	Kemala et al. (2022)
7	*Artemisia absinthium*	Silver nanoparticles	*c. albicans*	In vitro	0.7 µg/mL	NA	Rodríguez-Torres et al. (2019)
			C. parapsilosis				
			I. orientalis				
8	*Azadirachta indica*	Silver nanoparticles	*C. tropicalis*	In vitro	4.12 µg/mL	Inhibit fungal adherence	Al Aboody (2019)
		Copper nanoparticles	*Alternaria mali, Diplodia seriata* and *Botryosphaeria dothidea*	In vitro	0.5 µg/mL	Damage the DNA	Ahmad et al. (2020)
9	*Juniperus phoenicea*	Titanium dioxide nanoparticles	*A. niger* and *P. digitatum*	In vitro	40 µL/mL	Effect on cell membrane	Al Masoudi et al. (2023)
10	*Achillea millefolium*	Copper oxide nanoparticles	*A. flavus, M. canis* and *G. glabrata*	In vitro	50 µg/mL	Rupturing the cell walls	Rabiee et al. (2020)

11	*Ziziphus Mauritiana*	Silver chloride nanoparticles	*A. niger*	In vitro	60 μg/mL	Inhibit protein synthesis	Kabir et al. (2020)
12	Lemongrass and clove oil	Silica nanoparticles	*Gaeumannomyces tritici*	In vitro	7.9 μg/mL	Destroy cell proteins	Sattary et al. (2020)
13	*Nigrospora oryzae*	Silver nanoparticles	*Fusarium* spp.	In vitro	NA	NA	Dawoud et al. (2021)
14	Grape pomace	Biopolymeric nanoparticles	*C. albicans*	In vitro	50 μg/mL	Inhibit biofilm	Simonetti et al. (2019)
15	*Ulva lactuca*	Silver nanoparticles	*F. oxysporum*	In vitro	80 μg/mL	NA	Sahayaraj et al. (2019)
16	*Melia azedarach*	Selenium nanoparticles	*F. mangiferae*	In vitro	300 μg/mL	Altered cell shape	Shahbaz et al. (2023)
17	*Lycopersicon esculentum*	Silver nanoparticles	*Candida species*	In vitro	32 μg/mL	Inhibit biofilm formation	Choi et al. (2019)
18	*Zingiber officinale* and *Thymus vulgaris*	Silver nanoparticles	*C. albicans*	In vitro	0.7 μg/mL	Interrupting the membrane integrity	Mohammadi et al. (2019)
19	*Calotropis gigantean*	Zinc oxide nanoparticles	*A. paracistic* and *F. solani*	In vitro	25 μg/mL	NA	Farooq et al. (2022)
20	*Pulicaria vulgaris*	Silver chloride nanoparticles	*C. glabrata* and *C. albicans*	In vitro	60 μg/mL	NA	Sharifi-Rad and Pohl (2020)

13.6 Conclusion

Finally, plant-based nanoparticles synthesized using green technologies provide a convincing answer to the problems faced by fungal diseases. Their noteworthy antifungal qualities, low cost and eco-friendliness make them an attractive choice for a variety of applications, including nanomedicine, agriculture and cosmetics. Characterization approaches have offered useful insights into the characteristics of nanoparticles, assisting in their optimization. The antifungal activity mechanisms of these nanoparticles have considerable potential for revolutionizing fungal infection therapy. Further formulation refinement, study of combination medicines, targeted delivery tactics, biodegradable possibilities, clinical trials and scaled-up production methods will pave the path for their greater usage in healthcare as research progresses. In conclusion, plant-based nanoparticles have the potential to enhance global public health outcomes in the battle against fungal diseases.

References

Abass S, Parveen R, Irfan M, Jan B, Husain SA, Ahmad S (2022) Synergy based extracts of medicinal plants: Future antimicrobials to combat multidrug resistance. Curr Pharm Biotechnol:23. https://doi.org/10.2174/1389201023666220126115656

Abdallah BM, Ali EM (2021) Green synthesis of silver nanoparticles using the lotus lalambensis aqueous leaf extract and their anti-candidal activity against oral candidiasis. ACS Omega. https://doi.org/10.1021/acsomega.0c06009

Abdallah BM, Ali EM (2022) Therapeutic effect of green synthesized silver nanoparticles using erodium glaucophyllum extract against oral candidiasis: in vitro and in vivo study. Molecules. https://doi.org/10.3390/molecules27134221

Abomuti MA, Danish EY, Firoz A, Hasan N, Malik MA (2021) Green synthesis of zinc oxide nanoparticles using salvia officinalis leaf extract and their photocatalytic and antifungal activities. Biology. https://doi.org/10.3390/biology10111075

Ahmad H, Venugopal K, Bhat AH, Kavitha K, Ramanan A, Rajagopal K, Srinivasan R, Manikandan E (2020) Enhanced biosynthesis synthesis of copper oxide nanoparticles (CuO-NPs) for their antifungal activity toxicity against major phyto-pathogens of apple orchards. Pharm Res. https://doi.org/10.1007/s11095-020-02966-x

Ahmed S, Ahmad M, Swami BL, Ikram S (2016) A review on plants extract mediated synthesis of silver nanoparticles for antimicrobial applications: a green expertise. J Adv Res. https://doi.org/10.1016/j.jare.2015.02.007

Al Aboody MS (2019) Silver/silver chloride (Ag/AgCl) nanoparticles synthesized from Azadirachta indica lalex and its antibiofilm activity against fluconazole resistant Candida tropicalis. Artif Cells Nanomed Biotechnol. https://doi.org/10.1080/21691401.2019.1620257

Al Masoudi LM, Alqurashi AS, Abu Zaid A, Hamdi H (2023) Characterization and biological studies of synthesized titanium dioxide nanoparticles from leaf extract of Juniperus phoenicea (L.) growing in Taif Region, Saudi Arabia. Processes. https://doi.org/10.3390/pr11010272

Alharbi NS, Alsubhi NS, Felimban AI (2022) Green synthesis of silver nanoparticles using medicinal plants: characterization and application. J Radiat Res Appl Sci. https://doi.org/10.1016/j.jrras.2022.06.012

Ali EM, Abdallah BM (2020) Effective inhibition of candidiasis using an eco-friendly leaf extract of calotropis-gigantean-mediated silver nanoparticles. Nanomaterials (Basel). https://doi.org/10.3390/nano10030422

Al-Otibi F, Alkhudhair SK, Alharbi RI, Al-Askar AA, Aljowaie RM, Al-Shehri S (2021) The antimicrobial activities of silver nanoparticles from aqueous extract of grape seeds against pathogenic bacteria and fungi. Molecules. https://doi.org/10.3390/molecules26196081

Amooaghaie R, Saeri MR, Azizi M (2015) Synthesis, characterization and biocompatibility of silver nanoparticles synthesized from Nigella sativa leaf extract in comparison with chemical silver nanoparticles. Ecotoxicol Environ Saf. https://doi.org/10.1016/j.ecoenv.2015.06.025

Anand U, Carpena M, Kowalska-Góralska M, Garcia-Perez P, Sunita K, Bontempi E, Dey A, Prieto MA, Proćków J, Simal-Gandara J (2022) Safer plant-based nanoparticles for combating antibiotic resistance in bacteria: a comprehensive review on its potential applications, recent advances, and future perspective. Sci Total Environ. https://doi.org/10.1016/j.scitotenv.2022.153472

Anandalakshmi K, Venugobal J, Ramasamy V (2016) Characterization of silver nanoparticles by green synthesis method using Pedalium murex leaf extract and their antibacterial activity. Appl Nanosci. https://doi.org/10.1007/s13204-015-0449-z

Arsene MMJ, Jorelle ABJ, Sarra S, Viktorovna PI, Davares AKL, Ingrid NKC, Steve AAF, Andreevna SL, Vyacheslavovna YN, Carime BZ (2022) Short review on the potential alternatives to antibiotics in the era of antibiotic resistance. J Appl Pharm Sci. https://doi.org/10.7324/JAPS.2021.120102

Boral H, Metin B, Döğen A, Seyedmousavi S, Ilkit M (2018) Overview of selected virulence attributes in Aspergillus fumigatus, Candida albicans, Cryptococcus neoformans, Trichophyton rubrum, and Exophiala dermatitidis. Fungal Genet Biol 111:92–107

Bouz G, Doležal M (2021) Advances in antifungal drug development: an up-to-date mini review. Pharmaceuticals. https://doi.org/10.3390/ph14121312

Brown GD, Denning DW, Gow NAR, Levitz SM, Netea MG, White TC (2012) Hidden killers: human fungal infections. Sci Transl Med. https://doi.org/10.1126/scitranslmed.3004404

Carolin CF, Kumar PS, Saravanan A, Joshiba GJ, Naushad M (2017) Efficient techniques for the removal of toxic heavy metals from aquatic environment: a review. J Environ Chem Eng. https://doi.org/10.1016/j.jece.2017.05.029

Chahardoli A, Hajmomeni P, Ghowsi M, Qalekhani F, Shokoohinia Y, Fattahi A (2021) Optimization of quercetin-assisted silver nanoparticles synthesis and evaluation of their hemocompatibility, antioxidant, anti-inflammatory, and antibacterial effects. Global Chall. https://doi.org/10.1002/gch2.202100075

Choi JS, Lee JW, Shin UC, Lee MW, Kim DJ, Kim SW (2019) Inhibitory activity of silver nanoparticles synthesized using lycopersicon esculentum against biofilm formation in candida species. Nanomaterials (Basel). https://doi.org/10.3390/nano9111512

Chopra H, Bibi S, Singh I, Hasan MM, Khan MS, Yousafi Q, Baig AA, Rahman MM, Islam F, Emran TB, Cavalu S (2022) Green metallic nanoparticles: biosynthesis to applications. Front Bioeng Biotechnol. https://doi.org/10.3389/fbioe.2022.874742

Chowdhary A, Tarai B, Singh A, Sharma A (2020) Multidrug-resistant candida auris infections in critically ill coronavirus disease patients, India, April–July 2020. Emerg Infect Dis. https://doi.org/10.3201/eid2611.203504

Dawoud TM, Yassin MA, El-Samawaty ARM, Elgorban AM (2021) Silver nanoparticles synthesized by Nigrospora oryzae showed antifungal activity. Saudi J Biol Sci. https://doi.org/10.1016/j.sjbs.2020.12.036

Dikshit PK, Kumar J, Das AK, Sadhu S, Sharma S, Singh S, Gupta PK, Kim BS (2021) Green synthesis of metallic nanoparticles: applications and limitations. Catalysts. https://doi.org/10.3390/catal11080902

Dinesh S, Karthikeyan S, Arumugam P, Nadu T, Nadu T, Nadu T (2012) Biosynthesis of silver nanoparticles from Glycyrrhiza glabra root extract. Arch Appl Sci Res

Dutta T, Ghosh NN, Das M, Adhikary R, Mandal V, Chattopadhyay AP (2020) Green synthesis of antibacterial and antifungal silver nanoparticles using Citrus limetta peel extract: experimental and theoretical studies. J Environ Chem Eng. https://doi.org/10.1016/j.jece.2020.104019

Elangovan K, Elumalai D, Anupriya S, Shenbhagaraman R, Kaleena PK, Murugesan K (2015) Phyto mediated biogenic synthesis of silver nanoparticles using leaf extract of Andrographis echioides and its bio-efficacy on anticancer and antibacterial activities. J Photochem Photobiol B Biol. https://doi.org/10.1016/j.jphotobiol.2015.05.015

Farooq A, Khan UA, Ali H, Sathish M, Naqvi SAH, Iqbal S, Ali H, Mubeen I, Amir MB, Mosa WFA, Baazeem A, Moustafa M, Alrumman S, Shati A, Negm S (2022) Green chemistry based synthesis of zinc oxide nanoparticles using plant derivatives of Calotropis gigantea (giant milkweed) and its biological applications against various bacterial and fungal pathogens. Microorganisms. https://doi.org/10.3390/microorganisms10112195

Fissan H, Ristig S, Kaminski H, Asbach C, Epple M (2014) Comparison of different characterization methods for nanoparticle dispersions before and after aerosolization. Anal Methods. https://doi.org/10.1039/c4ay01203h

Garcia-Rubio R, de Oliveira HC, Rivera J, Trevijano-Contador N (2020) The fungal cell wall: candida, cryptococcus, and aspergillus species. Front Microbiol. https://doi.org/10.3389/fmicb.2019.02993

Ghannoum MA, Rice LB (1999) Antifungal agents: mode of action, mechanisms of resistance, and correlation of these mechanisms with bacterial resistance. Clin Microbiol Rev. https://doi.org/10.1128/cmr.12.4.501

Goswami L, Kim KH, Deep A, Das P, Bhattacharya SS, Kumar S, Adelodun AA (2017) Engineered nano particles: nature, behavior, and effect on the environment. J Environ Manage. https://doi.org/10.1016/j.jenvman.2017.01.011

Gunalan S, Sivaraj R, Rajendran V (2012) Green synthesized ZnO nanoparticles against bacterial and fungal pathogens. Prog Nat Sci Mater Int. https://doi.org/10.1016/j.pnsc.2012.11.015

Han HS, Koo SY, Choi KY (2022) Emerging nanoformulation strategies for phytocompounds and applications from drug delivery to phototherapy to imaging. Bioact Mater. https://doi.org/10.1016/j.bioactmat.2021.11.027

Hano C, Abbasi BH (2022) Plant-based green synthesis of nanoparticles: production, characterization and applications. Biomolecules. https://doi.org/10.3390/biom12010031

Hassanpour P, Hamishehkar H, Baradaran B, Mohammadi M, Shomali N, Spotin A, Hazratian T, Nami S (2020) An appraisal of antifungal impacts of nano-liposome containing voriconazole on voriconazole-resistant Aspergillus flavus isolates as a groundbreaking drug delivery system. Nanomed Res J. https://doi.org/10.22034/NMRJ.2020.01.010

Hawar SN, Al-Shmgani HS, Al-Kubaisi ZA, Sulaiman GM, Dewir YH, Rikisahedew JJ (2022) Green synthesis of silver nanoparticles from Alhagi graecorum leaf extract and evaluation of their cytotoxicity and antifungal activity. J Nanomater. https://doi.org/10.1155/2022/1058119

Hiba H, Thoppil JE (2022) Medicinal herbs as a panacea for biogenic silver nanoparticles. Bullet Natl Res Centre. https://doi.org/10.1186/s42269-021-00692-x

Hossain CM, Ryan LK, Gera M, Choudhuri S, Lyle N, Ali KA, Diamond G (2022) Antifungals and drug resistance. Encyclopedia. https://doi.org/10.3390/encyclopedia2040118

Hosseini Bafghi M, Zarrinfar H, Darroudi M, Zargar M, Nazari R (2022) Green synthesis of selenium nanoparticles and evaluate their effect on the expression of ERG3, ERG11 and FKS1 antifungal resistance genes in Candida albicans and Candida glabrata. Lett Appl Microbiol. https://doi.org/10.1111/lam.13667

Jadoun S, Arif R, Jangid NK, Meena RK (2021) Green synthesis of nanoparticles using plant extracts: a review. Environ Chem Lett. https://doi.org/10.1007/s10311-020-01074-x

Jia Z, Li J, Gao L, Yang D, Kanaev A (2023) Dynamic light scattering: a powerful tool for in situ nanoparticle sizing. Colloids Interfaces. https://doi.org/10.3390/colloids7010015

Kabir SR, Asaduzzaman AKM, Amin R, Haque AT, Ghose R, Rahman MM, Islam J, Amin MB, Hasan I, Debnath T, Chun BS, Zhao XD, Rahman Khan MK, Alam MT (2020) Zizyphus mauritiana fruit extract-mediated synthesized silver/silver chloride nanoparticles retain antimicrobial activity and induce apoptosis in MCF-7 cells through the Fas pathway. ACS Omega. https://doi.org/10.1021/acsomega.0c02878

Kemala P, Idroes R, Khairan K, Ramli M, Jalil Z, Idroes GM, Tallei TE, Helwani Z, Safitri E, Iqhrammullah M, Nasution R (2022) Green synthesis and antimicrobial activities of silver nanoparticles using Calotropis gigantea from Ie Seu-um geothermal area, Aceh Province, Indonesia. Molecules. https://doi.org/10.3390/molecules27165310

Khatoon N, Sharma Y, Sardar M, Manzoor N (2019) Mode of action and anti-Candida activity of Artemisia annua mediated-synthesized silver nanoparticles. J Mycol Med. https://doi.org/10.1016/j.mycmed.2019.07.005

Krithiga N, Rajalakshmi A, Jayachitra A (2015) Green synthesis of silver nanoparticles using leaf extracts of Clitoria ternatea and Solanum nigrum and study of its antibacterial effect against common nosocomial pathogens. J Nanosci. https://doi.org/10.1155/2015/928204

Lazcano-Ramírez HG, Garza-García JJO, Hernández-Díaz JA, León-Morales JM, Macías-Sandoval AS, García-Morales S (2023) Antifungal activity of selenium nanoparticles obtained by plant-mediated synthesis. Antibiotics. https://doi.org/10.3390/antibiotics12010115

Malatesta M (2021) Transmission electron microscopy as a powerful tool to investigate the interaction of nanoparticles with subcellular structures. Int J Mol Sci. https://doi.org/10.3390/ijms222312789

Mali SC, Dhaka A, Githala CK, Trivedi R (2020) Green synthesis of copper nanoparticles using Celastrus paniculatus Willd. leaf extract and their photocatalytic and antifungal properties. Biotechnol Rep. https://doi.org/10.1016/j.btre.2020.e00518

Mani-López E, Cortés-Zavaleta O, López-Malo A (2021) A review of the methods used to determine the target site or the mechanism of action of essential oils and their components against fungi. SN Appl Sci. https://doi.org/10.1007/s42452-020-04102-1

Mare AD, Ciurea CN, Man A, Mareş M, Toma F, Berţa L, Tanase C (2021) In vitro antifungal activity of silver nanoparticles biosynthesized with beech bark extract. Plan Theory. https://doi.org/10.3390/plants10102153

Miškovská A, Rabochová M, Michailidu J, Masák J, Čejková A, Lorinčík J, Maťátková O (2022) Antibiofilm activity of silver nanoparticles biosynthesized using viticultural waste. PLoS One. https://doi.org/10.1371/journal.pone.0272844

Mohamed NH, Ismail MA, Abdel-Mageed WM, Shoreit AAM (2014) Antimicrobial activity of latex silver nanoparticles using Calotropis procera. Asian Pac J Trop Biomed. https://doi.org/10.12980/APJTB.4.201414B216

Mohammadi M, Shahisaraee SA, Tavajjohi A, Pournoori N, Muhammadnejad S, Mohammadi SR, Poursalehi R, Hamid Delavari H (2019) Green synthesis of silver nanoparticles using Zingiber officinale and Thymus vulgaris extracts: characterisation, cell cytotoxicity, and its antifungal activity against Candida albicans in comparison to fluconazole. IET Nanobiotechnol. https://doi.org/10.1049/iet-nbt.2018.5146

Molina-Hernández JB, Scroccarello A, Della Pelle F, De Flaviis R, Compagnone D, Del Carlo M, Paparella A, Chaves López C (2022) Synergistic antifungal activity of catechin and silver nanoparticles on Aspergillus Niger isolated from coffee seeds. LWT. https://doi.org/10.1016/j.lwt.2022.113990

Mu L, Feng SS (2003) A novel controlled release formulation for the anticancer drug paclitaxel (Taxol®): PLGA nanoparticles containing vitamin E TPGS. J Control Release. https://doi.org/10.1016/S0168-3659(02)00320-6

Muthamil S, Devi VA, Balasubramaniam B, Balamurugan K, Pandian SK (2018) Green synthesized silver nanoparticles demonstrating enhanced in vitro and in vivo antibiofilm activity against Candida spp. J Basic Microbiol. https://doi.org/10.1002/jobm.201700529

Nami S, Aghebati-Maleki A, Aghebati-Maleki L (2021) Current applications and prospects of nanoparticles for antifungal drug delivery. EXCLI J. https://doi.org/10.17179/excli2020-3068

Naseer M, Aslam U, Khalid B, Chen B (2020) Green route to synthesize zinc oxide nanoparticles using leaf extracts of Cassia fistula and Melia azadarach and their antibacterial potential. Sci Rep. https://doi.org/10.1038/s41598-020-65949-3

Noah N (2018) Green synthesis: characterization and application of silver and gold nanoparticles. In: Green synthesis, characterization and applications of nanoparticles. https://doi.org/10.1016/B978-0-08-102579-6.00006-X

Osonga FJ, Kalra S, Miller RM, Isika D, Sadik OA (2020) Synthesis, characterization and antifungal activities of eco-friendly palladium nanoparticles. RSC Adv. https://doi.org/10.1039/c9ra07800b

Pal G, Rai P, Pandey A (2018) Green synthesis of nanoparticles: a greener approach for a cleaner future. In: Green synthesis, characterization and applications of nanoparticles. https://doi.org/10.1016/B978-0-08-102579-6.00001-0

Patil S, Muthusamy P (2020) A bio-inspired approach of formulation and evaluation of Aegle marmelos fruit extract mediated silver nanoparticle gel and comparison of its antibacterial activity with antiseptic cream. Eur J Integr Med. https://doi.org/10.1016/j.eujim.2019.101025

Rabiee N, Bagherzadeh M, Kiani M, Ghadiri AM, Etessamifar F, Jaberizadeh AH, Shakeri A (2020) Biosynthesis of copper oxide nanoparticles with potential biomedical applications. Int J Nanomedicine. https://doi.org/10.2147/IJN.S255398

Ramage G, Rajendran R, Sherry L, Williams C (2012) Fungal biofilm resistance. Int J Microbiol. https://doi.org/10.1155/2012/528521

Rasheed T, Bilal M, Iqbal HMN, Li C (2017) Green biosynthesis of silver nanoparticles using leaves extract of Artemisia vulgaris and their potential biomedical applications. Colloids Surf B: Biointerfaces. https://doi.org/10.1016/j.colsurfb.2017.07.020

Rodríguez-Torres MDP, Acosta-Torres LS, Díaz-Torres LA, Hernández Padrón G, García-Contreras R, Millán-Chiu BE (2019) Artemisia absinthium-based silver nanoparticles antifungal evaluation against three Candida species. Mater Res Express. https://doi.org/10.1088/2053-1591/ab1fba

Sahayaraj K, Rajesh S, Rathi JAM, Kumar V (2019) Green preparation of seaweed-based silver nano-liquid for cotton pathogenic fungi management. IET Nanobiotechnol. https://doi.org/10.1049/iet-nbt.2018.5007

Sattary M, Amini J, Hallaj R (2020) Antifungal activity of the lemongrass and clove oil encapsulated in mesoporous silica nanoparticles against wheat's take-all disease. Pestic Biochem Physiol. https://doi.org/10.1016/j.pestbp.2020.104696

Scorzoni L, de Paula e Silva ACA, Marcos CM, Assato PA, de Melo WCMA, de Oliveira HC, Costa-Orlandi CB, Mendes-Giannini MJS, Fusco-Almeida AM (2017) Antifungal therapy: new advances in the understanding and treatment of mycosis. Front Microbiol. https://doi.org/10.3389/fmicb.2017.00036

Shahbaz M, Akram A, Raja NI, Mukhtar T, Mehak A, Fatima N, Ajmal M, Ali K, Mustafa N, Abasi F (2023) Antifungal activity of green synthesized selenium nanoparticles and their effect on physiological, biochemical, and antioxidant defense system of mango under mango malformation disease. PLoS One. https://doi.org/10.1371/journal.pone.0274679

Sharifi-Rad M, Pohl P (2020) Synthesis of biogenic silver nanoparticles (Agcl-nps) using a pulicaria vulgaris gaertn. aerial part extract and their application as antibacterial, antifungal and antioxidant agents. Nanomaterials. https://doi.org/10.3390/nano10040638

Simonetti G, Palocci C, Valletta A, Kolesova O, Chronopoulou L, Donati L, Di Nitto A, Brasili E, Tomai P, Gentili A, Pasqua G (2019) Anti-Candida biofilm activity of pterostilbene or crude extract from non-fermented grape pomace entrapped in biopolymeric nanoparticles. Molecules. https://doi.org/10.3390/molecules24112070

Stetefeld J, McKenna SA, Patel TR (2016) Dynamic light scattering: a practical guide and applications in biomedical sciences. Biophys Rev. https://doi.org/10.1007/s12551-016-0218-6

Wunnoo S, Paosen S, Lethongkam S, Sukkurd R, Waen-ngoen T, Nuidate T, Phengmak M, Voravuthikunchai SP (2021) Biologically rapid synthesized silver nanoparticles from aqueous eucalyptus camaldulensis leaf extract: effects on hyphal growth, hydrolytic enzymes, and biofilm formation in Candida albicans. Biotechnol Bioeng. https://doi.org/10.1002/bit.27675

14 Green-Synthesized Nanoparticles: Antifungal Efficacy and Other Applications

Mostafa Mohammed Atiyah, M. S. Jisha,
and Smitha Vijayan

14.1 Introduction

Fungi play a crucial and multifaceted role in the natural world, and their importance can be observed in various aspects of life, including ecology, medicine, industry, and agriculture. Fungi are essential for the healthy growth of the majority of plants,; nevertheless, they can also cause diseases in humans, animals, and crops (Mohmand et al. 2011). Fungal growth in nature is facilitated by the filamentous growth, as well as the ability to secrete proteins and primary and secondary metabolites. These characteristics are used by the industry to make tiny chemical molecules, proteins, and, more recently, mycelium materials. These bio-based goods could be packaged and utilized as acoustic and thermal insulation and packaging (Jones et al. 2017). The goal of creating more environment-friendly manufacturing could be aided by fungi, which are a great option for producing a variety of goods like myco-leather and textiles, biofuels, construction materials, wastewater treatment, and sustainable meat alternatives (Pombeiro-Sponchiado et al. 2017; Poorniammal et al. 2021). In addition to being used in the industry and food, fungi are also used in medicine to

M. M. Atiyah
School of Biosciences, Mar Athanasios College for Advanced Studies, Thiruvalla, Kerala, India

Department of Biology, Thi-Qar Education Directorate, Thi-Qar, Iraq
e-mail: mostafa@macfast.ac.in

M. S. Jisha
School of Biosciences, Mahatma Gandhi University, Kottayam, Kerala, India
e-mail: jishams@mgu.ac.in

S. Vijayan (✉)
Department of Food Technology and Quality Assurance, Mar Athanasios College for Advanced Studies, Thiruvalla, Kerala, India
e-mail: smitha@macfast.org

N. Manzoor (ed.), *Advances in Antifungal Drug Development*,
https://doi.org/10.1007/978-981-97-5165-5_14

produce a wide range of compounds, such as antibiotics penicillin or cyclosporine, which is produced by *Tolypocladium inflatum*, and lovastatin, which is an anticholesterol medication made from *Aspergillus terreus* (Irum and Anjum 2012; Karwehl and Stadler 2016). Many chemical compounds identified from fungi may prove to be useful therapies for specific forms of cancer. The Chinese have long utilized a particular caterpillar fungus as a tonic (Mohmand et al. 2011).

Fungi are important in agriculture for a variety of reasons, including plant development and protection. For instance, mycorrhizal fungi form a mutualistic relationship with plant roots to increase the surface area of the root system, which enhances the plant's ability to absorb nutrients and this connection facilitates uptake of nutrients by plants that are not easily found in the soil, such as nitrogen and phosphorus (Smith and Read 2008; Eugenia and Florin 2010; Şesan et al. 2010). The endophytic fungi that grow on plant tissue are another important type of fungal species used in agriculture. Endophytic fungi and plant tissues have a complicated connection that involves regulating the plant defense mechanisms to prevent growth of phytopathogens but promote plant growth even in the face of biotic and abiotic stress (Galindo-Solís and Fernández 2022). However, in rare circumstances, certain fungal species have a less favorable effect on plant health, invading roots, leaves, and tissues to cause a variety of plant illnesses (Kaur et al. 2021). Although citric acid was originally extracted from lemon juice, fungi now produce over 95% of the world's supply of this acid, and approximately 80% of the food acidulants are utilized in the food business (Fig. 14.1) (Copetti 2019).

Skin infections can be brought on by fungus, viruses, bacteria, and parasites. The fact that fungal diseases affect the deeper layers of skin makes them more serious (Imtiaz et al. 2019). Fungi affect keratin tissue, which includes the skin, hair, and nails (Rai et al. 2017). *Trichophyton* species are the primary cause of a number of well-known severe skin illnesses, including *Tinea corporis*, *Tinea pedis*, *Tinea faciei*, *Tinea manuum*, *Tinea cruris*, and *Tinea barbae* (Wijesiri et al. 2018). The most common signs of fungal infections are itchy red areas, crusty patches, and hair loss (Abd Elaziz et al. 2020). Wearing clothing that is too tight or sharing a wardrobe, furniture, or changing with someone who is infected are two common circumstances that can result in fungal infection (Jain et al. 2010). Antifungal medications, mostly topical, oral, and intravenous, are used to treat a variety of fungal infections; however, compared to topical antifungal medications, oral antifungal medications are more harmful to human health. Azoles, echinocandins, and polyenes are a few examples of the various major groups of components included in regularly used antifungal medications (Nami et al. 2019). While polyenes directly bind to ergosterol and move inside the cell through the cell membrane by creating pores, echinocandins inhibit the synthesis of important polysaccharides (1,3-β-glucans) responsible for developing the cell wall. Through these pores, cellular organelles emerge and cause the fungal cells to die (Marek and Timmons 2019). Conventional drugs treat fungal infections, but they can have a variety of side effects at the application site, including burning, redness, and allergic responses (Girois et al. 2006). An infected foot ulcer in diabetic patients is a common source of potentially dreadful conditions such as gangrene and amputations (Abbas and Atiyah 2023). Fungal

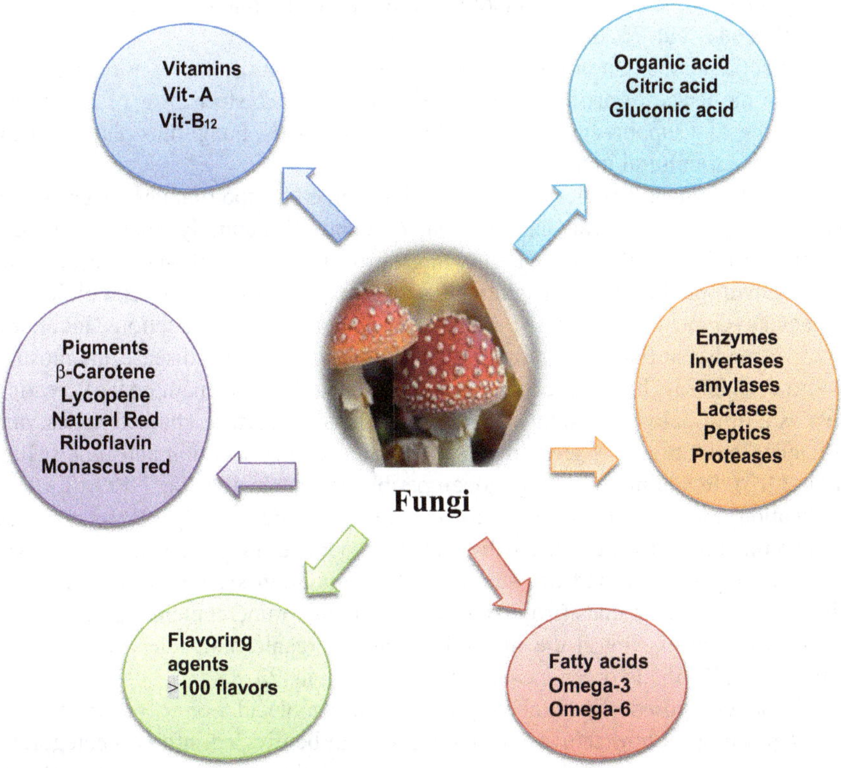

Fig. 14.1 Importance of fungi-based food ingredients

infections are extremely common in the tropical regions. Some are dangerous, and even deadly, as they produce a variety of injuries, ranging from superficial skin infection to invade the Internal device (systemic diseases) (Atiyah 2015).

Fungi can function as biological control agents to prevent plant diseases, for example, *Trichoderma* showed antagonistic effects against a variety of fungi that are harmful to plants (Lorito et al. 2010; Şesan 1986, 2017, 2003; Hermosa et al. 2013). Many biotechnology domains use *Trichoderma* species. They have great potential to protect against disease crops, promote plant growth, making them an exceptional addition to agriculture (Lorito et al. 2010; Lahlali et al. 2022). Many fungi are employed to manage insect pests, for example, *Beauveria bassiana* targets corn borer. *Verticillium lecanii* is effective against aphids and whiteflies, and *Metarhizium anisopliae* is used to suppress scarab larvae (Haritakun et al. 2010; Feng et al. 2023; Javed et al. 2019). The way the fungi work is that they infect the surface of the insect, which causes them to connect to its integuments. From there, they grow and multiply until they completely cover the insect (Yuvaraj and Ramasamy 2020). *Trichoderma* is the most studied group of filamentous fungi for its potential to biocontrol worm pests. Enzymes, space competition, and antibiosis (the synthesis of secondary metabolites) are the factors that cause harm to nematodes (Poveda et al. 2020).

The applications that are being shown have the ability to increase crop output, decrease the usage of artificial fertilizers and pesticides, steer clear of hazardous substances, and support sustainable agricultural methods. Fungi thus play a crucial role in the agricultural industry.

These days, nanoparticles (NPs) are widely valued for their numerous use in the biological, pharmaceutical, and medical domains. Structurally, they are hardly larger than 100 nm but have the ability to govern a wide range of medicines, including tiny hydrophobic and hydrophilic pharmaceuticals, vaccinations, and biological molecules (Gupta and Xie 2018). NPs have a wide range of applications, including targeted medicine delivery, tissue engineering scaffolds, and disease diagnostics (Mura et al. 2013). To achieve the greatest possible effect, the main goal of creating NPs as a drug delivery system is to regulate the size, surface characteristics, and efficient distribution of a given medication at a specified location and time (Mura et al. 2013). In addition to being biocompatible and biodegradable, NPs utilized in medication, should also have optimal mechanical qualities, timely release, and ease of manufacture. NPs are tracked by surface modification, get saved in the body through phagocytosis, and are stored in the circulation system (Mahapatro and Singh 2011). While metals (iron, gold, etc.) and other mineral elements are crucial to the structure of NPs in the mineral category, organic molecules make up the majority of NPs in the organic category (Erdoğar et al. 2018).

NPs are composed of natural or synthetic macromolecular or polymeric materials. Depending on how they are formed, they can be divided into two categories: nanospheres and nanocapsules. The medication contained in a core chamber of sac-like structures called nanocapsules, is enclosed in a polymeric coating. The medication and the polymer are either uniformly distributed or absorbed on the surface of a nanosphere, a matrix system. When polymers are used as nanoparticles, they are combined with medications that have therapeutic benefits for particular disorders, like cancer. Two methods are used to attach these NPs to nanocomposites: (1) The medications are contained in nanocarriers or (2) The medications are conjugated on the surface of the nanoparticles (Moritz and Geszke-Moritz 2015). The controlled release of macromolecules through polymers was first discovered by Langer and Folkman, and this discovery advanced the medication release method for antiangiogenic cancer treatments (Langer and Folkman 1976). An additional benefit of using polymer NPs as a medicinal nanocarrier is improved medication stability and solubility. Polymer NPs are therefore thought to be the most commonly used drug release mechanisms (Hickey et al. 2015).

14.2 Synthesis of Plant-Based Nanoparticles

Numerous simple, safe, affordable, reusable, and scalable green synthesis methods for NPs have been developed recently. Because of this, a number of biological systems—including bacteria, yeast, fungi, and plant extracts—are now widely used in green synthesis techniques to produce NPs (Nadeem et al. 2018; Jadoun et al. 2021). The three most crucial conditions for the green synthesis of nanoparticles are (1) choice of a safe stabilizing material, (2) an appropriate nontoxic reducing agent, and (3) a green or ecologically friendly solvent (water, ethanol, and their combinations are the most often utilized). In fact, a variety of synthetic approaches have also been employed to create nanoparticles, the most common ones being chemical, physical, and biosynthetic methods. Chemical processes typically include the use of poisonous and hazardous chemicals, which have a number of environmental concerns and are quite expensive (Nath and Banerjee 2013). Green synthesis offers a safe, environmentally acceptable, and biocompatible way to create NPs for a range of applications, including biological ones.

14.2.1 Medicinal Plants Used for the Synthesis of Nanoparticles and its enhanced efficacy

The qualities of medicinal plants stem from their abundant supply of secondary metabolites, which include flavonoids, alkaloids, phenolics, terpenoids, tannins, glycosides, quinones, steroids, and saponins (Van Wyk et al. 2009). But since they cannot penetrate the lipid bilayer of cells, majority of these chemicals have minimal absorption, which reduces their bioavailability and effectiveness (Singh et al. 2015). Numerous studies on these plants and their derivatives using phytochemical and pharmacological methods showed remarkable in vitro activity but less in vivo efficacy. An increasingly popular strategy in the literature to increase efficacy is to combine traditional medicinal herbs with nanotechnology (Majoumouo et al. 2020; Tavakoli et al. 2012; Khoobchandani et al. 2020). Among the newest methods with the potential to completely transform the pharmaceutical and medical industries is nanotechnology (Aboyewa et al. 2021; Khoobchandani et al. 2020). Using nanosystems in herbal medicine could be a useful strategy for getting rid of the restrictions connected to medicinal plants (Al-Yasiri et al. 2017). These plants can be used with newly developed technologies to create innovations that will improve the phytochemical distribution and effectiveness (Manisha et al. 2014). Various medicinal plants have apparently been employed to produce medicinal plants nanoparticles with better biocompatibility and biological activity. Green nanotechnology has shown considerable promise in this regard (Ahmed et al. 2016; Katas et al. 2019). Owing to the superior pharmacological properties of medicinal plant-mediated nanoparticles, it is now critical to concentrate on the ways in which traditional plants could assist the pharmaceutical industry, particularly in the area of drug discovery. Various endophytic microbes, such as bacteria, fungi, and actinomycetes, can be employed in the manufacture of nanoparticles made of various metals, including copper, zinc, gold, and silver. The most recent studies in this area are compiled in Table 14.1.

Plant-mediated synthesis of metal nanoparticles is becoming more and more significant due to its ease of use, quick rate of synthesis with a variety of morphologies, lack of complicated cell culture management, and environmental friendliness

Table 14.1 Biosynthesized nanoparticles from endophytes with their respective size and biological activity

Plants	Endophytes	Shape of NPs	Size	Types of NPs	Bioactivity	References
Coriandrum sativum	*Bacillus siamensis* C1	Spherical	25–50 nm	Silver	Antibacterial	Ibrahim et al. (2019)
Allium sativum	*Pseudomonas poae* CO	Spherical	19.8–44.9 nm	Silver	Antifungal	Ibrahim et al. (2020)
Mimosa pudica	*Aneurinibacillus migulanus*	Spherical, oval, cubic, triangular	~24.27 nm	Silver	Antibacterial	Mathew et al. (2010)
Coffea arabica	*Pseudomonas fluorescens 417*	Spherical	5–50 nm	Gold	Antibacterial	Syed et al. (2016)
Raphanus sativus	*Alternaria* sp.	Spherical	4–30 nm.	Silver	Antibacterial	Singh et al. (2017)
Taxus baccata	*Nemania* sp.	Spherical or ellipsoidal	5–70 nm	Silver	Antibacterial	Farsi and Farokhi (2018)
Erythrophleum fordii	*Alternaria tenuissima*	Spherical	15–45 nm.	Zinc oxide	Antimicrobial, anticancer, and antioxidant	Abdelhakim et al. (2020)
Chonemorpha fragrans	*Fusarium solani*	Spindle	40–45 nm	Gold	Anticancer	Clarance et al. (2020)
Cinnamomum zeylanicum	*Lasiodiplodia theobromae*	Spherical to oval	~76 nm	Silver	Antibacterial	Adnan (2020)
Chiliadenus montanus	*Trichoderma atroviride*	Spherical	10–15 nm	Silver	Antibacterial	Abdel-Azeem et al. (2020)
Madhuca longifolia	*Pestalotia* sp.	Angular	<40 nm	Silver	Antibacterial	Verma et al. (2016)
Pinus densiflora	*Talaromyces purpureogenus*	Round to triangle	~25 nm	Silver	Antimicrobial and anticancer	Hu et al. (2019)
Ocimum tenuiflorum	*Exserohilum rostrata*	Spherical	10–15 nm	Silver	Antibacterial, anti-inflammatory, and antioxidant	Bagur et al. (2020)
Borszczowia aralocaspica	*Isoptericola* SYSU 333150	Spherical	11–40 nm	Silver	Antibacterial	Dong et al. (2017)

Oxalis corniculate	*Streptomyces zaomyceticus Oc-5*	Spherical	~78 nm	Copper	Antimicrobial, antioxidant, and anticancer	Hassan et al. (2019)
Mentha longifolia	*Streptomyces* sp.	Spherical	2.3–85 nm	Silver	Antimicrobial	El-Gamal et al. (2018)
Convolvulus arvensis	*Streptomyces capillispiralis* Ca-1	Spherical	3.6–59 nm	Copper	Antimicrobial and insecticidal	Saad et al. (2018)
Ocimum sanctum	*Streptomyces coelicolor*	Spherical and ellipsoidal	~25 nm	*Magnesium*	Antimicrobial	El-Moslamy (2018)

(Ghosh et al. 2012). The manufacture of metal nanoparticles using "green" techniques is gaining popularity; biomass or plant extracts have been successfully used as reducing agents (Singh et al. 2010). There have also been reports of the creation of nanoparticles from certain therapeutic herbs (Karnani and Chowdhary 2013).

14.2.2 Possible Mechanism of Plant-Based Nanoparticles

Two predominant methodologies are used to synthesize nanoparticles: the top-down approach and the bottom-up approach (Kharissova et al. 2013). The bottom-up method uses biopolymers and smaller active chemicals found in plant extract and the secretions of microorganisms to synthesize nanoparticles. On the other hand, in the bottom-down technique, the majority of NPs in the top-down approach are size-reducing byproducts of several photochemical reactions (Abinaya et al. 2018). Compared to other methods, the plant-based green synthesis methodology is more commonly used since plants are easier to find and safer for the environment than synthesis mediated by microorganisms and fungi. Both require an extended period of cell culture (Xu et al. 2018; Rafique et al. 2017). Active terpenoids, alkaloids, phenols, tannins, and vitamin components found in plant extract have been shown to have medical and environmental benefits (Madkour 2018). For the manufacture of NPs based on bacteria, two approaches have been reported: the extracellular method and the intracellular method (Meeuwissen 2013). According to studies, the following processes lead to the creation of NPs: initially, the metal ion is trapped inside or on the surface of the bacterial cell. Secondly, when the bacterial enzymes are involved, this ion experiences a reduction process (Qamar and Ahmad 2021).

14.2.3 Endophytic Microorganisms as a Source of Nanoparticles

The synthesis of metal nanoparticles is increasingly utilizing biological methods. One such method involves the use of saprophytic microorganisms, such as fungi and bacteria, which can convert metal ions from their surroundings into metallic nanoparticles through the action of enzymes and secondary metabolites produced by cell activity. This process yields more stable and suitable-sized nanoparticles (Soliman et al. 2018). The use of endophytic microbes has become a new field of study for green manufactured nanoparticles. This area of study has the potential to reveal new avenues for the development of innovative nanomaterials with a wide range of applications at the nexus of biology and nanotechnology (Joshi et al. 2017). Various endophytic microbes, such as bacteria, fungi, and actinomycetes, can be employed in the manufacture of nanoparticles made of various metals, including copper, zinc, gold, and silver.

14.2.3.1 Nanoparticles Synthesized Using Endophytic Bacteria

To combat the toxicity of metal ions, certain endophytic bacteria have evolved a unique defense mechanism. This mechanism is based on the precipitation of metal ions at the nanoscale scale, which results in the production of nanoparticles (Iravani et al. 2014). Certain endophytic bacteria were shown to be able to grow and survive in the presence of high quantities of metal ions. Because of their exceptional capacity to convert metal ions into nanoparticles, bacteria may be a viable option for the manufacture of nanoparticles (Syed et al. 2019). *Bacillus siamensis* C1 isolated from *Coriandrum sativum* and *Pseudomonas poae* CO from *Allium sativum* had the ability to manufacture spherical silver nanoparticles and showed promise as antibacterial, antibiofilm, and antifungal agents (Ibrahim et al. 2019, 2020). The endophytic bacteria *Pseudomonas fluorescens*, isolated from the plant *Coffea arabica*, has been used to successfully manufacture gold nanoparticles with spherical forms and a size range of 5–50 nm. A panel of clinically relevant pathogens is susceptible to the bactericidal activity of the produced gold nanoparticles (Syed et al. 2016). The strain *Aneurinibacillus migulanus*, isolated from the surface-sterilized inner leaf segment of *Mimosa pudica*, is used by the same author, for the biosynthesis of silver nanoparticles with a variety of forms, including spherical, oval, cubic, and triangular shapes (Syed et al. 2016). Using the dynamic light scattering (DLS) method, the average particle size was found to be 24.27 nm. Their bactericidal effect suggests intriguing efficacy against both gram-positive and gram-negative pathogenic bacteria. The bacteria *Pseudomonas aeruginosa*, which is regarded as clinically significant, had the highest level of activity.

14.2.3.2 Nanoparticles Synthesized Using Endophytic Fungi

Due to their capacity for metal toleration, absorption, and accumulation, endophytic fungi have garnered increased interest in the creation of metallic nanoparticles in recent years (Namvar et al. 2015). Fungi, in contrast to other microbes, are effective catalysts for the synthesis of metallic nanoparticles of any kind and have a number of benefits, including: simple isolation from plants or soil than uncommon bacteria and actinomycetes, which need to be isolated using specific enrichment techniques (Hu et al. 2019). They release copious amounts of extracellular enzymes and metabolites, which help reduce metal ions into nanoparticles. Also, they are simple to scale up, given their quick rate of a rapid growth. Since most fungi grow in a wide range of pH, temperature, and NaCl, it is easier to adjust the culture conditions and generate homogenous nanoparticles (Casagrande and De Lima 2019). The endophytic fungus *Fusarium solani* was isolated from the plant *Chonemorpha fragrans*, which is utilized in the production of gold nanoparticles (Clarance et al. 2020). The produced nanomaterials revealed strong peak plasmon band between 510 and 560 nm, pink-ruby red hues, and needle- and flower-like structures with spindle shape morphology. The gold-based nanoparticles had lethal effects on human breast cancer cells (MCF-7) (IC_{50}: 1.3 ± 0.5 μg/mL) and cervical cancer cells (HeLa) (IC_{50}: 0.8 ± 0.5 μg/mL). Zinc oxide nanoparticles were made by using the culture filtrate of the endophytic fungus *Alternaria tenuissima*, isolated from *Erythrophleum fordii* (Abdelhakim et al. 2020). The biosynthesized nanoparticles had a spherical form, with a diameter ranging from 15 to 45 nm, and strong antioxidant, antibacterial, and anticancer properties (Bagur et al. 2020). The endophyte *Exserohilum rostrata* was isolated from the plant *Ocimum tenuiflorum*. This strain was utilized to produce spherical silver nanoparticles that ranged in size from 10 to 15 nm and demonstrated notable antimicrobial activity as well as other biological characteristics like antioxidant and anti-inflammatory activities.

14.2.3.3 Nanoparticles Synthesized Using Endophytic Actinomycetes

Actinomycetes are gram-positive bacteria with a high G + C (guanine-cytosine) content and are a part of the Actinobacteria phylum, one of the largest taxonomic groups in the Bacteria domain (Barka et al. 2016; Messaoudi et al. 2019). These bacteria are distinguished by their ability to produce a diverse array of beneficial secondary metabolites. Approximately 70–80% of secondary metabolites that are currently used in medicine, such as immunosuppressives, antibiotics, antifungals, anticancer agents, insecticides, and antivirals, have been identified and isolated from various actinomycetes, especially those belonging to the *Streptomyces* genus (Hassan et al. 2019; El-Gamal et al. 2018). *Streptomyces zaomyceticus* Oc-5 and *Streptomyces capillispiralis* Ca-1 have been isolated from *Oxalis corniculata* and *Convolvulus arvensis*, respectively. The production of copper nanoparticles, which showed varying biological activities, such as antibacterial, antioxidant, and anticancer properties as well as insecticidal properties, was carried out using both strains. In another study, an uncommon actinobacterium was employed to manage the disease brought on by *Staphylococcus warneri*, which poses a serious risk to human health (Dong et al. 2017). Utilizing the strain Isoptericola SYSU 333150 that was isolated from the plant *Borszczowia aralocaspica*, the researchers were able to produce spherical nanoparticles with a size range of 11–40 nm that showed antimicrobial activity against the pathogen *S. warneri* through photo-irradiation with sunlight exposure for varying periods of time. Numerous research attest to the attractiveness of endophytic microorganism-produced nanoparticles with varying metallic natures, sizes, and shapes, since they demonstrate a range of biological activities, such as antibacterial, cytotoxic, anti-inflammatory, and antioxidant properties (Farsi and Farokhi 2018; Adnan 2020; Abdel-Azeem et al. 2020).

14.2.4 Plant-Based Metal Nanoparticles

The manufacture of metal nanoparticles using "green" techniques is gaining popularity; biomass or extracts from various plants have been successfully used as reducing agents (Singh et al. 2010). Numerous studies have been conducted in this field showing immense application of metal nanoparticles for the management of phytopathogenic fungi. Metals like Ag, Cu, Fe, Zn, Se, Ni, and Pd have demonstrated exceptional antifungal activities. Hence, a thorough and critical evaluation of recent developments in the application of metal nanoparticles on phytopathogenic fungi is required.

14.2.4.1 Silver (Ag) Nanoparticles

Numerous plant and fungal extracts have been employed in biological systems to synthesize silver nanoparticles. A recent study performed synthesis and comparison of the antifungal evaluation of biogenic silver nanoparticles formed by using *Momordica charantia* and *Psidium guajava* leaf extract (Nguyen et al. 2020). Green synthesis of silver nanoparticles using *Melia azedarach* leaf extract and their antifungal activities were studied both in vitro and in vivo (Jebril et al. 2020). Biological synthesis of silver nanoparticles and evaluation of antibacterial and antifungal properties of silver and copper nanoparticles have been reported (Jafari et al. 2015). Studies have optimized the biosynthesis and antifungal activity of *Osmanthus fragrans* leaf extract-mediated silver nanoparticles (Huang et al. 2017). There are reports on the many plant and fungal extracts that have been utilized to create silver nanoparticles (Ege et al. 2020). These have been produced by a variety of chemical techniques, including microemulsion, sol-gel, and chemical reduction (Elgorban et al. 2016; Shanmugam et al. 2015; Mendes et al. 2014; Tarazona et al. 2019; Shivamogga Nagaraju et al. 2020; Kriti et al. 2020). Physical techniques like the irradiation process and high-voltage arc discharge have been employed, but to a lesser extent (Kasprowicz et al. 2010; Gorczyca et al. 2015; Luan and Xo 2018). These techniques have allowed for the synthesis of Ag nanoparticles with exceptional antifungal capabilities. Furthermore, since biological synthesis is ecologically friendly, they offer an extra advantage. It is also noteworthy that a number of commercial Ag nanoparticles have demonstrated exceptional antifungal qualities when tested for their ability to prevent the growth of phytopathogenic fungus.

14.2.4.2 Copper (Cu) Nanoparticles

The majority of the produced Cu nanoparticles are spherical and have demonstrated exceptional antifungal properties in this form. Many reports claim that spherical nanoparticles have the best chance of crossing the fungal membranes and reaching the enzymes to begin cellular inhibition at a fast rate (Ouda 2014; Bramhanwade et al. 2016). The first investigation on Cu nanoparticles against fungal pathogens was published by Giannousi et al. (2013) and has been regarded as a promising antifungal therapy option fever since (Elmer et al. 2018; Elmer and White 2018). The primary techniques for synthesizing Cu nanoparticles to combat infection have been described. Chemical reduction and hydrothermal synthesis are two techniques of chemical synthesis (Van Viet et al. 2016; Maqsood et al. 2020; Seku et al. 2018); Nemati et al. 2015; Hashim et al. 2019; Hermida-Montero et al. 2019). Conversely, biological synthesis, which uses various plant extracts, is frequently employed due to its ease, negligible toxicity and lack of environmental damage (RajeshKumar and Rinitha 2018; Asghar et al. 2018; Jagana et al. 2017; Mali et al. 2020; Hasanin et al. 2021). Furthermore, the ability of commercial nanoparticles to prevent phytopathogenic fungus has also been tested. These particles are efficient and simple to obtain (Malandrakis et al. 2019; Aleksandrowicz-Trzcińska et al. 2018; Zalewska et al. 2016; Ouda 2014).

14.2.4.3 Biopolymer-Based Nanoparticles

Plant cells function as little factories, producing a wide range of biochemicals and materials needed for their development, reproduction, protection from pests and diseases, and observation and interaction with their surroundings. Polymers that are widely used by people in daily life, such as cellulose, hemicellulose, lignin, and starch, are naturally produced by plants. Most of the natural rubber used in tires, tubing, gloves, and other consumer goods worldwide is produced by a single plant called *Hevea brasiliensis*, sometimes known as the Para rubber tree (van Beilen and Poirier 2012). By adding genes encoding enzyme activities that transform an endogenous plant metabolite into a polymeric structure that is unique to the plant, the spectrum of biomaterials produced by plants can be increased. This tactic has made it possible for plants to produce polyhydroxyalkanoates, cyanophycin, collagen, elastin, and silk (Schubert et al. 1988; Slater et al. 1988). Considerable efforts have been involved over the years to enhance production of these innovative biopolymers from plants. As a case study, consider the synthesis of poly[(R)-3-hydroxybutyrate] (PHB). PHB is a type of polyhydroxyalkanoate with potential applications in the plastic and other industries. Research has shown that PHB synthesis occurs in both oilseed and biomass crops, making it a prime candidate for the combined production of polymers and energy (Petrasovits et al. 2012).

14.3 Mechanism of Antifungal Action

The plant-based nanoparticles exhibit strong antifungal activity against a broad range of phytopathogenic fungi. Their application provides a unique method of controlling the phytopathogenic fungi in agriculture (Arciniegas-Grijalba et al. 2017; Medda et al. 2015). The antifungal effectiveness depends on a number of variables, including the nanoparticle size distribution, shape, composition, crystallinity, agglomeration, and surface chemistry (Koduru et al. 2018; Kasana et al. 2017). Nanoparticles, prefer a specific surface area-to-volume ratio, which enhances their antifungal activity (Alghuthaymi et al. 2015). Additionally, it has been shown that the synthesis pathway may be crucial to the antifungal action because the surfactants and metal precursors may not be easily extracted always. Consequently, the synthesis residues have the ability to alter the surface chemistry of nanoparticles, which in turn affects their antifungal activity (Alghuthaymi et al. 2015). The mechanisms of antifungal activity can vary based on the nanoparticle composition, size, shape, and surface properties, as well as the specific fungal species being targeted. It is important to note that research in this field is ongoing, and the understanding of green nanoparticle-based antifungal mechanisms continues to evolve (Fig. 14.2). Further research and clinical trials are necessary to validate the effectiveness of these approaches in treating fungal infections.

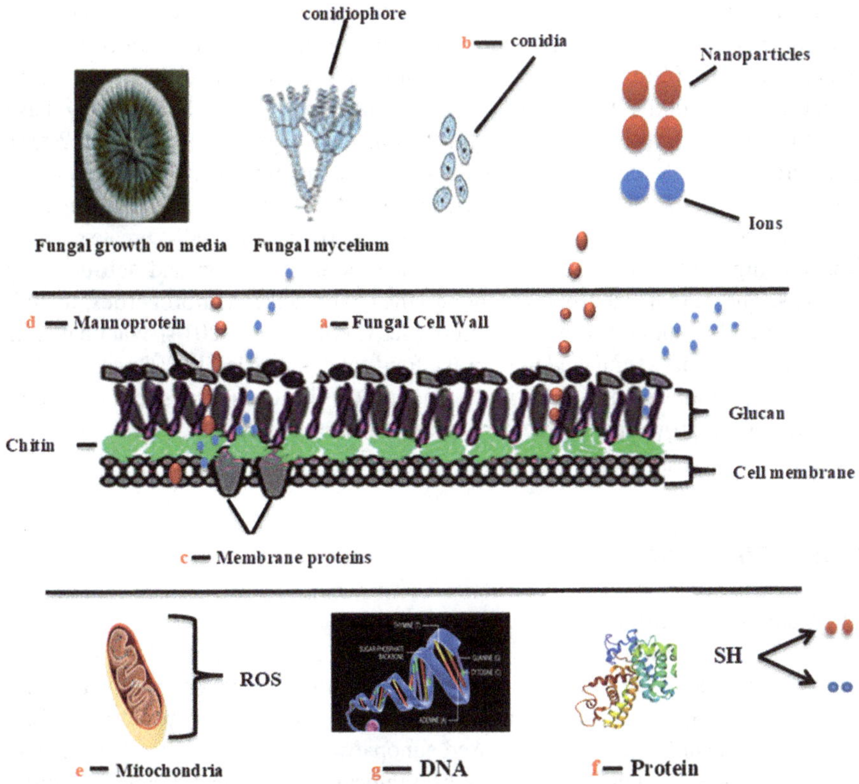

Fig. 14.2 This is an illustration of the possible mechanisms of action of metal nanoparticles on phytopathogenic fungi. (a) Ions are released by nanoparticles and bind to certain protein groups, which affect the function of essential membrane proteins and interfere with cell permeability. (b) The nanoparticles inhibit the germination of the conidia and suppress their development. (c) Nanoparticles and released ions disrupt electron transport, protein oxidation, and alter membrane potential. (d) They also interfere with protein oxidative electron transport. (e) They affect the potential of the mitochondrial membrane by increasing the levels of transcription of genes in response to oxidative stress (ROS). (f) ROS induces the generation of reactive oxygen species, triggering oxidation reactions catalyzed by the different metallic nanoparticles, causing severe damage to proteins, membranes, and deoxyribonucleic acid (DNA), interfering with nutrient absorption. (g) The ions of the nanoparticles have a genotoxic effect that destroys DNA, therefore causing cell death (Huerta-García et al. 2014; Mikhailova 2020; Zhao et al. 2018; Kumari et al. 2017; Rana et al. 2020)

14.4 Other Applications of Plant-Based Nanoparticles

Molecular nanomaterials find wide-ranging industrial applications across multiple sectors (Qiu Zhao et al. 2003). Nanomaterials have use in many sectors based on the various significant features of molecular nanoparticles, including size, shape, and surface characteristics. These applications span various sectors, from electronics to healthcare and environmental remediation. Nanoparticles are used in electronic components, such as conductive inks for printed electronics and semiconductors (Cui and Lieber 2001). They are widely employed as catalysts in chemical reactions to increase reaction rates and efficiency (Tauster and Fung 1978). They have immense applications in drug delivery, imaging, and diagnostics (Peer et al. 2007). They are also employed to remove pollutants and contaminants from water and air. They play a role in improving the efficiency and performance of batteries, supercapacitors, and solar cells (Dunn et al. 2011). Nanoparticles are used to create specialized coatings with enhanced properties, such as anticorrosion and self-cleaning coatings. They can be incorporated into textiles to enhance their properties, including antimicrobial and UV-protective properties (Petkova et al. 2014). Nanoparticles are used in food packaging and to improve food safety and shelf life (Duncan 2011). Thus, besides biomedical applications, nanoparticles have numerous applications as discussed here providing in-depth insights into the research and development in these areas.

14.5 Conclusion

The study on the antifungal efficacy of plant-based nanoparticles reveals promising potential for the development of novel and eco-friendly antifungal agents. These nanoparticles, derived from plant extracts, demonstrate significant antifungal activity, offering a natural and sustainable alternative to conventional synthetic antifungal agents. The utilization of plant-based nanoparticles not only shows efficiency in inhibiting fungal growth but also presents a range of potential applications, including agriculture, pharmaceuticals, and cosmetics. Further research and development in this area could lead to the creation of effective, environment-friendly antifungal solutions that have minimal adverse effects on human health and the environment. Overall, the findings of this study underscore the value of harnessing nature's resources for combatting fungal infections, paving the way for innovative and sustainable approaches in the field of antifungal therapy.

References

Abbas HH, Atiyah MM (2023) Anti-fungal activities of aqueous and alcoholic leaf extracts of *Moringa oleifera* Lam. on *Candida albicans* isolated from diabetic foot infections. AIP Conf Proc 2414(1)

Abd Elaziz D, Abd El-Ghany M, Meshaal S, El Hawary R, Lotfy S, Galal N, Ouf SA, Elmarsafy A (2020) Fungal infections in primary immunodeficiency diseases. Clin Immunol 219:108553

Abdel-Azeem A, Nada AA, O'Donovan A, Thakur VK, Elkelish A (2020) Mycogenic silver nanoparticles from endophytic Trichoderma atroviride with antimicrobial activity. J Renew Mater 8(2):171–185

Abdelhakim HK, El-Sayed ER, Rashidi FB (2020) Biosynthesis of zinc oxide nanoparticles with antimicrobial, anticancer, antioxidant and photocatalytic activities by the endophytic Alternaria tenuissima. J Appl Microbiol 128(6):1634–1646

Abinaya M, Vaseeharan B, Divya M, Sharmili A, Govindarajan M, Alharbi NS, Kadaikunnan S, Khaled JM, Benelli G (2018) Bacterial exopolysaccharide (EPS)-coated ZnO nanoparticles showed high antibiofilm activity and larvicidal toxicity against malaria and Zika virus vectors. J Trace Elem Med Biol 45:93–103

Aboyewa JA, Sibuyi NRS, Meyer M, Oguntibeju OO (2021) Gold nanoparticles synthesized using extracts of cyclopia intermedia, commonly known as honeybush, amplify the cytotoxic effects of doxorubicin. Nano 11(1):132

Adnan M (2020) Synthesis, characterization and applications of endophytic fungal nanoparticles. Inorganic Nano-Metal Chem 51(2):280–287

Ahmed S, Ahmad M, Swami BL, Ikram S (2016) Green synthesis of silver nanoparticles using Azadirachta indica aqueous leaf extract. J Radiat Res Appl Sci 9(1):1–7

Al-Yasiri AY, Khoobchandani M, Cutler CS, Watkinson L, Carmack T, Smith CJ, Kuchuk M, Loyalka SK, Lugão AB, Katti KV (2017) Mangiferin functionalized radioactive gold nanoparticles (MGF-198 AuNPs) in prostate tumor therapy: green nanotechnology for production, in vivo tumor retention and evaluation of therapeutic efficacy. Dalton Trans 46(42):14561–14571

Aleksandrowicz-Trzcińska M, Szaniawski A, Olchowik J, Drozdowski S (2018) Effects of copper and silver nanoparticles on growth of selected species of pathogenic and wood-decay fungi in vitro. For Chron 94(2):109–116

Alghuthaymi MA, Almoammar H, Rai M, Said-Galiev E, Abd-Elsalam KA (2015) Myconanoparticles: synthesis and their role in phytopathogens management. Biotechnol Biotechnol Equip 29(2):221–236

Arciniegas-Grijalba PA, Patiño-Portela MC, Mosquera-Sánchez LP, Guerrero-Vargas JA, Rodríguez-Páez JE (2017) ZnO nanoparticles (ZnO-NPs) and their antifungal activity against coffee fungus Erythricium salmonicolor. Appl Nanosci 7:225–241

Asghar MA, Zahir E, Shahid SM, Khan MN, Asghar MA, Iqbal J, Walker G (2018) Iron, copper and silver nanoparticles: green synthesis using green and black tea leaves extracts and evaluation of antibacterial, antifungal and aflatoxin B1 adsorption activity. LWT 90:98–107

Atiyah MM (2015) Isolation of the fungal samples from the patient's skin in Vijayawada City Hospital, Andhra Pradesh, India. Int J Curr Microbiol App Sci 4(6):287–291

Bagur H, Poojari CC, Melappa G, Rangappa R, Chandrasekhar N, Somu P (2020) Biogenically synthesized silver nanoparticles using endophyte fungal extract of Ocimum tenuiflorum and evaluation of biomedical properties. J Clust Sci 31:1241–1255

Barka EA, Vatsa P, Sanchez L, Gaveau-Vaillant N, Jacquard C, Klenk H-P, Clément C, Ouhdouch Y, van Wezel GP (2016) Taxonomy, physiology, and natural products of Actinobacteria. Microbiol Mol Biol Rev 80(1):1–43

Bramhanwade K, Shende S, Bonde S, Gade A, Rai M (2016) Fungicidal activity of Cu nanoparticles against Fusarium causing crop diseases. Environ Chem Lett 14:229–235

Casagrande MG, De Lima R (2019) Synthesis of silver nanoparticles mediated by fungi: a review. Front Bioeng Biotechnol 7:287

Clarance P, Luvankar B, Sales J, Khusro A, Agastian P, Tack J-C, Al Khulaifi MM, Al-Shwaiman HA, Elgorban AM, Syed A (2020) Green synthesis and characterization of gold nanoparticles using endophytic fungi Fusarium solani and its in-vitro anticancer and biomedical applications. Saudi J Biol Sci 27(2):706–712

Copetti MV (2019) Fungi as industrial producers of food ingredients. Curr Opin Food Sci 25:52–56

Cui Y, Lieber CM (2001) Functional nanoscale electronic devices assembled using silicon nanowire building blocks. Science 291(5505):851–853

Dong Z-Y, Narsing Rao MP, Xiao M, Wang H-F, Hozzein WN, Chen W, Li W-J (2017) Antibacterial activity of silver nanoparticles against Staphylococcus warneri synthesized using endophytic bacteria by photo-irradiation. Front Microbiol 8:1090

Duncan TV (2011) Applications of nanotechnology in food packaging and food safety: barrier materials, antimicrobials and sensors. J Colloid Interface Sci 363(1):1–24

Dunn B, Kamath H, Tarascon J-M (2011) Electrical energy storage for the grid: a battery of choices. Science 334(6058):928–935

Ege E, Kurtay G, Karaca B, Büyük İ, Gökdemir FŞ, Sumer A (2020) Green synthesis of silver nanoparticles from Phaseolus vulgaris L. extracts and investigation of their antifungal activities. Hacettepe J Biol Chem 49(1):11–23
El-Gamal MS, Salem SS, Abdo AM (2018) Biosynthesis, characterization, and antimicrobial activities of silver nanoparticles synthesized by endophytic Streptomyces sp. J Biotechnol 56:69–85
El-Moslamy SH (2018) Bioprocessing strategies for cost-effective large-scale biogenic synthesis of nano-MgO from endophytic Streptomyces coelicolor strain E72 as an anti-multidrug-resistant pathogens agent. Sci Rep 8(1):3820
Elgorban AM, El-Samawaty AE-RM, Yassin MA, Sayed SR, Adil SF, Elhindi KM, Bakri M, Khan M (2016) Antifungal silver nanoparticles: synthesis, characterization and biological evaluation. Biotechnol Biotechnol Equip 30(1):56–62
Elmer W, De La Torre-Roche R, Pagano L, Majumdar S, Zuverza-Mena N, Dimkpa C, Gardea-Torresdey J, White JC (2018) Effect of metalloid and metal oxide nanoparticles on Fusarium wilt of watermelon. Plant Dis 102(7):1394–1401
Elmer W, White JC (2018) The future of nanotechnology in plant pathology. Annu Rev Phytopathol 56:111–133
Erdoğar N, Akkın S, Bilensoy E (2018) Nanocapsules for drug delivery: an updated review of the last decade. Recent Pat Drug Deliv Formul 12(4):252–266
Eugenia ŞT, Florin O (2010) Trichoderma viride Pers.-experimental model for biological and biotechnological investigations of mycromyceta with importance in obtaining plant protection bioproducts. J Plant Dev 17(1)
Farsi M, Farokhi S (2018) Biosynthesis of antibacterial silver nanoparticles by endophytic fungus Nemania sp. isolated from Taxus baccata L.(Iranian yew). *Zahedan*. J Res Med Sci 20(6)
Feng M, Zhang Y, Coates BS, Du Q, Gao Y, Li L, Yuan H, Sun W, Chang X, Zhou S (2023) Assessment of Beauveria bassiana for the biological control of corn borer, Ostrinia furnacalis, in sweet maize by irrigation application. BioControl 68(1):49–60
Galindo-Solís JM, Fernández FJ (2022) Endophytic fungal terpenoids: natural role and bioactivities. Microorganisms 10(2):339
Ghosh S, Patil S, Ahire M, Kitture R, Gurav DD, Jabgunde AM, Kale S, Pardesi K, Shinde V, Bellare J (2012) Gnidia glauca flower extract mediated synthesis of gold nanoparticles and evaluation of its chemocatalytic potential. J Nanobiotechnol 10:1–9
Giannousi K, Avramidis I, Dendrinou-Samara C (2013) Synthesis, characterization and evaluation of copper based nanoparticles as agrochemicals against Phytophthora infestans. RSC Adv 3(44):21,743–21,752
Girois SB, Chapuis F, Decullier E, Revol BGP (2006) Adverse effects of antifungal therapies in invasive fungal infections: review and meta-analysis. Eur J Clin Microbiol Infect Dis 25:138–149
Gorczyca A, Pociecha E, Kasprowicz M, Niemiec M (2015) Effect of nanosilver in wheat seedlings and Fusarium culmorum culture systems. Eur J Plant Pathol 142:251–261
Gupta R, Xie H (2018) Nanoparticles in daily life: applications, toxicity and regulations. J Environ Pathol Toxicol Oncol 37(3):209
Haritakun R, Sappan M, Suvannakad R, Tasanathai K, Isaka M (2010) An antimycobacterial cyclodepsipeptide from the entomopathogenic fungus Ophiocordyceps communis BCC 16475. J Nat Prod 73(1):75–78
Hasanin M, Al Abboud MA, Alawlaqi MM, Abdelghany TM, Hashem AH (2021) Ecofriendly synthesis of biosynthesized copper nanoparticles with starch-based nanocomposite: antimicrobial, antioxidant, and anticancer activities. Biol Trace Elem Res 200:1–14
Hashim AF, Youssef K, Abd-Elsalam KA (2019) Ecofriendly nanomaterials for controlling gray mold of table grapes and maintaining postharvest quality. Eur J Plant Pathol 154:377–388
Hassan SE-D, Fouda A, Radwan AA, Salem SS, Barghoth MG, Awad MA, Abdo AM, El-Gamal MS (2019) Endophytic actinomycetes Streptomyces spp mediated biosynthesis of copper oxide nanoparticles as a promising tool for biotechnological applications. JBIC J Biol Inorg Chem 24:377–393

Hermida-Montero LA, Pariona N, Mtz-Enriquez AI, Carrión G, Paraguay-Delgado F, Rosas-Saito G (2019) Aqueous-phase synthesis of nanoparticles of copper/copper oxides and their antifungal effect against Fusarium oxysporum. J Hazard Mater 380:120850

Hermosa R, Rubio MB, Cardoza RE, Nicolás C, Monte E, Gutiérrez S (2013) The contribution of Trichoderma to balancing the costs of plant growth and defense. Int Microbiol 16(2):69–80

Hickey JW, Santos JL, Williford J-M, Mao H-Q (2015) Control of polymeric nanoparticle size to improve therapeutic delivery. J Control Release 219:536–547

Hu X, Saravanakumar K, Jin T, Wang M-H (2019) Mycosynthesis, characterization, anticancer and antibacterial activity of silver nanoparticles from endophytic fungus Talaromyces purpureogenus. Int J Nanomedicine 14:3427–3438

Huang W, Chen X, Duan H, Bi Y, Yu H (2017) Optimized biosynthesis and antifungal activity of Osmanthus fragrans leaf extract-mediated silver nanoparticles. Int J Agric Biol 19:668–672

Huerta-García E, Pérez-Arizti JA, Márquez-Ramírez SG, Delgado-Buenrostro NL, Chirino YI, Iglesias GG, López-Marure R (2014) Titanium dioxide nanoparticles induce strong oxidative stress and mitochondrial damage in glial cells. Free Radic Biol Med 73:84–94

Ibrahim E, Fouad H, Zhang M, Zhang Y, Qiu W, Yan C, Li B, Mo J, Chen J (2019) Biosynthesis of silver nanoparticles using endophytic bacteria and their role in inhibition of rice pathogenic bacteria and plant growth promotion. RSC Adv 9(50):29293–29299

Ibrahim E, Zhang M, Zhang Y, Hossain A, Qiu W, Chen Y, Wang Y, Wu W, Sun G, Li B (2020) Green-synthesization of silver nanoparticles using endophytic bacteria isolated from garlic and its antifungal activity against wheat Fusarium head blight pathogen Fusarium graminearum. Nanomaterials 10(2):219

Imtiaz N, Niazi MBK, Fasim F, Khan BA, Bano SA, Shah GM, Badshah M, Menaa F, Uzair B (2019) Fabrication of an original transparent PVA/gelatin hydrogel: in vitro antimicrobial activity against skin pathogens. Int J Polym Sci 2019:1–11

Iravani S, Korbekandi H, Mirmohammadi SV, Zolfaghari B (2014) Synthesis of silver nanoparticles: chemical, physical and biological methods. Res Pharm Sci 9(6):385

Irum W, Anjum T (2012) Production enhancement of Cyclosporin 'A'by Aspergillus terreus through mutation. Afr J Biotechnol 11(7):1736–1743

Jadoun S, Arif R, Jangid NK, Meena RK (2021) Green synthesis of nanoparticles using plant extracts: a review. Environ Chem Lett 19:355–374

Jafari A, Pourakbar L, Farhadi K, Gholizad LM, Goosta Y (2015) Biological synthesis of silver nanoparticles and evaluation of antibacterial and antifungal properties of silver and copper nanoparticles. Turk J Biol 39(4):556–561

Jagana D, Hegde YR, Lella R (2017) Green nanoparticles: a novel approach for the management of banana anthracnose caused by Colletotrichum musae. Int J Curr Microbiol Appl Sci 6(10):1749–1756

Jain A, Jain S, Rawat S (2010) Emerging fungal infections among children: a review on its clinical manifestations, diagnosis, and prevention. J Pharm Bioallied Sci 2(4):314

Javed K, Javed H, Mukhtar T, Qiu D (2019) Efficacy of Beauveria bassiana and Verticillium lecanii for the management of whitefly and aphid. Pak J Agric Sci 56(3)

Jebril S, Jenana RKB, Dridi C (2020) Green synthesis of silver nanoparticles using Melia azedarach leaf extract and their antifungal activities: in vitro and in vivo. Mater Chem Phys 248:122898

Jones M, Huynh T, Dekiwadia C, Daver F, John S (2017) Mycelium composites: a review of engineering characteristics and growth kinetics. J Bionanosci 11(4):241–257

Joshi CG, Danagoudar A, Poyya J, Kudva AK, Dhananjaya BL (2017) Biogenic synthesis of gold nanoparticles by marine endophytic fungus-Cladosporium cladosporioides isolated from seaweed and evaluation of their antioxidant and antimicrobial properties. Process Biochem 63:137–144

Karnani RL, Chowdhary A (2013) Biosynthesis of silver nanoparticle by eco-friendly method. Indian J Nanosci 1(1):25–31

Karwehl S, Stadler M (2016) Exploitation of fungal biodiversity for discovery of novel antibiotics. In: Stadler M, Dersch P (eds) How to overcome the antibiotic crisis: facts, challenges, technologies and future perspectives. Springer, pp 303–338

Kasana RC, Panwar NR, Kaul RK, Kumar P (2017) Biosynthesis and effects of copper nanoparticles on plants. Environ Chem Lett 15:233–240

Kasprowicz MJ, Kozioł M, Gorczyca A (2010) The effect of silver nanoparticles on phytopathogenic spores of Fusarium culmorum. Can J Microbiol 56(3):247–253

Katas H, Lim CS, Azlan AYHN, Buang F, Busra MFM (2019) Antibacterial activity of biosynthesized gold nanoparticles using biomolecules from Lignosus rhinocerotis and chitosan. Saudi Pharm J 27(2):283–292

Kaur C, Mishra Y, Mishra V, Saraogi GK, Tambuwala MM (2021) Recent advancement and biomedical applications of fungal metabolites. In: Singh J, Gehlot P (eds) New and future developments in microbial biotechnology and bioengineering. Elsevier, pp 47–67

Kharissova OV, Dias HVR, Kharisov BI, Pérez BO, Pérez VMJ (2013) The greener synthesis of nanoparticles. Trends Biotechnol 31(4):240–248

Khoobchandani M, Katti KK, Karikachery AR, Thipe VC, Srisrimal D, Dhurvas Mohandoss DK, Darshakumar RD, Joshi CM, Katti KV (2020) New approaches in breast cancer therapy through green nanotechnology and nano-ayurvedic medicine–pre-clinical and pilot human clinical investigations. Int J Nanomedicine Volume 15:181–197

Koduru JR, Kailasa SK, Bhamore JR, Kim K-H, Dutta T, Vellingiri K (2018) Phytochemical-assisted synthetic approaches for silver nanoparticles antimicrobial applications: a review. Adv Colloid Interf Sci 256:326–339

Kriti A, Ghatak A, Mandal N (2020) Inhibitory potential assessment of silver nanoparticle on phytopathogenic spores and mycelial growth of Bipolaris sorokiniana and Alternaria brassicicola. Int J Curr Microbiol Appl Sci 9(3):692–699

Kumari M, Shukla S, Pandey S, Giri VP, Bhatia A, Tripathi T, Kakkar P, Nautiyal CS, Mishra A (2017) Enhanced cellular internalization: a bactericidal mechanism more relative to biogenic nanoparticles than chemical counterparts. ACS Appl Mater Interfaces 9(5):4519–4533

Lahlali R, Ezrari S, Radouane N, Kenfaoui J, Esmaeel Q, El Hamss H, Belabess Z, Barka EA (2022) Biological control of plant pathogens: a global perspective. Microorganisms 10(3):596

Langer R, Folkman J (1976) Polymers for the sustained release of proteins and other macromolecules. Nature 263(5580):797–800

Lorito M, Woo SL, Harman GE, Monte E (2010) Translational research on Trichoderma: from 'omics to the field. Annu Rev Phytopathol 48:395–417

Luan LQ, Xo DH (2018) In vitro and in vivo fungicidal effects of γ-irradiation synthesized silver nanoparticles against Phytophthora capsici causing the foot rot disease on pepper plant. J Plant Pathol 100(2):241–248

Madkour LH (2018) Biogenic–biosynthesis metallic nanoparticles (MNPs) for pharmacological, biomedical and environmental nanobiotechnological applications. Chron Pharm Sci J 2(1):384–444

Mahapatro A, Singh DK (2011) Biodegradable nanoparticles are excellent vehicle for site directed in-vivo delivery of drugs and vaccines. J Nanobiotechnol 9:1–11

Majoumouo MS, Sharma JR, Sibuyi NRS, Tincho MB, Boyom FF, Meyer M (2020) Synthesis of biogenic gold nanoparticles from Terminalia mantaly extracts and the evaluation of their in vitro cytotoxic effects in cancer cells. Molecules 25(19):4469

Malandrakis AA, Kavroulakis N, Chrysikopoulos CV (2019) Use of copper, silver and zinc nanoparticles against foliar and soil-borne plant pathogens. Sci Total Environ 670:292–299

Mali SC, Dhaka A, Githala CK, Trivedi R (2020) Green synthesis of copper nanoparticles using Celastrus paniculatus Willd. leaf extract and their photocatalytic and antifungal properties. Biotechnol Reports 27:e00518

Manisha DR, Alwala J, Kudle KR, Rudra MPP (2014) Biosynthesis of silver nanoparticles using flower extracts of Catharanthus roseus and evaluation of its antibacterial efficacy. World J Pharm Pharm Sci 3:877–885

Maqsood S, Qadir S, Hussain A, Asghar A, Saleem R, Zaheer S, Nayyar N (2020) Antifungal properties of copper nanoparticles against Aspergillus niger. Sch Int J Biochem 3:87–91

Marek CL, Timmons SR (2019) Antimicrobials in pediatric dentistry. In: Nowak AJ et al (eds) Pediatric dentistry. Elsevier, pp 128–141

Mathew L, Chandrasekaran N, Mukherjee A (2010) Biomimetic synthesis of nanoparticles: science, technology & applicability. In: Mukherjee A (ed) Biomimetics learning from nature. IntechOpen

Medda S, Hajra A, Dey U, Bose P, Mondal NK (2015) Biosynthesis of silver nanoparticles from Aloe vera leaf extract and antifungal activity against Rhizopus sp. and Aspergillus sp. Appl Nanosci 5:875–880

Meeuwissen SA (2013) Manipulating polymersomes: control over functionality and morphology. Nijmegen:[Sn]

Mendes JE, Abrunhosa L, Teixeira JAd, De Camargo ER, De Souza CP, Pessoa JDC (2014) Antifungal activity of silver colloidal nanoparticles against phytopathogenic fungus (*Phomopsis* sp.) in soybean seeds. Int Sch Sci Res Innov 8(9)

Messaoudi O, Sudarman E, Bendahou M, Jansen R, Stadler M, Wink J (2019) Kenalactams A–E, polyene macrolactams isolated from Nocardiopsis CG3. J Nat Prod 82(5):1081–1088

Mikhailova EO (2020) Silver nanoparticles: mechanism of action and probable bio-application. J Funct Biomater 11(4):84

Mohmand AQK, Kousar MW, Zafar H, Bukhari KT, Khan MZ (2011) Medical importance of fungi with special emphasis on mushrooms. ISRA Med J 3(1):1–44

Moritz M, Geszke-Moritz M (2015) Recent developments in the application of polymeric nanoparticles as drug carriers. Adv Clin Exp Med 24(5):749–758

Mura S, Nicolas J, Couvreur P (2013) Stimuli-responsive nanocarriers for drug delivery. Nat Mater 12(11):991–1003

Nadeem M, Tungmunnithum D, Hano C, Abbasi BH, Hashmi SS, Ahmad W, Zahir A (2018) The current trends in the green syntheses of titanium oxide nanoparticles and their applications. Green Chem Lett Rev 11(4):492–502

Nami S, Aghebati-Maleki A, Morovati H, Aghebati-Maleki L (2019) Current antifungal drugs and immunotherapeutic approaches as promising strategies to treatment of fungal diseases. Biomed Pharmacother 110:857–868

Namvar F, Moniri M, Tahir M, Azizi S, Mohamad R (2015) Nanoparticles biosynthesized by fungi and yeast: a review of their preparation, properties, and medical applications. Molecules (Basel, Switzerland) 20(9):16,540–16,565

Nath D, Banerjee P (2013) Green nanotechnology–a new hope for medical biology. Environ Toxicol Pharmacol 36(3):997–1014

Nemati A, Shadpour S, Khalafbeygi H, Ashraf S, Barkhi M, Soudi MR (2015) Efficiency of hydrothermal synthesis of nano/microsized copper and study on in vitro antifungal activity. Mater Manuf Process 30(1):63–69

Nguyen DH, Vo TNN, Nguyen NT, Ching YC, Hoang Thi TT (2020) Comparison of biogenic silver nanoparticles formed by Momordica charantia and Psidium guajava leaf extract and antifungal evaluation. PLoS One 15(9):e0239360

Ouda SM (2014) Antifungal activity of silver and copper nanoparticles on two plant pathogens, Alternaria alternata and Botrytis cinerea. Res J Microbiol 9(1):34

Peer D, Karp JM, Hong S, Farokhzad OC, Margalit R, Langer R (2007) Nanocarriers as an emerging platform for cancer therapy. Nat Nanotechnol 2(12):751–760

Petkova P, Francesko A, Fernandes MM, Mendoza E, Perelshtein I, Gedanken A, Tzanov T (2014) Sonochemical coating of textiles with hybrid ZnO/chitosan antimicrobial nanoparticles. ACS Appl Mater Interfaces 6(2):1164–1172

Petrasovits LA, Zhao L, McQualter RB, Snell KD, Somleva MN, Patterson NA, Nielsen LK, Brumbley SM (2012) Enhanced polyhydroxybutyrate production in transgenic sugarcane. Plant Biotechnol J 10(5):569–578

Pombeiro-Sponchiado SR, Sousa GS, Andrade JCR, Lisboa HF, Gonçalves RC (2017) Production of melanin pigment by fungi and its biotechnological applications. Melanin 1(4):47–75

Poorniammal R, Prabhu S, Dufossé L, Kannan J (2021) Safety evaluation of fungal pigments for food applications. J Fungi 7(9):692

Poveda J, Abril-Urias P, Escobar C (2020) Biological control of plant-parasitic nematodes by filamentous fungi inducers of resistance: trichoderma, mycorrhizal and endophytic fungi. Front Microbiol 11:992

Qamar SUR, Ahmad JN (2021) Nanoparticles: mechanism of biosynthesis using plant extracts, bacteria, fungi, and their applications. J Mol Liq 334:116040

Qiu Zhao Q, Boxman A, Chowdhry U (2003) Nanotechnology in the chemical industry–opportunities and challenges. J Nanopart Res 5:567–572

Rafique M, Sadaf I, Rafique MS, Tahir MB (2017) A review on green synthesis of silver nanoparticles and their applications. Artif Cells Nanomed Biotechnol 45(7):1272–1291

Rai M, Ingle AP, Pandit R, Paralikar P, Gupta I, Anasane N, Dolenc-Voljč M (2017) Nanotechnology for the treatment of fungal infections on human skin. In: Kon K, Rai M (eds) The microbiology of skin, soft tissue, bone and joint infections. Elsevier, pp 169–184

RajeshKumar S, Rinitha G (2018) Nanostructural characterization of antimicrobial and antioxidant copper nanoparticles synthesized using novel Persea americana seeds. OpenNano 3:18–27

Rana A, Yadav K, Jagadevan S (2020) A comprehensive review on green synthesis of nature-inspired metal nanoparticles: mechanism, application and toxicity. J Clean Prod 272:122880

Saad EL, Salem SS, Fouda A, Awad MA, El-Gamal MS, Abdo AM (2018) New approach for antimicrobial activity and bio-control of various pathogens by biosynthesized copper nanoparticles using endophytic actinomycetes. J Radiat Res Appl Sci 11(3):262–270

Schubert P, Steinbüchel A, Schlegel H (1988) Cloning of the Alcaligenes eutrophus genes for synthesis of poly-beta-hydroxybutyric acid (PHB) and synthesis of PHB in Escherichia coli. J Bacteriol 170(12):5837–5847

Seku K, Ganapuram BR, Pejjai B, Kotu GM, Narasimha G (2018) Hydrothermal synthesis of copper nanoparticles, characterization and their biological applications. Int J Nano Dimension 9(1):7–14

Şesan TE (1986) Studiul biologic al speciilor de ciuperci antagoniste faţă de unii patogeni cu produc micoze la plante [Biological study of fungi species antagonistic towards some phytopathogens]. ICEBiol 1:89

Şesan TE (2003) Sustainable management of gray mold (Botrytis spp.) of horticultural crops. Adv Plant Dis Manag Res Signpost 37:121–152

Şesan TE (2017) Trichoderma spp. applications in agriculture and horticulture. Editura Universitatii din Bucuresti, Bucharest, Romania

Şesan TE, Oancea F, Toma C, Matei G-M, Matei S, Chira F, Chira D, Fodor E, Mocan C, Ene M (2010) Approaches to the study of mycorrhizas in Romania. Symbiosis 52:75–85

Shanmugam C, Gunasekaran D, Duraisamy N, Nagappan R, Krishnan K (2015) Bioactive bile salt-capped silver nanoparticles activity against destructive plant pathogenic fungi through in vitro system. RSC Adv 5(87):71174–71182

Shivamogga Nagaraju R, Holalkere Sriram R, Achur R (2020) Antifungal activity of Carbendazim-conjugated silver nanoparticles against anthracnose disease caused by Colletotrichum gloeosporioides in mango. J Plant Pathol 102:39–46

Singh A, Jain D, Upadhyay MK, Khandelwal N, Verma HN (2010) Green synthesis of silver nanoparticles using Argemone mexicana leaf extract and evaluation of their antimicrobial activities. Dig J Nanomater Bios 5(2):483–489

Singh S, Krishna THA, Kamalraj S, Kuriakose GC, Valayil JM, Jayabaskaran C (2015) Phytomedicinal importance of Saraca asoca (Ashoka): an exciting past, an emerging present and a promising future. Curr Sci 109:1790–1801

Singh T, Jyoti K, Patnaik A, Singh A, Chauhan R, Chandel SS (2017) Biosynthesis, characterization and antibacterial activity of silver nanoparticles using an endophytic fungal supernatant of Raphanus sativus. J Genet Eng Biotechnol 15(1):31–39

Slater SC, Voige WH, Dennis D (1988) Cloning and expression in Escherichia coli of the Alcaligenes eutrophus H16 poly-beta-hydroxybutyrate biosynthetic pathway. J Bacteriol 170(10):4431–4436

Smith SE, Read D (2008) The symbionts forming arbuscular mycorrhizas. In: Smith SE, Read D (eds) Mycorrhizal symbiosis. Academic, pp 13–41

Soliman H, Elsayed A, Dyaa A (2018) Antimicrobial activity of silver nanoparticles biosynthesised by Rhodotorula sp. strain ATL72. Egypt J Basic Appl Sci 5(3):228–233

Syed B, Prasad MNN, Satish S (2019) Synthesis and characterization of silver nanobactericides produced by Aneurinibacillus migulanus 141, a novel endophyte inhabiting Mimosa pudica L. Arab J Chem 12(8):3743–3752

Syed B, Prasad NMN, Satish S (2016) Endogenic mediated synthesis of gold nanoparticles bearing bactericidal activity. J Microsc Ultrastruct 4(3):162–166

Tarazona A, Gómez JV, Mateo EM, Jiménez M, Mateo F (2019) Antifungal effect of engineered silver nanoparticles on phytopathogenic and toxigenic Fusarium spp. and their impact on mycotoxin accumulation. Int J Food Microbiol 306:108259

Tauster SJ, Fung SC (1978) Strong metal-support interactions: occurrence among the binary oxides of groups IIA–VB. J Catal 55(1):29–35

Tavakoli J, Miar S, Zadehzare MM, Akbari H (2012) Evaluation of effectiveness of herbal medication in cancer care: a review study. Iran J Cancer Prevent 5(3):144

van Beilen JB, Poirier Y (2012) Plants as factories for bioplastics and other novel biomaterials. In: Altman A, Hasegawa PM (eds) Plant biotechnology and agriculture. Elsevier, pp 481–494

Van Wyk B, Van Oudtshoorn B, Gericke N (2009) Turning folklore into an ethnomedicinal catalogue medicinal plants of South Africa. Briza, South Africa

Verma SK, Gond SK, Mishra A, Sharma VK, Kumar J, Singh DK (2016) Biofabrication of antibacterial and antioxidant silver nanoparticles (Agnps) by an endophytic fungus Pestalotia Sp. isolated from Madhuca Longifolia. J Nanomater Mol Nanotechnol 5(4):3

Van Viet P, Nguyen HT, Cao TM, Van Hieu L (2016) Fusarium antifungal activities of copper nanoparticles synthesized by a chemical reduction method. J Nanomater 2016:2016

Wijesiri N, Yu Z, Tang H, Zhang P (2018) Antifungal photodynamic inactivation against dermatophyte Trichophyton rubrum using nanoparticle-based hybrid photosensitizers. Photodiagn Photodyn Ther 23:202–208

Xu C, Guan S, Wang S, Gong W, Liu T, Ma X, Sun C (2018) Biodegradable and electroconductive poly (3, 4-ethylenedioxythiophene)/carboxymethyl chitosan hydrogels for neural tissue engineering. Mater Sci Eng C 84:32–43

Yuvaraj M, Ramasamy M (2020) Role of fungi in agriculture. In: Biostimulants in plant science, pp 1–12

Zalewska ED, Machowicz-Stefaniak Z, Król ED (2016) Antifungal activity of nanoparticles against chosen fungal pathogens of caraway. Acta Scientiarum Polonorum Hortorum Cultus 15(6):121–137

Zhao X, Zhou L, Riaz Rajoka MS, Yan L, Jiang C, Shao D, Zhu J, Shi J, Huang Q, Yang H (2018) Fungal silver nanoparticles: synthesis, application and challenges. Crit Rev Biotechnol 38(6):817–835

Metal Nanoparticles: Management and Control of Phytopathogenic Fungi

15

Juned Ali, Danish Alam, Rubia Noori, Shazia Faridi, and Meryam Sardar

15.1 Introduction

Plant diseases have significant ecological effects in addition to endangering the world's food supply. Plant diseases are caused by infection from bacteria, viruses, viriods, nematodes, mycoplasmas, and fungi (plant pathogen). These diseases can harm trees, decorative plants, and crops and negatively impact plant health, growth, and production (Chaloner et al. 2021) and also destroy crops and jeopardize people's livelihoods (Jang and Seo 2023). Several types of plant pathogens have been listed, along with the associated diseases in Table 15.1. Comprehending plant disease biology, spread, and control is vital (Mwangi et al. 2023) for preventing and managing these diseases. Research and innovation are critical components in creating long-term remedies that safeguard plants and maintain the integrity of ecosystems around the globe. Protecting the health and vitality of our crops, gardens, and ecosystems requires a never-ending war against plant diseases (Ahmed and Yadav 2023; Mukhtar et al. 2023). This relentless struggle assumes paramount significance, particularly within the agricultural domain, where the imperative of ensuring food security is crucial. The endeavor to counter plant pathogens unfolds across multiple fronts, including cutting-edge breeding methods that produce plant types resistant to disease, scientific research that expands our knowledge of plant–pathogen interactions, and the prudent application of pesticides to control outbreaks (Lahlali et al. 2022). Within this perpetual conflict, promoting an understanding of the detrimental impacts that plant diseases impose on the realms of economy, ecology, and nutrition is just as crucial as cultivating resilient and sustainable farming practices (Fletcher et al. 2020).

J. Ali · D. Alam · R. Noori · S. Faridi · M. Sardar (✉)
Enzyme Technology Lab, Department of Biosciences, Jamia Millia Islamia, New Delhi, India
e-mail: alijuned7786@gmail.com; danish.ngl@gmail.com; rubiarn08@gmail.com; faridishazia10@gmail.com; msardar@jmi.ac.in

N. Manzoor (ed.), *Advances in Antifungal Drug Development*,
https://doi.org/10.1007/978-981-97-5165-5_15

Table 15.1 Plant pathogens and their associated diseases

Plant pathogens	Associated infections
Bacteria	Bacterial blight and bacterial wilt
Fungi	Rusts, powdery mildew, rot
Viruses	Mosaic, yellowing, and necrosis
Oomycetes	Late blight in potatoes and downy mildew in grapes

Fig. 15.1 (**a**) Rust disease: reddish-brown pustules on leaves; (**b**) Powdery mildew disease: white, powdery material on leaves; (**c**) Blight disease: wilted and brown leaves; (**d**) Canker disease: brown lesions on leaves

Plant diseases are a broad category of illnesses that impact different parts of the plant, such as leaves, stems, and roots. Fungal plant pathogens are attributed to the emergence of the following prevalent diseases (Fig. 15.1):

(a) **Rust:** Numerous fungal species of order *Pucciniales* are responsible for rust infections, which appear as orange to reddish-brown pustules on plant leaves, stems, and fruit. They damage various plants, such as roses, wheat, and soybeans. Several well-known rust diseases, including cedar-apple, coffee, soybean, and wheat rust, can cause severe problems for their host plants. Farmers and gardeners battle constant rust disease management, needing alertness and proactive steps to reduce the effects of disease and safeguard plant life (Shafi et al. 2022; Heller 2020).

(b) **Powdery mildew**: Numerous fungal species of order *Erysiphales* are responsible for powdery mildew infections, characterized by a white, powdery material that appears on the surfaces of leaves, stems, and flowers. Powdery mildew can weaken plants, lower their output, and increase their susceptibility to other stresses and illnesses, yet it is rarely fatal. Prompt diagnosis and suitable treatment are crucial to safeguard plant health and preserve the aesthetic appeal of gardens and landscapes (Seethapathy et al. 2022; Vielba-Fernández et al. 2020).

(c) **Blight**: Various fungal and bacterial species can produce blight, a plant disease that can harm a broad range of plants, including decorative plants, trees, vegetables, and fruit trees. The quick and fast wilting, browning, and death of plant tissues that characterize blight can result in significant damage and crop losses. Blight illnesses come in several forms, each brought on by a unique pathogen. Notable instances are fire blight in pears and potatoes. Diseases caused by blight can have severe ecological and economic effects, lowering agricultural yields and ruining landscapes. Particularly in agricultural and horticultural contexts, efficient management and control techniques are crucial to preventing or lessening the impact of blight diseases (Jindo et al. 2021; Guarnaccia et al. 2020).

(d) **Late Blight**: Late blight is a destructive disease mainly affecting potatoes and tomatoes (Solanaceae family) and is caused by the oomycete pathogen *Phytophthora infestans*. It drives fast foliar deterioration. Proactive management and attentive observation are crucial for safeguarding tomato and potato harvests and other vulnerable Solanaceae plants because late blight may be highly damaging and spread quickly. Although this infamous disease has been lessened in its effects by advancements in fungicide technology, resistant cultivars, and modern agricultural methods, producers still face difficulty managing it (Ivanov et al. 2021; Majeed et al. 2020).

(e) **Canker**: Specific, frequently submerged lesions or sores on the stems, branches, or trunks of woody plants are the hallmark of a particular kind of plant disease called canker. Many pathogens, including bacteria and fungi, are usually responsible for these lesions, which can damage and even kill afflicted plants. Gardeners, arborists, and forest managers face the issue of canker diseases. The health of woody plants must be preserved, and the financial burden of these diseases must be reduced via prompt detection and suitable care (Holland et al. 2021; Li et al. 2019).

Effectively regulating fungal infections in plants is crucial for a multitude of reasons. Throughout the plant world, fungi threaten food security, ecological balance, and agricultural production because of their capacity to cause deadly diseases. A worldwide food supply and economic stability may be affected if fungal infections are allowed to grow unchecked. These infections also have a negative impact on biodiversity and the state of the environment by compromising the health of natural ecosystems, forests, and beautiful landscapes. Fungi can also produce toxins that endanger the health of both plants and animals. Reducing the need for chemical treatments, which may have adverse effects on the environment, is another goal in the fight against fungal infections, in addition to protecting our green areas and crop production. Controlling and avoiding fungal infections is imperative to ensure a robust, sustainable, and peaceful cohabitation between plants and the natural environment. Among the various approaches available for combating fungal diseases in plants, utilization of nanoparticles emerges as a novel and ground-breaking strategy for managing fungal diseases (Verma et al. 2023).

15.2 Nanoparticles

Nanoparticles (NPs) are tiny particles ranging from 1 to 100 nm. Due to their small size and ability to infiltrate fungal cells rapidly, they may rupture their membranes and obstruct vital biological processes. Drugs can be dissolved, encapsulated, or attached using a nanoparticle matrix. NPs have the potential to be next-generation antimicrobial molecules. While their antibacterial abilities have been extensively explored, the application of nanoparticles as antifungal agents has not garnered as much attention as their antibacterial abilities noted by Huang et al. (2023). Nonetheless, a variety of characteristics of NPs may make them effective antimycotic agents. For example, NPs can attack microorganisms through various mechanisms due to their high specific surface area, high degree of tunable nature, and distinct physical and chemical characteristics (Begum et al. 2022). Antimicrobial resistance can be prevented or markedly delayed by targeting bacteria with NPs that exhibit multiple simultaneous mechanisms of action (Chao et al. 2022). It is a promising line of research to determine whether this tactic could prevent the emergence of fungal resistance. Qayyum and Khan (2016) have demonstrated that NPs can both penetrate biofilms and inhibit the formation of biofilms, making them a potentially useful tool for treating or preventing fungal biofilm infections. Their large surface area also makes it possible for effective drug loading and controlled release, which increases the efficacy of the antifungal agents and reduces their toxicity to the host organism. Another advantage of NPs is targeted drug delivery, as they can precisely deliver therapeutic drugs to the site of infection. This novel approach, which is becoming more essential in medical mycology and public health, has the potential to produce more effective, less toxic, and patient-friendly treatments for fungal infections.

15.3 Nanoparticles Classification

Nanoparticles are categorized depending on their size, composition, origin, and structure.

1. **Carbon-based NPs:** These NPs typically consist of carbon and can take the form of hollow tubes, ellipsoids, or spheres. Carbon-based NPs include fullerenes (C60), graphene (Gr), carbon nanotubes (CNTs), carbon nanofibers, carbon black, and carbon onions. With the exception of carbon black, the three primary synthesizing techniques for these carbon-based materials are chemical vapor deposition (CVD), arc discharge, and laser ablation (Kumar and Kumbhat 2016).

2. **Organic-based NPs:** NPs primarily derived from organic matter are classified as organic-based NPs; carbon-based and inorganic-based NPs are not included in this class. Some structures that can be made from organic NPs are dendrimers, micelles, liposomes, and polymer NPs. These are made by using weak interactions between molecules to facilitate their self-assembly and design (Jeevanandam et al. 2018). According to Gokarna et al. (2014), organic molecules with a 100 nm size are responsible for the production of organic NPs. Micelles and liposomes are examples of NPs with a hollow interior that are biodegradable, nontoxic, and sensitive to electromagnetic radiation (heat and light). They are better options for delivering drugs due to their unique characteristics. While considerations such as size, composition, surface shape, and other factors are significant, the efficacy and potential applications of nanoparticles hinge on the specific drug they carry, their stability, and whether they deliver the drug through an adsorbed drug system or an entrapped drug (Jeevanandam et al. 2018). Organic nanoparticles find extensive application in biomedicine, particularly in drug delivery systems, owing to their effective performance and the capability to be injected into specific anatomical sites, a technique referred to as targeted drug delivery (Tiwari et al. 2012).
3. **Inorganic-based NPs:** Metal and metal oxide NPs are examples of inorganic-based NPs. Metals like Au or Ag NPs, metal oxides like TiO_2 and ZnO-NPs, and semiconductors like silicon and ceramics can all be synthesized from these NPs.

15.4 Synthesis of Nanoparticles

NPs can be synthesized via chemical, biological, or physical approaches. Physical techniques use a top-down methodology and fabricate NPs from bulk materials. The process typically consists of two basic steps: rapid, controlled condensation after the bulk material has evaporated to produce the desired particle size. Three methods can be used to evaporate the material: high-voltage arc discharge, laser ablation, and γ-irradiation (Lara et al. 2018). In chemical methods (bottom-up approach), high-valence states of elements are typically converted to zero-valence NPs through various chemical reactions. Solvothermal synthesis, thermal decomposition, chemical precipitation, sonochemical methods, sol-gel methods, chemical reduction, and catalytic reduction are some of the chemical ways to make NPs (Khan et al. 2022). Biological synthesis is classified as a bottom-up method. The primary way that biological synthesis differs from standard chemical reactions is that it employs biological products as stabilizers and reducing agents. These biological products can be further divided into filtrates of bacteria and fungi, as well as sources from plants and animals. According to Nguyen et al. (2022), plant sources comprise leaf, seed, and flower extracts from various plant species, while animal sources include things like cow's milk (Lee et al. 2013). Researchers are becoming interested in biological synthesis, which is a very common fabrication technique for the creation of antifungal NPs. The mild reaction conditions of biological products make them a better option for use than industrial chemicals. Biological synthesis techniques have produced a wide variety of NPs, such as silver, gold, zinc oxide, copper oxide, selenium, zirconium oxide, platinum, etc. (Huang et al. 2023). It has been possible to synthesize both organic and inorganic NPs with antimycotic qualities; each has advantages and disadvantages of its own. According to Arciniegas-Grijalba et al. (2017), size and shape may affect cell uptake and other factors, which in turn may affect the antimicrobial activity of NPs. By optimizing certain aspects of the nanoparticle preparation process, such as reagent types or concentrations, reaction temperature, and reagent addition rate, particles with a range of sizes and shapes can be synthesized. For instance, varying the reducing agent's concentration allowed for the fabrication of Se NPs in various sizes, with lower concentrations constructing larger SeNPs (Huang et al. 2019). Other studies have changed other variables to vary the particle size and shape. Peng et al., for instance, fabricated different sizes of NPs by varying the temperature while maintaining constant reagent concentrations (Peng et al. 2010). Important elements that may influence how NPs interact with cells and fungi are surface characteristics like hydrophobicity and surface charge. Additionally, the stabilizing agents on NPs may interact directly with fungi and cells. Functionalizing NPs with antifungal chemicals can produce synergistic effects (Gibała et al. 2021). The surface hydrophobicity of the materials may prevent fungi from growing on the surface or within their structure (Benkovicová et al. 2019). Microbes typically have a lipid layer and a peptidoglycan-based negative-charge cell membrane. Therefore, the positive net surface charge can enhance the interaction between NPs and microbes. It is known that the presence of a positive net charge can increase NPs' antibacterial activity (Huang et al. 2020). Moreover, this may strengthen the antifungal effect. Through ligand engineering, it is simple to modify the surface characteristics of NPs to produce particles with novel and emergent properties. Consequently, a key factor in synthesizing NPs with strong antifungal efficacy is surface chemical modification with various chemicals.

15.5 Metal and Metal Oxide NPs as Potent Antifungal Agents

NPs without a carbon atom are known as inorganic NPs. Inorganic NPs are composed of metals or metal oxides.

15.5.1 Metal Nanoparticles

Metal-based nanoparticles can be synthesized by either constructive or destructive methods. Nanoparticles of nearly all metals, including aluminum (Al), cadmium (Cd), cobalt (Co), copper (Cu), gold (Au), iron (Fe), lead (Pb), silver (Ag), and zinc (Zn), can be synthesized using different approaches. Due to their large surface area to volume ratio, metal nanoparticles have great electrical, thermal, catalytic, UV-visible, and antibacterial properties (Khan et al. 2022).

15.5.1.1 Silver Nanoparticles (AgNPs)

Silver nanoparticles (AgNPs) are the inorganic NPs that have been studied the most for antifungal and antibacterial applications. They are utilized as antimicrobial agents in a variety of commercial products, such as medical device coatings and wound dressings, since they provide broad-spectrum antibacterial action against microbes (Li et al. 2020). AgNPs have been proved effective against different types of fungi, including *Aspergillus fumigatus*, *Aspergillus brasiliensis*, *Aspergillus oryzae*, *Penicillium chrysogenum*, *Bipolaris sorokiniana*, *Magnaporthe grisea*, *Candida albicans*, *Candida auris*, *Cryptococcus* sp., *Candida glabrata*, *Chaetomium globosum*, *Cladosporium cladosporioides*, *Mortierella alpine*, *Penicillium brevicompactum*, *Stachybotrys chartarum*, and *Rhizoctonia solani* (Nejad et al. 2016). Crucially, it has been demonstrated that the antifungal efficacy of silver nanoparticles (AgNPs) extends to fungal species that display resistance to pharmaceutical drugs. According to Gibała et al. (2021), AgNPs are generally believed to release soluble ions to carry out their antimicrobial action. Various studies have revealed that the choice of particle stabilizer and particle size influences the antifungal activity of AgNPs. For instance, ionic silver and AgNPs that were not stabilized were used as controls to test the antifungal effects of different AgNPs stabilized with surfactants or polymers (Panáček et al. 2009). Every AgNP used in this study was spherical and had a mean size of 25 nm. It was found that the nonstabilized AgNPs had an MIC of 0.21–0.42, 1.68, and 0.84 μg/mL against *Candida albicans*, *Candida parapsilosis*, and *Candida tropicalis*, respectively. AgNPs stabilized with surfactant or polymer exhibited greater antifungal efficacy against the same fungal strains, as evidenced by their respective lower MICs of 0.052–0.21, 0.84, and 0.42 μg/mL. Monteiro et al. synthesized AgNPs in a range of sizes (5, 10, and 60 nm) using ammonia and polyvinylpyrrolidone as stabilizers (Monteiro et al. 2012). AgNPs were more effective overall than many antifungal agents, including amphotericin B, fluconazole, ketoconazole, natamycin, griseofulvin, and itraconazole. For example, AgNPs were better at killing *Candida albicans* and *Candida tropicalis* than amphotericin B (Xue et al. 2016; Mallmann et al. 2015). In one study, AgNPs inhibited the formation of biofilms and destroyed pre-existing biofilms. Despite their remarkable antifungal properties, most of the aforementioned studies have solely examined the antifungal efficacy of AgNPs.

15.5.1.2 Gold Nanoparticles (AuNPs)

Gold NPs (AuNPs) have not got as much experimental attention as AgNPs when it comes to their potential use in antifungal applications. According to Eskandari-Nojedehi et al. (2018), AuNPs synthesized by bottom-up approach (biological method) were mostly used in antifungal studies. These methods involved using biological products as reducing agents, like plant extracts, bacterial filtrate, or fungal filtrate. Their size ranges from 20 to 300 nm, and they have demonstrated antifungal activity against a variety of fungal species, such as *Aspergillus flavus*, *Aspergillus niger*, *Candida albicans*, *Fusarium oxysporum*, *Microphyton gypseum*, *Aspergillus terreus*, and *Trichophyton rubrum* (Gopal et al. 2013). Most of these studies used the qualitative agar diffusion assay for antifungal testing, which makes quantitative evaluations and comparisons challenging. The MIC_{80} of AuNPs (25 and 30 nm, synthesized using different reducing agents) against *Candida* species were found to be 16–128 μg/mL (Wani and Ahmad 2013). This study and certain other studies (Ahmad et al. 2013) show that smaller AuNPs may be more effective against fungal infections. Dananjaya et al. synthesized AuNPs (16–23 nm) using a polysaccharide solution extracted from cyanobacteria as the reducing agent. After being treated with AuNPs, these particles changed shape significantly. They had an MIC and MCF of 32 μg/mL and 64 μg/mL, respectively, against *Candida albicans*. The antifungal action of AuNPs is believed to be achieved through membrane disruption and enzyme deactivation (Dananjaya et al. 2020). The transmembrane protein H^+-ATPase is another possible interaction target for them. ATP powers this enzyme and changes the energy from breaking down ATP into differences in the electrical potential across different biological membranes. Proton gradients power secondary transport mechanisms necessary for the fungal cell to grow. AuNPs can interact with H^+-ATPase and stop it from working properly. This can affect the fungal metabolism and kill it (Wani and Ahmad 2013).

15.5.1.3 Copper Nanoparticles (CuNPs)

CuNPs are small particles of copper that are said to be antifungal. They can kill a variety of different types of fungi, including *Aspergillus niger*, *Aspergillus flavus*, *Candida albicans*, *Candida tropicalis*, *Fusarium equiseti*, *Fusarium oxysporum*, *Fusarium culmorum*, *Phoma destructive*, and *Stachybotrys chartarum* (Bramhanwade et al. 2016). In contrast to AgNPs, AuNPs, and ZnONPs, antifungal investigations on CuNPs have been documented with considerable superficiality. Most of the CuNPs in these reports have particle sizes between 2 and 350 nm and have been synthesized using chemical reduction techniques. The agar diffusion method used in most of these studies for antifungal tests has inherent limitations, as previously discussed. CuNPs were demonstrated to have a low minimum inhibitory concentration (MIC) and a strong antifungal effectiveness against a range of fungus species. For example, MIC of 7 μg/mL was observed for ultrafine CuNPs (2–4 nm) produced by chemical reduction against *Corticium salmonicolor* (Cao et al. 2014). Using chemical reduction, carboxymethylated chitosan-stabilized CuNPs (4–15 nm) showed an MIC of 3.9 μg/mL against *Candida tropicalis* (Tantubay et al. 2015). The MIC of CuNPs (9–34 nm) produced by thermal degradation against *Stachybotrys chartarum* was around 2.5 μg/mL.

15.5.1.4 Selenium Nanoparticles (SeNPs)

Over the past 10 years, there has been a growing body of research on the antimicrobial activity of selenium NPs (SeNPs), with encouraging findings, especially for antibacterial applications (Mirza et al. 2022). SeNPs also exhibit antifungal activity, as shown by tests *on Aspergillus flavus*, *Aspergillus fumigatus*, *Aspergillus niger*, *Candida albicans*, *Candida glabrata*, *Candida krusei*, *Candida tropicalis*, and *Trichophyton rubrum* (Bafghi et al. 2021). It's interesting to note that the SeNPs utilized in a few of these reports were produced inside of bacteria. For example, SeNPs made by the Gram-negative bacteria *Klebsiella pneumoniae* had minimum inhibitory concentrations (MICs) of 250 µg/mL against *Aspergillus niger* and 2000 µg/mL against *Candida albicans* (67). Their sizes ranged from 90 to 320 nm. The Gram-positive bacteria Bacillus species Msh-1 made SeNPs between 120 and 140 nm in size. These had MICs of 100 µg/mL against *Aspergillus fumigatus* and 70 µg/mL against *Candida albicans* (Shakibaie et al. 2015). The researchers used the Gram-positive *Bacillus mycoides* and the Gram-negative *Stenotrophomonas maltophilia* bacteria to make SeNPs. The average particle size was 161 ± 52 nm for the *Bacillus mycoides* and 171 ± 35 nm for the *Stenotrophomonas maltophilia.* In relation to *Candida albicans*, these SeNPs demonstrated MICs of 512 and 256 µg/mL, respectively. SeNPs' antibiofilm activity was also documented. When used at 50 and 400 µg/mL, the biosynthesized SeNPs stopped *Candida albicans* biofilm formation by about 60% and 80%, respectively (Cremonini et al. 2016).

15.5.2 Metal Oxide Nanoparticles

Over the past few decades, metal oxides have drawn increasing attention from researchers. Metal oxides are ionic compounds made up of a combination of positive metallic ions and negative oxygen ions. The electrostatic interactions between the positive metal ions and the negative oxygen ions are what give rise to strong and stable ionic interactions (Devan et al. 2012). For example, when exposed to oxygen at room temperature, iron NPs (Fe) are easily transformed into iron oxide (Fe_2O_3) NPs, significantly increasing their reactivity compared to iron NPs. The purpose of synthesizing these oxide-based NPs is to alter the characteristics of their metal-based equivalents. Metal oxide NPs are frequently synthesized in order to benefit from their increased reactivity and efficiency. Among the most commonly synthesized oxides are silicon dioxide (SiO_2), titanium oxide (TiO_2), zinc oxide (ZnO), and aluminum oxide (Al_2O_3) (Khan et al. 2022). These NPs exhibit remarkable properties when compared to their metal analogs.

15.5.2.1 Zinc Oxide Nanoparticles (ZnONPs)

The most studied metal oxide NPs for antifungal applications are ZnONPs. They are frequently mentioned in the literature because zinc compounds are the primary ingredient in fungicides that are frequently used in agriculture. ZnONPs have been found to have antifungal activity against a broad range of fungal species (Ali et al. 2021), including *Alternaria alternata*, *Rhizopus stolonifera*, *Mucor plumbeus*, *Fusarium oxysporum* (Wani and Shah 2012), *Sclerotium rolfsii*, *Alternaria solani* (Surendra et al. 2016), *Aspergillus aculeatus*, *Aspergillus brasiliensis* (Vlad et al. 2012), *Aspergillus flavus*, *Trichoderma harzianum*, *Aspergillus nidulans* (Gunalan et al. 2012), *Aspergillus fumigatus*, *Aspergillus niger* (Senthilkumar and Sivakumar 2014), *Botrytis cinereal*, *Penicillium expansum* (He et al. 2011), *Candida albicans* (Pillai et al. 2020), *Erythricium salmonicolor* (Arciniegas-Grijalba et al. 2017), *Fusarium graminearum* (Dimkpa et al. 2013), *Trichophyton mentagrophytes*, *Microsporum canis* (Tiwari et al. 2016), *Fusarium* sp. (Sharma et al. 2010), and *Pythium debaryanum* (Sharma et al. 2011). ZnONPs were found to exhibit superior antifungal properties in comparison to silica and other metal oxide NPs. Karimiyan et al. compared several metal oxide NPs' antifungal activity against *Candida albicans* and observed that ZnONPs (20 nm), MgONPs (40 nm), SiO_2NPs (10 nm), and CuONPs (60 nm) were the most successful. On a mass basis, all metal oxide NPs that were investigated were far less effective than the antifungal drug amphotericin B (Karimiyan et al. 2015).

15.5.2.2 Titanium Dioxide Nanoparticles (TiO_2-NPs)

TiO_2-NPs have antifungal qualities against a range of fungal species, including *Fusarium oxysporum* and *Candida albicans* (Mukherjee et al. 2020). TiO_2-NPs, with sizes ranging from 70 to 100 nm, synthesized through the hydrolysis of titanium tetrachloride ($TiCl_4$), demonstrated a notable reduction in the formation of biofilms of *C. albicans*. This effect was observed in strains resistant to fluconazole (Haghighi et al. 2013). The photocatalytic activity of TiO_2-NPs is thought to be the source of their antimicrobial activity. Darbari et al. tested this theory by putting TiO_2-NPs on branched CNTs and found that the antifungal effects of light clearly caused the photodegradation of *C. albicans* biofilms in the presence of visible light. The authors hypothesized that because the CNTs mixed with TiO_2-NPs had a smaller band gap than pure TiO_2, visible light would make the CNT/TiO_2 interface more excited, which would create an electron hole in TiO_2. According to Darbari et al. (2011), the photogenerated holes may cause OH radicals to be produced on the surface of the microorganism, which could harm the cells. However, the majority of the time, UV light (between 2% and 3% of sunlight) is used as an irradiation source for TiO_2-NPs, which restricts their ability to combat microbes. TiO_2-NPs doped with fluorine and nitrogen can use visible light to make reactive oxygen species (ROS). This gives them a good level of antifungal activity when they are exposed to visible light (Mukherjee et al. 2020).

15.5.2.3 Iron Oxide Nanoparticles

Iron oxide NPs, such as magnetite (Fe_3O_4) and maghemite (γ-Fe_2O_3), have gained prominence recently due to their demonstrated value as safe, adaptable nanomaterials for biomedical applications. Their low cytotoxicity, biodegradability, and ability to modify their active surfaces with biocompatible coatings make them highly favored. The predominant focus of research on iron oxide nanoparticles (NPs) has revolved around their magnetic properties, extensively employed for applications such as targeted drug delivery, medical imaging, and various therapeutic and fabrication tasks (Jafari et al. 2019). However, recent research has also looked into the possible antibacterial and antifungal uses of these NPs. They made Fe_2O_3 nanoparticles (10–30 nm) with tannic acid as a stabilizing and reducing agent. They showed MIC values of 16–63 μg/mL against *Penicillium chrysogenum*, *Aspergillus niger*, *Cladosporium herbarum*, *Alternaria alternata*, and *Trichothecium roseum* (Parveen et al. 2018). According to Seddighi et al. (2017), Fe_3O_4 NPs demonstrated some noteworthy antifungal activity against *Candida* species, with MIC and MFC values of 62.5–500 μg/mL and 500–1000 μg/mL, respectively. Further research is necessary because the antifungal mechanisms of iron oxide NPs are still not fully understood.

15.6 Antifungal Mechanism of Inorganic Nanoparticles

Researchers have described the antifungal mechanisms of inorganic NPs in a number of reports (Fig. 15.2). These mechanisms include the release of mycotoxic ions (Li et al. 2017), damage to membranes, proteins, DNA, and other essential cellular components, excess production of reactive oxygen species (ROS), and depletion of ATP (Huang et al. 2023). The positively charged surface of AgNPs (Ag+/AgNPs) interacts electrostatically with the microorganism's negatively charged cell membrane. This induces alterations in the cellular structure, leading to the breakdown of the cell wall and cytoplasmic membrane, consequently resulting in the leakage of intracellular components. This process inflicts damage upon proteins, DNA, and other cellular contents. Additionally, silver nanoparticles (AgNPs) stimulate the production of reactive oxygen species (ROS), culminating in mitochondrial dysfunction and eventual cell death (Li et al. 2017; Prasher et al. 2018). In addition to the specific mechanisms mentioned above, the NPs appealing antifungal activity may also benefit from their distinctively small diameters. Due to their substantial surface area-to-volume ratio, NPs can adeptly envelop microorganisms, curbing their access to oxygen. This presents a supplementary avenue for inducing damage to fungal entities (Abdeen et al. 2013).

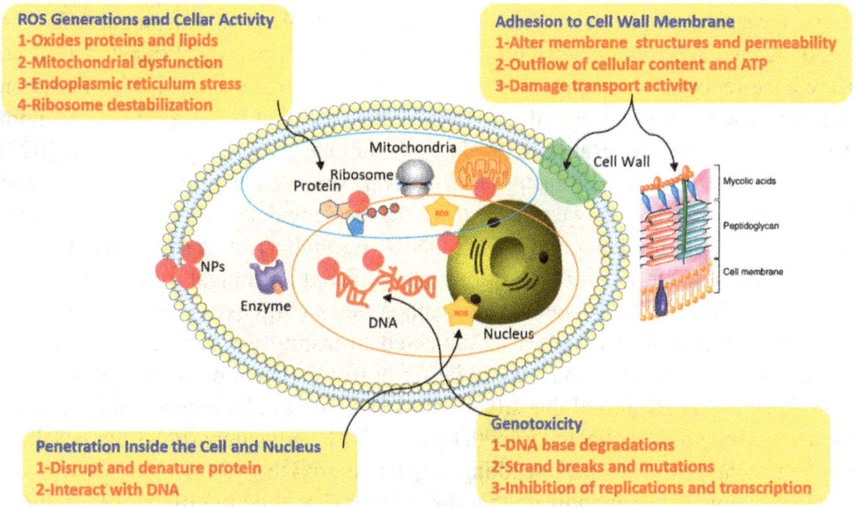

Fig. 15.2 Antifungal mechanism of metal oxide nanoparticles (This is an Open Access article distributed under the terms of the Creative Commons Attribution License (http://creativecommons.org/licenses/ by/2.0), which permits unrestricted use, distribution, and reproduction in any medium provided the original workis properly cited. Copyright@ Elsevier.) (Alghuthaymi et al. 2021)

15.7 Application of Nanoparticles Against Plant Pathogens

Research on the application of nanotechnology in agriculture is now being done in the areas of plant chemical delivery, water and seed control, nanobarcoding, gene transfer, controlled release of agrochemicals, and nanosensors. Numerous scientists have synthesized nanoparticles for protective coatings (protectants) or precise delivery of an active ingredient by encapsulation and adsorption (carrier). Using nanotechnology, a new generation of fungicides with various active components for controlling fungal diseases in plants may be developed (Kutawa et al. 2021; Maluin et al. 2020).

15.7.1 Nanoparticles as Protectants

The direct use of nanoparticles as antimicrobial agents has been proven successful against various soil-borne diseases. It can protect soil, seeds, roots, and leaves from pests and pathogens such as fungi, bacteria, and viruses. When nanoparticles enter a plant system, they either directly combat the pathogen or act as elicitor molecules, triggering local and systemic defense responses. The most thoroughly investigated nanoparticles are those made of metals such as gold, silver, titanium, zinc, and copper oxide, which are also known to have antifungal, antibacterial, and antiviral characteristics (Kim et al. 2018). Several nanoparticles have been documented for their utility as protectants, as outlined in Table 15.2. The antifungal effect of biogenically produced silver nanoparticles (AgNPs) against soil-borne phytopathogens *Rhizoctonia solani*, *Sclerotium rolfsii*, *Sclerotinia sclerotiorum*, and *Fusarium oxysporum* was investigated by Kaman and Dutta (2019). The antifungal effect of AgNP against *Sclerotium rolfsii* in wheat plants was reported by Desai et al. (2021). They have reported the suppression of mycelial growth and sclerotial germination and concluded that plant growth and disease management were optimal at 50 ppm AgNP concentration (Desai et al. 2021). Another investigation by Zaki et al. (2022) found that soil-borne pathogens *Rhizoctonia solani*, *Fusarium fujikuroi*, and *Macrophomina phseolina* were resistant to the antifungal activity of mycogenically produced zinc oxide nanoparticles (ZnONPs) through *Trichoderma* spp. Both on cotton seedlings and in vitro conditions, a significant antifungal impact was noted (Zaki et al. 2022). Magnesium oxide nanoparticles (MgONPs) and copper oxide nanoparticles (CuONPs) were also subjected to testing for their efficacy against *Phytophthora nicotianae* and *Thielaviopsis basicola* (Chen et al. 2022). Alvarez-Carvajal et al. (2020) found that silver-chitosan nanoparticles considerably reduced *Fusarium oxysporum* mycelial growth by up to 70% and nanoparticles did not show any detrimental effects on the seedling vegetative growth.

Microscopic findings indicate that the mechanism of action of AgNPs includes fungal cell disintegration, layer separation of the hyphal wall, leading to hyphal collapse subsequent death. SEM and EDS analysis reveal a direct interaction between nanoparticles and fungal cells, including contact and accumulation of AgNP inside fungal cells, the formation of micropores or fissures in the fungal cell wall, and the

Table 15.2 Role of metallic nanoparticles as protectant and carrier against fungal pathogens

Metal nanoparticles	Plant species	Target pathogen	Effect/mode of action	References
As protectant				
AgNPs	Mustard (*Brassica juncea*)	*Sclerotinia sclerotiorum*	Inhibition of sclerotial formation, hyphal growth and myceliogenic germination of sclerotia	Tomah et al. (2020)
AgNPs	Cotton (*Gossypium herbaceum*)	*Fusarium fujikuroi*, *Rhizoctonia solani* and *Macrophomina phseolina*	Mycelial growth reduction and cotton seedlings illness	Zaki et al. (2022)
AgNPs	Cereals, pulses and vegetables	*Rhizoctonia solani*, *Fusarium oxysporum*, *Sclerotium rolfsii* and *Sclerotinia sclerotiorum*	Inhibition of mycelial growth at 100 ppm of AgNPs	Kaman and Dutta (2019)
Capped AgNPs	Vegetables	*Sclerotinia sclerotiorum*	Inhibition of sclerotia germination and mycelial growth	Guilger-Casagrande et al. (2021)
AgNPs	Soybean (*Glycine max*)	*Phomopsis* sp., Soybean seeds	Inhibition of pathogen was seen at a concentration of 270 and 540 ppm	Mendes et al. (2014)
AgNPs	Strawberry (*Fragaria ananassa*)	*Macrophomina phaseolina* and *Fusarium solani*	Broad spectrum antagonism against *M. phaseolina* (67.05%) and *F. solani* (83.05%)	Paola et al. (2018)
MgONPs	Potato (*Solanum tuberosum*)	*Phytophthora infestans*	Cell membrane distortion, metabolic pathways disruption oxidative stress, and membrane transport activity	Wang et al. (2022)
MgONPs	Tobacco (*Nicotiana tabacum*)	*Thielaviopsis basicola* and *Phytophthora nicotianae*	Fungal growth inhibition, spore germination and impediment of sporangium development	Chen et al. (2020)

(continued)

Table 15.2 (continued)

Metal nanoparticles	Plant species	Target pathogen	Effect/mode of action	References
CuONPs	Tobacco (*Nicotiana tabacum*)	*Phytophthora nicotianae*	Increase in control efficacy and suppression of tobacco black shank disease without inducing phytotoxicity at 100 mg L^{-1} CuONPs	Chen et al. (2022)
CuONPs	Tomato and Eggplant	*Fusarium*	Increase of yield by 33% and 34% in tomato and eggplant	Elmer and White (2016)
ZnONPs	–	*Botrytis cinerea* and *Penicillium expansum*	Cellular functions hampered causing deformation in fungal hyphae	He et al. (2011)
Ag-doped TiO_2 NPs	–	*Fusarium solani* and *Venturia inaequalis*	Photocatalysis causing cell death and arresting the production of toxic pigment	Boxi et al. (2016)
TiO_2 NPs	Cowpea	Curvularia	Increased cowpea yield by 8.74–36.11%	Owolade and Ogunleti (2008)
CeO_2 NPs	*Solanum lycopersicum*	*Fusarium oxysporum* f. sp. *lycopersici*	Suppress wilting and improved the chlorophyll content in tomato plants	Adisa et al. (2018)
As carriers				
AgNPs and Fluconazole	Pulses, Vegetables	*Phoma glomerata*, *Phoma herbarum*, *Fusarium semitectum*, *Trichoderma* sp., and *Candida albicans*	Enhanced antifungal activity of fluconazole	Gajbhiye et al. (2009)
Alginate–Arabic gum and SiO_2 NPs and TiO_2 NPs	Cucumber (*Cucumis sativus*)	*Pythium aphanidermatum*	95% reduction in damping-off disease and showed more potential effects on increasing plant growth traits under green-house condition	Saberi Riseh et al. (2022)

(continued)

Table 15.2 (continued)

Metal nanoparticles	Plant species	Target pathogen	Effect/mode of action	References
Alginate bentonite coating with TiO_2 NPs	Bean (*Phaseolus vulgaris* L.)	*Rhizoctonia solani*	90% inhibition of pathogen as compared to 60% inhibition by free Vru 1 and vegetative growth parameters in bean were significantly enhanced	Saberi-Rise and Moradi-Pour (2020)
Ferbam and AuNPs	Tea leaves	–	–	Hou et al. (2016)
Tebuconazole, Propineb, Fludioxonil with AgNPs	–	*Bipolaria maydis*	–	Huang et al. (2018)

production of lamellar fragments (Tomah et al. 2020). The underlying mechanisms implicated in the antifungal impact were direct contact, adsorption of nanoparticles by fungal hypha, and alterations in cell morphology, as illustrated through SEM, TEM, and EDS (Chen et al. 2020).

15.7.2 Nanoparticles as Carriers

Nanotechnology can contribute to meeting the demands of sustainable agriculture by reducing the reliance on chemical fungicides and pesticides in the environment. Antimicrobial agent's effectiveness is increased by using nanoparticles as a carrier molecule that enables slow, timely, and targeted release of active substances in the environment. The consistent and delayed release of active substances is necessary to protect the crop during the growing season (Khan et al. 2011). Smaller nanoparticles with higher kinetic stability, optical transparency, and lower viscosity can function as more effective delivery systems (Xu et al. 2010). The solubility, wettability, bioavailability, and dispersion of chemical fungicides and pesticides can all be improved by using nanoformulations as carriers for active antimicrobial agents. Nanomaterials have unique qualities such as solubility, thermal stability, permeability, crystallinity, and biodegradability that are crucial for the development of nanopesticides. Nanoencapsulates, nanowires, nanotubes, nanocages, and nanoemulsions are typical nano-based delivery methods that may be vitally significant in plant protection techniques. Agrochemicals can be encapsulated using interactive surfaces with nanoencapsulates, primarily nanoclay (Bordes et al. 2009). Several nanoparticles have been reported as fungicide carriers and are summarized in Table 15.2.

The methods of controlled release and encapsulation are currently playing a significant part in regulating the use of various pesticides, herbicides, and fertilizers in agriculture across the globe. The most promising way of protecting host plants from insect or pathogen is undoubtedly nanoencapsulation, which involves encapsulating the nanomaterial. Since nanocapsules are typically made of polymers, this technique aids in the formulation of nanocide, which contains pesticides inside the shell (Fig. 15.3) (Elmer and White 2018).

According to Santiago et al. (2019), chitosan nanoparticles (ChNp) that had been loaded with AgNP exhibited antibacterial activity against the pathogen that causes tomato bacterial wilt, *Ralstonia solanacearum*, and hypothesized that they would be an effective substitute for bactericides or chemical antibiotics (Santiago et al. 2019). An investigation was carried out by Qian et al. (2011) to induce gradual release of the chemicals using nanosized calcium carbonate which carries the active components. Most of the research has been done to produce nanoencapsulated fungicides, but there are few literatures on encapsulating antibiotics and antiviral medicines to fight phytopathogens. Consequently, it is proposed that further investigation is warranted in this domain (Dutta et al. 2023).

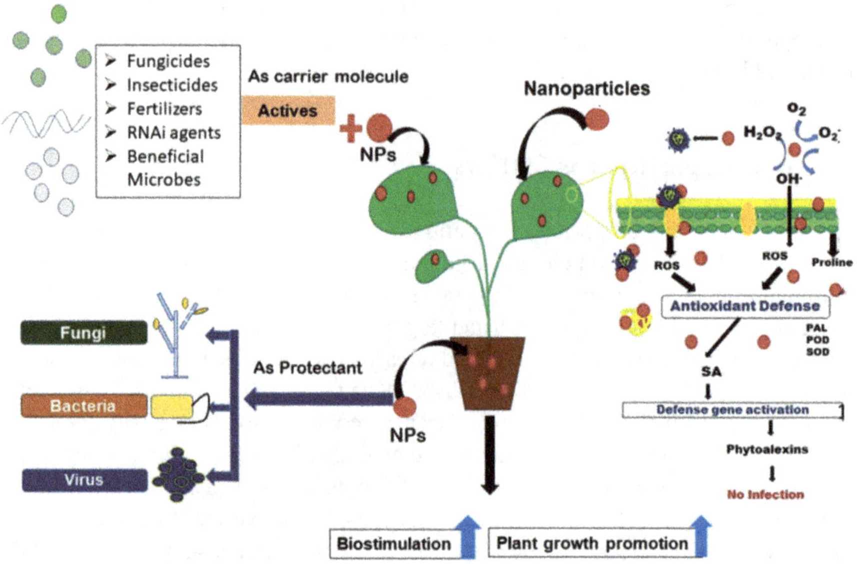

Fig. 15.3 Schematic representation of different approaches of nanoparticles for management of soil-borne plant pathogen (This is an Open Access article distributed under the terms of the Creative Commons Attribution (https://creativecommons.org/licenses/by/4.0/), which permits unrestricted use, distribution, and reproduction in any medium provided the original work is properly cited; Dutta et al. 2023)

15.7.3 Nanoparticles Inducing Plant Defense Mechanism

Induction of plant defense mechanisms stands as a crucial element within any comprehensive disease management strategy. This involves the stimulation and enhancement of inherent mechanisms within plants that serve to resist, counteract, or minimize the impact of pathogenic threats thereby contributing to a more resilient and sustainable approach to plant health management. One aspect of the plant defense response to various stresses is the generation of ROS, which prevents the spread of pathogens and sets off local and systemic defense reactions, such as the release of pathogenesis-related (PR) proteins. By raising or lowering oxidative stress, nanoparticles change the cellular redox balance (Tan et al. 2018). According to earlier research, nanoparticles can either promote the generation of ROS or decrease the oxidative burst by producing secondary metabolites and antioxidant enzymes such as superoxide dismutase (SOD), ascorbate peroxidase (APX), glutathione-S-transferase (GST), dehydroascorbate reductase (DHAR), catalase (CAT), and peroxidase (POX) depending on the needs of the host plant (Soares et al. 2018). According to Abdelrhim et al. (2021), silicon dioxide nanoparticles (SiO_2 NPs) were found to activate wheat seedlings' innate defense mechanisms and antioxidant systems in response to the parasite *R. solani*. Using proteome analysis, SOD, APX, and GST were discovered in higher concentrations in AgNP treated *Oryza sativa* roots (Mirzajani et al. 2014). Additionally, although inhibiting glutathione reductase (GR) and DHAR, these nanoparticles significantly boosted the activity of SOD and APX in *Pisum sativum* L. seedlings (Tripathi et al. 2017). Catalase (CAT), another enzyme that protects cells from oxidative damage, was significantly elevated in wheat roots subjected to CuONPs (Dimkpa et al. 2012). The accumulation of H_2O_2 in maize plants growing in soil supplemented with cerium dioxide nanoparticles (CeO_2NPs) increased with concentration (Zhao et al. 2012). Abdelaziz et al. (2022) demonstrated that ZnO NPs enhanced *F. oxysporum* infected eggplant healing by increasing metabolic and morphological markers. As demonstrated in a few studies, the induction of antioxidant machinery by nanoparticles may promote plant development as long as a dangerous level of ROS is not reached in the cells (Burman et al. 2013).

The balance between the defenses and growth of plants is controlled by crosstalk of different plant hormones and the balance of plant hormones has been shown to be affected by nanoparticles (Rastogi et al. 2017). Abscisic acid and indole-3-acetic acid, which are the stress-signaling molecules, accumulation in strawberry plants treated with selenium nanoparticles (SeNPs) was reported by Zahedi et al. (2019). Azhar et al. (2021) investigated phytohormone signaling when several metallic nanoparticles (ZnO, SiO_2, and ZnO/SiO_2 composite NPs) were exposed to *Arabidopsis*. The impact of CuONPs on the soil-borne pathogen *Gibberella fujikuroi* that causes Bakanae disease in rice plants was investigated by Shang et al. (2020).

PR proteins are a crucial element of a plant's immune system and are utilized as diagnostic biomarkers in plant defense signaling pathways (Ali et al. 2018). Treatment of healthy tobacco plants (*Nicotiana benthamiana*) with SiO_2 NPs and ZnO NPs increased the expression of the SA-inducible PR1 and PR2 genes, while treatment with magnetite nanoparticles (Fe_3O_4 NPs) had a similar effect. Therefore, the altered concentrations of phytohormones and PR proteins in plants exposed to nanoparticles imply activation of defense system (Cai et al. 2020). Grodetskaya et al. (2022) investigated the impact of CuONPs on soil-borne pathogen *Fusarium avenaceum* and *Fusarium oxysporum*. CuONPs dramatically reduced *F. avenaceum* infection while having no effect on *F. oxysporum*.

15.7.4 Nanoparticles as Biosensors for Pathogen Detection

Any sanitation or quarantine procedure depends on the ability to identify infectious or infested material before it is introduced into the greenhouse, field, state, or nation. For this method to work, pathogen detection needs to be done quickly. Nanotechnology makes significant advancements in this field of diagnostics by enabling faster and more sensitive pathogen probes. Nanoparticles can be employed as rapid diagnostic instruments to identify disease-causing bacterial, fungal, nematodal, and viral pathogens (Boonham et al. 2008; Yao et al. 2009). While nanotechnology has made considerable strides in human health over the past decade, there has been a notable scarcity of publications addressing its application for the diagnosis of plant diseases.

The improvement in plant pathogen diagnosis by nanotechnology has been briefly described (Ferreira et al. 2017; Jain 2005). Biosensors are nanoanalytical devices that emit an electrical signal by integrating a biological sensing element into a physicochemical transducer when it comes in contact with the target analyte (a pathogen). The biosensor can be loaded sufficiently to cause an increase in electrical signal in proportion to pathogen density. A digital output is produced as a result of a biomolecular interaction. Carbon nanomaterials (carbon nanotubes and graphene) and metalloid or metal oxide nanoparticles have also been used to construct biosensors. The production of biosensors with various nanoparticle and nanostructure types has now become easier because of breakthrough in nanotechnology (Fang and Ramasamy 2015). Although the methods have been pursued in medicine and water purification for more than 10 years, plant pathology has recently begun to focus on the advancement in the usage of super paramagnetic iron oxide nanoparticles to aid in pathogen identification (Li et al. 2011). Magnetic nanoparticles bind to DNA and biological tissue, making it easier to identify and/or extract these materials (Ahmadov et al. 2014). After being used on *F. oxysporum*, it was discovered that super paramagnetic iron oxide nanoparticles were distinct from quantum dots. Nano-Au functionalized with single-stranded oligonucleotides was employed in a new nanobiosensor to detect *R. solanacearum* genomic DNA in agricultural soil (Khaledian et al. 2017). Numerous articles from the preceding decade have demonstrated the capacity of antibody-based biosensors to detect plant pathogens (Chartuprayoon et al. 2013; Lin et al. 2014). Singh and colleagues reported a nano-Au-based dipstick for quick detection of Karnal bunt disease of wheat when grown in the field (Singh et al. 2014, 2010). Wang et al. (2010) employed indirect stimulation to build a sensitive electrochemical sensor using a modified gold electrode with copper nanoparticles to detect the harmful fungus *Sclerotinia sclerotiorum* (Wang et al. 2010). To advance this field, further research on related sensors and sensing systems is imperative for the identification of pathogens, their metabolites, or the monitoring of physiological changes induced by infections in plants. It is anticipated that these integrated techniques will soon be applied on-site, enabling precise detection of various soil-borne diseases.

In addition to single biosensor focused on a particular pathogen, research is being done to create nanochip microarrays with several fluorescent oligo probes to identify small nucleotide variations in fungi, bacteria and viruses that cause plant diseases (López et al. 2009). With this method, yields and earnings could rise while agrochemical inputs are reduced (Bergeson 2010). Once portable gadgets with biosensors are developed, nonresearchers could diagnose on the spot (Kashyap et al. 2016). In order to detect soil-borne diseases in situ, efforts must be focused on developing portable, affordable, effective, and hand-carried nanodevices (Alonso-Lomilloa et al. 2010).

15.8 Conclusion

In conclusion, the exploration of metal and metal oxide nanoparticles as fungicides presents a promising approach with significant implications for both the agricultural and industrial sectors. The distinctive physicochemical properties of these nanoparticles, characterized by their small size and high surface area, contribute to heightened antifungal activity. Nanoparticles are effective against a wide range of fungal pathogens, highlighting their potential to revolutionize current approaches to fungal disease management. Their ability to disrupt key cellular processes in fungi and the potential for targeted delivery minimizes the impact on nontarget organisms and ecosystems. Further investigations into the mechanisms of action, synergistic effects with existing fungicides, and optimization of nanoparticle formulations will be essential. Additionally, assessing the ecological implications of widespread nanoparticle use and developing strategies for responsible application are vital for ensuring the sustainability of this technology. Metal and metal oxide nanoparticles offer a promising solution for combating fungal infections, opening new possibilities for sustainable and efficient fungicide applications. As we move forward, interdisciplinary collaboration and a holistic approach to research and development will be vital in harnessing the full potential of these nanomaterials for a more resilient and productive agricultural and industrial future.

References

Abdeen S, Isaac RR, Geo S, Sornalekshmi S, Rose A, Praseetha P (2013) Evaluation of antimicrobial activity of biosynthesized iron and silver nanoparticles using the fungi fusarium oxysporum and actinomycetes sp. on human pathogens. Nano Biomed Eng 5(1):39–45

Abdelaziz AM, Salem SS, Khalil AM, El-Wakil DA, Fouda HM, Hashem AH (2022) Potential of biosynthesized zinc oxide nanoparticles to control *Fusarium* wilt disease in eggplant (*Solanum melongena*) and promote plant growth. Biometals 35(3):601–616

Abdelrhim AS, Mazrou YS, Nehela Y, Atallah OO, El-Ashmony RM, Dawood MF (2021) Silicon dioxide nanoparticles induce innate immune responses and activate antioxidant machinery in wheat against *Rhizoctonia solani*. Plan Theory 10(12):2758

Adisa IO, Reddy Pullagurala VL, Rawat S, Hernandez-Viezcas JA, Dimkpa CO, Elmer WH et al (2018) Role of cerium compounds in Fusarium wilt suppression and growth enhancement in tomato (*Solanum lycopersicum*). J Agric Food Chem 66(24):5959–5970

Ahmad T, Wani IA, Lone IH, Ganguly A, Manzoor N, Ahmad A, Ahmed J, Al-Shihri AS (2013) Antifungal activity of gold nanoparticles prepared by solvothermal method. Mater Res Bull 48(1):12–20

Ahmadov IS, Ramazanov MA, Sienkiewicz A, Forro L (2014) Uptake and intracellular trafficking of superparamagnetic iron oxide nanoparticles (spions) in plants. Dig J Nanomater Biostruct 9(3):1149–1157

Ahmed I, Yadav PK (2023) Plant disease detection using machine learning approaches. Expert Syst 40(5):e13136

Alghuthaymi MA, Rajkuberan C, Rajiv P, Kalia A, Bhardwaj K, Bhardwaj P, Abd-Elsalam KA et al (2021) Nanohybrid antifungals for control of plant diseases: current status and future perspectives. J Fungi 7(1):48

Ali S, Ganai BA, Kamili AN, Bhat AA, Mir ZA, Bhat JA et al (2018) Pathogenesis-related proteins and peptides as promising tools for engineering plants with multiple stress tolerance. Microbiol Res 212:29–37

Ali J, Mazumder JA, Perwez M, Sardar M (2021) Antimicrobial effect of ZnO nanoparticles synthesized by different methods against food borne pathogens and phytopathogens. Mater Today Proc 36:609–615

Alonso-Lomilloa MA, Domínguez-Renedoa O, Ferreira-Gonçalves L, Arcos-Martínez MJ (2010) Sensitive enzyme-biosensor based on screen-printed electrodes for ochratoxin A. Biosens Bioelectron 25:1333–1337

Alvarez-Carvajal F, Gonzalez-Soto T, Armenta-Calderón AD, Méndez Ibarra R, Esquer-Miranda E, Juarez J, Encinas-Basurto D (2020) Silver nanoparticles coated with chitosan against *Fusarium oxysporum* causing the tomato wilt. Biotecnia 22(3):73–80

Arciniegas-Grijalba P, Patiño-Portela M, Mosquera-Sánchez L, Guerrero-Vargas J, Rodríguez-Páez J (2017) ZnO nanoparticles (ZnO-NPs) and their antifungal activity against coffee fungus Erythricium salmonicolor. Appl Nanosci 7(5):225–241

Azhar BJ, Noor ASMA, Zulfiqar ALVEENA, Zeenat A, Ahmad SHAKEEL, Chishti I et al (2021) Effect of ZnO, SiO_2 and composite nanoparticles on *Arabidopsis thaliana* and involvement of ethylene and cytokinin signalling pathways. Pak J Bot 53(2):437–446

Bafghi MH, Darroudi M, Zargar M, Zarrinfar H, Nazari R (2021) Biosynthesis of selenium nanoparticles by Aspergillus flavus and Candida albicans for antifungal applications. Micro Nano Lett 16(14):656–669

Begum T, Follett PA, Mahmud J, Moskovchenko L, Salmieri S, Allahdad Z, Lacroix M (2022) Silver nanoparticles-essential oils combined treatments to enhance the antibacterial and antifungal properties against foodborne pathogens and spoilage microorganisms. Microb Pathog 164:105411

Benkovicová M, Kisová Z, Bucková M, Majková E, Šiffalovic P, Pangallo D (2019) The antifungal properties of super-hydrophobic nanoparticles and essential oils on different material surfaces. Coatings 9(3):176

Bergeson LL (2010) Nanosilver pesticide products: what does the future hold? Environ Q Manag 19(4):73–82

Boonham N, Glover R, Tomlinson J, Mumford R (2008) Exploiting generic platform technologies for the detection and identification of plant pathogens. In: Collinge DB, Munk L, Cooke BM (eds) Sustainable disease management in a European context. Springer, pp 355–363

Bordes P, Pollet E, Avérous L (2009) Nano-biocomposites: biodegradable polyester/nanoclay systems. Prog Polym Sci 34(2):125–155

Boxi SS, Mukherjee K, Paria S (2016) Ag doped hollow TiO_2 nanoparticles as an effective green fungicide against *Fusarium solani* and *Venturia inaequalis* phytopathogens. Nanotechnology 27(8):085103

Bramhanwade K, Shende S, Bonde S, Gade A, Rai M (2016) Fungicidal activity of Cu nanoparticles against Fusarium causing crop diseases. Environ Chem Lett 14(2):229–235

Burman U, Saini M, Kumar P (2013) Effect of zinc oxide nanoparticles on growth and antioxidant system of chickpea seedlings. Toxicol Environ Chem 95(4):605–612

Cai L, Cai L, Jia H, Liu C, Wang D, Sun X (2020) Foliar exposure of Fe3O4 nanoparticles on *Nicotiana benthamiana*: evidence for nanoparticles uptake, plant growth promoter and defense response elicitor against plant virus. J Hazard Mater 393:122415

Cao VD, Nguyen PP, Khuong VQ, Nguyen CK, Nguyen XC, Dang CH, Tran NQ (2014) Ultrafine copper nanoparticles exhibiting a powerful antifungal/killing activity against Corticium salmonicolor. Bull Korean Chem Soc 35(9):2645–2648

Chaloner TM, Gurr SJ, Bebber DP (2021) Plant pathogen infection risk tracks global crop yields under climate change. Nat Clim Chang 11(8):710–715

Chao D, Dong Q, Chen J, Yu Z, Wu W, Fang Y, Liu L, Dong S (2022) Highly efficient disinfection based on multiple enzyme-like activities of Cu3P nanoparticles: a catalytic approach to impede antibiotic resistance. Appl Catal B 304:121017

Chartuprayoon N, Rheem Y, Ng JC, Nam J, Chen W, Myung NV (2013) Polypyrrole nanoribbon based chemiresistive immunosensors for viral plant pathogen detection. Anal Methods 5(14):3497–3502

Chen J, Wu L, Lu M, Lu S, Li Z, Ding W (2020) Comparative study on the fungicidal activity of metallic MgO nanoparticles and macroscale MgO against soilborne fungal phytopathogens. Front Microbiol 11:365

Chen JN, Wu LT, Song K, Zhu YS, Ding W (2022) Nonphytotoxic copper oxide nanoparticles are powerful "nanoweapons" that trigger resistance in tobacco against the soil-borne fungal pathogen *Phytophthora nicotianae*. J Integr Agric 21(11):3245–3262

Cremonini E, Zonaro E, Donini M, Lampis S, Boaretti M, Dusi S, Melotti P, Lleo MM, Vallini G (2016) Biogenic selenium nanoparticles: characterization, antimicrobial activity and effects on human dendritic cells and fibroblasts. Microb Biotechnol 9(6):758–771

Dananjaya S, Thao NT, Wijerathna H, Lee J, Edussuriya M, Choi D, Kumar RS (2020) In vitro and in vivo anticandidal efficacy of green synthesized gold nanoparticles using Spirulina maxima polysaccharide. Process Biochem 92:138–148

Darbari S, Abdi Y, Haghighi F, Mohajerzadeh S, Haghighi N (2011) Investigating the antifungal activity of TiO 2 nanoparticles deposited on branched carbon nanotube arrays. J Phys D Appl Phys 44(24):245401

Desai P, Jha A, Markande A, Patel J (2021) Silver nanoparticles as a fungicide against soil-borne *sclerotium rolfsii*: a case study for wheat plants. In: Biobased nanotechnology for green applications, Springer Nature Switzerland AG, pp 513–542

Devan RS, Patil RA, Lin JH, Ma YR (2012) One-dimensional metal-oxide nanostructures: recent developments in synthesis, characterization, and applications. Adv Funct Mater 22:3326–3370

Dimkpa CO, McLean JE, Latta DE, Manangón E, Britt DW, Johnson WP et al (2012) CuO and ZnO nanoparticles: phytotoxicity, metal speciation, and induction of oxidative stress in sand-grown wheat. J Nanopart Res 14:1–15

Dimkpa CO, Latta DE, McLean JE, Britt DW, Boyanov MI, Anderson AJ (2013) Fate of CuO and ZnO nano-and microparticles in the plant environment. Environ Sci Technol 47(9):4734–4742

Dutta P, Kumari A, Mahanta M, Upamanya GK, Heisnam P, Borua S et al (2023) Nanotechnological approaches for management of soil-borne plant pathogens. Front Plant Sci 14:1136233

Elmer WH, White JC (2016) The use of metallic oxide nanoparticles to enhance growth of tomatoes and eggplants in disease infested soil or soilless medium. Environ Sci Nano 3(5):1072–1079

Elmer W, White JC (2018) The future of nanotechnology in plant pathology. Annu Rev Phytopathol 56:111–133

Eskandari-Nojedehi M, Jafarizadeh-Malmiri H, Rahbar-Shahrouzi J (2018) Hydrothermal green synthesis of gold nanoparticles using mushroom (Agaricus bisporus) extract: physico-chemical characteristics and antifungal activity studies. Green Process Synth 7(1):38–47

Fang Y, Ramasamy RP (2015) Current and prospective methods for plant disease detection. Biosensors 5(3):537–561

Ferreira MAM, Filipe JA, Coelho M, Chavaglia J (2017) Nanotechnology applications in industry and medicine. Nanotechnology applications in industry and medicine, 2:31–50

Fletcher J, Gamliel A, Gullino ML, McKirdy SJ, Smith GR, Stack JP (2020) A fresh look at graduate education in plant pathology in a changing world: global needs and perspectives. J Plant Pathol 102:609–618

Gajbhiye M, Kesharwani J, Ingle A, Gade A, Rai M (2009) Fungus-mediated synthesis of silver nanoparticles and their activity against pathogenic fungi in combination with fluconazole. Nanomedicine 5(4):382–386

Gibała A, Zeliszewska P, Gosiewski T, Krawczyk A, Duraczynska D, Szaleniec J, Szaleniec M, Ocwieja M (2021) Atibacterial and antifungal properties of silver nanoparticles-effect of a surface-stabilizing agent. Biomol Ther 11(10):1481

Gokarna A, Parize R, Kadiri H, Nomenyo K, Patriarche G, Miska P, Lerondel G (2014) Highly crystalline urchin-like structures made of ultra-thin zinc oxide nanowires. RSC Adv 4(88):47234–47239

Gopal JV, Thenmozhi M, Kannabiran K, Rajakumar G, Velayutham K, Rahuman AA (2013) Actinobacteria mediated synthesis of gold nanoparticles using Streptomyces sp. VITDDK3 and its antifungal activity. Mater Lett 93:360–362

Grodetskaya TA, Evlakov PM, Fedorova OA, Mikhin VI, Zakharova OV, Kolesnikov EA et al (2022) Influence of copper oxide nanoparticles on gene expression of birch clones in vitro under stress caused by phytopathogens. Nanomaterials 12(5):864

Guarnaccia V, Martino I, Tabone G, Brondino L, Gullino ML (2020) Fungal pathogens associated with stem blight and dieback of blueberry in northern Italy. Phytopathol Mediterr 59(2):229–245

Guilger-Casagrande M, Germano-Costa T, Bilesky-José N, Pasquoto-Stigliani T, Carvalho L, Fraceto LF et al (2021) Influence of the capping of biogenic silver nanoparticles on their toxicity and mechanism of action towards *Sclerotinia sclerotiorum*. J Nanobiotechnol 19:53. https://doi.org/10.1186/s12951-021-00797-5

Gunalan S, Sivaraj R, Rajendran V (2012) Green synthesized ZnO nanoparticles against bacterial and fungal pathogens. Prog Nat Sci Mater Int 22(6):693–700

Haghighi F, Roudbar Mohammadi S, Mohammadi P, Hosseinkhani S, Shipour R (2013) Antifungal activity of TiO2 nanoparticles and EDTA on Candida albicans biofilms. Infect Epidemiol Microbiol 1(1):33–38

He L, Liu Y, Mustapha A, Lin M (2011) Antifungal activity of zinc oxide nanoparticles against *Botrytis cinerea* and *Penicillium expansum*. Microbiol Res 166(3):207–215. https://doi.org/10.1016/j.micres.2010.03.003

Heller A (2020) Host-parasite interaction during subepidermal sporulation and pustule opening in rust fungi (Pucciniales). Protoplasma 257(3):783–792

Holland LA, Trouillas FP, Nouri MT, Lawrence DP, Crespo M, Doll DA et al (2021) Fungal pathogens associated with canker diseases of almond in California. Plant Dis 105(2):346–360

Hou R, Zhang Z, Pang S, Yang T, Clark JM, He L (2016) Alteration of the nonsystemic behavior of the pesticide ferbam on tea leaves by engineered gold nanoparticles. Environ Sci Technol 50:6216–6223

Huang W, Wang C, Duan H, Bi Y, Wu D, Du J, Yu H (2018) Synergistic antifungal effect of biosynthesized silver nanoparticles combined with fungicides. Int J Agric Biol 20:1225–1229

Huang T, Holden JA, Heath DE, O'Brien-Simpson NM, O'Connor AJ (2019) Engineering highly effective antimicrobial selenium nanoparticles through control of particle size. Nanoscale 11:14937–14951

Huang T, Holden JA, Reynolds EC, Heath DE, O'Brien-Simpson NM, O'Connor AJ (2020) Multifunctional antimicrobial polypeptide-selenium nanoparticles combat drug-resistant bacteria. ACS Appl Mater Interfaces 12(50):55696–55709

Huang T, Li X, Maier M, O'Brien-Simpson NM, Heath DE, O'Connor AJ (2023) Using inorganic nanoparticles to fight fungal infections in the antimicrobial resistant era. Acta Biomater 158:56

Ivanov AA, Ukladov EO, Golubeva TS (2021) Phytophthora infestans: an overview of methods and attempts to combat late blight. J Fungi 7(12):1071

Jafari J, Han XL, Palmer J, Tran PA, O'Connor AJ (2019) Remote control in formation of 3D multicellular assemblies using magnetic forces. ACS Biomater Sci Eng 5(5):2532–2542

Jain KK (2005) Nanotechnology in clinical laboratory diagnostics. Clin Chim Acta 358(1–2):37–54

Jang H, Seo S (2023) Interaction of microbial pathogens with plants: attachment to persistence. In: The produce contamination problem. Academic, pp 13–45

Jeevanandam J, Barhoum A, Chan YS, Dufresne A, Danquah MK (2018) Review on nanoparticles and nanostructured materials: history, sources, toxicity and regulations. Beilstein J Nanotechnol 9(1):1050–1074

Jindo K, Evenhuis A, Kempenaar C, Pombo Sudré C, Zhan X, Goitom Teklu M, Kessel G (2021) Holistic pest management against early blight disease towards sustainable agriculture. Pest Manag Sci 77(9):3871–3880

Kaman PK, Dutta P (2019) Synthesis, characterization and antifungal activity of biosynthesized silver nanoparticle. Indian Phytopathol 72:79–88

Karimiyan A, Najafzadeh H, Ghorbanpour M, Hekmati-Moghaddam SH (2015) Antifungal effect of magnesium oxide, zinc oxide, silicon oxide and copper oxide nanoparticles against Candida albicans. Zahedan J Res Med Sci 17(10):e2179

Kashyap PL, Rai P, Sharma S, Chakdar H, Kumar S et al (2016) Nanotechnology for the detection and diagnosis of plant pathogens. In: Ranjan S, Dasgupta N, Lichtfouse E (eds) Nanoscience in food and agriculture, 2nd edn. Springer, New York, pp 253–276

Khaledian S, Nikkhah M, Shams-bakhsh M, Hoseinzadeh S (2017) A sensitive biosensor based on gold nanoparticles to detect *Ralstonia solanacearum* in soil. J Gen Plant Pathol 83:231–239

Khan MR, Majid S, Mohidin FA, Khan N (2011) A new bioprocess to produce low cost powder formulations of biocontrol bacteria and fungi to control fusarial wilt and root-knot nematode of pulses. Biol Control 59(2):130–140

Khan Y, Sadia H, Ali Shah SZ, Khan MN, Shah AA, Ullah N, Ullah MF, Bibi H, Bafakeeh OT, Khedher NB, Eldin SM (2022) Classification, synthetic, and characterization approaches to nanoparticles, and their applications in various fields of nanotechnology: a review. Catalysts 12(11):1386

Kim DY, Kadam A, Shinde S, Saratale RG, Patra J, Ghodake G (2018) Recent developments in nanotechnology transforming the agricultural sector: a transition replete with opportunities. J Sci Food Agric 98(3):849–864

Kumar N, Kumbhat S (2016) Essentials in nanoscience and nanotechnology. John Wiley & Sons

Kutawa AB, Ahmad K, Ali A, Hussein MZ, Abdul Wahab MA, Adamu A et al (2021) Trends in nanotechnology and its potentialities to control plant pathogenic fungi: a review. Biology 10(9):881

Lahlali R, Ezrari S, Radouane N, Kenfaoui J, Esmaeel Q, El Hamss H et al (2022) Biological control of plant pathogens: a global perspective. Microorganisms 10(3):596

Lara HH, Guisbiers G, Mendoza J, Mimun LC, Vincent BA, Lopez-Ribot JL, Nash KL (2018) Synergistic antifungal effect of chitosan-stabilized selenium nanoparticles synthesized by pulsed laser ablation in liquids against Candida albicans biofilms. Int J Nanomedicine 13:2697

Lee KJ, Park SH, Govarthanan M, Hwang PH, Seo YS, Cho M, Lee WH, Lee JY, Kamala-Kannan S, Oh BT (2013) Synthesis of silver nanoparticles using cow milk and their antifungal activity against phytopathogens. Mater Lett 105:128–131

Li XM, Xu G, Liu Y, He T (2011) Magnetic Fe3O4 nanoparticles: synthesis and application in water treatment. Nanosci Nanotechnol Asia 1(1):14–24

Li J, Sang H, Guo H, Popko JT, He L, White JC, Dhankher OP, Jung G, Xing B (2017) Antifungal mechanisms of ZnO and Ag nanoparticles to Sclerotinia homoeocarpa. Nanotechnology 28(15):155101

Li P, Liu W, Zhang Y, Xing J, Li J, Feng J et al (2019) Fungal canker pathogens trigger carbon starvation by inhibiting carbon metabolism in poplar stems. Sci Rep 9(1):10,111

Li X, Huang T, Heath DE, O'Brien-Simpson NM, O'Connor AJ (2020) Antimicrobial nanoparticle coatings for medical implants: design challenges and prospects. Biointerphases 15(6):060801

Lin HY, Huang CH, Lu SH, Kuo IT, Chau LK (2014) Direct detection of orchid viruses using nanorod-based fiber optic particle plasmon resonance immunosensor. Biosens Bioelectron 51:371–378

López MM, Llop P, Olmos A, Marco-Noales E, Cambra M, Bertolini E (2009) Are molecular tools solving the challenges posed by detection of plant pathogen bacteria and viruses? Curr Issues Mol Biol 11(1):13–46

Majeed A, Muhammad Z, Inayat N, Siyar S, Wani SH (2020) Reducing the use of fungicides sprays with natural compounds and biocontrol agents for late blight management: a review. J Plant Sci Res 36:185

Mallmann EJJ, Cunha FA, Castro BN, Maciel AM, Menezes EA, Fechine PBA (2015) Antifungal activity of silver nanoparticles obtained by green synthesis. Rev Inst Med Trop São Paulo 57(2):165–167

Maluin FN, Hussein MZ, Azah Yusof N, Fakurazi S, Idris AS, Zainol Hilmi NH, Jeffery Daim LD (2020) Chitosan-based agronanofungicides as a sustainable alternative in the basal stem rot disease management. J Agric Food Chem 68(15):4305–4314

Mendes J, Abrunhosa L, Teixeira J, Camargo E, Souza C, Pessoa J (2014) Antifungal activity of silver colloidal nanoparticles against phytopathogenic fungus (*Phomopsis* sp.) in soybean seeds. Int J Biol Vet Agric Food Eng 8:928–933

Mirza K, Naaz F, Ahmad T, Manzoor N, Sardar M (2022) Development of cost-effective, eco-friendly selenium nanoparticle-functionalized cotton fabric for antimicrobial and antibiofilm activity. Fermentation 9(1):18

Mirzajani F, Askari H, Hamzelou S, Schober Y, Römpp A, Ghassempour A, Spengler B (2014) Proteomics study of silver nanoparticles toxicity on *Oryza sativa* L. Ecotoxicol Environ Saf 108:335–339

Monteiro D, Silva S, Negri M, Gorup L, De Camargo E, Oliveira R, Barbosa DdB, Henriques M (2012) Silver nanoparticles: influence of stabilizing agent and diameter on antifungal activity against Candida albicans and Candida glabrata biofilms. Lett Appl Microbiol 54(5):383–391

Mukherjee K, Acharya K, Biswas A, Jana NR (2020) TiO 2 nanoparticles co-doped with nitrogen and fluorine as visible-light-activated antifungal agents. ACS Appl Nano Mater 3(2):2016–2025

Mukhtar T, Vagelas I, Javaid A (2023) New trends in integrated plant disease management. Front Agron 4:1104122

Mwangi RW, Mustafa M, Charles K, Wagara IW, Kappel N (2023) Selected emerging and reemerging plant pathogens affecting the food basket: A threat to food security. J Agric Food Res 14:100827

Nejad MS, Bonjar GHS, Khatami M, Amini A, Aghighi S (2016) In vitro and in vivo antifungal properties of silver nanoparticles against Rhizoctonia solani, a common agent of rice sheath blight disease. IET Nanobiotechnol 11(3):236–240

Nguyen NT, Nguyen LM, Nguyen TT, Nguyen TT, Nguyen DT, Tran TV (2022) Formation, antimicrobial activity, and biomedical performance of plant-based nanoparticles: a review. Environ Chem Lett 20(4):2531–2571

Owolade O, Ogunleti D (2008) Effects of titanium dioxide on the diseases, development and yield of edible cowpea. J Plant Prot Res 48:329

Panáček A, Kolář M, Večeřová R, Prucek R, Soukupová J, Kryštof V, Hamal P, Zbořil R, Kvítek L (2009) Antifungal activity of silver nanoparticles against Candida spp. Biomaterials 30(31):6333–6340

Paola RR, Benjamín VS, Daniel GM, Vianey MT (2018) Antifungal effects of silver phytonanoparticles from *Yucca shilerifera* against strawberry soil-borne pathogens: *Fusarium solani* and

Macrophomina phaseolina. Mycobiology 46(1):47–51. https://doi.org/10.1080/12298093.2018.1454011

Parveen S, Wani AH, Shah MA, Devi HS, Bhat MY, Koka JA (2018) Preparation, characterization and antifungal activity of iron oxide nanoparticles. Microb Pathog 115:287–292

Peng S, McMahon JM, Schatz GC, Gray SK, Sun Y (2010) Reversing the size-dependence of surface plasmon resonances. Proc Natl Acad Sci 107(33):14530–14534

Pillai AM, Sivasankarapillai VS, Rahdar A, Joseph J, Sadeghfar F, Rajesh K, Kyzas GZ (2020) Green synthesis and characterization of zinc oxide nanoparticles with antibacterial and antifungal activity. J Mol Struct 1211:128107

Prasher P, Singh M, Mudila H (2018) Green synthesis of silver nanoparticles and their antifungal properties. BioNanoScience 8:254–263

Qayyum S, Khan AU (2016) Nanoparticles vs. biofilms: a battle against another paradigm of antibiotic resistance. MedChemComm 7(8):1479–1498

Qian K, Shi T, Tang T, Zhang S, Liu X, Cao Y (2011) Preparation and characterization of nano-sized calcium carbonate as controlled release pesticide carrier for validamycin against *Rhizoctonia solani*. Microchim Acta 173:51–57

Rastogi A, Zivcak M, Sytar O, Kalaji HM, He X, Mbarki S, Brestic M (2017) Impact of metal and metal oxide nanoparticles on plant: a critical review. Front Chem 5:78

Saberi Riseh R, Moradi Pour M, Ait Barka E (2022) A Novel route for double-layered encapsulation of *Streptomyces fulvissimus* Uts22 by alginate–Arabic gum for controlling of *Pythium aphanidermatum* in cucumber. Agronomy 12(3):655

Saberi-Rise R, Moradi-Pour M (2020) The effect of Bacillus subtilis Vru1 encapsulated in alginate–bentonite coating enriched with titanium nanoparticles against *Rhizoctonia solani* on bean. Int J Biol Macromol 152:1089–1097

Santiago TR, Bonatto CC, Rossato M, Lopes CA, Lopes CA, Mizubuti ESG, Silva LP (2019) Green synthesis of silver nanoparticles using tomato leaf extract and their entrapment in chitosan nanoparticles to control bacterial wilt. J Sci Food Agric 99(9):4248–4259

Seddighi NS, Salari S, Izadi AR (2017) Evaluation of antifungal effect of iron-oxide nanoparticles against different Candida species. IET Nanobiotechnol 11(7):883–888

Seethapathy P, Sankaralingam S, Pandita D, Pandita A, Loganathan K, Wani SH et al (2022) Genetic diversity analysis based on the virulence, physiology and regional variability in different isolates of powdery mildew in pea. J Fungi 8(8):798

Senthilkumar S, Sivakumar T (2014) Green tea (Camellia sinensis) mediated synthesis of zinc oxide (ZnO) nanoparticles and studies on their antimicrobial activities. Int J Pharm Pharm Sci 6(6):461–465

Shafi U, Mumtaz R, Shafaq Z, Zaidi SMH, Kaifi MO, Mahmood Z, Zaidi SAR (2022) Wheat rust disease detection techniques: a technical perspective. J Plant Dis Prot 129(3):489–504

Shakibaie M, Mohazab NS, Mousavi SAA (2015) Antifungal activity of selenium nanoparticles synthesized by Bacillus species Msh-1 against Aspergillus fumigatus and Candida albicans. Jundishapur J Microbiol 8(9):e26381

Shang H, Ma C, Li C, White JC, Polubesova T, Chefetz B, Xing B (2020) Copper sulfide nanoparticles suppress *Gibberella fujikuroi* infection in rice (Oryza sativa L.) by multiple mechanisms: contact-mortality, nutritional modulation and phytohormone regulation. Environ Sci Nano 7(9):2632–2643

Sharma D, Rajput J, Kaith B, Kaur M, Sharma S (2010) Synthesis of ZnO nanoparticles and study of their antibacterial and antifungal properties. Thin Solid Films 519(3):1224–1229

Sharma D, Sharma S, Kaith BS, Rajput J, Kaur M (2011) Synthesis of ZnO nanoparticles using surfactant free in-air and microwave method. Appl Surf Sci 257(22):9661–9672

Singh S, Singh M, Agrawal VV, Kumar A (2010) An attempt to develop surface plasmon resonance based immunosensor for Karnal bunt (*Tilletia indica*) diagnosis based on the experience of nano-gold based lateral flow immuno-dipstick test. Thin Solid Films 519:1156–1159

Singh S, Gupta AK, Gupta S, Gupta S, Kumar A (2014) Surface Plasmon Resonance (SPR) and cyclic voltammetry based immunosensor for determination of teliosporic antigen and diagnosis of Karnal Bunt of wheat using anti-teliosporic antibody. Sensors Actuators B Chem 191:866–873

Soares C, Pereira R, Fidalgo F (2018) Metal-based nanomaterials and oxidative stress in plants: current aspects and overview. In: Faisal M et al (eds) Phytotoxicity of nanoparticles. Springer, pp 197–227

Surendra T, Roopan SM, Al-Dhabi NA, Arasu MV, Sarkar G, Suthindhiran K (2016) Vegetable peel waste for the production of ZnO nanoparticles and its toxicological efficiency, antifungal, hemolytic, and antibacterial activities. Nanoscale Res Lett 11(1):546

Tan BL, Norhaizan ME, Liew WPP, Sulaiman Rahman H (2018) Antioxidant and oxidative stress: a mutual interplay in age-related diseases. Front Pharmacol 9:1162

Tantubay S, Mukhopadhyay SK, Kalita H, Konar S, Dey S, Pathak A, Pramanik P (2015) Carboxymethylated chitosan-stabilized copper nanoparticles: a promise to contribute a potent antifungal and antibacterial agent. J Nanopart Res 17(6):243

Tiwari JN, Tiwari RN, Kim KS (2012) Zero-dimensional, one-dimensional, two-dimensional and three-dimensional nanostructured materials for advanced electrochemical energy devices. Prog Mater Sci 57(4):724–803

Tiwari N, Pandit R, Gaikwad S, Gade A, Rai M (2016) Biosynthesis of zinc oxide nanoparticles by petals extract of Rosa indica L., its formulation as nail paint and evaluation of antifungal activity against fungi causing onychomycosis. IET Nanobiotechnol 11(2):205–211

Tomah AA, Alamer ISA, Li B, Zhang JZ (2020) Mycosynthesis of silver nanoparticles using screened *Trichoderma* isolates and their antifungal activity against *Sclerotinia sclerotiorum*. Nanomaterials 10(10):1955

Tripathi DK, Singh S, Singh S, Srivastava PK, Singh VP, Singh S et al (2017) Nitric oxide alleviates silver nanoparticles (AgNps)-induced phytotoxicity in *Pisum sativum* seedlings. Plant Physiol Biochem 110:167–177

Verma G, Mishra KK, Raigond B, Jeevan B, Singh AK, Paschapur AU, Kant L (2023) Nanodiagnostics: tool for diagnosis of plant pathogens. In: Mishra KK, Kumar S (eds) Biotic stress management of crop plants using nanomaterials. CRC Press, pp 113–123

Vielba-Fernández A, Polonio Á, Ruiz-Jiménez L, de Vicente A, Pérez-García A, Fernández-Ortuño D (2020) Fungicide resistance in powdery mildew fungi. Microorganisms 8(9):1431

Vlad S, Tanase C, Macocinschi D, Ciobanu C, Balaes T, Filip D, Gostin I, Gradinaru L (2012) Antifungal behaviour of polyurethane membranes with zinc oxide nanoparticles. Dig J Nanomater Biostruct 7:51–58

Wang Z, Wei F, Liu SY, Xu Q, Huang JY et al (2010) Electrocatalytic oxidation of phytohormone salicylic acid at copper nanoparticles-modified gold electrode and its detection in oilseed rape infected with fungal pathogen *Sclerotinia sclerotiorum*. Talanta 80:1277–1281

Wang ZL, Zhang X, Fan GJ, Que Y, Xue F, Liu YH (2022) Toxicity effects and mechanisms of MgO nanoparticles on the oomycete pathogen *Phytophthora infestans* and its host *Solanum tuberosum*. Toxics 10:553. https://doi.org/10.3390/toxics10100553

Wani IA, Ahmad T (2013) Size and shape dependant antifungal activity of gold nanoparticles: a case study of Candida. Colloids Surf B Biointerfaces 101:162–170

Wani A, Shah M (2012) A unique and profound effect of MgO and ZnO nanoparti- cles on some plant pathogenic fungi. J Appl Pharm Sci 2(3):4

Xu L, Liu Y, Bai R, Chen C (2010) Applications and toxicological issues surrounding nanotechnology in the food industry. Pure Appl Chem 82(2):349–372

Xue B, He D, Gao S, Wang D, Yokoyama K, Wang L (2016) Biosynthesis of silver nanoparticles by the fungus Arthroderma fulvum and its antifungal activity against genera of Candida, Aspergillus and Fusarium. Int J Nanomedicine 11:1899

Yao KS, Li SJ, Tzeng KC, Cheng TC, Chang CY, Chiu CY et al (2009) Fluorescence silica nanoprobe as a biomarker for rapid detection of plant pathogens. Adv Mater Res 79:513–516

Zahedi SM, Abdelrahman M, Hosseini MS, Hoveizeh NF, Tran LSP (2019) Alleviation of the effect of salinity on growth and yield of strawberry by foliar spray of selenium-nanoparticles. Environ Pollut 253:246–258

Zaki SA, Ouf SA, Abd-Elsalam KA, Asran AA, Hassan MM, Kalia A, Albarakaty FM (2022) Trichogenic silver-based nanoparticles for suppression of fungi involved in damping-off of cotton seedlings. Microorganisms 10(2):344

Zhao L, Peng B, Hernandez-Viezcas JA, Rico C, Sun Y, Peralta-Videa JR et al (2012) Stress response and tolerance of *Zea mays* to CeO2 nanoparticles: cross talk among H2O2, heat shock protein, and lipid peroxidation. ACS Nano 6(11):9615–9622

Phytosynthesized Nanoparticles: Antifungal Activity and Mode of Action 16

Kainat Mirza, Danish Alam, and Meryam Sardar

16.1 Introduction

Fungal infections (FIs) can cause a broad spectrum of disorders in people, from superficial infections to invasive FIs that can be fatal (Fisher et al. 2022). An estimated 1.5 million deaths and 13 million fungal infections occur globally each year; the majority of those afflicted have weakened immune systems (Bongomin et al. 2017; Rayens and Norris 2022). Conventionally four types of antifungal agents (AFAs) have been approved for treatment: (1) azoles, (2) polyenes, (3) echinocandins, and (4) 5-flucytosine. But, fungi are resistant to chemical attacks and treatment failure is often the result. This could be due to factors like antifungal drug properties (pharmacokinetics, pharmacodynamics, and drug–drug interactions), antifungal tolerance, host immune defects, and antifungal resistance. Antifungal drug resistance is a growing global concern in both space and time (Fisher et al. 2022). This includes newly emerging species that are resistant to multiple antifungal drugs, like the yeast *Candida auris* (Rhodes and Fisher 2019) as well as novel resistant variants of previously susceptible pathogens (like the ubiquitous mold *Aspergillus fumigatus*) (Verweij et al. 2020). The US CDC in 2019 placed both of these pathogens on its Antimicrobial Resistance (AMR) Threat List, as a result of rising public health burdens (Fisher et al. 2022).

Due to growing concern of antifungal resistance and restrictions of existing AFAs, international funding agencies have added antifungal resistance to their research programs. For example, the resistance of various non-*Candida albicans* spp. to echinocandins and azoles and the emergence of azole-resistant *Aspergillus fumigatus* (*A. fumigatus*) have been observed (Rhodes and Fisher 2019; Verweij et al. 2020). It could be due to exposure to AFAs. Furthermore, several fungal

K. Mirza · D. Alam · M. Sardar (✉)
Department of Biosciences, Jamia Millia Islamia, New Delhi, India
e-mail: kainatmirza09@gmail.com; danish.ngl@gmail.com; msardar@jmi.ac.in

N. Manzoor (ed.), *Advances in Antifungal Drug Development*,
https://doi.org/10.1007/978-981-97-5165-5_16

pathogen species have already become resistant to currently available AFAs or have shown reduced susceptibility to them.

In order to combat drug resistance there is a need for alternative AFAs which do not elicit resistance as well as are nontoxic. Nowadays nanotechnology is emerging as a promising field to combat drug resistance. Nanotechnology has improved the efficacy and safety of conventional therapeutics; for example, it targets directly on specific site of infection (Mirza et al. 2022; Noori et al. 2023). It was observed that nanotechnology based AFAs, which can be termed as nanomedicine, show multiple modes of action against fungus, which is beneficial for controlling the development of resistance in microbes (Mirza et al. 2022). Recently there is a growing interest in metallic nanoparticles (MNPs) owing to their small sizes, shapes and high surface area to volume ratio (Madkhali 2023). In this chapter, the various modes of synthesis and recent advances in green-synthesized MNPs using plant extracts as reductants and stabilizers have been systematically summarized. In addition, the different nanoparticles (NPs) and their formulations used for the treatment of various fungal infections in humans are discussed.

16.2 Biological Significance of Fungi

Fungi are heterotrophic eukaryotes, either unicellular or multicellular, that reproduce sexually, asexually, or both. They can live in association with bacteria, plants, and animals. They also have cell walls. Since they are eukaryotic, fungi and their host cells have many characteristics in common (Kumar et al. 2019).

16.2.1 Unique Features

- Certain species are unicellular yeasts that divide by fission or budding. When exposed to different environmental conditions, dimorphic fungus can transition between a yeast phase and a hyphal phase.
- A chitin-glucan complex makes up the fungal cell wall. Although plants also have glucans and arthropod exoskeletons include chitin, fungi are the only species that have both structural components together in their cell wall. The cellulose found in plant and oomycete cell walls are absent from the fungal cell walls (Gow et al. 2017).

16.2.2 Interaction of Fungi with Human

Fungal interaction with humans can be both beneficial and detrimental.

16.2.3 Beneficial Effects of Fungi

- They act as decomposers, resulting in the recycling of carbon and nutrients.
- They act as factories for biosynthesis of food, medicines, alcohol, antibiotics, and acids.
- They can be used as food supplement for human beings like *Agaricus* and *Marchella* (Kumar et al. 2019).
- *Penicillium* molds are used to mature a variety of cheese and as an important ingredient in bread like *Saccharomyces cerevisiae*, also known as baker's yeast.
- Antibiotics (like penicillin and cephalosporins) that either kill or stop the growth of other dangerous diseases are naturally produced by fungi.
- Used as model organisms for biochemical and genetic studies such as *Neurospora*, *Saccharomyces,* and *Aspergillus* (Kumar et al. 2019).

16.2.4 Detrimental Effects of Fungi

- They can spoil food, cloth, paper, and lumber.
- Causative agent for various diseases and allergies in animals as well as human (Kurup et al. 2000).

16.3 Fungal Infections in Human

They can infect the lungs or other regions of the body, although they typically affect the skin, hair, nails, or mucous membranes. They can get past host defenses. Individuals who are receiving immunosuppressive drugs, cancer therapies, or HIV/AIDS patients are more vulnerable to FIs. To infect people, fungi must meet certain requirements:

1. the ability to digest and absorb elements of human tissues,
2. the ability to grow in the human body temperature (37 °C or above),
3. the ability to infect the tissues involved in their pathogenesis,
4. the ability to form capsule (mucopolysaccharide), that can inhibit phagocytosis of the yeast,
5. should possess morphological diversity (Dasgupta et al. 2016) (such as yeasts, hyphae, spherules, and sclerotic bodies) to survive in various tissue environments,
6. the ability to generate toxic enzymes (phenyl oxidase, keratinase) that affect host tissues and some fungi can inhibit phagosome–lysosome fusion (Kumar et al. 2019).

16.3.1 Upsurge in Fungal Infections

- Damaged Anatomical Barriers—Mucosa and skin serve as the body's first line of defense against infections. Mucus, acidic pH, enzymes, and other antimicrobial secretions shield the anatomical barriers from infection in turn. Fungi look for entry points into the body during surgery and when indwelling catheters are used (Bajpai et al. 2019; Sardar and Mirza 2023). Furthermore, certain viral infections, burns, radiation therapy, and chemotherapy can all harm skin. Additionally, certain mucosal lesions give the fungi access to the blood and tissues.
- Weakened Immune System—Immune system serves as the second line of defense in healthy individuals. The likelihood of FIs increases during suppressed or weaken immune system such as HIV. Some cancers are also a cause of immunodeficiency (Bajpai et al. 2019). Immune system can be suppressed by some drugs/medicines.
- Changing Climate—Using fossil fuels increases greenhouse gases in the atmosphere, which causes climate change on Earth (Bajpai et al. 2019). Consequently, there is a change of the prevalence and distribution of fungal infections. Climatic change, such as changes in temperature and humidity affects pathogen sporulation and dispersal which can be beneficial for certain pathogens may also introduce new potential vectors (Bajpai et al. 2019). Global warming is also a reason for the evolution of heat-tolerant fungi that can withstand high temperatures.
- Nosocomial Infections—Patients in long-term acute care, skilled nursing, and hospital settings are susceptible to diseases known as nosocomial infections or healthcare-associated illnesses (HAIs). Some HAIs are fungal infections, like aspergillosis, candidemia, and mucormycosis. Many patients recovering from surgeries or being treated for illnesses like cancer, or taking certain medications have weakened immune systems. Use of medical devices, like catheters and ventilators, also create opportunities for fungi and other pathogens to enter and infect the body (Sardar and Mirza 2023). Thus, patients can develop invasive infections, severe infections that affect the blood, heart, brain, eyes, or other parts of the body. It was reported that the length of ICU stay affects enhances the risk of fungal infections (Bajpai et al. 2019).
- Antibiotic Exposure—Overuse of antibiotics either due to prevailing disease condition or unnecessary use by patients has resulted in resistant fungal infections and new variants of fungal strains (Mirza et al. 2022).
- Ecological Factors—Numerous environmental conditions have the potential to spread these spores to healthy people. Household pets, small rodents, reptiles, and exotic animal species such as parrots may transmit fungal infections to humans. Soil and decaying vegetation are also among the most common sources of fungi (Bajpai et al. 2019). Other common sources of fungi are household dust, building material, tobacco, ornamental plants, food, and water. Spores resemble plant seeds and are the reproductive form of fungi. Spores are very resistant to outside stimuli like heat, cold, or drugs and they may last longer periods of time in unfavorable environments. Construction, poor or insufficient air handling, water leaks, dust accumulation, and moisture buildup are all favorable conditions

for spore dispersal. Furthermore, there are an ever-increasing number of periodic constructions and renovations at the hospital premises resulting in dust contamination as well as possible fungal spore dispersal (Kanamori et al. 2015).

16.4 Type of Infections Caused by Fungal Pathogens

There are many ways in which fungal infections can be distinguished. They are: (1) Primary mycoses—whether the infection arises in a healthy host or (2) Opportunistic mycoses—whether the host has an impaired immune system due to any previous medical conditions. It can be further classified based on whether the infection has spread into the bloodstream (i.e., systemic or disseminated mycoses) or is localized to the outer layers of epithelia (i.e., superficial mycoses) (Sullivan et al. 2011).

16.4.1 Classification of Fungal Diseases Based on the Site of Infection in the Human Body

Superficial mycoses: caused by fungi infecting the outer layers of epithelia, i.e., skin or hair surface.

Cutaneous mycoses or dermatomycoses: superficial infection of skin, nails, or hair. No living tissue is infected here.

Subcutaneous mycoses: spreads beneath the skin and affects subcutaneous, bone, and connective tissue. This can occur in healthy individuals and includes sporotrichosis, phaeohyphomycosis, chromoblastomycosis, hyalohyphomycosis, and eumycoticmycetoma (Kumar et al. 2019).

Systemic or deep mycoses: can penetrate internal organs and spread throughout the body. This form is frequently lethal (Table 16.1).

16.4.2 Classification of Fungal Infections Based on the Affected Organs of the Human Body

Fungal infections of the skin: It is easy to get fungal spores into skin, especially if the skin is broken. The areas which are more prone to skin infections are the skin of the feet, fingers, scalp, mouth, and vagina. Fungus proliferates in hot and humid places between toes, armpits, and genitalia. Tinea and *Candida* are commonly associated with toe infection. Based on causative fungal agents, skin infections are classified as: dermatophytosis, yeast infections (including *Pityriasis versicolor*, candidiasis and other yeast infections) and mold infections (Rai et al. 2017).

Dermatophytosis (tinea)—It is caused by dermatophytes and infect the keratinized tissues such as the stratum corneum of the epidermis, hairs, and nails. There are number of stages in the pathogenesis of tinea which begins with the contact with infective fungal spores, adherence to the superficial cells, and invasion of keratin layers by the secretion of the keratinases and initiation of inflammation. The enzyme

Table 16.1 Overview of fungal diseases and their treatment

Fungal diseases	Most common fungal species	Symptoms	Site of infection	Antifungal therapy	References
Opportunistic invasive mycoses					
Aspergillosis	*Aspergillus fumigatus*	Fever; dyspnea; cough; subacute invasive pulmonary aspergillosis; cavitary tuberculosis	Lungs (mainly)	Polyenes, echinocandins, and azoles	Latgé and Chamilos (2019)
Candidiasis	*Candida albicans*	Fever; polymyalgia; chills; polyarthralgia; retinal exudate; candidemia; tenosynovitis; septic arthritis, meningitis; endophthalmitis; infective endocarditis	Esophagus; vagina; mouth; and throat	Polyenes, echinocandins, and azoles	Du et al. (2021)
Cryptococcosis	*Cryptococcus neoformans*	Fever; headache; signs of meningeal irritation; memory loss; lethargy; altered mentation; cranial neuropathies	Lung and CNS	Polyenes, azoles, and 5-fluorocytosine	Molloy et al. (2018)
Pneumocystis	*Pneumocystis jirovecii*	Fever; cough; difficulty breathing; chest pain; chills; fatigue (tiredness)	Lungs	Trimethoprim-sulfamethoxazole	Gingerich et al. (2021)
Endemic systemic mycoses					
Blastomycosis	*Blastomyces dermatitidis*	Fever; night sweats; anorexia; malaise; weight loss; hemoptysis; persistent cough	Lungs; CNS; Joints; skin and Bones	Polyenes and azoles	McBride et al. (2017)
Coccidioidomycosis	*Coccidioides immitis*	Arthralgias; fever; pleuritic pain; cough; profound fatigue; headache; meningitis	Lungs (mainly); skin; bones; liver; brain; heart and the membranes that protect the brain and spinal cord (meninges)	Polyenes and azoles	Garcia Garcia et al. (2015)
Histoplasmosis	*Histoplasma capsulatum*	Fever; headache; weakness; malaise; dry cough; sweats; pleuritic chest pain; dyspnea	Lungs (mainly); adrenal glands; liver; spleen	Azoles and polyenes	Lakhani et al. (2019)

Paracoccidioidomycosis	*Paracoccidioides brasiliensis*	Malaise; fever; dyspnea; cough; pulmonary hypertension; pulmonary bullae; pulmonary fibrosis	Lungs (mainly); lymph nodes; skin lesions; mouth and throat	Polyenes and azoles	Peçanha et al. (2022)
Penicilliosis	*Penicillium marneffei*	Skin lesions; lymphadenopathy; anemia; weight loss; fever	Lungs; liver; skin and mouth	Azoles and polyenes	Wang and Deng (2021)
Superficial mycoses					
Pityriasis versicolor	*Malassezia globosa*	Psoriasis on the head and neck; worsening of atopic dermatitis	Skin	Azoles	Nagpal and Kaur (2021)
Ringworm (e.g., tinea capitis)	*Trichosporon/ Microsporon*	Alopecia; severe itchiness; scaly rashes	Skin	Azoles, allylamines, and griseofulvin	Degreef (2008)
Athlete's foot (e.g., tinea pedis)	*Trichosporon/ Microsporon*	Scaly rash; itching; stinging	Skin	Azoles; Allylamines	Degreef (2008)

keratinase degrades the tough keratin into low molecular weight components that serve as food for dermatophytes. Symptoms include erythema and intense itching. Worldwide, *Trichophyton (T.) rubrum*, a common dermatophyte, is the causative fungal pathogen for athlete's foot, jock itch, fungal infection of nail, and ringworm (Nagpal and Kaur 2021; Rai et al. 2017).

Pityriasis Versicolor—After dermatophytosis, the next most common fungal infection is Pityriasis versicolor (also known as tinea versicolor) mostly caused by yeast *Malassezia globosa*. These yeasts are lipophilic and reside in the seborrheic parts of the skin. Symptoms include psoriasis on the head and neck and worsening of atopic dermatitis (Nagpal and Kaur 2021; Rai et al. 2017).

Candidiasis—The causative agent of this disease are *Candida* species which are opportunistic yeasts. They become pathogenic mainly in a weakened immune system. Conditions like immunosuppression, diabetes, cushing's syndrome, obesity, wet conditions, and other hormonal diseases are some of the other causes for candidiasis. Candidiasis mostly occurs in inguinal folds, skin folds in axillae, skin beneath the breasts in women, hands, fingernails, genital and perigenital areas, and mucous membranes. Candidiasis is caused mainly by *Candida albicans* but other non-albicans species, such as *C. glabrata, C. parapsilosis, C. tropicalis, C. krusei,* and *C. Guilliermondii*, are also responsible for the disease (Nagpal and Kaur 2021; Rai et al. 2017).

Mold Infections—They are less likely to cause infections and characterized as nondermatophytic filamentous opportunistic fungi. Generally, are not pathogenic in healthy hosts, but are sometimes isolated from toenails and feet. The symptoms are like tinea pedis, tinea unguium, and tinea manuum. These infections are common in tropical zones. In some geographical areas *Aspergillus niger* is the main causative agent for mold infections (Nagpal and Kaur 2021; Rai et al. 2017).

16.4.3 Fungal Infections of the Mouth

Under normal circumstances, mouth serves as a niche for fungus along with other microbes, living harmoniously. But in an immune compromised host undergoing any form of chemotherapy or on immunosuppressant treatments, i.e., after organ transplant or AIDS, these fungi can elicit infections. Onset of AIDS can be marked by fungal infection of the mouth. Oral thrush, a fungal infection of the mouth caused by *C. albicans*, is quite common in children and older people. Symptoms include thick white wipeable plaques (like coagulated milk) and are usually asymptomatic. These plaques can be present anywhere in the mouth including the hard palate, tongue, and buccal mucosa (Awaad et al. 2017).

16.4.4 Fungal Infections of the Respiratory Tract

In immunocompromised individuals, *Candida* usually causes throat infection termed as *Candida* esophagitis (Nassar et al. 2018). Sometimes *Candida*,

Cryptococcus, and *Aspergillus* infect the lower respiratory tract (Shailaja et al. 2004) examples are pulmonary aspergillosis and systemic candidiasis.

16.4.5 Fungal Infection in the Urogenital Tract

The most prevalent vaginal fungal infections are vulvovaginal candidiasis (VVC) (Jaeger et al. 2013). The various symptoms are burning, soreness, itching, vaginal discharge and pain during intercourse and/or urine (Jaeger et al. 2013; S. Kumar et al. 2019). Men, too, can acquire yeast infections. This is more frequent in uncircumcised men. Yeast infection in males includes itching, red rash on the penis, and burning on the top region of the penis. In women, VVC is mainly due to an imbalance in the vaginal flora. When the levels of female reproductive hormones (estrogen and progesterone) are high in the later stages of the luteal cycle, then occurrence of VVC is more. Generally, the mucous membrane protects itself by keeping a slightly acidic pH. Any imbalance in this pH either due to overuse of soap or any other reason can raise the chances of fungal infections. It is also reported that intrauterine devices for contraception also increase the chances of bacterial and fungal infections. However, VVC rarely causes any serious complications in women of reproductive age but recurrent infections can decrease the quality of life leading to depression. VVC can be cured using a single dose of fluconazole or applying clotrimazole cream 2–3 times a day (Kumar et al. 2019).

16.4.6 Fungemia

This is a condition in which fungi are found in human blood. *Candida*, *Aspergillus*, *Saccharomyces*, *Cryptococcus*, and *Histoplasma* are the most common fungal pathogens responsible for Fungemia (S. Kumar et al. 2019). It is common in immunocompromised persons.

16.5 Emergence of Multidrug-Resistant (MDR) Fungi

MDR refers to resistance against a range of drugs that are different from one another in terms of target and structure. Biedler and Riehm (1970) reported that cell lines resistant to Actinomycin D and Vinca alkaloids showed cross-resistance to other drugs. Resistance might be innate, randomly acquired, or nosocomial. It is caused by patient's irregular usage of the medication and reinfection with different infectious strains. Table 16.2 shows the list of important AFAs available and their mode of action on fungus. In order to achieve the goal of combating drug resistance, it is therefore necessary to be one step ahead of the pathogen and constantly develop new targets and strategies for fighting infection.

Table 16.2 List of currently available antifungal drugs which are available for the treatment of common fungal infections

Antifungal class and drugs	Biological effect	Mode of action
Azoles *Fluconazole, itraconazole, ketoconazole*	Mostly fungistatic	Inhibit the fungal cytochrome P450-dependent enzyme 14-α-lanosterol demethylase encoded by ERG11 (*Candida* spp.) or Cyp51A (*Aspergillus* spp.) thereby inhibiting ergosterol synthesis.
Echinocandins *Caspofungin, Micafungin*	Fungicidal against *Candida* spp., but fungistatic against *Aspergillus* spp.	Target 1,3-β-d-glucan synthase, thus preventing production of cell wall 1,3-β-d-glucan. Thus, it compromises the fungal cell wall stability and synthesis.
Polyenes *Amphotericin B, Nystatin*	Fungicidal	Target ergosterol directly, induces pore formation and extracts sterols from fungal cell membranes.
Analogues of pyrimidine *Flucytosine*	Fungicidal against *Cryptococcus* spp.	Inhibits DNA and RNA synthesis.
Allylamines *Terbinafine*	Fungicidal	Inhibit ergosterol synthesis by interfering with squalene epoxidase

16.5.1 Mechanisms of Drug Resistance in Fungi

The pace at which resistance develops in fungi depends upon the rate of mutation and size of population. There are various ways through which resistance is acquired. Some of them are as follow:

1. Change in drug–target interaction.
2. Mutation in the target binding site, for example, mutation of the genes encoding lanosterol demethylase for azoles or β-glucan synthase for echinocandins,
3. Increase in expression of the target available,
4. Change in the effective drug concentration by increase in drug efflux activity for intracellular drugs such as azoles, or the inhibition of prodrug activation for flucytosine (Fisher et al. 2022).
5. A recent finding showed role of heat shock protein (Hsp90) in drug resistance in fungi. Hsp90 can change the link between genotype and phenotype of fungi (Fisher et al. 2022). Figure 16.1 depicts the various ways by which the fungus develops resistance.

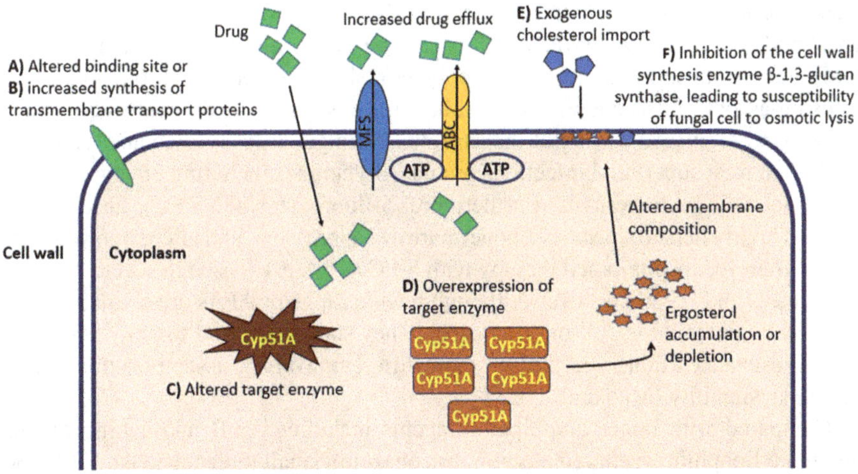

Fig. 16.1 Graphical representation at the cellular level of the principal antifungal mechanisms of action relevant in clinical practice. Molecular mechanism of resistance to antifungal therapy occur mainly through alterations in the cellular binding site (A) or increased synthesis of transmembrane transport proteins (B), which both lead to reduced intracellular accumulation. In addition, alteration (C) or overexpression (D) of the enzyme targeted by the drug can cause ergosterol intracellular accumulation and consequent altered composition and permeability of the cell wall. Import of exogenous cholesterol (E) can alter the cell wall composition and permeability resulting in inadequate intracellular drug concentration. Finally, inhibition of the cell wall synthesis enzyme β-1,3-glucan synthase (F) can lead to fungal cell susceptibility to osmotic lysis. ABC, ATP-binding cassette transporter family; MFS, major facilitator superfamily of transporters (This is an Open Access article distributed under the terms of the Creative Commons Attribution License (http://creativecommons.org/licenses/by/2.0), which permits unrestricted use, distribution, and reproduction in any medium provided the original work is properly cited from Vitiello, A., F. Ferrara, M. Boccellino, A. Ponzo, C. Cimmino, E. Comberiati, A. Zovi, S. Clemente and M. Sabbatucci. "Antifungal Drug Resistance: An Emergent Health Threat." Biomedicines **11**(4): 1063. Copyright (2023), MDPI.)

16.6 Existing Antifungal Drugs and Their Limitations

Despite the rise in the economic burden and stress caused by FIs and MDR, the list of currently available AFAs is limited. One of the reasons for this is the lack of antifungal targets which can be utilized to develop AFAs, due to the fungus being a eukaryote similar to humans. Hence, AFAs are far less compared to antibacterials. Since the 1990s, the pace at which AFAs are developing has slowed down. After echinocandins, two decades have already passed for any new class of drugs to be introduced in the market. To treat invasive FIs, four classes of AFAs have been approved by Food and Drug Administration (FDA namely, echinocandins, azoles, flucytosine, and polyenes (Wall and Lopez-Ribot 2020). But they have limited use owing to the side effects and unpleasant properties. In clinical practice, azoles are widely used; the major problem is that they interact with drugs that are substrates of cytochrome p450 (CYP450) enzymes. It is these interactions which can elicit resistance to azoles and off target toxicity (Mohd-Assaad et al. 2016). Also, polyenes bind to fungal ergosterol, which is structurally similar to mammalian cholesterol due to this, AmB (a major polyene) shows severe nephrotoxicity and infusion-associated reactions. As a result, AmB dosage is strictly controlled, and it is frequently replaced by one of the triazole antifungal drugs (voriconazole). Other AFAs that treat superficial infections include allylamines including onychomycosis of the toenails or fingernails. Furthermore, 5-fluorocytosine (5-FC), an effective AFA, is highly hepatoxic, causes bone-marrow suppression, and elicits fungal resistance. Therefore, a combined therapy with 5-FC and AmB is used in severe cases of candidiasis and cryptococcosis. Although a wide range of AFAs are available their therapeutic potential is still unsatisfactory. They show increased toxicity, and antifungal resistance (Table 16.3). They show different levels of tissue penetration and oral bioavailability (Madkhali 2023).

Compared with larger amphipathic agents including AmB and echinocandins and more lipophilic agents (including itraconazole), small molecules like 5-FC and voriconazole exhibit improved tissue penetration. AFAs, echinocandins and AmB can accumulate in tissues and show slow drug metabolism (Felton et al. 2014).

16.6.1 Potential Measures to Be Taken

1. Identify novel fungal antigens for vaccine development.
2. Discover new targets for AFAs.
3. Evaluate existing drugs for their possible antifungal actions.
4. Encapsulate or alter existing AFAs to improve their effectiveness. This can be achieved using Nanotechnology. Therefore, in the subsequent section, we will discuss the mode of synthesis of NPs (specifically green synthesis) and antifungal activity in detail.

Table 16.3 Sensitivity of existing AFAs to pathogenic fungal pathogens in clinical practice (High: High; Medium: Intermediate; Low: Low sensitivity)

Common fungal pathogens encountered in clinical practice	Sensitivity to AFAs			
	Triazoles		Echinocandins	Polyene
	Fluconazole	*Itraconazole*	*Anidulafungin Caspofungin Micafungin*	*Amphotericin B*
A. fumigatus, *A. hiratsukae* *A. lentulus*	L	L	H	L
Candida spp.	Species specific	M/H	M/H	M (but sensitive to *C. krusei, C. lusitania*)
Fusarium spp.	L	L	L	M
Fusarium solani	L	L	L	M
Lamentospora prolificans	L	L	L	L
Mucorales	L	L	L	M

16.7 Nanotechnology

It is the branch of science in which atoms and molecules are designed and produced at nanoscale, i.e., having one or more dimensions of the order of 100 nanometers (100 millionth of a millimeter) or less. The new materials thus formed has very different properties and new effects compared to the same materials made at larger sizes (Mirza et al. 2022). The reason is the high surface to volume ratio of nanoparticles compared to larger particles. In the next section synthesis of NPs were discussed in detail.

16.7.1 Synthesis of NPs

Currently, there are two methods used to synthesize NPs: "top-down" and "bottom-up" (Fig. 16.2). The top-down method is the most commonly used approach, where the synthesis starts with breaking bulk material into pieces which are then converted

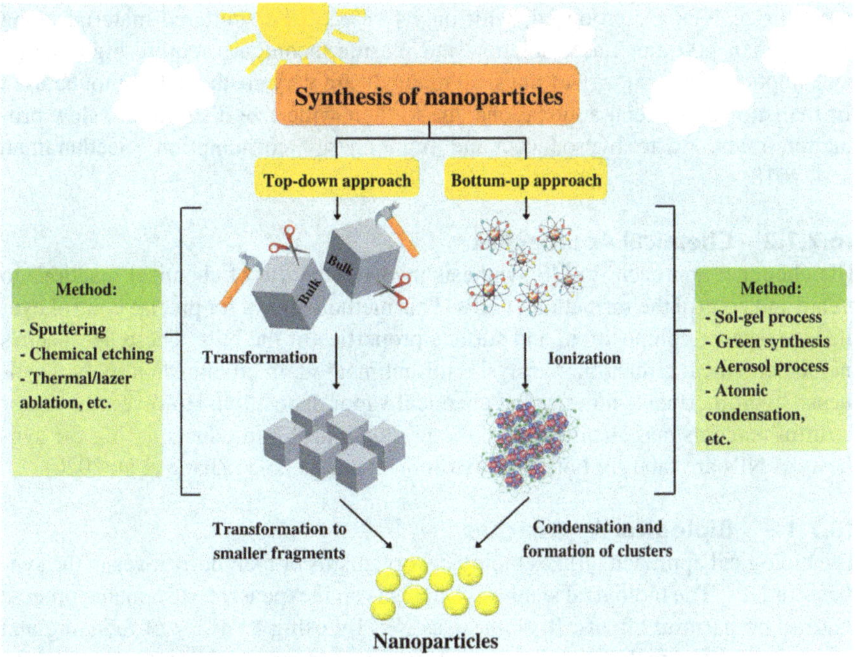

Fig. 16.2 Synthesis of nanoparticles by top-down and bottom-up methods (This is an Open Access article distributed under the terms of the Creative Commons Attribution License (http://creativecommons.org/licenses/by/2.0), which permits unrestricted use, distribution, and reproduction in any medium provided the original work is properly cited from Nguyen, N. T. T., L. M. Nguyen, T. T. T. Nguyen, T. T. Nguyen, D. T. C. Nguyen and T. V. Tran. "Formation, antimicrobial activity, and biomedical performance of plant-based nanoparticles: a review." Environmental Chemistry Letters **20**(4): 2531–2571. Copyright (2022), Springer)

into NPs. Using the bottom-up approach, different types of NPs are synthesized by assembling atoms and molecules. The top-down approach's most useful advantage is that it speeds up the production of NPs in large quantities (Karunakaran et al. 2023). The primary advantage of the bottom-up approach is that it produces NPs with a higher specific surface area and defined crystallographic properties. It is not possible to synthesize the defined shape using a top-down method. Nonetheless, waste components can be removed and fewer amounts of NPs can be produced by employing a bottom-up strategy (Krukowski et al. 2018). In general, there are three methods for the synthesis of NPs: chemical, physical, and biological approaches.

16.7.1.1 Physical Approaches

In physical approaches, temperature, pressure, and energy are all used to obtain NPs. These methods often rely on the principles of physics and typically involve the manipulation of physical properties to generate NPs with desired characteristics. Physical methods such as sputtering, deposition, ball milling, and plasma-based techniques are used in the synthesis of NPs (Dhand et al. 2015). In the majority of these techniques, the synthesis rate of metal NPs is slow. For instance, ball milling techniques yield 50% or less of NPs (Yadav et al. 2012). While sputtering, larger particle-size NPs are produced, with only 6–8% of the sputtered material being smaller than 100 nm. Laser ablation and plasma techniques require high energy consumption. The majority of physical methods are very costly and cannot be used for realistic commercial applications due to their wide size distribution, slow production rates, waste byproducts, and high energy consumption (Seetharaman et al. 2018).

16.7.1.2 Chemical Approaches

The chemical approach for NP synthesis involves the use of chemical reactions to create and control the formation of NPs. This method allows for precise control over the size, shape, composition, and surface properties of the NPs. These approaches include chemical reduction, pyrolysis, sol-gel method, microemulsion, polyol synthesis, hydrothermal synthesis, and chemical vapor deposition. However, the use of harmful reagents and chemicals and the production of byproducts during the synthesis of NPs are fatal for both the environment and people (Zhang et al. 2020).

16.7.1.3 Biological Approaches

The biological approach utilizes biological organisms or their derivatives in the synthesis of NPs. The biological synthesis of NPs is an inexpensive, sustainable process that has no harmful effects. It synthesizes NPs by using a variety of reducing and stabilizing agents, including microbes, plants, and some natural agents. The demand for biological approaches for the synthesis of NPs is increasing because of their low cost, great stability, and lack of toxicity. The biological synthesis approach was a safe and environment-friendly strategy to work with NPs that didn't harm individuals or the environment. The traditional approach might be effective with a large amount of well-defined size and shape NPs. When synthesizing NPs, the biological method offers several benefits over chemical and physical approaches, such as simplicity, low cost, and minimal waste (Khan et al. 2022; Ying et al. 2022). The various ways in which NPs can be synthesized are:

1. The utilization of secondary metabolites and enzymes present in bacteria, viruses, and fungi was a sustainable process used by these organisms to form NPs. These kinds of organisms provide the raw materials for the synthesis and modification of more ordered NPs (Nair et al. 2022). But maintaining and culturing microbes is expensive, time consuming, and laborious.
2. Using plants is another biological approach for producing NPs. Using plant extracts may be a simpler method to synthesize NPs than microbes. Moreover, plant extracts have the potential to reduce metallic ions, which could facilitate the formation and stabilization of metallic NPs (Mirza et al. 2020; Naikoo et al. 2021; Zhang et al. 2020). Plant extracts contain a variety of ingredients that may stabilize and facilitate the formation of NPs, including proteins, polysaccharides, amino acids, and phytochemicals like flavonoids, alkaloids, tannins, and polyphenols. Therefore, in this chapter, we have described the plant-based synthesis of NPs and their antifungal activity in detail.

16.7.2 Plant-Based Synthesis of NPs

For the synthesis of NPs, plant parts are utilized, including leaves, stems, flowers, bark, roots, fruits, vegetables, and shoots (Hano and Abbasi 2021; Muddapur et al. 2022). For a long time, plant parts have been used for treating a wide range of serious illnesses and are utilized as efficient cancer medications (Mirza et al. 2020) (Fig. 16.3). Plants are also a rich source of reducing and stabilizing agents. Plants or their extracts, especially flavonoids have been utilized in the bioremediation of metals. Therefore, several studies have proposed to synthesize different metallic NPs using extracted secondary metabolites, like proteins, sugars, and phenolic components. Quercetin, one of the plant metabolite flavonoids, was used to biosynthesize copper and silver NPs in micelle suspension, which have a higher surface area and better antibacterial activity (Karunakaran et al. 2023). The different NPs made from *Chrysophyllum oliviforme* plants have diameters between 20 and 50 nm and less polydispersity, which shows that the in vivo synthesis process worked well (Anju Varghese et al. 2015). The bioconversion of silver ions was observed in the *Veronica amygdalina* extract. A 15 nm size of silver NPs was developed more quickly. Terpenoids like geraniol, linalool, and citronellal that are found in geranium leaf extracts may be able to reduce the silver ions in the body (Aisida et al. 2019).

Metallic NPs were synthesized from various plant parts. Metal NPs, including titanium oxide, magnetite, nickel, platinum, zinc, copper, selenium, gold, and silver, were synthesized using natural resources. Metal NPs in a range of sizes and forms are produced using plant parts such as the stem, root, fruit, seed, callus, peel, leaves, and flower. The amount of plant extract in the reaction medium and the wide range of metal concentrations, that can affect the biosynthesis reaction, can change the size and shape of the NPs (Kuppusamy et al. 2016).

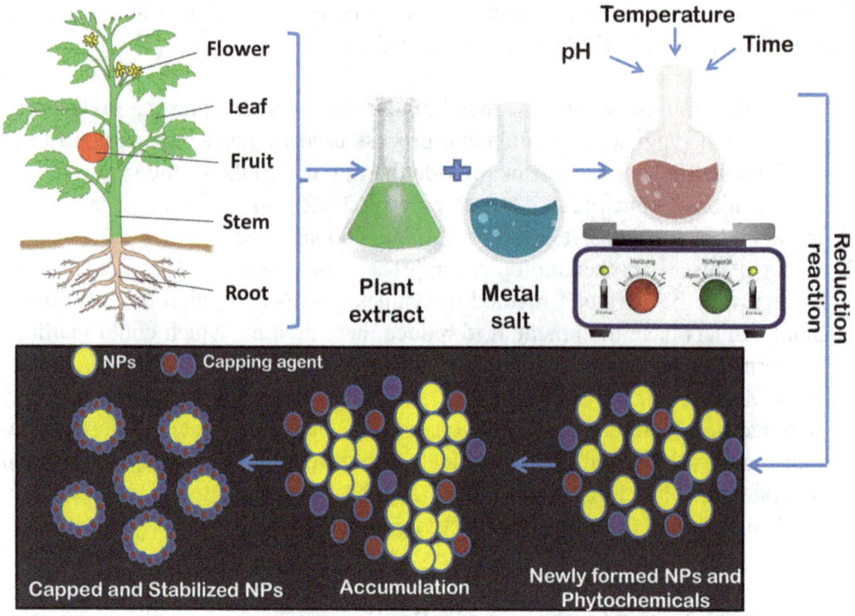

Fig. 16.3 Nanoparticles synthesized from the reducing and capping agents present in plant parts

The various functional groups, namely the carboxyl, amine, and phenolic compounds involved in the reduction of silver ions, are present in the stem portion of the plant extract. Vanaja et al. (2013) synthesized silver NPs using stem extract of *Coleus aromaticus*. The carboxylic and amine groups capped the silver nanoparticles and protect from aggregation (Vanaja et al. 2013). In another study, the methanolic extract of the *Callicarpa maingayi* stem was utilized by Shameli et al. (2012) to synthesize silver NPs (Shameli et al. 2012). The aldehyde group found in plant extracts is primarily responsible for reducing silver ions into metallic Ag NPs. Biosynthesized silver nanoparticles can kill microbes by attaching to their outer cell walls and messing up the lipoproteins. Ultimately, cell division was halted, which resulted in cell death. Plant extracts of *H. abyssinica* were used to synthesize Cu NPs, as reported by Murthy et al. (2020) (Murthy et al. 2020). It is possible for strong antioxidants like tannins, phenolic compounds, and anthraquinone glycosides to reduce Cu $(NO_3)_2$ salt solution into CuO NPs. These biomolecules are essential in enveloping and stabilizing the Cu NPs following the reduction. The phytochemicals found in plant extracts play a significant role in the formation of other metal oxide NPs, such as ZnO NPs. ZnO NP fusion with the help of biomolecules from *M. sativa* extract was investigated by Król et al. (2019). Because of zinc's coordination chemistry, the zinc aqua complex can exchange water molecules with protein ligands during binding. However, via particular metal ion binding sites, flavonoid (such as quercetin, rutin, and galangin) can chelate with Zn^{2+} ions, and as a result, ZnO NPs are synthesized.

16.7.2.1 Fruit as a Source for the Synthesis of NPs

Baldea et al. (2020) investigated the biological effects of the Au NPs as well as the stability of synthesized Au NPs using water extracts of Cornus mas fruit. It was found that at pH 7.5, $HAuCl_4$ could be reduced to Au NPs and that Au NPs are selectively toxic to hypertrophic keratinocytes. The potential application of Au NPs in the treatment of diseases associated with oral dysplasia was indicated by their good biocompatibility with normal gingival fibroblasts. According to Sathishkumar et al. (2016), Au NPs were synthesized by reducing $HAuCl_4$ with fruit extracts of *Couroupita guianensis* at 70 °C and pH 7 after 60 minutes of reaction (Sathishkumar et al. 2016). The obtained Au NPs were negatively charged, nonagglomerated, and averaged 26 ± 11 nm in size. Additionally, they were coated with a layer of polyphenolic compounds. Furthermore, it was discovered that the prepared Au NPs exhibited strong hemocompatibility and antioxidant abilities. Using mangosteen polyphenols from the aqueous extracts of *Garcinia mangostana* and *L. pericarp* as reducing and stabilizing agents, respectively, Chen (2014) prepared hydrophilic Au NPs with sizes ranging from 9 to 23 nm (Chen 2014). Jiang et al. (2016) prepared Ag NPs by reducing silver nitrate using hawthorn fruit aqueous and ethanolic extracts as stabilizing and reducing agents, respectively. The Ag NPs produced from the aqueous extracts were smaller, more uniform, and exhibited stronger antibacterial activity, according to a comparative analysis (Jiang et al. 2016). Based on extracts of *Eugenia stipitata* McVaugh fruits, Kumar et al. (2016) investigated the feasibility of synthesizing Ag NPs, and the results were satisfactory. Furthermore, infrared spectroscopy showed that the synthesis of Ag NPs was associated with malic acid, citric acid, and carotenoids in the extracts. The synthesized Ag NPs were spherical in shape, 15–45 nm in size, and exhibited strong antioxidant effects (Kumar et al. 2016). A phytochemical called isoleucine acid is found in *L. acidissima* fruit extract and is responsible for the formation of highly stable MgO NPs (Nijalingappa et al. 2019). In this study, the formation of the isoleucine-MgO complex occurred subsequent to the binding of isoleucine acid to magnesium nitrate. Subsequently, the complex was heated to a high temperature (500–800 °C) in order to form MgO NPs that exhibited high stability and good dispersion (Nijalingappa et al. 2019).

16.7.2.2 Seed as a Source for the Synthesis of NPs

Fenugreek seed extract contains high concentrations of flavonoids and other naturally occurring bioactive substances like lignin, saponin, and vitamins. When combined with other strong reducing agents, fenugreek seed extract works as an excellent surfactant to efficiently reduce chloroauric acid. Along with the functional groups C=N and C=C, the extract from the seeds also contains the carboxylic COO- group. As described by Mittal et al. (2013), the functional group of metabolites acts as a surfactant of the particles, and the flavonoids are capable of stabilizing the electrostatic stabilization of Au NPs (Mittal et al. 2013). The *Macrotyloma uniflorum* aqueous extract speeds up the rate at which silver ions are reduced. This might occur as a result of the extract's caffeic acid content. The caffeic acid reduction reaction consequently occurred in less than a minute. Cu NPs were synthesized by reducing copper sulfate using aqueous extracts of dried *Ziziphus spina-christi* (L.) Willd. seed powder combined with starch as a stabilizer, as reported by Khani et al. (2018). This resulted in spherical Cu NPs with a particle size of 5–20 nm. The resultant Cu NPs exhibited concentration-dependent inhibitory activity against *Staphylococcus aureus* and *Escherichia coli*. Furthermore, it was discovered that the Cu NPs had a strong adsorption effect on crystalline violet. In 2016, Sajadi et al. produced CuNPs-Fe_3O_4 composites at 60 °C using aqueous extracts of *Silybum marianum* L. seeds in combination with ferric chloride hexahydrate and copper chloride dihydrate. The composites have sizes ranging from 8.5 to 60 nm and good stability lasting longer than 2 months (Sajadi et al. 2016).

16.7.2.3 Leaves as a Source for the Synthesis of NPs

It was reported that plant leaf extract was utilized as a mediator in the synthesis of NPs. Numerous plant leaf extracts, including those of *Centella asiatica, Murraya koenigii*, and *Alternanthera sessilis*, have been studied. It was recently reported that *P. nigrum* leaf extracts have been found to contain longumine and piper longminine, which function as capping agents for the formation of silver NPs and could improve their cytotoxic effects on tumor cells. *P. nigrum* leaves contain a significant bioactive compound that aids in the sustainable production of NPs. Ag NPs are useful in treating a wide range of cancers and terrible illnesses (Jacob et al. 2012). *Psidium guajava* leaf extract was used to synthesize Se NPs, which were then in situ coated on alkali-activated cotton fabric to provide antibacterial and antifungal effects (Mirza et al. 2022). The potential of leaf extract of *Eupatorium odoratum* for the synthesis of Ag/Ag_2O was examined by Elemike et al. (2017). The findings demonstrated that the constituents of the leaf extract act as both reducing and stabilizing agents (Elemike et al. 2017). The primary function of polyphenols and flavonoids in *A. vulgaris* leaves as a bio-capping agent for Au NP synthesis was demonstrated by Sundararajan and Kumari (2017). The presence of dihydrokaempferol, caffeic acid, quercetin, and kaempferol in *P. undulata* extract demonstrated as metal-reducing agents during the formation of Au and Ag NPs, according to a different study (Khan et al. 2022). According to research published in 2019 by Chandraker et al., *A. conyzoides* leaves contain a variety of compounds that can reduce, cap, and stabilize Ag NPs, including alkaloids, flavonoids, terpenoids, saponins, and tannins (Chandraker et al. 2019). Guo et al. (2014) used ethanol extracts of dried powdered vine tea leaves to prepare Au NPs by reducing chloroauric acid ($HAuCl_4$). Additionally, they looked at how reaction parameters like temperature, pH, and extract amount affected the physicochemical characteristics of Au NPs. They discovered that high temperatures were better for synthesizing small-sized Au NPs and that low temperatures were more stable for Au NPs. Alkaline conditions or excessive vine tea extracts were the causes of Au NPs aggregation (Guo et al. 2014). According to Tao et al. (2019), aqueous extracts of aloe vera leaves were used to reduce $HAuCl_4$ to create spherical and highly crystalline Au NPs with particle sizes between 20 and 60 nm. The Au NPs were found to be highly stable and less prone to oxidation and agglomeration while protected by the extracts (Jing et al. 2019). Pang et al. 2020 investigated that *Youngia japonica* leaf extracts could be used to construct Ag NPs at 60 °C and 40 min of reaction time. The resulting Ag NPs had an average particle size of 20 nm, with the majority of them being spherical. These Ag NPs significantly suppressed the bacteria that were isolated from the stem ends of cut lilies and are anticipated to be applied to the freshness treatment of cut flowers following the reaction (Pang et al. 2020). By reducing silver nitrate with green tea aqueous extracts, Wang et al. (2013) synthesized spherical Ag NPs with a particle size of 30–40 nm at 40 °C for a duration of 2 h.

16.7.2.4 Flower as a Source for the Synthesis of NPs

As demonstrated by Ghosh et al. (2012), *Gnidia glauca* flower aqueous extracts could reduce $HAuCl_4$ and synthesize Au NPs in less than 20 min. The final Au NPs were mostly about 10 nm in size and came in spherical, triangular, and hexagonal shapes. Additionally, it was found that the Au NPs significantly assisted in the reduction of 4-nitrophenol with $NaBH_4$ to produce 4-aminophenol (Ghosh et al. 2012). Zangeneh and Zangeneh (2020) synthesized spherical Au NPs with a particle size of 15–45 nm by reducing $HAuCl_4 \cdot 3H_2O$ with water extracts from *Hibiscus sabdariffa* flowers. It is interesting to note that the as-synthesized Au NPs could greatly increase anti-inflammatory cytokines while decreasing proinflammatory cytokines. Moreover, Au NPs exhibited no appreciable cytotoxicity toward endothelial cells, similar to daunorubicin (Zangeneh and Zangeneh 2020). Dinesh et al. (2020) reduced copper acetate monohydrate by applying aqueous extracts of *Hibiscus rosasinensis* flowers, resulting in square Cu NPs measuring 0.115–1.1 µm in a 30-min reaction facilitated by acoustic waves. In the meantime, it was found that the Cu NPs could produce free radicals to accelerate the breakdown of lovastatin and 5-fluorouracil (Dinesh et al. 2020). In order to produce Cu NPs, Pinto et al. (2019) reduced copper chloride dihydrate at 120 °C in ethanol, oleamide, and oleic acid systems, respectively, using varying concentrations of aqueous extracts of *Eucalyptus globulus* bark. At the same time, the phytoactive substances in the extracts were attached to Cu NPs, which made them more conductive and antioxidant (Pinto et al. 2019).

16.7.3 Factors Influencing the Synthesis of NPs

Variations in pH, temperature, reaction time, and reactant concentration can be applied to modify the green synthesis of NPs. These parameters have mainly recognized the effect of environmental factors on NP synthesis, and it is possible that these elements will be crucial in optimizing the synthesis of metallic NPs (Zhang et al. 2020).

16.7.3.1 Temperature

Globally, different kinds of studies and reports are being carried out to understand how temperature affects NPs. The main factor that determines the size, shape, and degree of synthesis of the NPs is temperature. Different shapes (such as triangles, octahedral platelets, spheres, and rods) and the size of the NPs synthesized can be varied in response to temperature. The formation of nucleation centers is facilitated by an increase in the reaction response rate with temperature (Rana et al. 2020).

16.7.3.2 pH

The formation of NPs is significantly influenced by the medium pH reaction. The pH increase automatically increased the number of nucleation centers, which is crucial for promoting the formation of metal NPs. It is known that pH plays a crucial role in determining the size and shape of the NPs (Singh et al. 2020).

16.7.3.3 Reaction Time

The shape, size, and yield of NPs produced through green synthesis are primarily determined by the reaction time, which is a significant factor (Mughal and Hassan 2022; Roy et al. 2022). Along with temperature and pH, the most significant factor influencing the structural morphology of NPs is reaction time. Karade et al. (2018) noted that reaction time plays an important role in the synthesis of magnetic NPs.

16.7.4 Characterization of NPs

The characterization of NPs can be categorized based on the analysis of physical and chemical instruments, such as Fourier transform infrared spectroscopy (FT-IR), UV/Vis spectroscopy, scanning electron microscopy (SEM), X-ray diffraction (XRD), atomic force microscopy (AFM), dynamic light scattering (DLS), surface-enhanced remain spectroscopy (SERS), atomic absorption spectroscopy (AAS), energy dispersive spectroscopy (EDS), X-ray photoelectron spectroscopy (XPS), and high angle annular dark-field (HAADF) (Habeeb Rahuman et al. 2022).

16.7.4.1 FTIR

FTIR analysis involves passing infrared red light through the sample, allowing some of the rays to be absorbed by the material and the remaining rays to pass through it. The absorption or transmission as a function of wavelength provided by the spectrum describes the sample materials (Rozali et al. 2023). FTIR analysis reveals the various functional groups present in NPs and can be correlated with plant metabolite functional groups. It gives a glimpse of the role of biomolecules in the reduction of NPs (silver nitrate to silver). FTIR analysis is a suitable, economical, straightforward, and noninvasive technique (Naganthran et al. 2022).

16.7.4.2 UV-Vis Spectrophotometry

UV-Vis spectrophotometry is used to analyze the absorption maxima of NPs. The NPs synthesized have characteristic wavelength maxima, which could be used as a fingerprint for a particular NP. Typically, the wavelength range set to monitor NP synthesis is from 300 to 800 nm. Strong absorption results in a point spectrum in the observable region for the metallic NPs produced under specific salt conditions (Khan et al. 2022). According to prior study findings, NPs with sizes between 1 and

100 nm should be classified according to their absorption of wavelengths between 300 and 800 nm wavelength range (Begum et al. 2022).

16.7.4.3 Scanning Electron Microscope (SEM)

SEM can be used to characterize NPs. The purpose of this instrumentation investigation is to determine the distribution, morphology, size, and shape of synthesized NPs (Habeeb Rahuman et al. 2022). The SEM analysis evaluated how the morphological structure of the microbes changed both before and after they were treated with NPs. According to earlier research, detectable alterations in bacterial cell shape and holes made by NPs in the cell wall have been utilized as markers of the antimicrobial activity of NPs (Nahari et al. 2022; Wu et al. 2022).

16.7.4.4 X-Ray Diffraction (XRD)

With XRD, material atomic structures can be examined. To distinguish between the qualitative and quantitative levels of materials, this approach is useful. The size and structure of crystalline NPs were identified and verified using XRD analysis (Habeeb Rahuman et al. 2022). The Debye-Scherrer formula was used to analyze the particle dimension of nanomaterials from XRD data (Alaallah et al. 2023).

16.7.4.5 Atomic Force Microscopy (AFM)

The size, shape, and outside region of synthesized NPs have been confirmed and classified using atomic force microscopy (Habeeb Rahuman et al. 2022).

16.7.4.6 Transmission Electron Microscopy (TEM)

At the nanoscale level, the material's crystal structure and size of particles have been defined and confirmed by transmission electron microscopy (TEM) (Habeeb Rahuman et al. 2022).

16.7.4.7 Annular Dark Field Imaging (HAADF)

Annular dark-field imaging (HAADF) showed how NPs interacted with bacteria and gave information about the size distribution of NPs that interacted with each type of bacteria (Habeeb Rahuman et al. 2022).

16.7.4.8 Intracranial Pressure (ICP)

Intracranial pressure (ICP) spectrometry can determine the metal concentration of NPs. Metal NP concentrations are experimentally determined using inductively coupled plasma mass spectrometry (ICP-MS) and inductively coupled plasma emission spectroscopy (ICP-ES) (Habeeb Rahuman et al. 2022).

16.8 Antifungal Activity of NPs

A wide range of nanomaterials have been researched for their potential as AFAs, including biodegradable copolymeric and polymeric-based structures, lipid-based nanosystems, metallic nanocomposites, and NPs. Due to the small size they can facilitate transport of available drugs via different routes including nasal, intraocular, and oral routes (Du et al. 2021). Due to the different mode of action of NPs against fungus, it also reduces the risk of antifungal resistance. Therefore, plant-based synthesis of NPs is an inexpensive and environment-friendly method for antifungal applications. Green-synthesized MNPs release metal ions which generate reactive oxygen species and free radicals to kill fungal species (Lipovsky et al. 2011). Nanoparticles penetrate the cell wall of fungus leaking out intracellular components. They specifically destroy cell membranes by interfering with oxidative stress and redox balance (Kumari et al. 2019). NPs interact with the sulfur moieties or phosphorus of DNA, thus preventing DNA replication, inhibiting the growth of microbes, and causing cell death (Shanmugam et al. 2016). Apart from this, they can cause enzyme inhibition and disruption, protein denaturation and damage, denature ribosome, and inhibition of ribosome-synthesized proteins. In addition NPs are capable of inhibiting antioxidant enzymes like glutathione-producing enzymes (GHS), decreasing the chances of survival in fungi (Sun et al. 2018). The various ways through which NPs interact with fungi is shown in Fig. 16.4.

Antifungal activity of several NPs synthesized from plant extract is shown in Table 16.4. Vijayan et al. (2018) synthesized Ag and Au nanoparticles from *S. nodiflora* leaf extract and compared their antifungal activity. The results showed that Ag NPs showed large zone of inhibition (ZOI) about 12.9 mm, and 10.9 mm, compared to Au NPs (~ 9 mm) against *Aspergillus* spp. and *Penicillium* spp., respectively (Vijayan et al. 2018). This could be due to the smaller particle size of Ag nanoparticles (19.4 nm) than that of Au nanoparticles (22.01 nm), making them easier to penetrate and destroy fungal cells. In a study, Qasim Nasar et al. (2019) fabricated Ag NPs with *S. quettense* extract and obtained considerably larger particle size between 48.40 and 55.35 nm compared with that by Vijayan et al. (2018) (Qasim Nasar et al. 2019; Vijayan et al. 2018). Still, the ZOI against *Aspergillus* was much wider at 13.2 mm compared to Vijayan et al. (2018). Thus, green-synthesized Ag nanoparticles from *S. quettense* extract showed better antifungal activity compared to *S. nodiflora*. This could be due to the significant amount of flavonoid and phenolic compounds found in the plant extracts of *S. quettense* (40–56 μg per mg dried extract). As a result, the antifungal properties of metallic nanoparticles are dependent not only on particle size but also on phytochemicals present in plant extracts during the synthesis.

Jebril et al. (2020) synthesized Ag nanoparticles from *M. azedarach* leaf extracts and checked its antifungal performance. As a result, the percentage of inhibition of Ag NPs against *V. dahliae* was 18%, 33%, and 51% at doses of 20, 40, and 60 ppm, respectively. The higher the concentration of Ag NPs, the better the antifungal action (Jebril et al. 2020).

Aside from precious NPs such as Au and Ag other NPs of ZnO, CuO and Se could be used as antifungal agents against *C. albicans, C. tropicalis,* and *F. oxysporum*. Rajapriya et al. (2020) have green-synthesized ZnO NPs from *C. scolymus* (globe artichoke) leaf extract. The minimum inhibitory concentration of ZnO NPs against *C. albicans* and *C. tropicalis* was ~100 and 0.35 μg/ mL, respectively. The findings suggest higher antifungal activity against *C. tropicalis* (approximately 300

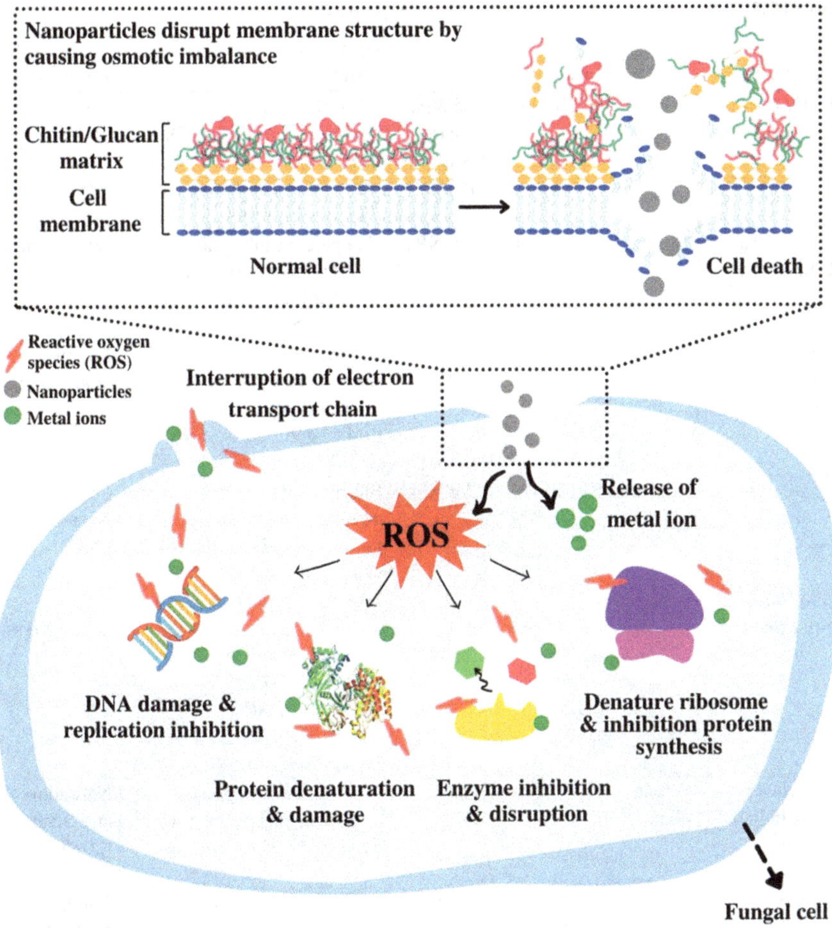

Fig. 16.4 Antifungal mechanism of metallic nanoparticles produced using plants. The reactive oxygen species are responsible for protein damage, DNA replication inhibition, enzyme inhibition, and so forth, causing fungal cell death. Abbreviations: ROS, reactive oxygen species; DNA, deoxyribonucleic acid (This is an Open Access article distributed under the terms of the Creative Commons Attribution License (http://creativecommons.org/licenses/by/2.0), which permits unrestricted use, distribution, and reproduction in any medium provided the original work is properly cited from Nguyen, N. T. T., L. M. Nguyen, T. T. T. Nguyen, T. T. Nguyen, D. T. C. Nguyen and T. V. Tran. "Formation, antimicrobial activity, and biomedical performance of plant-based nanoparticles: a review." Environmental Chemistry Letters **20**(4): 2531–2571. Copyright (2022), Springer)

times) than that against *C. albicans* (Rajapriya et al. 2020). In another study, Narendhran and Sivaraj (2016) produced ZnO NPs from *L. aculeata* extract for antifungal purposes. The inhibition zones for *A. flavus* and *F. oxysporum* were found to be 21 and 19 mm, respectively (Narendhran and Sivaraj 2016). Khanderao et al. demonstrated promising antifungal efficacy of green CuO NPs using *Moringa oleifera* leaves extract (Pagar et al. 2020). In an investigation by Kainat et al. (2022),

Table 16.4 Antifungal activity of plant-based nanoparticles

Plant name	Part	Size (nm)	NPs	Optimal conditions and results	References
Psidium guajava (Guava)	Leaf	17 (XRD)	Se NPs coated cotton fabric	Inhibition zone (mm) against *C. albicans* (35), *C. tropicalis* (35), *C. glabrata* (45); 100% reduction in *C. albicans* biofilm	Mirza et al. (2022)
M. azedarach (Chinaberry)	Leaf	23 (SEM)	Ag NPs	The percent inhibition (%) at the concentration of 60 µg/mL against: *V. dahliae* (51)	Jebril et al. (2020)
C. scolymus (Globe artichoke)	Leaf	65.9 (SEM)	ZnO NPs	Minimum inhibition concentration (MIC_{50}) (µg/mL) against: *C. albicans* (>100) and *C. tropicalis* (0.35)	Rajapriya et al. (2020)
S. quettense (Podlech)	Whole plant	53 (SEM)	Ag NPs	Inhibition zone (mm) against *A. niger* (13.2), *A. fumigatus* (12), *A. flavus* (10), *Mucor* spp. (11)	Qasim Nasar et al. (2019)
S. nodifora (Nodeweed)	Leaf	Ag NPs—19.4 Au NPs—22 (TEM)	Ag NPs, Au NPs	Inhibition zone (mm) against fungal: +Ag NPs: *Aspergillus* spp. (12.9), *Penicillium* spp. (10.9) + Au NPs: *Aspergillus* spp. (9) *Penicillium* spp. (9)	Vijayan et al. (2018)
L. aculeata (West Indian lantana)	Whole plant	12 (TEM)	ZnO NPs	Inhibition zone (mm) against at 100 µg/mL: *A. flavus* (21), *F. oxysporum* (19)	Narendhran and Sivaraj (2016)
N. arbor-tristis (Night-flowering jasmine)	Flower	12–32 (TEM)	ZnO NPs	Minimum inhibition concentration (µg/mL) against: *A. alternata* (64), *A. niger* (16), *B. cinerea* (128), *F. oxysporum* (64), *P. expansum* (128)	Jamdagni et al. (2018)
S. arvensis (Wild mustard)	Seed	14 (TEM)	Ag NPs	The percent inhibition (%) at the concentration of 2.5–40 µg/mL against *N. parvum* (15–83)	Khatami et al. (2015)

authors have in situ coated green-synthesized Se NPs from guava leaf extract on cotton fabric for antibacterial and antifungal performance. The results showed much larger ZOI against *Candida* spp. compared to bacterial spp. (*E. coli, S. aureus, K. pneumonia*). The Se NPs coated fabric was able to eliminate 100% biofilm formed on the fabric as compared to uncoated fabric (Fig. 16.5). The results showed Se NPs coated fabric can be used as undergarments for women suffering from VVCs and other skin related fungal infections (Mirza et al. 2022).

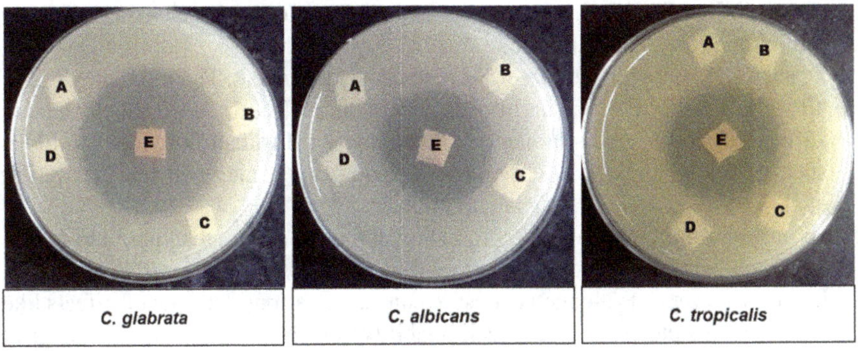

Fig. 16.5 Agar diffusion assay of (1) *C. glabrata,* (2) *C. albicans,* and (3) *C. tropicalis* showing ZOI around the Se NPs coated cotton fabric E, where A, B, C, and D represent controls: sodium selenite salt, guava extract, uncoated fabric, and activated fabric samples, respectively (This is an Open Access article distributed under the terms of the Creative Commons Attribution License (http://creativecommons.org/licenses/by/2.0), which permits unrestricted use, distribution, and reproduction in any medium provided the original work is properly cited from Mirza, K., F. Naaz, T. Ahmad, N. Manzoor and M. Sardar. "Development of Cost-Effective, Ecofriendly Selenium Nanoparticle-Functionalized Cotton Fabric for Antimicrobial and Antibiofilm Activity." Fermentation **9**(1):18. Copyright (2022), MDPI.)

Overall, green-synthesized NPs from plant extracts are emerging as a promising nanomedicine for treating a wide range of fungal infections (Nguyen et al. 2022).

16.9 Limitations of Metal Nanoparticles

Although MNPs exhibit various advantages as antifungal agents but the acute cytotoxicity and biocompatibility is always an issue. The toxicity of MNPs is dependent on various factors.

1. **Size**—NPs have size in nanoscale. If a particle is smaller in size than the extent at which it shows antifungal activity is comparatively higher but at the same time it can harm human cells. Contradictory results have been observed in this regard. In few studies it is shown that the toxicity of Au NPs is size dependent. As the size mainly decides the endocytosis of NPs into the cell (Madkhali 2023). It was reported that 1.4 nm Au NPs exhibit more cytotoxicity compared to larger Au NPs (up to 15 nm). Conversely, in another study Au NPs of 45 nm exhibited higher toxicity than smaller Au NPs (Madkhali 2023).
2. **Shape**—The cytotoxicity of NPs is shape dependent. For example, it was reported that spherical nanoparticles (approx. 40 nm) were more toxic than larger nanospheres. Also, nanorods are more toxic than spherical ZnO NPs (Madkhali 2023).
3. **Physicochemical properties**—NPs of different metal have varying toxicity. For example Se and Ag NPs are promising antifungal agents. But in terms of bio-

compatibility, Se NPs are favored as Se is present as a trace element in human (Attarilar et al. 2020; Mirza et al. 2022).

4. **Dosage**—Antifungal activity of MNPs is dose dependent. The more the dose of NPs, the more will be its antifungal performance. But the dose should remain sufficiently low to elicit human cell cytotoxicity (Madkhali 2023). Several NPs such as Ag NPs show dose dependent toxicity in human cell lines (Attarilar et al. 2020).
5. **Exposure time**—Time duration to which tissues or organ are exposed to NPs is also responsible for toxicity. Lesser the time of exposure lesser will be the toxicity and vice versa. Prolonged exposure can cause various detrimental effects like increased apoptosis, decrease cell viability and loss of membrane integrity (Attarilar et al. 2020).

It has been reported in several studies that NPs can cause oxidative stress, inflammation, apoptosis, and DNA damage and can interfere with cell metabolism. Intravenous administration of 35 nm TiO_2 NPs caused pregnancy complications in pregnant mouse models (Madkhali 2023). Various in vitro and in vivo studies revealed the genotoxicity and cytotoxicity of NPs while some showed little to no toxic effects. Due to the conflicting results it is essential to conduct extensive research on the toxicity profile, dosage and various other aspects of green-synthesized NPs. Also the various NPs synthesized haven't undergone clinical trials, restricting their commercialization and public usage.

16.10 Possible Directions for Further Research

Although, there is a shortage of AFAs and the rate at which drug resistant fungi are developing is high but following steps could be taken:

1. Presently researchers are focusing on structural modification of new AFAs to improve their safety and efficacy; but, extra efforts are required to identify new targets and develop NP-based AFAs.
2. Various studies have demonstrated the likely of NPs as AFAs; nevertheless, more research with their hybrids is needed. Moreover, the mechanisms of action of NPs need to be studied (Madkhali 2023).
3. The toxicity profile of NPs needs to be addressed unbiased and internationally agreed with in vivo toxicological models. For the toxicity studies the design, validation, and adaptation of such models need to be carefully considered.

16.11 Conclusion

Traditional AFAs should be broad spectrum and show multiple mode of action on fungal pathogen and sustained antimicrobial properties. In addition, these agents should have excellent biocompatibility and stability. Due to an upsurge in resistant strains to conventional AFAs, these AFAs are on a back fall. In this regard, plant-based NPs have a promising future as a potential alternative therapeutic against the fungus to overcome antifungal resistance. NPs can exert their antifungal effects against fungal pathogens by targeting several moieties of fungus simultaneously, which hinder the development of resistant strains. Despite this, NPs have a narrow therapeutic window; therefore, extensive physicochemical characterization during the early stages of therapeutics development is required. Extensive in vivo studies are also needed during preclinical and clinical studies for successful pharmaceutical development to avert any failures in the late phases of drug development.

References

Aisida SO, Ugwu K, Akpa PA, Nwanya AC, Nwankwo U, Botha SS, Ejikeme PM, Ahmad I, Maaza M, Ezema FI (2019) Biosynthesis of silver nanoparticles using bitter leave (Veronica amygdalina) for antibacterial activities. Surf Interfaces 17:100359

Alaallah NJ, Abd Alkareem E, Ghaidan A, Imran NA (2023) Eco-friendly approach for silver nanoparticles synthesis from lemon extract and their anti-oxidant, anti-bacterial, and anti-cancer activities. J Turkish Chem Soc Sect A Chem 10(1):205–216

Anju Varghese, R., Anandhi, P., Arunadevi, R., Boovisha, A., Sounthari, P., Saranya, J., Parameswari, K., & Chitra, S. (2015). Satin leaf (Chrysophyllum oliviforme) extract mediated green synthesis of silver nanoparticles: antioxidant and anticancer activities

Attarilar S, Yang J, Ebrahimi M, Wang Q, Liu J, Tang Y, Yang J (2020) The toxicity phenomenon and the related occurrence in metal and metal oxide nanoparticles: a brief review from the biomedical perspective. Front Bioeng Biotechnol 8:822

Awaad AS, Al-Mudhayyif HA, Al-Othman MR, Zain ME, El-Meligy RM (2017) Amhezole, a novel fungal secondary metabolite from Aspergillus terreus for treatment of microbial mouth infection. Phytother Res 31(3):395–402

Bajpai VK, Khan I, Shukla S, Kumar P, Rather IA, Park Y-H, Huh YS, Han Y-K (2019) Invasive fungal infections and their epidemiology: measures in the clinical scenario. Biotechnol Bioprocess Eng 24:436–444

Baldea I, Florea A, Olteanu D, Clichici S, David L, Moldovan B, Cenariu M, Achim M, Suharoschi R, Danescu S (2020) Effects of silver and gold nanoparticles phytosynthesized with Cornus mas extract on oral dysplastic human cells. Nanomedicine 15(1):55–75

Begum T, Follett PA, Mahmud J, Moskovchenko L, Salmieri S, Allahdad Z, Lacroix M (2022) Silver nanoparticles-essential oils combined treatments to enhance the antibacterial and antifungal properties against foodborne pathogens and spoilage microorganisms. Microb Pathog 164:105411

Biedler JL, Riehm H (1970) Cellular resistance to actinomycin D in Chinese hamster cells in vitro: cross-resistance, radioautographic, and cytogenetic studies. Cancer Res 30(4):1174–1184

Bongomin F, Gago S, Oladele RO, Denning DW (2017) Global and multi-national prevalence of fungal diseases—estimate precision. J Fungi 3(4):57

Chandraker SK, Lal M, Shukla R (2019) DNA-binding, antioxidant, H 2 O 2 sensing and photocatalytic properties of biogenic silver nanoparticles using Ageratum conyzoides L. leaf extract. RSC Adv 9(40):23,408–23,417

Chen L (2014) Study on biosynthesis and spectral property of gold nanoparticles in the extracts of mangosteen (Garcinia mangostana L) pericarp. Chem Res Appl 26(1):74–80

Dasgupta A, Fuller KK, Dunlap JC, Loros JJ (2016) Seeing the world differently: variability in the photosensory mechanisms of two model fungi. Environ Microbiol 18(1):5–20

Degreef H (2008) Clinical forms of dermatophytosis (ringworm infection). Mycopathologia 166(5–6):257–265

Dhand C, Dwivedi N, Loh XJ, Ying ANJ, Verma NK, Beuerman RW, Lakshminarayanan R, Ramakrishna S (2015) Methods and strategies for the synthesis of diverse nanoparticles and their applications: a comprehensive overview. RSC Adv 5(127):105,003–105,037

Dinesh GK, Pramod M, Chakma S (2020) Sonochemical synthesis of amphoteric Cu0-nanoparticles using Hibiscus rosa-sinensis extract and their applications for degradation of 5-fluorouracil and lovastatin drugs. J Hazard Mater 399:123035

Du W, Gao Y, Liu L, Sai S, Ding C (2021) Striking back against fungal infections: the utilization of nanosystems for antifungal strategies. Int J Mol Sci 22(18):10104

Elemike EE, Onwudiwe DC, Ekennia AC, Sonde CU, Ehiri RC (2017) Green synthesis of Ag/Ag2O nanoparticles using aqueous leaf extract of Eupatorium odoratum and its antimicrobial and mosquito larvicidal activities. Molecules 22(5):674

Felton T, Troke PF, Hope WW (2014) Tissue penetration of antifungal agents. Clin Microbiol Rev 27(1):68–88

Fisher MC, Alastruey-Izquierdo A, Berman J, Bicanic T, Bignell EM, Bowyer P, Bromley M, Brüggemann R, Garber G, Cornely OA (2022) Tackling the emerging threat of antifungal resistance to human health. Nat Rev Microbiol 20(9):557–571

Garcia Garcia SC, Salas Alanis JC, Flores MG, Gonzalez Gonzalez SE, Vera Cabrera L, Ocampo Candiani J (2015) Coccidioidomycosis and the skin: a comprehensive review. An Bras Dermatol 90:610–619

Ghosh S, Patil S, Ahire M, Kitture R, Gurav DD, Jabgunde AM, Kale S, Pardesi K, Shinde V, Bellare J (2012) Gnidia glauca flower extract mediated synthesis of gold nanoparticles and evaluation of its chemocatalytic potential. J Nanobiotechnol 10:1–9

Gingerich AD, Norris KA, Mousa JJ (2021) Pneumocystis pneumonia: immunity, vaccines, and treatments. Pathogens 10(2):236

Gow NA, Latge J-P, Munro CA (2017) The fungal cell wall: structure, biosynthesis, and function. Microbiol Spectr 5(3):funk-0035-2016. https://doi.org/10.1128/microbiolspec

Guo Q, Fu Z, Dong C (2014) Biosynthesis of gold nanoparticles using vine tea powder extracts. Chin J Appl Chem 31(7):841–846

Habeeb Rahuman HB, Dhandapani R, Narayanan S, Palanivel V, Paramasivam R, Subbarayalu R, Thangavelu S, Muthupandian S (2022) Medicinal plants mediated the green synthesis of silver nanoparticles and their biomedical applications. IET Nanobiotechnol 16(4):115–144

Hano C, Abbasi BH (2021) Plant based green synthesis of nanoparticles: production, characterization and applications, vol 12. MDPI, p 31

Jacob J, Mukherjee T, Kapoor S (2012) A simple approach for facile synthesis of Ag, anisotropic Au and bimetallic (Ag/Au) nanoparticles using cruciferous vegetable extracts. Mater Sci Eng C 32(7):1827–1834

Jaeger M, Plantinga TS, Joosten LA, Kullberg B-J, Netea MG (2013) Genetic basis for recurrent vulvo-vaginal candidiasis. Curr Infect Dis Rep 15:136–142

Jamdagni P, Khatri P, Rana J-S (2018) Green synthesis of zinc oxide nanoparticles using flower extract of Nyctanthes arbor-tristis and their antifungal activity. J King Saud Univ Sci 30(2):168–175

Jebril S, Jenana RKB, Dridi C (2020) Green synthesis of silver nanoparticles using Melia azedarach leaf extract and their antifungal activities: in vitro and in vivo. Mater Chem Phys 248:122898

Jiang Y, Li F, Liu C, Xu H, Sun H, Wang L (2016) Biosynthezied silver nanopaticles using hawthorn fruit extract and their antibacterial activity against four common aquatic pathogens. Oceanologia et Limnologia Sinica/Hai Yang Yu Hu Chao 47(1):253–260

Jing T, Zhengwei F, Chunfa D, Xiangjie W, Xiuzhi Y (2019) Green synthesis and characterization of monodisperse gold nanoparticles using aloe vera leaf extract. Rare Metal Mater Eng 48(11):3470–3475

Kanamori H, Rutala WA, Sickbert-Bennett EE, Weber DJ (2015) Review of fungal outbreaks and infection prevention in healthcare settings during construction and renovation. Clin Infect Dis 61(3):433–444

Karade V, Dongale T, Sahoo SC, Kollu P, Chougale A, Patil P, Patil P (2018) Effect of reaction time on structural and magnetic properties of green-synthesized magnetic nanoparticles. J Phys Chem Solids 120:161–166
Karunakaran G, Sudha KG, Ali S, Cho E-B (2023) Biosynthesis of nanoparticles from various biological sources and its biomedical applications. Molecules 28(11):4527
Khan F, Shahid A, Zhu H, Wang N, Javed MR, Ahmad N, Xu J, Alam MA, Mehmood MA (2022) Prospects of algae-based green synthesis of nanoparticles for environmental applications. Chemosphere 293:133571
Khani R, Roostaei B, Bagherzade G, Moudi M (2018) Green synthesis of copper nanoparticles by fruit extract of Ziziphus spina-christi (L.) Willd.: application for adsorption of triphenylmethane dye and antibacterial assay. J Mol Liq 255:541–549
Khatami M, Pourseyedi S, Khatami M, Hamidi H, Zaeifi M, Soltani L (2015) Synthesis of silver nanoparticles using seed exudates of Sinapis arvensis as a novel bioresource, and evaluation of their antifungal activity. Bioresour Bioprocess 2:1–7
Król A, Railean-Plugaru V, Pomastowski P, Buszewski B (2019) Phytochemical investigation of Medicago sativa L. extract and its potential as a safe source for the synthesis of ZnO nanoparticles: the proposed mechanism of formation and antimicrobial activity. Phytochem Lett 31:170–180
Krukowski S, Lysenko N, Kolodziejski W (2018) Synthesis and characterization of nanocrystalline composites containing calcium hydroxyapatite and glycine. J Solid State Chem 264:59–67
Kumar B, Smita K, Debut A, Cumbal L (2016) Extracellular green synthesis of silver nanoparticles using Amazonian fruit Araza (Eugenia stipitata McVaugh). Trans Nonferrous Metals Soc China 26(9):2363–2371
Kumar S, Jain T, Banerjee D (2019) Fungal diseases and their treatment: a holistic approach. In: Pathogenicity and drug resistance of human pathogens: mechanisms and novel approaches, pp 111–134
Kumari M, Giri VP, Pandey S, Kumar M, Katiyar R, Nautiyal CS, Mishra A (2019) An insight into the mechanism of antifungal activity of biogenic nanoparticles than their chemical counterparts. Pestic Biochem Physiol 157:45–52
Kuppusamy P, Yusoff MM, Maniam GP, Govindan N (2016) Biosynthesis of metallic nanoparticles using plant derivatives and their new avenues in pharmacological applications–an updated report. Saudi Pharmaceutical Journal 24(4):473–484
Kurup VP, Shen H-D, Banerjee B (2000) Respiratory fungal allergy. Microbes Infect 2(9):1101–1110
Lakhani P, Patil A, Majumdar S (2019) Challenges in the polyene-and azole-based pharmacotherapy of ocular fungal infections. J Ocul Pharmacol Ther 35(1):6–22
Latgé J-P, Chamilos G (2019) Aspergillus fumigatus and Aspergillosis in 2019. Clin Microbiol Rev 33(1):00140-00118. https://doi.org/10.1128/cmr
Lipovsky A, Nitzan Y, Gedanken A, Lubart R (2011) Antifungal activity of ZnO nanoparticles—the role of ROS mediated cell injury. Nanotechnology 22(10):105101
Madkhali OA (2023) A comprehensive review on potential applications of metallic nanoparticles as antifungal therapies to combat human fungal diseases. Saudi Pharmaceutical Journal 101733:101733
McBride JA, Gauthier GM, Klein BS (2017) Clinical manifestations and treatment of blastomycosis. Clin Chest Med 38(3):435–449
Mirza K, Sharma A, Naved T, Akhtar J, Sarwat M (2020) New technologies and interventions for the standardization of Unani drugs. International Journal of Unani and Integrative Medicine 4(1):22–25
Mirza K, Naaz F, Ahmad T, Manzoor N, Sardar M (2022) Development of cost-effective, eco-friendly selenium nanoparticle-functionalized cotton fabric for antimicrobial and Antibiofilm activity. Fermentation 9(1):18
Mittal AK, Chisti Y, Banerjee UC (2013) Synthesis of metallic nanoparticles using plant extracts. Biotechnol Adv 31(2):346–356

Mohd-Assaad N, McDonald BA, Croll D (2016) Multilocus resistance evolution to azole fungicides in fungal plant pathogen populations. Mol Ecol 25(24):6124–6142

Molloy SF, Kanyama C, Heyderman RS, Loyse A, Kouanfack C, Chanda D, Mfinanga S, Temfack E, Lakhi S, Lesikari S (2018) Antifungal combinations for treatment of cryptococcal meningitis in Africa. N Engl J Med 378(11):1004–1017

Muddapur UM, Alshehri S, Ghoneim MM, Mahnashi MH, Alshahrani MA, Khan AA, Iqubal SS, Bahafi A, More SS, Shaikh IA (2022) Plant-based synthesis of gold nanoparticles and theranostic applications: a review. Molecules 27(4):1391

Mughal SS, Hassan SM (2022) Comparative study of AgO nanoparticles synthesize via biological, chemical and physical methods: A review. American Journal of Materials Synthesis and Processing 7(2):15–28

Murthy H, Desalegn T, Kassa M, Abebe B, Assefa T (2020) Synthesis of green copper nanoparticles using medicinal plant hagenia abyssinica (brace) JF. Gmel leaf extract: Antimicrobial properties *Journal of nanomaterials* 2020

Naganthran A, Verasoundarapandian G, Khalid FE, Masarudin MJ, Zulkharnain A, Nawawi NM, Karim M, Che Abdullah CA, Ahmad SA (2022) Synthesis, characterization and biomedical application of silver nanoparticles. Materials 15(2):427

Nagpal M, Kaur M (2021) Nanomaterials for skin antifungal therapy: an updated review. Journal of Applied Pharmaceutical Science 11(1):015–025

Nahari MH, Al Ali A, Asiri A, Mahnashi MH, Shaikh IA, Shettar AK, Hoskeri J (2022) Green synthesis and characterization of iron nanoparticles synthesized from aqueous leaf extract of Vitex leucoxylon and its biomedical applications. Nano 12(14):2404

Naikoo GA, Mustaqeem M, Hassan IU, Awan T, Arshad F, Salim H, Qurashi A (2021) Bioinspired and green synthesis of nanoparticles from plant extracts with antiviral and antimicrobial properties: A critical review. J Saudi Chem Soc 25(9):101304

Nair GM, Sajini T, Mathew B (2022) Advanced green approaches for metal and metal oxide nanoparticles synthesis and their environmental applications. Talanta Open 5:100080

Narendhran S, Sivaraj R (2016) Biogenic ZnO nanoparticles synthesized using L. aculeata leaf extract and their antifungal activity against plant fungal pathogens. Bull Mater Sci 39:1–5

Nassar Y, Eljabbour T, Lee H, Batool A (2018) Possible risk factors for candida esophagitis in immunocompetent individuals. Gastroenterology Res 11(3):195

Nguyen NTT, Nguyen LM, Nguyen TTT, Nguyen TT, Nguyen DTC, Tran TV (2022) Formation, antimicrobial activity, and biomedical performance of plant-based nanoparticles: a review. Environ Chem Lett 20(4):2531–2571

Nijalingappa T, Veeraiah M, Basavaraj R, Darshan G, Sharma S, Nagabhushana H (2019) Antimicrobial properties of green synthesis of MgO micro architectures via Limonia acidissima fruit extract. Biocatal Agric Biotechnol 18:100991

Noori R, Ali J, Mirza K, Sardar M (2023) Nanoparticles mimicking oxidase activity and their application in synthesis of neurodegenerative therapeutic drug L-DOPA. ChemistrySelect 8(5):e202203808

Pagar K, Ghotekar S, Pagar T, Nikam A, Pansambal S, Oza R, Sanap D, & Dabhane H (2020) Antifungal activity of biosynthesized CuO nanoparticles using leaves extract of Moringa oleifera and their structural characterizations. Asian J of Nanoscience and Materials, 3(1):15–23

Pang Z, Li H, Liu J, Yu G, He S (2020) Nano-silver preparation using leaf extracts of Youngia japonica and its in hibitory effects on growth of bacteria from cut lily stemends. Northern Horticulture 14:103–109

Peçanha PM, Peçanha-Pietrobom PM, Grão-Velloso TR, Rosa Júnior M, Falqueto A, Gonçalves SS (2022) Paracoccidioidomycosis: what we know and what is new in epidemiology, diagnosis, and treatment. Journal of fungi 8(10):1098

Pinto RJ, Lucas JM, Silva FM, Girão AV, Oliveira FJ, Marques PA, Freire CS (2019) Bio-based synthesis of oxidation resistant copper nanowires using an aqueous plant extract. J Clean Prod 221:122–131

Qasim Nasar M, Zohra T, Khalil AT, Saqib S, Ayaz M, Ahmad A, Shinwari ZK (2019) Seripheidium quettense mediated green synthesis of biogenic silver nanoparticles and their theranostic applications. Green Chemistry Letters and Reviews 12(3):310–322

Rai M, Ingle A, Pandit R, Paralikar P, Gupta I, Anasane N, Dolenc-Voljč M (2017) Nanotechnology for the treatment of fungal infections on human skin *the microbiology of skin, soft tissue, bone and joint infections*. Elsevier, pp 169–184

Rajapriya M, Sharmili SA, Baskar R, Balaji R, Alharbi NS, Kadaikunnan S, Khaled JM, Alanzi KF, Vaseeharan B (2020) Synthesis and characterization of zinc oxide nanoparticles using Cynara scolymus leaves: enhanced hemolytic, antimicrobial, antiproliferative, and photocatalytic activity. J Clust Sci 31:791–801

Rana A, Yadav K, Jagadevan S (2020) A comprehensive review on green synthesis of nature-inspired metal nanoparticles: mechanism, application and toxicity. J Clean Prod 272:122880

Rayens E, Norris KA (2022) Prevalence and healthcare burden of fungal infections in the United States, 2018. Paper presented at the Open forum infectious diseases, vol 9

Rhodes J, Fisher MC (2019) Global epidemiology of emerging Candida auris. Curr Opin Microbiol 52:84–89

Roy A, Pandit C, Gacem A, Alqahtani MS, Bilal M, Islam S, Hossain MJ, Jameel M (2022) Biologically derived gold nanoparticles and their applications. Bioinorg Chem Appl 2022

Rozali NL, Azizan KA, Singh R, Jaafar SNS, Othman A, Weckwerth W, Ramli US (2023) Fourier transform infrared (FTIR) spectroscopy approach combined with discriminant analysis and prediction model for crude palm oil authentication of different geographical and temporal origins. Food Control 146:109509

Sajadi SM, Nasrollahzadeh M, Maham M (2016) Aqueous extract from seeds of Silybum marianum L. as a green material for preparation of the cu/Fe3O4 nanoparticles: a magnetically recoverable and reusable catalyst for the reduction of nitroarenes. J Colloid Interface Sci 469:93–98

Sardar M, Mirza K (2023) Affinity immobilization and affinity layers *biocatalyst immobilization*. Elsevier, pp 269–290

Sathishkumar G, Jha PK, Vignesh V, Rajkuberan C, Jeyaraj M, Selvakumar M, Jha R, Sivaramakrishnan S (2016) Cannonball fruit (Couroupita guianensis, Aubl.) extract mediated synthesis of gold nanoparticles and evaluation of its antioxidant activity. J Mol Liq 215:229–236

Seetharaman PK, Chandrasekaran R, Gnanasekar S, Chandrakasan G, Gupta M, Manikandan DB, Sivaperumal S (2018) Antimicrobial and larvicidal activity of eco-friendly silver nanoparticles synthesized from endophytic fungi Phomopsis liquidambaris. Biocatal Agric Biotechnol 16:22–30

Shailaja V, Pai L, Mathur D, Lakshmi V (2004) Prevalence of bacterial and fungal agents causing lower respiratory tract infections in patients with human immunodeficiency virus infection. Indian J Med Microbiol 22(1):28–33

Shameli K, Bin Ahmad M, Jaffar Al-Mulla EA, Ibrahim NA, Shabanzadeh P, Rustaiyan A, Abdollahi Y, Bagheri S, Abdolmohammadi S, Usman MS (2012) Green biosynthesis of silver nanoparticles using Callicarpa maingayi stem bark extraction. Molecules 17(7):8506–8517

Shanmugam C, Sivasubramanian G, Parthasarathi B, Baskaran K, Balachander R, Parameswaran V (2016) Antimicrobial, free radical scavenging activities and catalytic oxidation of benzyl alcohol by nano-silver synthesized from the leaf extract of Aristolochia indica L.: a promenade towards sustainability. Appl Nanosci 6:711–723

Singh A, Gautam PK, Verma A, Singh V, Shivapriya PM, Shivalkar S, Sahoo AK, Samanta SK (2020) Green synthesis of metallic nanoparticles as effective alternatives to treat antibiotics resistant bacterial infections: A review. Biotechnology Reports 25:e00427

Sullivan DJ, Moran GP, Coleman DC (2011) Fungal infections of humans. In: Fungi: biology and applications, pp 257–278

Sun Q, Li J, Le T (2018) Zinc oxide nanoparticle as a novel class of antifungal agents: current advances and future perspectives. J Agric Food Chem 66(43):11209–11220

Sundararajan B, Kumari BR (2017) Novel synthesis of gold nanoparticles using Artemisia vulgaris L. leaf extract and their efficacy of larvicidal activity against dengue fever vector Aedes aegypti L. J Trace Elem Med Biol 43:187–196

Vanaja M, Rajeshkumar S, Paulkumar K, Gnanajobitha G, Malarkodi C, Annadurai G (2013) Phytosynthesis and characterization of silver nanoparticles using stem extract of Coleus aromaticus. Int J Mater Biomater Appl 3(1):1–4

Verweij PE, Lucas JA, Arendrup MC, Bowyer P, Brinkmann AJ, Denning DW, Dyer PS, Fisher MC, Geenen PL, Gisi U (2020) The one health problem of azole resistance in Aspergillus fumigatus: current insights and future research agenda. Fungal Biol Rev 34(4):202–214

Vijayan R, Joseph S, Mathew B (2018) Eco-friendly synthesis of silver and gold nanoparticles with enhanced antimicrobial, antioxidant, and catalytic activities. IET Nanobiotechnol 12(6):850–856

Wall G, Lopez-Ribot JL (2020) Current antimycotics, new prospects, and future approaches to antifungal therapy. Antibiotics 9(8):445

Wang Y, Deng K (2021) Environmental risk factors for Talaromycosis hospitalizations of HIV-infected patients in Guangzhou, China: case crossover study. Front Med 8:731188

Wang Y, Chen H, Lan R (2013) Preparation of nano-silver with green tea extract. Environ Sci Technol 36(12):122–125

Wu X, Fang F, Zhang B, Wu JJ, Zhang K (2022) Biogenic silver nanoparticles-modified forward osmosis membranes with mitigated internal concentration polarization and enhanced antibacterial properties. NPJ Clean Water 5(1):41

Yadav TP, Yadav RM, Singh DP (2012) Mechanical milling: a top down approach for the synthesis of nanomaterials and nanocomposites. Nanosci Nanotechnol 2(3):22–48

Ying S, Guan Z, Ofoegbu PC, Clubb P, Rico C, He F, Hong J (2022) Green synthesis of nanoparticles: current developments and limitations. Environ Technol Innov 26:102336

Zangeneh MM, Zangeneh A (2020) Novel green synthesis of Hibiscus sabdariffa flower extract conjugated gold nanoparticles with excellent anti-acute myeloid leukemia effect in comparison to daunorubicin in a leukemic rodent model. Appl Organomet Chem 34(1):e5271

Zhang D, Ma X-L, Gu Y, Huang H, Zhang G-W (2020) Green synthesis of metallic nanoparticles and their potential applications to treat cancer. Front Chem 8

Antifungal Efficacy of Plant-Based Nanoparticles as a Putative Tool for Antifungal Therapy

17

Sradhanjali Mohapatra, Nazia Hassan, Mohd. Aamir Mirza, and Zeenat Iqbal

17.1 Introduction

Fungal infection is one of the most common public health concerns and the number has increased in the past few decades (Fisher et al. 2012). This infection may be categorized into superficial (affecting skin) and systemic infections (invasion of internal organs). Superficial infections are less dangerous than invasive forms but are contagious and should not be neglected. The most common fungal species which are responsible for this infection may be summarized as *Aspergillus*, *Candida*, *Cryptococcus*, and *Pneumocystis*. Millions of people suffer from this infection every year throughout the world. The reason may be attributed to the common use of traditional drugs as antifungal therapy and the increased resistance of some species of fungi toward drugs frequently used for this infection (Arastehfar et al. 2020; Rodriguez-Tudela et al. 2008). Negligence of the disease may further add burden to this problem. Additionally, the rate of infection is higher in the case of immunosuppressed patients suffering from conditions like AIDS, organ transplant, chemotherapy, etc. These above facts demand development of novel antifungal drugs involving wide spectrum structural classes, which can act on new targets with minimal adverse effects.

Plants or plant-based products may be the perfect solution to this problem. Standardized plant extracts or pure phytoconstituents may possibly be used for antifungal therapy as they have chemical diversity against widely different targets. Plant-derived medicines are preferred over synthetic ones due to their accessibility, low toxicity, efficient therapeutic use, affordability and higher

S. Mohapatra · N. Hassan · M. A. Mirza · Z. Iqbal (✉)
Department of Pharmaceutics, School of Pharmaceutical Education and Research, Jamia Hamdard, New Delhi, India
e-mail: sibanee@gmail.com; naziazahidhasan@gmail.com; aamir_pharma@yahoo.com; zeenatiqbal@jamiahamdard.ac.in

N. Manzoor (ed.), *Advances in Antifungal Drug Development*,
https://doi.org/10.1007/978-981-97-5165-5_17

degree of patient acceptance. Traditional uses of medicinal plants include the treatment of both human and animal mycoses, and they are regarded as a valuable resource for the development of novel antifungal medications. Numerous secondary metabolites found in plants, such as terpenoids, saponins, and alkaloids that have antifungal properties, have been studied abundantly (Kumar Mishra et al. 2020).

Currently, nanotechnology is one of the rapidly emerging areas and plays an undeniably significant role in the field of medicine, and technology. It is a multidisciplinary area which has been widely exploited in health care for developing nanomedicines of proficient use. They have a size range of 1–100 nm which is advantageous in pharmaceutical formulations. They have the ability to alter and improve the pharmacokinetic and pharmacodynamic characteristics of the drugs. They are biocompatible and have the capacity to improve the solubility and stability of the drugs. These characteristics of nanoformulations make them more appropriate for therapeutic use. They can effectively target systemic circulation with less harmful effects by reducing dose and dose frequency. Further, nanoparticulate formulations are also effective for topical infections (Gupta et al. 2022).

Presently, plant-based nanocarriers are widely investigated for their safety and efficacy (Parveen et al. 2022). They are widely implicated in the pharmaceutical field for a variety of diseases including fungal infections (Mohapatra et al. 2022). Nowadays herbal extracts are exploited as potential means for producing nanoformulations in a safer and cost-effective way. Green synthesis of nanoparticles is an approach which is used to generate several nanomedicines in treating fungal infections. Due to its simplicity of use and the variety of plants available, plant-based nanoparticle green synthesis is now regarded as the gold standard among these green biological techniques. Here, plant extracts, algae, fungi, viruses, or bacteria are utilized for the synthesis of nanoparticles. It offers many benefits such as simple and rapid preparation procedure, biocompatibility, cost-effectiveness and scalability. Metals such as, silver, gold, copper, zinc, titanium, iron etc. are used for the preparation of plant-based nanoparticles specifically for fungal infections. Metal nanoparticles themselves possess antifungal infection as reported in several research (Lotfali et al. 2020; Wani and Ahmad 2013; Ibarra-Laclette et al. 2022; He et al. 2011; Seddighi et al. 2017). Additionally, the nanoencapsulation approach is also used which is popular for encapsulating extracts or bioactive molecules derived from plant and are widely used for the management of different ailments including fungal infection (Dureja et al. 2022).

The focus of this chapter was the compilation of research done in the field of plant-based nanoparticles having antifungal efficacy, which will be a future prospect for conducting research in this novel emerging area to provide better antifungal therapy. This compilation would draw the attention of researchers in this budding area for conducting more investigation to curb fungal diseases through an effective and environmentally friendly manner.

17.2 Nanoparticles Encapsulating Herbal Extract for Antifungal Efficacy

It is undeniable that plant extracts are a source of many active constituents that offer several properties such as antibiotic, antioxidant, antihypertensive, antibacterial, antifungal, antiviral, insecticide, hypoglycemic, anticancer, and many others. Incorporation of nanotechnology brings evolution in this area by developing formulations which enhance their action many fold while maintaining safety. Plant extract-loaded nanocarriers, by virtue of their unique characteristics, mask unpleasant taste and aromas, protect sensitive phytochemicals from biological besides environmental factors, increase bioavailability, facilitate passage across the biological barriers, provide targeted/controlled delivery of phytochemicals to specific organs, etc. Presently, it is used as a novel strategy for ecological and sustainable production. Biopolymer encapsulation has been preferably exploited for the incorporation of bio constituents for their controlled release and action.

In the realm of innovative antifungal therapies, the encapsulation of herbal extracts within various modern nanoparticle systems has emerged as a promising approach. Utilizing diverse nanoparticle formulations, like polymer-based, lipid-based, and emulsion-based carriers, encapsulated herbal extracts offer a multifaceted solution in combating fungal infections. Polymeric nanoparticles, with their adjustable properties and controlled release mechanisms, enable the sustained delivery of herbal antifungal compounds to targeted sites, enhancing efficacy and minimizing potential side effects. Lipid-based nanoparticles, owing to their biocompatibility and structural resemblance to cell membranes, facilitate enhanced penetration and absorption of herbal extracts, optimizing their antifungal activity. Emulsion-based nanoparticles, through their stability and capacity for incorporating both hydrophilic and hydrophobic herbal components, further enhance the solubility and bioavailability of the antifungal agents. This amalgamation of herbal extracts with diverse nanoparticle systems showcases a promising frontier in developing efficient and targeted antifungal therapies, potentially revolutionizing the treatment landscape for fungal infections. Table 17.1 conscripts list of nanocarriers encapsulating herbal extract showing antifungal action.

Table 17.1 Nanoparticles encapsulating herbal extract for antifungal action

S. No.	Type of nanocarriers	Biological source	Active constituents	Other excipients	Method of preparation	Particle/ vesicle size	Evaluation	Targeted pathogenic species	References
1	Solid lipid nanoparticles	*Zataria multiflora*	Essential oil	0.03% essential oil in 5% of lipid phase, tween 80, and Poloxamer 188 (2.5% w/v) in aqueous phase	High shear homogenization and ultra sound technique	255.5 ± 3 nm	In vitro	*Aspergillus ochraceus, Aspergillus niger, Aspergillus flavus, Alternaria solani, Rhizoctonia solani*, and *Rhizopus stolonifer*	Nasseri et al. (2016)
2	Chitosan nanoparticles	*Cymbopogon martinii*	Essential oil	Chitosan, tween 80, TPP	Homogenization under magnetic stirring	455–480 nm	In vitro	*Fusarium graminearum*	Kalagatur et al. (2018)
3	Chitosan nanoparticles	*Carum copticum* essential oil & *Peganum harmala* extract	Essential oil & extract	Chitosan in an aqueous acetic acid, tween 80, TPP	Homogenization under magnetic stirring	100 nm	In vitro	*Alternaria alternata*	Izadi et al. (2021)
4	Chitosan nanoparticles	*Mentha longifolia*	Leaf extract	Chitosan, tween 80, TPP	Homogenization under magnetic stirring followed by centrifugation	157 nm	In-vitro	Against mycelium growth of *A. niger*	El-Aziz et al. (2018)
5	Chitosan nanoparticles	Thyme extract	Leaf extract (hydro alcoholic mixture)	Chitosan, acetic acid, TPP	Ionotropic gelation technique	75.07 nm	In vitro	*Macrophomina phaseolina*	Tabarestani (2022b)
6	Chitosan nanoparticles	*Carum copticum* L.	Essential oils of oregano & lemongrass	Oregano oil and lemongrass		100–200 nm	In vitro	*C. albicans*	Ashraf et al. (2022)

7	Chitosan nanocapsules	*Trichoderma harzianum*	Extract	Chitosan, tripolyphosphate	Ionic gelation method	77.91 nm	In vitro	*Macrophomina phaseolina*	Tabarestani (2022a)
8	Chitosan nanoparticles	s-nitrosomercaptosuccinic acid	–	Chitosan	Ionic gelation method	241.69 ± 18.95 nm	In vitro	*C. albicans*, *C. glabrata*, *C. krusei* and *C. parapsilosis*	Moron (2020)

17.3 Metal Nanoparticles as Novel Antifungal Agents

Metal nanoparticles display antifungal activity owing to their intrinsic antimicrobial nature (Asghari et al. 2016). They display unique physical and chemical properties. Their size, charge, porosity, and hydrophobicity may change depending on the synthesis process, and have an impact on particle stability, biocompatibility, biodistribution, circulation time, cellular localization etc. Although the exact mechanism of action of metal nanoparticles exhibiting antifungal activity is not known, the possible hypothesis for their mechanism can be suggested to include direct uptake of nanoparticles, release of ions or reactive oxygen species (ROS), impairment of cell wall/membrane through accumulation, inhibition of enzymatic activity, declining of ATP levels, DNA, protein, gene expression regulation, and mitochondrial dysfunction (Don et al. 2016).

It has been reported that plants are used for the synthesis of a vast number of metallic nanoparticles. Plant-mediated synthesis can be done either intracellularly (inside the plant), extracellularly (using plant extracts), or by using individual phytochemicals. Plants have the capacity to accrue metals and then convert them incrementally into nanoparticles intracellularly. On the other hand, bioactive substances present inside the plant help in the reduction of the metal. This difference in the stabilizing and reducing potential of biomolecules is determinant of the variation in size, shape, etc. The formation of gold nanoparticles inside the living plant, alfalfa in $AuCl_4$ rich environment and accumulation inside *Medicago sativa* and *Brassica juncea* plants in aqueous solutions of $KAuCl_4$ medium have been reported by Gardea-Torresdey et al. (Gardea-Torresdey et al. 2002) and Bali and Harris (Bali and Harris 2010), respectively. However, most of the research has been focused on the synthesis of metallic nanoparticles from their extract obtained from different parts of the plant such as root, stem, leaves, flower, fruit, seed, seed coat, latex, etc. Table 17.2 enlists metal nanoparticles synthesized from various plants.

Table 17.2 Metal nanoparticles synthesized from diverse plants possessing antifungal activity

S. No	Type of nanoparticles	Name of the plant	Part of the plant	Use of plant material	Size of nanoparticles	Reference
1	Silver	*Lippia citriodora*	Leaf extract (Aqueous)	Reducing agent	15–30 nm	Cruz et al. (2010)
2	Silver	*Cucumis prophetarum*	Leaf extract (Aqueous)	Reducing agent & stabilizing agent	90 nm	Hemlata et al. (2020)
3	Silver	*Catharanthus roseus*	Leaf extract (Aqueous)	Reducing agent & stabilizing agent	35–55 nm	Ponarulselvam et al. (2012)
4	Silver	*Phyllanthus emblica*	Fruit extract	Capping/stabilizing agents	19.8–92.8 nm	Masum et al. (2019)
5	Gold	Bayberry tannin	Tannin	Reducing agent & stabilizing agent	1.8 ± 0.3 nm	Huang et al. (2010)
6	Gold	Tea leaves	Leaf extract	Reducing agent		Sharma et al. (2012)
7	Gold	*Morinda citrifolia*	Root extract	Reducing agent	12.17–38.26 nm	Suman et al. (2014)
8	Gold	*Garcinia mangostana*	Aqueous peel extract	Reducing agent	32.96 ± 5.25 nm	Xin Lee et al. (2016)
9	Gold	*Pelargonium graveolens* & its endophytic fungus	Leaf extract	Reducing & capping agents	510–560 nm	Shankar et al. (2003)
10	Gold	*Punica granatum* (Pomegranate)	Fruit extract	Reducing agent	10–50 nm	Basavegowda et al. (2013)
11	Iron	*Camellia sinensis*	Leaf extract	Reducing & capping agent	5–15 nm	Hoag et al. (2009)
12	Iron	*Phoenix dactylifera*	Leaf extract	Reducing agent	60–92 nm	Batool et al. (2021)
13	Iron	*Citrullus lanatus* rinds	Watermelon peel extract	Solvent, capping, & reducing agent	2–20 nm	Prasad et al. (2016)
14	Iron	*Sesbania grandiflora*	Leaf Extract	Reducing agent	25–60 nm	Rajendran and Sengodan (2017)
15	Zinc oxide	*Myristica fragrans*	Fruit extracts	Stabilizing & capping	66 nm	Faisal et al. (2021)

(continued)

Table 17.2 (continued)

S. No	Type of nanoparticles	Name of the plant	Part of the plant	Use of plant material	Size of nanoparticles	Reference
16	Zinc oxide	*Hibiscus subdariffa*	Leaf extract	Stabilizing	12–46 nm	Bala et al. (2015)
17	Zinc oxide	*Raphanus sativus* var. Longipinnatus	Leaf extract	Capping/stabilizing agents	209 nm	Umamaheswari et al. (2021)
18	Zinc oxide	*Coriandrum sativum*	Leaves extract	Stabilizing & reducing agent	Around 100 nm	Ukidave and Ingale (2022)
19	Zinc oxide	*Sesbania grandiflora*	Leaf extract	Reducing agent	15–35 nm	Rajendran and Sengodan (2017)

17.4 Green Synthesis of Nanoparticles in Curbing Fungal Infection

To overcome the toxicity and environmental concerns associated with some of the methods involved in the preparation of nanoparticles, green synthesis has emerged as a viable alternative. Green chemistry is a dependable, efficient, environmentally safe, and biologically appropriate method that uses a variety of plant materials rich in phytochemicals, microbes, marine algae, etc. However, nanomaterials produced from microbes are slow having large size distribution. Since long, plants have been used as reducing agents for different chemical reactions, without knowing their specific mechanism of action. This has led to increased attention and evolution of green synthesis.

Methods used to synthesize nanoparticles can be broadly classified as physical, chemical and biological (green synthesis). Physical and chemical methods involve high energy, temperature, pressure radiation, harmful chemicals, complex equipment and synthesis conditions, etc., whereas green synthesis exploits the use of plant parts and other microorganisms with lesser energy consumption and avoids the use of toxic and harmful reagents. Green synthesis has several benefits over other processes, including the fact that it is pollution-free, nontoxic, environment friendly, cost-effective, and more sustainable. However, there are a few concerns regarding the raw material extraction, concentration, pH of the medium, temperature conditions, reaction times, and product quality.

17.4.1 Green Synthesis of Metallic Nanoparticles

This process involves the following steps:

1. Preparation of plant extract
2. Addition of the extract to the metal precursor solution ($AgNO_3$, $HAuCl_4$, H_2PtCl_6, $FeCl_3{\cdot}6H_2O$, Cu $(NO_3)_2{\cdot}3H_2O$, $(NiNO_3)_2{\cdot}6H_2O$, and Na_2SeO_3 are commonly used for the synthesis of Ag, Au, Pt, Fe, Cu, Ni, and Se nanoparticles, respectively) containing the salts of respective metals
3. Synthesis of metal nanoparticles
   (a) Metal ions are reduced to metal atoms, and the reduced metal atoms are then successively nucleated.
   (b) Merging of small adjacent nanoparticles into larger size particles occurs with increase in thermodynamic stability.
   (c) When the nanoparticles receive their final shape, the process is finished.

Thus, the active biomolecules present in the extract play an important role in the reduction and stabilization of metal ions in the solution (Makarov et al. 2014).

17.4.2 Types of Metal Nanoparticles Exhibiting Antifungal Activity

Different types of metal nanoparticles exhibiting antifungal activity can be listed as:

1. Silver nanoparticles
2. Gold nanoparticles
3. Iron nanoparticles
4. Zinc Oxide nanoparticles
5. Other metal nanoparticles

17.4.2.1 Silver Nanoparticles

Among all the metallic nanoparticles, the silver nanoparticles are one of the most significant and fascinating due to their special qualities like catalytic activity and stability. They have a multifaceted role in the biomedical field. They have evolved as a new class of antimicrobial agents against pathogenic microbes and play a positive role as anticancer agents owing to their antiproliferative and targeting action (Dikshit et al. 2021). At present, they are being investigated as candidates for carrying a wide range of antifungal activity due to recent improvements in their properties like biocompatibility and stability through surface modifications. Research has reported that silver nanoparticles, in combination with clinically approved antifungal drugs, showed improved efficacy with negligible cytotoxicity (Jia and Sun 2021; Hussain et al. 2019). Currently, these particles are synthesized using green methods for increased biological activity with reduced toxicity and cost.

17.4.2.2 Gold Nanoparticles

Preparation of gold nanoparticles is based on the reduction of gold ions using reducing agents such as sodium borohydride and sodium citrate (Ghosh et al. 2008) followed by the prevention of growth and particle agglomeration of the newly formed nanoparticles with the help of protecting agents. Hence, the shape and size of the nanoparticles during synthesis can be controlled by the using proper reducing agents, protection agents and their concentration, along with the synthesis conditions, such as temperature, pH, stirring methods and time. However, these components (reducing and stabilizing agents) are derived from natural sources such as plants, bacteria, algae, fungi, etc. in case of green synthesis (Vankar and Bajpai 2010; Das and Marsili 2010). These nanoparticles are widely used for their anticancer properties, induce oxidative stress, and destroy cancerous cells. It displays substantial antimicrobial activity [32, 33] owing to strong cytotoxicity to several microorganisms including the fungi.

17.4.2.3 Iron Oxide Nanoparticles

These are made by precipitating them in isobutanol, which also serves as a surfactant, along with ammonium hydroxide and sodium hydroxide. Soon after they are formed, they must be surface coated with different moieties to minimize the accumulation. These particles have hydrophobic surfaces with a large surface-area-to-volume ratio, strong magnetic attraction among particles and van der Waals forces, which are responsible for their aggregation. As these are unstable in aqueous medium and may be sequestered by macrophages inside the blood, they can be subsequently rejected by the reticuloendothelial system. Further, large concentration of iron in blood is also toxic to living being. Iron oxides are found in three forms: magnetite (Fe_2O_3), maghemite (γ-Fe_2O_3), and hematite (α-Fe_2O_3). Superparamagnetism is exhibited by nanoparticles made of ferromagnetic materials with a size of less than 10–20 nm. Magnetic iron oxide nanoparticles have received a lot of attention due to their surface area and volume ratio, low toxicity, superparamagnetic characteristics, and separation methodology. They are of particular interest in biomedical applications for protein immobilization, such as thermal therapy, drug delivery, and diagnostic magnetic resonance imaging (Ali et al. 2016; Xu et al. 2014). The said nanoparticles have wide applications in anticancer strategy (hyperthermia), where they destroy tissues by generating heat in the adjoining area of cancer cells. Due to the reactivity of nanoscale iron and its high surface area to volume ratio, these are also employed in the fight against environmental pollution (Saif et al. 2016).

17.4.2.4 Zinc Oxide Nanoparticles

Among several synthesized inorganic metal oxides, zinc oxide nanoparticles are of great interest because they are easy to prepare, inexpensive and safe. Zinc oxide is a metal oxide that the US FDA has designated as GRAS (generally recognized as safe) (Kashid et al. 2011). Because of their remarkable optical, physical, and antimicrobial properties, zinc oxide nanoparticles have a lot of potential to advance agriculture, field of electronics and biomedical systems (Anbuvannan et al. 2015; Jayaseelan et al. 2012). Research has proved that Zinc oxide nanoparticles have antifungal activity against dermatophytes and other pathogenic fungi, such as *Candida* and *Aspergillus*. Furthermore, it also shows synergistic antifungal activity with common antifungal drugs, which is helpful in reducing the dose and hence lessen the associated toxicity and increases their efficacy (Agarwal et al. 2017).

Table 17.3 enlists several nanoparticles produced by green synthesis showing antifungal action.

Table 17.3 Nanoparticles produced by green synthesis, their source, and antifungal action

S. No.	Biological source	Natural extract/ compound	Types of nanoparticles	Biological activity	Type of study & fungal strains tested	Reference
1	*Moringa Olifera*	Leaf & flower extract	Silver nanoparticles	Significantly decreases fungal growth at lower concentrations & appears as a potent fungistatic agent	In vitro studies against plant pathogenic fungi *Pestalotiopsis mangiferae* (isolated from infected coconut palm)	Jenish et al. (2022)
2	*Brassica rapa* L.	Leaf extract	Silver nanoparticles	Antifungal action	In vitro studies against wood-degrading fungal pathogens: *Gloeophyllum abietinum*, *G. trabeum*, *Chaetomium globosum*, and *Phanerochaete sordida*	Narayanan and Park (2014)
3	*Mentha pulegium*	Leaf extracts (Aqueous)	Silver nanoparticles	Antifungal action	In vitro studies against fluconazole-resistant *Candida albicans*	Abd Kelkawi et al. (2017)
4	*Lotus lalambensis* Schweinf	Leaf extracts (Aqueous)	Silver nanoparticles	Inhibited morphogenesis in *C. albicans*; suppressed adhesion & formation of biofilms; causes 80% decrease in the production of antioxidant enzymes in *C. albicans*	In vitro studies against fluconazole-resistant *C. albicans*	Abdallah and Ali (2021)
5	*Glycosmis pentaphylla*	Fruit extract	Silver nanoparticles	Antifungal action	In vitro studies against *Alternaria alternata*, *Fusarium moniliforme, Colletotrichum lindemuthianum*, and *Candida glabrata*	Dutta et al. (2022)
6	-	Rutin	Silver nanoparticles	Skin candidiasis	In vitro studies against *C. albicans*	Alqarni et al. (2022)
7	-	Naringenin	Silver nanoparticles	Antifungal action	In vitro studies against *C. albicans* and *C. glabrata*	Katta et al. (2023)

8	Aloe vera extract	Leaf extract	Fluconazole-loaded silver nanoparticles	Antifungal action	In vitro studies against clinical *C. albicans* strains and a reference strain of *C. albicans*	Arsène et al. (2023)
9	Neem extract		Fluconazole-loaded silver nanoparticles	A synergistic antifungal effect of fluconazole & neem extract was observed	In vitro studies against *C. albicans*	Chauhan et al. (2022)
10	*Nyctanthes arbortristis*	Flower extract (aqueous)	Zinc oxide nanoparticle	Antifungal action	In vitro studies against *A. alternate*, *A. niger*, *B. cinerea*, *F. oxysporum*, and *P. expansum*	Jamdagni et al. (2018)
11	*Amphipterygium glaucum* & *Calendula officinalis*	Leaves & flowers respectively	Selenium Nanoparticles	Antifungal action	Antifungal agents against two commercially relevant plant pathogenic fungi, *Fusarium oxysporum* and *Colletotrichum gloeosporioides*	Lazcano-Ramírez et al. (2023)
12	Aloe vera	Leaf extract	Selenium nanoparticle	Antifungal action	Decreases the radial mycelial growth of *Colletotrichum coccodes* and *Penicillium digitatum*	Fardsadegh and Jafarizadeh-Malmiri (2019)
13	*Aconitum heterophyllum*	Leaf extract	Gold nanoparticles	Antibiofilm and photocatalytic activity	*P. auruginosa*, *B. subtilis*, and *C. albicans*	Rasool et al. (2022)
14	*Aspalathus linearis* tea	Leaf extract (aqueous)	Gold nanoparticles	Demonstrates substantial increase in the antifungal action of pristine antibiotic agents	4 *Aspergillus* species	Thipe et al. (2015)
15	*Euphorbia prostrata* & *Pelargonium graveolens* extracts	Aerial parts collected at the flowering stage	Mn-Ni@Fe_3O_4-NPs & Mn:Fe $(OH)_3$-NPs	Antifungal action	*C. albicans*	Dashamiri et al. (2018)
16	*Rosmarinus officinalis*	Leaf extract	Palladium nanoparticles	Antifungal activity	*Candida parapsilolis*, *C. albicans*, *C. glabrata*, and *Candida krusei*	Rabiee et al. (2020)

(continued)

Table 17.3 (continued)

S. No.	Biological source	Natural extract/ compound	Types of nanoparticles	Biological activity	Type of study & fungal strains tested	Reference
17	*Argemone mexicana L*	Plant extract	Gold & silver nanoparticles	Extract as well as the silver nanoparticles act synergistically to increase inhibition in the growth of resistant bacteria & fungus	*C. albicans*	Téllez-de-Jesús et al. (2021)
18	*Capparis decidua*	Fruit powder	Selenium Nanoparticles	Antifungal activity	*Aspergillus fumigatus* and *C. albicans*	Ali et al. (2020)
19	*Bryophyllum pinnatum (lam.) and Polyalthia longifolia (Sonn.)*	Leaves	Copper oxide nanoparticles	Antifungal activity	*Galactomyces geotrichum*	Hanisha et al. (n.d.)

17.5 Limitation of Green Synthesis

"Plant-based Nanoparticles" present a promising avenue for novel antifungal therapies; however, it also comes with certain limitations which can be described as follows:

Limited Research: Despite the potential of plant-based nanoparticles for antifungal efficacy, the research in this specific area may still be relatively limited. The field is relatively new, and there may not be enough comprehensive studies to fully understand the safety, long-term effects, and overall efficacy of these nanoparticles for treating fungal infections.

Standardization of Plant Extracts: Plant extracts used for nanoparticle synthesis can vary in their chemical composition due to factors like geographical location, climate, and harvesting methods. Therefore, achieving consistent and standardized extracts with reproducible antifungal properties may be challenging.

Delivery Challenges: While nanotechnology enhances drug delivery, ensuring the targeted delivery of plant-based nanoparticles to specific fungal sites in the body can be a challenge. Proper targeting is essential to maximize the therapeutic benefits and reduce potential side effects.

Development Costs: The development and production of nanoparticles can be costly, which might hinder their widespread availability and affordability, especially in resource-limited settings where fungal infections are prevalent.

Regulatory Hurdles: Introducing new therapies, especially involving nanoparticles, in the medical field requires extensive regulatory approval and safety assessments. Meeting the stringent requirements set by regulatory authorities may delay the clinical translation of plant-based nanoparticles for antifungal treatment.

Drug Resistance: As with any antifungal treatment, the risk of fungal strains developing resistance to plant-based nanoparticles is a concern. Continuous monitoring and research are necessary to assess the potential for the emergence of resistant strains.

Interaction with Other Medications: If plant-based nanoparticles are used alongside conventional antifungal drugs or other medications, potential drug interactions need to be thoroughly studied to avoid adverse effects or reduced efficacy.

Patient Acceptance: Acceptance and adherence to novel therapies can be influenced by factors such as patient awareness, cultural beliefs, and personal preferences. Hence, patient acceptance may affect the successful implementation of nanoparticle-based antifungal therapies.

Nanoparticle Toxicity: Although plant-based nanoparticles are generally considered biocompatible, there could still be concerns regarding their toxicity at the cellular level. The interaction between nanoparticles and human cells or tissues is complex, and potential adverse effects need to be thoroughly investigated.

Scaling Up Production: The large-scale production of plant-based nanoparticles may pose technical challenges and require sophisticated manufacturing processes, impacting their accessibility for widespread use.

Therefore, it is crucial to address these limitations through further research, preclinical, and clinical studies. Moreover, collaboration between different scientific disciplines to unlock the full potential of plant-based nanoparticles as effective antifungal agents needs to be explored.

17.6 Conclusion and Future Prospective

Plant-based nanoparticles offer an innovative approach in the quest for effective antifungal therapies. Fungal infections continue to be a significant global health concern, and the rise of drug-resistant fungal strains underscores the urgency for new and improved treatments. The integration of nanotechnology with plant extracts holds great potential in addressing the limitations of conventional antifungal drugs. Plant-based nanoparticles combine the diverse array of active constituents found in plants with the unique advantage of nanoscale drug delivery. Terpenoids, saponins, alkaloids, and other plant secondary metabolites have demonstrated antifungal properties and can be encapsulated within nanoparticles to enhance their efficacy and targeted delivery. Additionally, green synthesis is also evolved as an eco-friendly and safe method for synthesizing nanocarriers in limiting fungal diseases.

However, despite the promises of this emerging field, there are certain limitations that must be addressed. Limited research, potential nanoparticle toxicity, standardization of plant extracts, delivery challenges, and regulatory hurdles all require careful consideration and extensive investigation. Additionally, the cost of development and production, the risk of drug resistance, potential drug interactions, and patient acceptance are critical factors that need to be studied and managed. To fully unlock the potential of plant-based nanoparticles as antifungal agents, interdisciplinary collaboration between researchers, pharmacologists, clinicians, and regulatory bodies is essential. Rigorous preclinical and clinical studies are needed to thoroughly evaluate the safety, efficacy, and long-term effects of these nanoparticles in treating fungal infections. Standardization and optimization of nanoparticle synthesis processes should be pursued to ensure reproducibility and scalability for widespread implementation. Synthesis of plant-based nanocarriers demand an extension of laboratory work to industrial scale with elucidation of phytochemicals used in the synthesis by different advance tools and techniques. Further, it is essential to characterize the exact mechanisms involved in the entire process of synthesis as well as inhibition of fungal species. Overcoming these challenges offers the prospect of revolutionary advancements in antifungal therapy. Plant-based nanoparticles have the potential to provide a safer, more effective, and affordable alternative to existing antifungal treatments, benefiting millions of people affected by fungal infections worldwide. As research in this field continues to expand, the development of targeted, biocompatible, and sustainable antifungal formulations may become a reality, significantly improving the outcomes for patients with fungal infections.

References

Abd Kelkawi AH, Kajani AA, Bordbar AK (2017) Green synthesis of silver nanoparticles using Mentha pulegium and investigation of their antibacterial, antifungal and anticancer activity. IET Nanobiotechnol. https://doi.org/10.1049/iet-nbt.2016.0103

Abdallah BM, Ali EM (2021) Green synthesis of silver nanoparticles using the lotus lalambensis aqueous leaf extract and their anti-Candidal activity against oral candidiasis. ACS Omega. https://doi.org/10.1021/acsomega.0c06009

Agarwal H, Venkat Kumar S, Rajeshkumar S (2017) A review on green synthesis of zinc oxide nanoparticles—an eco-friendly approach. Resour-Effic Technol. https://doi.org/10.1016/j.reffit.2017.03.002

Ali A, Zafar H, Zia M, ul Haq I, Phull AR, Ali JS, Hussain A (2016) Synthesis, characterization, applications, and challenges of iron oxide nanoparticles. Nanotechnol Sci Appl. https://doi.org/10.2147/NSA.S99986

Ali SJ, Preetha S, Lavanya MJ, Prathap L, Rajeshkumar S (2020) Antifungal activity of selenium nanoparticles extracted from Capparis decidua Fruit against Candida albicans. J Evol Med Dent Sci. https://doi.org/10.14260/jemds/2020/533

Alqarni MH, Foudah AI, Alam A, Salkini MA, Muharram MM, Labrou NE, Kumar P (2022) Development of gum-acacia-stabilized silver nanoparticles gel of rutin against Candida albicans. Gels. https://doi.org/10.3390/gels8080472

Anbuvannan M, Ramesh M, Viruthagiri G, Shanmugam N, Kannadasan N (2015) Synthesis, characterization and photocatalytic activity of ZnO nanoparticles prepared by biological method. Spectrochimica Acta A Mol Biomol Spectrosc. https://doi.org/10.1016/j.saa.2015.01.124

Arastehfar A, Gabaldón T, Garcia-Rubio R, Jenks JD, Hoenigl M, Salzer HJF, Ilkit M, Lass-Flörl C, Perlin DS (2020) Drug-resistant fungi: an emerging challenge threatening our limited antifungal armamentarium. Antibiotics. https://doi.org/10.3390/antibiotics9120877

Arsène MMJ, Viktorovna PI, Alla M, Mariya M, Nikolaevitch SA, Davares AKL, Yurievna ME, Rehailia M, Gabin AA, Alekseevna KA, Vyacheslavovna YN, Vladimirovna ZA, Svetlana O, Milana D (2023) Antifungal activity of silver nanoparticles prepared using Aloe vera extract against Candida albicans. Vet World. https://doi.org/10.14202/vetworld.2023.18-26

Asghari F, Jahanshiri Z, Imani M, Shams-Ghahfarokhi M, Razzaghi-Abyaneh M (2016) Antifungal nanomaterials: synthesis, properties, and applications. In: Nanobiomaterials in antimicrobial therapy: applications of nanobiomaterials. https://doi.org/10.1016/B978-0-323-42864-4.00010-5

Ashraf H, Gul H, Jamil B, Saeed A, Pasha M, Kaleem M, Khan AS (2022) Synthesis, characterization, and evaluation of the antifungal properties of tissue conditioner incorporated with essential oils-loaded chitosan nanoparticles. PLoS One. https://doi.org/10.1371/journal.pone.0273079

Bala N, Saha S, Chakraborty M, Maiti M, Das S, Basu R, Nandy P (2015) Green synthesis of zinc oxide nanoparticles using Hibiscus subdariffa leaf extract: Effect of temperature on synthesis, anti-bacterial activity and anti-diabetic activity. RSC Adv. https://doi.org/10.1039/c4ra12784f

Bali R, Harris AT (2010) Biogenic synthesis of Au nanoparticles using vascular plants. Ind Eng Chem Res. https://doi.org/10.1021/ie101600m

Basavegowda N, Sobczak-Kupiec A, Fenn RI, Dinakar S (2013) Bioreduction of chloroaurate ions using fruit extract Punica granatum (Pomegranate) for synthesis of highly stable gold nanoparticles and assessment of its antibacterial activity. Micro Nano Lett. https://doi.org/10.1049/mnl.2013.0137

Batool F, Iqbal MS, Khan SUD, Khan J, Ahmed B, Qadir MI (2021) Biologically synthesized iron nanoparticles (FeNPs) from Phoenix dactylifera have anti-bacterial activities. Sci Rep. https://doi.org/10.1038/s41598-021-01374-4

Chauhan P, Shivani Rana RR, Bassi P (2022) Green synthesis of metal nanoparticles for development of antifungal topical gel formulation. IJPPR 24

Cruz D, Falé PL, Mourato A, Vaz PD, Luisa Serralheiro M, Lino ARL (2010) Preparation and physicochemical characterization of Ag nanoparticles biosynthesized by Lippia citriodora (Lemon Verbena). Colloids Surf B: Biointerfaces. https://doi.org/10.1016/j.colsurfb.2010.06.025

Das SK, Marsili E (2010) A green chemical approach for the synthesis of gold nanoparticles: characterization and mechanistic aspect. Rev Environ Sci Biotechnol. https://doi.org/10.1007/s11157-010-9188-5

Dashamiri S, Ghaedi M, Naghiha R, Salehi A, Jannesar R (2018) Antibacterial, anti fungal and E. coli DNA cleavage of Euphorbia prostrata and Pelargonium graveolens extract and their combination with novel nanoparticles. Braz J Pharm Sci. https://doi.org/10.1590/s2175-97902018000417724

Dikshit PK, Kumar J, Das AK, Sadhu S, Sharma S, Singh S, Gupta PK, Kim BS (2021) Green synthesis of metallic nanoparticles: applications and limitations. Catalysts. https://doi.org/10.3390/catal11080902

Don MM, San CY, Jeevanandam J (2016) Antimicrobial properties of nanobiomaterials and the mechanism. In: Nanobiomaterials in antimicrobial therapy: applications of nanobiomaterials. https://doi.org/10.1016/B978-0-323-42864-4.00008-7

Dureja H, Murthy NN, Wich PR, Dua K (2022) Drug delivery systems for metabolic disorders. https://doi.org/10.1016/C2021-0-01991-X

Dutta T, Chowdhury SK, Ghosh NN, Chattopadhyay AP, Das M, Mandal V (2022) Green synthesis of antimicrobial silver nanoparticles using fruit extract of Glycosmis pentaphylla and its theoretical explanations. J Mol Struct. https://doi.org/10.1016/j.molstruc.2021.131361

El-Aziz ARMA, Al-Othman MR, Mahmoud MA, Shehata SM, Abdelazim NS (2018) Chitosan nanoparticles as a carrier for Mentha longifolia extract: Synthesis, characterization and antifungal activity. Curr Sci. https://doi.org/10.18520/cs/v114/i10/2116-2122

Faisal S, Jan H, Shah SA, Shah S, Khan A, Akbar MT, Rizwan M, Jan F, Wajidullah, Akhtar N, Khattak A, Syed S (2021) Green synthesis of zinc oxide (ZnO) nanoparticles using aqueous fruit extracts of Myristica fragrans: their characterizations and biological and environmental applications. ACS Omega. https://doi.org/10.1021/acsomega.1c00310

Fardsadegh B, Jafarizadeh-Malmiri H (2019) Aloe vera leaf extract mediated green synthesis of selenium nanoparticles and assessment of their in vitro antimicrobial activity against spoilage fungi and pathogenic bacteria strains. Green Process Synth. https://doi.org/10.1515/gps-2019-0007

Fisher MC, Henk DA, Briggs CJ, Brownstein JS, Madoff LC, McCraw SL, Gurr SJ (2012) Emerging fungal threats to animal, plant and ecosystem health. Nature. https://doi.org/10.1038/nature10947

Gardea-Torresdey JL, Parsons JG, Gomez E, Peralta-Videa J, Troiani HE, Santiago P, Yacaman MJ (2002) Formation and growth of Au nanoparticles inside Live Alfalfa Plants. Nano Lett. https://doi.org/10.1021/nl015673+

Ghosh P, Han G, De M, Kim CK, Rotello VM (2008) Gold nanoparticles in delivery applications. Adv Drug Deliv Rev. https://doi.org/10.1016/j.addr.2008.03.016

Gupta V, Mohapatra S, Mishra H, Farooq U, Kumar K, Ansari MJ, Aldawsari MF, Alalaiwe AS, Mirza MA, Iqbal Z (2022) Nanotechnology in cosmetics and cosmeceuticals—a review of latest advancements. Gels. https://doi.org/10.3390/gels8030173

Hanisha R, Udayakumar R, Selvayogesh S, Keerthivasan P, Gnanasekaran R (n.d.) Anti fungal activity of green synthesized copper nanoparticles using plant extract of Bryophyllum Pinnatum (Lam.) and Polyalthia Longifolia (Sonn.). Biosci Biotechnol Res Asia

He L, Liu Y, Mustapha A, Lin M (2011) Antifungal activity of zinc oxide nanoparticles against Botrytis cinerea and Penicillium expansum. Microbiol Res. https://doi.org/10.1016/j.micres.2010.03.003

Hemlata, Meena PR, Singh AP, Tejavath KK (2020) Biosynthesis of silver nanoparticles using Cucumis prophetarum aqueous leaf extract and their antibacterial and antiproliferative activity against cancer cell lines. ACS Omega. https://doi.org/10.1021/acsomega.0c00155

Hoag GE, Collins JB, Holcomb JL, Hoag JR, Nadagouda MN, Varma RS (2009) Degradation of bromothymol blue by "greener" nano-scale zero-valent iron synthesized using tea polyphenols. J Mater Chem. https://doi.org/10.1039/b909148c

Huang X, Wu H, Liao X, Shi B (2010) One-step, size-controlled synthesis of gold nanoparticles at room temperature using plant tannin. Green Chem. https://doi.org/10.1039/b918176h

Hussain MA, Ahmed D, Anwar A, Perveen S, Ahmed S, Anis I, Shah MR, Khan NA (2019) Combination therapy of clinically approved antifungal drugs is enhanced by conjugation with silver nanoparticles. Int Microbiol. https://doi.org/10.1007/s10123-018-00043-3

Ibarra-Laclette E, Blaz J, Pérez-Torres CA, Villafán E, Lamelas A, Rosas-Saito G, Ibarra-Juárez LA, García-ávila C d J, Martínez-Enriquez AI, Pariona N (2022) Antifungal effect of copper nanoparticles against Fusarium kuroshium, an obligate symbiont of Euwallacea kuroshio ambrosia beetle. J Fungi. https://doi.org/10.3390/jof8040347

Izadi M, Moosawi Jorf SA, Nikkhah M, Moradi S (2021) Antifungal activity of hydrocolloid nano encapsulated Carum copticum essential oil and Peganum harmala extract on the pathogenic fungi Alternaria alternata. Physiol Mol Plant Pathol. https://doi.org/10.1016/j.pmpp.2021.101714

Jamdagni P, Khatri P, Rana JS (2018) Green synthesis of zinc oxide nanoparticles using flower extract of Nyctanthes arbor-tristis and their antifungal activity. J King Saud Univ Sci. https://doi.org/10.1016/j.jksus.2016.10.002

Jayaseelan C, Rahuman AA, Kirthi AV, Marimuthu S, Santhoshkumar T, Bagavan A, Gaurav K, Karthik L, Rao KVB (2012) Novel microbial route to synthesize ZnO nanoparticles using Aeromonas hydrophila and their activity against pathogenic bacteria and fungi. Spectrochimica Acta A Mol Biomol Spect. https://doi.org/10.1016/j.saa.2012.01.006

Jenish A, Ranjani S, Hemalatha S (2022) Moringa oleifera nanoparticles demonstrate antifungal activity against plant pathogenic fungi. Appl Biochem Biotechnol. https://doi.org/10.1007/s12010-022-04007-2

Jia D, Sun W (2021) Silver nanoparticles offer a synergistic effect with fluconazole against fluconazole-resistant Candida albicans by abrogating drug efflux pumps and increasing endogenous ROS. Infect Genet Evol. https://doi.org/10.1016/j.meegid.2021.104937

Kalagatur NK, Nirmal Ghosh OS, Sundararaj N, Mudili V (2018) Antifungal activity of chitosan nanoparticles encapsulated with Cymbopogon martinii essential oil on plant pathogenic fungi Fusarium graminearum. Front Pharmacol. https://doi.org/10.3389/fphar.2018.00610

Kashid MN, Kiwi-Minsker L, Gehr R, Chen D, Moreau M, Kikutani Y et al (2011) Peracetic acid (CAS No. 79-21-0) and its equilibrium solutions. Chem Eng Sci. ISSN-0733-6339-40

Katta C, Shaikh AS, Bhale N, VGS SJ, Kaki VR, Dikundwar AG, Singh PK, Shukla R, Mishra K, Madan J (2023) Naringenin-capped silver nanoparticles amalgamated gel for the treatment of cutaneous candidiasis. AAPS PharmSciTech

Kumar Mishra K, Deep Kaur C, Kumar Sahu A, Panik R, Kashyap P, Prasad Mishra S, Dutta S (2020) Medicinal plants having antifungal properties. In: Medicinal plants - use in prevention and treatment of diseases. https://doi.org/10.5772/intechopen.90674

Lazcano-Ramírez HG, Garza-García JJO, Hernández-Díaz JA, León-Morales JM, Macías-Sandoval AS, García-Morales S (2023) Antifungal activity of selenium nanoparticles obtained by plant-mediated synthesis. Antibiotics. https://doi.org/10.3390/antibiotics12010115

Lotfali E, Toreyhi H, Sharabiani KM, Fattahi A, Soheili A, Ghasemi R, Keymaram M, Rezaee Y, Iranpanah S (2020) Comparison of antifungal properties of gold, silver, and selenium nanoparticles against amphotericin b-resistant candida glabrata clinical isolates. Avicenna J Med Biotechnol. https://doi.org/10.18502/ajmb.v13i1.4578

Makarov VV, Love AJ, Sinitsyna OV, Makarova SS, Yaminsky IV, Taliansky ME, Kalinina NO (2014) "Green" nanotechnologies: synthesis of metal nanoparticles using plants. Acta Nat. https://doi.org/10.32607/20758251-2014-6-1-35-44

Masum MI, Siddiqa MM, Ali KA, Zhang Y, Abdallah Y, Ibrahim E, Qiu W, Yan C, Li B (2019) Biogenic synthesis of silver nanoparticles using phyllanthus emblicafruit extract and its inhibitory action against the pathogen acidovorax oryzaestrain RS-2 of rice bacterial brown stripe. Front Microbiol. https://doi.org/10.3389/fmicb.2019.00820

Mohapatra S, Iqubal A, Ansari MJ, Jan B, Zahiruddin S, Mirza MA, Ahmad S, Iqbal Z (2022) Benefits of black cohosh (Cimicifuga racemosa) for women health: an up-close and in-depth review. Pharmaceuticals. https://doi.org/10.3390/ph15030278

Moron G (2020) Antifungal activity of loaded chitosan nanoparticles with S-nitrosomercaptosuccinic acid against Candida sp. J Nanosci Curr Res

Narayanan KB, Park HH (2014) Antifungal activity of silver nanoparticles synthesized using turnip leaf extract (Brassica rapa L.) against wood rotting pathogens. Eur J Plant Pathol. https://doi.org/10.1007/s10658-014-0399-4

Nasseri M, Golmohammadzadeh S, Arouiee H, Jaafari MR, Neamati H (2016) Antifungal activity of zataria multiflora essential oil-loaded solid lipid nanoparticles in-vitro condition. Iran J Basic Med Sci. https://doi.org/10.22038/ijbms.2016.7824

Parveen R, Mohapatra S, Ahmad S, Husain SA (2022) Amalgamation of nanotechnology for delivery of bioactive constituents in solid tumors. Curr Drug Deliv. https://doi.org/10.2174/1567201819666220425093102

Ponarulselvam S, Panneerselvam C, Murugan K, Aarthi N, Kalimuthu K, Thangamani S (2012) Synthesis of silver nanoparticles using leaves of Catharanthus roseus Linn. G. Don and their antiplasmodial activities. Asian Pac J Trop Biomed. https://doi.org/10.1016/S2221-1691(12)60100-2

Prasad C, Gangadhara S, Venkateswarlu P (2016) Bio-inspired green synthesis of Fe3O4 magnetic nanoparticles using watermelon rinds and their catalytic activity. Appl Nanosci (Switzerland). https://doi.org/10.1007/s13204-015-0485-8

Rabiee N, Bagherzadeh M, Kiani M, Ghadiri AM (2020) Rosmarinus officinalis directed palladium nanoparticle synthesis: investigation of potential anti-bacterial, anti-fungal and Mizoroki-Heck catalytic activities. Adv Powder Technol. https://doi.org/10.1016/j.apt.2020.01.024

Rajendran SP, Sengodan K (2017) Synthesis and characterization of zinc oxide and iron oxide nanoparticles using Sesbania grandiflora leaf extract as reducing agent. J Nanosci. https://doi.org/10.1155/2017/8348507

Rasool MA, Gautam GK, Panda DP, Sahu DC (2022) Preparation, characterisation and antifungal activity of gold nanoparticles prepared with fresh extract of Aconitum heterophyllum leaves. Res J Pharma Technol. https://doi.org/10.52711/0974-360X.2022.00544

Rodriguez-Tudela JL, Alcazar-Fuoli L, Cuesta I, Alastruey-Izquierdo A, Monzon A, Mellado E, Cuenca-Estrella M (2008) Clinical relevance of resistance to antifungals. Int J Antimicrob Agents. https://doi.org/10.1016/S0924-8579(08)70010-4

Saif S, Tahir A, Chen Y (2016) Green synthesis of iron nanoparticles and their environmental applications and implications. Nano. https://doi.org/10.3390/nano6110209

Seddighi NS, Salari S, Izadi AR (2017) Evaluation of antifungal effect of iron-oxide nanoparticles against different Candida species. IET Nanobiotechnol. https://doi.org/10.1049/iet-nbt.2017.0025

Shankar SS, Ahmad A, Pasricha R, Sastry M (2003) Bioreduction of chloroaurate ions by geranium leaves and its endophytic fungus yields gold nanoparticles of different shapes. J Mater Chem. https://doi.org/10.1039/b303808b

Sharma RK, Gulati S, Mehta S (2012) Preparation of gold nanoparticles using tea: a green chemistry experiment. J Chem Educ. https://doi.org/10.1021/ed2002175

Suman TY, Radhika Rajasree SR, Ramkumar R, Rajthilak C, Perumal P (2014) The green synthesis of gold nanoparticles using an aqueous root extract of Morinda citrifolia L. Spectrochimica Acta A Mol Biomol Spect. https://doi.org/10.1016/j.saa.2013.08.066

Tabarestani MS (2022a) Nanoencapsulation of thyme extract with antifungal and antioxidant effects. J Nanostruct. https://doi.org/10.22052/JNS.2022.01.002

Tabarestani MS (2022b) Evaluation of antifungal effect of biodegradable nano encapsulated extract of Trichoderma harzianum. Irarian Plant Prot Res

Téllez-de-Jesús DG, Flores-Lopez NS, Cervantes-Chávez JA, Hernández-Martínez AR (2021) Antibacterial and antifungal activities of encapsulated Au and Ag nanoparticles synthesized using Argemone mexicana L extract, against antibiotic-resistant bacteria and Candida albicans. Surf Interfaces. https://doi.org/10.1016/j.surfin.2021.101456

Thipe VC, Njobeh PB, Mhlanga SD (2015) Optimization of commercial antibiotic agents using gold nanoparticles against toxigenic Aspergillus spp. In: Materials today: proceedings. https://doi.org/10.1016/j.matpr.2015.08.044

Ukidave VV, Ingale LT (2022) Green synthesis of zinc oxide nanoparticles from Coriandrum sativum and their use as fertilizer on Bengal gram, Turkish gram, and green gram plant growth. Int J Agron. https://doi.org/10.1155/2022/8310038

Umamaheswari A, Prabu SL, John SA, Puratchikody A (2021) Green synthesis of zinc oxide nanoparticles using leaf extracts of Raphanus sativus var. Longipinnatus and evaluation of their anticancer property in A549 cell lines. Biotechnol Rep. https://doi.org/10.1016/j.btre.2021.e00595

Vankar PS, Bajpai D (2010) Preparation of gold nanoparticles from Mirabilis Jalapa flowers. Indian J Biochem Biophys

Wani IA, Ahmad T (2013) Size and shape dependant antifungal activity of gold nanoparticles: a case study of Candida. Colloids Surf B Biointerfaces. https://doi.org/10.1016/j.colsurfb.2012.06.005

Xin Lee K, Shameli K, Miyake M, Kuwano N, Bt Ahmad Khairudin NB, Bt Mohamad SE, Yew YP (2016) Green synthesis of gold nanoparticles using aqueous extract of Garcinia mangostana fruit peels. J Nanomater. https://doi.org/10.1155/2016/8489094

Xu J, Sun J, Wang Y, Sheng J, Wang F, Sun M (2014) Application of iron magnetic nanoparticles in protein immobilization. Molecules. https://doi.org/10.3390/molecules190811465

Part V

Plant-Based Chemical Derivatives as Antifungals

Antifungal Efficacy of Natural Product-Based Chemical Derivatives

18

Hari Madhav and Nasimul Hoda

18.1 Introduction

The emergence of resistance to current mainline antimicrobial drugs that target pathogenic parasites, viruses, bacteria, and fungi underlines the urgent need for the development of new antimicrobial agents (Aldholmi et al. 2019). When it comes to fungus, invasive and chronic fungal infections can be extremely expensive, especially for hospitalized and immunocompromised individuals. Although an array of fungal species may result in infections, over 90% of fungal-related fatalities are caused by invasive pathogens from the genera *Aspergillus*, *Cryptococcus*, *Candida*, and *Pneumocystis* (Heard et al. 2021). The excessive use of antifungal drugs developed drug resistance within the pathogenic fungi and made them nonsusceptible to traditional drugs (Arif et al. 2009).

Since the dawn of civilization, herbs have been utilized as remedies to facilitate human health (Bhuiyan et al. 2020). They are excellent when searching for novel treatments because they offer an abundance of therapeutic molecules with a wide range of chemical and functional diversity. In recent years, studies on naturally occurring compounds with antifungal properties have significantly increased (Zhang et al. 2023). However, even though natural products contain antifungal characteristics that are frequently preferable to those of commercial pharmaceuticals, it is still very difficult to employ them therapeutically (Heard et al. 2021). A potential approach to address this issue is the structural diversification of natural compounds that have a high sensitivity to fungal strains. Thus far, natural product identification has yielded several significantly efficacious antifungal agents, including polyenes, echinocandins, and pyrimidine analogues, as well as optimized molecular building

H. Madhav · N. Hoda (✉)
Drug Design and Synthesis Laboratory, Department of Chemistry, Jamia Millia Islamia, New Delhi, India
e-mail: harimadhavgautam@gmail.com; nhoda@jmi.ac.in

N. Manzoor (ed.), *Advances in Antifungal Drug Development*,
https://doi.org/10.1007/978-981-97-5165-5_18

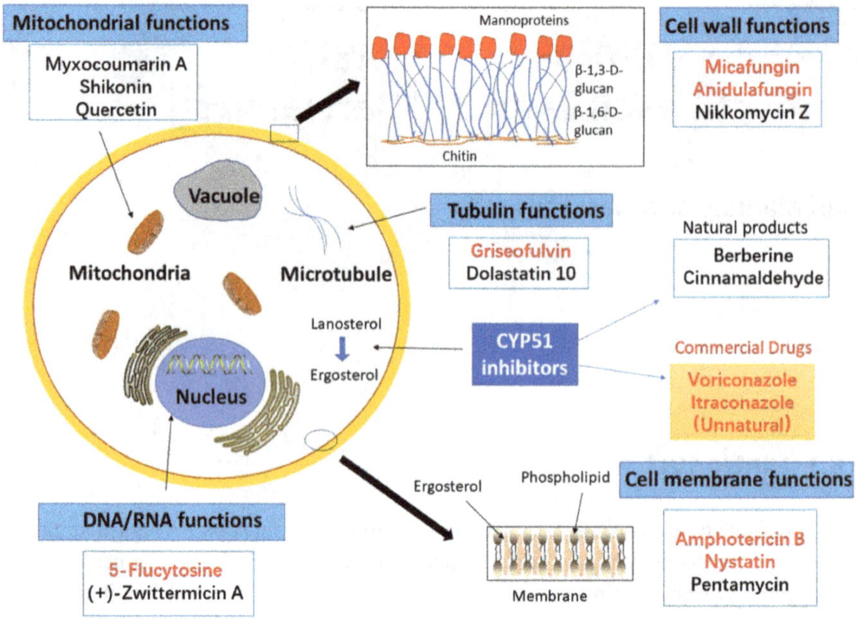

Fig. 18.1 Some fungal cell targets and their related commercially accessible pharmaceuticals (Adapted from (Zhang et al. 2023) licensed under CC BY-NC-ND 4.0)

blocks with antifungal activity (Roemer and Krysan 2014). Many natural compounds and their derived analogues have demonstrated significant antifungal effectiveness against a variety of fungal cell targets such as cell walls, cell membranes, etc. (Garcia-Rubio et al. 2020) as described in Fig. 18.1.

Fungal drug targets such as glucans, chitin, mannoproteins, etc. are cell wall polymers, enzymes and pathways like topoisomerases, nucleases, elongation factors, myristoylation, ergosterol biosynthesis (cell membrane), DNA, protein kinases and protein phosphatases that are involved in signal transduction pathways are some examples (Wills et al. 2000). There is a class of antifungals, viz., azoles that potentially inhibit lanosterol-14α-demethylase to prevent ergosterol biosynthesis and destroy cell membrane integrity (Bardal et al. 2011; Onyewu et al. 2003). The activity of 1,3-*β*-D-glucan synthase is significantly inhibited by echinocandins which results in the arrest of cell wall synthesis and consequently cause cell wall rapture (Farhadi et al. 2020; Lima et al. 2019). Furthermore, the biosynthetic pathway responsible for the molecular level regular expression of DNA or RNA is inhibited by pyrimidine analogues and marked as potential antifungal agents (Sanglard 2002). Hence, the present chapter is focused on detailed insight into the significance of natural products and natural product-derived molecules for the development of antifungal agents. The detail about different natural products and their derivatives as antifungals approved by the US Food and Drug Administration (FDA) or in the pipeline is discussed in the subsequent sections.

18.1.1 Clinically Approved Antifungal Natural Products

Antifungal drug development faces significant challenges due to the cellular and metabolic similarities between mammalian and fungal cells. These similarities also limit the availability of potential drug targets. At present, antifungal drugs are based on three major molecular classes such as azoles, polyenes, and echinocandins which too are based on natural products. Echinocandins, often referred to as cyclic lipopeptides with an N-linked acyl lipid side chain, are naturally occurring substances formed from fungi that function as noncompetitive inhibitors of 1,3-β-D-glucan synthase, hence influencing the formation of fungal cell walls (Grover 2010). This cellular component is lacking in human cells; hence, they show a relatively low toxicity profile. The semisynthetic derivatives of echinocandin B, pneumocandin B_0, and acylated cyclic hexapeptide FR901379 such as anidulafungin (**1**), caspofungin (**2**), and micafungin (**3**), respectively, are approved by the FDA to treat systemic fungal infections (Fig. 18.2) (Hashimoto 2009; Heard et al. 2021; Letscher-Bru 2003; Onishi et al. 2000). Recently, echinocandin rezafungin (**4**) and triterpenoid ibrexafungerp (**5**) were approved by the FDA for the treatment of fungal infections (Phillips et al. 2023; Syed 2023).

Another class of natural product's polyenes falls under the category of bacteria-derived macrocyclic lactones which include nystatin (**6**), amphotericin B (**7**), and natamycin (**8**) (Fig. 18.3) (Serhan et al. 2014).

Pentamycin (**9**) is another polyene natural product that got approval in Switzerland for the treatment of fungal infections (Zhou et al. 2019). Polyenes showed oscillating conjugated bonds within their macrolide cyclic structure. When polyenes bind to sterols in cell membranes—such as ergosterol in fungal cells—channels are formed in the membrane, causing the membrane to leak (Kristanc et al. 2019). Griseofulvin (**10**) is a polyketide natural product that targets β-tubulin and inhibits microtubule aggregation (Fig. 18.4) (Mazu et al. 2016).

Further, plant-derived unsaturated fatty acid undecylenic acid (**11**) inhibits morphogenesis to show antifungal efficacy and also possibly inhibits fatty acid biosynthesis (Heard et al. 2021). Similarly, some other antifungal drugs such as allylamines like terbinafine (**12**) and naftifine (**13**), benzylamine like butenafine (**14**), and thiocarbamate tetralin class liranaftate (**15**) also got approval which was developed by mimicking natural products (Heard et al. 2021). These drugs act through the inhibition of squalene epoxidase which resulted in the inhibition of ergosterol biosynthesis and accumulation of squalene which made fungal cell walls weaker (Ryder 1992).

18.1.2 Antifungal Natural Products in Clinical Trials

Through the past few decades, treatment with antifungal drugs has undergone a major transformation. Antifungal chemotherapy has changed as a result of the discovery of azoles that are effective systemically and, lastly, the entry of echinocandins. For more severe invasive fungal infections, the prior preference for amphotericin B formulations has given way to more efficient and typically less toxic drugs. For

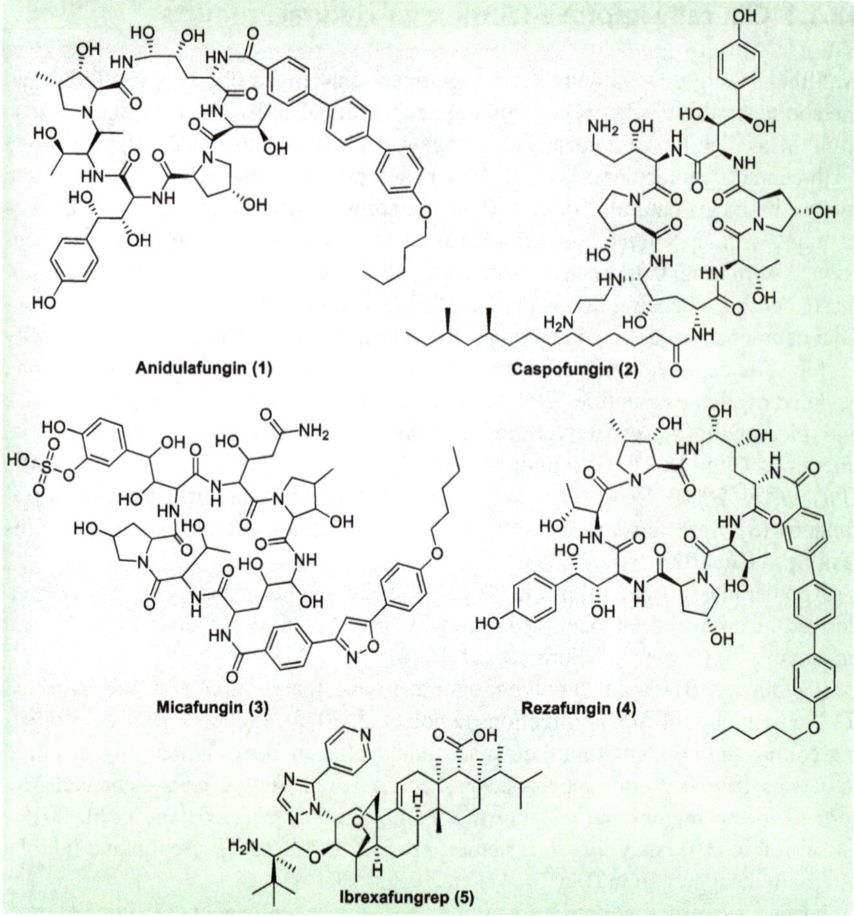

Fig. 18.2 Natural product-based antifungal drugs of the Echinocandin family

instance, the major types of invasive candidiasis are now treated with echinocandins alternatively as the first course of action, and the mold-active azoles have almost entirely supplanted amphotericin B as the treatment commonly used for invasive aspergillosis (McCarty and Pappas 2021). Drug toxicity, drug–drug interactions, and failure of oral bioavailability frequently make it difficult to use invasive antifungal medicines (IFDs). Rising of antifungal resistance and new infections caused by new pathogens, a number of those that have become resistant to existing antifungal drugs, novel therapeutic alternatives are therefore required.

Clinical studies to cure invasive fungal infections have attracted a lot of interest and challenge during the past few decades. Some of the most cited clinical studies involving anti-infectives have compared diverse therapies for diseases like invasive *aspergillosis*, *candidemia*, endemic *mycoses*, and *cryptococcal meningitis* (Pappas 2014). Currently, natural product-based antifungal drug development is focused on

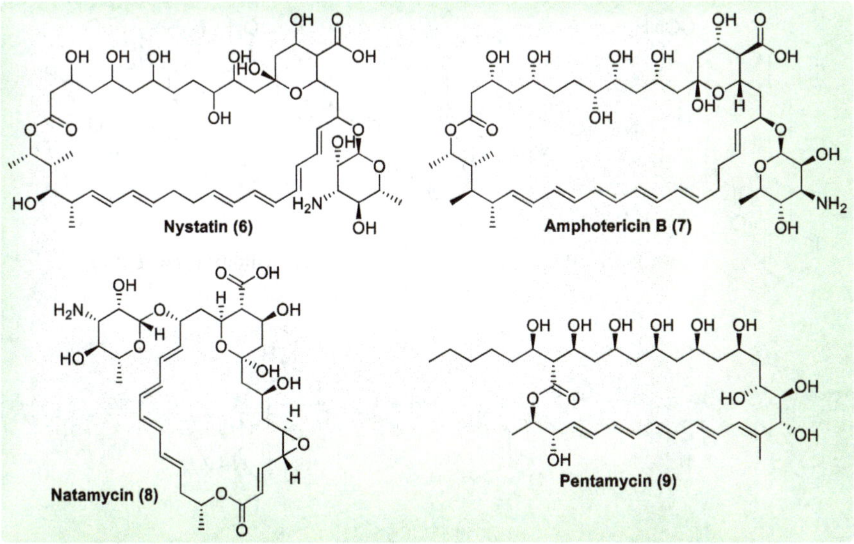

Fig. 18.3 Natural product-based antifungal drugs of the Polyene family

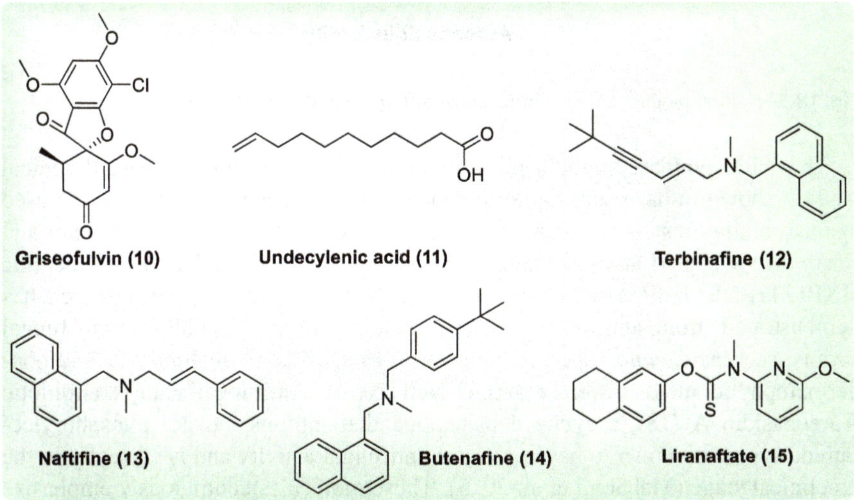

Fig. 18.4 Natural product-based antifungal drugs of other different families

the identification of new antifungals selective toward fungal targets over human eukaryotic targets (Chen et al. 2014; Perfect 2017). Multiple optimistic antifungals with distinct ways of therapeutic action are currently in the clinical development stage (Allen et al. 2015; Rauseo et al. 2020). Natural antifungals in clinical trials include cyclic hexapeptide VL 2397 (another name: ASP2397) (**16**) which is produced by the fungal strain MF-347833 (Fig. 18.5) (Nakamura et al. 2017).

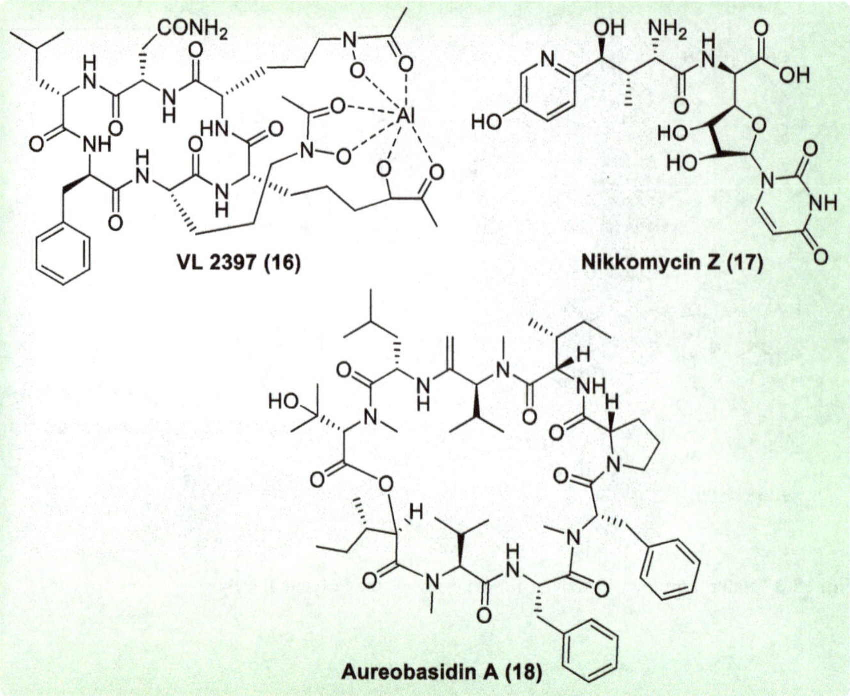

Fig. 18.5 Natural product-based antifungals of different families in clinical studies

The chitin synthase inhibitor Nikkomycin Z (**17**) is currently in phase II clinical studies, shown to have substantial therapeutic advantages in mammals when used against ubiquitous, dimorphic fungi such as *Coccidioides*, *Histoplasma*, and *Blastomyces* spp. (Larwood 2020; Poester et al. 2023). Similar to this, peptide HXP124 (PDB: 1MR4), which is presently in the first phase of clinical research, has demonstrated strong antifungal efficacy over a variety of potentially serious fungal strains, such as *Candida* species, *Cryptococcus* species, dermatophytes, and non-dermatophytic molds (Mercer and O'Neil 2020). Antifungal natural antibiotic Aureobasidin A (**18**), a cyclic depsipeptide that inhibits inositol phosphorylceramide synthase shown to have potential antifungal activity and is currently in the preclinical stage (Maharani et al. 2015). These molecules continuously emphasize the significance of natural products in antifungal drug discovery.

18.2 Natural Product Derivatives as Antifungals

Natural products are significant sources of therapeutically important chemical scaffolds and medicines. Simplifying natural product structures without sacrificing their biological function is difficult, though. The structural and stereochemical complexity of NPs is a key characteristic that makes synthesis challenging and restricts future research into the chemical space surrounding NPs. Moreover, the complex structures of NPs make it more challenging to achieve the proper drug-likeness, such as favorable physicochemical properties and optimal pharmacokinetic (PK)/pharmacodynamic (PD) profiles. The structural modification of NPs and the elucidation of the structural requirements for biological activity represent a possible strategy to tackle these problems. In the consecutive section, we have discussed the diversification of different natural products for the development of potential antifungal agents.

18.2.1 Polyene Derivatives

Amphotericin B (AmB), a polyene macrolide, is the most commonly used therapy for systemic fungal infection due to its wide range of fungicidal activity. The association of polyene macrolides to ergosterol in the fungal cell membrane is thought to be the mechanism of action of these compounds, since it causes pores to form and ions to flow out, ultimately leading to the death of the fungal cell. Therefore, it encouraged medicinal chemists to design and develop new polyene derivatives as new antifungal agents. In this context, Preobrazhenskaya et al. (Preobrazhenskaya et al. 2010) reported semisynthetic derivatives of the heptene nystatin analogue 28, 29-didehydronystatin A_1 (S44HP) to afford new antifungals. The authors of the study tested newly developed analogues in vitro against yeasts and fungi followed by the most active compound in vivo. According to the investigation, the analogue **19** (Fig. 18.6) had MIC values of 0.5, 0.5, 1, and 1 μg/mL for *Candida albicans* ATCC 14053, *Cryptococcus humicolus* ATCC 9949, *Aspergillus niger* ATCC 16404, and *Fusarium oxysporum* VKM F-140, respectively.

The outcomes demonstrated that in the mouse model of candidosis sepsis on a leucopenic history, the newly identified molecules exhibited superior chemotherapeutic action compared to AmB. Further, this group reported the development and detailed SAR of new analogues of polyene antibiotics of the AmB group (Tevyashova et al. 2013). The results of the study highlighted that the antifungal activity of these polyene antibiotics was significantly influenced by diversifications at the C-7 to C-10 polyol region. The research effort produced a pair of the most effective molecules, **20** and **21**, which had superior toxicological qualities and strong antifungal activity. These compounds appeared competitive to AmB in both in vitro and in vivo studies.

Kim et al. described the biosynthesis of polyene **22** derived from nystatin-like *Pseudonocardia* polyene (NPP) A1, which is a distinct tetraene antifungal macrolide that contains disaccharides and is generated by *Pseudonocardia autotrophica* (Kim et al. 2018). The NPP A1 biosynthetic pathway gene's particular polyketide enoyl reductase (ER) domain was altered to produce polyene **22**, which represented the heptene version of NPP A1. The research focused on the in vivo cytotoxicity and bioavailability of the drug in addition to its effectiveness against *C. albicans*. The findings demonstrated that polyene **22**'s antifungal activity is on par with that of AmB, the most effective heptane antifungal antibiotic. Similarly, Sheng et al.

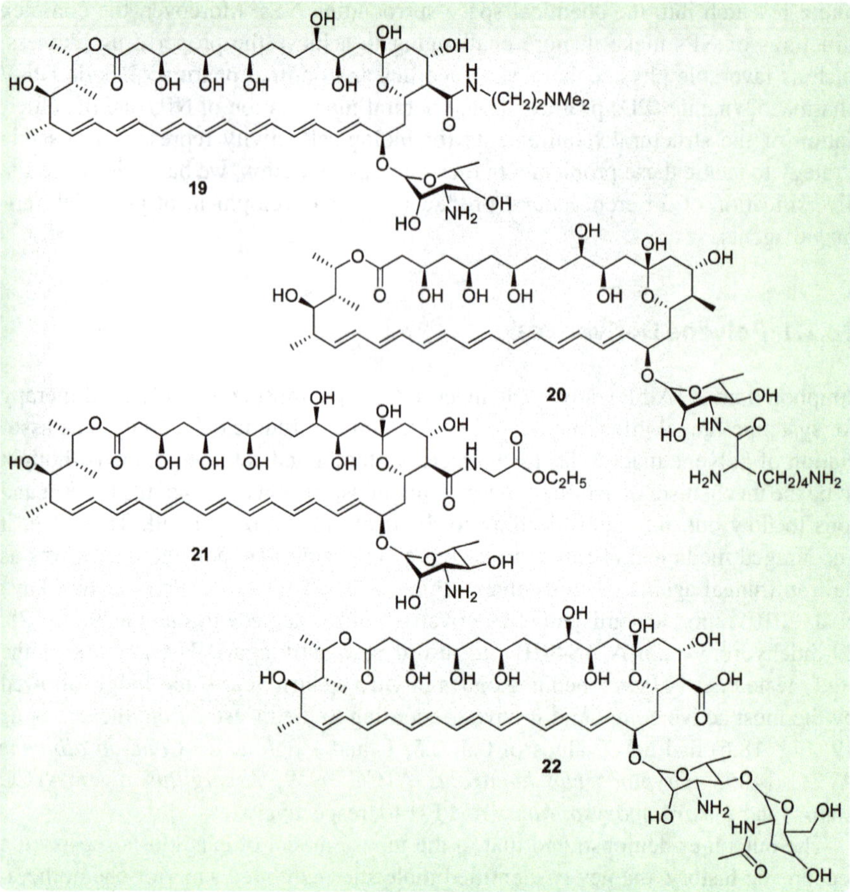

Fig. 18.6 Chemical structures of natural product-derived antifungals **19–22**

developed derivatives of polyene tetramycin B produced by *Streptomyces hygrospinosus* var. *beijingensis* CGMCC 4.1123 to improve pharmacological properties based on pathway engineering (Sheng et al. 2020). The authors performed genetic engineering to achieve 12-decarboxy-12-methyl tetramycin B (**23**, Fig. 18.7) as the derivative of tetramycin B.

The results emphasized that derivative **23** showed improved antifungal potency against *Saccharomyces cerevisiae* and *Rhodotorula glutinis* as compared to tetramycin B. Likewise, Tevyashova et al. reported the semisynthetic amide derivatives of polyene antibiotic natamycin (Tevyashova et al. 2023b). Investigations into the antifungal properties of amide analogues were conducted against filamentous fungi and a group of clinical strains of *C. auris*. The results demonstrated that semisynthetic natamycines were more efficient against fungal strains as compared to the initial antibiotic. Amide **24** displayed an MIC value of 2 μg/mL while the parent natamycin displayed an MIC value of 8 μg/mL. Additionally, **25**'s large lipophilic

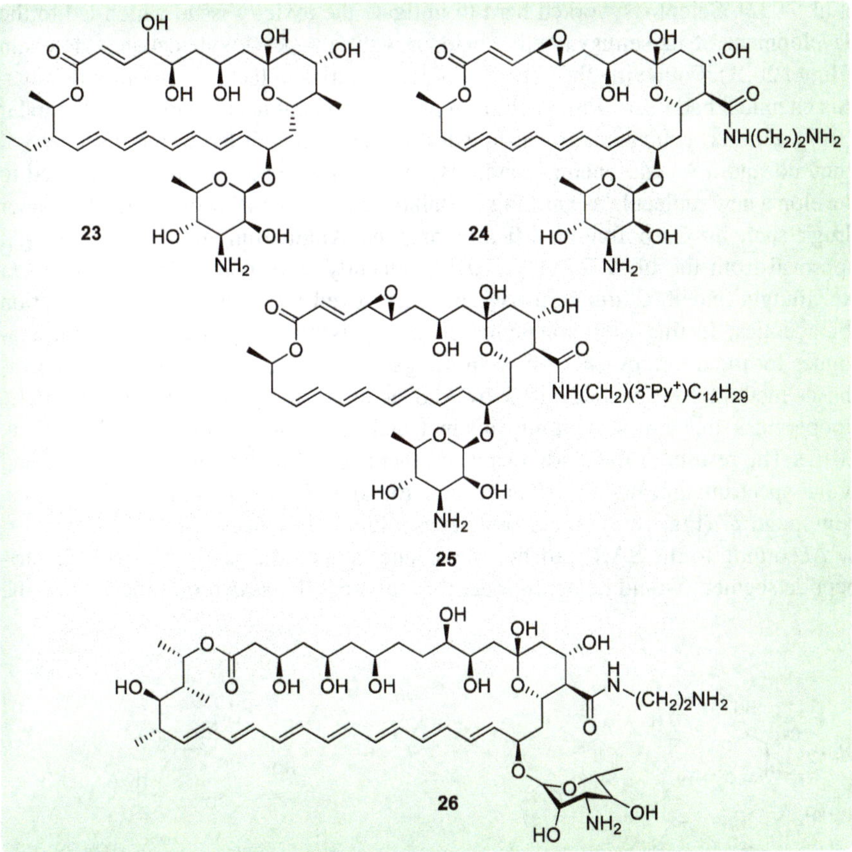

Fig. 18.7 Chemical structures of natural product-derived antifungals **23–26**

side chains demonstrated superior antifungal potential and particularly high EI in vitro, although it was more vulnerable to HPF. The in vivo results highlighted that amide **24** was efficient on a mouse candidemia model with a larger LD_{50}/ED_{50} ratio in comparison to AmB. Later, this group explored polyenes AmB and Nystatin A_1 to develop their new amide derivatives and performed a comparative antifungal study in vitro (Tevyashova et al. 2023a). The study achieved a potent compound **26** with an MIC value of 0.5 μg/mL against *C. parapsilosis* ATCC 22019. These studies opened the window of possibilities for the development of polyene derivatives-based new antifungal agents.

18.2.2 Echinocandin Derivatives

Echinocandin B was the first antifungal drug discovered in 1974 which showed good antifungal properties, but it also showed a strong hemolytic effect (Szymański et al. 2022). Scientists worked hard to mitigate the toxicity issue which led to the development of its semisynthetic analogue with a 4-octyloxybenzoate side chain (Hüttel 2021). Following this, because of its toxicity and lack of solubility in water, this chemical had been eliminated from phase II clinical trials (Patil and Majumdar 2017). Further, echinocandin antifungal drug development led to the discovery of pneumocandin A_0 and pneumocandin B_0, of which pneumocandin B_0 was used to develop a new molecule as antifungal (Balkovec et al. 2014). Moreover, a few other drugs such as Caspofungin, Micafungin, and Anidulafungin of this class got approval from the FDA (Li et al. 2018). Recently, two more echinocandins, i.e., Rezafungin and Ibrexafungerp have been approved by FDA for fungal infection therapeutics. In this view, medicinal chemists also developed echinocandin analogues for the development of new antifungal agents. Moreover, the complete synthesis methodology was devised by Yao et al. to synthesize antifungal cyclic lipopeptides that are structurally distinct and resemble caspofungin (Yao et al. 2012). The results of the study displayed that cyclic lipopeptides were potent and broad-spectrum antifungals. When it came to *Aspergillus fumigatus* or *C. albicans*, compound **27** (Fig. 18.8) outperformed caspofungin in vitro.

According to the SAR statistics, the cyclic lipopeptide scaffold's "left" lipotripeptide segment would be appropriate for a hydrophilic structural motif, while the

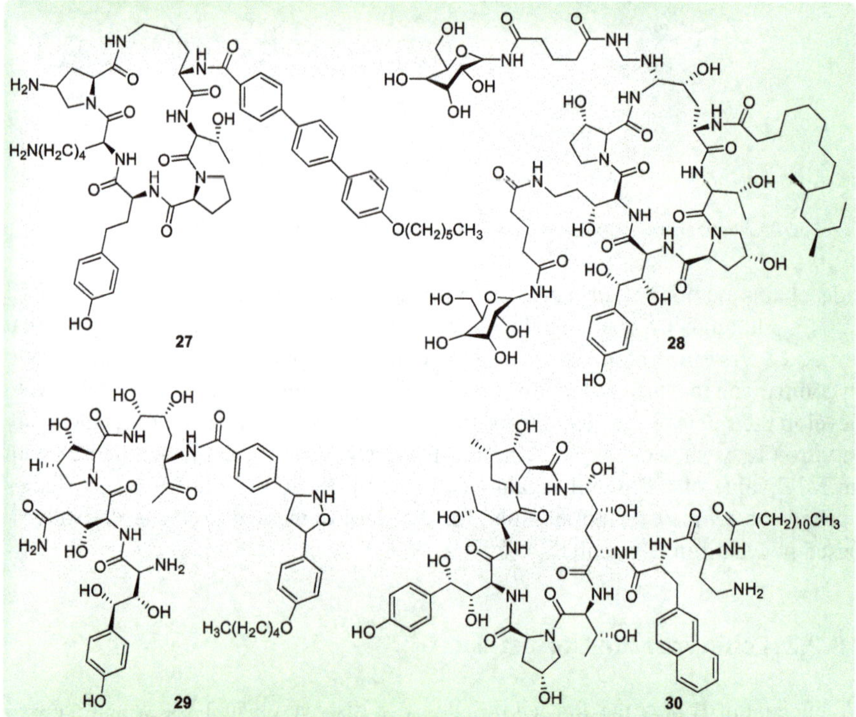

Fig. 18.8 Chemical structures of natural product-derived antifungals **27–30**

"right" lipotripeptide region had been chosen as a hydrophobic core. Guo et al synthesized glycosylated derivatives of the cyclic peptide fungicide Caspofungin as new antifungal agents (Guo et al. 2012). The study delivered β-D-glucopyranosyl derivative **28** as a potent and broad-spectrum antifungal. Later, Singh et al. (Singh et al. 2013) synthesized pentyloxyl-diphenylisoxazoloyl derivatives of pneumocandins and echinocandins and evaluated their antifungal activity. The results highlighted that pneumocandin derivative **29** was the most potent among others. The study also demonstrated the *β*-glucan synthase inhibition potential of **29**, demonstrating that **29** inhibits the caspofungin-resistant β-glucan synthase enzymes with IC_{50} values comparable to those found for the wild-type enzyme. Moreover, to improve the solubility and minimize the toxicity of echinocandin B, Zhu et al. synthesized N-acylated analogues as new antifungal agents (Zhu et al. 2020). The investigations demonstrated that analogue **30** was obtained with comparable antifungal efficacy and no hemolytic effect. The results concluded that compound **30** showed comparable fungicidal effectiveness, much higher solubility, and lower toxicity with anidulafungin and has potential for further development. Recently, anisofungin and rezafungin were structurally manipulated by Logviniuk et al. through the precise elimination of the benzylic hydroxy group of the nonproteinogenic amino acid 3S,4S-dihydroxy-L-homotyrosine (Logviniuk et al. 2022). The effect of the removal of selective benzylic alcohol on fungal activity is given by schematic representation, as shown in Fig. 18.9. According to the authors, this alteration allowed them to combat a wide range of echinocandin-resistant *Candida* strains once again effectively.

According to the study, the dehydroxylated echinocandins bound to the amino acids of the Fks proteins, which are anticipated to be close to the L-homotyrosine residue of *Candida* strains with mutations in the FKS1 and/or FKS2 genes of the original strains. This study demonstrated that chemical alteration of a selective group of echinocandin drugs may restore their efficacy in resistant strains and this strategy might be useful for the identification of new antifungal agents.

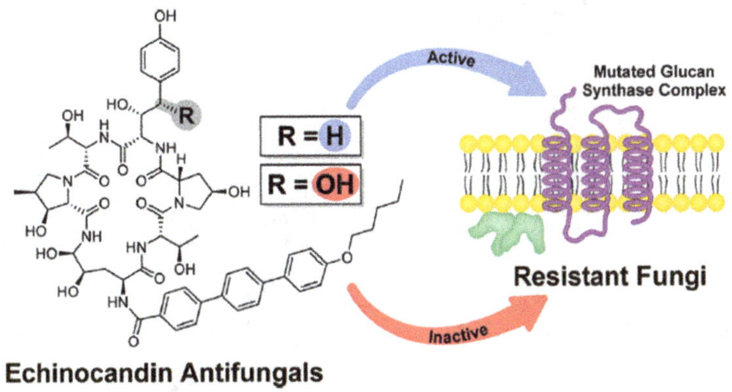

Fig. 18.9 The effect of selective benzylic alcohol on antifungal efficacy on resistant strain (Adapted from (Logviniuk et al. 2022) is licensed under CC-BY 4.0)

18.2.3 Marine-Derived Natural Product Derivatives

Marine species are excellent sources of biologically active compounds because they synthesize secondary metabolites to sustain themselves in the water. Marine algae have been used industrially for farming edible species and producing agar, carrageenan, and alginate (Hu et al. 2011). Marine-derived natural products exhibited a wide range of bioactivities that could be useful as drugs for humans. Pharmaceuticals derived from aquatic environments have shown remarkable promise in the treatment of a broad range of illnesses, from acute to chronic ailments (Bhatia et al. 2022). Therefore, multiple studies were focused on the diversification of marine-derived natural products to afford potential antifungal agents. In this view, Liu et al. (Liu et al. 2020a) in silico designed derivatives of marine natural product chlorotetaine with different amino acids followed by chemical synthesis as new antifungal agents. A unique aquatic strain of *Bacillus amyloliquefaciens* ZJU-2011 produced broth that contained chlorotetaine. The results highlighted that the derivative with lysine (**31**, Fig. 18.10) demonstrated potential in vitro antifungal activities against *C. krusei* CBS573 and *C. parapsilosis* JCM1785 with MIC values of 1.47 and 1.54 μg/mL, respectively.

Further, two novel tetrahydrofuran derivatives, designated as aspericacids A (**32**) and B, were isolated and characterized by Liu et al. by using metabolites generated by the sponge-associated *Aspergillus* sp. LS78 (Liu et al. 2020b). The in vitro antifungal results demonstrated that **32** showed a moderate MIC value of 50 μg/mL

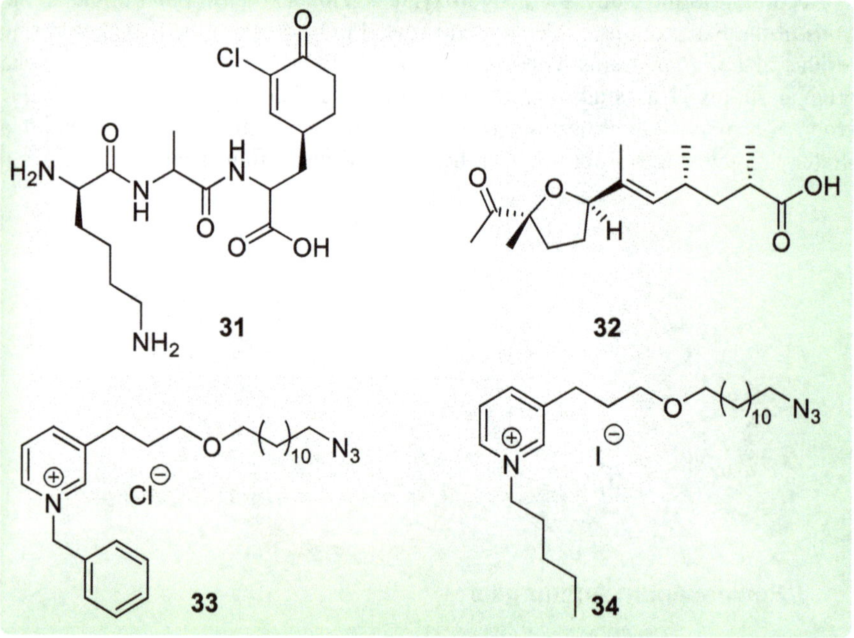

Fig. 18.10 Chemical structures of natural product-derived antifungals **31–34**

against both *C. albicans* and *C. neoformans*. The synthetic derivatives of marine alkaloids were reported by Andrade et al. (Andrade et al. 2021) and studied against *Candida* spp. by performing in vitro and in vivo assays. The results described that derivatives **33** and **34** were potentially active against *C. albicans*, *C. glabrata*, and *C. tropicalis* with an MIC value of 7.8 mg/mL and showed promising antifungal efficacy against *C. krusei* with an MIC value of 3.9 mg/mL. Compound **33** was found to have no severe toxic effects in vivo and to have a considerable effect on the fungal burden of *C. albicans* on the kidney and spleen of animals suffering from intra-abdominal candidiasis (IAC). Recently, Liu et al. synthesized a series of derivatives of marine-derived natural product Streptochlorin through acylation and oxidative annulation to develop new antifungals (Liu et al. 2023b).

18.2.4 Other Natural Product Derivatives

An azaoxoaporphine alkaloid designated sapangine (**35**) was taken from the stem bark of *Cananga odorata* and demonstrated wide-spectrum antifungal activity (Muhammad et al. 2001). Hence, Jiang et al. reported two new antifungal lead compounds **36** and **37** through a structural simplification approach on **35**, shown in Fig. 18.11 (Jiang et al. 2015).

The study demonstrated that the analogues exhibited broad-spectrum antifungal activity with MIC_{80} values of 0.5 to 0.125 mg/mL including fluconazole-resistant isolates. The study could be useful for further antifungal drug development. The study of Silva et al. disclosed the mechanism of action of broad-spectrum antifungal natural product Berberine (**38**, Fig. 18.12) (da Silva et al. 2016).

Compound **38** is an isoquinoline alkaloid belonging to the protoberberine family that was extracted from the stem bark, rhizomes, and roots of natural herbs, including *Phellodendron amurense*, *Hydrastis canadensis*, *Berberis aristata*, and *Berberis aquifolium*. The study demonstrated that after 24 and 72 h of berberine treatment, the MIC values for fluconazole-resistant *Candida* and *Cryptococcus neoformans* strains were equal to 8 and 16 μg/mL, respectively. The study highlighted that **38** resulted in DNA damage, changes to the functionality of the mitochondrial and plasma membranes, and cellular death, most likely through apoptosis.

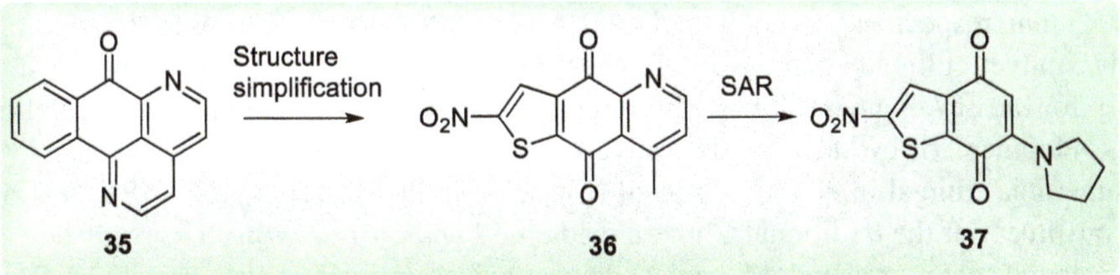

Fig. 18.11 Structural simplification approach to afford natural product-derived antifungals **36** and **37**

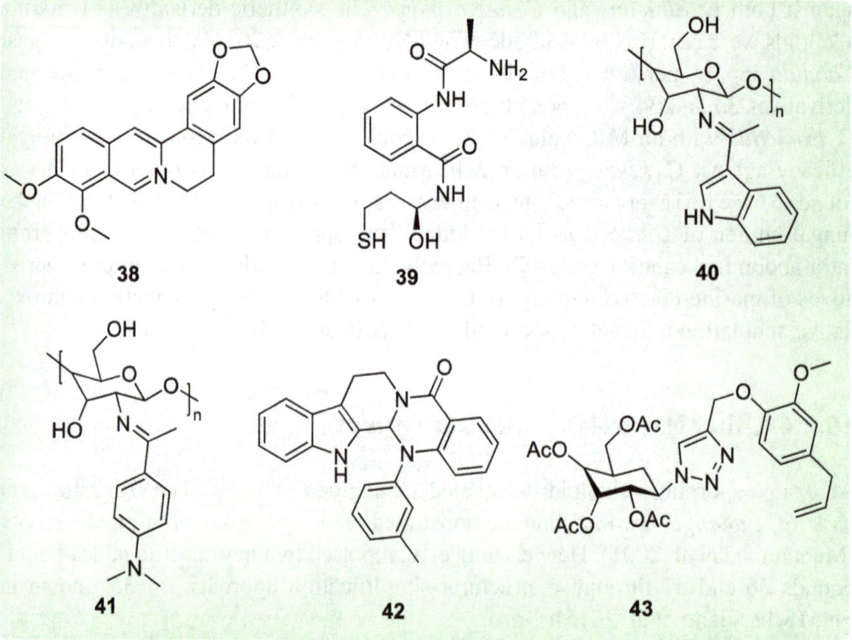

Fig. 18.12 Chemical structures of natural product-derived antifungals **38–43**

Ibrahim et al. isolated and identified a new antifungal natural product Fusarithioamide A (**39**) from ethanolic extract of the endophytic fungus *Fusarium chlamydosporium* from the leaves of *Anvillea garcinia* (Burm.f.) DC. (Asteraceae) (Ibrahim et al. 2016). Compound **39** exhibited antibacterial and antifungal potential against an array of microorganisms. The study demonstrated that **39** showed a zone of inhibition (ZOI) of 16.2 mm against *C. albicans* as compared to clotrimazole, 18.5 mm. The compound was selectively cytotoxic toward BT-549 and SKOV-3 cell lines with IC_{50} values of 0.4 and 0.8 μM and could be further diversified into potential antifungal and cytotoxic agents.

Hassan et al. synthesized phenolic Schiff-base derivatives of chitosan with indole-3-carboxaldehyde (**40**) and 4-dimethylaminobenzaldehyde (**41**) to boost the antimicrobial properties of chitosan (Hassan et al. 2018). The study indicated that ZOI for the synthesized analogues **40** and **41** against *C. albicans* was at 15.8 and 15.5 mm, respectively as compared to 11.4 mm for the pristine Chitosan. Therefore, the study highlighted that the development of Schiff bases of chitosan might be useful for the development of new antifungals. The study of Liang et al. dealt with the development of evodiamine derivative **42** as potential lead antifungal agents for superficial fungal infection treatment (Liang et al. 2022). Evodiamine (EVO) is a constituent of the traditional Chinese medicine *Evodia rutaecarpa*. The study demonstrated that compound **42** displayed remarkable antifungal efficacy with MIC_{100} values of 2, 38, and 38 μg/mL against *C. albicans*, *T. rubrum*, and *T. mentagrophytes*, respectively. These results signify the importance of evodiamine for further antifungal drug development. Further, using the sugar moiety, Goswami et al. create conjugate compounds that act as antifungal agents against *Aspergillus fumigatus* by combining eugenol and isoeugenol, two secondary metabolites of the plant *Myristica fragrans* (Goswami et al. 2022). The results demonstrated that two conjugates **43** and **44** (Fig. 18.13) showed remarkable antifungal efficacy with IC_{50} values of 5.42 and 9.39 μM, respectively. It was noted that both the quantity of conidia and the melanin hydrophobicity correlated with cell walls are inhibited by these analogues.

Berdzik et al. reported the development of gramine-based bioconjugates by in silico design followed by chemical synthesis using click chemistry as antifungal agents (Berdzik et al. 2023). The authors performed molecular docking studies to predict their binding with CYP51 (PDB ID: 5V5Z) of *C. albicans* and showed that indole-1,4-disubstituted-1,2,3-triazole conjugates may be the potential inhibitors of lanosterol 14α-demethylases to show antifungal properties. Two new analogues, **45** and **46** of natural product enfumafungin were reported from *Hormonema carpetanum* as antifungal agents by Cheng et al. The study disclosed that analogues **45** and **46** were active against multiple fungal strains. The MIC_{80} values were 32 and 16 μg/mL observed for analogues **45** and **46**, respectively against *C. glabrata* 537. The results encourage medicinal chemists to develop further analogues of enfumafungin for antifungal drug development (Cheng et al. 2023).

18.2.5 Natural Products Inspired Analogues

Multiple studies were pointed out in the literature focused on the development of new antifungals based on the scaffolds of natural products like quinoline, quinazoline, β-carboline, phytoalexins, etc. (Chen et al. 2022; Chung et al. 2015; Pedras et al. 2011; Singh et al. 2016). Here, we underlined a few recent studies related to the development of new antifungals based on these scaffolds. In this connection, Ji et al. designed new chitin synthase inhibitors as antifungal agents based on quinolone followed by chemical synthesis (Ji et al. 2019). The study came forward with multiple analogues with potential antifungal efficacy against different fungal strains. The study highlighted derivative **47** (Fig. 18.14) as the most potent chitin synthase

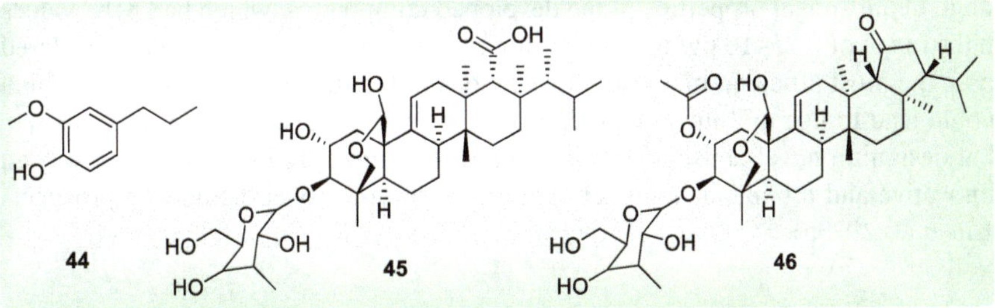

Fig. 18.13 Chemical structures of natural product-derived antifungals **44**–**46**

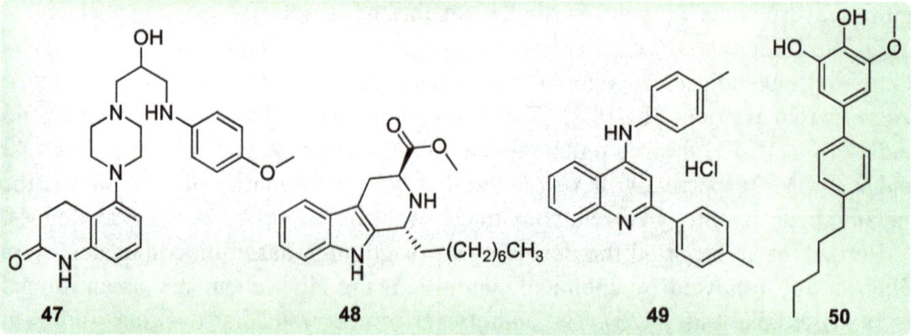

Fig. 18.14 Chemical structures of natural product-derived antifungals **47–50**

inhibitor with an IC_{50} value of 0.10 mM as compared with Polyoxin B (IC_{50} = 0.18 mM).

The authors investigated the antifungal efficacy against *C. albicans*, *Cryptococcus neoformans*, and *Aspergillus flavus* with an MIC of 32 mg/mL which was similar to the Polyoxin B. Singh et al. synthesized and reported new derivatives of C-1 alkylated tetrahydro-β-carboline keeping the aim to afford new antifungal lead (Singh et al. 2020). The derivatives were marked as broad-spectrum antifungals that were found active against an array of fungal strains. Lead compound **48** with an n-alkyl chain of eight carbons was most active against *C. glabrata* and *C. kefyr* with an MIC of 9.7 μg/mL. The derivatives of quinoline and quinazolines as potent antifungals were also reported by Qin et al. The authors evaluated the derivatives against four fungi which revealed that few derivatives exhibited antifungal efficacy moderate to excellent. The most promising compound **49** of the study demonstrated MIC values of 16, 8, 4, and 8 μg/mL against *C. tropicalis* ATCC 1369, *C. neoformans* ATCC 32045, *A. fumigatus* AS 3.3913, and *C. albicans* ATCC 10231, respectively. The study's mechanistic investigations proved that the chemicals' antifungal action was not brought on by the rupture of the fungal membrane. These findings imply that 4-aminoquinolines might be exciting and novel options for future research into antifungal medication development (Qin et al. 2022).

Recently, to discover new antifungals inspired by natural products, Liu et al. synthesized new biphenyl compounds inspired by naturally occurring phytoalexins as potent antifungal agents. The research's antifungal outcomes showcased the wide range of antifungal properties of the developed compounds, which had MIC values in the range of 0.25–16 μg/mL. According to the authors, compound **50**, considered to be the most efficient, may cause disruptions in the fungal cell membrane, which would lead to its rapid fungicidal effect. The aforementioned findings demonstrate that designing novel antifungal candidates with natural product inspiration was an innovative and promising approach that might provide novel therapeutic prospects (Liu et al. 2023a).

18.3 Conclusion and Future Perspectives

In conclusion, the escalating resistance to existing antimicrobial drugs, including those targeting fungi, poses a serious global health threat. Fungal infections, particularly those caused by invasive pathogens such as *Aspergillus*, *Cryptococcus*, *Candida*, and *Pneumocystis*, have become increasingly challenging to treat due to the emergence of drug-resistant strains. This necessitates the exploration of alternative therapeutic strategies, and natural products have emerged as a promising avenue for the development of novel antifungal agents. The significance of natural products in antifungal drug development is evident from the examples of clinically approved drugs such as echinocandins, polyenes, and azoles. These drugs have demonstrated efficacy by targeting various fungal cell components like cell walls, cell membranes, and biosynthetic pathways, providing a foundation for further exploration. However, challenges persist in drug development, given the cellular and metabolic similarities between mammalian and fungal cells. Ongoing clinical trials explore new antifungal agents with distinct mechanisms of action, providing optimism for the development of first-in-class treatments. These clinical studies reflect a dynamic effort to address emerging issues in invasive fungal infections, including resistance to existing drugs and infections caused by new pathogens.

The chapter highlights the structural diversification of natural compounds emerges as a key strategy for overcoming challenges in therapeutic application. This approach is exemplified by the development of polyene and echinocandin derivatives, as well as the exploration of marine-derived natural product derivatives. Structural modifications aim to improve drug solubility, reduce toxicity, and enhance efficacy, paving the way for the development of next-generation antifungal therapies. Beyond direct derivatives, natural products also inspire the design of analogues with potential antifungal properties. The development of compounds based on quinolones, tetrahydro-β-carbolines, quinolines, and quinazolines showcases the versatility and adaptability of natural product scaffolds for antifungal drug discovery.

Looking ahead, the exploration of natural products and their derivatives continues to hold great promise in the development of antifungal agents. The dynamic landscape of research, including in silico design, chemical synthesis, and biological evaluation, emphasizes the multifaceted approach required to meet the challenges posed by fungal infections. The pursuit of novel antifungal candidates inspired by natural products is not only a scientifically sound strategy but also a beacon of hope in addressing the pressing global issue of antimicrobial resistance. As research in this field progresses, it is anticipated that a deeper understanding of natural compounds and their derivatives will lead to the discovery of effective and innovative solutions to combat fungal infections in the future.

Acknowledgments The author **HM** wishes to acknowledge the Indian Council of Medical Research (ICMR), Government of India, for providing financial assistance through Senior Research Fellowship (SRF) with file no. 45/02/2020-Nan/BMS.

References

Aldholmi M, Marchand P, Ourliac-Garnier I, Le Pape P, Ganesan A (2019) A decade of antifungal leads from natural products: 2010–2019. Pharmaceuticals 12(4):182

Allen D, Wilson D, Drew R, Perfect J (2015) Azole antifungals: 35 years of invasive fungal infection management. Expert Rev Anti-Infect Ther 13(6):787–798

Andrade JT, Lima WG, Sousa JF, Saldanha AA, De Sá NP, Morais FB et al (2021) Design, synthesis, and biodistribution studies of new analogues of marine alkaloids: potent in vitro and in vivo fungicidal agents against *Candida* spp. Eur J Med Chem 210:113048

Arif T, Bhosale JD, Kumar N, Mandal TK, Bendre RS, Lavekar GS et al (2009) Natural products—antifungal agents derived from plants. J Asian Nat Prod Res 11(7):621–638

Balkovec JM, Hughes DL, Masurekar PS, Sable CA, Schwartz RE, Singh SB (2014) Discovery and development of first in class antifungal caspofungin (CANCIDAS®)—a case study. Nat Prod Rep 31(1):15–34

Bardal SK, Waechter JE, Martin DS (2011) Infectious diseases. Appl Pharmacol Elsevier:233–291

Berdzik N, Jasiewicz B, Ostrowski K, Sierakowska A, Szlaużys M, Nowak D et al (2023) Novel gramine-based bioconjugates obtained by click chemistry as cytoprotective compounds and potent antibacterial and antifungal agents. Nat Prod Res 1–7

Bhatia S, Makkar R, Behl T, Sehgal A, Singh S, Rachamalla M et al (2022) Biotechnological innovations from ocean: transpiring role of marine drugs in management of chronic disorders. Molecules 27(5):1539

Bhuiyan FR, Howlader S, Raihan T, Hasan M (2020) Plants metabolites: possibility of natural therapeutics against the COVID-19 pandemic. Front Med 7:7

Chen X, Ren B, Chen M, Liu M-X, Ren W, Wang Q-X et al (2014) ASDCD: antifungal synergistic drug combination database. Coste AT, editor. PLoS One 9(1):e86499

Chen J, Wang Y, Luo X, Chen Y (2022) Recent research progress and outlook in agricultural chemical discovery based on quinazoline scaffold. Pestic Biochem Physiol 184:105122

Cheng Z, Wu W, Liu Y, Chen S, Li H, Yang X et al (2023) Natural Enfumafungin analogues from Hormonema carpetanum and their antifungal activities. J Nat Prod 86(10):2407–2413

Chung P-Y, Bian Z-X, Pun H-Y, Chan D, Chan AS-C, Chui C-H et al (2015) Recent advances in research of natural and synthetic bioactive quinolines. Future Med Chem 7(7):947–967

Farhadi Z, Farhadi T, Hashemian SM (2020) Virtual screening for potential inhibitors of β(1,3)-D-glucan synthase as drug candidates against fungal cell wall. J Drug Assess 9(1):52–59

Garcia-Rubio R, de Oliveira HC, Rivera J, Trevijano-Contador N (2020) The fungal cell wall: Candida, Cryptococcus, and Aspergillus species. Front Microbiol 10

Goswami L, Gupta L, Paul S, Vermani M, Vijayaraghavan P, Bhattacharya AK (2022) Design and synthesis of eugenol/isoeugenol glycoconjugates and other analogues as antifungal agents against Aspergillus fumigatus. RSC Med Chem 13(8):955–962

Grover N (2010) Echinocandins: a ray of hope in antifungal drug therapy. Indian J Pharmacol 42(1):9

Guo J, Hu H, Zhao Q, Wang T, Zou Y, Yu S et al (2012) Synthesis and antifungal activities of glycosylated derivatives of the cyclic peptide fungicide caspofungin. ChemMedChem 7(8):1496–1503

Hashimoto S (2009) Micafungin: a sulfated echinocandin. J Antibiot (Tokyo) 62(1):27–35

Hassan MA, Omer AM, Abbas E, Baset WMA, Tamer TM (2018) Preparation, physicochemical characterization and antimicrobial activities of novel two phenolic chitosan Schiff base derivatives. Sci Rep 8(1):11416

Heard SC, Wu G, Winter JM (2021) Antifungal natural products. Curr Opin Biotechnol 69:232–241

Hu G-P, Yuan J, Sun L, She Z-G, Wu J-H, Lan X-J et al (2011) Statistical research on marine natural products based on data obtained between 1985 and 2008. Mar Drugs 9(4):514–525

Hüttel W (2021) Echinocandins: structural diversity, biosynthesis, and development of antimycotics. Appl Microbiol Biotechnol 105(1):55–66

Ibrahim SRM, Elkhayat ES, Mohamed GAA, Fat'hi SM, Ross SA (2016) Fusarithioamide A, a new antimicrobial and cytotoxic benzamide derivative from the endophytic fungus Fsarium chlamydosporium. Biochem Biophys Res Commun 479(2):211–216

Ji Q, Deng Q, Li B, Li B, Shen Y (2019) Design, synthesis and biological evaluation of novel 5-(piperazin-1-yl)quinolin-2(1H)-one derivatives as potential chitin synthase inhibitors and antifungal agents. Eur J Med Chem 180:204–212

Jiang Z, Liu N, Hu D, Dong G, Miao Z, Yao J et al (2015) The discovery of novel antifungal scaffolds by structural simplification of the natural product sampangine. Chem Commun 51(78):14648–14651

Kim H-J, Han C-Y, Park J-S, Oh S-H, Kang S-H, Choi S-S et al (2018) Nystatin-like Pseudonocardia polyene B1, a novel disaccharide-containing antifungal heptaene antibiotic. Sci Rep 8(1):13584

Kristanc L, Božič B, Jokhadar ŠZ, Dolenc MS, Gomišček G (2019) The pore-forming action of polyenes: from model membranes to living organisms. Biochim Biophys Acta Biomembr 1861(2):418–430

Larwood DJ (2020) Nikkomycin Z—ready to meet the promise? J Fungi 6(4):261

Letscher-Bru V (2003) Caspofungin: the first representative of a new antifungal class. J Antimicrob Chemother 51(3):513–521

Li Y, Lan N, Xu L, Yue Q (2018) Biosynthesis of pneumocandin lipopeptides and perspectives for its production and related echinocandins. Appl Microbiol Biotechnol 102(23):9881–9891

Liang Y, Zhang H, Zhang X, Peng Y, Deng J, Wang Y et al (2022) Discovery of evodiamine derivatives as potential lead antifungal agents for the treatment of superficial fungal infections. Bioorg Chem 127:105981

Lima SL, Colombo AL, de Almeida Junior JN (2019) Fungal cell wall: emerging antifungals and drug resistance. Front Microbiol 10:2573

Liu XH, Zhao JF, Wang T, Bin WM (2020a) Design, identification, antifungal evaluation and molecular modeling of chlorotetaine derivatives as new anti-fungal agents. Nat Prod Res 34(12):1712–1720

Liu Y, Ding L, Zhang Z, Yan X, He S (2020b) New antifungal tetrahydrofuran derivatives from a marine sponge-associated fungus aspergillus sp. LS78. Fitoterapia 146:104677

Liu J-C, Yang J, Lei S-X, Wang M-F, Ma Y-N, Yang R (2023a) Natural phytoalexins inspired the discovery of new biphenyls as potent antifungal agents for treatment of invasive fungal infections. Eur J Med Chem 261:115842

Liu J-R, Gao Y, Jin B, Guo D, Deng F, Bian Q et al (2023b) Design, synthesis, antifungal activity, and molecular docking of Streptochlorin derivatives containing the nitrile group. Mar Drugs 21(2):103

Logviniuk D, Jaber QZ, Dobrovetsky R, Kozer N, Ksiezopolska E, Gabaldón T et al (2022) Benzylic dehydroxylation of echinocandin antifungal drugs restores efficacy against resistance conferred by mutated glucan synthase. J Am Chem Soc 144(13):5965–5975

Maharani R, Sleebs BE, Hughes AB (2015) Macrocyclic N-methylated cyclic peptides and depsipeptides, pp 113–249

Mazu TK, Bricker B, Flores-Rozas H, Ablordeppey S (2016) The mechanistic targets of antifungal agents: an overview. Mini-reviews. Med Chem 16(7):555–578

McCarty TP, Pappas PG (2021) Antifungal Pipeline. Front Cell Infect Microbiol 11:11

Mercer DK, O'Neil DA (2020) Innate inspiration: antifungal peptides and other immunotherapeutics from the host immune response. Front Immunol 11:11

Muhammad I, Dunbar DC, Takamatsu S, Walker LA, Clark AM (2001) Antimalarial, cytotoxic, and antifungal alkaloids from Duguetia hadrantha. J Nat Prod 64(5):559–562

Nakamura I, Yoshimura S, Masaki T, Takase S, Ohsumi K, Hashimoto M et al (2017) ASP2397: a novel antifungal agent produced by Acremonium persicinum MF-347833. J Antibiot 70(1):45–51

Onishi J, Meinz M, Thompson J, Curotto J, Dreikorn S, Rosenbach M et al (2000) Discovery of novel antifungal (1,3)-β-d-glucan synthase inhibitors. Antimicrob Agents Chemother 44(2):368–377

Onyewu C, Blankenship JR, Del Poeta M, Heitman J (2003) Ergosterol biosynthesis inhibitors become fungicidal when combined with calcineurin inhibitors against Candida albicans, Candida glabrata, and Candida krusei. Antimicrob Agents Chemother 47(3):956–964

Pappas PG (2014) Antifungal clinical trials and guidelines: what we know and do not know. Cold Spring Harb Perspect Med 4(11):a019745–a019745

Patil A, Majumdar S (2017) Echinocandins in antifungal pharmacotherapy. J Pharm Pharmacol 69(12):1635–1660

Pedras MSC, Yaya EE, Glawischnig E (2011) The phytoalexins from cultivated and wild crucifers: chemistry and biology. Nat Prod Rep 28(8):1381

Perfect JR (2017) The antifungal pipeline: a reality check. Nat Rev Drug Discov 16(9):603–616

Phillips NA, Rocktashel M, Merjanian L (2023) Ibrexafungerp for the treatment of vulvovaginal candidiasis: design, development and place in therapy. Drug Des Devel Ther 17:363–367

Poester VR, Munhoz LS, Stevens DA, Melo AM, Trápaga MR, Flores MM et al (2023) Nikkomycin Z for the treatment of experimental sporotrichosis caused by Sporothrix brasiliensis. Mycoses 66(10):898–905

Preobrazhenskaya MN, Olsufyeva EN, Tevyashova AN, Printsevskaya SS, Solovieva SE, Reznikova MI et al (2010) Synthesis and study of the antifungal activity of new mono- and disubstituted derivatives of a genetically engineered polyene antibiotic 28,29-didehydronystatin A1 (S44HP). J Antibiot 63(2):55–64

Qin T-H, Liu J-C, Zhang J-Y, Tang L-X, Ma Y-N, Yang R (2022) Synthesis and biological evaluation of new 2-substituted-4-amino-quinolines and -quinazoline as potential antifungal agents. Bioorg Med Chem Lett 72:128877

Rauseo AM, Coler-Reilly A, Larson L, Spec A (2020) Hope on the horizon: novel fungal treatments in development. Open Forum Infect Dis 7(2):ofaa016

Roemer T, Krysan DJ (2014) Antifungal drug development: challenges, unmet clinical needs, and new approaches. Cold Spring Harb Perspect Med 4(5):a019703–a019703

Ryder NS (1992) Terbinafine: mode of action and properties of the squalene epoxidase inhibition. Br J Dermatol 126(s39):2–7

Sanglard D (2002) Clinical relevance of mechanisms of antifungal drug resistance in yeasts. Enferm Infecc Microbiol Clin 20(9):462–470

Serhan G, Stack CM, Perrone GG, Morton CO (2014) The polyene antifungals, amphotericin B and nystatin, cause cell death in Saccharomyces cerevisiae by a distinct mechanism to amphibian-derived antimicrobial peptides. Ann Clin Microbiol Antimicrob 13(1):18

Sheng Y, Ou Y, Hu X, Deng Z, Bai L, Kang Q (2020) Generation of tetramycin B derivative with improved pharmacological property based on pathway engineering. Appl Microbiol Biotechnol 104(6):2561–2573

da Silva AR, de Andrade Neto JB, da Silva CR, de Sousa Campos R, Costa Silva RA, Freitas DD et al (2016) Berberine antifungal activity in fluconazole-resistant pathogenic yeasts: action mechanism evaluated by flow cytometry and biofilm growth inhibition in Candida spp. Antimicrob Agents Chemother 60(6):3551–3557

Singh SB, Herath K, Kahn JN, Mann P, Abruzzo G, Motyl M (2013) Synthesis and antifungal evaluation of pentyloxyl-diphenylisoxazoloyl pneumocandins and echinocandins. Bioorg Med Chem Lett 23(11):3253–3256

Singh D, Devi N, Kumar V, Malakar CC, Mehra S, Rattan S et al (2016) Natural product inspired design and synthesis of β-carboline and γ-lactone based molecular hybrids. Org Biomol Chem 14(34):8154–8166

Singh R, Jaisingh A, Maurya IK, Salunke DB (2020) Design, synthesis and bio-evaluation of C-1 alkylated tetrahydro-β-carboline derivatives as novel antifungal lead compounds. Bioorg Med Chem Lett 30(3):126869

Syed YY (2023) Rezafungin: first approval. Drugs 83(9):833–840

Szymański M, Chmielewska S, Czyżewska U, Malinowska M, Tylicki A (2022) Echinocandins—structure, mechanism of action and use in antifungal therapy. J Enzyme Inhib Med Chem 37(1):876–894

Tevyashova AN, Olsufyeva EN, Solovieva SE, Printsevskaya SS, Reznikova MI, Trenin AS et al (2013) Structure-antifungal activity relationships of polyene antibiotics of the amphotericin B group. Antimicrob Agents Chemother 57(8):3815–3822

Tevyashova A, Efimova S, Alexandrov A, Omelchuk O, Ghazy E, Bychkova E et al (2023a) Semisynthetic amides of amphotericin B and nystatin A1: A comparative study of in vitro activity/toxicity ratio in relation to selectivity to ergosterol membranes. Antibiotics 12(1):151

Tevyashova AN, Efimova SS, Alexandrov AI, Ghazy ESMO, Bychkova EN, Solovieva SE et al (2023b) Semisynthetic amides of polyene antibiotic Natamycin. ACS Infect Dis 9(1):42–55

Wills EA, Redinbo MR, Perfect JR, Del PM (2000) New potential targets for antifungal development. Emerg Ther Targets 4(3):265–296

Yao J, Liu H, Zhou T, Chen H, Miao Z, Dong G et al (2012) Total synthesis and structure–activity relationships of caspofungin-like macrocyclic antifungal lipopeptides. Tetrahedron 68(14):3074–3085

Zhang C-W, Zhong X-J, Zhao Y-S, Rajoka MSR, Hashmi MH, Zhai P et al (2023) Antifungal natural products and their derivatives: a review of their activity and mechanism of actions. Pharmacol Res - Mod Chin Med 7:100262

Zhou S, Song L, Masschelein J, Sumang FAM, Papa IA, Zulaybar TO et al (2019) Pentamycin biosynthesis in Philippine Streptomyces sp. S816: cytochrome P450-catalyzed installation of the C-14 hydroxyl group. ACS Chem Biol 14(6):1305–1309

Zhu B, Dong Y, Ma J, Chen M, Ruan S, Zhao W et al (2020) The synthesis and activity evaluation of N-acylated analogs of echinocandin B with improved solubility and lower toxicity. J Pept Sci 26(11):e3278

Therapeutic Potential of Phytochemicals and Their Derivatives as Antifungal Candidates: Recent Discovery and Development

19

Kashish Azeem, Iram Irfan, Mohd. Shakir, Diwan S. Rawat, and Mohammad Abid

19.1 Introduction: The Fungal Menace

Fungal infections have gained significant attention as a growing threat in the landscape of infectious diseases. Despite their often-subtle presentation, fungi can cause severe illnesses in immunocompromised individuals and those with underlying health conditions. The increasing prevalence of fungal infections can be attributed to factors such as climate change, global travel, and the widespread use of immunosuppressive therapies (Oryan and Akbari 2016; Seagle et al. 2021; Polgreen and Polgreen 2017; Casadevall et al. 2019; Ashraf et al. 2020). The urgent need to combat these infections has driven the importance of antifungal agents to the forefront of medical and scientific concerns. The emergence of drug-resistant fungi complicates treatment regimens, demanding novel antifungal strategies (Arastehfar et al. 2020a, b). Fungal pathogens possess a remarkable ability to adapt and evolve, leading to the rapid emergence of drug-resistant strains. Frequent and prolonged use of antifungal drugs in both clinical and agricultural settings has exerted selective pressure on fungi, favoring the survival and propagation of resistant strains. *Candida*

K. Azeem · I. Irfan · M. Abid (✉)
Medicinal Chemistry Laboratory, Department of Biosciences, Jamia Millia Islamia, New Delhi, India
e-mail: rs.kazeem@jmi.ac.in; iramirfanchem@gmail.com; mabid@jmi.ac.in

Mohd. Shakir
Medicinal Chemistry Laboratory, Department of Biosciences, Jamia Millia Islamia, New Delhi, India

Department of Chemistry, University of Delhi, University of Delhi, Delhi, India
e-mail: shakirsalim7534@gmail.com

D. S. Rawat
Department of Chemistry, University of Delhi, University of Delhi, Delhi, India
e-mail: dsrawat@chemistry.du.ac.in

N. Manzoor (ed.), *Advances in Antifungal Drug Development*,
https://doi.org/10.1007/978-981-97-5165-5_19

albicans is a significant fungal pathogen known for causing persistent infections that are difficult to treat, rapid development of drug-resistant strains, and its capability to survive and multiply within macrophages. Study supports that intramacrophage *C. glabrata* serves as a reservoir for persistent and drug-resistant infections, making it challenging to treat effectively (Arastehfar et al. 2023). Drug-resistant fungal infections have profound clinical implications, particularly for immunocompromised patients and individuals with underlying health conditions. In hospitals and healthcare facilities, patients undergoing chemotherapy, organ transplantation, or immunosuppressive therapies are particularly vulnerable to invasive fungal infections. Opportunistic fungi exploit weakened immune defenses, leading to invasive candidiasis and other severe infections with high morbidity and mortality rate (Lionakis 2023; Kangabam and Nethravathy 2023; Loh and Lam 2023). Beyond immunocompromised patients, allergic bronchopulmonary aspergillosis (ABPA) presents a recurring challenge for asthmatics and those with compromised respiratory health (Patel et al. 2019; Dhooria et al. 2020; Barry et al. 2020; Gothe et al. 2020; Maleki et al. 2020). ABPA is a condition caused by a hypersensitivity reaction to the *Aspergillus fungus*. It is rarely reported in patients without asthma or cystic fibrosis. However, in a case study of ABPA in a patient with a history of chronic cocaine use and tuberculosis (but no history of asthma or cystic fibrosis), the individual presented with progressively worsening dyspnea and experienced a significant weight loss (Ayoubi et al. 2019). *Candida auris* is resistant to various antifungal medications, including azoles, polyenes, and echinocandins. Patients who have undergone organ transplant, are on immunosuppressive agents, have diabetes, recent antibiotic use, catheter use, and prolonged hospital or nursing home stays have been found more susceptible to *C. auris* infections (Sanyaolu et al. 2022). Moreover, neglected tropical fungal diseases inflict substantial burdens in regions with limited healthcare access (Rodrigues and Nosanchuk 2021; Lim et al. 2022). In addition, there are various cases found for paracoccidioidomycosis (a deep mycosis endemic to Latin America) (Griffiths et al. 2019), talaromycosis (penicilliosis) (an invasive mycosis endemic in tropical and subtropical Asia) (Narayanasamy et al. 2021), and fungal keratitis (a severe eye infection involving the cornea) (Brown et al. 2022), which are called to be accepted as neglected tropical diseases due to their endemic nature. Beyond healthcare settings, drug resistance presents significant challenges in managing neglected tropical fungal diseases. These infections, prevalent in low-resource regions, often lack effective treatments, and the emergence of resistance further exacerbates the burden on vulnerable populations. As many of these neglected diseases disproportionately affect marginalized communities, the impact of drug resistance on healthcare equity is deeply concerning. Furthermore, the intrinsic genetic plasticity of fungi allows for rapid acquisition and dissemination of resistance mechanisms (Hokken et al. 2019). The rise of resistance compromises the effectiveness of conventional antifungal treatments, leading to treatment failures, prolonged hospital stays, increased morbidity, and elevated mortality rates. In response to the escalating threat of fungal infections, there is a pressing need to develop effective antifungal agents.

Phytochemicals exhibit a wide range of antifungal activities, capable of targeting various fungal pathogens (Lagrouh et al. 2017). They have been shown to inhibit the growth and development of common fungal species such as *C. albicans, Aspergillus fumigatus, Cryptococcus neoformans*, and *Trichophyton rubrum*. Phytochemicals encompass a wide variety of chemical classes, such as alkaloids, flavonoids, terpenoids, polyphenols, and essential oils. Each class possesses distinct antifungal mechanisms, including disruption of fungal cell membranes, inhibition of key enzymes, interference with biofilm formation, and modulation of cell signaling pathways (Fig. 19.1) (Cushnie and Lamb 2005; Chen et al. 2023; Hussain et al. 2021). For instance, *Calotropis gigantean* crude extracts have shown promising antifungal activity against various pathogenic fungal species including *C. albicans, Aspergillus niger, Aspergillus ochraceus, Aspergillus ustus*, and *Rizopus oryzae* (Parvin et al. 2014). One study showed that phytochemicals *Cirsium englerianum, Rumex abyssinicus, Discopodium penninervium*, and *Lippia adoensis* exhibited impressive antifungal activity against *C. albicans* and *T. mentagrophytes*. Among them, *R. abyssinicus* demonstrated the most potent antifungal effect, with MIC (Minimum Inhibitory Concentration) values ranging from 32 to 64 μg/mL (Kebede et al. 2021). Combining phytochemicals with conventional antifungal drugs has demonstrated synergistic effects, which enhanced the overall antifungal activity and potentially reducing the development of drug resistance in fungi (Alves et al. 2014; Pinmai et al. 2008; Essid et al. 2017; Muthamil et al. 2018). This characteristic makes them attractive candidates for further therapeutic development. Ongoing research on natural products has revealed novel sources of phytochemicals with potent antifungal properties. These discoveries provide valuable insights into previously untapped reservoirs of potential drug leads.

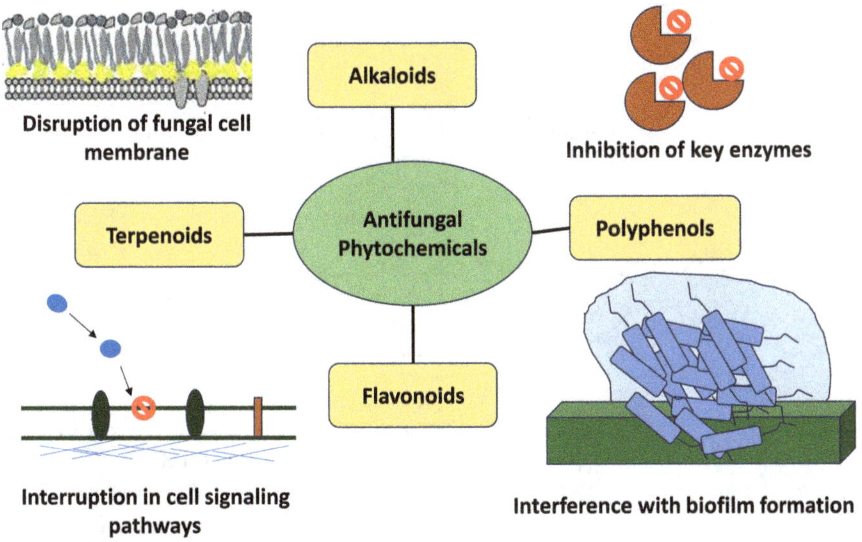

Fig. 19.1 Phytochemicals, including alkaloids, flavonoids, terpenoids, polyphenols, and essential oils, exhibit diverse antifungal mechanisms, targeting fungal cell membranes, enzymes, biofilm formation, and cell signaling pathways

19.2 Traditional Antifungal Agents

There are several groups of antifungal agents (Table 19.1, Fig. 19.2) that have been used to combat the fungal diseases and infections since many decades. Cells of both yeast and human, however, are eukaryotic and that is why these antifungal agents target both, causing mild to severe side effects. Moreover, the fungistatic nature of most of these agents and the rise of clinical resistance over the period of time limit their success. Nevertheless, till the new drug or new approach come into the action, these traditional antifungals are in the forefront of the treatment of the fungal diseases. Each group has its unique mechanism of action which against different fungi type. The structure of antifungal agents plays a crucial role in determining their antifungal activity and also in the development of new drug with better efficacy. These structures dictate target specificity, influencing affinity for fungal components like ergosterol or β-1,3-glucan. Membrane permeability, enzyme inhibition,

Table 19.1 Chemical structures of some common and widely used antifungal agents

Sr. no.	Group	Antifungal agents	Mechanism of action	References
1	Polyene	Amphotericin B Nystatin	Work by binding to the fungal cell membrane and disrupting its structure, leading to cell death	Carolus et al. (2020)
		Flucytosine (5-FC)	Often used in combination with other antifungal drugs to treat systemic fungal infections. It is converted into fluorouracil within the fungal cell, leading to the disruption of RNA and protein synthesis, ultimately causing cell death	Delma et al. (2021)
2	Allylamins	Terbinafine (oral antifungal)	Work by interfering with fungal cell division and preventing the growth of new fungal cells	Du et al. (2021)
3	Azoles	Ketoconazole Fluconazole Itraconazole Voriconazole Posaconazole	Work by inhibiting an enzyme called cytochrome P450-dependent enzyme lanosterol 14α-demethylase, which is essential for the synthesis of ergosterol, a crucial component of the fungal cell membrane	Emami et al. (2017)
4	Echinocandins	Caspofungin Micafungin Anidulafungin	Work by inhibiting the synthesis of beta-glucan, a key component of the fungal cell wall, leading to cell wall disruption and fungal cell death	Lima et al. (2019)
5	Topical antifungals (creams, lotions and ointment)	Clotrimazole, miconazole, econazole, ciclopirox	Cause leakage of folin-positive substances (amino acids and peptides) and potassium ions from the fungal cells	Gupta and Plott (2004) Subissi et al. (2010)

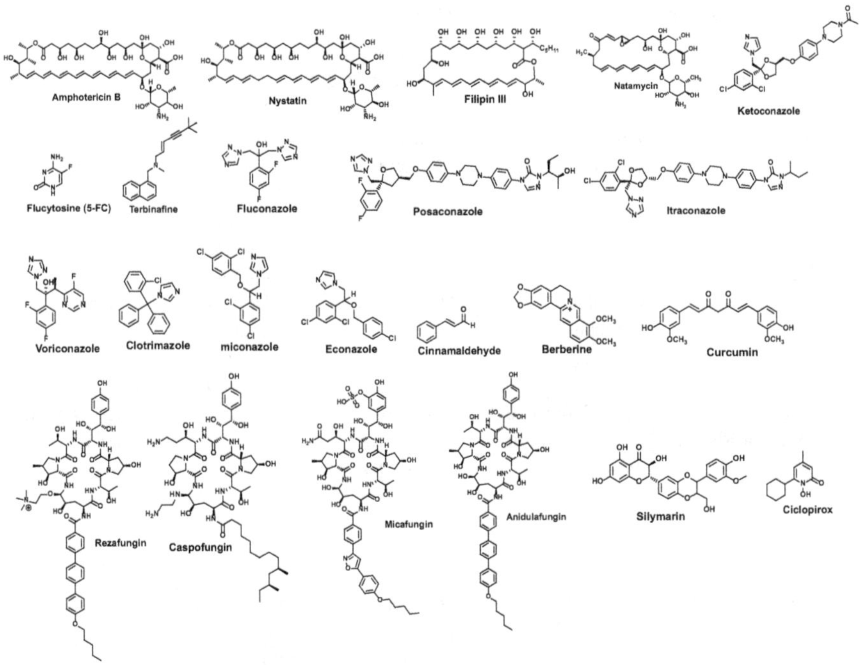

Fig. 19.2 Chemical structures of some commonly used antifungals

and metabolic disruption are determined by the design of these structures, impacting the agents' ability to penetrate fungal cells and disrupt essential processes. Additionally, structural considerations address the avoidance of resistance and influence pharmacokinetics, affecting bioavailability. Rational design, considering structure–activity relationships, is crucial for optimizing efficacy, minimizing side effects, and combating a broad spectrum of fungal infections, guiding the development of new antifungal drugs and combination therapies.

19.3 Phytochemicals: Antifungal Arsenal

Phytochemicals, often referred to as secondary metabolites, are naturally occurring bioactive compounds found in plants. These compounds are not directly involved in the essential functions of growth, development, and reproduction but serve vital roles in plant defense against various stresses, including microbial pathogens, insects, and environmental challenges. Plants synthesize a vast array of phytochemicals to protect themselves from potential threats in their surroundings. These compounds are found in various plant parts, such as leaves, stems, roots, fruits, and flowers, and their concentrations can vary depending on factors like plant species, growth conditions, and environmental stimuli (Velu et al. 2018). Table 19.2 represents important groups of medicinal phytochemicals with their biological properties.

Table 19.2 Important groups of phytochemical antifungals

Sr. no.	Phytochemicals group	Biological activities	Compounds with antifungal properties	Plant sources	References
1	Alkaloids	Antifungal, antiviral, insecticide	Caffeine Nicotine	*Coffea* spp. Tobacco leaves *(Nicotiana tabacum)*	Alamgir and Alamgir (2017)
2	Flavonoids	Antioxidant, pollinator attractants	Quercetin Catechins	Onions (*Allium cepa)* Green tea (*Camellia sinensis*)	Pistelli and Giorgi (2012)
3	Terpenoids (isoprenoids)	Plant defense and communication, antifungal	Thymol Limonene Saponins	Thyme (*Thymus vulgaris*) Citrus fruits (orange) Brazilian Caatinga	Barros et al. (2023)
4	Polyphenols	Antioxidant, antimicrobial	Resveratrol Curcumin Tannin	Grapes (*Vitis vinifera*) turmeric (*Curcuma longa*)	Rathod et al. (2023)
5	Essential oils	Antifungal	Tea tree oil Eucalyptus oil	*Melaleuca alternifolia* *Eucalyptus* species	Noumi et al. (2011)
7	Cyanogenic glycosides	Antifungal	Amygdalin Prunasin	Bitter almonds Cherry laurel	Bartnik and Facey (2024)

19.4 Phytochemical Antifungal Agents and Their Mechanism of Action

Several mechanisms contribute to the antifungal properties of phytochemicals. Phytochemicals can target and disrupt the structure and integrity of fungal cell membranes. These compounds insert themselves into the fungal membrane, leading to leakage of cellular contents and ultimately causing cell death. For instance, isoquercetin, a flavonoid isolated from *Aster yomena,* perturbs the lipid bilayer of *C. albicans* without causing hemolysis, leading to the membrane damage, thereby, increasing permeability (Yun et al. 2015).

Inhibition of Key Enzymes: Some phytochemicals inhibit specific enzymes that are essential for fungal growth and survival. By interfering with these enzymes' activities, the phytochemicals disrupt vital metabolic processes and impede ability of the fungus to thrive (Ramakrishnan et al. 2016). Notable antifungal agents such as echinocandins and azole drugs have demonstrated potential as inhibitors of β-1,3-glucan synthase and lanosterol 14-α-demethylase, respectively. Azole drugs, in particular, have played a crucial therapeutic role for over two decades, given the limited availability of antifungal medications. However, the extensive use of triazoles has contributed to the emergence of drug-resistant *Candida* sp. and *Aspergillus* sp. Consequently, the exploration of antifungal agents with novel drug targets stands as a long-term objective for numerous research groups. Carrying such researches ahead, a recent study showed that mixtures of phytochemical (including naringenin, cinnamic acid, quercetin, curcumin) effectively prevented the fungal growth by inactivating the lignocellulytic enzymes produced by *Aspergillus niger* (Mondal et al. 2022). These phytochemicals functioned as natural preservatives being safer than the fungicides. Seaweeds such as phlorotannins, can increase the activity of mitochondria dehydrogenases inhibiting the dimorphic transition of *C. albicans*, leading to the formation of pseudo-hyphae with reduced ability to adhere to the epithelial cells (Lopes et al. 2013). This reduces the fungal virulence and the ability to attack host cells.

Interference with Fungal Biofilm Formation: Fungal biofilms are structured communities of fungi that adhere to surfaces and are highly resistant to antifungal treatments. Phytochemicals can inhibit the formation of fungal biofilms or disperse preformed biofilms, making the fungi more susceptible to antifungal agents (Shankar Raut and Mohan Karuppayil 2016). Plant-derived alkaloid, piperine, demonstrated a concentration-dependent antibiofilm and antihyphal activity against *C. albicans* (Priya and Pandian 2020). It inhibited hyphal extension and swapping hyphal phase to yeast forms yet under filamentation-inducing circumstances by downregulating the expression of several biofilm-related and hyphal-specific genes. Another study showed that treatment with terpenoids camphor and eucalyptol disrupted the ability of *C. albicans* strains to form biofilm in vitro (Ivanov et al. 2021).

Generation of Reactive Oxygen Species (ROS): Phytochemicals can induce the production of ROS within fungal cells. ROS are highly reactive molecules that cause oxidative stress, damaging cellular components and leading to fungal cell death. Apigenin, from *Aster yomen,* a flowering plant, reduced the viability of fungal cells, caused mitochondrial ROS accumulation and lipid peroxidation in *C. albicans* (Lam et al. 2020). The overproduction of mitochondrial ROS disrupted the intracellular redox homeostasis and reduces the glutathione level in fungal cells resulting in oxidative stress.

Modulation of Fungal Virulence Factors: Some phytochemicals can suppress the expression of virulence factors in fungi. Virulence factors are molecules that contribute to the ability of the fungus to infect and cause disease in the host. By reducing the expression of these factors, the phytochemicals can attenuate the pathogenicity of the fungi. Extracts of *Anadenanthera colubrina* modulated the virulence enzymes of *C. albicans* which downregulated the gene expressions of the SAP-1, PLB-1(Maia et al. 2021). This caused decreased fungal cell growth with minimal toxicity against the host cells.

Alteration of Fungal Cell Signaling: Phytochemicals may interfere with fungal cell signaling pathways, disrupting communication between cells and compromising their ability to coordinate essential functions for survival and growth. Polyphenols from various seaweeds were shown to inhibit the growth of *Candida* spp. by reducing the violacein production essential for their quorum sensing activity (Nazzaro et al. 2019).

Efflux Pump Inhibition: Fungal cells can develop resistance to antifungal agents by employing efflux pumps that expel these compounds from within the cell. Certain phytochemicals can inhibit these efflux pumps, making the fungus more susceptible to conventional antifungal treatments. Camphor downregulated the level of CDR2, a gene encoding efflux pumps, interfering with the efflux pump activity in *Candida* (Ivanov et al. 2021).

19.5 Difference in Mechanism of Antifungal Action of Phytochemicals and Conventional Drugs

The mechanisms by which phytochemicals exert their antifungal effects can differ from those of conventional antifungal drugs. Phytochemicals are naturally occurring compounds found in plants, while conventional antifungal drugs are synthetic chemicals created in laboratories. The natural origin of phytochemicals may offer some advantages, such as potentially lower toxicity and a broader range of biological activities. Phytochemicals often have multiple mechanisms of action (Chen and Liu 2018). They can target various aspects of fungal biology, making it harder for fungi to develop resistance. In contrast, conventional antifungal drugs typically have a specific mode of action, such as targeting a particular enzyme or cellular structure in the fungus. Conventional antifungal drugs are generally developed to target specific fungal species or groups. Some antifungal drugs are effective against a wide range of fungi (broad spectrum), while others are more targeted (narrow-spectrum). Phytochemicals, on the other hand, can exhibit a broader spectrum of activity, often being effective against multiple types of fungi due to their diverse mechanisms of action. Fungi can develop resistance to both conventional antifungal drugs and phytochemicals. However, because phytochemicals typically have multiple modes of action, the risk of developing resistance may be lower compared to single-target conventional drugs. Phytochemicals are generally considered safe for human consumption when derived from edible plants. However, some may still have potential side effects or interactions with certain medications. Conventional antifungal drugs are carefully evaluated for safety and efficacy before clinical use. These drugs undergo rigorous testing and regulatory approval processes before they can be prescribed to patients. In contrast, phytochemicals from plants are often considered as potential leads for drug development and require further research and refinement before they can be used as mainstream antifungal treatments. Conventional antifungal drugs are widely available in pharmacies and hospitals, and many have well-established dosing regimens and treatment guidelines. Phytochemicals, especially those from traditional medicinal plants, may vary in their availability, and their efficacy and safety may not be as extensively studied as conventional drugs (Boukhatem and Setzer 2020).

19.6 Recent Discoveries of Antifungal Phytochemicals

Alkaloids are nitrogen-containing compounds with diverse pharmacological activities. Many alkaloids have been found to possess antifungal properties. One prominent example is berberine, a benzylisoquinoline alkaloid found in various plants such as *Berberis* spp., *Coptis chinensis*, and *Hydrastis canadensis*. Berberine has demonstrated broad-spectrum antifungal activity against *Candida albicans* and other pathogenic fungi by disrupting cell membranes and inhibiting fungal growth (Lam et al. 2016; Zhou et al. 2022). Terpenoids, or isoprenoids, represent the largest class of phytochemicals, characterized by their multiple isoprene units. These compounds play essential roles in plant defense against pathogens, including fungi (Toffolatti et al. 2021). Terpenoids with antifungal activity have been found to trigger mitochondrial dysfunction in *Saccharomyces cerevisiae* (Haque et al. 2016). Thymol, a monoterpene phenol found in *Thymus vulgaris* (thyme), exhibits potent antifungal activity against various fungi, including *Candida* species, by disrupting cell membranes and inhibiting spore formation (Qi et al. 2023). Flavonoids are a diverse group of polyphenolic compounds known for their antioxidant, antimicrobial and antimicrobial properties. For example, Quercetin, a flavonol found in onions, apples, and tea, shows antifungal effects against *Candida* species by disrupting fungal cell membranes and inhibiting virulence factors (Smiljković et al. 2019). Polyphenols are a broad class of compounds with multiple phenolic rings and diverse biological activities. Several polyphenols have shown antifungal properties. Epigallocatechin gallate (EGCG), a catechin polyphenol abundant in green tea, exhibits antifungal effects against *Candida* species by disrupting cell membranes and inhibiting fungal growth. Phenolic acids are aromatic compounds with at least one phenolic group. Some phenolic acids have demonstrated antifungal activities. Caffeic acid, found in coffee beans and other plant sources, has been shown to inhibit the growth of various pathogenic fungi, including *Aspergillus fumigatus*. Essential oils are volatile compounds derived from plants, often rich in terpenoids and phenolic compounds. Many essential oils have potent antifungal properties (Tariq et al. 2019). For instance, eucalyptus oil from Eucalyptus species and tea tree oil, obtained from *Melaleuca alternifolia*, contains terpenoids like terpinen-4-ol and alpha-terpineol, which exert antifungal effects against *Candida* and dermatophyte fungi (Yasin et al. 2021). Essential oils are volatile compounds derived from plants, often rich in terpenoids and phenolic compounds. Lectins among the essential oils are proteins or glycoproteins that can agglutinate cells and have been found to exhibit antifungal activity. Lectins from various plant sources have shown potential antifungal effects and putative use in the development of novel active principles against various fungal infections (Del Rio et al. 2020). Saponins are other groups of glycosides known for their surfactant properties and diverse biological activities, including antifungal activity. Saponins extracted from pericarp of the fruit *Sapindus mukorossi* were identified as fungicidal against two important fungal pathogens—*Venturia inaequalis* and *Botrytis cinerea* in vitro (Porsche et al. 2018). Saponins extracted and purified from the *Chamaecostus cuspidatus* are shown to function as effective antifungals and antibiofilm agent against *Candida albicans,* causing profound changes in the fungal cells morphology (Barros et al. 2023). Recently, tea saponins form *Camelliea oleifera* showed significant antifungal activity against the cell membrane structure of *C. albicans* leading to the leakage of cell contents and inhibiting the growth of mycelium. It further reduced the cell adhesion and aggregation and thus effectively inhibited the biofilm formation (Yu et al. 2022). Cyanogenic glycosides release hydrogen cyanide upon hydrolysis, and some of them have shown antifungal properties. Examples include amygdalin found in bitter almonds and prunasin in cherry laurel. Silymarin is a mixture of flavonolignans extracted from milk thistle (Silybum marianum) (Yun and Lee 2017). Studies have shown its antifungal activity against various fungal species, including *Candida* and *Aspergillus*. Silymarin was found to disrupt fungal cell membranes and inhibit fungal growth (Janeczko and Kochanowicz 2019). Cinnamaldehyde is a major compound found in cinnamon bark oil. Research has demonstrated its potent antifungal properties against common fungal pathogens like *C. albicans* Chen et al. 2019. Recently, it was shown to inhibit ergosterol biosynthesis in *Fusarium sambucinum* leading to the disruption of cell membrane integrity (Wei et al. 2020). Curcumin is a polyphenol found in turmeric (*Curcuma longa*). It exhibits antifungal effects against different fungi, including *Candida* species (Hussain et al. 2022). Its antifungal activity is attributed to its interference with fungal cell signaling and modulation of gene expression (Chen et al. 2016).

19.7 Beyond Tradition: Modern Approaches to Antifungal Drug Development

Antifungal drug development has been an important area of research to combat fungal infections, which can range from mild superficial infections to life-threatening systemic diseases. It is important to note that the development of new antifungal drugs is a complex and lengthy process that requires rigorous testing for safety and efficacy. In recent years, several modern approaches have emerged to tackle the challenges posed by fungal pathogens which are discussed below.

Combination Therapies: Just like in bacterial infections, combination therapies involving two or more antifungal agents are being investigated. This approach aims to enhance efficacy, prevent drug resistance, and broaden the spectrum of activity against different fungal species. One study assessed the effectiveness of a combination of berberine and amphotericin B, an antifungal agent, against dual-species biofilms of *Candida albicans* and *Staphylococcus aureus* (Gao et al. 2021). The results exhibited the potential of berberine and amphotericin B for treating the dual biofilms related infections and also presented the molecular basis for the efficacy of combinatorial treatment. A recent study showed the results which suggest that a mixture of cinnamaldehyde and eugenol can be a possible treatment for the gastrointestinal and vulvovaginal candidiasis as it led to a rapid loss of viable fungal cells (Saracino et al. 2022).

Nanotechnology: Nanoparticles have shown promise as drug delivery systems for antifungal agents. Nanocarriers can improve drug solubility, stability, and specifically target infected tissues, thereby reducing side effects and improving treatment outcomes. Phytocomponents such as flavonoids, phenolic acid, terpenoids, and organic acid and terpenoids are the possible bioreducing and stabilizing agents for the synthesis of metal-based nanoparticles as antifungal agents (Cruz-Luna et al. 2021; Sánchez-López et al. 2020). For instance, gold and silver nanoparticles have exhibited diminished fungal growth in *C. albicans* and *S. cerevisiae* by developing inhibition zones, the latter one being the most active (Khatoon et al. 2018). Recently, a study demonstrated that nanocurcumin caused a significant reduction in number of *Candida* colonies, showing a good antifungal effect against oral candidiasis (Anwar et al. 2023).

Immunomodulation*:* Enhancing the host immune response against fungal infections is another avenue for drug development. Immunomodulatory agents can boost the host immune defenses, aiding in the clearance of fungal pathogens. Immunomodulating therapy that uses host immune components including CT-cells, cytokine, natural killer cells, dendritic cells, monoclonal antibodies, can improve the antifungal treatment and enhance the efficacy of the antifungal agents (Ademe 2020). In general, such therapies are safer with lesser risk of resistance and broad spectrum of activity. However, this approach is still under careful exploration.

Repurposing Drugs: Studies show that several non-antifungal drugs, like antiarrhythmic, antipsychotic, antidepressant and nonsteroidal anti-inflammatory drugs have demonstrated antifungal activity (Zhang et al. 2021). Repurposing these drugs can expedite the drug development process since their safety profiles are already established. However, further investigations including transcriptome analysis, molecular techniques and animal model experiments. There is an interconnectedness of human, animal and environmental health suggesting that drug repurposing, as opposed to the traditional combination therapy, offers a potential strategy to circumvent issues like drug antagonism, multidrug interactions and cytotoxicity (Kim et al. 2020).

Natural Products: Researchers continue to investigate natural sources like plants, marine organisms, and microorganisms for antifungal compounds. These natural products often offer diverse chemical structures and may provide new leads for drug development (Atanasov et al. 2021). Natural products often possess a complex structure. Various strategies are employed to obtain analogues with comparatively simpler structures for antifungal drug development.

Genomics and Proteomics: Advances in genomics and proteomics have enabled a better understanding of fungal pathogens and their drug resistance mechanisms (Cowen et al. 2015; Peraman et al. 2021). By delving into the genetic and proteomic makeup of these pathogens, valuable insights can be gained into the specific molecular pathways and cellular processes that contribute to their ability to resist drugs (Ball et al. 2020). This in-depth understanding serves as a foundation for the development of targeted therapies. Armed with knowledge about the genetic and protein-level intricacies of fungal resistance, new therapies can be designed and tailored that specifically address these resistance mechanisms, potentially leading to more effective and precise treatments for fungal infections (Robbins and Cowen 2022).

CRISPR-Cas9 Technology: The advent of gene editing tools, such as CRISPR-Cas9, represents a groundbreaking advance in identifying genes associated with drug resistance in fungi (Javed et al. 2023; Krappmann 2016). By utilizing these tools, genetic material of fungal organisms can be precisely manipulated, allowing for the identification of specific genes responsible for their resistance to drugs (Getahun et al. 2022). This targeted approach enables a more accurate understanding of the molecular mechanisms behind drug resistance.

19.8 Combating Antifungal Resistance Through Multifaceted Approach

Antifungal resistance is a significant challenge in the treatment of fungal infections. It occurs when fungi develop mechanisms to evade the effects of antifungal drugs, rendering them less effective or completely ineffective. Overcoming antifungal resistance requires a multifaceted approach that targets various aspects of resistance development. Implementing antifungal stewardship programs in healthcare settings can help ensure the appropriate use of antifungal drugs. This includes optimizing drug dosages, minimizing unnecessary use, and promoting adherence to treatment regimens, which can reduce the emergence of resistance. In some cases, combining antifungal drugs with non-antifungal agents that have synergistic effects can improve the overall treatment efficacy and prevent resistance development (Moraes and Ferreira-Pereira 2019). For instance, combination of non-antifungal drugs auranofin and pentamidine displayed synergistic inhibitory effect against both drug-susceptible and drug-resistant *C. albicans* and biofilm as well (Lin et al. 2022). Regular surveillance of antifungal resistance patterns is crucial to detect emerging resistance early (Fisher et al. 2022). This helps healthcare providers adjust treatment strategies and improve patient outcomes. In-depth research on the genetic and molecular mechanisms of antifungal resistance is vital to identify potential drug targets and develop specific strategies to counter resistance (Lee et al. 2020). Adjuvants are substances that enhance the effects of antifungal drugs. Combining antifungal agents with adjuvants, such as plant-derived phenols and terpenoids can lead to improved drug efficacy and reduced resistance development (Zacchino et al. 2017). Environmental factors, such as excessive use of antifungals in agriculture or hospitals, can contribute to the development of resistance. Implementing measures to reduce antifungal usage in nonclinical settings can help mitigate resistance development. Overall, a comprehensive and multidisciplinary approach is necessary to overcome antifungal resistance. By combining drug development, improved treatment strategies, and vigilant surveillance, we can better manage fungal infections and preserve the effectiveness of existing antifungal agents.

19.9 Conclusion

Phytochemicals offer several advantages as antifungal agents, making them attractive candidates for combating fungal infections. They have demonstrated significant antifungal activity against a wide range of fungal pathogens. Plants are renewable resources, and the extraction of phytochemicals can be done sustainably. Utilizing these compounds as antifungal agents aligns with eco-friendly practices and reduces the dependence on nonrenewable resources. The rise of drug-resistant fungal strains has become a significant challenge in healthcare. Phytochemicals offer a potential solution as they can target different pathways in fungi, making it harder for the pathogens to develop resistance. Their use may provide new avenues for combating drug-resistant fungi. Phytochemicals can often be produced at a lower cost compared to synthetic drugs, making them potentially more affordable and accessible, especially in regions with limited resources. Some phytochemicals not only possess antifungal properties but also exhibit other beneficial biological activities, such as antioxidant, anti-inflammatory, and immunomodulatory effects. This multifunctionality can contribute to improved overall health outcomes during fungal infections. These modern approaches hold promise in improving the management of fungal infections and addressing the growing concern of antifungal resistance.

19.10 Future Prospects

The scarcity of new antifungal classes and the emergence of drug-resistant strains necessitate innovation in the field of antifungal drug discovery. Effective antifungal agents are vital to improve patient outcomes, especially in vulnerable populations where fungal infections have high mortality rates. Addressing the challenges posed by fungal infections requires a comprehensive approach. Research efforts focus on unraveling the molecular mechanisms of fungal pathogenesis and identifying potential drug targets. Preclinical studies play a crucial role in the discovery and development of new antifungal compounds. Bridging the gap between laboratory findings and clinical applications, well-designed clinical trials are essential for evaluating the safety and efficacy of antifungal therapies in human subjects. The escalating prevalence of fungal infections demands a concerted effort to strengthen our arsenal of antifungal agents. Advancements in research, coupled with effective clinical implementation, are paramount to combating fungal diseases successfully. Collaborative endeavors among researchers, clinicians, and pharmaceutical industries are necessary to translate scientific discoveries into tangible solutions. Empowering the future requires a resilient commitment to confront the growing impact of fungal infections, ensuring improved patient outcomes and a healthier global population. To address the urgency of drug resistance, it is imperative to pursue innovative approaches to antifungal therapies. Research efforts must focus on discovering new antifungal compounds with novel mechanisms of action. Combination therapies, employing synergistic effects of multiple antifungal agents, offer potential solutions to combat resistance effectively. Furthermore, promoting prudent antifungal drug use, infection control measures, and surveillance programs are vital to curbing the spread of drug resistance.

References

Ademe M (2020) Immunomodulation for the treatment of fungal infections: opportunities and challenges. Front Cell Infect Microbiol 10:469

Alamgir ANM, Alamgir ANM (2017) Medicinal, non-medicinal, biopesticides, color-and dye-yielding plants; secondary metabolites and drug principles; significance of medicinal plants; use of medicinal plants in the systems of traditional and complementary and alternative medicines (CAMs). In: Therapeutic use of medicinal plants and their extracts: volume 1: pharmacognosy, pp 61–104

Alves CT, Ferreira IC, Barros L, Silva S, Azeredo J, Henriques M (2014) Antifungal activity of phenolic compounds identified in flowers from north eastern Portugal against Candida species. Future Microbiol 9(2):139–146

Anwar SK, Elmonaem SNA, Moussa E, Aboulela AG, Essawy MM (2023) Curcumin nanoparticles: the topical antimycotic suspension treating oral candidiasis. Odontology 111(2):350–359

Arastehfar A, Gabaldón T, Garcia-Rubio R, Jenks JD, Hoenigl M, Salzer HJ et al (2020a) Drug-resistant fungi: an emerging challenge threatening our limited antifungal armamentarium. Antibiotics 9(12):877

Arastehfar A, Lass-Flörl C, Garcia-Rubio R, Daneshnia F, Ilkit M, Boekhout T et al (2020b) The quiet and underappreciated rise of drug-resistant invasive fungal pathogens. J Fungi 6(3):138

Arastehfar A, Daneshnia F, Cabrera N, Penalva-Lopez S, Sarathy J, Zimmerman M et al (2023) Macrophage internalization creates a multidrug-tolerant fungal persister reservoir and facilitates the emergence of drug resistance. Nat Commun 14(1):1183

Ashraf N, Kubat RC, Poplin V, Adenis AA, Denning DW, Wright L et al (2020) Re-drawing the maps for endemic mycoses. Mycopathologia 185:843–865

Atanasov AG, Zotchev SB, Dirsch VM, Supuran CT (2021) Natural products in drug discovery: advances and opportunities. Nat Rev Drug Discov 20(3):200–216

Ayoubi N, Jalali S, Kapadia N (2019) A case of allergic bronchopulmonary aspergillosis (ABPA) in a patient with a history of cocaine use and tuberculosis. Case Rep Med 2019:1

Ball B, Langille M, Geddes-McAlister J (2020) Fun (gi) omics: advanced and diverse technologies to explore emerging fungal pathogens and define mechanisms of antifungal resistance. MBio 11(5):10–1128

Barros DB, Nascimento NS, Sousa AP, Barros AV, Borges YWB, Silva WMN et al (2023) Antifungal activity of terpenes isolated from the Brazilian Caatinga: a review. Braz J Biol 83:e270966

Barry J, Gadre A, Akuthota P (2020) Hypersensitivity pneumonitis, allergic bronchopulmonary aspergillosis and other eosinophilic lung diseases. Curr Opin Immunol 66:129–135

Bartnik M, Facey P (2024) Glycosides. In: Pharmacognosy. Academic, pp 103–165

Boukhatem MN, Setzer WN (2020) Aromatic herbs, medicinal plant-derived essential oils, and phytochemical extracts as potential therapies for coronaviruses: future perspectives. Plan Theory 9(6):800

Brown L, Kamwiziku G, Oladele RO, Burton MJ, Prajna NV, Leitman TM, Denning DW (2022) The case for fungal keratitis to be accepted as a neglected tropical disease. J Fungi 8(10):1047

Carolus H, Pierson S, Lagrou K, Van Dijck P (2020) Amphotericin B and other polyenes—discovery, clinical use, mode of action and drug resistance. J Fungi 6(4):321

Casadevall A, Kontoyiannis DP, Robert V (2019) On the emergence of Candida auris: climate change, azoles, swamps, and birds. MBio 10(4):10–1128

Chen H, Liu RH (2018) Potential mechanisms of action of dietary phytochemicals for cancer prevention by targeting cellular signaling transduction pathways. J Agric Food Chem 66(13):3260–3276
Chen J, He ZM, Wang FL, Zhang ZS, Liu XZ, Zhai DD, Chen WD (2016) Curcumin and its promise as an anticancer drug: An analysis of its anticancer and antifungal effects in cancer and associated complications from invasive fungal infections. Eur J Pharmacol 772:33–42
Chen L, Wang Z, Liu L, Qu S, Mao Y, Peng X et al (2019) Cinnamaldehyde inhibits Candida albicans growth by causing apoptosis and its treatment on vulvovaginal candidiasis and oropharyngeal candidiasis. Appl Microbiol Biotechnol 103:9037–9055
Chen Y, Xing M, Chen T, Tian S, Li B (2023) Effects and mechanisms of plant bioactive compounds in preventing fungal spoilage and mycotoxin contamination in postharvest fruits: a review. Food Chem 415:135787
Cowen LE, Sanglard D, Howard SJ, Rogers PD, Perlin DS (2015) Mechanisms of antifungal drug resistance. Cold Spring Harb Perspect Med 5(7):a019752
Cruz-Luna AR, Cruz-Martínez H, Vásquez-López A, Medina DI (2021) Metal nanoparticles as novel antifungal agents for sustainable agriculture: current advances and future directions. J Fungi 7(12):1033
Cushnie TT, Lamb AJ (2005) Antimicrobial activity of flavonoids. Int J Antimicrob Agents 26(5):343–356
Del Rio M, de la Canal L, Regente M (2020) Plant antifungal lectins: mechanism of action and targets on human pathogenic fungi. Curr Protein Peptide Sci 21(3):284–294
Delma FZ, Al-Hatmi AM, Brüggemann RJ, Melchers WJ, de Hoog S, Verweij PE, Buil JB (2021) Molecular mechanisms of 5-fluorocytosine resistance in yeasts and filamentous fungi. J Fungi 7(11):909
Dhooria S, Sehgal IS, Muthu V, Agarwal R (2020) Treatment of allergic bronchopulmonary aspergillosis: from evidence to practice. Future Microbiol 15(5):365–376
Du W, Gao Y, Liu L, Sai S, Ding C (2021) Striking back against fungal infections: the utilization of nanosystems for antifungal strategies. Int J Mol Sci 22(18):10104
Emami S, Tavangar P, Keighobadi M (2017) An overview of azoles targeting sterol 14α-demethylase for antileishmanial therapy. Eur J Med Chem 135:241–259
Essid R, Hammami M, Gharbi D, Karkouch I, Hamouda TB, Elkahoui S et al (2017) Antifungal mechanism of the combination of Cinnamomum verum and Pelargonium graveolens essential oils with fluconazole against pathogenic Candida strains. Appl Microbiol Biotechnol 101(18):6993–7006
Fisher MC, Alastruey-Izquierdo A, Berman J, Bicanic T, Bignell EM, Bowyer P et al (2022) Tackling the emerging threat of antifungal resistance to human health. Nat Rev Microbiol 20(9):557–571
Gao S, Zhang S, Zhang S (2021) Enhanced in vitro antimicrobial activity of amphotericin B with berberine against dual-species biofilms of Candida albicans and Staphylococcus aureus. J Appl Microbiol 130(4):1154–1172
Getahun YA, Ali DA, Taye BW, Alemayehu YA (2022) Multidrug-resistant microbial therapy using antimicrobial peptides and the CRISPR/Cas9 system. Vet Med Res Rep 13:173–190
Gothe F, Schmautz A, Häusler K, Tran NB, Kappler M, Griese M (2020) Treating allergic bronchopulmonary aspergillosis with short-term prednisone and itraconazole in cystic fibrosis. J Allergy Clin Immunol Pract 8(8):2608–2614
Griffiths J, Lopes Colombo A, Denning DW (2019) The case for paracoccidioidomycosis to be accepted as a neglected tropical (fungal) disease. PLoS Negl Trop Dis 13(5):e0007195
Gupta AK, Plott T (2004) Ciclopirox: a broad-spectrum antifungal with antibacterial and anti-inflammatory properties. Int J Dermatol 43(S1):3–8
Haque E, Irfan S, Kamil M, Sheikh S, Hasan A, Ahmad A et al (2016) Terpenoids with antifungal activity trigger mitochondrial dysfunction in Saccharomyces cerevisiae. Microbiology 85:436–443

Hokken MW, Zwaan BJ, Melchers WJG, Verweij PE (2019) Facilitators of adaptation and antifungal resistance mechanisms in clinically relevant fungi. Fungal Genet Biol 132:103254

Hussain AY, Hussein HJ, Al-Rubaye AF (2021) Antifungal efficacy of the crude flavonoid, terpenoid, and alkaloid extracted from Myrtus communis L. against Aspergillus species isolated from stored medicinal plants seeds in the Iraqi markets. J Biotechnol Res Center 15(2):73–80

Hussain Y, Alam W, Ullah H, Dacrema M, Daglia M, Khan H, Arciola CR (2022) Antimicrobial potential of curcumin: therapeutic potential and challenges to clinical applications. Antibiotics 11(3):322

Ivanov M, Kannan A, Stojković DS, Glamočlija J, Calhelha RC, Ferreira IC et al (2021) Camphor and eucalyptol—Anticandidal spectrum, antivirulence effect, efflux pumps interference and cytotoxicity. Int J Mol Sci 22(2):483

Janeczko M, Kochanowicz E (2019) Silymarin, a popular dietary supplement shows anti–Candida activity. Antibiotics 8(4):206

Javed MU, Hayat MT, Mukhtar H, Imre K (2023) CRISPR-Cas9 system: a prospective pathway toward combatting antibiotic resistance. Antibiotics 12(6):1075

Kangabam N, Nethravathy V (2023) An overview of opportunistic fungal infections associated with COVID-19. *3*. Biotech 13(7):231

Kebede T, Gadisa E, Tufa A (2021) Antimicrobial activities evaluation and phytochemical screening of some selected medicinal plants: a possible alternative in the treatment of multidrug-resistant microbes. PLoS One 16(3):e0249253

Khatoon UT, Rao GN, Mohan MK, Ramanaviciene A, Ramanavicius A (2018) Comparative study of antifungal activity of silver and gold nanoparticles synthesized by facile chemical approach. J Environ Chem Eng 6(5):5837–5844

Kim JH, Cheng LW, Chan KL, Tam CC, Mahoney N, Friedman M et al (2020) Antifungal drug repurposing. Antibiotics 9(11):812

Krappmann S (2016) CRISPR-Cas9, the new kid on the block of fungal molecular biology. Med Mycol myw097

Lagrouh F, Dakka N, Bakri Y (2017) The antifungal activity of Moroccan plants and the mechanism of action of secondary metabolites from plants. J Mycol Med 27(3):303–311

Lam P, Kok SHL, Lee KKH, Lam KH, Hau DKP, Wong WY et al (2016) Sensitization of Candida albicans to terbinafine by berberine and berberrubine. Biomed Rep 4(4):449–452

Lam PL, Wong RM, Lam KH, Hung LK, Wong MM, Yung LH et al (2020) The role of reactive oxygen species in the biological activity of antimicrobial agents: an updated mini review. Chem Biol Interact 320:109023

Lee Y, Puumala E, Robbins N, Cowen LE (2020) Antifungal drug resistance: molecular mechanisms in Candida albicans and beyond. Chem Rev 121(6):3390–3411

Lim W, Verbon A, van de Sande W (2022) Identifying novel drugs with new modes of action for neglected tropical fungal skin diseases (fungal skinNTDs) using an Open Source Drug discovery approach. Expert Opin Drug Discov 17(6):641–659

Lima SL, Colombo AL, de Almeida Junior JN (2019) Fungal cell wall: emerging antifungals and drug resistance. Front Microbiol 10:492317

Lin J, Xiao X, Liang Y, Zhao H, Yu Y, Yuan P et al (2022) Repurposing non-antifungal drugs auranofin and pentamidine in combination as fungistatic antifungal agents against C. albicans. Front Cell Infect Microbiol 12:1065962

Lionakis MS (2023) Exploiting antifungal immunity in the clinical context. In: Seminars in immunology, vol 67. Academic, p 101,752

Loh JT, Lam KP (2023) Fungal infections: immune defense, immunotherapies and vaccines. Adv Drug Deliv Rev 196:114775

Lopes G, Pinto E, Andrade PB, Valentao P (2013) Antifungal activity of phlorotannins against dermatophytes and yeasts: approaches to the mechanism of action and influence on Candida albicans virulence factor. PLoS One 8(8):e72203

Maia CMDA, Pasetto S, Nonaka CFW, Costa EMMDB, Murata RM (2021) Yeast-host interactions: Anadenanthera colubrina modulates virulence factors of C. albicans and inflammatory response in vitro. Front Pharmacol 12:629778

Maleki M, Mortezaee V, Hassanzad M, Mahdaviani SA, Poorabdollah M, Mehrian P et al (2020) Prevalence of allergic bronchopulmonary aspergillosis in cystic fibrosis patients using two different diagnostic criteria. Eur Ann Allergy Clin Immunol 52(3):104–111
Mondal S, Santra S, Uddin H, Pal K, Halder SK, Chattopadhyay S, Mondal KC (2022) Application of phytochemicals to combat fungal pathogens of pulses: An approach toward inhibition of fungal propagation and Invasin activity. J Agric Food Chem 70(25):7662–7673
Moraes DC, Ferreira-Pereira A (2019) Insights on the anticandidal activity of non-antifungal drugs. J Mycol Med 29(3):253–259
Muthamil S, Balasubramaniam B, Balamurugan K, Pandian SK (2018) Synergistic effect of quinic acid derived from Syzygium cumini and undecanoic acid against Candida spp. biofilm and virulence. Front Microbiol 9:417399
Narayanasamy S, Dat VQ, Thanh NT, Ly VT, Chan JFW, Yuen KY et al (2021) A global call for talaromycosis to be recognised as a neglected tropical disease. Lancet Glob Health 9(11):e1618–e1622
Nazzaro F, Fratianni F, d'Acierno A, De Feo V, Ayala-Zavala FJ, Gomes-Cruz A et al (2019) Effect of polyphenols on microbial cell-cell communications. In: Quorum sensing. Academic, pp 195–223
Noumi E, Snoussi M, Hajlaoui H, Trabelsi N, Ksouri R, Valentin E, Bakhrouf A (2011) Chemical composition, antioxidant and antifungal potential of Melaleuca alternifolia (tea tree) and Eucalyptus globulus essential oils against oral Candida species. J Med Plants Res 5(17):4147–4156
Oryan A, Akbari M (2016) Worldwide risk factors in leishmaniasis. Asian Pac J Trop Med 9(10):925–932
Parvin S, Kader MA, Chouduri AU, Rafshanjani MAS, Haque ME (2014) Antibacterial, antifungal and insecticidal activities of the n-hexane and ethyl-acetate fractions of methanolic extract of the leaves of Calotropis gigantea Linn. J Pharmacogn Phytochem 2(5):47–51
Patel AR, Patel AR, Singh S, Singh S, Khawaja I (2019) Treating allergic bronchopulmonary aspergillosis: a review. Cureus 11(4)
Peraman R, Sure SK, Dusthackeer VA, Chilamakuru NB, Yiragamreddy PR, Pokuri C et al (2021) Insights on recent approaches in drug discovery strategies and untapped drug targets against drug resistance. Future J Pharm Sci 7:1–25
Pinmai K, Chunlaratthanabhorn S, Ngamkitidechakul C, Soonthornchareon N, Hahnvajanawong C (2008) Synergistic growth inhibitory effects of Phyllanthus emblica and Terminalia bellerica extracts with conventional cytotoxic agents: doxorubicin and cisplatin against human hepatocellular carcinoma and lung cancer cells. World J Gastroenterol: WJG 14(10):1491
Pistelli L, Giorgi I (2012) Antimicrobial properties of flavonoids. In: Dietary phytochemicals and microbes, pp 33–91
Polgreen PM, Polgreen EL (2017) Emerging and re-emerging pathogens and diseases, and health consequences of a changing climate. Infect Dis 40
Porsche FM, Molitor D, Beyer M, Charton S, André C, Kollar A (2018) Antifungal activity of saponins from the fruit pericarp of Sapindus mukorossi against Venturia inaequalis and Botrytis cinerea. Plant Dis 102(5):991–1000
Priya A, Pandian SK (2020) Piperine impedes biofilm formation and hyphal morphogenesis of Candida albicans. Front Microbiol 11:528306
Qi X, Zhong S, Schwarz P, Chen B, Rao J (2023) Mechanisms of antifungal and mycotoxin inhibitory properties of Thymus vulgaris L. essential oil and their major chemical constituents in emulsion-based delivery system. Ind Crop Prod 197:116575
Ramakrishnan J, Rathore SS, Raman T (2016) Review on fungal enzyme inhibitors–potential drug targets to manage human fungal infections. RSC Adv 6(48):42,387–42,401
Rathod NB, Elabed N, Punia S, Ozogul F, Kim SK, Rocha JM (2023) Recent developments in polyphenol applications on human health: a review with current knowledge. Plan Theory 12(6):1217
Robbins N, Cowen LE (2022) Genomic approaches to antifungal drug target identification and validation. Ann Rev Microbiol 76:369–388

Rodrigues ML, Nosanchuk JD (2021) Fungal diseases as neglected pathogens: a wake-up call to public health officials. In: Advances in clinical immunology, medical microbiology, COVID-19, and big data. Jenny Stanford Publishing, pp 399–411

Sánchez-López E, Gomes D, Esteruelas G, Bonilla L, Lopez-Machado AL, Galindo R et al (2020) Metal-based nanoparticles as antimicrobial agents: an overview. Nano 10(2):292

Sanyaolu A, Okorie C, Marinkovic A, Abbasi AF, Prakash S, Mangat J et al (2022) Candida auris: an overview of the emerging drug-resistant fungal infection. Infect Chemother 54(2):236

Saracino IM, Foschi C, Pavoni M, Spigarelli R, Valerii MC, Spisni E (2022) Antifungal activity of natural compounds vs. candida spp.: a mixture of cinnamaldehyde and eugenol shows promising in vitro results. Antibiotics 11(1):73

Seagle EE, Williams SL, Chiller TM (2021) Recent trends in the epidemiology of fungal infections. Infect Dis Clin 35(2):237–260

Shankar Raut J, Mohan Karuppayil S (2016) Phytochemicals as inhibitors of Candida biofilm. Curr Pharm Des 22(27):4111–4134

Smiljković M, Kostić M, Stojković D, Glamočlija J, Soković M (2019) Could flavonoids compete with synthetic azoles in diminishing Candida albicans infections? A comparative review based on in vitro studies. Curr Med Chem 26(14):2536–2554

Subissi A, Monti D, Togni G, Mailland F (2010) Ciclopirox: recent nonclinical and clinical data relevant to its use as a topical antimycotic agent. Drugs 70:2133–2152

Tariq S, Wani S, Rasool W, Shafi K, Bhat MA, Prabhakar A et al (2019) A comprehensive review of the antibacterial, antifungal and antiviral potential of essential oils and their chemical constituents against drug-resistant microbial pathogens. Microb Pathog 134:103580

Toffolatti SL, Maddalena G, Passera A, Casati P, Bianco PA, Quaglino F (2021) Role of terpenes in plant defense to biotic stress. In: Biocontrol agents and secondary metabolites. Woodhead Publishing, pp 401–417

Velu G, Palanichamy V, Rajan AP (2018) Phytochemical and pharmacological importance of plant secondary metabolites in modern medicine. In: Bioorganic phase in natural food: an overview, pp 135–156

Wei J, Bi Y, Xue H, Wang Y, Zong Y, Prusky D (2020) Antifungal activity of cinnamaldehyde against Fusarium sambucinum involves inhibition of ergosterol biosynthesis. J Appl Microbiol 129(2):256–265

Yasin M, Younis A, Javed T, Akram A, Ahsan M, Shabbir R et al (2021) River tea tree oil: composition, antimicrobial and antioxidant activities, and potential applications in agriculture. Plan Theory 10(10):2105

Yu Z, Wu X, He J (2022) Study on the antifungal activity and mechanism of tea saponin from Camellia oleifera cake. Eur Food Res Technol 248:1–13

Yun DG, Lee DG (2017) Assessment of silibinin as a potential antifungal agent and investigation of its mechanism of action. IUBMB Life 69(8):631–637

Yun J, Lee H, Ko HJ, Woo ER, Lee DG (2015) Fungicidal effect of isoquercitrin via inducing membrane disturbance. Biochim Biophys Acta BBA Biomembr 1848(2):695–701

Zacchino SA, Butassi E, Di Liberto M, Raimondi M, Postigo A, Sortino M (2017) Plant phenolics and terpenoids as adjuvants of antibacterial and antifungal drugs. Phytomedicine 37:27–48

Zhang Q, Liu F, Zeng M, Mao Y, Song Z (2021) Drug repurposing strategies in the development of potential antifungal agents. Appl Microbiol Biotechnol 105:5259–5279

Zhou Y, Yang CJ, Luo XF, Li AP, Zhang SY, An JX et al (2022) Design, synthesis, and biological evaluation of novel berberine derivatives against phytopathogenic fungi. Pest Manag Sci 78(10):4361–4376

Bioactive Heterocyclic Analogs as Antifungal Agents: Recent Advances and Future Aspects

20

Mohd Danish Ansari, Nouman, Rabiya Mehandi, Manish Rana, and Rahisuddin

20.1 Introduction

Contagious fungal infections or mycoses brought about by different pathogenic growths, have arisen as a huge worldwide wellbeing concern, influencing people across different socioeconomics and topographical districts (Sharma and Nonzom 2021). The frequency of these diseases has increased lately, especially among immune compromised patients, bringing about increase in death rates (Perfect and Casadevall 2006). By and large, the advancements in antifungal medications have lingered behind that of antibacterial and antiviral compounds, somewhat because of the underlying similarities between the microbe and human host cells, making drug configuration difficult. In the domain of clinical sciences, the constant fight against microbes has prodded intensified examination into the advancement of antifungal medications (Gupte et al. 2002). Among the different procedures utilized, using bioactive heterocycles as antimicrobials has made tremendous progress (Kathiravan et al. 2012). These perplexing ring structures, frequently containing nitrogen, oxygen, sulfur, or different other components, have inborn natural properties that can be used for therapeutic purposes (Alvarez-Builla and Barluenga 2011; Taylor et al.

M. D. Ansari
Laboratory of Green Synthesis, Department of Chemistry, University of Allahabad, Allahabad, India
e-mail: danishansari93@gmail.com

Nouman · R. Mehandi · Rahisuddin (✉)
Molecular and Biophysical Research Lab (MBRL), Department of Chemistry, Jamia Millia Islamia, New Delhi, India
e-mail: Noumancho786@gmail.com; Merabiyamehandi333@gmail.com; rahisuddin@jmi.ac.in

M. Rana
Department of Chemistry, Ramjas College, University of Delhi, Delhi, India
e-mail: manishrana@ramjas.du.ac.in

N. Manzoor (ed.), *Advances in Antifungal Drug Development*, https://doi.org/10.1007/978-981-97-5165-5_20

2016). Lately, noteworthy steps have been made in the designing of bioactive heterocycles and assessment of their antifungal properties (Blokhina et al. 2021; Ming et al. 2017; Ryu et al. 2009). These advances have not just improved how we might interpret microbial pathogenesis and medication systems, but have additionally helped in the preparation of novel treatment modalities.

When pathogenic fungi present a huge danger to the wellbeing of a large population worldwide, understanding new developments and future possibilities of improvement in medications are of central significance (Yang et al. 2022). This section helps in understanding the antifungal mechanism of action of bioactive heterocycles and their potential to be used as novel antifungal drugs (Table 20.1). Recent years have seen significant advancements in the assessment of antifungal compounds established as bioactive heterocycles. Scientists have modified the primary heterocyclic rings of compounds to make novel composites with powerful antifungal properties (Mehandi et al. 2021). For example, pyrazole derivatives have been shown to display promising inhibitory impact on microorganisms by inhibiting the basic enzymatic pathways (Qiao et al. 2019; Mehandi et al. 2023; Rana et al. 2023a, b). Likewise, imidazole-based particles have shown competence in disrupting robust biofilms and fundamental cell membrane properties (Rani et al. 2013). These developments demonstrate the synthesis of bioactive heterocycles with improved antifungal properties. While improvement in the antifungal efficacy holds huge commitment, a few difficulties must be defeated for fruitful their clinical

Table 20.1 Some heterocyclic antifungal drugs and their mechanism of action (Nami et al. 2019; Nigam 2015)

Antifungal drugs	Mechanism of action	Chemical class
Griseofulvin	• Inhibits fungal cell mitosis (sliding of microtubules) and nucleic acid synthesis	Benzofuran
Caspofungin, micafungin and Anidulafungin	• Inhibits (1,3)-β-D-glucan synthase	Echinocandins
Ciclopirox	• Inhibits transport of amino acids and nutrients	Hydroxypyridone derivative
Fluocytosine	• Inhibits thymidylate synthetase	Antimetabolite
• *First-generation*: ketoconazole, clotrimazole, econazole, bifonazole. • *Second-generation*: fluconazole, Itraconazole, ravuconazole, posaconazole, voriconazole, isavuconazole, terconazole and albaconazole.	• Inhibits fungal cytochrome P450 (sterol 14α-demethylase or CYP51) • Inhibits ergosterol biosynthesis	Azoles
AmB (Amphotericin B) and Nystatin	• Alters membrane barrier function and ergosterol levels	Polyenes
Amorolfine	• Inhibits cholesterol isomerase and sterol reductase; depletes ergosterol and causes accumulation of ignosterol	Morpholines

interpretation (Fisher et al. 2018). One main difficulty is increasing drug resistance, which highlights the requirement for methodologies that limits this problem, like using drug combinations and developing novel drug delivery frameworks. Moreover, working on the selectivity and pharmacokinetic properties of bioactive heterocycles remains a critical goal. Analysts are investigating new methodologies like synthesis of nanoparticles to upgrade the adequacy and accuracy of antifungal treatments (Di Santo et al. 2005; El-Sayed et al. 2022).

A constant and significant threat to human health and life is posed by fungi (Garibotto et al. 2010). These microbes are ubiquitous, being present in humans, animals, air, soil, water, decaying organic matters and plants. By breaking down organic materials into simpler forms, fungi and bacteria play a critical role in maintaining the ecosystem. A large number of fungi do not cause disease, but become pathogenic in people with a weakened immune system. Some disorders that lower host immunity include different cancers, diabetes, trauma, excessive iron intake, taking antibiotics and steroids, malnutrition, other infections etc. (Qadir et al. 2022). Fungal infections in people may be divided into three categories: mycoses (infections), toxic manifestations of fungal toxins and allergic reactions to fungal proteins (Kathiravan et al. 2012). The three primary pathogens that cause fungal infections are still *Candida* spp., *Aspergillus* spp. and *Cryptococcus neoformans* (Richardson and Lass-Flörl 2008). Based on the manner in which they work, there are four major classes of clinically available antifungal drugs: echinocandins, such as caspofungin and micafungin (Denning 2003), polyenes, for example nystatin and amphotericin B (Richardson and Lass-Flörl 2008), antimetabolites, like 5-fluorocytosine (Moudgal and Sobel 2010), and azoles, like fluconazole, voriconazole and

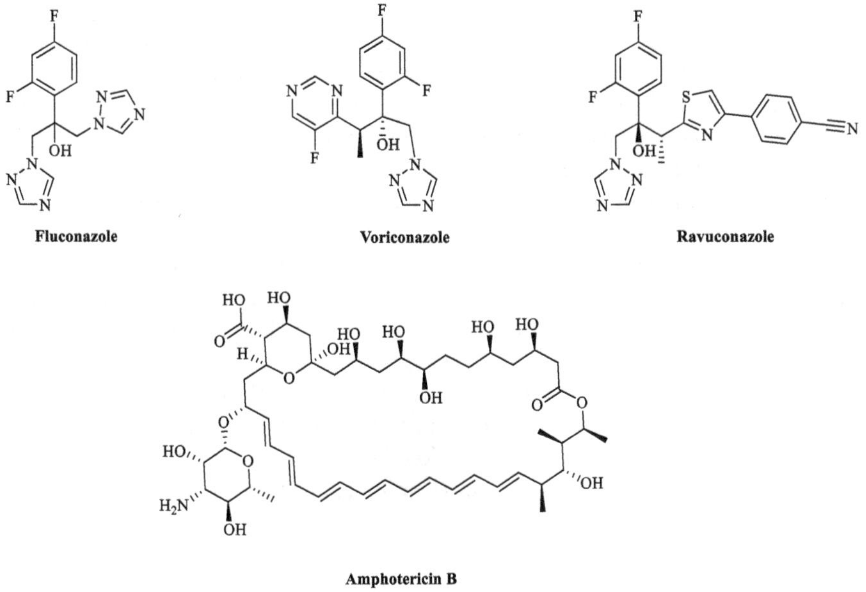

Fig. 20.1 Some triazole derivatives and polyene antifungal drugs

ravuconazole etc. (Fig. 20.1) (Allen et al. 2015). Many cases of black fungus have been reported in India during COVID-19 (Qadir et al. 2022). The chance of getting black fungus within 6 weeks is highest for patients receiving COVID treatment, according to a top neurosurgeon at AIIMS, New Delhi, India. Ampho B 89 was frequently used for the treatment of black fungus (Gallis et al. 1990; Jeong et al. 2019; Pagano et al. 2004; Prakash et al. 2019).

Eventually, the combination of state-of-the-art science, multidisciplinary cooperation and the inborn capability of bioactive heterocycles holds the way to reshaping the scene of antifungal therapeutics. As we proceed on this journey from the current status to future advancements, there is immense potential for improved viability, less resistance and further enhanced patient outcomes resulting in a new era of antifungal drug development.

20.2 Heterocycles in the Service of Humankind

Heterocyclic compounds, a fascinating class of organic compounds, have profoundly altered science and technology, evolving into substances that have used in a wide range of fields (Liu 2001; Reddy et al. 2004). These substances, which are identified by their ring structures, comprising atoms from at least two distinct elements, have significantly advanced the fields of materials sciences, agriculture, pharmacology and several other disciplines (Abdel-Hafez 2008; Eicher et al. 2013).

20.2.1 Heterocycles in Pharmaceuticals and Medicines

Heterocycles have excellent prospects for drug development because of their amazing structural diversity. Because of their capacity to alter a variety of biological targets, including receptors, enzymes and transporters, medicines covering a variety of therapeutic fields have been developed. Pyridines, pyrimidines, pyrazoles, imidazoles and thiazoles are a few of the most common heterocyclic systems (Aneja et al. 2011; Ansari et al. 2022; Nakamoto et al. 2010). Because of these properties, people in pharmaceutical sciences can create customized molecules with features useful for the treatment of a range of ailments. Numerous important medications that treat a variety of diseases rely on heterocyclic compounds as their foundation. Their complex structures and functional groups offer a design framework or scaffold for designing medications with increased specificity and activity. For instance, the antiviral medication Remdesivir, which has a heterocyclic core, played a crucial role in our fight against the COVID-19 pandemic. The capacity of the drug to block viral RNA polymerase as its mode of action, highlights the significance of heterocycles in resolving several health issues (Gordon et al. 2020).

20.2.2 Key Contributions in Drug Discovery

20.2.2.1 Antiviral and Antibiotic Agents

Several drugs, that are antiviral and antibacterial agents, have been developed based on heterocyclic molecules. Heterocyclic moieties are essential for the action of quinolones like ciprofloxacin, an antibiotic. **Ritonavir** and **nelfinavir** are used to treat HIV infection and AIDS while nucleoside analogs like acyclovir inhibit herpesvirus (Bardsley-Elliot and Plosker 2000; O'Brien and Campoli-Richards 1989; Walmsley et al. 2002). Penicillin G, piperacillin, cefadroxil, VK and cefradine are examples of antibiotics used against bacterial infections (Fig. 20.2) (Boyd 1982; De Marco and Salgado 2017).

20.2.2.2 Anticancer Medications

Cancer is the leading cause of mortality around the world. According to the World Health Organization (WHO) (2020), more than 9 million people died from cancer in 2018 across the globe. Cancer is a broad category of disorders that can affect neighboring tissues and cause aberrant cells to develop uncontrollably. Radiation therapy, chemotherapy and surgery are the primary therapies. However, for majority of tumor types, there is no cancer treatment that is 100% successful, and a few heterocyclic mixtures have proven to be vital to disease treatment. The pyrimidine analogs, cytarabine and gemcitabine, are utilized in leukemia and pancreatic disease treatment, respectively (Kindler et al. 2010; Reese and Schiller 2013). Moreover, the imidazole-containing compound, **imatinib** has changed the therapy of chronic myeloid leukemia (Ramchandren and Schiffer 2009). **Axitinib** has been approved for the treatment of renal cell carcinoma (Wu et al. 2015), **ponatinib** for Philadelphia chromosome-positive acute lymphoblastic leukemia and chronic myeloid leukemia, while olaparib has been approved for advanced ovarian cancer (Fig. 20.2) (Ledermann et al. 2012).

20.2.2.3 Central Nervous System Drugs

In comparison to other diseases, developing therapies for illnesses of the central nervous system (CNS) involve tough efforts. CNS drug candidates usually have a 30% higher development cost than most other drug discovery domains and a 50% lower possibility of being commercialized than other medication prospects (Kesselheim et al. 2015). Compounds with heterocyclic rings are prevalent in drugs that target the CNS. Since many years ago, the market has been flooded with morpholine derivatives and benzodiazepines like diazepam, doxapram, phendimetrazine and reboxetine which have been used as prominent CNS drugs (Fig. 20.2) (Camerman and Camerman 1972; Lenci et al. 2021).

20.2.2.4 Antihypertensive Drugs

Because of alterations in lifestyle and rise in stress, hypertension has become a major problem in people these days. Antihypertensive drugs are used to avoid

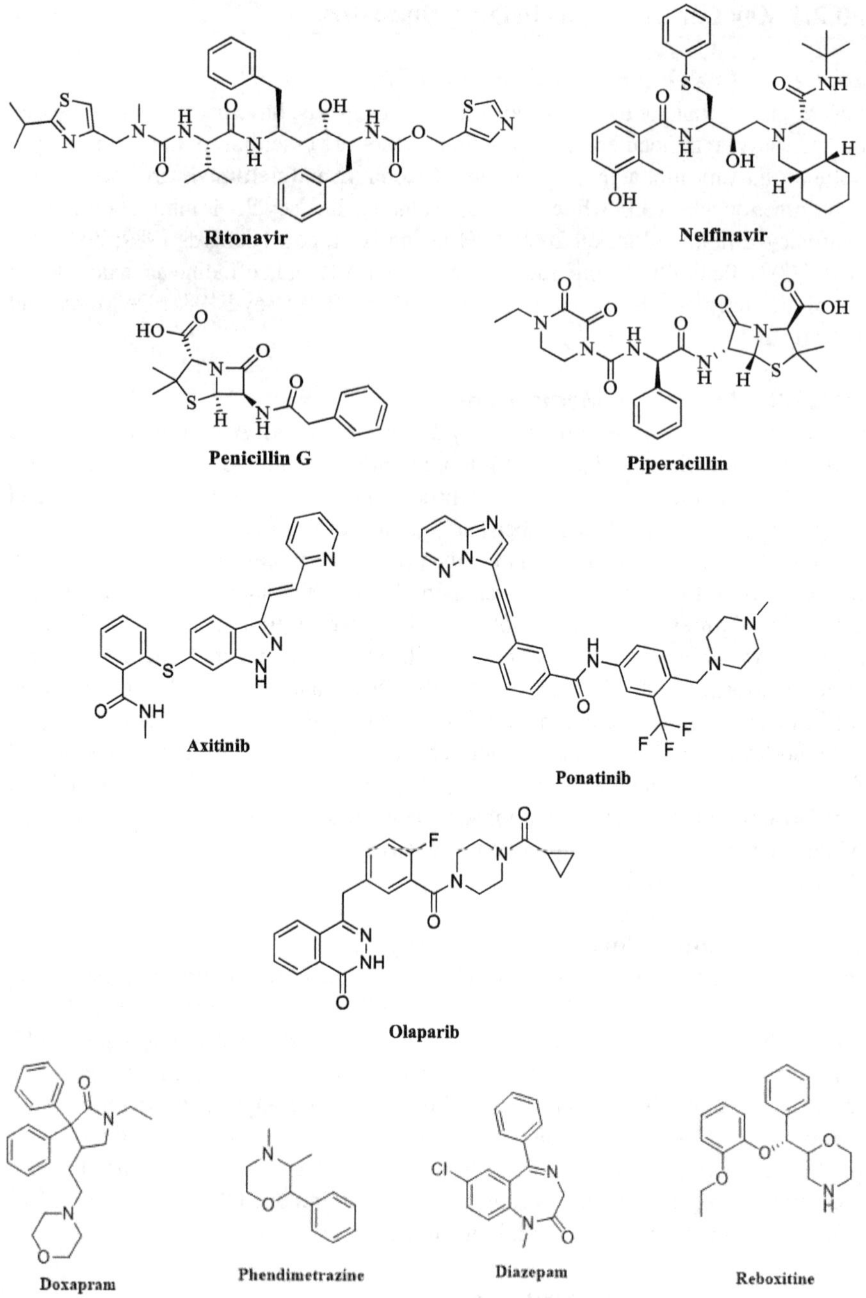

Fig. 20.2 Some biologically active N and O containing heterocyclic scaffolds

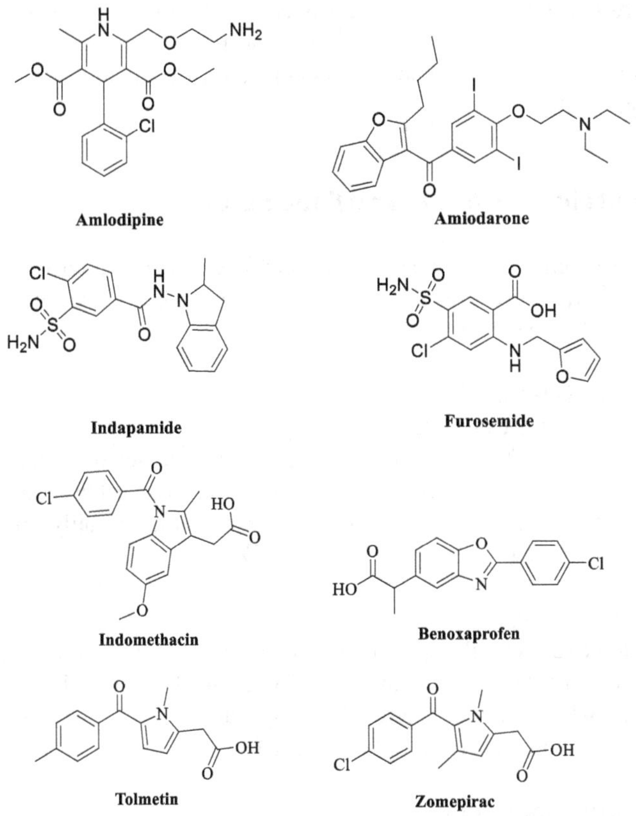

Fig. 20.2 (continued)

cardiovascular problems linked to increased blood pressure. These medications exist in a variety of different forms and all of them work differently to lower blood pressure. While some drain the body of extra water and salt, others relax and enlarge the capillaries or lower the pulse rate. Heterocycles play a crucial role as constituents of cardiovascular drugs, including beta-blockers (**propranolol**), calcium channel blockers (**amlodipine**) and antiarrhythmics (**amiodarone**) (Hamilton et al. 2020; Harrison et al. 1988; Kumar et al. 2003). Indapamide and furosemide are utilized for the treatment of both decompensated heart failure and hypertension (Fig. 20.2) (Anavekar et al. 1979; Stason et al. 1966).

20.2.2.5 Anti-Inflammatory Agents

Anti-inflammatory substances reduce swelling and inflammations. These types of drugs make up about half of analgesics and alleviate pain by reducing inflammation as opposed to opioids which affect the CNS to block pain signaling to the brain. Nonsteroidal anti-inflammatory drugs (NSAIDs) like ibuprofen**, indomethacin** and **celecoxib** contain heterocyclic rings in their structures (Bancos et al. 2009; Kumar et al. 2013). **Benoxaprofen**, **tolmetin** and **zomepirac** are mostly used to decrease levels of hormones that cause pain, soreness, swelling and stiffness in illnesses such as rheumatoid arthritis and osteoarthritis, and especially juvenile rheumatoid arthritis (Astbury 1994) (Fig. 20.2).

20.2.3 Advancements in Agriculture and Food Security

The heterocyclic analogs assume a critical part in different fields, including farming and food security. This section discusses some other ways by which heterocycles are engaged in farming and food security.

20.2.3.1 Pesticides and Herbicides

Numerous pesticides and herbicides utilized in farming contain heterocyclic mixtures as dynamic components. These mixtures are intended to target unwanted growth or weeds while limiting damage to the crops. Pyrethroids, for instance, are a class of engineered pesticides containing heterocyclic rings that are usually utilized in horticulture to control bugs (Ravula and Yenugu 2021).

20.2.3.2 Fungicides

Heterocyclic mixtures are likewise utilized in the improvement of crop yield using fungicides, which are fundamental for shielding crops from fungal diseases. Drugs like triazoles and azoles, which contain heterocyclic rings, are utilized to control different disease-causing microorganisms (Paul et al. 2010).

20.2.3.3 Plant Growth Regulators

Heterocycles are utilized as controllers of plant growth to upgrade crop yield. For example, cytokinins, a class of plant growth regulator that controls cell division and plant development, contain heterocyclic rings (Schäfer et al. 2015).

20.2.3.4 Crop Protection Research

In the last 20 years, 70% of all agrochemicals introduced in the market had at least one heterocyclic moiety (Lamberth and Dinges 2012). Analysts are persistently investigating new heterocyclic compounds for their true capacity in crop security. This includes the union and testing of novel mixtures that can battle rising horticultural difficulties, like pesticide-resistant organisms and new diseases (Lamberth 2013).

20.2.3.5 Sustainable Agriculture

Heterocyclic compounds may also be utilized in economical horticultural practices, like natural cultivation. Natural pesticides and composts frequently depend on regular heterocyclic substances to limit the ecological effect of farming. Heterocyclic mixtures assume a significant part in present day horticulture by adding to the advancement of powerful practices that are harmless to the ecosystem and give answers for crop security, soil improvement and upgraded crop yields. These are all fundamental for food security. (Ahrens et al. 2013).

20.2.4 Use of Heterocycles for Environmental Sustainability and Renewable Energy

Heterocyclic compounds also have important applications in the field of environmental sustainability and renewable energy. Some of the several ways in which heterocycles are involved include:

20.2.4.1 Solar Cells

Organic photovoltaic (OPV) cells, which are a sort of solar cells, frequently integrate heterocyclic moiety as a feature of their light-retaining and charge-shipping materials. These mixtures can upgrade the proficiency and cost adequacy of sustainable power age. For instance, polythiophene derivatives, which contain heterocyclic thiophene rings, are commonly used as electron-donating materials in OPV cells due to their excellent electron mobility and absorption properties (Chung et al. 2016).

20.2.4.2 Battery Innovation

A few heterocycles are being explored for their true energy capacity, especially in the improvement of cutting-edge batteries. They can work on the exhibition and security of battery-powered batteries, making them more appropriate for environment-friendly power frameworks. Heterocyclic compounds are commonly used as electrolyte solvents and additives in lithium-ion batteries. For example, cyclic carbonates like EC (ethylene carbonate) and PC (propylene carbonate), which contain heterocyclic carbonate rings, are essential components of the electrolyte in many lithium-ion batteries (Xu 2004).

20.2.4.3 Hydrogen Storage

Exploration is continuous to track down productive strategies for putting away and delivering hydrogen as a perfect energy transporter. Certain heterocyclic molecules can act as hydrogen stockpiling materials, which is essential for the improvement of hydrogen-based sustainable power frameworks. Certain heterocyclic compounds, such as graphene-like materials containing pyridine, have been studied for their physisorption-based hydrogen storage capabilities. These materials can adsorb hydrogen through weak van der Waals interactions (Wang et al. 2016).

20.2.4.4 Biofuels

Heterocycles can be found in several biofuels and biofuel precursors. They can be found in the various cosmetics of biodiesel, bioethanol and other useful substances, contributing to lower emissions of substances that deplete the ozone layer in comparison to nonrenewable energy sources. Algae-based biofuels often contain heterocyclic compounds, such as fatty acid methyl esters (FAMEs) and lipids, as major constituents. These heterocycles are derived from the cellular structure of microalgae and serve as feedstocks for biodiesel production (Eibler et al. 2017).

20.2.4.5 Environmental Sensors

Heterocyclic compounds are used in the development of sensors and detectors for monitoring environmental parameters (Andersson et al. 2006). These sensors can be employed to detect pollutants, greenhouse gases and other environmental factors critical for sustainability. Heterocyclic compounds like 8-hydroxyquinoline and its derivatives have been employed as fluorescent sensors for metal ion detection in environmental samples (Farruggia et al. 2006). These compounds can selectively bind to specific metal ions, producing distinct fluorescent signals that enable the quantification of metal concentrations.

20.2.4.6 Waste-to-Energy

Heterocyclic compounds are also involved in waste-to-energy technologies (Zhou and Wang 2020). Some of these compounds can assist in the breakdown of organic waste materials into biogas or other forms of renewable energy. Heterocyclic compounds can be part of the organic fraction of municipal solid waste (OFMSW) or agricultural waste. Anaerobic digestion of OFMSW and similar waste streams produce biogas, which consists of methane and carbon dioxide. Heterocyclic compounds contribute to the overall organic content that can be converted into biogas.

20.2.4.7 Water Purification

Certain heterocyclic compounds can be part of water purification processes, helping to remove contaminants and make water safer for consumption or ecosystem health. For example, heterocyclic polymers like polypyrrole have been investigated for their ability to eliminate pollutants, including heavy metals, from water through electrochemical methods (Mahmud et al. 2016).

20.2.4.8 Carbon Capture

Research is going on to develop materials that can capture and store carbon dioxide (CO_2) emission from industrial processes and power plants (Benson and Orr 2008). Heterocyclic compounds are among the chemicals being explored for this purpose (Bhavsar et al. 2023). Metal-organic frameworks (MOFs), a class of porous materials containing heterocyclic ligands, and zeolitic imidazolate frameworks (ZIFs) hold immense potential for gas storage and separation (Liu et al. 2012; Phan et al. 2009; Wang et al. 2008). N-heterocyclic carbenes (NHCs) have gained a lot of attention lately because of their capacity to react stoichiometrically and reversibly in solution with CO_2 (Duong et al. 2004; Van Ausdall et al. 2009). Such heterocyclic frameworks offer an adaptable base for research into the applicability of including Lewis fundamental functioning in a heterocycle for CO_2 separations. Due to pi-electron transfer, from the nearby heteroatoms (X) into the vacant *p*-orbital of the carbene carbon, the suitably functionalized singlet carbenes are stable and manageable (Arduengo III et al. 1991; Dixon and Arduengo III 1991).

20.2.5 Use of Heterocycles for Catalysis and Green Chemistry

Heterocycles, which are organic molecules with at least one noncarbon atom in their ring structures, have become useful tools in the domains of catalysis and green chemistry. They are extremely helpful in fostering ecologically friendly and sustainable chemical processes due to their distinctive chemical features, which include a variety of electronic configurations and functional groups. These examine the crucial function of heterocyclic compounds in green chemistry and catalysis, emphasizing their advantages in lowering environmental impact and promoting sustainable practices.

20.2.5.1 Catalysis with Heterocycles

Heterocycles play a crucial role in catalytic processes, allowing chemical reactions to proceed more quickly while improving selectivity and efficiency. The use of heterocyclic ligands in transition metal catalysis is one of the notable instances. A variety of transformations in organic synthesis are made possible by the stable complexes formed by compounds produced from pyridine, imidazole and other compounds, among others, and transition metals (Phukan et al. 2009; Tseberlidis et al. 2017).

20.2.5.2 Green Solvents and Sustainable Reaction Media

Ionic liquids have become popular as green solvents because of their low volatility, nonflammability and capacity to dissolve a variety of organic and inorganic molecules. These ionic liquids usually contain heterocyclic ions in their compositions. These solvents improve worker safety in the lab and lessen the negative effects of chemical reactions on the environment (Hallett and Welton 2011; Pârvulescu and Hardacre 2007; Plechkova and Seddon 2008). According to the principles of green chemistry, heterocyclic-based solvents also contribute to energy-efficient microwave-assisted processes (Martínez-Palou 2010). The creation of environment-friendly solvents that are less risky to both human health and the environment relies on heterocyclic compounds, which also lessens the environmental impact of different chemical processes.

20.2.5.3 Renewable Feedstocks and Circular Economy

Heterocyclic compounds made from sustainable feedstocks, including biomass or agricultural waste, are replacing conventional petrochemical-based starting materials as viable substitutes. Utilizing these resources enable researchers to produce value-added compounds while reducing their dependency on fossil fuels, in line with the concepts of the circular economy (Sheldon 2014).

20.2.5.4 Reduction of Hazardous Byproducts

The focus of green chemistry is on reducing or eliminating waste and toxic byproducts from chemical operations. In order to promote more sustainable practices, heterocyclic compounds are typically used in reactions that create less waste products (Anastas and Warner 1998).

20.2.5.5 Sustainable Polymers and Materials

In order to crack the problem of plastic waste and advance the field of material science toward a more sustainable future, heterocyclic monomers are essential in the synthesis of sustainable polymers and biodegradable plastics (Ali and Kaneko 2019).

20.2.6 Academic Research and Beyond

Heterocyclic chemistry continues to be a fertile ground for academic research, enabling the synthesis of complex molecules and the discovery of novel chemical transformations. The Buchwald–Hartwig amination, a cornerstone of modern synthetic chemistry, employs heterocyclic substrates for carbon-nitrogen bond formation.

20.2.6.1 Medicinal Chemistry and Drug Discovery

Heterocycles play a crucial role in medicinal chemistry, helping to create drugs that treat a variety of illnesses (Jassas et al. 2023). Heterocyclic scaffolds are included in several medications, including tamoxifen, aspirin and penicillin (Clardy et al. 2009; Jordan 2003; Lichterman 2004). To find novel medicinal treatments with improved potency, selectivity and less adverse effects, researchers are still investigating the synthesis and modification of heterocyclic molecules.

20.2.6.2 Materials Science and Nanotechnology

Heterocyclic compounds are crucial in the construction of innovative materials with specialized features, according to material science. They play main role in the creation of organic electronics, conductive polymers and photovoltaic materials used in solar cells (Nitha et al. 2021). The use of heterocyclic compounds as building blocks for the production of nanoparticles and nanomaterials has also been observed in nanotechnology (Wang et al. 2021).

20.2.6.3 Organic Synthesis and Methodology

The advancement of organic synthesis techniques and heterocycles are at the forefront of science. Their incorporation into complex compounds is frequently a crucial step in the creation of medications, sophisticated materials and natural goods. Synthetic methods are becoming more effective and ecologically benign as heterocycle synthesis strategies development (Arora et al. 2012).

20.2.6.4 Beyond Research

Heterocycles are used in a variety of sectors outside of academics. They are employed in the production of explosives, flavoring agents, perfumes and dyes. Their flexibility and versatility make them essential in a variety of economic areas (Shimotori et al. 2020; Wu et al. 2021).

20.3 Recent Advances in Heterocyclic Drug Discovery

The fungal kingdom is spread over the earth, adapting to almost every climatic and substrate situation (Wang et al. 2021). Fungal infections have increased significantly in recent years. Antifungal chemotherapeutic treatments are ineffective in current medical settings resulting in serious human ailments and mortality (Fisher et al. 2022). Heterocyclic compounds are becoming more and more important in the field of medicinal chemistry as they exhibit a wide range of physiological actions (Jampilek 2019; Kabi and Uzzaman 2022; Qadir et al. 2022). The derivatives of indole, imidazole, thiazole, pyridine, oxadiazole, quinazoline, etc. are particularly appealing among heterocycles containing N- and S-atoms (Tyrell and Quinn 2010; De et al. 2021). The productivity and quality of agricultural crops has significantly reduced because of the persistent attack by fungal pathogens. One of India's most important crops, maize, is deteriorating day by day because of fungal strains, primarily *Fusarium*, *Aspergillus* and *Penicillium* species (Ahmad et al. 2020; Oke et al. 2021; Green et al. 2022).

Extensive use of antifungal drugs and long-term therapies lead to toxicity and drug resistance. Thus, there is an urgent need for effective antifungal agents with fewer adverse effects for the cure of mycoses (De Oliveira et al. 2023). To prevent cross-resistance to currently available therapeutics, new drugs should preferably have chemical features that clearly differ from those of existing agents and with alternative mechanisms of action in order to combat the rapidly developing drug resistance (Cui et al. 2022; Bajpai et al. 2019). Due to wide range of chemical properties and biological activities, heterocyclic compounds, which often contain one or more hetero-atoms in the ring structure (nitrogen, oxygen or sulfur), have been significant in the development of new drugs. The following section mentions some developments in the search for heterocyclic drugs:

(a) **Computational Approaches:** Computational methods, such as molecular docking, molecular dynamics simulations and machine learning algorithms, have become increasingly important in virtual screening and designing of novel heterocyclic compounds. These tools can predict the binding affinity and selectivity of compounds, thus expediting the drug discovery process.
(b) **Fragment-Based Drug Design:** Fragment-based drug design (FBDD) has recently gained popularity. It includes the screening of small fragments that bind to a target protein and then growing them into larger heterocyclic compounds. This strategy may result in the identification of potent and targeted therapeutic candidates.
(c) **Diversity-Oriented Synthesis:** Researchers have been exploring new synthetic methodologies to access diverse libraries of heterocyclic compounds. Diversity-oriented synthesis (DOS) is a strategy that aims to create structurally diverse compounds for screening against various biological targets. This approach helps identify novel drug candidates.
(d) **Biological Target Classes:** Heterocyclic compounds have been designed and tested against a wide range of biological target classes, including G-protein coupled receptors (GPCRs), protein kinases, enzymes and ion channels involved in specific disease pathways. Tailoring heterocyclic structures for these targets has led to the development of targeted therapies.
(e) **Antiviral and Antimicrobial Agents:** The discovery of heterocyclic compounds with antiviral and antimicrobial properties has been a major focus, especially in the context of emerging infectious diseases. Researchers have

been actively searching for compounds effective against viruses like HIV, hepatitis and SARS-CoV-2, as well as against antibiotic-resistant bacteria.

(f) **Fragment-Based Covalent Inhibitors:** In addition to noncovalent binding, the design of covalent inhibitors containing heterocyclic warheads has gained attention. These compounds can form irreversible bonds with their target proteins, enhancing potency and selectivity.

(g) **Central Nervous System (CNS) Drug Discovery:** Heterocyclic compounds have been explored for CNS drug discovery, including compounds for neurodegenerative diseases and psychiatric disorders. Key considerations have been optimizing blood-brain barrier penetration and selectivity for specific neurotransmitter receptors.

(h) **Natural Product-Inspired Heterocycles:** Natural products, such as alkaloids and alkaloid-like compounds, have served as valuable inspiration for the design of heterocyclic drug candidates. These compounds often have complex structures with diverse biological activities.

(i) **Green Chemistry and Sustainability:** There has been a growing emphasis on green chemistry principles in heterocyclic drug discovery, aiming for more sustainable and environment-friendly synthesis processes.

A series of Schiff base derivatives have been synthesized via condensation of substituted maleimides with *p*-toluene sulfonyl hydrazide. The antifungal activity of all these compounds was screened against two harmful fungal strains, including *Aspergillus niger* and *Candida albicans*. Antimicrobial screening data analysis revealed that the evaluated drugs were moderate to excellent inhibitors. It is important to note that derivative **1** (Fig. 20.3) showed more potent antifungal activity than did the regular drug, fluconazole (Bhagare et al. 2020). The thiazolyl-pyrazoline derivatives were synthesized by Altintop et al. and evaluated for their antifungal results against pathogenic molds and yeasts. The Ames assay was used to evaluate the genotoxicity of the most active antifungal derivative. The cytotoxicity of the derivatives was also investigated against human lung adenocarcinoma A549 and mouse embryonic fibroblast NIH/3T3 cells. Due to its notable inhibitory effect on *Candida zeylanoides*, which includes low cytotoxicity against NIH/3T3 cells, no mutagenic effect and low MIC value (250 μg/mL) compared to ketoconazole (MIC 1/4 250 μg/mL), derivative **2** (Fig. 20.3) was chosen as the most promising anticandidal derivative (Altıntop et al. 2015).

As antifungal agents, the fluconazole-based compounds containing 1,2,3-triazole were designed and synthesized. Their antifungal activities were screened *in-vitro* by measuring the minimum inhibitory concentrations (MICs). Compound **3** was initiated to be more potent against fungal pathogens (*Candida*) than the standard drugs fluconazole and amphotericin B. Furthermore, the compound was evaluated *in vivo* against *Candida albicans* in Swiss mice. At a lower dose (0.001 mg/mL), compound **3** was reported to reduce the fungal load in mice by 97.4% without having a significant proliferative impact (Aher et al. 2009). A series of imidazolyl derivatives were synthesized to treat cryptococcal infections. Researchers designed an arrangement of the imidazolyl chromanone oxime ether fragment from oxiconazole and

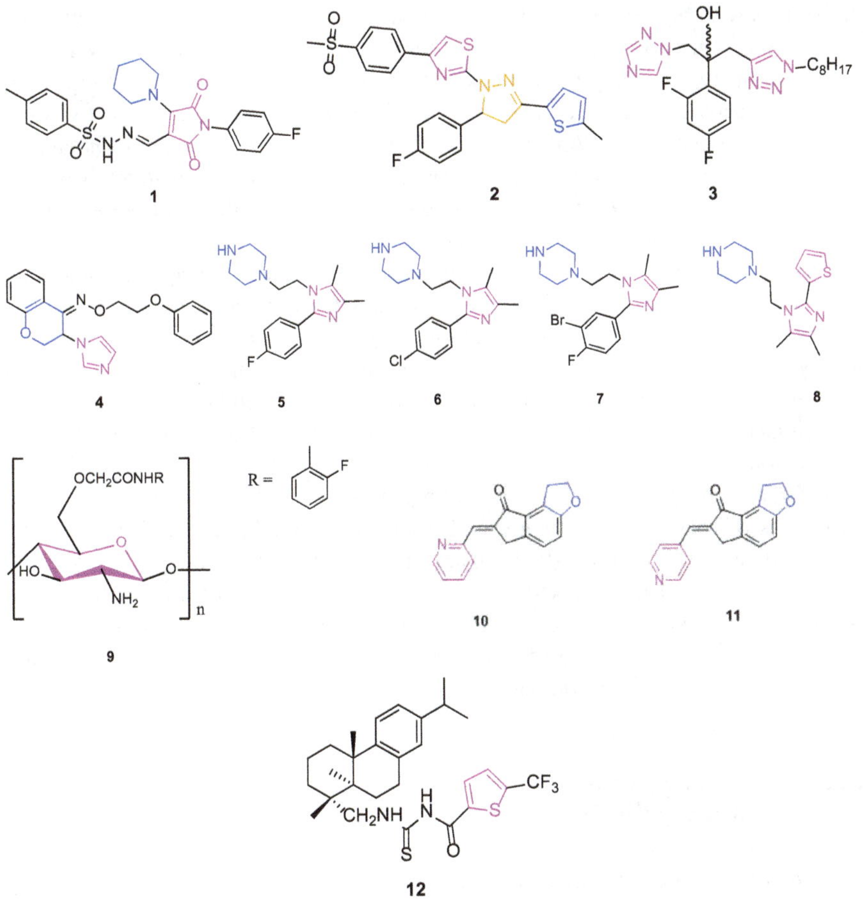

Fig. 20.3 Structures of molecules that belong to the different classes of antifungal drugs

2-phenoxyethyl ether moiety instead of the benzyl ether moiety in omoconazole. The synthesized compounds were screened against *Exophiala dermatitidis, Candida albicans, Aspergillus fumigatus* and *Cryptococcus gattii* for their *in vitro* activities with itraconazole and fluconazole as standard. All the compounds showed potent activity against *C. gattii* with MIC value ≤4 μg/mL. The compound **4** (Fig. 20.3) (MIC 0.5 μg/mL) was the most active derivative (Chandrika and Sharma 2020).

A highly useful synthesis of piperazine derivatives was done. All the synthesized compounds were evaluated for *in vitro* antimicrobial activities. The compounds **5**, **6**, **7** and **8** (Fig. 20.3) showed excellent antibacterial activity against bacterial strains (*E. coli, S. aureus, P. aeruginosa* and *S. typhi*) and fungal strains (*Rhizopus* sp., *C. albicans, A. flavus* and *A. niger*) with zone of inhibition 14–18 mm and 16–22 mm, respectively (Rajkumar et al. 2014). By adding aminobenzene and heterocyclic compounds to carboxymethyl chitosan, Mi et al. created a number of new chitosan derivatives with benzenoid moiety. Their antifungal activity against *C. lagenarium* and *P. asparagi*, two plant pathogenic fungi, was screened in vitro by evaluating reduction in hyphae length and showed that all the products exhibited excellent antifungal activity. The derivative **9** (Fig. 20.3) exhibited excellent activity, at 94.73% and 87.65% (1 mg/ml) against *Colletotrichum lagenarium* and *Phomopsis asparagi*, respectively. Besides, the cytotoxicity of chitosan derivatives was also measured via CCK-8 *in-vitro* on L929 cells and observed that all the derivatives showed low cytotoxicity (Mia et al. 2020).

Scientists reported the expeditious synthesis of new antifungal agents containing 7-arylidene indanone moiety. The antifungal activity of the heterocyclic derivatives was evaluated against four fungal strains namely, *Aspergillus niger*, *Mucor mucedo*, *Rhizophus oryzae* and *Candida albicans*. Most of the compounds showed good inhibition, while the derivatives **10** and **11** (Fig. 20.3) showed large ZOIs against all three fungal strains. Furthermore, the findings using HL-60 cells showed very little to no cytotoxicity of the derivatives (Adole et al. 2021). To investigate the effect of thiophene heterocyclic ring as antifungal agent, Wu et al. designed and synthesized two series of rosin-based derivatives. The effect of thiophene-ring derivatives on *Valsa mali* demonstrated weak to high antifungal activity. The analog **12** (Fig. 20.3) had overall the best results, with an EC_{50} value that was lower than carbendazim (standard) at 0.184 mg/L. Compound **12** was similarly effective against *V. mali* , leading to noticeably increased cell membrane permeability, altered shape and lower pectinase activity in comparison to the control group (Wu et al. 2020).

20.4 Novel Approaches for Antifungal Heterocyclic Compounds

A novel series of heterocyclic azole derivatives were designed and synthesized by Cao et al. with the purpose of developing antifungal agents. The structures of the derivatives were confirmed by means of ^{1}H, ^{13}C-NMR, IR, mass spectral and elemental analysis. The heterocyclic derivatives were screened for antifungal activity against *Candida* spp., *Cryptococcus* spp. and *Aspergillus* spp. Most of the target derivatives exhibited excellent antifungal activities. The most potent compound was **13** (Fig. 20.4) (MIC = 0.25 μg/mL) that exhibited the highest antifungal activity (Cao et al. 2015).

To investigate more active heterocyclic derivatives with antifungal activity, a series of 6-perfluoropropanyl derivatives of quinoline were designed by Fang et al., via a two-step reaction and characterized using mass spectrometry and ^{1}H- NMR spectroscopy. All the compounds were further evaluated for their *in vivo* antifungal activity against fungal strains. At 100 mg/L, the majority of them exhibited excellent *in vivo* antifungal efficacy against *P. oryzae*. Compound **14** (Fig. 20.4), which is very active at 10 mg/L, was the best among them (Fang et al. 2017). A group of heterocyclic compounds was synthesized for their potential as antifungal agents. All

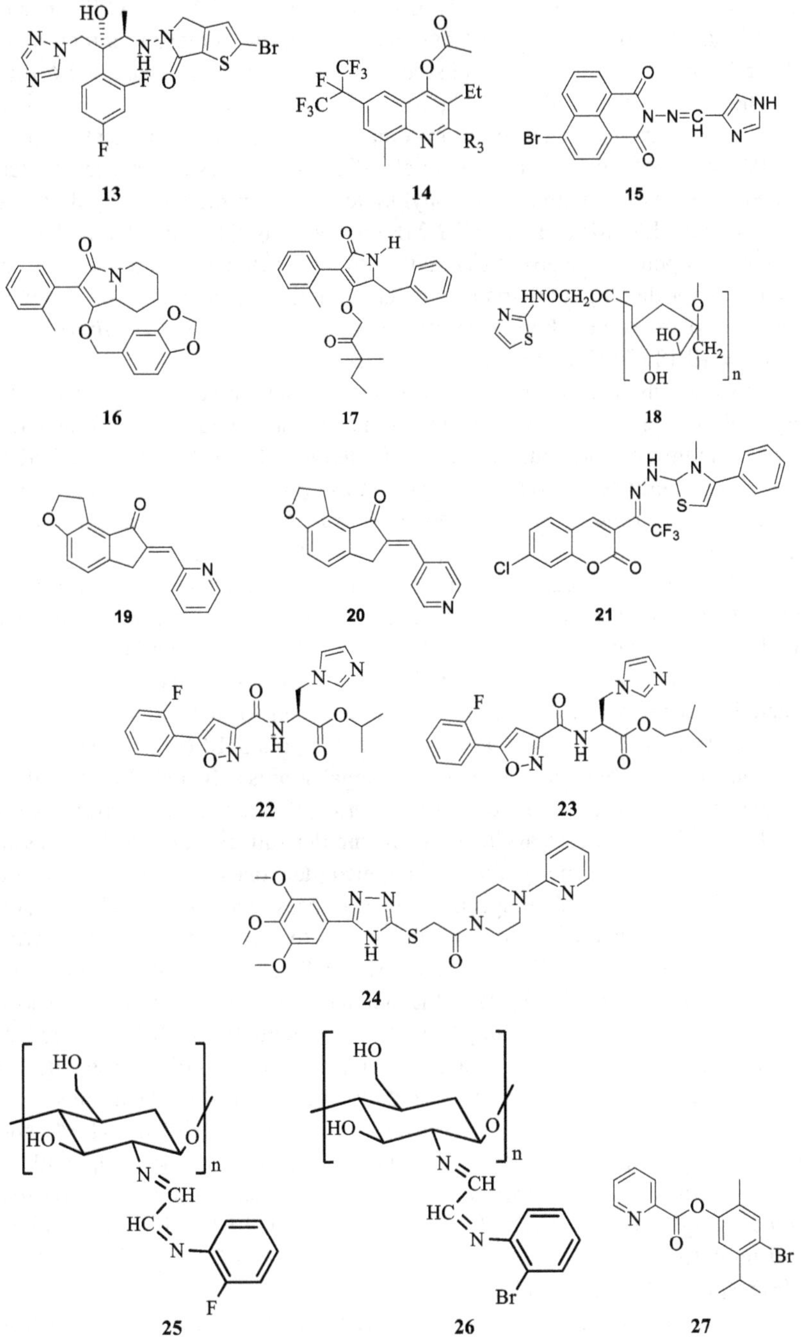

Fig. 20.4 Structures of representative antifungal heterocyclic compounds

the derivatives were characterized by UV-visible, FT-IR, ^{1}H, ^{13}C-NMR spectroscopy and mass spectrometry. The compounds were further screened for their *in vitro* antifungal activity against the fungal strains: *C. albicans, C. glabrata* and *C. tropicalis.* The results revealed that the heterocyclic analog **15** (Fig. 20.4) exhibited significant activity against fungal strains in comparison to the standard drug fluconazole with 200 μg/ml MIC value (Nouman et al. 2023). Two series of heterocyclic tetramic acid derivatives containing piperonyl moiety for their high antifungal potential were developed. ESI-MS, ^{1}H and ^{13}C-NMR data were used for elucidating the structure of the compounds. Twelve different typical plant pathogenic fungi were used as test subjects for the target compounds potential antifungal activity. The compound **16** (Fig. 20.4) showed the highest antifungal activity against *Phytophthora capsici* with $IC_{50} = 19.13$ μM (Wang et al. 2019a, b, c).

A series of amino acid-oriented tetramic acid derivatives was established by Wang et al. in an effort to discover new promising compounds for antifungal activity. The structures of compounds were confirmed by ^{1}H and ^{13}C NMR and ESI-MS spectroscopic methods. Six typical fungal strains prevalent in agriculture, including *P. adianticola, F. graminearum, A. Nees, M. oryzae, G. theae-sinensis* and *S. sclerotiorum,* were used to test for their antifungal activity. Of all the tested compounds, **17** (Fig. 20.4) exhibited highest antifungal inhibition rate against *S. sclerotiorum* as compared with hymexazol (Wang et al. 2019a, b, c). Mi et al. developed and created a number of new heterocyclic inulin derivatives by attaching amino heterocyclic moieties to carboxymethyl inulin. FT-IR and ^{1}H NMR spectroscopy was used to characterize them. All the compounds were screened for their antifungal activity against *Colletotrichum l.* and *Botrytis c.* The compound **18** (Fig. 20.4) exhibited highest antifungal activity against both the fungal strains (Mi et al. 2021). Utilizing a solvent-free microwave technique, a new series of heterocyclic derivatives were created by utilizing 7-heteroarylidene indanone derivatives from easily accessible heterocyclic aldehydes. The synthesized targeted indanone derivatives were characterized by FTIR, ^{1}H, ^{13}C-NMR spectroscopy and mass spectrometry. All derivatives were further investigated for their antifungal activity against fungal strains *Mucor mucido, Rhizophus oryzae, Candida albicans* and *Aspergillus niger.* Comparing the compounds **19** and **20** (Fig. 20.4) to the standard fluconazole, the results showed that they have outstanding antifungal inhibitory potentiality with MIC = 15.62 to 31.25 μg/mL (Adole et al. 2021). Yang et al. reported a series of heterocyclic compounds containing -CF_3 group, synthesized through solvent-free conditions in one-pot reaction. The derivatives were characterized by using IR, HRMS, ^{1}H, ^{13}C-NMR and X-ray single crystal diffraction. The antifungal activity of the compounds was further examined against *Fusarium graminearum, Moniliforme* and *Curvularia lunata.* Among all the synthesized derivatives, **21** (Fig. 20.4) exhibited the highest rate of inhibitor against *F. graminearum* and *C. lunata* at 0.5 mg/mL of 89% and 93.4%, respectively (Yang et al. 2021).

A potent innovative series of heterocyclic compounds was synthesized and tested for *in vitro* antifungal efficacy. Most of the target substances effectively inhibited both *C. albicans* and *Cryptococcus neoformans.* The 2-F substituted analogs, **22** and **23** (Fig. 20.4) showed a preponderance *in vitro* antifungal strength against

Candida spp., *C. neoformans*, *A. fumigatus* and fluconazole-resistant *C. albicans* strains. The values were better than or close to the reference drugs voriconazole and fluconazole (Zhao et al. 2017). Heterocyclic derivatives containing amide moiety of 1,2,4-triazole were synthesized and characterized using NMR spectroscopy, mass spectrometry and elemental analysis. They were also evaluated for antifungal activity against tested strains. The active analog **24** (Fig. 20.4) showed highest antifungal strength against *Botrytis cinerea* with 93.6% inhibition rate at a concentration of 50 μg/mL (Wu et al. 2019).

Chitosan heterocyclic analogs with active imines of halogenated aromatic compounds were synthesized by Lijie et al. They were characterized by FTIR, ^{13}C-NMR spectroscopy and elemental analysis. All derivatives were screened for their *in vitro* antifungal activity against *B. cinereal*, *Fusarium oxysporum f.* sp. *cucumerinum* and *Fusarium oxysporum f.* sp. *niveum* using hyphae measurements. The results showed that chitosan derivatives **25** and **26** (Fig. 20.4) exhibited antifungal activity at 1.0 mg/mL (Wei et al. 2021). Wang et al. developed heterocyclic derivatives with the aim of obtaining different units for carvacrol and thymol esters. The derivatives were characterized by ^{13}C, ^{1}H-NMR and UV-spectroscopy and *in vitro* antifungal activity was screened against various fungal strains, by assessing the mycelium growth rate. The results revealed that compound **27** (Fig. 20.4) showed excellent antifungal activity against *Botrytis cinerea* (Wang et al. 2019a, b, c).

20.5 Antifungal Activity of Heterocyclic Scaffolds

According to the study of antifungal activity, the compounds of spiro-[benzoxazine-piperidin]-one demonstrated wide-spectrum antifungal activity as compared to fluconazole and polyoxin B. Combination testing revealed that these drugs exhibited either additive or synergistic effects when coupled with fluconazole. The compound **28** (Fig. 20.5) showed good antifungal activity with an MIC of 16 μg/ml toward fluconazole-resistant *C. albicans* (Xu et al. 2022a, b). The quinoline and quinazoline derivatives were synthesized and characterized with the goal of finding novel antifungal drugs. Following an evaluation of their antifungal properties against four invasive fungi, it became clear that few target analogs had moderate to exceptional inhibitory potencies. Compound **29** (Fig. 20.5), demonstrated potential antifungal activity and showed great stability in mouse plasma and a low risk of developing resistance (Qin et al. 2022).

The hydrophobic cleft of CYP51 was used to design the analogs of selenide and diselenide derivatives as effective antifungal drugs. The active compound **30** (Fig. 20.5) with low cytotoxic and hemolytic effect and showed remarkable action against the fluconazole-resistant strains C. *glabrata*, C. *albicans*, C. *parapsilosis* and *C. krusei*. Interestingly, compound **30** prevented fluconazole-resistant *C. albicans* from forming biofilms. The *in vivo* bioactivity of **30** is due to its exceptional pharmacological profile and investigation further revealed that this compound dramatically decreased kidney fungal load in mouse model during *C. albicans* infection,

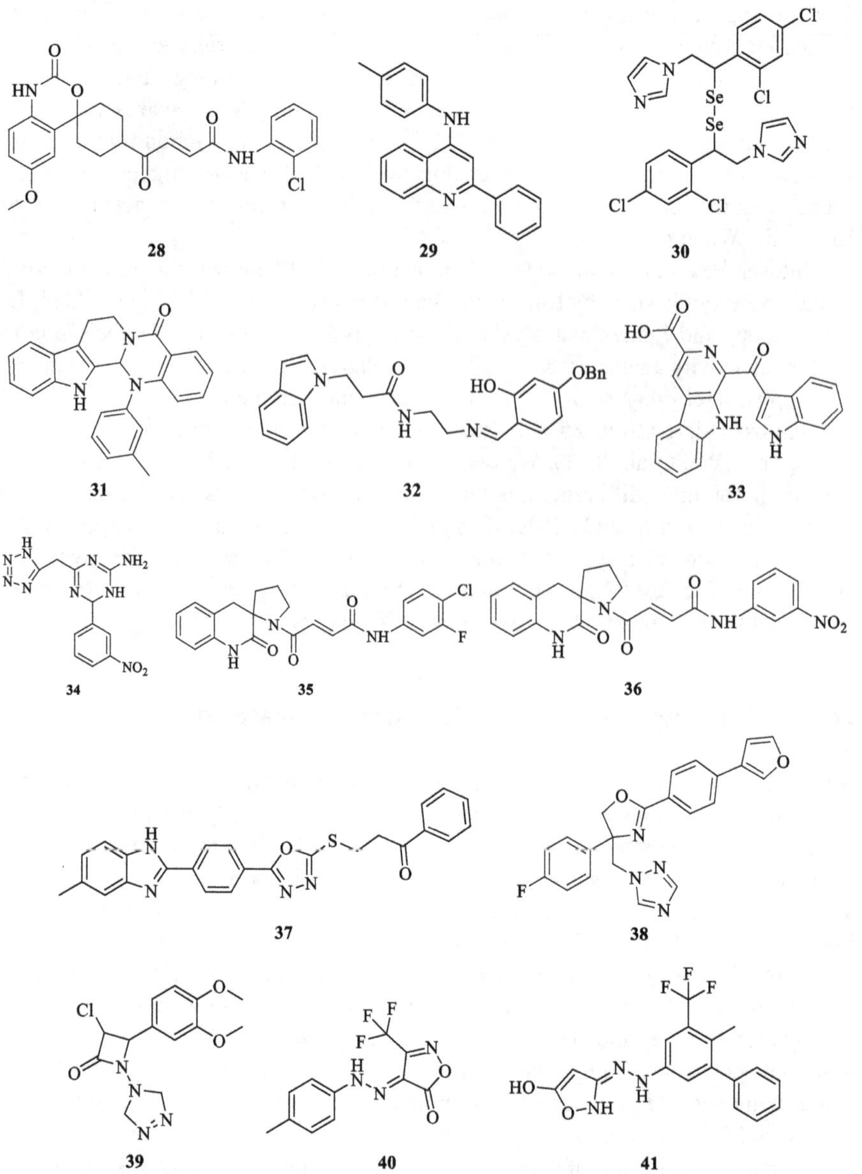

Fig. 20.5 Structural diversity of heterocyclic scaffolds with antifungal activity

besides showing only mild acute toxicity in mice. A docking analysis of *C. albicans* CYP51 revealed the manner in which compound **30** binds to the enzyme. The diselenide compound **30** can be developed further to potentially cure invasive fungal infections (Xu et al. 2022a, b).

Due to major threat from mycosis, particularly superficial fungal infections (SFIs), evodiamine (EVO), a potent ingredient in traditional Chinese medicinal plant, *Evodia rutaecarpa*, showed potent antifungal properties. A number of EVO analogs were synthesized and comprehensive assessment of fungicidal property was conducted in order to identify possible treatment agents against SFIs. Compound **31** (Fig. 20.5) showed good fungicidal activity against *T. rubrum*, *T. mentagrophytes* and *C. albicans* with MIC values of 38, 38 and 2 μg/mL, respectively. Additionally, when administered into guinea pigs with *T. rubrum* infection, compound **31** had outstanding antifungal effects, indicating that it is a desirable molecule and a lead derivative for the conception of antifungal medications. The compound also offered a wonderful, prospective treatment approach for fungal diseases (Liang et al. 2022). Drug resistance has been steadily developing in clinics due to the frequent use of azole antifungals. A number of indole and indoline compounds were prepared and their antifungal efficacy toward *C. albicans* was assessed. The results demonstrated that lead compound had a potent antifungal impact on *C. albicans*. The compound **32** (Fig. 20.5) showed IC_{50} of 18 μg/mL against *C. albicans* (Ma et al. 2022).

A group of significant heterocyclic natural compounds, known as carboline alkaloids, typically exhibit significant bioactivities. Pityriacitrin B belonging to the carboline alkaloid class, was obtained from Chinese Burkholderia species, throughout investigations for active compounds from natural products, indicating potential antifungal properties. Therefore, pityriacitrin compound was examined for its fungicidal activity toward *S. sclerotiorum*, *P. capsici*, *R. solani* and *B. cinerea* in order to develop the potential fungicidal agent. Compound **33** (Fig. 20.5) showed antifungal potential with MIC of 42.36, 42.55, 54.18 and 23.98 μg/mL against *P capsici*, *S. sclerotiorum*, *B. cinereal* and *R. solani*, respectively (Huang et al. 2022).

A series of 1,3,5-triazine derivatives was made in good yield by combining cyanamide, aromatic/heteroaromatic aldehydes and 5-amino-1,2,3,4-tetrazole in acetic acid. The antibacterial and antifungal properties of each heterocycle that was formed were examined. S-triazine and tetrazole were bound in a single, compact structure to develop a new series. When compared to fluconazole, the same compound showed remarkable efficacy against *C. albicans*. The MIC (μg/ml) values of compound **34** (Fig. 20.5) was found to be 0.1312, 0.05086 and 2.1851×10^{-4} against *E coli*, *P. aeruginosa* and *C. albicans*, respectively. Compound **34** showed significant *in vitro* inhibitory action against CYP51 protein as a potential antifungal target (Mekheimer et al. 2022).

The heterocyclic scaffolds were synthesized, and spectroscopic techniques were used to establish their structural details. Active derivatives **35** and **36** (Fig. 20.5) displayed fungicidal activity toward *C. albicans* as compared to fluconazole. The results demonstrated that antifungal properties with MIC values of 4, 4, 8 and 2 μg/mL for compound **35** and 4, 4, 2 and 8 μg/mL for **36** against *A. fumigatus*, *C. neoformans*, *C. albicans* and *A. flavus*, respectively. The antifungal efficacy of these compounds was due to the suppression of the formation of cell wall chitin. These findings showed that the synthetic compounds had both selective and potent antifungal activity for chitin synthase inhibitors (Wu et al. 2022).

The antifungal activity of heterocyclic derivatives was assessed against *C. glabrata, C. krusei, C. parapsilosis* and *C. albicans* using broth dilution method. Results showed that compound **37** (Fig. 20.5) had more inhibitory effect against *C. glabrata, C. albicans, C. krusei* and *C. parapsilopsis* with MIC values of 1.56, 1.56, 0.78 and 3.12 μg/mL, respectively than ketoconazole (MIC = 1.56 μg/mL). The 14-α-demethylase active site was the subject of molecular docking experiments, and estimated ADMET profiles were computed. Additionally, simulations lasting 100 ns were run on the protein–ligand complexes of CYP51-Compound **37**, and the results showed that the mean RMSD value was 0.23 nm. As a result of their close binding to CYP51 protein, lead compound exhibited conformational stability, according to simulation results (Cevik et al. 2021).

The dihydrooxazole analogs were synthesized to find antifungal agents which showed broad ranged and stable metabolism. With MIC values between 0.03 and 0.5 μg/mL against *C. albicans* and between 0.25 and 2 μg/mL for *C. neoformans* and *A. fumigatus*, all compounds demonstrated outstanding antifungal activity. Compound **38** (Fig. 20.5) exhibited antifungal results against *C. albicans* (I), *C. albicans* (II), *C. tropicalis, C. neoformans* and *A. fumigatus* with MIC of <0.03, <0.03, 0.125, 0.25 and 0.25 μg/mL, respectively. Compound **38** displayed adequate pharmacokinetic properties (Zhao et al. 2022). Goal of the study was to synthesize seven azomethine analogs of heterocyclic lactam derivatives. The molecular structures of the heterocyclic compounds were confirmed by elemental analysis, ^{1}H, ^{13}C NMR, UV-Vis and IR spectra. All compounds were tested for *in vitro* antifungal activity against four strains of maize fungi*: M. phaseolina, R. solani, F. verticillioides* and *D. maydis*. The ED_{50} of a β-lactam **39** (Fig. 20.5) was 10.84 μg/mL against *F. verticillioides* (Verma et al. 2022). The bulk of the examined strains of the *Trichophyton* family, *M. canis* and *E. floccosum* were inhibited by almost all isoxazolones and dihydroisoxazoles. The derivatives **40, 41** and **42** (Fig. 20.5) exhibited significant antifungal activity against all the above-mentioned strains (MIC = 0.78–25 μg/mL) and *C. albicans* (MIC = 31.2–62.5 μg/mL) (Elkina et al. 2022).

20.6 Conclusions

Tremendous progress has been made in the development of antifungal medications, based on bioactive heterocycles in recent years, opening up new possibilities for the treatment of fungal diseases. These advancements have been made possible by a better knowledge of fungal biology, the discovery of brand-new therapeutic targets and creative drug discovery techniques. In response to the rising problems of drug resistance and the lack of effective treatments for fungal infections, recent advances in this sector have led to the development of novel antifungal agents with greater efficacy and fewer adverse effects. Additionally, the inventory of bioactive heterocycles with antifungal potential has grown as a result of research into both natural and synthetic substances. Antifungal medication research should concentrate on a few important areas in the future. First, to find novel therapeutic targets and therapeutic strategies to combat resistance in strains, it is essential to continue studies on the processes that are responsible fungal disease and resistance. Second, the development of combination medicines that include bioactive heterocycles with current antifungal drugs may improve treatment effectiveness and lower the chance of resistance emergence. Third, for improved clinical results, medication delivery methods and formulations must be optimized to increase bioavailability and tissue targeting. Furthermore, collaborative work among chemists, microbiologists and pharmacologists will be crucial in quickening the development of therapeutically useful antifungal medications from research development. To bring these novel medicines to market, additional funding for research and development and regulatory assistance will be required.

References

Abdel-Hafez SH (2008) Selenium containing heterocycles: synthesis, anti-inflammatory, analgesic and anti-microbial activities of some new 4-cyanopyridazine-3 (2H) selenone derivatives. Eur J Med Chem 43:1971–1977

Adole VA, More RA, Jagdale BS, Pawar TB, Chobe SS, Shinde RA, Dhonnar SL, Koli PB, Patil AV, Bukane AR, Gacche RN (2021) Microwave prompted solvent-free synthesis of new series of heterocyclic tagged 7-arylidene indanone hybrids and their computational, antifungal, antioxidant, and cytotoxicity study. Bioorganic Chem 115:105259

Aher NG, Pore VS, Mishra NN, Kumar A, Shukla PK, Sharma A, Bhat MK (2009) Synthesis and antifungal activity of 1,2,3-triazole containing fluconazole analogues. Bioorg Med Chem Lett 19:759–763

Ahmad A, Arif MS, Yasmeen T, Riaz M, Rizwan M, Shahzad SM, Ali S, Riaz MA, Sarosh M (2020) Seasonal variations of soil phosphorus and associated fertility indicators in wastewater-irrigated urban aridisol. Chemosphere 239:124725

Ahrens H, Lange G, Müller T, Rosinger C, Willms L, van Almsick A (2013) 4-Hydroxyphenylpyruvate dioxygenase inhibitors in combination with safeners: solutions for modern and sustainable agriculture. Angew Chem Int Ed 52:9388–9398

Ali MA, Kaneko T (2019) Syntheses of aromatic/heterocyclic derived bioplastics with high thermal/mechanical performance. Ind Eng Chem Res 58:15958–15974

Allen D, Wilson D, Drew R, Perfect J (2015) Azole antifungals: 35 years of invasive fungal infection management. Expert Rev Anti-Infect Ther 13:787–798

Altıntop MD, Özdemir A, Turan-Zitouni G, Ilgın S, Atlı Ö, Demirel R, Kaplancıklı ZA (2015) A novel series of thiazolyl–pyrazoline derivatives: synthesis and evaluation of antifungal activity, cytotoxicity and genotoxicity. Eur J Med Chem 92:342–352

Alvarez-Builla J, Barluenga J (2011) Heterocyclic compounds: an introduction. In: Modern heterocyclic chemistry, pp 1–9

Anastas PT, Warner JC (1998) Green chemistry. Frontiers 640:850

Anavekar SN, Ludbrooke A, Louis WJ, Doyle AE (1979) Evaluation of indapamide in the treatment of hypertension. J Cardiovasc Pharmacol 1:389–394

Andersson JT, Hegazi AH, Roberz B (2006) Polycyclic aromatic sulfur heterocycles as information carriers in environmental studies. Anal Bioanal Chem 386:891–905

Aneja DK, Lohan P, Arora S, Sharma C, Aneja KR, Prakash O (2011) Synthesis of new pyrazolyl-2, 4-thiazolidinediones as antibacterial and antifungal agents. Organic Med Chem Lett 1:1–11

Ansari MD, Sagir H, Yadav VB, Verma A, Nazeef M, Shakya S, Siddiqui IR (2022) DFT analysis and synthesis of medicinally important pyrrolo [2, 3-d] pyrimidines by using thiamine hydrochloride as a recyclable organocatalyst in aqueous media. Polycycl Aromat Compd:1–16

Arduengo AJ III, Harlow RL, Kline M (1991) A stable crystalline carbene. J Am Chem Soc 113:361–363

Arora P, Arora V, Lamba HS, Wadhwa D (2012) Importance of heterocyclic chemistry: a review. Int J Pharm Sci Res 3:2947

Astbury C (1994) Non-steroidal anti-inflammatory drugs. Mechanisms and clinical uses. Ann Rheum Dis 53:719

Bajpai VK, Khan I, Shukla S, Kumar P, Rather IA, Park YH, Huh YS, Han YK (2019) Invasive fungal infections and their epidemiology: measures in the clinical scenario. Biotechnol Bioprocess Eng 24:436–444

Bancos S, Bernard MP, Topham DJ, Phipps RP (2009) Ibuprofen and other widely used non-steroidal anti-inflammatory drugs inhibit antibody production in human cells. Cell Immunol 258:18–28

Bardsley-Elliot A, Plosker GL (2000) Nelfinavir: an update on its use in HIV infection. Drugs 59:581–620

Benson SM, Orr FM (2008) Carbon dioxide capture and storage. MRS Bull 33:303–305

Bhagare AM, Aher JS, Gaware MR, Lokhande DD, Kardel AV, Bholay AD, Dhayagude AC (2020) Novel Schiff bases derived from N-aryl maleimide derivatives as an effective antimicrobial agent: theoretical and experimental approach. Bioorg Chem 103:104129

Bhavsar A, Hingar D, Ostwal S, Thakkar I, Jadeja S, Shah M (2023) The current scope and stand of carbon capture storage and utilization a comprehensive review. Case Stud Chem Environ Eng 8:100368

Blokhina SV, Sharapova AV, Ol'khovich MV, Doroshenko IA, Levshin IB, Perlovich GL (2021) Synthesis and antifungal activity of new hybrids thiazolo [4, 5-d] pyrimidines with (1H-1, 2, 4) triazole. Bioorg Med Chem Lett 40:127944

Boyd DB (1982) Theoretical and physicochemical studies on β-lactam antibiotics. In: Penicillins and cephalosporins. Elsevier, pp. 437–545

Camerman A, Camerman N (1972) Stereochemical basis of anticonvulsant drug action. II. Molecular structure of diazepam. J Am Chem Soc 94:268–272

Cao X, Xu Y, Cao Y, Wang R, Zhou R, Chu W, Yang Y (2015) Design, synthesis, and structure-activity relationship studies of novel thienopyrrolidone derivatives with strong antifungal activity against *aspergillus fumigates*. Eur J Med Chem 102:471–476

Cevik UA, Celik I, Işık A, Pillai RR, Tallei TE, Yadav R, Özkay Y, Kaplancıkl ZA (2021) Synthesis, molecular modeling, quantum mechanical calculations and ADME estimation studies of benzimidazole-oxadiazole derivatives as potent antifungal agents. J Mol Struct 1252:132095

Chandrika KVSM, Sharma S (2020) Promising antifungal agents: a minireview. Bioorganic Med Chem 28:115398

Chung C-L, Chen C-Y, Kang H-W, Lin H-W, Tsai W-L, Hsu C-C, Wong K-T (2016) A–D–A type organic donors employing coplanar heterocyclic cores for efficient small molecule organic solar cells. Org Electron 28:229–238

Clardy J, Fischbach MA, Currie CR (2009) The natural history of antibiotics. Curr Biol 19:R437–R441

Cui X, Wang L, Lu Y, Yue C (2022) Development and research progress of anti-drug resistant fungal drugs. J Infect Public Health 15:986–1000

De Marco BA, Salgado HRN (2017) Characteristics, properties and analytical methods of cefadroxil: a review. Crit Rev Anal Chem 47:93–98

De Oliveira HC, Bezerra BT, Rodrigues ML (2023) Antifungal development and the urgency of minimizing the impact of fungal diseases on public health. ACS Bio Med Chem Au 3:137–146

De A, Sarkar S, Majee A (2021) Recent advances on heterocyclic compounds with antiviral properties. Chem Heterocycl Compd 57:410–416

Denning DW (2003) Echinocandin antifungal drugs. Lancet 362:1142–1151

Di Santo R, Tafi A, Costi R, Botta M, Artico M, Corelli F, Forte M, Caporuscio F, Angiolella L, Palamara AT (2005) Antifungal agents. 11. N-substituted derivatives of 1-[(aryl)(4-aryl-1 H-pyrrol-3-yl) methyl]-1 H-imidazole: synthesis, anti-candida activity, and QSAR studies. J Med Chem 48:5140–5153

Dixon DA, Arduengo AJ III (1991) Electronic structure of a stable nucleophilic carbene. J Phys Chem 95:4180–4182

Duong HA, Tekavec TN, Arif AM, Louie J (2004) Reversible carboxylation of N-heterocyclic carbenes. Chem Commun:112–113

Eibler D, Hammerschick T, Buck L, Vetter W (2017) Up to 21 different sulfur-heterocyclic fatty acids in rapeseed and mustard oil. J Am Oil Chem Soc 94:893–903

Eicher T, Hauptmann S, Speicher A (2013) The chemistry of heterocycles: structures, reactions, synthesis, and applications. John Wiley & Sons

Elkina NA, Shchegolkov EV, Burgart YV, Agafonova NA, Perminova AN, Gerasimova NA, Makhaeva GF, Rudakova EV, Kovaleva NV, Boltneva NP, Serebryakova OG, Borisevich SS, Evstigneeva NP, Zilberberg NV, Kungurov NV, Saloutin VI (2022) Synthesis and biological evaluation of polyfluoroalkyl-containing 4-arylhydrazinylidene-isoxazoles as antifungal agents with antioxidant activity. Russ Chem Bull 254:109935

El-Sayed A, Abu-Bakr S, Swelam S, Khaireldin N, Shoueir K, Khalil A (2022) Applying nanotechnology in the synthesis of benzimidazole derivatives: a pharmacological approach. Biointerface Res Appl Chem 12:992–1005

Fang YM, Zhang RR, Shen ZH, Wu HK, Tan CX, Weng JQ, Xu TM, Liu XH (2017) Synthesis, antifungal activity, and SAR study of some new 6-perfluoropropanyl quinoline derivatives. Molecules 55:240–245

Farruggia G, Iotti S, Prodi L, Montalti M, Zaccheroni N, Savage PB, Trapani V, Sale P, Wolf FI (2006) 8-Hydroxyquinoline derivatives as fluorescent sensors for magnesium in living cells. J Am Chem Soc 128:344–350

Fisher MC, Hawkins NJ, Sanglard D, Gurr SJ (2018) Worldwide emergence of resistance to antifungal drugs challenges human health and food security. Science 360:739–742

Fisher MC, Alastruey-Izquierdo A, Berman J, Bicanic T, Bignell EM, Bowyer P, Bromley M, Brüggemann R, Garber G, Cornely OA, Gurr SJ, Harrison TS, Kuijper E, Rhodes J, Sheppard DC, Warris A, White PL, Xu J, Zwaan B, Verweij PE (2022) Tackling the emerging threat of antifungal resistance to human health. Nat Rev Microbiol 20:557–571

Gallis HA, Drew RH, Pickard WW (1990) Amphotericin B: 30 years of clinical experience. Rev Infect Dis 12:308–329

Garibotto FM, Garro AD, Masman MF, Rodríguez AM, Luiten PG, Raimondi M, Zacchino SA, Somlai C, Penke B, Enriz RD (2010) New small-size peptides possessing antifungal activity. Bioorg Med Chem 18:158–167

Gordon CJ, Tchesnokov EP, Woolner E, Perry JK, Feng JY, Porter DP, Götte M (2020) Remdesivir is a direct-acting antiviral that inhibits RNA-dependent RNA polymerase from severe acute respiratory syndrome coronavirus 2 with high potency. J Biol Chem 295:6785–6797

Green ER, Fakhoury JN, Monteith AJ, Pi H, Giedroc DP, Skaar EP (2022) Bacterial hydrophilins promote pathogen desiccation tolerance. Cell Host Microbe 30:975–987

Gupte M, Kulkarni P, Ganguli B (2002) Antifungal antibiotics. Appl Microbiol Biotechnol 58:46–57

Hallett JP, Welton T (2011) Room-temperature ionic liquids: solvents for synthesis and catalysis. 2. Chem Rev 111:3508–3576

Hamilton D, Nandkeolyar S, Lan H, Desai P, Evans J, Hauschild C, Choksi D, Abudayyeh I, Contractor T, Hilliard A (2020) Amiodarone: a comprehensive guide for clinicians. Am J Cardiovasc Drugs 20:549–558

Harrison MR, Smith MD, Nissen SE, Grayburn PA, DeMaria AN (1988) Use of exercise Doppler echocardiography to evaluate cardiac drugs: effects of propranolol and verapamil on aortic blood flow velocity and acceleration. J Am Coll Cardiol 11:1002–1009

Huang D, Zhang Z, Li Y, Liu F, Huang W, Min Y, Wang K, Yang J, Cao C, Gong Y, Ke S (2022) Carboline derivatives based on natural pityriacitrin as potential antifungal agents. Phytochem Lett 48:100–105

Jampilek J (2019) Heterocycles in medicinal chemistry. Molecules 24:3839

Jassas RS, Naeem N, Sadiq A, Mehmood R, Alenazi NA, Al-Rooqi MM, Mughal EU, Alsantali RI, Ahmed SA (2023) Current status of N-, O-, S-heterocycles as potential alkaline phosphatase inhibitors: a medicinal chemistry overview. RSC Adv 13:16413–16452

Jeong W, Keighley C, Wolfe R, Lee WL, Slavin MA, Kong DCM, Chen S-A (2019) The epidemiology and clinical manifestations of mucormycosis: a systematic review and meta-analysis of case reports. Clin Microbiol Infect 25:26–34

Jordan VC (2003) Tamoxifen: a most unlikely pioneering medicine. Nat Rev Drug Discov 2:205–213

Kabi E, Uzzaman M, A (2022) Review on biological and medicinal impact of heterocyclic compounds, Results Chem 4:100606

Kathiravan MK, Salake AB, Chothe AS, Dudhe PB, Watode RP, Mukta MS, Gadhwe S (2012) The biology and chemistry of antifungal agents: a review. Bioorg Med Chem 20:5678–5698

Kesselheim AS, Hwang TJ, Franklin JM (2015) Two decades of new drug development for central nervous system disorders. Nat Rev Drug Discov 14:815

Kindler HL, Niedzwiecki D, Hollis D, Sutherland S, Schrag D, Hurwitz H, Innocenti F, Mulcahy MF, O'Reilly E, Wozniak TF (2010) Gemcitabine plus bevacizumab compared with gemcitabine plus placebo in patients with advanced pancreatic cancer: phase III trial of the Cancer and Leukemia Group B (CALGB 80303). J Clin Oncol 28:3617

Kumar KA, Ganguly K, Mazumdar K, Dutta NK, Dastidar SG, Chakrabarty AN (2003) Amlodipine: a cardiovascular drug with powerful antimicrobial property. Acta Microbiol Pol 52:285–292

Kumar V, Kaur K, Gupta GK, Gupta AK, Kumar S (2013) Developments in synthesis of the anti-inflammatory drug, celecoxib: a review. Recent Patents Inflamm Allergy Drug Discov 7:124–134

Lamberth C (2013) Heterocyclic chemistry in crop protection. Pest Manag Sci 69:1106–1114

Lamberth C, Dinges J (2012) The significance of heterocycles for pharmaceuticals and agrochemicals. In: Bioactive heterocyclic compound classes: agrochemicals, pp 1–20

Ledermann J, Harter P, Gourley C, Friedlander M, Vergote I, Rustin G, Scott C, Meier W, Shapira-Frommer R, Safra T (2012) Olaparib maintenance therapy in platinum-sensitive relapsed ovarian cancer. N Engl J Med 366:1382–1392

Lenci E, Calugi L, Trabocchi A (2021) Occurrence of morpholine in central nervous system drug discovery. ACS Chem Neurosci 12:378–390

Liang Y, Zhang H, Zhang X, Peng Y, Deng J, Wang Y, Li R, Liu L, Wang Z (2022) Discovery of evodiamine derivatives as potential lead antifungal agents for the treatment of superficial fungal infections. Bioorg Chem 127:105981

Lichterman BL (2004) Aspirin: the story of a wonder drug. BMJ 329:1408

Liu R-S (2001) Synthesis of oxygen heterocycles via alkynyltungsten compounds. Pure Appl Chem 73:265–269

Liu Y, Wang ZU, Zhou H-C (2012) Recent advances in carbon dioxide capture with metal-organic frameworks. Greenhouse Gases Sci Technol 2:239–259

Ma J, Jiang Y, Zhuang X, Chen H, Shen Y, Mao Z, Rao G, Wang R (2022) Discovery of novel indole and indoline derivatives against Candida albicans as potent antifungal agents. Bioorg Med Chem Lett 71:128826

Mahmud HNME, Huq AO, Binti Yahya R (2016) The removal of heavy metal ions from wastewater/aqueous solution using polypyrrole-based adsorbents: a review. RSC Adv 6:14,778–14,791

Martínez-Palou R (2010) Microwave-assisted synthesis using ionic liquids. Mol Divers 14:3–25

Mehandi R, Arif R, Rana M, Ahmedi S, Sultana R, Khan MS, Maseet M, Khanuja M, Manzoor N, Rahisuddin NN (2021) Synthesis, characterization, DFT calculation, antifungal, antioxidant, CT-DNA/pBR322 DNA interaction and molecular docking studies of heterocyclic analogs. J Mol Struct 1245:131248

Mehandi R, Twala C, Ahmedi S, Fatima A, Ul Islam K, Rana M, Sultana R, Manzoor N, Javed S, Haque MM, Iqbal J, Rahisuddin, Nishat N (2023) 1,3,4-oxadiazole derivatives: synthesis, characterization, antifungal activity, DNA binding investigations, TD-DFT calculations, and molecular modelling. J Biomol Struct Dyn:1–33

Mekheimer RA, Rahma GEDA, Elmonem MA, Yahia R, Hisham M, Hayallah AM, Mostafa SM, Abo-Elsoud FA, Sadek KU (2022) New s-Triazine/Tetrazole conjugates as potent antifungal and antibacterial agents: design, molecular docking and mechanistic study. Pharmaceuticals 1267:133615

Mi Y, Zhang J, Han X, Tan W, Miaoa Q, Cui J, Li Q, Guo Z (2021) Modification of carboxymethyl inulin with heterocyclic compounds: synthesis, characterization, antioxidant and antifungal activities. Int J Biol Macromol 181:572–581

Mia Y, Zhanga J, Chena Y, Suna X, Tana W, Lia Q, Guo Z (2020) New synthetic chitosan derivatives bearing benzenoid/heterocyclic moieties with enhanced antioxidant and antifungal activities. Carbohydr Polym 249:116

Ming LS, Jamalis J, Al-Maqtari HM, Rosli MM, Sankaranarayanan M, Chander S, Fun H-K (2017) Synthesis, characterization, antifungal activities and crystal structure of thiophene-based heterocyclic chalcones. Chemical Data Collections 9:104–113

Moudgal V, Sobel J (2010) Antifungals to treat Candida albicans. Expert Opin Pharmacother 11:2037–2048

Nakamoto K, Tsukada I, Tanaka K, Matsukura M, Haneda T, Inoue S, Murai N, Abe S, Ueda N, Miyazaki M (2010) Synthesis and evaluation of novel antifungal agents-quinoline and pyridine amide derivatives. Bioorg Med Chem Lett 20:4624–4626

Nami S, Mohammadi R, Vakili M, Khezripour K, Mirzaei H, Morovati H (2019) Fungal vaccines, mechanism of actions and immunology: a comprehensive review. Biomed Pharmacother 109:333–344

Nigam PK (2015) Antifungal drugs and resistance: current concepts. Our Dermatol Online 6:212

Nitha PR, Soman S, John J (2021) Indole fused heterocycles as sensitizers in dye-sensitized solar cells: an overview. Mater Adv 2:6136–6168

Nouman, Rana M, Ahmedi S, Arif R, Manzoor N, Rahisuddin (2023) 4-Bromo-1,8-naphthalimide derivatives as antifungal agents: synthesis, characterization, DNA binding, molecular docking, antioxidant and ADMET studies, Polycycl Aromat Compd, 44:748–772

O'Brien JJ, Campoli-Richards D (1989) Acyclovir: an updated review of its antiviral activity, pharmacokinetic properties and therapeutic efficacy. Drugs 37:233–309

Oke OE, Uyanga VA, Iyasere OS, Oke FO, Majekodunmi BC, Logunleko MO, Abiona JA, Nwosu EU, Abioja MO, Daramola JO, Onagbesan OM (2021) Environmental stress and livestock productivity in hot-humid tropics: alleviation and future perspectives. J Therm Biol 100:103077

Pagano L, Offidani M, Fianchi L, Nosari A, Candoni A, Picardi M, Corvatta L, D'Antonio D, Girmenia C, Martino P (2004) Mucormycosis in hematologic patients. Haematologica 89:207–214

Pârvulescu VI, Hardacre C (2007) Catalysis in ionic liquids. Chem Rev 107:2615–2665

Paul PA, McMullen MP, Hershman DE, Madden LV (2010) Meta-analysis of the effects of triazole-based fungicides on wheat yield and test weight as influenced by Fusarium head blight intensity. Phytopathology 100:160–171

Perfect JR, Casadevall A (2006) Fungal molecular pathogenesis: what can it do and why do we need it? Molecular principles of fungal pathogenesis, pp 1–11

Phan A, Doonan CJ, Uribe-Romo FJ, Knobler CB, O'keeffe M, Yaghi OM (2009) Synthesis, structure, and carbon dioxide capture properties of zeolitic imidazolate frameworks

Phukan M, Borah KJ, Borah R (2009) Henry reaction in environmentally benign methods using imidazole as catalyst. Green Chem Lett Rev 2:249–253

Plechkova NV, Seddon KR (2008) Applications of ionic liquids in the chemical industry. Chem Soc Rev 37:123–150

Prakash H, Ghosh AK, Rudramurthy SM, Singh P, Xess I, Savio J, Pamidimukkala U, Jillwin J, Varma S, Das A (2019) A prospective multicenter study on mucormycosis in India: epidemiology, diagnosis, and treatment. Med Mycol 57:395–402

Qadir T, Amin A, Sharma PK, Jeelani I, Abe H (2022) A review on medicinally important heterocyclic compounds. Open Med Chem J 16. https://doi.org/10.2174/18741045-v16-e2202280

Qiao L, Zhai Z-W, Cai P-P, Tan C-X, Weng J-Q, Han L, Liu X-H, Zhang Y-G (2019) Synthesis, crystal structure, antifungal activity, and docking study of difluoromethyl pyrazole derivatives. J Heterocyclic Chem 56:2536–2541

Qin TH, Liu JC, Zhang JY, Tang LX, Ma YN, Yang R (2022) Synthesis and biological evaluation of new 2-substituted-4-amino-quinolines and -quinazoline as potential antifungal agents. Bioorg Med Chem Lett 72:128877

Rajkumar R, Kamaraj A, Krishnasamy K (2014) Synthesis, spectral characterization and biological evaluation of novel 1-(2-(4,5-dimethyl-2-phenyl1H-imidazol-1-yl)ethyl)piperazine derivatives. J Saudi Chem Soc 18:735–743

Ramchandren R, Schiffer CA (2009) Dasatinib in the treatment of imatinib refractory chronic myeloid leukemia. Targets Ther, Biologics, pp 205–214

Rana M, Fatima A, Siddiqui N, Ahmedi S, Dar SH, Manzoor N, Javed S, Rahisuddin. (2023a) Carbothioamide based Pyrazoline derivative: synthesis, single crystal structure, DFT/TD-DFT, Hirshfeld surface analysis and biological studies. Polycycl Aromat Compd 43(7):6181–6201

Rana M, Hungyo H, Parashar P, Ahmad S, Mehandia R, Tandonc V, Razad K, Assirie MA, Alie TE, El-Bahyf ZM, Rahisuddin (2023b) Design, synthesis, X-ray crystal structures, anticancer, DNA binding, and molecular modelling studies of pyrazole–pyrazoline hybrid derivatives. RSC Adv 13:26766–26779

Rani N, Sharma A, Kumar Gupta G, Singh R (2013) Imidazoles as potential antifungal agents: a review. Mini Rev Med Chem 13:1626–1655

Ravula AR, Yenugu S (2021) Pyrethroid based pesticides–chemical and biological aspects. Crit Rev Toxicol 51:117–140

Reddy PVG, Kiran YBR, Reddy CS, Reddy CD (2004) Synthesis and antimicrobial activity of novel phosphorus heterocycles with exocyclic P–C link. Chem Pharm Bull 52:307–310

Reese ND, Schiller GJ (2013) High-dose cytarabine (HD araC) in the treatment of leukemias: a review. Curr Hematol Malig Rep 8:141–148

Richardson M, Lass-Flörl C (2008) Changing epidemiology of systemic fungal infections. Clin Microbiol Infect 14:5–24

Ryu C-K, Lee R-Y, Kim NY, Kim YH, Song AL (2009) Synthesis and antifungal activity of benzo [d] oxazole-4, 7-diones. Bioorg Med Chem Lett 19:5924–5926

Schäfer M, Brütting C, Meza-Canales ID, Großkinsky DK, Vankova R, Baldwin IT, Meldau S (2015) The role of cis-zeatin-type cytokinins in plant growth regulation and mediating responses to environmental interactions. J Exp Bot 66:4873–4884

Sharma B, Nonzom S (2021) Superficial mycoses, a matter of concern: global and Indian scenario-an updated analysis. Mycoses 64:890–908

Sheldon RA (2014) Green and sustainable manufacture of chemicals from biomass: state of the art. Green Chem 16:950–963

Shimotori Y, Hoshi M, Ogawa N, Miyakoshi T, Kanamoto T (2020) Synthesis, antibacterial activities, and sustained perfume release properties of optically active5-hydroxy-and 5-acetoxyalkanethioamide analogues. Heterocycl Commun 26:84–98

Stason WB, Cannon PJ, Heinemann HO, Laragh JH (1966) Furosemide: a clinical evaluation of its diuretic action. Circulation 34:910–920

Taylor AP, Robinson RP, Fobian YM, Blakemore DC, Jones LH, Fadeyi O (2016) Modern advances in heterocyclic chemistry in drug discovery. Org Biomol Chem 14:6611–6637. https://doi.org/10.1039/C6OB00936K

Tseberlidis G, Intrieri D, Caselli A (2017) Catalytic applications of pyridine-containing macrocyclic complexes. Eur J Inorg Chem 2017:3589–3603

Tyrell JA, Quinn LD (2010) Fundamentals of heterocyclic chemistry: importance in nature and in the synthesis of pharmaceuticals, 1st edn. Wiley, pp 8–98

Van Ausdall BR, Glass JL, Wiggins KM, Aarif AM, Louie J (2009) A systematic investigation of factors influencing the decarboxylation of imidazolium carboxylates. J Org Chem 74:7935–7942

Verma D, Sharma S, Sahni T, Kaur H, Kaur S (2022) Designing, antifungal and structure activity relationship studies of Azomethines and β-lactam derivatives of aza heterocyclic amines. J Indian Chem Soc 99:100587

Walmsley S, Bernstein B, King M, Arribas J, Beall G, Ruane P, Johnson M, Johnson D, Lalonde R, Japour A (2002) Lopinavir–ritonavir versus nelfinavir for the initial treatment of HIV infection. N Engl J Med 346:2039–2046

Wang B, Côté AP, Furukawa H, O'Keeffe M, Yaghi OM (2008) Colossal cages in zeolitic imidazolate frameworks as selective carbon dioxide reservoirs. Nature 453:207–211

Wang Z, Sun L, Xu F, Zhou H, Peng X, Sun D, Wang J, Du Y (2016) Nitrogen-doped porous carbons with high performance for hydrogen storage. Int J Hydrog Energy 41:8489–8497
Wang S, Bao L, Song D, Wang J, Cao X (2019a) Heterocyclic lactam derivatives containing piperonyl moiety as potential antifungal agents. Bioorg Med Chem Lett 29:126661
Wang S, Bao L, Song D, Wang J, Cao X, Ke S (2019b) Amino acid-oriented poly-substituted heterocyclic tetramic acid derivatives as potential antifungal agents. Eur J Med Chem 179:567–575
Wang K, Jiang S, Yang Y, Fan L, Su F, Ye M (2019c) Synthesis and antifungal activity of carvacrol and thymol esters with heteroaromatic carboxylic acids. Nat Prod Res 33:13
Wang Y, Chang J-P, Xu R, Bai S, Wang D, Yang G-P, Sun L-Y, Li P, Han Y-F (2021) N-heterocyclic carbenes and their precursors in functionalised porous materials. Chem Soc Rev 50:13559–13586
Wei L, Zhang J, Tan W, Wang G, Li Q, Dong F, Guo Z (2021) Antifungal activity of double Schiff bases of chitosan derivatives bearing active halogeno-benzenes. Int J Biol Macromol 179:292–298
World Health Organization (2020) Cancer: key facts
Wu P, Nielsen TE, Clausen MH (2015) FDA-approved small-molecule kinase inhibitors. Trends Pharmacol Sci 36:422–439
Wu W, Jiang Y, Fei Q, Du H, Yang (2019) Synthesis and antifungal activity of novel 1,2,4-triazole derivatives containing an amide moiety. J Heterocyclic Chem 57:1379–1386
Wu C, Tao P, Li J, Gao Y, Shang S, Song Z (2020) Antifungal application of pine derived products for sustainable forest resource exploitation. Ind Crop Prod 143:111892
Wu J, Jiang Y, Lian Z, Li H, Zhang J (2021) Computational design and screening of promising energetic materials: the coplanar family of novel heterocycle-based explosives. Int J Quantum Chem 121:e26788
Wu H, Du C, Xu Y, Liu L, Zhou X, Ji Q (2022) Design, synthesis, and biological evaluation of novel spiro[pyrrolidine-2,3′-quinolin]-2′-one derivatives as potential chitin synthase inhibitors and antifungal agents. Eur J Med Chem 233:114208
Xu K (2004) Nonaqueous liquid electrolytes for lithium-based rechargeable batteries. Chem Rev 104:4303–4418
Xu Y, Shen Y, Du C, Liu L, Wu H, Ji Q (2022a) Spiro[benzoxazine-piperidin]-one derivatives as chitin synthase inhibitors and antifungal agents: design, synthesis and biological evaluation. Eur J Med Chem 243:114723
Xu H, Mou Y, Guo ZR, Yan Z, An R, Wang X, Su X, Hou Z, Guo C (2022b) Discovery of novel selenium-containing azole derivatives as antifungal agents by exploiting the hydrophobic cleft of CYP51. Eur J Med Chem 243:114707
Yang G, Shi L, Pan Z, Wu L, Fan L, Wang C, Xua C, Liang J (2021) The synthesis of coumarin thiazoles containing a trifluoromethyl group and their antifungal activities. Arabian J Chem 14:102880
Yang L, Bo XW, Sun L, Zhang C, Hua Jin C (2022) SAR analysis of heterocyclic compounds with monocyclic and bicyclic structures as antifungal agents. ChemMedChem 17:e202200221
Zhao S, Zhang X, Wei P, Su X, Zhao L, Wu M, Hao C, Liu C, Zhao D, Cheng M (2017) Design, synthesis and evaluation of aromatic heterocyclic derivatives as potent antifungal agents. Eur J Med Chem 137:96–107
Zhao L, Sun Y, Yin W, Tian L, Sun N, Zheng Y, Zhang C, Zhao S, Su X, Zhao D, Cheng M (2022) Design, synthesis, and biological activity evaluation of 2-(benzo[b]thiophen-2-yl)-4-phenyl-4,5-dihydrooxazole derivatives as broad-spectrum antifungal agents. Eur J Med Chem 228:113987
Zhou C, Wang Y (2020) Recent progress in the conversion of biomass wastes into functional materials for value-added applications. Sci Technol Adv Mater 21:787–804

Part VI

Natural Products Derived from Microbes and Other Natural Sources

Bioactive Potential of *Streptomyces* Spp. Against Diverse Pathogenic Fungi

21

Harsha, Munendra Kumar, Prateek Kumar, Renu Solanki, and Monisha Khanna Kapur

21.1 Introduction

Around 1.5 billion individuals suffer from fungal infections estimated worldwide. Fungi are eukaryotic microorganisms including molds and yeasts that exist widely in nature. Pathogenic fungi, like *Aspergillus, Colletotrichum, Alternaria, Botrytis, Magnoporthe,* and *Cryptococcus* spp., threaten crop growth and human foodstuffs. The ability of fungi to successfully adapt to a wide range of spectra leads to the generation of mycotoxins (Van der Fels-Klerx and Camenzuli 2016). In the case of crops, a huge economic loss has been incurred during growth, storage, and transport due to mold rot, mildew, and blast diseases. This leads to a severe impact on our food production industry and our lives. Exposure of field crops to mycotoxins entails severe issues related to health and the environment. By triggering a variety of apoptotic and necrotic cell death pathways, the release of toxins by phytopathogenic fungi may obstruct seed germination and invade plant tissues (Pusztahelyi

Harsha · M. K. Kapur (✉)
Microbial Technology Laboratory, Acharya Narendra Dev College, University of Delhi, New Delhi, Delhi, India
e-mail: Harsha@zoology.du.ac.in; monishakhanna@andc.du.ac.in

M. Kumar
Department of Zoology, Rajiv Gandhi University, Doimukh, Arunachal Pradesh, India
e-mail: munendra.kumar@rgu.ac.in

P. Kumar
Department of Zoology, University of Allahabad, Prayagraj, Uttar Pradesh, India
e-mail: drprateekkumar@allduniv.ac.in

R. Solanki
Department of Zoology, Deen Dayal Upadhyaya, University of Delhi, New Delhi, India
e-mail: renu_slnk@ddu.ac.in

N. Manzoor (ed.), *Advances in Antifungal Drug Development*,
https://doi.org/10.1007/978-981-97-5165-5_21

et al. 2015). The most important crops in the field are maize, rice, and wheat; hence, exposure to mycotoxins might have a large negative impact on general public health (Jard et al. 2011; Rodrigues et al. 2019; Udovicki et al. 2018). Even though mycotoxicosis primarily affects underdeveloped nations, statistics from the last year revealed that modern industrialized nations too are affected by fungal infections (Bhatnagar et al. 2003; Milićević et al. 2020). Mold growth and spread are influenced by a variety of distinct factors, including social, economic, and environmental ones. Agriculture is commercially dependent on the use of chemical fungicides and pesticides. These substances are being overused, which has harmful side effects on human health and the environment (Nicolopoulou-Stamati et al. 2016; Alengebawy et al. 2021). This is because they are simple to use and inexpensive. The effects of excessive use of fungicides have been depicted in Table 21.1.

21.2 Increase in Multidrug Resistance (MDR) and Review of Resistance Mechanisms Against Commonly Used Antifungal Agents

The impact of microbial resistance to various antibiotics used in the management of severe fungal infections has gained significant global attention (Khushboo et al. 2022; Salam et al. 2023). The Centre for Disease Control and Prevention (CDC) reports that *Candida* and *Aspergillus* species have the highest levels of resistance to the antifungal medications currently in use (Table 21.2). The major threat of antibiotic resistance nationally is estimated at 2,868,700 infections each year by bacteria and fungi, and 35,900 deaths. However, the cases of resistance attained in several countries keep on increasing at a larger pace annually. An active person's immune system plays a greater role in spotting, reducing, and getting rid of fungal inflammation. Yet, the prevalence of fungal infections multiplies with diseases like cancer, AIDS, diabetes mellitus, and TB which are considered to be morbid. *Aspergillus* spp., *Cryptococcus* spp., and *Candida* spp. are examples of pathogenic fungi that can cause fungal infections and increase death rates. Moreover, more than 50% of mortality is due to the delayed prognosis of infection and recognition of causative fungi (Jampilek 2016; Fang et al. 2023). Polyenes, azoles, echinocandins, and pyrimidine analogs of 5-flucytosine are four systemic antifungal medicines. Azoles are a critical class of molecules for the management of fungal diseases in plants, animals, and humans as well as for the preservation of materials. The significant benefit of the usage of azoles in food security, animal health, and patient survival from life-threatening fungal infections are all due to successful treatment.

21.2.1 Azoles

Resistance is a heritable alteration in fungal sensitivity to a drug/fungicide that diminishes its effectiveness in controlling fungi. Increased risk of resistance to azoles was rare in the 1980s, but with AIDS epidemic in the 1990s, it was found that oral

Table 21.1 Effects of some commonly used fungicides on human health

S No.	Chemical fungicides	Active components in chemical fungicides	Mode of action	Implications on agriculture	Implications on human and animal	References
1.	Dithiocarbamates	Ziram Maneb Mancozeb, ferbam	Multisite inhibitor	Contact-dependent mobility	Developmental and teratogenic toxicity	Gikas et al. (2022), Caldas et al. (2001)
2.	Benzamides	Zoxamide	Anti-microtubule agent	Soil acidification	Nausea, vomiting, abdominal pain	Zhang et al. (2022), Petit et al. (2012)
3.	Phenylpyrroles	Fenpiclonil, fludioxonil	MAPK phosphorylation	Still unknown	Sweating, nausea, vomiting	Ghannoum et al. (2004), Brandhorst et al. (2019)
4.	Carbamates	Propamocarb	Defects in cell membrane permeabilization, Ach inhibitors	High risk of brain cancers in farmers	Carcinogenic	Zubrod et al. (2019)
5.	Phenylamides	Fluopyram	Inhibitor of RNA polymerization	Effect nutrient cycling, and homeostasis of soil	Giddiness, vomiting, nausea	Monkiedje and Spiteller (2002), Yang et al. (2011)
6.	Triazoles	Voriconazole	Inhibition of C-14 demethylase enzyme (DMI)	Increment of oxidative stress	Hypertrophy, vacuolization, and steatosis of hepatocytes	Roman et al. (2021)
7.	Phosphonate	Fosetyl-AI	Multisite inhibition	Formation of metal complex	Ataxia, depression, ptosis	Tudi et al. (2021)

Table 21.2 List of antifungal drugs against which different fungi attained resistance

Groups	Antifungal drugs	Example of resistant fungal species	Molecular mechanism	References
Azoles and triazoles	Epoxiconazole, Oxpoconazole, Clotrimazole, ketoconazole, Itraconazole, fluconazole	*Aspergillus* spp., *Candida albicans, Cryptococcus* spp.	14α-lanosterol demethylase (ERG11), alteration in the gene cyp51A (TR34/L98H), overexpression of genes coding for efflux transporter	Ruiz et al. (2021), Montoya et al. (2020)
Pyrimidine analog	Flucytosine, 5-fluorouracil, cytosine arabinoside	*Cryptococcus* sp., *aspergillus* spp.	Inhibitor of DNA/ RNA/protein synthesis	Kanafani and Perfect (2008)
Polyenes	Pimaricin, nystatin, amphotericin B	*Candida* sp., *botrytis cinerea*	LOF mutation in different genes in ergosterol biosynthetic pathway	Kim et al. (2020)
Echinocandins	Caspofungin, micafungin, anidulafungin, and rezafunzin	*Candida* spp., *aspergillus* spp.	Inhibition of β 1,3-D-glucan synthase enzyme	Al-Baqsami et al. 2020

candidiasis caused by *Candida albicans* occurred in 90% of all HIV-positive individuals. Ergosterol, a sterol unique to fungi, is analogous to cholesterol and sterols present in membranes of other eukaryotic cells. The sensitivity of fungal cells to a range of environmental stimuli may be greatly affected if there is any change in the quantity of ergosterol (Johnston et al. 2020). Ergosterol biosynthesis is specific to fungi and it becomes the primary target of antifungal drugs with minimal side effects on animals and plants. The principle mechanism of azoles is to target and inhibit the (ERG11) gene coding for lanosterol 14-demethylase, two cytochrome P450-dependent enzymes. The fungal pathogens which are naturally susceptible to azoles attain resistance by the following mechanisms. The ERG11 gene-encoded 14-lanosterol demethylase, which changes lanosterol into ergosterol in the membrane, is the target site of azoles (Ruiz et al. 2021; Suchodolski et al. 2021).

21.2.1.1 Increased Activity of Azole Efflux

Azole antifungals can be effluxed by either major facilitator superfamily (MFS) or ABC (ATP-binding cassette (ABC) transporters (Holmes et al. 2016; Toepfer 2023). The ABCG transporters Cdr1p and Cdr2p participate in multidrug resistance and transport phosphatidylethanolamine, phosphatidylcholine, and phosphatidylserine from the inner to the outer leaflet of the plasma membrane, thus affecting permeability. These efflux proteins from both the superfamily of transporters contain membrane-spanning domains, the difference lies in the source of energy each of them employs to drug extrusion. Efflux transporters are essential in eukaryotes to export toxins and fungicides out of the cell (Del Sorbo et al. 2000). Overexpression of the genes encoding these transporters can lead to resistance against azole antifungals as the intracellular concentration of a drug is reduced and leads to tolerance. In the case of MFS transporters such as CaMdr1p, the electrochemical gradient is utilized to translocate substrates (Keniya et al. 2015; Banerjee et al. 2021). *Saccharomyces cerevisiae* can acquire benomyl and methotrexate resistance because of the MFS transporter gene MDR1, also known as BENr. Azole-resistant clinical isolates of *C. albicans* frequently overexpress one or more efflux pumps, such as CaCdr1p, CaCdr2p, and CaMdr1p (Fling et al. 1991; Sheng et al. 2004), but azole-susceptible clinical isolates typically exhibit low-level constitutive expression of CaCdr1p (Holmes et al. 2008). In an animal model, CaMDR1 inactivation was found to significantly diminish *C. albicans* virulence (Becker et al. 1995; Prasad and Rawal 2014), although MRR1-dependent stimulation of CaMDR1 expression was later discovered not to enhance pathogenicity (Lohberger et al. 2014; Tyndall et al. 2016). High-level azole resistance is most frequently associated with CaCdr1p overexpression and to a lesser extent, CaCdr2p overexpression, while in certain isolates CaMdr1p is the sole overexpressed pump (Ivnitski-Steele et al. 2009; Rodrigues et al. 2019). Most often, mutations in the transcriptional regulator Mrr1p are the cause of this overexpression of CaMdr1p (Dunkel et al. 2008; Morschhäuser 2010). Fluconazole is a triazole medication, and overexpression of CaMdr1p imparts intermediate-level resistance to it (Hiller et al. 2006), but overexpression of CaCdr1p is linked to efflux of a larger spectrum of substrates. The structural and molecular differences between ABC and MFS pumps make this divergence in substrate specificity expected.

21.2.1.2 Overexpression of Genes

By overexpressing ERG11 gene, which causes a rise in the synthesis of the encoded enzyme, resistance is gained. For the same level of inhibition, higher enzyme concentrations call for higher medication concentrations. ERG11 expression is often elevated above normal in several *Candida* species in response to azole therapy (or other ergosterol biosynthesis inhibitors) (Henry et al. 2000; Hossain et al. 2022). Other ERG genes in the pathway are frequently included in this overexpression. In some circumstances, irreversible overexpression of ERG11 gene or the whole pathway might result in the resistant phenotype. There are essentially two ways that the ERG11 gene on chromosome 5 might become permanently upregulated in *C. albicans*: (1) by duplication, increasing the number of copies of the gene there; and (2) by gain-of-function (GOF) mutation in Upc2. The overexpression of several ergosterol biosynthesis genes, including the previously noted ERG11, is caused by GOF mutations in Upc2 (MacPherson et al. 2005). The ergosterol production pathway in *C. albicans* is primarily regulated by Upc2, a Zn2-Cys6 transcriptional factor. Constitutively active mutants of the two paralogs Upc2 (G888D) and Ecm22 (G790D) only cause sterol absorption in the model organism *S. cerevisiae* when the environment is aerobic (Crowley et al. 1998; David et al. 2015; Lewis et al. 1998; Shianna et al. 2001; Jordá et al. 2022). But in *C. albicans*, constitutively active mutants of Upc2 at the C-terminus

(Y642F, A643V/T, A646W, and G648D/S) or other regions (G304R and W478C) improve resistance to the azoles by direct ERG11 overexpression (Vasicek et al. 2014; Alves et al. 2014; Chen et al. 2011).

21.2.1.3 Minimal Amount of Azole Influx

Alteration in the plasma membrane composition affects membrane fluidity and asymmetry leading to decreased uptake of the drug (Park et al. 2002; Loeffler and Stevens 2003; Hernández Tasco et al. 2020). In the case of fluconazole-resistant *C. albicans*, a decreased amount of phosphatidylcholine to ethanolamine ratio in the plasma membrane may lead to reduced uptake of the drug (Kanafani and Perfect 2008).

21.2.1.4 Mechanism of Resistance of Azoles

According to AA locations G54, L98, and M220 where the most mutations are found. Tandem repetitions in the promoter are another component of cyp51A resistance mutations, which raise cyp51 expression. This is often combined with mutations in the cyp51A coding region itself (the most common is a mutation termed "TR/L98H" with a 36-base tandem repeat in the promoter and an L98H substitution in Cyp51A) (Verweij et al. 2009; Warrilow et al. 2013; Camps et al. 2012a, b; Mellado et al. 2007). Azole use in agriculture is usually linked to Cyp51A mutations (Fig. 21.1). Tebuconazole, a popular agricultural azole, generates a tandem repeat mutation in the cyp51A promoter, according to a study by Camps et al. (2012a, b).

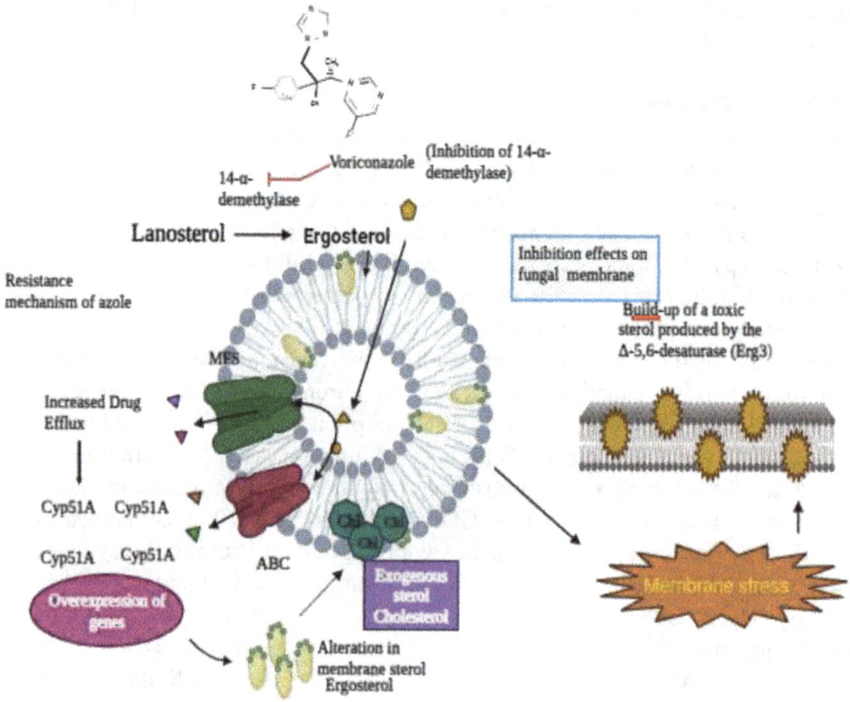

Fig. 21.1 Mode of action of commercially synthesized drug upon sterol of fungal membrane

21.2.2 Pyrimidine Analog

The synthetic structural analogs of DNA nucleotides employed are 5-Fluorocytosine (5-FC) and 5-Fluorouracil (5-FU). When taken orally, these molecules quickly disperse throughout the body owing to their highwater solubility and tiny size (Viswanathan et al. 2017). During clinical fungal infections, drugs enter the fungal cells quickly by certain transporters, such as cytosine permeases or pyrimidine transporters (Galocha et al. 2020). Another enzyme, uridine phosphoribosyl transferase (UPRT), transforms 5-FU into 5-fluorouracil monophosphate (5-FUMP). Then, 5-FUMP can either be transformed into 5-fluorouracil triphosphate, which substitutes UTP in fungal RNA and restricts protein synthesis, or into 5-fluorodeoxyuridine monophosphate, which inhibits thymidylate synthase, a critical enzyme for DNA synthesis, preventing the reproduction of fungal cells (Ahmed et al. 2022).

Resistance mechanism of FUR1 gene: 5-FU is then further processed by the product of the FUR1gene, a uracil phosphoribosyl transferase, and inhibits both DNA and protein synthesis. The mechanism of resistance is well understood in fungal pathogens like *Candida albicans*, where loss of function mutations in FCY1, FCY2, and FUR1 can mediate resistance (Fig. 21.2).

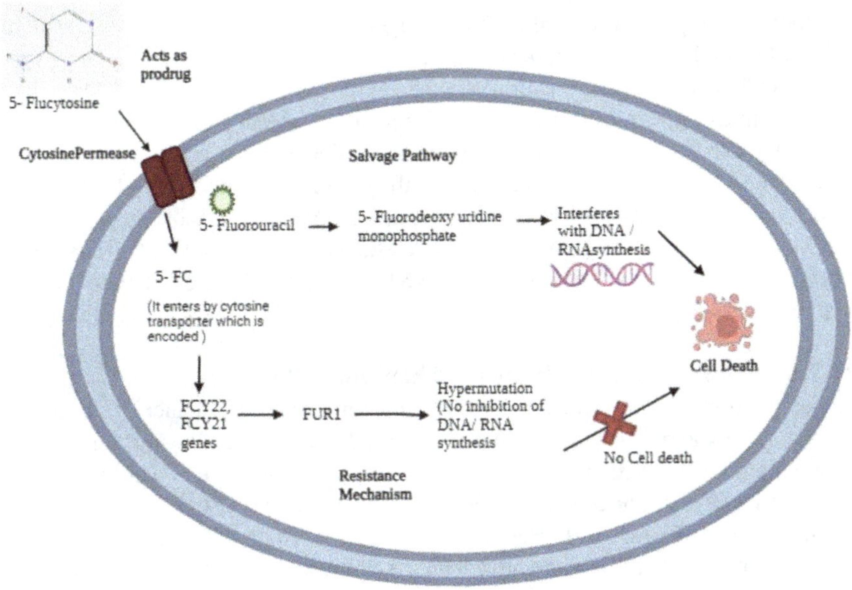

Fig. 21.2 Mechanism of action of pyrimidine analog on the fungal membrane

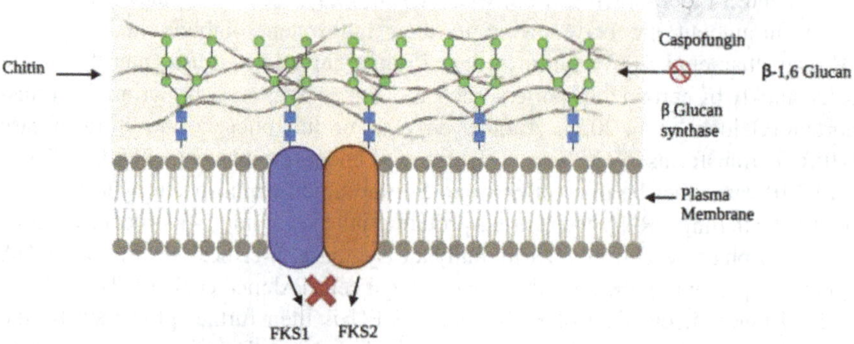

Fig. 21.3 Mechanism of action of echinocandin action on fungal membrane

21.2.3 Echinocandins

As a novel family of antifungal treatment, lipopeptide echinocandins work as non-competitive inhibitors of the glucan synthesis, operating upon the 1,3-D-glucan synthase. The major targets of this drug are the components of the cell wall, leading to a change in osmotic sensitivity, the development of false hyphae, a drop in sterol levels, and thicker cell walls are all effects of echinocandin therapy. Resistance to echinocandins is conferred by mutations in the FKS gene, which codes for the catalytic subunits of glucan synthase, in the case of *Candida* spp. (Johnson et al. 2011; Healey et al. 2018). According to reports, resistance develops in *C. glabrata* due to modifications in homologous areas of FKS1 and FKS2 (Fig. 21.3) (García-Cela et al. 2012; Arastehfar et al. 2020).

21.2.3.1 Resistance Mechanism of Echinocandins

The expression of FKS genes can influence echinocandin resistance. The mechanism of echinocandins is highly targeted and quiet apart from azole antifungal found in two highly conserved "hot spot" areas of FKS1. Limited amino acid changes in the Fks subunits of glucan synthase lead to echinocandin resistance, which causes clinical failures (Perlin 2011; Hernández Tasco et al. 2020). The minimum inhibitory concentration (MIC) values are enhanced as a result of the amino acid changes, which can reduce glucan synthase sensitivity by several folds (Rashed et al. 2013; Henry et al. 2011; Katiyar and Edlind 2009; David et al. 2015; Alastruey-Izquierdo and Martín-Galiano 2023).

21.2.4 Polyenes

Polyenes exert their effect by entering into the fungal membranes. The transmembrane potential is lost and cellular function is compromised by this method. Despite the fact that amphotericin B resistance among *Candida* strains is still uncommon, there have been recent reports of elevated MICs to amphotericin B among *C. krusei* and *C. glabrata* isolates (Tortorano et al. 2012). Additionally, *Trichosporon beigelii* (Walsh et al. 1991) and *Candida lusitaniae* (Pfaller and Barry 1994) both typically exhibit intrinsic polyene resistance. It has been challenging to identify isolates that are resistant to polyene (Rex et al. 1995). Compared to yeasts, filamentous fungi are more likely to attain resistance to polyenes. *Aspergillus terreus* is one of the *Aspergillus* species that is often resistant to amphotericin B (Sabatelli et al. 2006). Only 11.5% of *A. fumigatus* isolates were inhibited at a concentration of less than 1.0 g/ml, according to the SENTRY program, indicating a considerable rise in the incidence of polyene resistance among *Aspergillus* species (Messer et al. 2006). Large long-term investigations are missing, though. Invasive diseases have also been linked to uncommon *Aspergillus* species that are generally resistant to most antifungal medications, such as *Aspergillus ustus* and *Aspergillus lentulus*. Amphotericin B is often ineffective against other molds including *Scedosporium apiospermum, Scedosporium prolificans*, and *Fusarium* species (Sabatelli et al. 2006). There is no known resistance breaking point for polyenes. Amphotericin B resistance is typically identified by a MIC of less than 1.0 g/ml. Other sterols build up in the fungal membrane, when the ERG3 gene, which is involved in ergosterol production, is defective. In comparison to isolates that are vulnerable to polyene, *Candida* and *Cryptococcus* with high polyene resistance had comparatively low ergosterol contents (Zhu et al. 2022). Increased catalase activity may possibly play a role in amphotericin B resistance, resulting in a reduction in oxidative stress sensitivity (Fig. 21.4) (Sokol-Anderson et al. 1986).

21.3 Nurturing New Therapeutic Directions and Interventions

Researchers are showing interest in the identification of novel compounds synthesized by *Streptomyces* spp. The broad spectrum of actions with increased shelf life could be helpful in promoting plant development under field conditions which becomes a new insight for making use of technology. Many reports have suggested that the commercial biocontrol agents are easy to deliver, promote growth and stress resistance, and eventually increase plant biomass and yield (Masoudi et al. 2021).

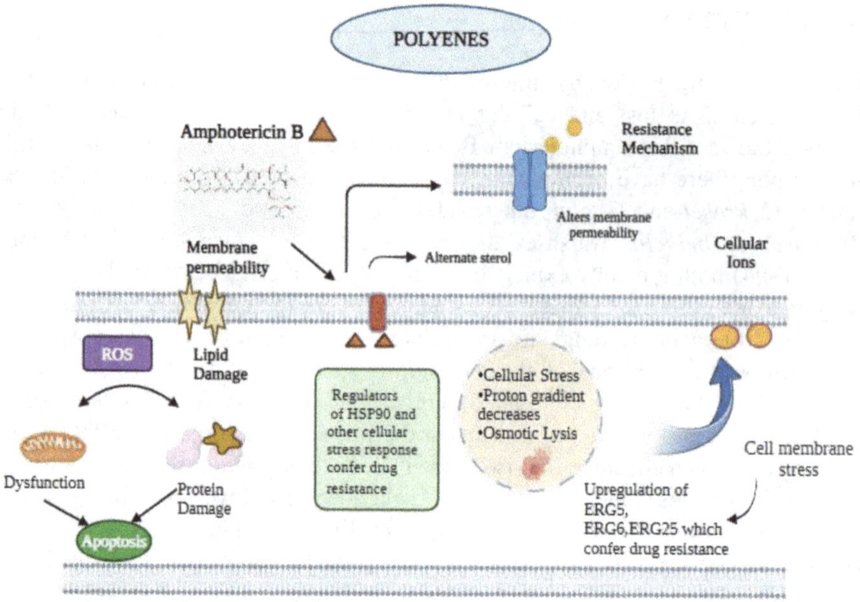

Fig. 21.4 Mode of action of polyenes upon fungal membrane and its cellular changes

21.4 Biofumigation Along with Volatile Organic Compounds from *Streptomyces* Spp. as an Alternative to Tackle Mycotoxins Production

The contamination of cereals with fungus is still a significant problem despite good manufacturing practices (GMP) and good agricultural practices (GAP) (Brandao et al. 2010; Pickova et al. 2021). *Aspergillus ochraceus* is one of the pollutants that, when exposed to the right environmental factors, has been demonstrated to create ochratoxin A (OTA), a serious hazard to both public health and food safety (da Rocha Neto et al. 2015; Durand et al. 2021). OTA has been described as teratogenic, carcinogenic, embryotoxic, genotoxic, neurotoxic (Toledo et al. 2015; Pei et al. 2021), and immunotoxic. Biological solutions are now viewed as a feasible alternative to reduce mycotoxin contamination in the context of the development of organic farming in order to limit the adverse effects of pesticides (Barral et al. 2017). The majority of volatile organic compounds (VOCs) of microbial origin are complex mixtures of low molecular weight substances that readily evaporate under ambient conditions of pressure and temperature (25 °C/1.01 105 psi). Due to their ability to preserve foods without causing organoleptic alterations, VOCs offer a particular benefit from a practical and homogeneous application standpoint (Passone and Etcheverry 2014). In the agriculture and food industries, VOCs from microbes are being explored as prospective substitute biocontrol agents since fungus contamination of cereal grains is still a serious issue on a global scale. According to research,

VOCs can shield cereals against fungi without coming into touch with them (Neuwald et al. 2018; Fenta et al. 2023). As a result, the cereals are preserved without significantly altering their organoleptic qualities. Due to their quick transition into the gaseous phase at room temperature, VOCs offer a significant advantage in terms of ease and homogeneity (Bluma et al. 2009).

21.5 Safer Approach of Using Biofertilizers and Biopesticides Derived from *Streptomyces* Spp.

The growth of agriculture comes with many problems due to increased challenges. The major one is the rise in demand for higher productivity owing to a scarcity of resources (Rosegrant et al. 2005). This in turn leads to dependence on over usage of chemical pesticides and fungicides for crop improvement to meet the global need for survival. When these compounds are overused, they start accumulating in nature and lead to eutrophication in water bodies which could be toxic to human health upon overexposure (Ahmed et al. 2022). To address this situation, biological control using microbial antagonists has been proposed as an alternative to chemical pesticides (Bardin et al. 2015). Actinomycetes belonging to the phylum Actinobacteria represent an important microbial community in the soil with increasing agriculture applications, especially in biological control, promoting plant growth and development (Yadav et al. 2018). These bacteria produce several bioactive compounds with diverse biological activities like antifungal, antibacterial, antihelminthic, and insecticidal. These secondary metabolites promote plant growth by protecting them from phytopathogens and helping in the management of agriculture (Yarzábal and Chica 2019).

A group studied the antagonistic activity of *Streptomyces* spp. AN090126 against various phytopathogenic bacteria and fungi (Charlet et al. 2022). They attempted various bioassays in which they reduced the wilting of tomatoes in a dose-dependent manner and also demonstrated synergistic interaction between *Pectobacterium carotovorum*, the causal agent of soft rot in kimchi cabbage (Human et al. 2016), isolated different *Streptomyces* from Sugarbush *Protea repens* and *Protea neriifolia*. They identified fungichromin and actiphenol compounds which showed prolific activity against variable fungal pathogens. According to Boukaew et al. (2022), the freeze-dried culture filtrate of *Streptomyces philanthi* RL-1-178 (DCF RL 1-178) showed antifungal and antiaflatoxigenic mechanism efficacy against two aflatoxigenic strains (*Aspergillus flavus* and *Aspergillus parasiticus*). According to the experiments conducted, they showed the antifungal properties of culture filtrate upon stored maize seeds and identified it as a bio-preservative. The antifungal compounds produced from *Streptomyces* species and commercial biopesticides are displayed in Tables 21.3 and 21.4, respectively.

Table 21.3 Antifungal compounds derived from *Streptomyces* spp.

Streptomyces spp.	Antibiotics	Bioactive compound	Type	References
S. Mediocidicus	Mediomycin B	Antifungal	Polyene	Mahajan et al. (2017)
S. Venezuela	Chloramphenicol	Antimicrobial	Macrolide	
S. Aureofaciens	Tetracycline	Antimicrobial	Macrolide	Quinn et al. (2021)
S. Noursei	Nystatin	Antifungal	Polyene macrolide	Brautaset et al. (2011)
S. Lomondensis	Lomofungin	Antifungal	Lomofungin	Luo et al. (2015)
S. Pristinaespiralis	Virginiamycin	Antimicrobial	Peptidolactones M and S	Dzhavakhiya et al. (2016)
S. Lincolnensis	Lincomycin	Antimicrobial	Lincosamides	Zhang et al. (2022)
S. Spectabilis	Spectinomycin	Antimicrobial	Aminocyclitol	
S. Vinaceus	Viomycin	Antituberculosis activity	Tuberactinomycin	
S. Rimosus	Oxytetracycline	Antimicrobial	Polyketide	Devi et al. (2023)

21.6 Molecular Docking as an Enigmatic Tool to Analyze Bond Interactions Between Enzymes and Ligands

Molecular docking is the most important technique in the field of pharmacodynamics and pharmacokinetics. It emphasizes detecting and analyzing active compounds present in extract or chemically characterized molecules (Zhu et al. 2018; Veras et al. 2016; Wang et al. 2019). To supplement the experimental approach, it virtually predicts the ideal energy-minimized posture of one molecule to another when they are linked together to create a stable complex. It involves two steps: (a) ligand conformation, position, and orientation prediction (within binding sites); (b) assessment of binding affinity (Muniaraj et al. 2018; Rizvi et al. 2013). In our previous studies, we have extracted the bioactive compounds from *Streptomyces* spp. and tested them on fungal pathogens (*Aspergillus flavus* and *Candida albicans*). The minimum inhibitory concentration of extract number 130, 169, 165 and 194 was determined to be 0.5–1.0 mg/ml, 0.5–1.0 mg/ml, 1.0–2.5 mg/ml, and 1.0–2.5 mg/ml. In silico studies (Chi_1389, Chi_897, ChiC_426, Chi_1731, Chi65_1884, and Chi_1686) with *Streptomyces* sp. strain 130 were used for molecular docking to check ligand-enzyme interaction (Kumar et al. 2019). dNAG (di N-Acetyl Glucosamine) was used as a ligand, which was initially converted into PDB format.

Autodock 4.2 tool was used for molecular docking (Rizvi et al. 2013; Patel et al. 2021). Single letter code of amino acid is used to represent the name of the amino acid and number next to the single letter code is used to denote the position of amino acid (e.g., M467). Bond interaction between enzymes and ligands was analyzed by

Table 21.4 Commercialized actinomycete-based biopesticides

Commercial name of biopesticide	Active compound	Target pests	Biocontrol mechanism	Country	References
Fungicides, nematicides, insecticides					
Rhizovit®	*Streptomyces rimosus*	*Phytophthora* Spp., *Rhizoctonia solani*, *Botrytis* spp.	Antibiosis	–	European Food Safety Authority and European Centre for Disease Prevention and Control (2015), Caldara et al. (2018)
Actinogrow®, Actinovat®	*Streptomyces lydicus*	*Erisphe* spp.	Hyperparasitism	European Union	Lichatowich (2007), Rahman et al. (2007), Machado et al. (2022)
Mycostop®	*Streptomyces griseoviridis* strain K61	*Alternaria* spp., *Rhizocotania solani*, *fusarium* spp.	Competition	Canada	Sun et al. (2017)
Avicta®	*Streptomyces avermitilis*	*Pratylenchus* spp., *Heterodera* spp., *M. Arenaria*	Antibiosis	Brazil, Argentina, South Africa	Sharma et al. (2019), Yang et al. (2015)
Tracer®	Spinosad and Spinosyn D from Saccharopolyspora	Lepidopterous larvae	Antibiosis	USA	
Milbeknock®	*Streptomyces hygroscopicus* (Milbemectin)	Spider mites	Nematicidal	Spain	

binding energy and inhibition coefficient (Mashraqi et al. 2023). The highest released binding energy (kcal/mol) represents the effective bond interaction (Wang et al. 2019). In the molecular docking, the binding energy of Chi_1389 and Chi65_1884 was −6.06 and − 6.30 kcal/mol, respectively (Oussou-Azo et al. 2020; Ngemenya et al. 2015; Hussain et al. 2020; Mashraqi et al. 2023). In the remaining protein models, Chi_426, ChiD_1731, and Chi_1686 showed the binding energy in the range of −5.0 to −6.0 kcal/mol. The application studies of the extract have also been done on spoiled food waste because the scientific community has a big focus on detoxifying the mycotoxins from grains and ingredients used in food preparation, but priority should be given to counter-attacking the direct source of mycotoxins through the food chain (Bryden 2012)

21.7 Nanotechnology-Based Antifungal Treatments to Battle Fungi

Nanomaterials (metal and polymeric nanoparticles) are used as potential therapeutic options for the most common invasive filamentous fungal infections. Nanoparticles (NPs) are a diverse class of particles with nanoscale dimensions ranging from 1 to 100 nm. These particles can have a variety of forms and morphologies, including cylindrical, spherical, tubular, conical, and spiral (León-Buitimea et al. 2021). Because they have several potential medicinal and diagnostic uses, NPs are now piquing the interest of the scientific community (Lee et al. 2017). Due to their low toxicity, ability to pass through a variety of biological barriers, and capability to covalently conjugate with hydrophobic or hydrophilic drugs and macromolecules, boosting solubility and stability, these materials are also seen as attractive therapeutic possibilities (Khan et al. 2019). Nanoparticles are characterized as organic or polymeric, inorganic (metallic), or carbon nanoparticles based on their structure, size, and chemical and physical characteristics. Organic NPs, such as liposomes, micelles, and dendrimers, are biodegradable, nontoxic, and heat and light-sensitive (Ealia and Saravanakumar 2017). Due to their great efficiency and injection in specified areas, this class of NPs is the top choice in the biomedical field, primarily for medication administration. Inorganic NPs include all particles made from metals and metal oxides. Cadmium (Cd), Aluminum (Al), Cobalt (Co), Copper (Cu), Gold (Au), Iron (Fe), Silver (Ag), and Zinc (Zn) are popular metal nanoparticle sources (Ijaz et al. 2022). Synthesis of NPs by actinomycetes especially *Streptomyces* might prove to be the preferred choice as a result of rapid growth, high biomass yield, and compatible safety of actinomycetes with the environment. By analyzing the size and dispersion of the biosynthesized gold nanoparticles (GNPs), HR-TEM revealed that GNPs with sizes between 5 and 50 nm were formed. The work demonstrates the quick and environmentally friendly synthesis of GNPs from *Streptomyces griseoruber*, and it is the first account to date of the catalytic activity of GNPs from actinomycetes (Ranjitha and Rai 2017). Researchers also have paid a lot of attention to various metal nanoparticles because of their peculiar physical, chemical, and biological characteristics. Metal nanoparticles have demonstrated their potential in a number of fields, including biotechnology, bioremediation of environmental pollutants, gene delivery for the treatment or prevention of genetic disorders, catalysis, imaging, particularly in medicine such as the delivery of antigens for vaccination, novel electronics, optics, and better drug delivery techniques. Environmentally friendly processes must be developed for the production of nanoparticles. The "bottom-up approach" and the "top-down approach" have been proposed as two key methods for the production of metal nanoparticles. The "top-down approach" concentrates on producing nanoparticles using bulk materials (larger ones) to regulate their assembly. However, starting at the molecular level and keeping meticulous control of molecular structure is how the "bottom-up approach" leads to the construction of larger and more sophisticated systems (Golinska et al. 2014; Das and Smita 2018). By merging these in vivo models with pharmacokinetic/pharmacodynamics models, dose studies estimating the risk of resistance developing and reducing the creation of resistance, fungal persistence, and tolerance should be made simpler.

21.8 Conclusion

There is an urgent need to develop new sources of antibiotics due to the growth of multi-resistant infections and the limited number of new medications being licensed for the health market. *Streptomyces* has made significant contributions to medicine during the last 80 years, not just through antibacterial antibiotics, but also through antifungal, antiparasitic, and anticancer chemicals. *Streptomyces* isolates from traditional medicine imply that these bacteria have been important in human health for far longer than previously assumed. This new *Streptomyces* source may also assist to restore the critically depleted stock of emergency antibiotics used to tackle multi-resistant infections, as well as providing the structural variety required for a new generation of innovative antibiotics. Furthermore, understanding their past uses is more than just a historical curiosity, as they may help us uncover critical variables in the complicated synthesis and/or use of antibiotics. Finally, to ensure the continued availability of this resource, it is critical to maintain the habitats and microbial genera associated with these *Streptomyces* spp. as well as to collect correct information and data on their prevalence, features, and characteristics.

Acknowledgments Acharya Narendra Dev College (ANDC), University of Delhi, is gratefully acknowledged for providing infrastructural facilities.

Contribution of Authors The paper was prepared by Harsha, who also carried out the literature search and data analysis. PK, MK, and RS provided critical revisions and assisted with the data analysis, and MKK oversaw the project.

References

Ahmed M, Sajid AR, Javeed A, Aslam M, Ahsan T, Hussain D et al (2022) Antioxidant, antifungal, and aphicidal activity of the triterpenoids spinasterol and 22, 23-dihydrospinasterol from leaves of Citrullus colocynthis L. Sci Rep 12(1):4910

Al-Baqsami ZF, Ahmad S, Khan Z (2020) Antifungal drug susceptibility, molecular basis of resistance to echinocandins and molecular epidemiology of fluconazole resistance among clinical Candida glabrata isolates in Kuwait. Sci Rep 10:6238. https://doi.org/10.1038/s41598-020-63240-z

Alastruey-Izquierdo A, Martín-Galiano AJ (2023) The challenges of the genome-based identification of antifungal resistance in the clinical routine. Front Microbiol 14:1134755

Alengebawy A, Abdelkhalek ST, Qureshi SR, Wang MQ (2021) Heavy metals and pesticides toxicity in agricultural soil and plants: ecological risks and human health implications. Toxics 9(3):42

Alves CT, Ferreira IC, Barros L, Silva S, Azeredo J, Henriques M (2014) Antifungal activity of phenolic compounds identified in flowers from north eastern Portugal against Candida species. Future Microbiol 9(2):139–146

Arastehfar A, Daneshnia F, Najafzadeh MJ, Hagen F, Mahmoudi S, Salehi M et al (2020) Evaluation of molecular epidemiology, clinical characteristics, antifungal susceptibility profiles, and molecular mechanisms of antifungal resistance of Iranian Candida parapsilosis species complex blood isolates. Front Cell Infect Microbiol 10:206

Banerjee A, Pata J, Sharma S, Monk BC, Falson P, Prasad R (2021) Directed mutational strategies reveal drug binding and transport by the MDR transporters of Candida albicans. J Fungi 2021(7):68

Bardin M, Ajouz S, Comby M, Lopez-Ferber M, Graillot B, Siegwart M, Nicot PC (2015) Is the efficacy of biological control against plant diseases likely to be more durable than that of chemical pesticides? Front Plant Sci 6:566

Barral B, Chillet M, Minier J, Léchaudel M, Schorr-Galindo S (2017) Evaluating the response to fusarium ananatum inoculation and antifungal activity of phenolic acids in pineapple. Fungal Biol 121(12):1045–1053

Becker JM, Henry LK, Jiang W, Koltin Y (1995) Reduced virulence of Candida albicans mutants affected in multidrug resistance. Infection and immunity, 63(11):4515–4518

Bhatnagar D, Ehrlich KC, Cleveland TE (2003) Molecular genetic analysis and regulation of aflatoxin biosynthesis. Appl Microbiol Biotechnol 61:83–93

Bluma R, Landa MF, Etcheverry M (2009) Impact of volatile compounds generated by essential oils on aspergillus section Flavi growth parameters and aflatoxin accumulation. J Sci Food Agric 89(9):1473–1480

Boukaew S, Cheirsilp B, Yossan S, Khunjan U, Petlamul W, Prasertsan P (2022) Utilization of palm oil mill effluent as a novel substrate for the production of antifungal compounds by Streptomyces philanthi RM-1-138 and evaluation of its efficacy in suppression of three strains of oil palm pathogen. J Appl Microbiol 132(3):1990–2003

Brandao LR, Medeiros AO, Duarte MC, Barbosa AC, Rosa CA (2010) Diversity and antifungal susceptibility of yeasts isolated by multiple-tube fermentation from three freshwater lakes in Brazil. J Water Health 8(2):279–289

Brandhorst TT, Kean IR, Lawry SM, Wiesner DL, Klein BS (2019) Phenylpyrrole fungicides act on triosephosphate isomerase to induce methylglyoxal stress and alter hybrid histidine kinase activity. Sci Rep 9(1):5047

Brautaset T, Sletta H, Degnes KF, Sekurova ON, Bakke I, Volokhan O et al (2011) New nystatin-related antifungal polyene macrolides with altered polyol region generated via biosynthetic engineering of Streptomyces noursei. Appl Environ Microbiol 77(18):6636–6643

Bryden WL (2012) Mycotoxin contamination of the feed supply chain: implications for animal productivity and feed security. Anim Feed Sci Technol 173(1–2):134–158

Caldas ED, Conceiçao MH, Miranda MCC, de Souza LCK, Lima JF (2001) Determination of dithiocarbamate fungicide residues in food by a spectrophotometric method using a vertical disulfide reaction system. J Agric Food Chem 49(10):4521–4525

Camps SM, Dutilh BE, Arendrup MC, Rijs AJ, Snelders E, Huynen MA et al (2012a) Discovery of a HapE mutation that causes azole resistance in aspergillus fumigatus through whole genome sequencing and sexual crossing. PLoS One 7(11):e50034

Camps SM, Rijs AJ, Klaassen CH, Meis JF, O'Gorman CM, Dyer PS et al (2012b) Molecular epidemiology of aspergillus fumigatus isolates harboring the TR34/L98H azole resistance mechanism. J Clin Microbiol 50(8):2674–2680

Charlet R, Le Danvic C, Sendid B, Nagnan-Le Meillour P, Jawhara S (2022) Oleic acid and palmitic acid from Bacteroides thetaiotaomicron and lactobacillus johnsonii exhibit anti-inflammatory and antifungal properties. Microorganisms 10(9):1803

Chen SCA, Slavin MA, Sorrell TC (2011) Echinocandin antifungal drugs in fungal infections: a comparison. Drugs 71:11–41

Crowley JH, Tove S, Parks LW (1998) A calcium-dependent ergosterol mutant of Saccharomyces cerevisiae. Curr Genet 34:93–99

da Rocha Neto AC, Maraschin M, Di Piero RM (2015) Antifungal activity of salicylic acid against Penicillium expansum and its possible mechanisms of action. Int J Food Microbiol 215:64–70

Das M, Smita SS (2018) Biosynthesis of silver nanoparticles using bark extracts of Butea monosperma (lam.) Taub. and study of their antimicrobial activity. Appl Nanosci 8:1059–1067

David A, Botías C, Abdul-Sada A, Goulson D, Hill EM (2015) Sensitive determination of mixtures of neonicotinoid and fungicide residues in pollen and single bumblebees using a scaled down QuEChERS method for exposure assessment. Anal Bioanal Chem 407:8151–8162

Del Sorbo G, Schoonbeek HJ, De Waard MA (2000) Fungal transporters involved in efflux of natural toxic compounds and fungicides. Fungal Genet Biol 30(1):1–15

Devi S, Verma J, Sohal SK, Manhas RK (2023) Insecticidal potential of endophytic Streptomyces sp. against Zeugodacus cucurbitae (Coquillett)(Diptera: Tephritidae) and biosafety evaluation. Toxicon 233:107246

Dunkel N, Liu TT, Barker KS, Homayouni R, Morschhäuser J, Rogers PD (2008) A gain-of-function mutation in the transcription factor Upc2p causes upregulation of ergosterol biosynthesis genes and increased fluconazole resistance in a clinical Candida albicans isolate. Eukaryot Cell 7(7):1180–1190

Durand C, Maubon D, Cornet M, Wang Y, Aldebert D, Garnaud C (2021) Can we improve antifungal susceptibility testing? Front Cell Infect Microbiol 11:720609

Dzhavakhiya V, Savushkin V, Ovchinnikov A, Glagolev V, Savelyeva V, Popova E et al (2016) Scaling up a virginiamycin production by a high-yield Streptomyces virginiae VKM ac-2738D strain using adsorbing resin addition and fed-batch fermentation under controlled conditions. 3 Biotech 6(2):240

Ealia SAM, Saravanakumar MP (2017) A review on the classification, characterisation, synthesis of nanoparticles and their application. In: IOP conference series: materials science and engineering, vol 263(3). IOP Publishing, p 032019

European Food Safety Authority, European Centre for Disease Prevention and Control (2015) The European Union summary report on trends and sources of zoonoses, zoonotic agents and food-borne outbreaks in 2013. EFSA J 13(1):3991

Fang W, Wu J, Cheng M, Zhu X, Du M, Chen C et al (2023) Diagnosis of invasive fungal infections: challenges and recent developments. J Biomed Sci 30(1):42

Fenta L, Mekonnen H, Kabtimer N (2023) The exploitation of microbial antagonists against post-harvest plant pathogens. Microorganisms 11(4):1044

Fling ME, Kopf J, Tamarkin A, Gorman JA, Smith HA, Koltin Y (1991) Analysis of a Candida albicans gene that encodes a novel mechanism for resistance to benomyl and methotrexate. Mol Gen Genet MGG 227:318–329

Galocha M, Costa IV, Teixeira MC (2020) Carrier-mediated drug uptake in fungal pathogens. Genes 11(11):1324

García-Cela E, Gil-Serna J, Marín S, Acevedo H, Patiño B, Ramos AJ (2012) Effect of preharvest anti-fungal compounds on aspergillus steynii and A. Carbonarius under fluctuating and extreme environmental conditions. Int J Food Microbiol 159(2):167–176

Ghannoum MA, Hossain MA, Long L, Mohamed S, Reyes G, Mukherjee PK (2004) Evaluation of antifungal efficacy in an optimized animal model of Trichophyton mentagrophytes-dermatophytosis. J Chemother 16(2):139–144

Golinska P, Wypij M, Ingle AP, Gupta I, Dahm H, Rai M (2014) Biogenic synthesis of metal nanoparticles from actinomycetes: biomedical applications and cytotoxicity. Appl Microbiol Biotechnol 98:8083–8097

Healey KR, Kordalewska M, Jiménez Ortigosa C, Singh A, Berrío I, Chowdhary A, Perlin DS (2018) Limited ERG11 mutations identified in isolates of Candida auris directly contribute to reduced azole susceptibility. Antimicrob Agents Chemother 62(10):10–1128

Henry T, Iwen PC, Hinrichs SH (2000) Identification of aspergillus species using internal transcribed spacer regions 1 and 2. J Clin Microbiol 38(4):1510–1515

Henry RS, Johnson WG, Wise KA (2011) The impact of a fungicide and an insecticide on soybean growth, yield, and profitability. Crop Prot 30(12):1629–1634

Hernández Tasco AJ, Ramírez Rueda RY, Alvarez CJ, Sartori FT, Sacilotto ACB, Ito IY et al (2020) Antibacterial and antifungal properties of crude extracts and isolated compounds from Lychnophora markgravii. Nat Prod Res 34(6):863–867

Hiller NJ, Day DV, Vance RJ (2006) Collective enactment of leadership roles and team effectiveness: a field study. Leadersh Q 17(4):387–397

Holmes AR, Lin YH, Niimi K, Lamping E, Keniya M, Niimi M et al (2008) ABC transporter Cdr1p contributes more than Cdr2p does to fluconazole efflux in fluconazole-resistant Candida albicans clinical isolates. Antimicrob Agents Chemother 52(11):3851–3862

Holmes AR, Cardno TS, Strouse JJ, Ivnitski-Steele I, Keniya MV, Lackovic K et al (2016) Targeting efflux pumps to overcome antifungal drug resistance. Future Med Chem 8(12):1485–1501

Hossain CM, Ryan LK, Gera M, Choudhuri S, Lyle N, Ali KA, Diamond G (2022) Antifungals Drug Resist Encycl 2(4):1722–1737

Human ZR, Moon K, Bae M, De Beer ZW, Cha S, Wingfield MJ et al (2016) Antifungal Streptomyces spp. associated with the infructescences of Protea spp. in South Africa. Front Microbiol 7:1657

Hussain A, Rashid MM, Akhtar N, Muin A, Ahmad G (2020) In-vitro evaluation of fungicides at different concentrations against Alternaria solani causing early blight of potato. Journal of Pharmacognosy and Phytochemistry 9(4):1874–1878

Ijaz I, Bukhari A, Gilani E, Nazir A, Zain H, Saeed R (2022) Green synthesis of silver nanoparticles using different plants parts and biological organisms, characterization and antibacterial activity. Environmental Nanotechnology, Monitoring & Management 18:100704

Ivnitski-Steele I, Holmes AR, Lamping E, Monk BC, Cannon RD, Sklar LA (2009) Identification of Nile red as a fluorescent substrate of the Candida albicans ATP-binding cassette transporters Cdr1p and Cdr2p and the major facilitator superfamily transporter Mdr1p. Anal Biochem 394(1):87–91

Jampilek J (2016) Potential of agricultural fungicides for antifungal drug discovery. Expert Opin Drug Discov 11(1):1–9

Jard G, Liboz T, Mathieu F, Guyonvarc'h, A., & Lebrihi, A. (2011) Review of mycotoxin reduction in food and feed: from prevention in the field to detoxification by adsorption or transformation. Food Addit Contam: Part A 28(11):1590–1609

Johnson ME, Katiyar SK, Edlind TD (2011) New Fks hot spot for acquired echinocandin resistance in Saccharomyces cerevisiae and its contribution to intrinsic resistance of Scedosporium species. Antimicrob Agents Chemother 55(8):3774–3781

Johnston EJ, Moses T, Rosser SJ (2020) The wide-ranging phenotypes of ergosterol biosynthesis mutants, and implications for microbial cell factories. Yeast 37(1):27–44

Jordá T, Martínez-Martín A, Martínez-Pastor MT, Puig S (2022) Modulation of yeast Erg1 expression and terbinafine susceptibility by iron bioavailability. Microb Biotechnol 15(11):2705–2716

Kanafani ZA, Perfect JR (2008) Resistance to antifungal agents: mechanisms and clinical impact. Clin Infect Dis 46(1):120–128

Katiyar SK, Edlind TD (2009) Role for Fks1 in the intrinsic echinocandin resistance of fusarium solani as evidenced by hybrid expression in Saccharomyces cerevisiae. Antimicrob Agents Chemother 53(5):1772–1778

Keniya MV, Fleischer E, Klinger A, Cannon RD, Monk BC (2015) Inhibitors of the Candida albicans major facilitator superfamily transporter Mdr1p responsible for fluconazole resistance. PLoS One 10(5):e0126350

Khan AU, Khan M, Khan MM (2019) Antifungal and antibacterial assay by silver nanoparticles synthesized from aqueous leaf extract of Trigonella foenum-graecum. BioNanoScience 9:597–602

Khushboo, Kumar P, Dubey KK, Usmani Z, Sharma M, Gupta VK (2022) Biotechnological and industrial applications of Streptomyces metabolites. Biofuels Bioprod Biorefin 16(1):244–264

Kim JH, Cheng LW, Chan KL, Tam CC, Mahoney N, Friedman M et al (2020) Antifungal drug repurposing. Antibiotics 9(11):812

Kumar M, Kumar P, Das P, Kapur MK (2019) Draft genome of Streptomyces sp. strain 130 and functional analysis of extracellular enzyme producing genes. Mol Biol Rep 46:5063–5071

Kuo SH, Lu PL, Chen YC, Ho MW, Lee CH, Chou CH, Lin SY (2021) The epidemiology, genotypes, antifungal susceptibility of Trichosporon species, and the impact of voriconazole on Trichosporon fungemia patients. J Formos Med Assoc 120(9):1686–1694

Lee T, Park D, Kim K, Lim SM, Yu NH, Kim S et al (2017) Characterization of bacillus amyloliquefaciens DA12 showing potent antifungal activity against mycotoxigenic fusarium species. Plant Pathol J 33(5):499

León-Buitimea A, Garza-Cervantes JA, Gallegos-Alvarado DY, Osorio-Concepción M, Morones-Ramírez JR (2021) Nanomaterial-based antifungal therapies to combat fungal diseases aspergillosis, Coccidioidomycosis, Mucormycosis, and candidiasis. Pathogens 10(10):1303

Lewis RE, Klepser ME, Pfaller MA (1998) Update on clinical antifungal susceptibility testing for Candida species. Pharmacotherapy 18(3):509–515

Lichatowich T (2007) The plant growth enhancing and biocontrol mechanisms of Streptomyces lydicus WYEC 108 and its use in nursery and greenhouse production. In: Riley LE, Dumroese RK, Landis TD (eds) National proceedings: forest and conservation nursery associations-2006. Proceedings RMRS-P-50. US Department of Agriculture, Forest Service, Rocky Mountain Research Station, Fort Collins, CO, pp 61–62. 50

Loeffler J, Stevens DA (2003) Antifungal drug resistance. Clin Infect Dis 36(Suppl. 1):S31–S41

Lohberger A, Coste AT, Sanglard D (2014) Distinct roles of Candida albicans drug resistance transcription factors TAC1, MRR1, and UPC2 in virulence. Eukaryot Cell 13(1):127–142

Luo Q, Hu H, Peng H, Zhang X, Wang W (2015) Isolation and structural identification of two bioactive phenazines from Streptomyces griseoluteus P510. Chinese Journal of Chemical Engineering 23(4):699–703

MacPherson S, Akache B, Weber S, De Deken X, Raymond M, Turcotte B (2005) Candida albicans zinc cluster protein Upc2p confers resistance to antifungal drugs and is an activator of ergosterol biosynthetic genes. Antimicrob Agents Chemother 49(5):1745–1752

Mahajan S, Tilak R, Kaushal SK, Mishra RN, Pandey SS (2017) Clinico-mycological study of dermatophytic infections and their sensitivity to antifungal drugs in a tertiary care center. Indian J Dermatol Venereol Leprol 83:436

Mashraqi A, Al Abboud MA, Ismail KS, Modafer Y, Sharma M, El-Shabasy A (2023) Correlation between antibacterial activity of two Artemisia sp. extracts and their plant characteristics. bioRxiv, 2023–05
Masoudi Y, van Rensburg W, Barnard-Jenkins B, Rautenbach M (2021) The influence of cellulose-type formulants on anti-candida activity of the tyrocidines. Antibiotics 10(5):597
Mellado E, Garcia-Effron G, Alcazar-Fuoli L, Melchers WJG, Verweij PE, Cuenca-Estrella M, Rodriguez-Tudela JL (2007) A new aspergillus fumigatus resistance mechanism conferring in vitro cross-resistance to azole antifungals involves a combination of cyp51A alterations. Antimicrob Agents Chemother 51(6):1897–1904
Messer LC, Laraia BA, Kaufman JS, Eyster J, Holzman C, Culhane J et al (2006) The development of a standardized neighborhood deprivation index. J Urban Health 83:1041–1062
Milićević D, Udovički B, Petrović Z, Janković S, Radulović S, Gurinović M, Rajković A (2020) Current status of mycotoxin contamination of food and feeds and associated public health risk in Serbia. Meat Technol 61(1):1–36
Monkiedje A, Spiteller M (2002) Effects of the phenylamide fungicides, mefenoxam and metalaxyl, on the microbiological properties of a sandy loam and a sandy clay soil. Biol Fertil Soils 35:393–398
Montoya MC, Beattie S, Alden KM, Krysan DJ (2020) Derivatives of the antimalarial drug mefloquine are broad-spectrum antifungal molecules with activity against drug-resistant clinical isolates. Antimicrob Agents Chemother 64(3):10–1128
Morschhäuser J (2010) Regulation of multidrug resistance in pathogenic fungi. Fungal Genet Biol 47(2):94–106
Muniaraj S, Subramanian V, Srinivasan P, Palani M (2018) In silico and in vitro studies on Lyngbya majuscula using against lung cancer cell line (A549). Pharm J 10(3):421
Neuwald DA, Saquet AA, Klein N (2018) Disorders during storage of fruits and vegetables. Postharvest Physiol Disord Fruits Vegetables:89–110
Ngemenya M, Metuge H, Mbah J, Zofou D, Babiaka S, Titanji V (2015) Isolation of natural product hits from Peperomia species with synergistic activity against resistant plasmodium falciparum strains. Eur J Med Plants 5(1):77–87
Nicolopoulou-Stamati P, Maipas S, Kotampasi C, Stamatis P, Hens L (2016) Chemical pesticides and human health: the urgent need for a new concept in agriculture. Front Public Health 4:148
Oussou-Azo AF, Nakama T, Nakamura M, Futagami T, Vestergaard MDCM (2020) Antifungal potential of nanostructured crystalline copper and its oxide forms. Nanomaterials 10(5):1003
Park RD, Jo KJ, Jo YY, Jin YL, Kim KY, Shim JH (2002) Variation of antifungal activities of chitosans on plant pathogens. J Microbiol Biotechnol 12(1):84–88
Passone MA, Etcheverry M (2014) Antifungal impact of volatile fractions of Peumus boldus and Lippia turbinata on aspergillus section Flavi and residual levels of these oils in irradiated peanut. Int J Food Microbiol 168:17–23
Patel A, Kumar A, Sheoran N, Kumar M, Sahu KP, Ganeshan P et al (2021) Antifungal and defense elicitor activities of pyrazines identified in endophytic pseudomonas putida BP25 against fungal blast incited by Magnaporthe oryzae in rice. J Plant Dis Protect 128:261–272
Pei X, Tekliye M, Dong M (2021) Isolation and identification of fungi found in contaminated fermented milk and antifungal activity of vanillin. Food Sci Human Wellness 10(2):214–220
Perlin DS (2011) Current perspectives on echinocandin class drugs. Future Microbiol 6(4):441–457
Petit AN, Fontaine F, Vatsa P, Clément C, Vaillant-Gaveau N (2012) Fungicide impacts on photosynthesis in crop plants. Photosynth Res 111:315–326
Pfaller MA, Barry AL (1994) Evaluation of a novel colorimetric broth microdilution method for antifungal susceptibility testing of yeast isolates. J Clin Microbiol 32(8):1992–1996
Pickova D, Ostry V, Toman J, Malir F (2021) Aflatoxins: history, significant milestones, recent data on their toxicity and ways to mitigation. Toxins 13(6):399
Prasad R, Rawal MK (2014) Efflux pump proteins in antifungal resistance. Front Pharmacol 5:202

Pusztahelyi T, Holb IJ, Pócsi I (2015) Secondary metabolites in fungus-plant interactions. Front Plant Sci 6:573

Quinn GA, Abdelhameed AM, Banat AM, Alharbi NK, Baker LM, Castro HC et al (2021) Streptomyces isolates from the soil of an ancient irish cure site, capable of inhibiting multi-resistant bacteria and yeasts. Appl Sci 11(11):4923

Rahman A, Bannigan A, Sulaman W, Pechter P, Blancaflor EB, Baskin TI (2007) Auxin, actin and growth of the Arabidopsis thaliana primary root. Plant J 50(3):514–528

Ranjitha VR, Rai VR (2017) Actinomycetes mediated synthesis of gold nanoparticles from the culture supernatant of Streptomyces griseoruber with special reference to catalytic activity. 3 Biotech 7:1–7

Rashed KN, Ćirić A, Glamočlija J, Calhelha RC, Ferreira IC, Soković M (2013) Antimicrobial activity, growth inhibition of human tumour cell lines, and phytochemical characterization of the hydromethanolic extract obtained from Sapindus saponaria L. aerial parts. BioMed Res Int 2013(1):659183

Rex JH, Rinaldi MG, Pfaller MA (1995) Resistance of Candida species to fluconazole. Antimicrob Agents Chemother 39(1):1–8

Rizvi ZF, Mukhtar R, Chaudhary MF, Zia M (2013) Antibacterial and antifungal activities of Lawsonia inermis, Lantana camara and Swertia angustifolia. Pak J Bot 45(1):275–278

Rodrigues AM, Eparvier V, Odonne G, Amusant N, Stien D, Houël E (2019) The antifungal potential of (Z)-ligustilide and the protective effect of eugenol demonstrated by a chemometric approach. Sci Rep 9(1):8729

Roman DL, Voiculescu DI, Filip M, Ostafe V, Isvoran A (2021) Effects of triazole fungicides on soil microbiota and on the activities of enzymes found in soil: A review. Agriculture 11(9):893

Rosegrant MW, Valmonte-Santos RA, Cline SA, Ringler C, Li W (2005) Water resources, agriculture and pasture: implications of growing demand and increasing scarcity. In: Grassland: a global resource. Proceedings 2005 international grassland congress, Dublin, Ireland, vol 26, pp 227–249

Ruiz-Baca E, Arredondo-Sánchez RI, Corral-Pérez K, López-Rodríguez A, Meneses-Morales I, Ayala-García VM, Martínez-Rocha AL (2021) Molecular mechanisms of resistance to antifungals in *Candida albicans*. Adv Candida Albicans 39:5772

Sabatelli F, Patel R, Mann PA, Mendrick CA, Norris CC, Hare R et al (2006) In vitro activities of posaconazole, fluconazole, itraconazole, voriconazole, and amphotericin B against a large collection of clinically important molds and yeasts. Antimicrob Agents Chemother 50(6):2009–2015

Salam MA, Al-Amin MY, Salam MT, Pawar JS, Akhter N, Rabaan AA, Alqumber MA (2023) Antimicrobial resistance: a growing serious threat for global public health. In: Healthcare, vol 11, no. 13. MDPI, p 1946

Sharma J, Rosiana S, Razzaq I, Shapiro RS (2019) Linking cellular morphogenesis with antifungal treatment and susceptibility in Candida pathogens. J Fungi 5(1):17

Sheng C, Zhang W, Zhang M, Song Y, Ji H, Zhu J et al (2004) Homology modeling of lanosterol 14α-demethylase of Candida albicans and aspergillus fumigatus and insights into the enzyme-substrate interactions. J Biomol Struct Dyn 22(1):91–99

Shianna KV, Dotson WD, Tove S, Parks LW (2001) Identification of a UPC2 homolog in Saccharomyces cerevisiae and its involvement in aerobic sterol uptake. J Bacteriol 183(3):830–834

Snelders E, van der Lee HAL, Kuijpers J, Rijs AJM, Varga J, Samson RA et al (2008) Emergence of azole resistance in aspergillus fumigatus and spread of a single resistance mechanism. PLoS Med 5(11):e219

Sokol-Anderson ML, Brajtburg J, Medoff G (1986) Amphotericin B-induced oxidative damage and killing of Candida albicans. J Infect Dis 154(1):76–83

Suchodolski J, Derkacz D, Bernat P, Krasowska A (2021) Capric acid secreted by saccharomyces boulardii influences the susceptibility of Candida albicans to fluconazole and amphotericin B. Sci Rep 11(1):6519

Sun W, Wang D, Yu C, Huang X, Li X, Sun S (2017) Strong synergism of dexamethasone in combination with fluconazole against resistant Candida albicans mediated by inhibiting drug efflux and reducing virulence. Int J Antimicrob Agents 50(3):399–405

Toepfer S (2023) Experimental combination therapy against azole-based multidrug resistant *Candida auris* (Doctoral dissertation, University of Otago)

Toledo AV, López SMY, Aulicino MB, Marino de Remes Lenicov AM, Balatti PA (2015) Antagonism of entomopathogenic fungi by bacillus spp. associated with the integument of cicadellids and delphacids. Int Microbiol 18

Tortorano AM, Prigitano A, Dho G, Grancini A, Passera M, Ecmm-Fimua Study Group (2012) Antifungal susceptibility profiles of Candida isolates from a prospective survey of invasive fungal infections in Italian intensive care units. J Med Microbiol 61(3):389–393

Tudi M, Daniel Ruan H, Wang L, Lyu J, Sadler R, Connell D, Phung DT (2021) Agriculture development, pesticide application and its impact on the environment. Int J Environ Res Public Health 18(3):1112

Tyndall JD, Sabherwal M, Sagatova AA, Keniya MV, Negroni J, Wilson RK et al (2016) Structural and functional elucidation of yeast lanosterol 14α-demethylase in complex with agrochemical antifungals. PLoS One 11(12):e0167485

Udovicki B, Audenaert K, De Saeger S, Rajkovic A (2018) Overview on the mycotoxins incidence in Serbia in the period 2004–2016. Toxins 10(7):279

Van der Fels-Klerx HJ, Camenzuli L (2016) Effects of milk yield, feed composition, and feed contamination with aflatoxin B1 on the aflatoxin M1 concentration in dairy cows' milk investigated using Monte Carlo simulation modelling. Toxins 8(10):290

Vasicek EM, Berkow EL, Flowers SA, Barker KS, Rogers PD (2014) UPC2 is universally essential for azole antifungal resistance in Candida albicans. Eukaryot Cell 13(7):933–946

Veras FF, Correa APF, Welke JE, Brandelli A (2016) Inhibition of mycotoxin-producing fungi by bacillus strains isolated from fish intestines. Int J Food Microbiol 238:23–32

Verweij PE, Gonzalez GM, Wiederhold NP, Lass-Flörl C, Warn P, Heep M et al (2009) In vitro antifungal activity of isavuconazole against 345 mucorales isolates collected at study centers in eight countries. J Chemother 21(3):272–281

Viswanathan P, Muralidaran Y, Ragavan G (2017) Challenges in oral drug delivery: a nano-based strategy to overcome. In: Nanostructures for oral medicine. Elsevier, pp 173–201

Walsh TJ, Lee JW, Kelly P, Bacher J, Lecciones J, Thomas V et al (1991) Antifungal effects of the nonlinear pharmacokinetics of cilofungin, a 1,3-beta-glucan synthetase inhibitor, during continuous and intermittent intravenous infusions in treatment of experimental disseminated candidiasis. Antimicrob Agents Chemother 35(7):1321–1328

Wang YS, Jia M, Zhang QL, Song XY, Yang DP (2019) Ab initio investigation of excited state dual hydrogen bonding interactions and proton transfer mechanism for novel oxazoline compound. Chinese Physics B 28(10):103105

Warrilow AG, Parker JE, Kelly DE, Kelly SL (2013) Azole affinity of sterol 14α-demethylase (CYP51) enzymes from Candida albicans and Homo sapiens. Antimicrobial agents and chemotherapy, 57(3):1352–1360

Yadav AN, Verma P, Kumar S, Kumar V, Kumar M, Sugitha TCK et al (2018) Actinobacteria from rhizosphere: molecular diversity, distributions, and potential biotechnological applications. In: New and future developments in microbial biotechnology and bioengineering. Elsevier, pp 13–41

Yang C, Hamel C, Vujanovic V, Gan Y (2011) Fungicide: modes of action and possible impact on nontarget microorganisms. Int Scholarly Res Notices 2011(1):130289

Yarzábal LA, Chica EJ (2019) Role of rhizobacterial secondary metabolites in crop protection against agricultural pests and diseases. In: New and future developments in microbial biotechnology and bioengineering. Elsevier, pp 31–53

Zhang S, Zhong R, Tang S, Han H, Chen L, Zhang H (2022) Baicalin alleviates short-term lincomycin-induced intestinal and liver injury and inflammation in infant mice. International Journal of Molecular Sciences 23(11):6072

Zhu F, Zhou YK, Ji ZL, Chen XR (2018) The plant ribosome-inactivating proteins play important roles in defense against pathogens and insect pest attacks. Front Plant Sci 9:146

Zhu GY, Chen Y, Wang SY, Shi XC, Herrera-Balandrano DD, Polo V, Laborda P (2022) Peel diffusion and antifungal efficacy of different fungicides in pear fruit: structure-diffusion-activity relationships. J Fungi 8(5):547

Zubrod JP, Bundschuh M, Arts G, Brühl CA, Imfeld G, Knäbel A et al (2019) Fungicides: an overlooked pesticide class? Environ Sci Technol 53(7):3347–3365

Antifungal Potential of Bioactive Compounds Derived from Microbes and Other Natural Sources: Challenges and Future Scope

22

Munendra Kumar, Kajal, Nargis Taranum, Khyati, Biji Balan, Prateek Kumar, and Amit Singh Dhaulaniya

22.1 Challenges in Currently Used Antifungal Drugs

Fungi are one of the most detrimental eukaryotic microorganisms on earth, contributing to the extinction of a variety of life forms (Fisher et al. 2020). More than 1.5 million people die from fungal infections each year in the world, which has become a significant occurrence (Deaguero et al. 2020). *Aspergillus, Candida,* and *Cryptococcus* are currently responsible for majority of fungal infections that result in human mortality and morbidity (Boral et al. 2018). Different anatomical areas, such as the skin, genitourinary, oral, lung, and gastrointestinal tract, all contain a variety of fungal species that are part of the normal flora and play a crucial role in maintaining human health (Rolling et al. 2020). But patients with an impaired immune system who have undergone transplant surgery, chemotherapy,

M. Kumar (✉)
Department of Zoology, Rajiv Gandhi University, Doimukh, Arunachal Pradesh, India
e-mail: munendra.kumar@rgu.ac.in

Kajal · N. Taranum · Khyati
School of Biological and Life Sciences, Galgotias University, Greater Noida, Uttar Pradesh, India
e-mail: mkajal230@gmail.com; nargistaranum98@gmail.com; khyati@galgotiasuniversity.edu.in

B. Balan
Department of Zoology, Dyal Singh College, University of Delhi, New Delhi, Delhi, India
e-mail: biji.balan619@gmail.com

P. Kumar
Department of Zoology, University of Allahabad, Prayagraj, Uttar Pradesh, India
e-mail: drprateekkumar@allduniv.ac.in

A. S. Dhaulaniya
Department of Zoology, Kirori Mal College, University of Delhi, New Delhi, Delhi, India
e-mail: amit.dhaulaniya@gmail.com

N. Manzoor (ed.), *Advances in Antifungal Drug Development*, https://doi.org/10.1007/978-981-97-5165-5_22

radiotherapy, hemodialysis, or received immunosuppressive medication are more susceptible to mycoses (Zhang et al. 2021). A significant challenge to disease control across several anthropogenic systems is posed by the quick spread of multi-drug-resistant pathogenic fungi and the more widely recognized threat of antibiotic-resistant bacteria (Fisher et al. 2018). The negative effects of fungi on human health are currently outpacing those of breast cancer or malaria in terms of global death rates, and they are at par with those of tuberculosis and HIV (Brown et al. 2012). Despite their prevalence, fungal infections have historically received much less attention than other types of infectious diseases.

22.1.1 Multiple Drug Resistance (MDR) as a Therapeutic Challenge

In the history of mankind, several deadly and infectious disease outbreaks have been seen at different times. According to the report of WHO, communicable diseases are one of the top causative reasons for deaths per year across the globe. Although some pathogens are unfortunately underestimated by health authorities, they are potent enough to cause several deaths and emerging diseases. Invasive fungal infections (IFIs) are majorly responsible for morbidity and mortality in humans. A recent report showed that every year invasive fungal infections cause almost 1.7 million deaths in immunocompromised and immunocompetent individuals such as elderly people, HIV/AIDS, cancer, and organ transplantation patients (Denning and Bromley 2015; Houšť et al. 2020; Pianalto and Alspaugh 2016). These patients have access to excessive antibiotic usage and immune-suppressing drugs enhance their susceptibility to the development of invasive fungal infections (Enoch et al. 2017; Lockhart and Guarner 2019). Moreover, Global Action Fund for Fungal Infections (GAFFI) estimated that each year approximately one million eyes become blind because of fungal keratitis (GAFFI 2018).

Fungi are the most diverse group of organisms; 5 million species of fungi are known till date; out of which, 300 species are responsible for causing various diseases in humans (Perfect 2017). Examples include *Candida albicans, Candida auris, Aspergillus fumigatus, Cryptococcus neoformans, Pneumocystis jirovecii, Histoplasma capsulatum, Taloromyces, Coccidioides immitis, Mucormycetes, Malassezia furfur, Blastomyces dermatitidis, Sporothrix* spp., *Fusarium,* and *Scedosporium* (Fernandez et al. 2020; Hasim and Coleman 2019; Schmiedel and Zimmerli 2016). *Candida* is a yeast that causes candidiasis and accounts for 750,000 cases. Aspergillosis is caused by *Aspergillus* spp. accounting for 300,000 cases and cryptococcosis caused by *Cryptococcus* accounts for 23,000 cases worldwide annually. Out of these, patients with aspergillosis have the highest mortality rate, i.e., 30–90% of invasive fungal infections (Houšť et al. 2020; Pianalto and Alspaugh 2016; Bongomin et al. 2017).

Recent therapeutic options for IFs infections are divided into five structural classes of antifungal drugs. These are polyenes, azoles, allylamines, pyrimidines, and echinocandins (Rauseo et al. 2020; Fernandez et al. 2020; Hasim and Coleman

2019; Houšť et al. 2020). The oldest class of antifungal agent is polyenes, in which amphotericin B binds to the ergosterol of fungal membranes, leading to permeability changes in the cell wall. They effectively work against *Aspergillus* spp., *Candida* spp., mycoses, *Fusarium* spp., *Cryptococcus* spp., and other fungi (Gray et al. 2012). The main limitation of polyenes is their toxicity. They have a tendency to bind with human cholesterol leading to toxicity of kidneys and liver, but lipid formulations in amphotericin B have overcome this problem (Hamil 2013; Birch and Sibley 2017). Another class of antifungal drugs is azoles, which is the widely prescribed class of drugs against fungal infections. Among many other azole drugs available, fluconazole, posaconazole, itraconazole, and voriconazole are mostly used in invasive fungal infections. Azoles inhibit the lanosterol 14α-demethylase enzyme which inhibits the synthesis of ergosterol, the most pivotal sterol in the fungal cell membrane (Meletiadis et al. 2007; Van Daele et al. 2019). Although azoles are well-accepted drugs, they have drawbacks such as interference with the metabolism of other drugs and show drug–drug interactions by interfering with the cytochrome P450 enzyme system. Pyrimidines interfere with DNA and RNA synthesis as its metabolism gets disrupted (Houst et al. 2020; Pianalto and Alspaugh 2016; Van Daele et al. 2019). Another class of antifungal drugs is allylamines which inhibit the enzyme squalene epoxidase involved in the pathway of ergosterol synthesis (Birnbaum 1990; Nowosielski et al. 2011). The most recent class of antifungal drugs is echinocandins, which inhibits 1,3-β-glucan synthesis. The 1,3-β-glucan is an important component of the fungal cell wall, and thus, these drugs alter the cell wall composition. Echinocandins show fungicidal and fungistatic activity against species of *Candida* and *Aspergillus*, respectively, but show no useful clinical response against *Cryptococcus* spp. Echinocandins showed significantly less drug–drug interactions and toxicity. Although they showed less resistance rate when compared to other antifungal drugs, however, recently few fungi such as non-albicans *Candida* spp. started developing resistance to echinocandins because of acquired mutations (Maligie and Selitrennikoff 2005).

Overuse of antimicrobial drugs leads to multidrug resistance against various microbial agents. It has dramatically increased in the last few decades and has become a global health concern. Multidrug resistance is the condition where microorganisms become insensitive toward various classes of antimicrobial drugs, even when the drugs are structurally not similar and have different molecular targets (Prestinaci et al. 2015). A very restricted range of antifungal agents are available in public for the treatment of chronic fungal diseases. The above-mentioned antifungal drugs are attaining resistance toward different fungal strains also, increasing cases of lower susceptibility of fungal strains. Further, a wide range of antifungal resistance has been seen in the case of recently isolated strains of fungi (Fig. 22.1). Resistance has been seen mainly in various fungal isolates such as *Candida* spp., *Aspergillus* spp., *Cryptococcus neoformans, Trichosporon beigelii, Scopulariopsis* spp., and *Pseudallescheria boydii* against drugs like polyene macrolides, azoles derivatives, 1,3-β-glucan synthase inhibitors, and RNA and DNA synthesis inhibitors (Loeffler et al. 2003). Reports suggest polyenes showed resistance against a wide range of isolates of fungal strains because of the modifications and remolding

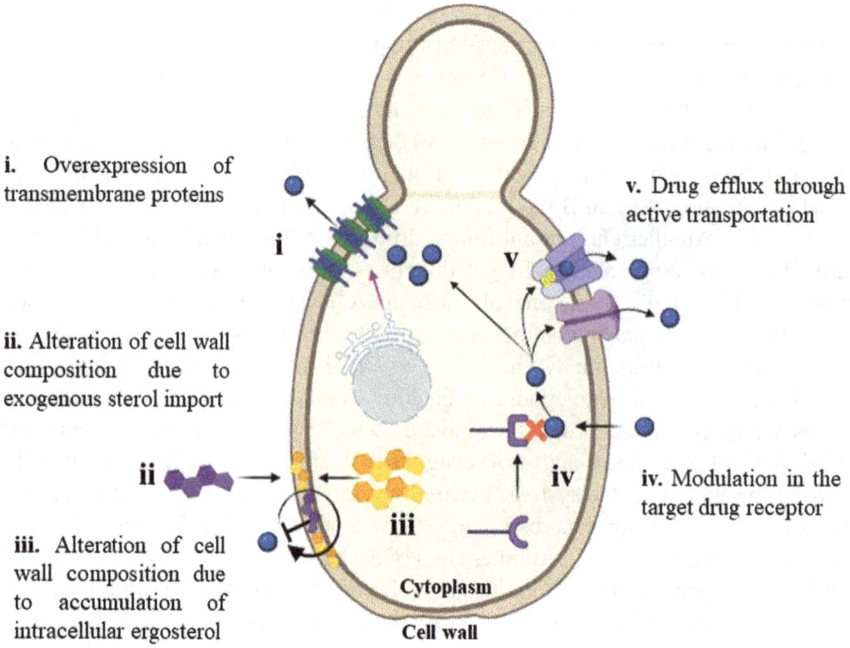

Fig. 22.1 Different mechanisms of antifungal drug resistance. (i) Overexpression of transmembrane transport proteins minimize the intracellular accumulation of drug molecules. (ii) & (iii) Import of exogenous sterol and accumulation of intracellular ergosterol consequent altered composition and permeability of cell wall, which shows resistance against the drug effect. (iv) Modulation in target drug receptors make the cell resistant against the antifungal drug. (v) Accumulated drug molecules are effluxed out through transporters like ABC (ATP-binding cassette transporter family) and MFS (major facilitator superfamily of transporters)

of the lipid composition of the plasma membrane, i.e., lack of certain isomerases which leads to ergosterol decline in the membrane (Kelly et al. 1994; Subden et al. 1977). This modification enables the binding affinity of AmB to the plasma membrane due to the unavailability of binding sites (Hitchcock et al. 1987). In the case of azoles, the resistance is possible due to potential molecular mechanisms. Progress of resistance to an azole drug causes further cross-resistance in other azole derivatives and sometimes the resistance is azole specific (Stevens and Stevens 1996). Laboratory data regarding echinocandin resistance is scarce and mainly based on laboratory mutant strains of *S. cerevisiae.* In this fungus, the β-glucan synthase complex is coded by two genes and regulated by another third gene. The mutations of one of the gene *FKS1*, leads to echinocandin resistance through changes in β-glucan synthase. Thus, echinocandins are no longer capable of entering the cytoplasm of fungal cells (Kurtz 1988). Thus, our limited antifungal agents and increasing resistance in various fungal strains, maximize the urge to utilize the current knowledge and clinical advancement which would guide in the development of new antifungal drugs and other treatment therapeutics.

22.1.2 Limited Spectrum and Fungistatic Challenges

Nystatin and polyenes, the first antifungal substances used in human healthcare, were found in the 1950s, whereas copper and sulfur fungicides were initially applied to treat agricultural diseases more than 150 years ago. As first-line therapies for fungal diseases in people and plants, systemic fungicides and antifungals are now often employed. The most clinically significant pathogens responsible for invasive fungal infections (IFIs) include fungi from the genera *Aspergillus*, *Candida*, *Pneumocystis*, and *Cryptococcus* (Arastehfar et al. 2020). Only three primary classes of antifungal medications—azoles, echinocandins, and polyenes—are used to treat invasive fungal infections, in contrast to the wide range of antibiotic drug classes that are available for use against bacterial diseases (Kaur and Nobile 2023). Azoles (fluconazole, posaconazole, voriconazole, itraconazole) are the most popularly prescribed antifungal medications used to treat superficial and systemic fungal infections, whereas polyenes (amphotericin B) are the most traditional class of antifungal medications used to treat severe systemic fungal infections, and echinocandins (micafungin and caspofungin) are the most recent class which is used to treat resistant fungal infections (Berman and Krysan 2020). Azoles inhibit the synthesis of ergosterol, via binding to *Cyp51A* in *Aspergillus* and *Erg11* in *Candida* species whereas Echinocandins interfere with the synthesis of 1,3-D-glucan, a significant structural component of the cell wall, by targeting the catalytic subunit of 1,3-D-glucan synthase, which is encoded by FKS genes. Last but not least, the binding of polyenes to ergosterol in the cell membrane leads to cell death due to the formation of pores in the membrane, which disrupts osmotic pressure (Arastehfar et al. 2020).

Antifungals come in two different types: fungicidal, which kills the fungal cell, and fungistatic, which stops the proliferation of the fungal cell but does not kill it. Fluconazole and echinocandins are some of the fungistatic drugs that work against *Candida* and *Aspergillus* species. However, the major problem with antifungals nowadays is the development of drug resistance or multidrug resistance. In recent years, there has been an upsurge in the multidrug-resistant fungus *Candida auris* worldwide. More importantly, some of these diseases are resistant to practically every type of antifungal medication currently available (Du et al. 2020). The epidemiological landscape of IFIs has changed as a result of the widespread use of antifungals; fungal species resistant to one or more antifungal classes are now more frequently seen in clinical settings and are linked to therapeutic failure (Romero et al. 2019; Latgé and Chamilos 2019). According to Chowdhary et al. (2020), 15 COVID-19 patients in New Delhi ICUs had secondary candidiasis, with *C. auris* accounting for two-thirds of cases and carrying a death rate of up to 60%. More worrisomely, in invasive aspergillosis, patients who contract azole-resistant *A. fumigatus* die with a mortality rate of 50–100% (Lestrade et al. 2019). A variety of techniques have been used to conduct antifungal therapies, including the synthesis of new substances, using organism extracts, drug repurposing, and an association between well-known antifungal medicines and non-antifungal agents (Robbins et al. 2016). Table 22.1 gives a list of some conventional antifungal drugs, their spectrum of activity, route of administration and mode of action. The search for

Table 22.1 Currently used antifungal drugs and their mechanism of action

Drugs	Spectrum of activity	Route of administration	Mechanism of action
Amphotericin B	Fungicidal	Topical, IV	Binds to ergosterol and disrupts fungal membrane
Nystatin	Fungicidal	Topical, oral suspension	Same as amphotericin B
Imidazoles • Econazole • Ketoconazole • Clotrimazole • Miconazole • Oxiconazole • Sulconazole	Fungistatic	Topical, oral	Inhibits 14-α-lanosterol demethylase, a fungal cytochrome P450
Triazoles • Fluconazole • Itraconazole • Posaconazole • Isavuconazole	Fungistatic	IV, oral Oral Oral IV, oral	Same as imidazoles
Allylamines • Naftifine • Terbinafine	Fungistatic and fungicidal	Topical Topical, Oral	Inhibition of squalene epoxidase
Flucytosine	Fungistatic or fungicidal	Oral	Inhibits nucleic acid synthesis
Griseofulvin	Fungistatic	Oral	Affects microtubule function, which inhibits mitosis
Nikkomycin Z	Fungicidal	IV	Inhibits fungal cell wall chitin synthesis
Amorolfine	Fungistatic and fungicidal	Topical	Reduces ergosterol accumulation in the fungal cell membrane by inhibiting D14 reductase and D7–D8 isomerase.
Ciclopirox	Fungistatic and fungicidal	Topical	Membrane instability
Tavaborole	Fungistatic and fungicidal	Topical	Inhibition of fungal protein synthesis
Echinocandins • Caspofungin • Anidulafungin • micafungin	Fungicidal	IV	Inhibits β-glucan synthesis which disrupts cell wall synthesis

antifungal drugs can be aided by traditional medicine. We can discover species that modern medicine has neglected by using the knowledge of plants employed as anti-infectives in systems of traditional medicine from all over the world. The plants used to make these tried-and-true traditional medicines may include bioactive substances that are useful against drug-resistant fungal diseases.

22.2 Medical Side Effects of Commercial Antifungal Drugs

22.2.1 Immunological Side Effects

The prevalence of fatal fungal infections has sharply increased during the past two decades as the number of AIDS, cancer, organ transplant, and immune-compromised patients has risen (Shukla et al. 2016). With an annual rise in the frequency of fungal infections, the development of antifungal therapies for prevention and cure is also taking place. As a part of antifungal therapies, the use of antifungal drugs is also on the rise. Although antifungal drugs have been proven effective against various fungal infections, the prolonged use of these drugs has shown adverse effects in many cases (Fig. 22.2). The well-known immunological adverse effects generally refer to cutaneous responses and bone marrow suppression (Labro 2012). Evidence for the immunomodulatory side effects of antifungal agents on various immune cells/systems is increasing. Amphotericin B has been known to cause infusion-related toxicity (fever and chills), suggested to be caused by innate immune cells producing proinflammatory cytokines and chemokines such as interleukin 1β, TNF-α, etc. (Houšť et al. 2020). Sau et al. (2003) hypothesized the release of inflammatory cytokines from cells expressing TLR2 and CD14. TLR4 mutant cells were less sensitive to amphotericin B activation than TLR4-normal cells resulting in less production of proinflammatory cytokines. Ries et al. (2019) investigated the effect of various antifungal drugs (a few azoles, caspofungin, micafungin, and amphotericin B), on effector functions (activation, degranulation, and phagocytosis, production of ROS, and IL-8) of neutrophils. Although they did not clarify the clinical significance except for liposomal amphotericin B, all other antifungal drugs were found involved in the modulation of one or other effector functions of neutrophils. Wheeler et al. (2016) investigated the immunological repercussions of intestinal fungal

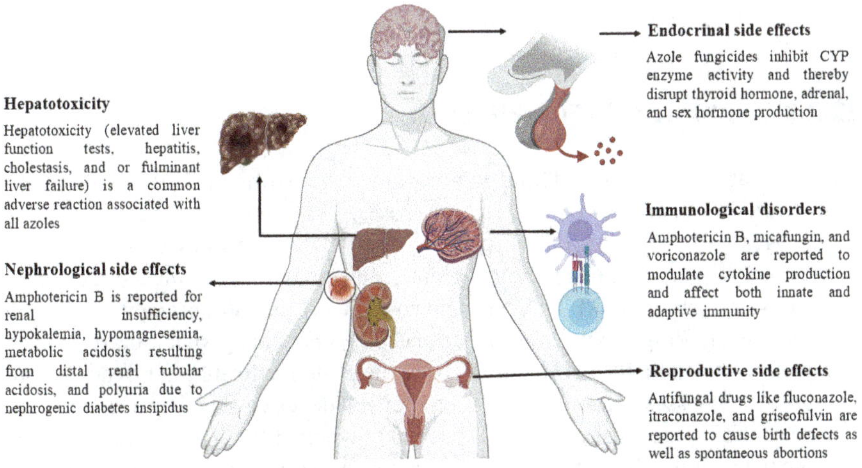

Fig. 22.2 Reported medical side effects of currently used antifungal drugs

dysbiosis. The researchers used an oral antifungal drug, fluconazole to disrupt the fungal community in mice and found that it worsened the outcome of colitis and allergic airway disease. Their study also suggests that the interactions between fungal and bacterial populations in the gut play a role in modulating immune homeostasis and can be affected adversely with the prolonged use of antifungal drugs such as fluconazole.

22.2.2 Endocrinal Side Effects

Antifungal medications used to prevent and/or cure mycoses may have detrimental effects on the endocrine and metabolic systems (Lionakis et al. 2008). Among all the classes of fungicides, azole fungicides are most concerning in terms of their adverse interaction with the endocrine system. Conazole may interact with the mammalian cytochrome P450 (CYP) system in addition to blocking fungus-specific enzymes (Juberg et al. 2006). As a result, it may have endocrine-disrupting effects by preventing certain enzymes from functioning in the steroid production pathway (Zhang et al. 2002; Trösken et al. 2004). This is especially concerning during pregnancy because sex hormones are essential for the development of the reproductive system (Draskau and Svingen 2022). Kjaerstad et al. (2010) studied the in vitro endocrine disruptive effect of imidazoles and triazoles. Although they all demonstrated the ability to act through a variety of routes, the disruption of steroid production appears to be the crucial mechanism. Their study also concluded that imidazoles are more potent endocrine disruptors than triazole. In a separate study conducted by Draskau et al. (2019), triticonazole and flusilazole, two commonly used agricultural azole fungicides, were examined for their ability to potentially affect reproductive parameters and their in-vitro antiandrogenic action. In the in-vitro study, both fungicides demonstrated potent androgen receptor (AR) antagonistic activity and impairment of steroid production with visible signs of shorter male anogenital distance (AGD) caused by triticonazole.

22.2.3 Nephrological Side Effects

Deoxycholate, liposomal, and lipid complexes are three different drug formulations that are used to treat different systemic mycoses. The primary distinction between them is nephrotoxicity, with the first being more toxic with rates as high as 80%. Acute renal failure, tubulopathies, and glomerulopathies are prominent clinical presentations of nephrotoxicity, which is defined as any renal harm brought on directly or indirectly by drugs (Kwiatkowska et al. 2021). For many years, deoxycholate formulation of Amphotericin B (AmBD) was the go-to medication for treating invasive fungal illnesses. However, nephrotoxicity side effects and infusion-related responses (IRRs) are frequently observed, making them difficult to utilize in clinical practice (Chai et al. 2013). Cases of tubulopathy are very common with amphotericin B. AmB has an antifungal impact because it can change the permeability of fungal cell membranes, which results in cell death. AmB can bind to cholesterol molecules in cellular membranes, changing the permeability of the membrane. The structure of amphotericin B enables it to mimic a false phospholipid, enabling membrane penetration. Within these membranes, it induces two significant detrimental outcomes: oxidative stress and ion disruption (Wang 2022). Acute tubular necrosis and tubular dysfunction may ensue from this impact, which can be harmful to renal tubular cells (Pazhayattil and Shirali 2014). The common clinical manifestations associated with AmB-induced tubular dysfunction include electrolyte disbalance (potassium and magnesium loss in urine), urine concentration issues, and renal tubular acidosis (Fanos and Cataldi 2000). Deoxycholate is also thought to be responsible for direct toxicity to the tubules (Faught et al. 2015). Chai et al. (2013) demonstrated that in patients getting AmBD therapy, cytokine levels change over time, which matters more than their exact levels. They found that high IL-6 after 1 week of treatment increases AKI risk by 2.74–6.93 times. Keeping IL-8 high also slightly raises nephrotoxicity risk. Caspofungin was the initial echinocandin approved for treating candidiasis and as a second option for aspergillosis. It hampers fungal cell walls by blocking 1,3-β-glucan production, affecting specific enzymes. It is now considered a suitable option for treating fungal infections in high-risk patients prone to amphotericin-induced AKI. While caspofungin is less kidney-damaging than amphotericin B according to clinical trials, lab studies suggest mild distal tubule damage as a possible AKI mechanism (Morales-Alvarez 2020).

22.3 Antifungal Compounds from Different Natural Sources

22.3.1 Antifungal Compounds from Microbes

A variety of natural products originating from microbes are being used to control the infection of pathogenic fungi. These natural products have a wide range of activities like antifungal, antibacterial, antitumor, etc. Two-thirds of antibiotics and half of the antifungal drugs like echinocandins, amphotericin B, fluconazole, voriconazole, and itraconazole are biosynthesized by the diverse species of bacteria and fungi. Echinocandin is a class of antifungal drugs composed of a semisynthetic lipid side chain with N-acyl lipid. These drugs are composed of three natural lipoproteins (caspofungin, micafungin, and anidulafungin) produced by the *Aspergillus* (Dongmo Fotsing and Bajaj 2024; Grover 2010; Spampinato and Leonardi 2013). Amphotericin B is a well-known antifungal drug obtained from microbes. It was extracted and purified in 1953 by Elizabeth L. Hazen and Rachel F. Brown from the secretion broth of the actinobacterium *Streptomyces nosodus*. Amphotericin B interacts and targets the ergosterol of the cell membrane of fungi. This binding irreversibly interferes with membrane integrity and results in cell death. Amphotericin B has long been the primary antifungal drug used to treat invasive and severe mycoses. It is used to treat a variety of fungal infections, including Mucormycosis, Aspergillosis, Blastomycosis, Candidiasis, and Leishmaniasis (Baginski and Czub 2009). Fluconazole: The active substance belongs to the triazole class of antifungals and is one of the most commonly used agents for this purpose. It interacts with the 14-α-demethylase enzyme, which is responsible for converting lanosterol into ergosterol. Ergosterol is an essential component of the fungal cellular membrane, and fluconazole thus inhibits the synthesis of ergosterol to improve cellular permeability. Additionally, the medication has other functions such as inhibition of endogenous respiration and inhibition of yeast formation (Pfaller et al. 2005; Govindarajan et al. 2023; Yüksekgönül et al. 2021; Govindarajan et al. 2022). Itraconazole: In order to treat a range of infections caused by fungi, it is used as an antifungal agent in the azole class of drugs. This drug works by inhibiting the growth of fungi, which are rare but can be damaging to those with a weakened immune system. The active metabolite of this drug is hydroxyitraconazole, and it has been shown to inhibit the formation of ergosterol. It is necessary for the metabolism of ergosterol in fungi, as well as cholesterol in mammals. This inhibition of the synthesis of ergosterol results in irregularities in the fungal membrane, increasing its permeability and disrupting the integrity thereby altering the activity of membrane-bound enzymes. 14-α-lanosterol demethylase, also known as 14LDM, is responsible for the formation of the ergosterol (Georgopapadakou and Walash 1996; Chen et al. 2018; Kurn and Wadhwa 2023). Flucytosine: When the drug 5-FC enters the fungal cell, it is metabolized by an enzyme called cytopermease. This enzyme is responsible for the conversion of 5-FC into 5-FU, which is the active form of the drug. 5-FU is then converted into its active form by the cytosine-deaminase enzyme, which does not exist in mammalian cells. Therefore, 5-FC cannot be metabolized in bacteria or human cells. 5-FC interacts with the RNA strand in a way similar to uracil. This interaction disrupts RNA synthesis and impairs protein synthesis in the fungal cells, resulting in DNA damage (Spampinato and Leonardi 2013; Padda and Parmar 2023). Voriconazole: This is a second generation triazole which has improved antifungal spectrum compared to older triazoles such as fluconazole. This active ingredient in Vfend (Pfizer) is an antifungal triazole medication used to treat severe invasive fungal infections. It is most commonly used to treat patients with immunocompromised conditions. This triazole is able to bind to sterol 14-α-demethylase (CYP51), also known as lanosterol demethylase. Lanosterol is demethylated by yeast and other fungi as part of the ergosterol synthesis pathway. Without sufficient ergosterol, the fungal cell membrane functions are disrupted and fungal cell growth is inhibited. When fungal growth is limited, the host immune system clears the invading organism (Ravikumar et al. 2007). Caspofungins: These agents are a new type of antifungals that work on the cell wall of fungi by blocking glucan synthesis. They work in combination with amphotericin B and triazoles, and acetate inhibits the production of β- and D-glucans, which are important for *Asperigillus* and *Candida* cell walls. β-glucan is not found in mammalian cells, but caspofungin has been shown to work against *Candida* and *Aspergillus* hyphae (Dongmo Fotsing and Bajaj 2024).

22.3.2 Natural Bioactive Compounds from Plants

22.3.2.1 Phenolic Compounds

Phenolic compounds are organic compounds with low molecular weight that are naturally found in plants. They can be found in plants that are vascular, like monocots or dicots, as well as plants that are not vascular, like fruits and vegetables (Ansari et al. 2013). In plants with vascular components, phenolic compounds can be found like rosmarinic acid in hornwort, agrestis, green algae, cereals, and other orchids. You can also find phenolic compounds in plants that do not have a vascular component. Ferns are a great source of phenolic compounds, like kaempferol (Adam 1999), which can be found in *Fernia molesta* and salvinia, and also in other plants like grapes and tomatoes (Ansari et al. 2013).

Phenolic compounds are not just found in plants—they can be found in animals too. Guaiacol, for example, is a compound found in the intestines of desert locust, *Schistocercera gregaria* (Choudhary et al. 2008). Unfortunately, the antifungal properties of some of these compounds are unknown, so it will take some more research to figure out how they work.

22.3.2.2 Flavonoid Compound

Flavones are made up of a single carbonyl group, and when you add a 3-hydroxy group, you get flavonol. One flavonoid, amentoflavone, from *Selaginella tamariscina* has been shown to have strong antifungal effects against a variety of fungal strains, and it has a very low hemolytic effect on humans. Another flavonoid from *Piper solmsianum* has been shown to have antifungal effects against all dermatophytes. A flavonoid from *Alpinia officinarum* rhizome has been shown to be very effective against a range of fungal diseases. Thus, flavonoids, azulenes, sesquiterpene, and essential oils from the plant have been shown to have immense antifungal benefits against dermatophytes. Isoflavan, a 2-hydroxymaackiain-derived flavonoid, has been demonstrated to possess antifungal properties. Other flavonoid derivatives, such those derived from scandenone, tiliroside, quercetin, kaempferol, and hispidulin have been demonstrated to inhibit the growth of a wide variety of human pathogens (Di Carlo et al. 1999).

22.3.2.3 Coumarins

Coumarins play an indirect negative role in the transmission of infections. An antifungal agent has been identified in the form of 1-tetrachloroethylene-8bH, 10bH-ethylene-7-en-8a, 12-olide from the micro-organism *Senecium poepigii*. Angelicin, a naturally occurring furanocoumarin, has been demonstrated to possess antifungal activities. Additionally, synthetic coumarins, as well as angelicin derivatives, have been identified as active against the fungal pathogens (*C. albicans*, *C. neoformans*, *S. cerevisiae*, and *A. niger*) (Pan et al. 2015; Zhang et al. 2017; Ramírez-Pelayo et al. 2019; Liang et al. 2017; Küpeli Akkol et al. 2020; Wittayapipath et al. 2020; Nofal et al. 2000).

22.3.2.4 Quinones

Quinones are two-ketone substitutions of aromatic rings, which inactivate proteins and interfere with their functions. These compounds bind to surface-exposed adhesin, cell wall polypeptides, and membrane-bound enzymes to form a complex that inactivates enzymes. Antifungal activity has been reported for four of the quinones: kigelinone (an o-quinone trimer), isopinnatal (an isopinnatal trimer), dihydro-α-lapachone (a dihydro-quinone nucleus), and lapachol (a lapachol trimer) (Arif et al. 2009). Another quinone, hopeanolin, found in the stem bark of *Hopea exalata*, has been identified as an antifungal agent. Additionally, a 2,6-dimethoxy-*p*-benzoquinone found in the heartwood of *Rheumatoida apiculata* has been reported to have antifungal activity (Lee et al. 2019; Wang et al. 2021).

22.3.2.5 Saponins

Glycosylated compounds, known as saponins, are widely distributed throughout the plant kingdom. There are three main categories of saponins: triterpenoids, steroids, and steroids. In the case of spirostanol, three saponins have been isolated from the roots, designated as spirostanol, sansevierin, and sanseviratin, and three steroidal (sensitizing) saponins. In particular, eight saponins were identified from *Tribulus terrestris*, two of which showed promising antifungal action against influenza-resistant strains of *Candida neoformans, Candida albicans, Candida clavulanicum, Candida fungalum, Candida tularemia*, and *Candida tropicalis*. Additionally, smilagenin, disporoside, and kalopanax, demonstrated antifungal efficacy against a panel of human pathogenic opportunistic fungi (*C. albicans*), and human pathogenic yeasts (Arif et al. 2009).

22.3.2.6 Xanthones

Xanthones are closely related to flavonoids and are biosynthesized by the phytochemicals that give them their name. Caledoxanthone E, isolated from the stem bark of *Calophyllum kaledonicum*, is reported to have potent antifungal properties. Xanthones after isoprenylation, such as xanthone C and xanthone W, show activity against *C. albicans* and *Coccidiomycosis australis*. The extract of *Securidaca longepedunculata* contains xanthones with a base structure of butenafine. Xanthone analogues are reported to be highly active against *Coccidia australis*, *Coccidia africanum, Coccidomycis australis*, *Coccidioides, C. neoformans,* and *penicillium* (Arif et al. 2009).

22.3.2.7 Alkaloids

Alkaloids like berberine and frangulanine have been found to have antifungal properties. Berberine is 32 times more active than other alkaloids when it comes to drug-resistant *C. albican* strains, while frangulanine and cinnamodial, both have strong antifungal activity against a variety of pathogenic fungi, as well as azole-resistant strains of Wangiella (Arif et al. 2009).

22.4 Chitinase-Based Therapeutics for Sustainable Management of Fungal Pathogens

22.4.1 Chitinases as Potential Antifungal Therapeutics

Chitinases produced from plants, fungi, and bacteria have the potential to replace traditional fungicides (Gebily et al. 2021; Vurukonda et al. 2018). Chitin plays a significant role in forming the structural building blocks of fungal cell walls. It is a linear polymer containing GlcNAc units (β-1,4-N-acetylglucosamine) and the second most abundant polysaccharide in nature, which can be readily broken down by extracellular chitinases (Singh and Arya 2019; Veliz et al. 2017; Beier and Bertilsson 2013). Chitinases secreted by *Streptomyces* exhibit high activity and stability over a wide range of pH and temperature conditions. Increasing the production of chitinase can be achieved through gene transformation into a compatible system (Subbanna et al. 2020; Kumar et al. 2019). The agricultural and biotechnological industries are actively working to enhance the production and stability of chitinases and other pesticidal compounds from *Streptomyces* spp. (Subbanna et al. 2020). Two primary approaches are being employed for this purpose. The first involves exploring geographically and environmentally diverse habitats to isolate *Streptomyces* spp. (Han et al. 2009; Kumar et al. 2019). The second approach focuses on modifying existing *Streptomyces* genes responsible for biopesticide production. Various comprehensive studies collectively demonstrated that chitinases have the potential to control various pathogenic fungi.

22.4.2 Challenges in Harnessing Chitinase for Antifungal Purposes

22.4.2.1 Unavailability of Stable Chitinase at the Industrial Scale

Different chitinases are identified from various sources like plants, bacteria, and fungi, but still, researchers are struggling with the production of stable chitinases at an industrial scale (Kumar et al. 2019). Several regulatory factors like poor stability, less production, and purification are limiting the application of chitinases (Kidibule et al. 2021). Poor stability of chitinases is a significant challenge observed in past years. Several factors like temperature, pH, proteolytic degradation, and chemical inhibitors play key roles in the stability of chitinases (Umar et al. 2021). As chitinases are produced by several organisms like plants, bacteria, fungi, and nonchordates and every organism has several chitinase-producing genes in their genome but still, any organism has not been found to produce an efficient scale of chitinase to fulfill the industrial demand (Abdel Wahab and Esawy 2022; Hasan et al. 2023). Apart from poor production, purification is also a limiting factor at large scale, which increases the production cost.

22.4.2.2 Chitinase Inhibitors

Chitinase inhibitors are substances that disrupt the activity of chitinase enzymes. In the recent development, a variety of synthetic and natural compounds are identified that can modulate the activity of chitinases. Compounds like Argifin, Argadin, allosamidin, and C2-decaffeine are notable antichitinases. While Argadin, Allosamidin, and Argifin are isolated from natural sources like bacteria and fungi, C2-decaffeine is a synthetic chitinase inhibitor (Hirose et al. 2010; Hartl et al. 2012). They impede the activity of chitinases to hydrolyze the chitin by binding to the active sites of enzymes. Allosamidin is majorly produced by *Streptomyces*, Argifin by *Serratia marcescens*, and Argadin is derived from the fungus *Trichoderma viridae* (Suzuki et al. 2014; Oyeleye and Normi 2018). Although, among these chitinase inhibitors, only Aragdin is produced from fungi, so only Aragdin can protect the fungi especially *Trichoderma viridae* from the hydrolyzing effect of chitinases in natural conditions. The significance of these chitinase inhibitors is ignorable with respect to the resistance of fungi against chitinases.

22.5 Conclusion

The development of drug-resistant fungi is a serious threat to mankind and its belongings. A significant challenge to disease control across several anthropogenic systems is posed by the quick spread of multidrug-resistant pathogenic fungi. The spreading of MDR pathogen is mainly caused by improper use of drugs; in the current situation, maximum available drug molecules are facing the problem of ineffectiveness against different fungi. The dominating effect of fungi on human health is correlated with patients of breast cancer, tuberculosis, HIV, COVID-19, etc. Few antifungal drugs (Azole) are fungistatic instead of fungicidal, which can only suppress the growth rate rather than directly killing the fungal cells. By utilizing these multiple factors, fungal strains maintain their structural and functional integrity against the susceptibility of drug effects. A variety of natural products originating from microbes are being used to control the infection of pathogenic fungi, but the above-mentioned factors are making an urgent need for the discovery of new bioactive compounds. Two primary approaches are being employed for the discovery of novel bioactive compounds. The first involves exploring geographically and environmentally diverse habitats to isolate novel bioactive compounds producing strains. The second approach focuses on modifying existing genes responsible for the production of bioactive compounds. In the 1980s and 1990s, the research community was very focused on chitinase enzyme as a bio alternative to antifungal drugs. This approach was very effective and could be exploited as a bio alternative but the discovery of stable chitinases and production at industrial scale has limited the scope of chitinases. More work on chitinases to scale up their production and stability enhancement may be the key factor in combating pathogenic fungi.

References

Abdel Wahab WA, Esawy MA (2022) Statistical, physicochemical, and thermodynamic profiles of chitinase production from local agro-industrial wastes employing the honey isolate *aspergillus Niger* EM77. Heliyon 8(10):e10869. https://doi.org/10.1016/j.heliyon.2022.e10869

Adam KP (1999) Phenolic constituents of the fern *Phegopteris connectilis*. Phytochemistry 52(5):929–934

Ansari MA, Anurag A, Fatima Z, Hameed S (2013) Natural phenolic compounds: a potential antifungal agent. Microb Pathog Strateg Combat Sci Technol Educ 1:1189–1195

Arastehfar A, Gabaldón T, Garcia-Rubio R, Jenks JD, Hoenigl M, Salzer HJ et al (2020) Drug-resistant fungi: an emerging challenge threatening our limited antifungal armamentarium. Antibiotics 9(12):877

Arif T, Bhosale JD, Kumar N, Mandal TK, Bendre RS, Lavekar GS, Dabur R (2009) Natural products–antifungal agents derived from plants. J Asian Nat Prod Res 11(7):621–638

Baginski M, Czub J (2009) Amphotericin B and its new derivatives-mode of action. Curr Drug Metab 10(5):459–469

Beier S, Bertilsson S (2013) Bacterial chitin degradation-mechanisms and ecophysiological strategies. Frontiers in microbiology 4:149. https://doi.org/10.3389/fmicb.2013.00149

Berman J, Krysan DJ (2020) Drug resistance and tolerance in fungi. Nat Rev Microbiol 18(6):319–331

Birch M, Sibley G (2017) Antifungal chemistry review. Editor(s): Samuel Chackalamannil, David Rotella, Simon E. Ward. Comprehensive Medicinal Chemistry III. Elsevier 703–716. https://doi.org/10.1016/B978-0-12-409547-2.12410-2

Birnbaum JE (1990) Pharmacology of the allylamines. Journal of the American Academy of Dermatology 23(4):782–785

Bongomin F, Gago S, Oladele RO, Denning DW (2017) Global and multi-national prevalence of fungal diseases—estimate precision. Journal of fungi 3(4):57

Boral H, Metin B, Döğen A, Seyedmousavi S, Ilkit M (2018) Overview of selected virulence attributes in *aspergillus fumigatus*, *Candida albicans*, Cryptococcus neoformans, Trichophyton rubrum, and Exophiala dermatitidis. Fungal Genet Biol 111:92–107

Brown GD, Denning DW, Gow NA, Levitz SM, Netea MG, White TC (2012) Hidden killers: human fungal infections. Sci Transl Med 4(165):165rv13

Chai LY, Netea MG, Tai BC, Khin LW, Vonk AG, Teo BW, Schlamm HT, Herbrecht R, Donnelly JP, Troke PF, Kullberg BJ (2013) An elevated pro-inflammatory cytokine response is linked to development of amphotericin B-induced nephrotoxicity. J Antimicrob Chemother 68(7):1655–1659. https://doi.org/10.1093/jac/dkt055

Chen K, Cheng L, Qian W, Jiang Z, Sun L, Zhao Y et al (2018) Itraconazole inhibits invasion and migration of pancreatic cancer cells by suppressing TGF-β/SMAD2/3 signalling. Oncol Rep 39(4):1573–1582

Choudhary MI, Naheed N, Abbaskhan A, Musharraf SG, Siddiqui H (2008) Phenolic and other constituents of fresh water fern *Salvinia molesta*. Phytochemistry 69(4):1018–1023

Chowdhary A, Tarai B, Singh A, Sharma A (2020) Multidrug-resistant Candida auris infections in critically ill coronavirus disease patients, India. Emerg Infect Dis 26(11):2694

Deaguero IG, Huda MN, Rodriguez V, Zicari J, Al-Hilal TA, Badruddoza AZM, Nurunnabi M (2020) Nano-vesicle based anti-fungal formulation shows higher stability, skin diffusion, biosafety and anti-fungal efficacy in vitro. Pharmaceutics 12(6):516

Denning DW, Bromley MJ (2015) How to bolster the antifungal pipeline. Science 347(6229):1414–1416

Di Carlo G, Mascolo N, Izzo AA, Capasso F (1999) Flavonoids: old and new aspects of a class of natural therapeutic drugs. Life Sci 65(4):337–353

Dongmo Fotsing LN, Bajaj T (2024) Caspofungin. In: StatPearls. StatPearls Publishing, Treasure Island, FL

Draskau MK, Svingen T (2022) Azole fungicides and their endocrine disrupting properties: perspectives on sex hormone-dependent reproductive development. Front Toxicol 4:883254. https://doi.org/10.3389/ftox.2022.883254

Draskau MK, Boberg J, Taxvig C, Pedersen M, Frandsen HL, Christiansen S, Svingen T (2019) In vitro and in vivo endocrine disrupting effects of the azole fungicides triticonazole and flusilazole. Environ Pollut (Barking, Essex: 1987) 255(Pt 2):113309. https://doi.org/10.1016/j.envpol.2019.113309

Du H, Bing J, Hu T, Ennis CL, Nobile CJ, Huang G (2020) *Candida auris*: epidemiology, biology, antifungal resistance, and virulence. PLoS Pathog 16(10):e1008921

Enoch DA, Yang H, Aliyu SH, Micallef C (2017) The changing epidemiology of invasive fungal infections. Human fungal pathogen identification: methods and:protocols 17-65

Fanos V, Cataldi L (2000) Amphotericin B-induced nephrotoxicity: a review. J Chemother (Florence, Italy) 12(6):463–470. https://doi.org/10.1179/joc.2000.12.6.463

Faught LN, Greff MJ, Rieder MJ, Koren G (2015) Drug-induced acute kidney injury in children. Br J Clin Pharmacol 80(4):901–909. https://doi.org/10.1111/bcp.12554

Fernandez de Ullivarri M, Arbulu S, Garcia-Gutierrez E, Cotter PD (2020) Antifungal peptides as therapeutic agents. Frontiers in Cellular and Infection Microbiology 10:105

Fisher MC, Hawkins NJ, Sanglard D, Gurr SJ (2018) Worldwide emergence of resistance to antifungal drugs challenges human health and food security. Science 360(6390):739–742

Fisher MC, Gurr SJ, Cuomo CA, Blehert DS, Jin H, Stukenbrock EH et al (2020) Threats posed by the fungal kingdom to humans, wildlife, and agriculture. MBio 11(3):10–1128

GAFFI (2018) Global Fund for Fungal Infections [Internet] Available from: https://www.gaffi.org

Gebily DAS, Ghanem GAM, Ragab MM, Ali AM, Soliman NEK, Abd El-Moity TH (2021) Characterization and potential antifungal activities of three Streptomyces spp. as biocontrol agents against Sclerotinia sclerotiorum (Lib.) de Bary infecting green bean. Egyptian Journal of Biological. Pest Control 31(1). https://doi.org/10.1186/s41938-021-00373-x

Georgopapadakou NH, Walsh TJ (1996) Antifungal agents: chemotherapeutic targets and immunologic strategies. Antimicrob Agents Chemother 40(2):279–291

Govindarajan A, Bistas KG, Ingold CJ, Aboeed A (2022) Fluconazole. In: StatPearls. StatPearls Publishing, Treasure Island, FL

Govindarajan A, Bistas KG, Ingold CJ, Aboeed A (2023) Fluconazole. In: StatPearls. StatPearls Publishing

Gray KC, Palacios DS, Dailey I, Endo MM, Uno BE, Wilcock BC, Burke MD (2012) Amphotericin primarily kills yeast by simply binding ergosterol. Proceedings of the National Academy of Sciences 109(7):2234–2239

Grover ND (2010) Echinocandins: a ray of hope in antifungal drug therapy. Indian J Pharmacol 42(1):9

Hamill RJ (2013) Amphotericin B formulations: a comparative review of efficacy and toxicity. Drugs 73:919–934

Han Y, Yang B, Zhang F, Miao X, Li Z (2009) Characterization of antifungal chitinase from marine Streptomyces sp. DA11 associated with South China sea sponge Craniella australiensis. Marine Biotechnology 11(1):132–140. https://doi.org/10.1007/s10126-008-9126-5

Hartl L, Zach S, Seidl-Seiboth V (2012) Fungal chitinases: diversity, mechanistic properties and biotechnological potential. Appl Microbiol Biotechnol 93(2):533–543. https://doi.org/10.1007/s00253-011-3723-3

Hasan I, Gai F, Cirrincione S, Rimoldi S, Saroglia G, Terova G (2023) Chitinase and insect meal in aquaculture nutrition: a comprehensive overview of the latest achievements. Aust Fish 8:607. https://doi.org/10.3390/fishes8120607

Hasim S, Coleman JJ (2019) Targeting the fungal cell wall: current therapies and implications for development of alternative antifungal agents. Future medicinal chemistry 11(08):869–883

Hirose T, Sunazuka T, Omura S (2010) Recent development of two chitinase inhibitors, Argifin and Argadin, produced by soil microorganisms. Proc Japan Acad Ser B Phys Biol Sci 86(2):85–102. https://doi.org/10.2183/pjab.86.85

Hitchcock CA, Barrett-Bee KJ, Russell NJ (1987) The lipid composition and permeability to azole of an azole-and polyeneresistant mutant of Candida albicans. Journal of Medical and Veterinary mycology 25(1):29–37

Houšť J, Spížek J, Havlíček V (2020) Antifungal drugs. Metabolites 10:106. https://doi.org/10.3390/metabo10030106

Juberg DR, Mudra DR, Hazelton GA, Parkinson A (2006) The effect of fenbuconazole on cell proliferation and enzyme induction in the liver of female CD1 mice. Toxicol Appl Pharmacol 214(2):178–187. https://doi.org/10.1016/j.taap.2006.01.017

Kaur J, Nobile CJ (2023) Antifungal drug-resistance mechanisms in Candida biofilms. Curr Opin Microbiol 71:102237

Kelly SL, Lamb DC, Taylor M, Corran AJ, Baldwin BC, Powderly WG (1994) Resistance to amphotericin B associated with defective sterol Δ 8→ 7 isomerase in a Cryptococcus neoformans strain from an AIDS patient. FEMS microbiology letters, 122(1-2):39–42

Kidibule PE, Costa J, Atrei A, Plou FJ, Lobato MF, Pogni R (2021) Production and characterization of chitooligosaccharides by the fungal chitinase Chit42 immobilized on magnetic nanoparticles and chitosan beads: selectivity, specificity and improved operational utility. RSC Adv 11:5529. https://doi.org/10.1039/d0ra10409d

Kjaerstad MB, Taxvig C, Nellemann C, Vinggaard AM, Andersen HR (2010) Endocrine disrupting effects in vitro of conazole antifungals used as pesticides and pharmaceuticals. Reprod Toxicol (Elmsford, N.Y.) 30(4):573–582. https://doi.org/10.1016/j.reprotox.2010.07.009

Kumar M, Kumar P, Das P, Kapur MK (2019) Draft genome of Streptomyces sp. strain 130 and functional analysis of extracellular enzyme producing genes. Mol Biol Rep 46(5):5063–5071. https://doi.org/10.1007/s11033-019-04960-y

Küpeli Akkol E, Genç Y, Karpuz B, Sobarzo-Sánchez E, Capasso R (2020) Coumarins and coumarin-related compounds in pharmacotherapy of cancer. Cancers 12(7):1959

Kurtz M (1988) New antifungal drug targets: A vision for the future. ASM news 64:31–39

Kurn H, Wadhwa R (2023) Itraconazole. In: StatPearls. StatPearls Publishing

Kwiatkowska E, Domański L, Dziedziejko V, Kajdy A, Stefańska K, Kwiatkowski S (2021) The mechanism of drug nephrotoxicity and the methods for preventing kidney damage. Int J Mol Sci 22(11):6109. https://doi.org/10.3390/ijms22116109

Labro M-T (2012) Immunomodulatory effects of antimicrobial agents. Part II: antiparasitic and antifungal agents. Expert Rev Anti-Infect Ther 10(3):341–357. https://doi.org/10.1586/eri.12.10

Latgé JP, Chamilos G (2019) *Aspergillus fumigatus* and aspergillosis in 2019. Clin Microbiol Rev 33(1):10–1128

Lee MH, Lapidus RG, Ferraris D, Emadi A (2019) Analysis of the mechanisms of action of naphthoquinone-based anti-acute myeloid leukemia chemotherapeutics. Molecules 24(17):3121

Lestrade PPA, Meis JF, Melchers WJG, Verweij PE (2019) Triazole resistance in aspergillus fumigatus: recent insights and challenges for patient management. Clin Microbiol Infect 25(7):799–806

Liang C, Ju W, Pei S, Tang Y, Xiao Y (2017) Pharmacological activities and synthesis of esculetin and its derivatives: a mini-review. Molecules 22(3):387

Lionakis MS, Samonis G, Kontoyiannis DP (2008) Endocrine and metabolic manifestations of invasive fungal infections and systemic antifungal treatment. Mayo Clin Proc 83(9):1046–1060. https://doi.org/10.4065/83.9.1046

Lockhart SR, Guarner J (2019) Emerging and reemerging fungal infections. In Seminars in diagnostic pathology 36(3):177–181. WB Saunders.

Loeffler J, Stevens DA (2003) Antifungal drug resistance. Clinical infectious diseases 36(1):S31–S41

Maligie MA, Selitrennikoff CP (2005) Cryptococcus neoformans resistance to echinocandins:(1, 3) β-glucan synthase activity is sensitive to echinocandins. Antimicrobial agents and chemotherapy 49(7):2851–2856

Meletiadis J, Antachopoulos C, Stergiopoulou T, Pournaras S, Roilides E, Walsh TJ (2007) Differential fungicidal activities of amphotericin B and voriconazole against Aspergillus species determined by micro broth methodology. Antimicrobial agents and chemotherapy 51(9):3329–3337

Morales-Alvarez MC (2020) Nephrotoxicity of antimicrobials and antibiotics. Adv Chronic Kidney Dis 27(1):31–37. https://doi.org/10.1053/j.ackd.2019.08.001

Nofal ZM, El-Zahar MI, Abd El-Karim SS (2000) Novel coumarin derivatives with expected biological activity. Molecules 5(2):99–113

Nowosielski M, Hoffmann M, Wyrwicz LS, Stepniak P, Plewczynski DM, Lazniewski M, Ginalski K, Rychlewski L (2011) Detailed mechanism of squalene epoxidase inhibition by terbinafine. Journal of chemical information and modeling 51(2):455–462

Oyeleye A, Normi YM (2018) Chitinase: diversity, limitations, and trends in engineering for suitable applications. Biosci Rep 38(4):BSR2018032300. https://doi.org/10.1042/BSR20180323

Padda IS, Parmar M (2023) Flucytosine. In: StatPearls. StatPearls Publishing, Treasure Island (FL). Available from: https://www.ncbi.nlm.nih.gov/books/NBK557607/

Pan L, Li XZ, Yan ZQ, Guo HR, Qin B (2015) Phytotoxicity of umbelliferone and its analogs: structure–activity relationships and action mechanisms. Plant Physiol Biochem 97:272–277

Pazhayattil GS, Shirali AC (2014) Drug-induced impairment of renal function. Int J Nephrol Renov Dis 7:457–468. https://doi.org/10.2147/IJNRD.S39747

Perfect JR (2017) The antifungal pipeline: a reality check. Nature reviews Drug discovery 16(9):603–616

Pfaller MA, Boyken L, Hollis RJ, Messer SA, Tendolkar S, Diekema DJ (2005) In vitro activities of anidulafungin against more than 2,500 clinical isolates of *Candida* spp., including 315 isolates resistant to fluconazole. J Clin Microbiol 43(11):5425–5427

Pianalto KM, Alspaugh JA (2016) New horizons in antifungal therapy. Journal of Fungi 2(4):26

Prestinaci F, Pezzotti P, Pantosti A (2015) Antimicrobial resistance: a global multifaceted phenomenon. Pathogens and global health 109(7):309–318

Ramírez-Pelayo C, Martínez-Quiñones J, Gil J, Durango D (2019) Coumarins from the peel of citrus grown in Colombia: composition, elicitation and antifungal activity. Heliyon 5(6):e01937

Rauseo AM, Coler-Reilly A, Larson L, Spec A (2020) Hope on the horizon: novel fungal treatments in development. In Open Forum Infectious Diseases 7(2). ofaa016. US: Oxford University Press

Ravikumar K, Sridhar B, Prasad KD, Bhujanga Rao AKS (2007) Voriconazole, an antifungal drug. Acta Crystallogr Sect E: Struct Rep Online 63(2):o565–o567

Ries F, Alflen A, Aranda Lopez P, Beckert H, Theobald M, Schild H, Teschner D, Radsak MP (2019) Antifungal drugs influence neutrophil effector functions. Antimicrob Agents Chemother 63(6):e02409–e02418. https://doi.org/10.1128/AAC.02409-18

Robbins N, Wright GD, Cowen LE (2016) Antifungal drugs: the current armamentarium and development of new agents. Microbiol Spectrum 4(5):4–5

Rolling T, Hohl TM, Zhai B (2020) Minority report: the intestinal mycobiota in systemic infections. Curr Opin Microbiol 56:1–6

Romero M, Messina F, Marin E, Arechavala A, Depardo R, Walker L et al (2019) Antifungal resistance in clinical isolates of aspergillus spp.: when local epidemiology breaks the norm. J Fungi 5(2):41

Sau K, Mambula SS, Latz E, Henneke P, Golenbock DT, Levitz SM (2003) The antifungal drug amphotericin B promotes inflammatory cytokine release by a toll-like receptor- and CD14-dependent mechanism. J Biol Chem 278(39):37561–37568. https://doi.org/10.1074/jbc.M306137200

Schmiedel Y, Zimmerli S (2016) Common invasive fungal diseases: an overview of invasive candidiasis, aspergillosis, cryptococcosis, and Pneumocystis pneumonia. Swiss medical weekly 146:w14281

Shukla PK, Singh P, Yadav RK, Pandey S, Bhunia SS (2016) Past, present, and future of antifungal drug development. In: Saxena A (ed) Communicable diseases of the developing world. Topics in medicinal chemistry, vol 29. Springer, Cham. https://doi.org/10.1007/7355_2016_4

Singh G, Arya SK (2019) Antifungal and insecticidal potential of chitinases: A credible choice for eco-friendly farming. Biocatalysis and Agricultural Biotechnology. https://doi.org/10.1016/j.bcab.2019.101289

Spampinato C, Leonardi D (2013) Candida infections, causes, targets, and resistance mechanisms: traditional and alternative antifungal agents. BioMed Res Int 2013(1):204237

Stevens DA, Stevens JA (1996) Cross-resistance phenotypes of fluconazole-resistant Candida species: results with 655 clinical isolates with different methods. Diagnostic microbiology and infectious disease 26(3-4):145–148

Subbanna ARNS, Stanley J, Rajasekhara H, Mishra KK, Pattanayak A, Bhowmick R (2020) Perspectives of microbial metabolites as pesticides in agricultural pest management. In: Merillon JM, Ramawat KG (eds) Co-evolution of secondary metabolites, reference series in phytochemistry. https://doi.org/10.1007/978-3-319-96397-6_44

Subden RE, Safe L, Morris DC, Brown RG, Safe S (1977) Eburicol, lichesterol, ergosterol, and obtusifoliol from polyene antibiotic-resistant mutants of Candida albicans. Canadian Journal of Microbiology 23(6):751–754

Suzuki S, Nagasawa H, Sakuda S (2014) Identification of the allosamidin-releasing factor in allosamidin-producing Streptomyces. J Antibiot 67:195–197. https://doi.org/10.1038/ja.2013.109

Trösken ER, Scholz K, Lutz RW, Völkel W, Zarn JA, Lutz WK (2004) Comparative assessment of the inhibition of recombinant human CYP19 (aromatase) by azoles used in agriculture and as drugs for humans. Endocr Res 30(3):387–394. https://doi.org/10.1081/erc-200035093

Umar AA, Hussaini AB, Yahayya J, Sani I, Aminu H (2021) Chitinolytic and antagonistic activity of *Streptomyces* isolated from Fadama soil against Phytopathogenic fungi. Trop Life Sci Res 32(3):25–38. https://doi.org/10.21315/tlsr2021.32.3.2

Van Daele R, Spriet I, Wauters J, Maertens J, Mercier T, Van Hecke S, Brüggemann R (2019) Antifungal drugs: what brings the future? Medical Mycology 57(3):S328–S343

Veliz EA, Martínez-Hidalgo P, Hirsch AM (2017) Chitinase-producing bacteria and their role in biocontrol. AIMS Microbiology 3(3):689–705. https://doi.org/10.3934/microbiol.2017.3.689

Vurukonda S, Giovanardi D, Stefani E (2018) Plant Growth Promoting and Biocontrol Activity of Streptomyces spp. as Endophytes. International journal of molecular sciences 19(4):952. https://doi.org/10.3390/ijms19040952

Wang J (2022) Influence of potassium ions on act of amphotericin B to the DPPC/Chol mixed monolayer at different surface pressures. Membranes 12(1):84. https://doi.org/10.3390/membranes12010084

Wang D, Wang XH, Yu X, Cao F, Cai X, Chen P et al (2021) Pharmacokinetics of anthraquinones from medicinal plants. Front Pharmacol 12:638993

Wheeler ML, Limon JJ, Bar AS, Leal CA, Gargus M, Tang J, Brown J, Funari VA, Wang HL, Crother TR, Arditi M, Underhill DM, Iliev ID (2016) Immunological consequences of intestinal fungal dysbiosis. Cell Host Microbe 19(6):865–873. https://doi.org/10.1016/j.chom.2016.05.003

Wittayapipath K, Yenjai C, Prariyachatigul C, Hamal P (2020) Evaluation of antifungal effect and toxicity of xanthyletin and two bacterial metabolites against Thai isolates of *Pythium insidiosum*. Sci Rep 10(1):4495

Yüksekgönül AÜ, Ertuğrul İ, Karagöz T (2021) Fluconazole-associated QT interval prolongation and Torsades de pointes in a paediatric patient. Cardiol Young 31(12):2035–2037

Zhang W, Ramamoorthy Y, Kilicarslan T, Nolte H, Tyndale RF, Sellers EM (2002) Inhibition of cytochromes P450 by antifungal imidazole derivatives. Drug Metab Dispos 30(3):314–318. https://doi.org/10.1124/dmd.30.3.314

Zhang S, Ma J, Sheng L, Zhang D, Chen X, Yang J, Wang D (2017) Total coumarins from *Hydrangea paniculata* show renal protective effects in lipopolysaccharide-induced acute kidney injury via anti-inflammatory and antioxidant activities. Front Pharmacol 8:872

Zhang Q, Liu F, Zeng M, Mao Y, Song Z (2021) Drug repurposing strategies in the development of potential antifungal agents. Appl Microbiol Biotechnol 105:5259–5279

Microbial and Plant Natural Products and Their Antifungal Targets

23

Prateek Kumar, Kapinder, Manish Sharma, Munendra Kumar, and Khyati

23.1 Antifungal Mode of Action of Natural Products on Cell Wall Targets

Like other microorganisms, fungus is also encircled in cell wall which provides protection to the cell from environment. Fungal cell wall is a pivotal organelle, plays significant role in biofilm generation and host pathogen communication (Borges-Walmsley et al. 2002; Netea et al. 2006; Heard et al. 2021). The important constituents of the fungal cell wall are glucans, chitin and glycoproteins. Chitin is crucial in terms of structural stability of cell wall found near to the plasma membrane (Saunders et al. 2010). Different fungal species and developmental stages show variations in composition of outer layer, which influences the ecology of the fungus. The inner cell wall is made up of β-(1,3) glucan covalently bonded with chitin (Latge 2007; Fleet 1991). This association of glucan and chitin lead to the formation of interchain hydrogen bonds which forms fibrous microfibrils that creates scaffolds surrounding the cell (Gow et al. 2017). Chitin and glucan association in cell wall of *Saccharomyces cerevisiae* and *Candida albicans* require specific enzymes and regulatory networks. Manipulations with the synthesis of the fungal cell wall lead to a defected cell wall which in turn is unable to protect the cell. Therefore, it is an

P. Kumar (✉) · Kapinder · M. Sharma
Department of Zoology, University of Allahabad, Prayagraj, Uttar Pradesh, India
e-mail: drprateekkumar@allduniv.ac.in; drkapinder@allduniv.ac.in; drmanishsharma@allduniv.ac.in

M. Kumar
Department of Zoology, Rajiv Gandhi University, Doimukh, Arunachal Pradesh, India
e-mail: munendra.kumar@rgu.ac.in

Khyati
School of Biological and Life Sciences, Galgotias University,
Greater Noida, Uttar Pradesh, India
e-mail: khyati@galgotiasuniversity.edu.in

N. Manzoor (ed.), *Advances in Antifungal Drug Development*,
https://doi.org/10.1007/978-981-97-5165-5_23

appealing target for various antifungal treatments by disrupting the normal formation of the structure of cell wall (Munro 2013). Certain challenges such as, evident similarities shared by fungal and mammalian cells at the cellular as well as metabolic levels and increased usage of antifungal agents which leads to drug resistance further complicate the antifungal therapeutics development. Recently, three major classes of clinically approved antifungal agents being utilized for treating fungal infections are azoles, polyenes and echinocandin. All of these, directly or indirectly involve in inhibiting the synthesis of components of fungal cell wall. Azoles, such as miconazole and econazole, ketoconazole, fluconazole and itraconazole, were proved to be successful toward human fungal infections. Widespread use of fluconazole has resulted in resistance to azoles which involves the novel compounds that are active against resistant strains (Ghannoum and Rice 1999; Fothergill et al. 1996). Polyenes, such as amphotericin B, interact with the sterols present in the fungal membrane, where annulus of eight amphotericin B make hydrophobic bonds with membrane sterols and this arrangement causes pores in the membrane where polyene residues modify permeability and pivotal constituent effusion (Kerridge 1980, 2020). Polyene also interacts with lipids and cause oxidative damage in *C. albicans* (Graybill et al. 1997; Titsworth and Grunberg 1973). Antifungal drug echinocandins, a semi synthetic lipopeptides interferes with the chitin synthesis pathway and inhibit β-(1,3) glucan synthase, blocking the synthesis of β-(1,3) glucan (Perlin 2020). Fungal infections are frequently transmitted and become highly resistant to several antifungal drugs. This creates a serious concern in front of our medical healthcare system, and there is utmost need to discover and develop new antifungal drugs against multidrug resistance of fungal pathogens. Nature has immense potential and many undiscovered molecules and microorganisms that can be a trustable source for developing relevant natural products and antibiotics. In the recent times, advancements in the research and technologies, serve many scientific approaches such as gene sequencing and genome mining tools, along with chemical and molecular pathways, this has prominently contributed in developing variable natural products. Cell wall is a necessary component of fungal cells as it provides protection from environment and appropriate conditions for survival and its development. Any type of damage to the cell wall leads to dead fungal cells and or reduced growth (Fig. 23.1). Constant research is going on to develop natural potent antifungal drugs which can interfere with the development of cell wall in fungal infections. In the following paragraphs we will discuss about the various natural known antifungal agents and their mode of action on the cell wall.

23.1.1 Inhibition of Glucan Synthase

The fungal cell wall has approximate thickness from 0.1 to 1.0 μm (Klis 1994; Klis et al. 2002). The major constituents of the cell wall of most of the fungi are β-(1,3) glucan and chitin, which are enclosed by a gel kind of substrate of galactomannoproteins and α (1–3)-glucans. β-(1,3) glucan composed of linear chain of glucose molecules which are the integral part for the cell wall structural stability and inside

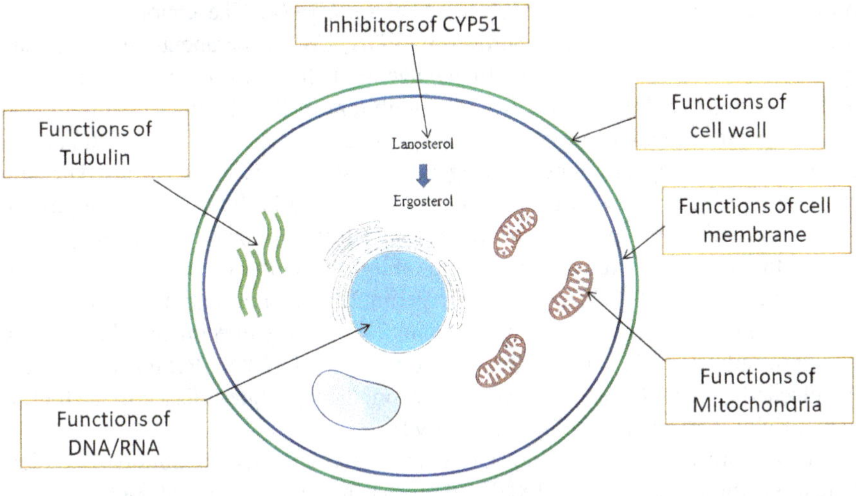

Fig. 23.1 Antifungal drug targets

Table 23.1 Natural products that impact (a) β-(1,3) glucan synthase and (b) chitin synthase

S. No.	Name of natural product	Origin/source
(a) Natural products that impact β-(1, 3) glucan synthase in fungal cell wall		
1.	Caspofungin	*Glarea lozoyensis* (Martos et al. 2012)
2.	Anidulafungin	*Aspergillus nidulans* (Martos et al. 2012; Ghannoum et al. 2020)
3.	Micafugin	*Coleophoma empetri* (Martos et al. 2012; Ghannoum et al. 2020; Marena et al. 2021)
(b) Natural products that target chitin synthase		
1.	Arnthrichitin	*H. oceanicum, A. phaeospermu-m* (Vicente et al. 2003)
2.	Nikkomycin Z	*Streptomyces tendae* (Fostel and Lartey 2000; Sass et al. 2021)
3.	Phellinsin A	*Phellinus* spp. (Vicente et al. 2003; Hwang et al. 2000)
4.	Polyoxin A, B and G	*Streptomyces cacaoi* (Lepesheva and Waterman 2007; Vicente et al. 2003)

layer of the two-layered cell wall has mainly comprised of glucan. In the cell wall these β-(1,3) glucan molecules form a network with β-(1,6) glucan polymer and chitin; this association of molecules interact with the adjacent polymer of adjacent molecules in other chain through hydrogen bonds and in return form a complex three-dimensional network of microfibrils (Fontaine et al. 1997; Latge 2007). Glucan synthase is the enzyme located in the plasma membrane and synthesizes β-(1,3) glucan. So, blocking the activity of this specific enzyme results in cell wall weakening (Fig. 23.1, Table 23.1) (Hartland et al. 1991). Because of these myriad variations in the components in this network, there arise variations among different groups of fungal pathogens.

In the 1970s, echinocandins were identified from *Aspergillus rugulosus* and *Aspergillus nidulans* in order to obtain drugs that cause lesser side effects than polyenes and azoles (Nyfeler and Keller-Schierlein 1974). These molecules are the recent group of drugs that got approved by FDA, US for therapeutic purposes and comprised of different types of antifungal agents such as casofungin, anidulafungin and micafungin. Echinocandins work as inhibitors of β-(1,3) glucan synthase and fungus specific metabolic processes (Odds et al. 2003). These are fungicidal agents against *Candida* spp. and fungistatic agent against *Aspergillus fumigatus.* They are also highly efficient toward azole resistant strains and biofilms. Caspofungins are the first echinocandins which were approved for usage, followed by micafungin and anidulafungin. These drugs are proved to inhibit the synthesis of β-(1,3) glucan to target fungal cell wall (Chen et al. 2011; Perlin 2011). After that, many echinocandins were introduced for clinical implementations. Family of echinocandins include pneumocandins, aculeacins, mulundocandin and WF11899. Echinocandins and papulacandins are fatty acid derivatives of cyclic hexapeptides, disaccharide β-(1,4) galactosylglucose, respectively (Tkacz 1992).

Mechanism of action: Echinocandins are noncompetitive inhibitors of β-(1,3) glucan synthase (Odds et al. 2003). Inhibition in the formation of β-(1,3) glucan leads to β-(1,6) and β-(1,3) glucan network decline as a result unstable and osmotically disbalanced fungal cell wall form, which ultimately lead to death of fungal cell and effectively decline the host cell damage (Kathiravan et al. 2012; Song and Stevens 2016). β-(1,3) glucan synthesis inhibition also influences the other constituents of cell wall, such as it causes ergosterol reduction and elevation of chitin concentration in the cell wall (Pfaller et al. 1989). *C. albicans* infection in murine model interferes with the interactions of cell wall β-(1,3) glucan and immune receptor dectin-1, by decreasing the efficiency of detection of these receptors by covering them with the layer of mannosylated glycoprotein (mannan) (Marakalala et al. 2013). This is known as masking of receptors, caspofungins causes cell wall changes and expose β-(1,3) glucan to dectin-1, as a result immune response of cells get activated in response to fungus (Marakalala et al. 2013; Eschenauer et al. 2007; Gonçalves et al. 2016).

The key difference between micafungin, caspofungin and anidulafungin lies in the composition of their side chains, where micafungin has a complex aromatic-substituent, caspofungin has fatty-acid side chain and anidulafungin has alkoxytriphenyl group (Eschenauer et al. 2007). These different modifications of side chains elevate the inhibition of glucan synthase synthesis and prevent the host cells from fungal infection (Eschenauer et al. 2007). Caspofungin long tail fatty-acid chain interaction with the cell membrane allows the interference in the activity of Fks1 enzyme. Fks1 gene mutations lead to the conformational modifications which ultimately results in lower affinity for echinocandins (Gonçalves et al. 2016; Balashov et al. 2006). Such mutations are responsible for the cross resistance to the agents of this class. Similar phenomenon has been seen in case of cdr2p gene, but at a much lesser extent (Schuetzer-Muehlbauer et al. 2003). Although echinocandins have comparatively fewer side effects than the other antifungal drugs available; there are certain drawbacks also, such as echinocandins metabolize in liver and does not reach the CNS and eyes, so it is not a preferred drug for such cases.

23.1.2 Inhibition of Chitin Synthesis

Chitin is a pivotal component of pathogenic fungal cell walls also present in exoskeletons of mosquitoes, sand flies, ticks and snails but absent in mammals and humans (Fig. 23.1, Table 23.1). Thus, chitin synthase proved to be an appealing target for the researchers for the development of antifungal therapeutics which are low in toxicity toward humans. Irrespective of its lower percentage dry weight in fungal cell wall, chitin contributes to the structural mechanical strength and integrity. It is a homopolymer of β-(1,4) linked *N*-acetylglucosamine, a polysaccharide found in all the fungal cell walls (Munro and Gow 2001; Latge 2007). These polysaccharide chains in fungi twist and fold back into antiparallel chains against the first chain. The folding creates intra chain hydrogen bonds which provide extra stiffness and create tougher microfibrils, making them extremely strong molecules. The 3D structure of these microfibrils are covalently bonded with the β-(1,3) glucans, another polysaccharide present in the fungal cell wall (Klis et al. 2006). In fungal cell wall different amounts of chitin is synthesized and upon deacetylated of this chitin, it becomes chitosan, by the enzyme chitin deacetylases. These chitin deacetylations provide elasticity and protection to the polymer from the host chitinases.

Chitin is biosynthesized by a family of enzymes called chitin synthases (CHS) placed in the plasma membrane (Roncero 2002). Family of CHSs belongs to glycosyltransferase family 2 which catalyze transfer of *N*-acetyl-d-glucosamine residue to the chitin chain from UDP-*N*-acetyl-d-glucosamine. The CHSs mainly consist of seven classes, whereas three typical classes are present and identified in yeast such as *C. albicans* and *S. cerevisiae* (Au-Young and Robbins 1990). The functional analysis of each CHS gene has been done by deletion of specific CHS strains. In vitro studies suggest that the Class I enzymes generate a minor amount of cell wall chitin. Under normal conditions, the strains with class I mutated CHS genes were found viable (Munro and Gow 2001). Class II enzymes also contribute very little in chitin formation in cell wall, but deletion studies of class II enzymes showed cell survival was severely affected and cell death occur by malformation of primary septum. Class III enzymes generating majority of chitin found in cell wall also contribute in formation of bud ring and cell wall maturation (Munro et al. 2001). Class V and VII chitin synthase enzymes have myosin motor like domains. Unlike other classes of CHSs, Class III, V, VI and VII were not identified from the yeast *Saccharomyces cerevisiae* and *C. albicans,* instead they have only been found from the filamentous fungi and few dimorphic fungi.

Initially, the compounds such as nikkomycins, polyoxins and plagiochin were discovered which affect chitin biosynthesis (Kathiravan et al. 2012). Polyoxins and nikkomycins are two closely related classes of natural products which are produced by *Streptomycetes*. These two are competitive inhibitors of chitin synthase. Nikkomycin and polyoxins are pyrimidine nucleosides which are connected with di- or tripeptide moiety having similarities in their structure with UDP-*N*-acetylglucosamine (substrate of the CHS). Therefore, the antifungal property of these compounds lies within the competitive inhibition of the catalytic site of chitin synthase enzyme. Nikkomycins have similar structure which causes higher binding affinity toward the catalytic site than *N*-acetylglucosamine (Kathiravan et al. 2012; Song and Stevens 2016). Nikkomycin derivatives have been exclusively used to cure the infections caused by *Epidemophyton floccosum, C. albicans, H. capsulation*, etc. Nikkomycin X and Z are nikkomycin derivatives which vary in nucleoside moiety. Studies showed that nikkomycin derivatives have varied inhibitory action toward different classes of chitin synthase as they inhibit CHSs class I and III, but not class II (Gaughran et al. 1994). Nikkomycin Z showed synergistic action on various group of fungi when combined with other antifungal agents such as fluconazole, ergosterol and echinocandins (Li and Rinaldi 1999; Chiou et al. 2001; Luque et al. 2003). Polyoxins and nikkomycin derivatives have similar mechanism of action on fungal cell wall by inhibiting chitin synthase (Kathiravan et al. 2012). Polyoxins A and B have been isolated from the purification of culture filtrates showing antifungal properties against phytopathogenic fungi *Pyricularia oryzae, Alternaria kikuchiana, Pellicularia filamentosa* and *Cochliobolus miyabeanus* (Chaudhary et al. 2013).

Phellinsin A is another naturally occurring antifungal agent, mainly screened for the inhibition of chitin synthase II of microbial origin, isolated from the cultural broth of fungus PL3. Phellinsin A has 2.5 times stronger inhibitory action than the other chitin inhibitors, although it shows weak suppression against chitin synthase I (Hwang et al. 2000). Another drug, Arthrichitin, showed poor action against *Candidia* and *Trychophyton* spp. and several other phytopathogens produced by *Arthrinium* spp. (Verekar et al. 2023).

23.2 Antifungal Mode of Action of Natural Products on Cell Membrane Targets

Natural products exhibit potent biological activities with precise target sites, therefore providing great pharmacological benefits. The novel plant derived compounds always have hidden expanding chemical space with unique chemical structures and membrane-binding properties (Pye et al. 2017). The main target for antimicrobial substances is usually the cell membrane (Fig. 23.1). The cell membrane of the fungus is made up of a variety of lipids from the sphingolipid, glycerophospholipid and sterol classes (Hannich et al. 2011; Marques et al. 2018). The ring structures of sterols, which are steroid alcohols with amphipathic lipids, are hard and compact. Ergosterol is the main component of the fungal cell membrane (Dupont et al. 2012). Drugs that alter the integrity of cell membranes are thought to be the most effective, and imidazole and triazoles, which target the formation of ergosterol, have proven to be the most effective till now.

23.2.1 Inhibition of Ergosterol Synthesis Pathway

The cytochrome P450 enzyme CYP51 (14 α-demethylase), which is encoded by the Erg11 gene, processes and matures ergosterol, the primary structural component of fungal cell membranes. It functions as a prospective therapeutic target location and engages in mono-oxygenated sterol production in vivo. If it is deactivated, it directly obstructs the formation of ergosterol in fungal cells (Lepesheva and Waterman 2007). As a result, fungal membrane is not correctly synthesized which leads to altered membrane morphology and ultimately reduced cell growth and death. Various drugs such as fluconazole, ketoconazole and itraconazole have high efficiency for this target (Parker et al. 2008).

Several natural products are found to be very effective against a large number of fungi which specifically target ergosterol biosynthesis in fungal cell membrane (Table 23.2; Fig. 23.1). The essential oil derived from *Thymus vulgaris, Origanum vulgare* and *O. majorana* consist of a monoterpenoid phenol named carvacrol that exhibited strong activity against fungi. *Ocimum sanctum* essential oil was found to be even more effective against *Candida* strains (Khan et al. 2010). The methyl chavicol and linalool present in the plant essential oil were found to damage the membrane as well as decreases ergosterol synthesis. By blocking ergosterol biosynthesis and, indirectly, cell wall formation, the eugenol and cinnamaldehyde compounds of *Syzygium aromaticum* and *Cinnamomum verum*, respectively, have anticandidal action (Mehmood et al. 1999; Khan et al. 2013). Similarly, cinnamaldehyde suppresses the growth of *Fusarium sambucinum*, probably via interfering with ergosterol production (Wei et al. 2020). Tomatidine (Goodridge et al. 2009), an antifungal compound derived from *Solanum lycopersicum*, was discovered to be effective against *Candida* spp. by targeting Erg6, a C-24 sterol methyl-transferase, an enzyme lacking in mammals (Dorsaz et al. 2017).

Melaleuca alternifolia essential oil contains terpinen-4-ol (Skepper et al. 2008) and 1,8-cineole (Friedman et al. 2020a, b) as active compounds against *Botrytis cinerea*, targeting cell membranes and organelles, respectively by lowering ergosterol concentration of cell membrane and are thought to be connected to CYP51 (Yu et al. 2015). Citronellol (Fostel and Lartey 2000) and geraniol (Arikan et al. 2002), on the other hand, harm *T. rubrum* cell walls and membranes by blocking ergosterol synthesis (Pereira et al. 2015). Moreover, antifungal activity of gallic acid is also achieved by the inhibition of ergosterol synthesis and CYP51 enzyme activity in *C. albicans* (Li et al. 2017). Furthermore, a methanolic extract of *Polyalthia longifolia var. pendula* leaves containing diterpene 1 as an active ingredient demonstrated antifungal activity against *C. albicans* and *C. neoformans* by modifying cell membrane permeability and producing intracellular ROS (Bhattacharya et al. 2015).

23.2.2 Inhibition of Sterols

It has also been discovered that natural compounds have a strong affinity for the sterols found on cell membranes (Table 23.2; Fig. 23.1). Amphotericin B, an

Table 23.2 Natural products that impact (a) ergosterol synthesis by interfering CYP51 (14 α-demethylase) and (b) fungal cell sterols

S. No.	Name of natural products	Origin/source
(a) Natural products as CYP51(14 α-demethylase) inhibitors		
1	Sanguinarine	*Macleaya cordata* (Hu et al. 2022)
2	Berberine	*Coptis chinensis* Franch (Xie et al. 2020)
3	Tomatidine	*Solanum lycopersicum* (Dorsaz et al. 2017)
4	Cinnamaldehyde	Cinnamon essential oil (Wei et al. 2020)
5	Citronellol	Citrus limon (L.) Burm. f. (Pereira Fde et al. 2015; Barrera 2008)
6	Geraniol	Thyme and cinnamon (Pereira et al. 2015; Barrera 2008; Singh et al. 2016)
7	Eugenol, methyl eugenol	*Eugenia caryophyllus* (Ahmad et al. 2010)
8	Gallic acid	*Punica granatum* L. (Li et al. 2017)
9	Linalool	*Eucalyptus* (Khan et al. 2010; Barrera 2008)
10	Menthol	Mint (Barrera 2008; Samber et al. 2015)
11	Bengazole A	Marine sponge *Jaspidae fijispong* (Jamison et al. 2019; Chandrasekhar and Sudhakar 2010)
(b) Natural products that bind to cell sterols		
1	Amphotericin B	*S. nodosus* (Ghannoum et al. 2020; Lemke et al. 2005; Chen et al. 2021)
2	Natamycin (Pimaricin)	*A. ficuum* (Brothers and Wyatt 2000; Ciesielski et al. 2016)
3	Nystatin	*S. noursei* (Lyu et al. 2016)
4	Pentamycin	*S. pentaticus* (Umezawa et al. 1958)
5	Carvacrol	Essential oil of *T. vulgaris* (Barrera 2008; Lima et al. 2013)
6	Magnolol	*Magnolia officinalis* (Li et al. 2021; Bchbchani et al. 2017)
7	Thymol	Thyme essential oil (Faria et al. 2011; Castro et al. 2015)
8	Citral	Citrus limon (L.) Burm. f. (Barrera 2008; Lima et al. 2012)
9	Fengycin	*B. subtilis* F-29–3 (Vanittanakom et al. 1986)
10	Iturin A	*B. subtilis* PY-1 (Quentin et al. 1982; Gong et al. 2006)

antifungal medication with a broad spectrum of action, attaches to the sterols in cell membranes, disturbs cell metabolism and enhances membrane permeability (Garcia-Effron 2020). It does, however, bind to sterols on human cell membranes, causing similar metabolic disturbance (Quentin et al. 1982). Nystatin, derived from *Streptomyces,* is a polyene antifungal and antibacterial that forms holes in membrane (Helrich et al. 2006; Marena et al. 2021). Similarly, Pentamycin, an antifungal polyene that shows affinity for sterols is derived from *Streptomyces pentaticus* (Helrich et al. 2006; Arana et al. 2009; Malova et al. 2015). Thymol (2-isopropyl-5-methyl phenol), a natural monoterpene with in vitro antifungal activity against *C. albicans*, *C. tropicalis* and *C. krusei* strains, was isolated from *Thymus vulgaris* (Isono et al. 1969). Thymol leads to the interference with ergosterol synthesis in the plasma membrane (Castro et al. 2015). Iturin A (Parker et al. 2008), isolated from a

Bacillus strain, has shown antifungal activity and an affinity for synthetic sterol-containing membranes (Quentin et al. 1982; Gong et al. 2006).

Polyene macrolide antibiotics are naturally occurring substances that attack membranes (Zotchev 2003). The bigger polyene antibiotics, such as amphotericin B and nystatin, generate barrel-stave transmembrane pores that collapse ion gradients and cause fungal death (de Kruijff and Demel 1974). Pimaricin specifically binds to ergosterol but does not alter fungal membrane (te Welscher et al. 2008); however, fungicidal activity elicits because ergosterol-dependent membrane proteins, including glucose and amino acid transporters are inhibited (te Welscher et al. 2012). A substance used to treat athlete's foot, holotoxin, is a saponin made from sea cucumber *Stichopus japonicas* (Shimada 1969) and is believed to target membrane sterols. Bicyclic peptides called theonellamides (TNMs) are extracted from marine sponges called *Theonella Swinhoei* (Youssef et al. 2014; Fukuhara et al. 2018). Theonegramide, theopalauamide and isotheopalauamide are compounds also isolated from *T. swinhoei* with minor modifications at specific amino acid side chains (Bewley and Faulkner 1994; Schmidt et al. 1998) exhibiting antifungal activity. The binding of physiologically significant fungal sterols is an important method that provides a direct route to a more effective antifungal treatment.

23.3 Inhibition of Lipid Metabolism

It is evident that lipids play vital roles in maintaining different cellular functions. Apart from being integral part of cellular membranes, they serve as major energy sources in the form of reserve materials. Some of the lipids also act as signaling molecules (Zhang et al. 2023; de Carvalho and Caramujo 2018), thus contributing in cellular communication. Therefore, lipids serve as a major target for many natural compounds by interfering with lipid biosynthesis and disruption of cell membranes (Table 23.3; Fig. 23.1). Some, like Aureobasidine A (Zhang et al. 2023; Teymuri et al. 2021), inhibit synthesis of inositol phosphoglycerides (Zhang et al. 2023; Lester and Dickson 1993) by inhibiting inositol phosphoglycerides synthase enzyme. The antifungal agent anthraquinone emodine disrupts lipid metabolism of fungal cells. The current trends in lipidomic technology gives a deep insight into further exploration of drug targets based on lipid metabolism.

23.4 Interference of DNA/RNA Function

The DNA/RNA functions are mainly enzyme dependent processes; thus, enzymes serve as main targets which ultimately disrupt vital cellular functions. Several important enzymes, including topoisomerases, are involved in the replication of DNA (Zhang et al. 2023; Stabb and Handelsman 1998) and polymerases that are drug targets for many antimicrobial agents. Some of the antimicrobial compounds act as transcription and translational inhibitors, thereby inhibiting protein synthesis (Table 23.3; Fig. 23.1) (Zhang et al. 2023; Galván Márquez et al. 2013). The

Table 23.3 Natural products that impact (a) lipid metabolism and (b) DNA/RNA function

S. No.	Name of natural product	Origin/source
(a) Natural products that impact the lipid metabolism		
1.	Emodin	*Rheum palmatum L.* (Zhang et al. 2023; Friedman et al. 2020a, b)
2.	(−)-(E)-dysidazirine	Marine sponge *Dysidea fragilis* (Zhang et al. 2023; Skepper et al. 2008; Molinski and Ireland 1988)
3.	Aureobasidine A	*A. pullulans* (Zhang et al. 2023; Teymuri et al. 2021)
(b) Natural products that impact the DNA/RNA function		
1.	Yatakemycin	*Streptomyces* sp. (Zhang et al. 2023; Igarashi et al. 2003)
2.	Chitosan	*A. Japonicus* (Zhang et al. 2023; Liu et al. 2019)
3.	Icofungipen	*Bacillus cereus* (Zhang et al. 2023; Hasenoehrl et al. 2006)
4.	(+)-Zwittermicin A	*Bacillus cereus* UW85 (Zhang et al. 2023; Silo-Suh et al. 1994)

compound parnafungin (Zhang et al. 2023; Hu et al. 2022) inhibits enzyme poly-adenosine polymerase and thus inhibits protein synthesis (Table 23.3).

23.5 Interference of Mitochondrial Functions

All eukaryotic cells possess mitochondria. This organelle performs the function of energy transduction and metabolism. The essential elements for the energy production are electron transport chain and respiratory chains (Guo et al. 2013; Xia et al. 2019; Zhang et al. 2023). Flaws in these chains may result in higher ROS levels and might cause oxidative stress within the cells, which can ultimately lead to cell death. Rosin-based amide derivatives form hydrogen bonds, conjugates and interact electrostatically with the target receptors. This is the reason why these derivatives are crucial for their antifungal efficacy (Tao et al. 2020; Zhang et al. 2023). Studies conducted in vitro have revealed that synthetic analogs of phloeodictines from marine sources exhibit strong fungicidal effects on a variety of fungal strains, including the drug-resistant ones (Li et al. 2011; Kumar et al. 2023; Zhang et al. 2023). During oxidative phosphorylation process in *Botrytis cinerea*, *p*-coumaric acid performs as an uncoupler (Morales et al. 2017). Quercetin performs its antifungal function by disrupting the fungal cell mitochondrial dysfunction (Kwun and Lee 2020; Zhang et al. 2023). The antifungal mode of action of shikonin is through excessive production of endogenous reactive oxygen species (ROS) and dysfunction of mitochondria (Shishodia and Shankar 2020). Owing to the fact that mitochondria are the powerhouse of eukaryotic cells, it serves as a key target for the preparation of antifungal medications (Table 23.4; Fig. 23.1). As a result, it is possible that specific fungal proteins involved in mitochondrial function could be used as targets for antifungal drug discovery (Xia et al. 2019; Zhang et al. 2023).

Table 23.4 Natural products that impact (a) mitochondrial functions and (b) tubulin functions

S. No.	Name of natural product	Origin/source
(a) Natural products that impact the fungal mitochondrial functions		
1.	*p*-Coumaric acid	Fruits, vegetables and cereals (Morales et al. 2017; Zhang et al. 2023)
2.	Myxocoumarin A	*S. aurantiaca* MYX-030 (Gulder et al. 2013; Zhang et al. 2023)
3.	Shikonin	*L. erythrorhizon* (Shishodia and Shankar 2020; Zhang et al. 2023)
4.	Quercetin	Food products (Gehrke 2013; Silva et al. 2014; Rocha 2019; Zhang et al. 2023)
(b) Natural products that impact the fungal tubulin functions		
1.	Griseofulvin	*Penicillium griseofulvum* (Zhang et al. 2023; Cardoso 2020)
2.	Dolastatin 10	Sea hare *Dolabella auriculari* (Zhang et al. 2023; Cardoso 2020)

23.6 Interference with Tubulin Functions

Maintenance of cell structure, cell morphogenesis and transduction of intracellular signals is performed by cellular microtubule proteins. These proteins are the most abundant components of eukaryotic cell microtubules (Goodson and Jonasson 2018). Microtubules are also important for transmitting signals, acting as carriers for cellular material transport and for transporting intracellular vesicles (Barlan and Gelfand 2017). An antifungal natural substance derived from *Dolabella uricularia,* dolastatin 10, inhibits tubulin production and guanosine triphosphate binding which is tubulin-dependent and prevents microtubule assembly, stops mitosis leading to cell death. The most widely utilized drugs are benzimidazoles, which are also among the most widely used commercial drugs that are sensitive to this target and are classified as natural products (Table 23.4) (Garcia-Bustos et al. 2019). Dermatophytosis and onychomycosis are both conditions that are commonly treated with the drug griseofulvin (Wordeman 2010). When the development of microtubule spindles is inhibited during metaphase, it affects the creation of cell walls and proteins in fungal cells. It also has neurotoxic and neurological adverse effects in people, such as headache, sleepiness and neurological weakening, because it specifically targets microtubule activity (Pizzorno and Shippy 2016). Antimitotic drugs that target tubulin interfere with microtubule dynamics by acting as tubulin-binding compounds, and these microtubule-targeted drugs are very effective against fungi (Table 23.4; Fig. 23.1). As a result, cutting-edge small molecules as tubulin-binding inhibitors for antifungal medications may someday be developed (Zhang et al. 2023).

23.7 Interference with Signaling Pathway

Drugs that target fungal cells specifically are those with a comparable mechanism of action (Table 23.5; Fig. 23.1). Fungal cells grow abnormally by affecting the signaling pathway (Agarwal et al. 2008; Huang et al. 2011; Zhang et al. 2023). Because many drugs have similar mechanisms of action, they can also affect how fungi communicate. This explains why they can more successfully destroy fungal cells by acting on signals that keep them functioning normally. This offers a practical solution to cure clinical disorders, symptoms as well as their underlying causes.

23.8 Interference with Transport Function

Presence of ions influences the metabolism of fungal cells. Apart from the presence of sodium and potassium ions that are involved in the maintenance of potential of plasma membranes, there is presence of ions of iron, calcium and other essential elements. A few medications, such as nonactin, trigger cell death in yeast cells by decreasing their ability to store potassium (Tebbets et al. 2013; Gao et al. 2021). Several other drugs, like guaiacol, damage cell membrane by destroying ion channels (Gao et al. 2021; Zhang et al. 2023). Ion channel proteins are involved in nearly all physiological processes, such as secretion, cell division and electrolyte and water balance. Ion channels are therefore excellent candidates for drug discovery due to their various physiological activities (Table 23.5; Fig. 23.1). Furthermore, if we compare these targets to other target classes it was found that ion channel proteins are very challenging and complex targets for initial drug discovery phase.

23.9 Conclusion

The antifungal arsenal can be greatly improved by using natural products. Our review of the literature revealed a number of unique natural compounds that have been claimed to have promising antifungal action. According to the distribution of the sources of natural products, microbes (fungi and bacteria) and plants are the primary sources of antifungal natural products. The antifungal targets for natural products include the enzymes involved in glucan and chitin synthesis in the fungal

Table 23.5 Natural products that impact (a) signaling pathway and (b) fungal transport function

S. No.	Name of natural product	Origin/source
(a) Natural products that impact the fungal signaling pathway		
1.	Sampangine	Annonaceae family (Zhang et al. 2023; Huang et al. 2011; Agarwal et al. 2008)
(b) Natural products that impact the fungal transport function		
1.	Guaiacol	Guaiac resin (Zhang et al. 2023; Gao et al. 2021)
2.	Dinactin, nonactin, Trinactin, Monactin	*Streptomyces* sp. YIM56295 (Zhang et al. 2023; Tebbets et al. 2013)

cell walls and ergosterol synthesis in the cell membranes. Furthermore, interference with lipid metabolism, DNA/RNA function, mitochondrial function, tubulin function, signaling pathway and transport function are the other important targets for antifungal natural products.

Acknowledgments Authors are thankful to Department of Zoology, University of Allahabad, Prayagraj, Uttar Pradesh, India and School of Biological and Life Sciences, Galgotias University, Greater Noida, Uttar Pradesh, India.

Declaration of Competing InterestThe authors declare that they have no competing interest.

References

Au-Young J, Robbins PW (1990) Isolation of a chitin synthase gene (CHS1) from *Candida albicans* by expression in *Saccharomyces cerevisiae*. Mol Microbiol 4(2):197–207

Arikan S et al (2002) In vitro synergy of caspofungin and amphotericin B against *as-pergillus* and *fusarium spp*. Antimicrob Agents Chemother 46(1):245–247

Agarwal AK et al (2008) Role of heme in the antifungal activity of the aza-oxoaporphine alkaloid sampangine. Eukaryot Cell 7(2):387–400. https://doi.org/10.1128/EC.00323-07

Arana DM et al (2009) The role of the cell wall in fungal pathogenesis. Microb Biotechnol 2(3):308–320

Ahmad A et al (2010) Evolution of ergosterol biosynthesis inhibitors as fungicidal against *Candida*. Microb Pathog 48(1):35–41

Bewley CA, Faulkner DJ (1994) Theonegramide, an antifungal glycopeptide from the Philippine lithistid sponge *Theonella swinhoei*. J Org Chem 59:4849–4852

Brothers AM, Wyatt RD (2000) The antifungal activity of natamycin toward molds isolated from commercially manufactured poultry feed. Avian Dis 44(3):490–497

Balashov SV, Park S, Perlin DS (2006) Assessing resistance to the echinocandin antifungal drug caspofungin in *Candida albicans* by profiling mutations in FKS1. Antimicrob Agents Chemother 50(6):2058–2063

Barrera L (2008) Antifungal activity of essential oils and their compounds on the growth of *fusarium sp*. isolate from papaya (*Carica papaya*). Rev Científica UDO Agrícola 8

Bhattacharya AK, Chand HR, John J, Deshpande MV (2015) Clerodane type diterpene as a novel antifungal agent from *Polyalthia longifolia var. pendula*. Eur J Med Chem 94:1–7

Barlan K, Gelfand VI (2017) Microtubule-based transport and the distribution, tethering and organization of organelles. Cold Spring Harb Perspect Biol 9(5). https://doi.org/10.1101/cshperspect.a025817

Behbehani J et al (2017) The natural compound magnolol affects growth, biofilm formation, and ultrastructure of oral *Candida* isolates. Microb Pathog 113:209–217

Borges-Walmsley DMI, Chen Xinhua, Shu Adrian R, Walmsley (2002) The pathobiology of Paracoccidioides brasiliensis Trends in Microbiology 10(2):80–87 https://doi.org/10.1016/S0966-842X(01)02292-2

Cardoso SRS (2020) Characterizing Naturally-Occurring Entomopathogenic Fungi in Reproductive Females of Atta spp. Int J Agric Res 9(1)

Chiou CC, Mavrogiorgos N, Tillem E, Hector R, Walsh TJ (2001) Synergy, pharmacodynamics, and time-sequenced ultrastructural changes of the interaction between nikkomycin Z and the echinocandin FK463 against *aspergillus fumigatus*. Antimicrob Agents Chemother 45(12):3310–3321

Chandrasekhar S, Sudhakar A (2010) Total synthesis of bengazole a. Org Lett 12(2):236–238

Chen SCA, Slavin MA, Sorrell TC (2011) Echinocandin antifungal drugs in fungal infections: a comparison. Drugs 71:11–41

Chaudhary PM, Tupe SG, Deshpande MV (2013) Chitin synthase inhibitors as antifungal agents. Mini-Rev Med Chem 13(2):222–236

Ciesielski F et al (2016) Recognition of membrane sterols by polyene antifungals amphotericin B and Natamycin, a (13) C MAS NMR study. Front Cell Dev Biol 4(57):57

Chen XF et al (2021) Antifungal susceptibility profiles and resistance mechanisms of clinical *Diutina catenulata* isolates with high MIC values. Front Cell Infect Microbiol 11:739496

de Kruijff B, Demel RA (1974) Polyene antibiotic-sterol interactions in membranes of Acholeplasma laidlawii cells and lecithin liposomes. Molecular structure of the polyene antibiotic-cholesterol complexes. Biochim Biophys Acta 339:57–70

Dupont S, Lemetais G, Ferreira T, Cayot P, Gervais P, Beney L (2012) Ergosterol biosynthesis: a fungal pathway for life on land? Evolution 66:2961–2968

Castro RD et al (2015) Antifungal activity and mode of action of thymol and its synergism with nystatin against *Candida* species involved with infections in the oral cavity: an in vitro study. BMC Complement Altern Med 15:417

Dorsaz S et al (2017) Identification and mode of action of a plant natural product targeting human fungal pathogens. Antimicrob Agents Chemother 9:61

de Carvalho CCCR, Caramujo MJ (2018) The various roles of fatty acids. Moecules (Basel, Switzerland) 23(10):2583. https://doi.org/10.3390/molecules23102583

Eschenauer G, De Pestel DD, Carver PL (2007) Comparison of echinocandin antifungals. Ther Clin Risk Manag 3(1):71–97

Fleet GH (1991) Cell walls. Yeasts 4:199–277

Fontaine T, Mouyna I, Hartland RP, Paris S, Latge JP (1997) From the surface to the inner layer of the fungal cell wall. Biochem Soc Trans 25(1):194–199

Fothergill AW, Sutton DA, Rinaldi MG (1996) An in vitro head-to-head comparison of Schering 56592, amphotericin B, fluconazole, and Itraconazole against a spectrum of filamentous fungi, abstr. F89. In: Program and abstracts of the 36th interscience conference on antimicrobial agents and chemother. American Society for Microbiology, Washington, DC, p 115

Fostel JM, Lartey PA (2000) Emerging novel antifungal agents. Drug Discov Today 5(1):25–32

Faria NC et al (2011) Enhanced activity of antifungal drugs using natural phenolics against yeast strains of *Candida* and *Cryptococcus*. Lett Appl Microbiol 52(5):506–513

Fukuhara K, Takada K, Watanabe R, Suzuki T, Okada S, Matsunaga S (2018) Colony-wise analysis of a Theonella swinhoei marine sponge with a yellow interior permitted the isolation of Theonellamide I. J Nat Prod 81:2595–2599

Friedman M et al (2020a) The inhibitory activity of anthraquinones against pathogenic protozoa, bacteria, and fungi and the relationship to structure. Molecules 13:25

Friedman M, Xu A, Lee R, Nguyen DN, Phan TA, Hamada SM, Panchel R, Tam CC, Kim JH, Cheng LW et al (2020b) The inhibitory activity of anthraquinones against pathogenic protozoa, bacteria, and fungi and the relationship to structure. Molecules 25(13):3101. https://doi.org/10.3390/molecules25133101

Gaughran JP, Lai MH, Kirsch DR, Silverman SJ (1994) Nikkomycin Z is a specific inhibitor of *Saccharomyces cerevisiae* chitin synthase isozyme Chs3 in vitro and in vivo. J Bacteriol 176(18):5857–5860

Gehrke N (2013) Oxidative damage of DNA confers resistance to cytosolic nuclease TREX1 degradation and potentiates STING-dependent immune sensing. Immunity 39(3):482–495. https://doi.org/10.1016/j.immuni.2013.08.004

Graybill JR, Montalbo E, Kirkpatrick WR, Revankar S, Rinaldi M, Patterson T (1997) *Candida albicans*: less predictable than we may think, abstr. O49. In: 13th international conference for human and animal mycology

Ghannoum MA, Rice LB (1999) Antifungal agents: mode of action, mechanisms of resistance, and correlation of these mechanisms with bacterial resistance. Clin Microbiol Rev 12(4):501–517

Gong M et al (2006) Study of the antifungal ability of Bacillus subtilis strain PY-1 in vitro and identification of its antifungal substance (iturin a). Acta Biochim Biophys Sin Shanghai 38(4):233–240

Goodridge HS, Wolf AJ, Underhill DM (2009) Beta-glucan recognition by the innate immune system. Immunol Rev 230(1):38–50

Guo C et al (2013) Oxidative stress, mitochondrial damage and neurode-generative diseases. Neural Regen Res 8(21):2003–2014. https://doi.org/10.3969/j.issn.1673-5374.2013.21.009

Galván Márquez I, Akuaku J, Cruz I, Cheetham J, Golshani A, Smith ML (2013) Disruption of protein synthesis as antifungal mode of action by chitosan. Int J Food Microbiol 164(1):108–112. https://doi.org/10.1016/j.ijfoodmicro.2013.03.025

Gonçalves SS, Souza ACR, Chowdhary A, Meis JF, Colombo AL (2016) Epidemiology and molecular mechanisms of antifungal resistance in *Candida* and *Aspergillus*. Mycoses 59(4):198–219

Gow NA, Latge JP, Munro CA (2017) The fungal cell wall: structure, biosynthesis, and function. Microbiol Spectr 5(3):10–1128

Goodson HV, Jonasson EM (2018) Microtubules and microtubule- associated proteins. Cold Spring Harb Perspect Biol 10(6). https://doi.org/10.1101/cshperspect.a022608

Garcia-Bustos JF, Sleebs BE, Gasser RB (2019) An appraisal of natural products active against parasitic nematodes of animals. Parasit Vectors 12(1):306. https://doi.org/10.1186/s13071-019-3537-1

Garcia-Effron G (2020) Rezafungin-mechanisms of action, susceptibility and resistance: similarities and differences with the other Echinocandins. J Fungi (Basel) 6(4):262

Ghannoum M et al (2020) Ibrexafungerp: a novel Oral triterpenoid antifungal in development for the treatment of *Candida auris* infections. Antibiotics (Basel) 9(9):539

Gao T et al (2021) The antioxidant guaiacol exerts fungicidal activity against fungal growth and deoxynivalenol production in *fusarium graminearum*. Front Microbiol 12:762844. https://doi.org/10.3389/fmicb.2021.762844

Gulder TA et al (2013) Isolation, structure elucidation and total synthesis of lajollamide A from the marine fungus Asteromyces cruciatus. Mar Drugs 10:2912–2935. https://doi.org/10.3390/md10122912

Hartland RP, Emerson GW, Sullivan PA (1991) A secreted β-glucan-branching enzyme from *Candida albicans*. Proc R Soc Lond Ser B Biol Sci 246(1316):155–160

Hwang EI et al (2000) Phellinsin a, a novel chitin synthases inhibitor produced by *Phellinus* sp. PL3. J Antibiot 53(9):903–911

Helrich CS, Schmucker JA, Woodbury DJ (2006) Evidence that nystatin channels form at the boundaries, not the interiors of lipid domains. Biophys J 91(3):1116–1127

Hasenoehrl A et al (2006) In vitro activity and in vivo efficacy of icofungipen (PLD-118), a novel oral antifungal agent, against the pathogenic yeast *Candida albicans*. J Antimicrob Agents 50(9):3011–3018. https://doi.org/10.1128/AAC.00254-06

Hannich JT, Umebayashi K, Riezman H (2011) Distribution and functions of sterols and sphingolipids. Cold Spring Harb Perspect Biol 3:1–14

Huang Z et al (2011) Sampangine inhibits heme biosynthesis in both yeast and human. Eukaryot Cell 10(11):1536–1544. https://doi.org/10.1128/EC.05170-11

Heard SC, Wu G, Winter JM (2021) Antifungal natural products. Curr Opin Biotechnol 69:232–241

Hu Z et al (2022) Sanguinarine, isolated from *Macleaya cordata*, exhibits potent anti-fungal efficacy against *Candida albicans* through inhibiting ergosterol synthesis. Front Microbiol 13:908461

Isono K, Asahi K, Suzuki S (1969) Studies on polyoxins, antifungal antibiotics. The structure of polyoxins. J Am Chem Soc 91(26):7490–7505

Igarashi Y et al (2003) Yatakemycin, a novel antifungal antibiotic produced by *Streptomyces* sp. TP-A0356. J Antibiot 56(2):107–113. https://doi.org/10.7164/antibiotics.56.107

Jamison MT et al (2019) Synergistic anti-candida activity of Bengazole a in the presence of Bengamide a dagger. Mar Drugs 17(2):102

Kerridge D (1980) The plasma membrane of *Candida albicans* and its role in the action of antifungal drugs. Eukaryotic Microbial Cell 103

Klis FM (1994) Cell wall assembly in yeast. Yeast 10(7):851–869

Klis FM, Mol P, Hellingwerf K, Brul S (2002) Dynamics of cell wall structure in *Saccharomyces cerevisiae*. FEMS Microbiol Rev 26(3):239–256

Klis FM, Boorsma A, De Groot PW (2006) Cell wall construction in *Saccharomyces cerevisiae*. Yeast 23(3):185–202

Khan A, Ahmad A, Akhtar F, Yousuf S, Xess I, Khan LA, Manzoor N (2010) *Ocimum sanctum* essential oil and its active principles exert their antifungal activity by disrupting ergosterol biosynthesis and membrane integrity. Res Microbiol 161:816–823

Khan M, Ahmad I, Cameotra SS (2013) Phenyl aldehyde and propanoids exert multiple sites of action towards cell membrane and cell wall targeting ergosterol in *Candida albicans*. AMB Express 3:54

Kathiravan MK et al (2012) The biology and chemistry of antifungal agents: a review. Bioorg Med Chem 20(19):5678–5698

Kwun MS, Lee DG (2020) Quercetin-induced yeast apoptosis through mitochondrial dysfunction under the accumulation of magnesium in *Candida albicans*. Fungal Biol 124(2):83–90. https://doi.org/10.1016/j.funbio.2019.11.009

Kerridge D (2020) The protoplast membrane and antifungal drugs. In: F. protoplasts. CRC Press, pp 135–169

Kumar P, Kumar M, Kundu A et al (2023) Chemical profiling of *Streptomyces* sp. for detection of potential pharmaceutical molecules. Biologia 78:3275. https://doi.org/10.1007/s11756-023-01485-5

Lester RL, Dickson RC (1993) Sphingolipids with inositolphosphate-containing head groups. Adv Lipid Res 26:253–274

Li RK, Rinaldi MG (1999) In vitro antifungal activity of nikkomycin Z in combination with fluconazole or itraconazole. Antimicrob Chemother 43(6):1401–1405

Luque JC, Clemons KV, Stevens DA (2003) Efficacy of micafungin alone or in combination against systemic murine aspergillosis. Antimicrob Chemother 47(4):1452–1455

Lemke A, Kiderlen AF, Kayser O (2005) Amphotericin B. Appl. Microbiol Biotechnol 68(2):151–162

Latge JP (2007) The cell wall: a carbohydrate Armour for the fungal cell. Mol Microbiol 66(2):279–290

Lepesheva GI, Waterman MR (2007) Sterol 14alpha-demethylase cytochrome P_{450} (CYP51), a P_{450} in all biological kingdoms. Biochim Biophys Acta 1770(3):467–477

Li XC et al (2011) Natural product based 6-hydroxy-2,3,4,6-tetrahydropyrrolo[1,2-*a*]pyrimidinium scaffold as a new antifungal template. ACS Med Chem Lett 2(5):391–395. https://doi.org/10.1021/ml200020h

Lima IO et al (2012) Anti-*Candida albicans* effectiveness of citral and in-vestigation of mode of action. Pharm Biol 50(12):1536–1541

Lima IO et al (2013) Antifungal activity and mode of action of carvacrol against *Candida albicans* strains. J Essential Oil Res 25(2):138–142

Lyu X et al (2016) Efficacy of nystatin for the treatment of oral candidiasis: a sys-tematic review and meta-analysis. Drug Des Dev Ther 10:1161–1171

Li ZJ et al (2017) Antifungal activity of gallic acid in vitro and in vivo. Phytother Res 31(7):1039–1045

Liu H et al (2019) Synergistic effect of natural antifungal agents for postharvest diseases of blackberry fruits. J Sci Food Agric 99(7):3343–3349. https://doi.org/10.1002/jsfa.9551

Li H et al (2021) Design, synthesis, and structure-activity relationship studies of Magnolol derivatives as antifungal agents. J Agric Food Chem 69(40):11781–11793

Molinski TF, Ireland CM (1988) Dysidazirine, a cytotoxic azacyclopropene from the marine sponge *Dysidea fragilis*. J Org Chem 53(9):2103–2105. https://doi.org/10.1021/jo00244a049

Mehmood Z, Ahmad I, Mohammad F, Ahmad S (1999) Indian medicinal plants: a potential source for anticandidal drugs. Pharm Biol 3:237–242

Munro CA, Gow NAR (2001) Chitin synthesis in human pathogenic fungi. Med Mycol 39(1):41–53

Munro CA, Winter K, Buchan A, Henry K, Becker JM, Brown AJ, Bulawa CE, Gow NA (2001) Chs1 of Candida albicans is an essential chitin synthase required for synthesis of the septum and for cell integrity. Mol Microbiol 39(5):1414–1426

Martos AI et al (2012) Evaluation of disk diffusion method compared to broth microdilution for antifungal susceptibility testing of 3 echinocandins against *aspergillus* spp. Diagn Microbiol Infect Dis 73(1):53–56

Marakalala MJ, Vautier S, Potrykus J, Walker LA, Shepardson KM, Hopke A, Mora-Montes HM, Brown GD (2013) Differential adaptation of *Candida albicans* in vivo modulates immune recognition by dectin-1. PLoS Pathog 9(4):e1003315

Munro CA (2013) Chitin and glucan, the yin and yang of the fungal cell wall, implications for antifungal drug discovery and therapy. Adv Appl Microbiol 83:145–172

Malova IO et al (2015) Natamycin—antimycotic of polyene macrolides class with un-usual properties, vol 91. Vestn Dermatol Venerol, p 161

Morales J, Mendoza L, Cotoras M (2017) Alteration of oxidative phosphory-lation as a possible mechanism of the antifungal action of p-coumaric acid against *Botrytis cinerea*. J Appl Microbiol 123(4):969–976. https://doi.org/10.1111/jam.13540

Marques JT, Marinho HS, de Almeida RFM (2018) Sphingolipid hydroxylation in mammals, yeast and plants an integrated view. Prog Lipid Res 71:18–42

Marena GD et al (2021) Biological properties and analytical methods for mica-fungin: a critical review. Crit Rev Anal Chem 51(4):312–328

Netea MG et al (2006) Immune sensing of Candida albicans requires cooperative recognition of mannans and glucans by lectin and Toll-like receptors. Journal of Clinical Investigation 116(6):1642–1650. https://doi.org/10.1172/JCI27114

Nyfeler R, Keller-Schierlein W (1974) Metabolites of microorganisms. 143. Echinocandin B, a novel polypeptide-antibiotic from aspergillus nidulans var. echinulatus: isolation and structural components. Helv Chim Acta 57(8):2459–2477

Odds FC, Brown AJ, Gow NA (2003) Antifungal agents: mechanisms of action. Trends Microbiol 11(6):272–279

Pfaller M, Riley J, Koerner T (1989) Effects of cilofungin (LY121019) on carbohydrate and sterol composition of *Candida albicans*. Eur J Clin Microbiol Infect Dis 8:1067–1070

Parker JE et al (2008) Differential azole antifungal efficacies contrasted using a sac-charomyces cerevisiae strain humanized for sterol 14 alpha-demethylase at the homologous locus. Antimicrob Agents Chemother 52(10):3597–3603

Perlin DS (2011) Current perspectives on echinocandin class drugs. Future Microbiol 6(4):441–457

Pereira FO et al (2015) Antifungal activity of geraniol and citronellol, two monoterpenes alcohols, against *Trichophyton rubrum* involves inhibition of ergosterol biosynthesis. Pharm Biol 53(2):228–234

Pizzorno J, Shippy A (2016) Is mold toxicity really a problem for our patients? Part 2-nonrespiratory conditions. Integr Med (Encinitas) 15(3):8–14

Pye CR, Bertin MJ, Lokey RS, Gerwick WH, Linington RG (2017) Retrospective analysis of natural products provides insights for future discovery trends. Proc Natl Acad Sci 114:5601–5606

Perlin DS (2020) Cell wall-modifying antifungal drugs. In: The fungal cell wall: an armour and a weapon for human fungal pathogens, pp 255–275

Quentin MJ et al (1982) Action of peptidolipidic antibiotics of the iturin group on erythrocytes. Biochim Biophys Acta 684(2):207–211

Rocha I (2019) Seed Coating: A Tool for Delivering Beneficial Microbes to Agricultural Crops Frontiers in Plant Science 10. https://doi.org/10.3389/fpls.2019.01357

Roncero C (2002) The genetic complexity of chitin synthesis in fungi. Curr Genet 41:367–378

Shimada S (1969) Antifungal steroid glycoside from sea cucumber. Science 163:1462

Silo-Suh LA et al (1994) Biological activities of two fungistatic antibiotics produced by *Bacillus cereus* UW85. Appl Environ Microbiol 60(6):2023–2030. https://doi.org/10.1128/aem.60.6.2023-2030.1994

Schmidt EW, Bewley CA, Faulkner DJ (1998) Theopalauamide, a bicyclic glycopeptide from filamentous bacterial symbionts of the lithistid sponge *Theonella swinhoei* from Palau and Mozambique. J Org Chem 63:1254–1258

Stabb EV, Handelsman J (1998) Genetic analysis of zwittermicin a resistance in Escherichia coli: effects on membrane potential and RNA polymerase. Mol Microbiol 27(2):311–322. https://doi.org/10.1046/j.1365-2958.1998.00678.x

Schuetzer-Muehlbauer M, Willinger B, Krapf G, Enzinger S, Presterl E, Kuchler K (2003) The Candida albicans Cdr2p ATP-binding cassette (ABC) transporter confers resistance to caspofungin. Mol Microbiol 48(1):225–235

Skepper CK, Dalisay DS, Molinski TF (2008) Synthesis and antifungal ac-tivity of (−)-(Z)-dysidazirine. Org Lett 10(22):5269–5271

Samber N et al (2015) Synergistic anti-candidal activity and mode of action of *Mentha piperita* essential oil and its major components. Pharm Biol 53(10):1496–1504

Saunders M, Glenn AE, Kohn LM (2010) Exploring the evolutionary ecology of fungal endophytes in agricultural systems: using functional traits to reveal mechanisms in community processes Abstract. Evolutionary Applications 3(5-6):525–537. https://doi.org/10.1111/eva.2010.3.issue-5-6 https://doi.org/10.1111/j.1752-4571.2010.00141.x

Silva IR et al (2014) Diversity of arbuscular mycorrhizal fungi along an environmental gradient in the Brazilian semiarid. Appl Soil Ecol 84:166–175. https://doi.org/10.1016/j.apsoil.2014.07.008

Singh S, Fatima Z, Hameed S (2016) Insights into the mode of action of anticandidal herbal monoterpenoid geraniol reveal disruption of multiple MDR mechanisms and virulence attributes in *Candida albicans*. Arch Microbiol 198(5):459–472

Song JC, Stevens DA (2016) Caspofungin: pharmacodynamics, pharmacokinetics, clinical uses and treatment outcomes. Crit Rev Microbiol 42(5):813–846

Shishodia SK, Shankar J (2020) Proteomic analysis revealed ROS-mediated growth inhibition of *aspergillus terreus* by shikonin. J Proteome 224:103849. https://doi.org/10.1016/j.jprot.2020.103849

Sass G, Larwood DJ, Martinez M, Shrestha P, Stevens DA (2021) Efficacy of nikkomycin Z in murine CNS coccidioidomycosis: modelling sustained-release dosing. J Antimicrob Chemother 76(10):2629–2635

Titsworth E, Grunberg E (1973) Chemotherapeutic activity of 5-fluorocytosine and amphotericin B against Candida albicans in mice. Antimicrob Agents Chemother 4(3):306–308

Tkacz JS (1992) Emerging targets in antibacterial and antifungal chemotherapy. In: Georgopapadakou NH, Sutcliffe JA (eds) 606pp

te Welscher YM, ten Napel HH, Balague MM, Souza CM, Riezman H, de Kruijff B, Breukink E (2008) Natamycin blocks fungal growth by binding specifically to ergosterol without permeabilizing the membrane. J Biol Chem 283:6393–6401

te Welscher YM, van Leeuwen MR, de Kruijff B, Dijksterhuis J, Breukink E (2012) Polyene antibiotic that inhibits membrane transport proteins. Proc Natl Acad Sci USA 109(28):11156–11159

Tebbets B et al (2013) Identification of antifungal natural products via sac-charomyces cerevisiae bioassay: insights into macrotetrolide drug spectrum, potency and mode of action. Med Mycol 51(3):280–289. https://doi.org/10.3109/13693786.2012.710917

Tao P et al (2020) Antifungal application of rosin derivatives from renewable pine resin in crop protection. J Agric Food Chem 68(14):4144–4154. https://doi.org/10.1021/acs.jafc.0c00562

Teymuri M, Shams-Ghahfarokhi M, Razzaghi-Abyaneh M (2021) Inhibitory effects and mechanism of antifungal action of the natural cyclic depsipeptide, aureobasidin a against *Cryptococcus neoformans*. Bioorg Med Chem 41:128013. https://doi.org/10.1016/j.bmcl.2021.128013

Umezawa S et al (1958) A new antifungal antibiotic, pentamycin. J Antibiot (Tokyo) 11(1):26–29

Vanittanakom N et al (1986) Fengycin–a novel antifungal lipopeptide antibiotic pro-duced by *Bacillus subtilis* F-29-3. J Antibiot (Tokyo) 39(7):888–901

Vicente MF, Basilio A, Cabello A, Peláez F (2003) Microbial natural products as a source of antifungals. Clin Microbial Infect 9(1):15–32

Verekar SA, Gupta MK, Deshmukh SK (2023) Discovery of bioactive metabolites from the genus Arthrinium. In: Fungi and fungal products in human welfare and biotechnology, pp 257–287

Wordeman L (2010) How kinesin motor proteins drive mitotic spindle function: lessons from molecular assays. Semin Cell Dev Biol 21(3):260–268. https://doi.org/10.1016/j.semcdb.2010.01.018

Wei J et al (2020) Antifungal activity of cinnamaldehyde against *fusarium sambucinum* involves inhibition of ergosterol biosynthesis. J Appl Microbiol 129(2):256–265

Xia M et al (2019) Communication between mitochondria and other organelles: a brand-new perspective on mitochondria in cancer. Cell Biosci 9(27):27. https://doi.org/10.1186/s13578-019-0289-8

Xie Y, Liu X, Zhou P (2020) In vitro antifungal effects of Berberine against *Candida spp*, in: planktonic and biofilm conditions. Drug Des Devel Ther 14:87–101

Youssef DT, Shaala LA, Mohamed GA, Badr JM, Bamanie FH, Ibrahim SR (2014) Theonellamide G, a potent antifungal and cytotoxic bicyclic Glycopeptide from the Red Sea marine sponge *Theonella swinhoei*. Mar Drugs 12:1911–1923

Yu D et al (2015) Antifungal modes of action of tea tree oil and its two characteristic components against Botrytis cinerea. J Appl Microbiol 119(5):1253–1262

Zotchev SB (2003) Polyene macrolide antibiotics and their applications in human therapy Curr. Med Chem 10:211–223

Zhang CW et al (2023) Antifungal natural products and their derivatives: a review of their activity and mechanism of actions. Pharmacol Res Modern C Med 7:100262. https://doi.org/10.1016/j.prmcm.2023.100262

Part VII

Toxicology of Natural Antifungals and Other Applications

Toxicology of Antifungal and Antiviral Drugs

24

Sarika Bano, Saiema Ahmedi, Nikhat Manzoor, and Sanjay Kumar Dey

24.1 Introduction

Fungi are a diverse group of eukaryotic microorganisms that can exist in various forms, including yeasts, molds, and mushrooms. They are widespread in the environment, and many species are commensals in the human body. While most fungi are harmless, some have the potential to cause infections when the balance between the host immune defenses and the fungal virulence factors is disrupted. Fungal infections can be broadly categorized into superficial, cutaneous, subcutaneous and systemic infections, with varying levels of severity and tissue involvement (Bonamigo and Dornelles 2023). Superficial fungal infections typically affect the skin, hair, and nails, while systemic infections can invade multiple organ systems and pose life-threatening risks (Kaushik et al. 2015). Understanding the type of fungal infection is crucial in selecting the appropriate antifungal therapy. Fungal infections remain a significant global health concern, and advancing antifungal strategies is pivotal in mitigating their impact on individuals and healthcare systems worldwide. Antifungal drugs have revolutionized the management of fungal infections, offering hope to patients affected by these infections. Understanding the mechanisms of action and classes of antifungal and antiviral drugs is essential for healthcare providers when making treatment decisions. Despite the availability of multiple antifungal drugs, several challenges persist in managing fungal infections. However, the emergence of antifungal resistance and the challenges associated with

Sarika Bano and Saiema Ahmedi contributed equally to this work.

S. Bano · S. K. Dey (✉)
Laboratory for Structural Biology of Membrane Proteins, Dr. B.R. Ambedkar Center for Biomedical Research, University of Delhi, New Delhi, Delhi, India
e-mail: sarika.bano11@gmail.com; sdey@acbr.du.ac.in

S. Ahmedi · N. Manzoor
Department of Biosciences, Jamia Millia Islamia, New Delhi, Delhi, India
e-mail: saiemaahmedi1995@gmail.com; nmanzoor@jmi.ac.in

N. Manzoor (ed.), *Advances in Antifungal Drug Development*, https://doi.org/10.1007/978-981-97-5165-5_24

therapy call for continued research and innovation in the field of medical mycology. Resistance to antifungals is a growing concern, especially among certain *Candida* and *Aspergillus* species (Houšť et al. 2020). Additionally, drug interactions, adverse effects, and the need for prolonged treatment can complicate therapy. Antifungal drugs are sometimes used in combination with other medications to treat complex medical conditions (Ben-Ami and Kontoyiannis 2021). For example, in the treatment of invasive fungal infections in immunocompromised patients, a combination of antifungal agents may be employed to improve efficacy and reduce the risk of resistance (Johnson et al. 2004). The choice of antifungal agent depends on the type and severity of the infection, as well as individual patient factors. Furthermore, the increasing prevalence of immunocompromised patients, such as those with HIV/AIDS or undergoing chemotherapy, underscores the importance of effective antifungal therapy. Antifungal infections can be prevalent in certain regions and are considered neglected diseases.

Similarly, a virus is a microscopic organism that needs its host body to reproduce on its own. The genetic material of the viruses is either deoxyribonucleic acid (DNA) or ribonucleic acid (RNA), and it can be single- or double-stranded (Courouble et al. 2021; Dey and Dey 2020; Dey et al. 2022; Farasati Far et al. 2023; Rizwan et al. 2022; Shahjahan et al. 2023; Yadav et al. 2022). Human viral infections are the cause of millions of deaths worldwide and are linked to conditions like hepatitis, influenza, herpes simplex, HIV/AIDS, the common cold and others (Evans 2013). While each virus species has its own unique process of infection, they all infect their animal hosts by taking similar steps. (i) The first stage of a virus infection is attachment and entry. Initially, the virus's glycoproteins bind to receptor or coreceptor molecules on the host cell membrane, allowing the virus to enter the cell through endocytosis (Soderstrom 2014). (ii) Virus uncoating: the viral components (genetic material and proteins) are released into the host cytosol when the host cell enzymes break down the capsid of the internalized virus. (iii) Viral genome transcription and replication: the viral genome, which is made up of DNA or RNA, is carried into the nucleus where it is transcribed and replicated to create many copies of the genome and messenger RNA (mRNA) molecules, respectively (Rice et al. 1985). Depending on the type of genome, like, DNA/RNA, single-stranded or double-stranded, positive or negative sense, different replication mechanisms apply. Within the cytoplasm, the RNA genome is capable of replication. (iv) Protein synthesis: Using the machinery of the host cell, the viral mRNAs are translated into structural and regulatory proteins in the cytoplasm. (v) Assembly: When the viral genome replicates and expresses properly, it produces the components required for the progeny virus to survive after being released from the host cell. As a result, all the elements required to create new viruses are combined. (vi) Release: By lysis of the host cell or budding, the assembled progeny viruses are released into the extracellular fluid. The host cell dies during the lysis process, but it might not during budding (Thomas et al. 1965). Antiviral drugs are approved for each step or the other to combat viral disorders as detailed in this book chapter. Major toxicities exerted by them are also explained herein. Due to their contagious nature, viruses force scientists to create antiviral drugs that inhibit one or more of the aforementioned stages of the viral life cycle, thereby limiting the virus's ability to survive and spread. An antiviral medication is a small- or large-molecule, synthetic or natural agent that can lessen virally-induced infectious disease. The US Food and Drug Administration (FDA) authorized idoxuridine in 1963 as the first antiviral medication to treat herpes simplex infections (De Clercq and Li 2016). The growing need for innovative approaches to develop antiviral agents is highlighted by the rise in chronic viral infections like HIV, HCV, HBV, and others, as well as the emergence of new viruses like coronaviruses that cause severe acute respiratory syndrome (SARS). Specifically, the global AIDS epidemic of the 1980s, which was linked to the human immunodeficiency virus (HIV), boosted efforts toward advancement of basic and therapeutic science. This resulted in the creation of antiviral inhibitors that could combat other viruses in addition to HIV (Dieffenbach and Fauci 2011). Unfortunately, most of them do pose some toxicity to human host at certain conditions which has also been explained in this chapter.

24.2 Toxicity of Antifungal Drugs

24.2.1 Classes of Antifungal Drugs

The primary and most obvious reason for using antifungal drugs is to treat fungal infections. Fungal infections can affect various parts of the body, including the skin, nails, respiratory tract, and internal organs. Without appropriate treatment, these infections can cause discomfort, pain, and potentially life-threatening complications. Antifungal drugs are categorized into several classes, each with its own mechanism of action and spectrum of activity. These include azoles, polyenes, echinocandins, allylamines, and pyrimidine analogs (Nicola et al. 2019). Azoles, such as fluconazole and itraconazole, are commonly used for the treatment of various fungal infections (Robbins et al. 2016). Polyenes, like amphotericin B, are effective against a broad range of fungi and are particularly useful in severe systemic infections (Mount et al. 2018). Echinocandins, represented by caspofungin, target the fungal cell wall and are valuable options in cases of invasive candidiasis. Allylamines like terbinafine are predominantly used to treat dermatophyte infections, while pyrimidine analogs, including flucytosine, inhibit nucleic acid synthesis and are used in combination therapy for systemic fungal infections (Bonamigo and Dornelles 2023). Most of them are summarized in Table 24.1.

24.2.2 Mechanism of Action of Antifungal Drugs

24.2.2.1 Azoles

Azoles represent a significant class of antifungal drugs that have revolutionized the treatment of fungal infections. Azoles inhibit the biosynthesis of ergosterol, a vital component of the fungal cell membranes. This disruption in ergosterol production compromises membrane integrity, rendering the fungal cell more permeable and

Table 24.1 Antifungal drugs and their mode of action

Class	Examples	Route of administration	Mode of action	Side effects	References
Azoles	Fluconazole, itraconazole, voriconazole, posaconazole, and ketoconazole	Oral, topical application	Target an enzyme lanosterol 14α-demethylase, which is crucial in the conversion of lanosterol to ergosterol	Hepatotoxicity, headache, fever, and vomiting sensation	Chandrasekaran et al. (2019), Van Daele et al. (2019)
Polyenes	Amphotericin B and nystatin	Topical application	Binding with ergosterol and disruption of fungal cell membrane leading to the formation of membrane pores	Nephrotoxicity, neurotoxicity, muscle, and joint pain	Kovacic and Cooksy (2012)
Echinocandins	Caspofungin, micafungin, and anidulafungin	Topical application	Target fungal cell wall by inhibiting the synthesis of beta-1,3-D-glucan	Gastrointestinal effects, diarrhea, vomiting, and nausea	Perlin (2015)
Allylamines	Terbinafine, naftifine, and butenafine	Topical application, oral	Target an enzyme squalene epoxidase, which is crucial for synthesis of ergosterol	Skin reactions, itching, redness at the site of application, and liver function abnormalities	Vanreppelen et al. (2023)
Nucleotide analogs	Flucytosine	Oral, intravenous	Disrupting the synthesis of pyrimidine analogs impede fungal DNA and RNA replication	Bone marrow suppression, decrease in RBC, and anemia	Trivedi et al. (2022)

vulnerable to damage. Azoles target a specific enzyme, lanosterol 14α-demethylase, which is crucial in the conversion of lanosterol to ergosterol (Como and Dismukes 1994). By interfering with this enzyme, azoles impede the production of ergosterol, leading to membrane instability and cell death. Notable members of this class include fluconazole, itraconazole, voriconazole, posaconazole, and ketoconazole (Van Daele et al. 2019). Fluconazole is widely used for treating superficial and systemic candidiasis. Voriconazole, with its broad-spectrum activity, is essential in the treatment of invasive candidiasis while ketoconazole, although less commonly used today, was historically employed in various fungal infections, including dermatophytosis and candidiasis (Como and Dismukes 1994).

24.2.2.2 Polyenes

Polyenes are a crucial class of antifungal drugs that have played a significant role in the treatment of fungal infections for decades and well known for their effectiveness against a broad spectrum of fungal pathogens, polyenes offer a unique mechanism of action that sets them apart from other antifungal agents. Their unique mechanism of action involves binding to and disrupting the fungal cell membrane. Specifically, they interact with ergosterol, a vital component of the fungal cell membrane, leading to the formation of membrane pores (Kovacic and Cooksy 2012). These pores result in increased permeability and the leakage of essential cellular components, ultimately leading to fungal cell death (Carolus et al. 2020). The selective targeting of ergosterol distinguishes polyenes from other antifungal classes. The polyene class includes several members, with amphotericin B and nystatin being the most notable. Amphotericin B, a versatile and potent polyene antifungal, is used to treat a wide range of fungal infections, including systemic mycoses such as aspergillosis and candidiasis (Kovacic and Cooksy 2012). Nystatin, on the other hand, is primarily employed topically, often in the treatment of oral and cutaneous fungal infections. Both polyenes are integral components of antifungal therapy, each with its unique applications.

24.2.2.3 Echinocandins

Echinocandins constitute a vital class of antifungal drugs that have transformed the treatment of invasive fungal infections. They specifically target the fungal cell wall by inhibiting the enzyme glucan synthase and inhibit the synthesis of β-1,3-D-glucan, a critical component responsible for maintaining cell wall integrity (Chen et al. 2011). By interfering with this process, echinocandins weaken the cell wall, making the fungal cells more susceptible to osmotic stress and ultimately leading to their death (Perlin 2015). Echinocandins include three main members: caspofungin, micafungin, and anidulafungin (Grover 2010). While they share a common mechanism of action, these agents exhibit some variations in their clinical applications and pharmacokinetics. Caspofungin is a broad-spectrum echinocandin employed in the treatment of invasive aspergillosis, candidiasis, and other fungal infections (Perlin 2015). Micafungin is primarily used for candidiasis, and anidulafungin finds utility in the treatment of esophageal candidiasis (Grover 2010).

24.2.2.4 Allylamines

Allylamines owe their antifungal properties to their distinct mechanism of action. Unlike many other antifungal classes, which primarily inhibit ergosterol synthesis, allylamines target an enzyme called squalene epoxidase. Squalene epoxidase is a crucial component in the biosynthesis of ergosterol, a fundamental constituent of the fungal cell membrane (Stütz 1987). By inhibiting this enzyme, allylamines disrupt the conversion of squalene to lanosterol, leading to an accumulation of squalene within the fungal cell (Petranyi et al. 1987). This disruption destabilizes the fungal cell membrane, resulting in increased membrane permeability and the leakage of vital cellular components. Ultimately, the fungal cell is unable to maintain its structural integrity. Allylamines includes Terbinafine, Naftifine, and Butenafine, etc. (Vanreppelen et al. 2023). Terbinafine is one of the most well-known and widely used allylamines. It is available in various formulations, including topical creams and ointments for skin infections and oral tablets for more severe cases like onychomycosis (fungal nail infections) (Hammoudi Halat et al. 2022). Naftifine is primarily available as a topical cream or gel for the treatment of various dermatophyte infections, such as athlete's foot, jock itch, and ringworm. Butenafine is another allylamine primarily used in topical formulations for the treatment of fungal skin infections. It is effective against dermatophytes and is often used for athlete's foot (Hammoudi Halat et al. 2022).

24.2.2.5 Pyrimidine Analogs

Pyrimidine analogs exert their antifungal effects through a distinctive mechanism of action. They interfere with fungal nucleic acid synthesis by inhibiting enzymes involved in the biosynthesis of pyrimidine nucleotides, such as cytosine and thymine (Basha and Goudgaon 2021). By disrupting the synthesis of these essential building blocks, pyrimidine analogs impede fungal DNA and RNA replication (Trivedi et al. 2022). This interference with nucleic acid synthesis ultimately leads to the inhibition of fungal growth and reproduction (Trivedi et al. 2022). Pyrimidine analogs can be categorized into different classes based on their chemical structure and mechanism of action. Some well-known classes of pyrimidine analogs include-Flucytosine, and 5-Fluorouracil (Basha and Goudgaon 2021). Flucytosine is a pyrimidine analog used primarily in combination therapy for the treatment of systemic fungal infections, particularly those caused by *Candida* and *Cryptococcus* species (Chandrasekaran et al. 2019). While 5-fluorouracil is primarily known for its use in cancer chemotherapy, it also exhibits antifungal properties. It is occasionally employed in the treatment of superficial fungal infections (Basha and Goudgaon 2021).

24.2.3 Major Toxicities Associated with Various Classes of Antifungal Drugs

24.2.3.1 Azoles: Hepatotoxicity and Drug–Drug Interactions

Azoles, a widely used class of antifungal drugs, are known for their potential hepatotoxicity (Fig. 24.1). These drugs can lead to elevated liver enzymes,

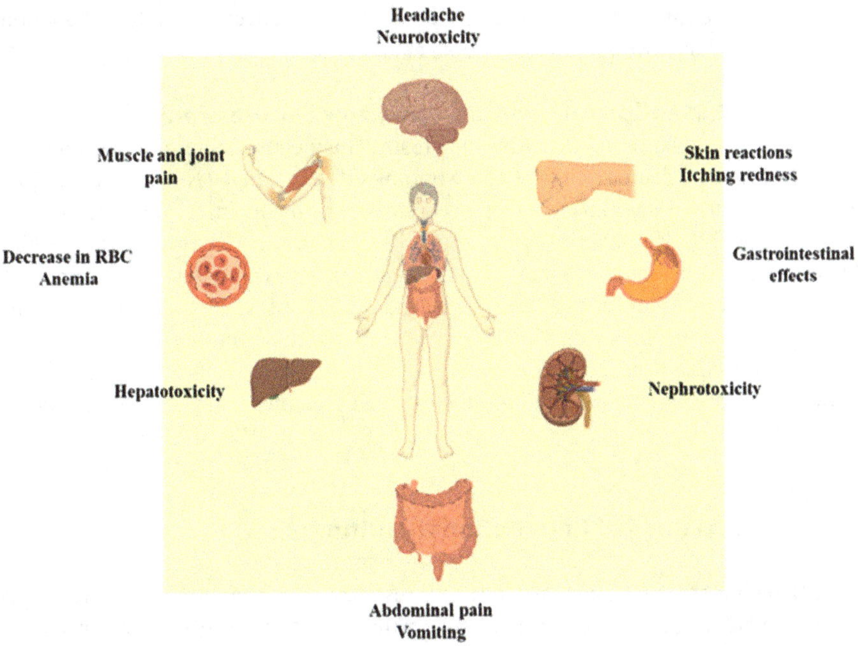

Fig. 24.1 Major toxicities caused by antifungal drugs on human body

hepatocellular damage and, in rare cases, severe hepatotoxicity (Ahmedi et al. 2022; Tverdek et al. 2016). Additionally, azoles are susceptible to various drug–drug interactions, necessitating caution and careful monitoring when coadministered with other medications (Nicola et al. 2019).

24.2.3.2 Polyenes: Nephrotoxicity and Neurotoxicity

Polyenes, such as amphotericin B, are renowned for their potent antifungal properties. However, they are also infamous for their significant toxicity, particularly nephrotoxicity and neurotoxicity. Nephrotoxicity manifests as impaired renal function, leading to electrolyte imbalances, reduced glomerular filtration rate and acute kidney injury (Carolus et al. 2020). Neurotoxicity, though less common, can result in various central nervous system symptoms (Kovacic and Cooksy 2012).

24.2.3.3 Echinocandins: Gastrointestinal Effects

Echinocandins, such as caspofungin, exhibit a more favorable safety profile. However, they are associated with gastrointestinal side effects, including nausea, vomiting, and diarrhea (Grover 2010). These adverse effects, while generally mild, can impact patient compliance and overall well-being.

24.2.3.4 Allylamines: Skin Reactions and Liver Function Abnormalities

Allylamines are primarily used in the treatment of superficial fungal infections. They can lead to localized skin reactions, including itching and redness at the site of application (Petranyi et al. 1987). Additionally, isolated cases of liver function abnormalities have been reported (Nicola et al. 2019).

24.2.3.5 Nucleotide Analogs: Bone Marrow Suppression

Nucleotide analogs, represented by flucytosine, may cause bone marrow suppression, leading to a decrease in red and white blood cell counts (Nicola et al. 2019). This can result in anemia and increased susceptibility to infections (Basha and Goudgaon 2021). The severity of these toxicities can vary and may be influenced by factors such as the patient's health and concomitant medications. Balancing the potential benefits of antifungal treatment with the risk of adverse effects is a complex task that necessitates careful patient selection and diligent monitoring. As the field of antifungal drug development advances, efforts are focused on minimizing toxicity and tailoring treatments through targeted therapies and personalized medicine approaches.

24.2.4 Limitations of Conventional Antifungals

Fungal pathogens can adapt and become resistant to commonly used antifungal agents, reducing the effectiveness of treatment. This resistance is a growing concern, particularly in healthcare settings. Another limitation is the narrow spectrum of many antifungal drugs. They are often designed to target specific fungal species, so the choice of antifungal treatment relies heavily on accurate diagnosis (Espinel-Ingroff 2022). Misdiagnosis can result in ineffective treatment. Antifungal drugs can also be associated with significant toxicities, particularly when used over extended periods or at high doses (Kaur et al. 2023). This can limit their use, especially in patients with underlying health conditions. The potential for drug–drug interactions further complicates treatment. In addition, some antifungal drugs can be costly, posing a barrier to access for certain patients, especially in regions with limited healthcare resources (Trivedi et al. 2022). Patient compliance with antifungal regimens, which often require long-term treatment, can be challenging. Incomplete treatment can lead to recurrent infections. The emergence of new and drug-resistant fungal pathogens is an ongoing challenge, necessitating the development of novel antifungal agents (Vanreppelen et al. 2023). These limitations underscore the need for continued research and innovation in the field of antifungal drugs to address these challenges and improve the efficacy and safety of antifungal therapies.

24.2.5 Plant-Based Natural Antifungals

Plants serve as a rich source of bioactive secondary metabolites, consists a wide array of compounds such as tannins, terpenoids, saponins, alkaloids, flavonoids, and various others, which have been documented to possess in vitro antifungal properties (Table 24.2) (Cardoso et al. 2012). Plant kingdom offers valuable lead compounds with novel structures, and there has been a growing interest in conducting extensive investigations of tropical plant species. Consequently, research on natural products and their derivatives has gained momentum in recent years, primarily because of their significance in drug discovery (Manzoor 2019). Numerous molecules exhibiting antifungal activity against various fungal strains have been discovered in plants, carrying substantial importance for human applications. These molecules can directly serve as the basis for the development of enhanced antifungal compounds. Many phytochemicals exhibit broad-spectrum antifungal activity, meaning they can target a wide range of fungal species (Ahmedi et al. 2022). This versatility can be particularly useful when the specific fungal pathogen is unknown. Fungi may be less prone to developing resistance to phytochemicals compared to some synthetic antifungal drugs. This can be advantageous in preventing treatment failures due to resistant strains. Table 24.2 consists of list of some medicinal plants with their antifungal property.

Table 24.2 List of a few medicinal plants with their antifungal property

Medicinal plants constituents	Bioactive components	Antifungal activity against	References
Tea tree oil	Terpinen-4-ol, γ-terpinene and α-terpinene	Candida, trichophyton, dermatophytes, *Saccharomyces cerevisiae*	Hammer et al. (2004)
Garlic	Allicin	Penicillium, aspergillus, candida, cryptococcus	Kuete (2017), Wan et al. (2019)
Oregano oil	Carvacrol and thymol	Candida, aspergillus, sclerotinia, cryptococcus	Ahmed and Hussain (2013), Manohar et al. (2001), Soylu et al. (2010)
Coconut oil	Lauric acid	Candida, dermatophytes	Abdullah et al. (2020)
Grapefruit seed extract	Trans-ferulic acid, rosmarinic acid, and trans-2-hydroxycinnamic acid	Botrytis, candida	Tsutsumi-Arai et al. (2019)
Neem	Azadirachtin, nimbolinin, nimbin, and nimbidol	Dermatophytes, microsporum, aspergillus, candida	Ospina et al. (2015)
Cinnamon	Eugenol, cinnamaldehyde, and cinnamic acid	Rhizopus, aspergillus, penicillium, candida	Shreaz et al. (2016), Xing et al. (2010)
Echinacea	Alkylamides and caffeic acid derivatives	Candida, *Alternaria*	Ismail et al. (2022), Katarina et al. (2012)
Aloe Vera	Hyaluronic acid	Rhizopus, aspergillus	Medda et al. (2015)
Turmeric	Curcumin	*Fusarium,* dermatophytes, aspergillus, *Helminthosporium oryzae, fusarium*	Akter et al. (2018), Chowdhury et al. (2008)

24.3 Toxicity of Antiviral Drugs

Depending on the particular medication and how it works, antiviral medications with FDA approval can have varying degrees of toxicity. Common side effects associated with antiviral medications include nausea, headache, fatigue, and gastrointestinal disturbances (Fig. 24.2). Serious side effects are possible, especially in people with pre-existing medical conditions or after extended use (Tompa et al. 2021). It is important to remember that antiviral medications have different toxicity profiles, and before receiving FDA approval, each medication must pass stringent testing to determine its safety and effectiveness. When prescribing antiviral medications, medical professionals carefully balance the risks and benefits, taking into account the patient's general health as well as the specific viral infection being treated. To manage and mitigate any potential toxicity issues, patients and healthcare professionals must communicate and monitor on a regular basis.

According to recent update September 2021, several antiviral drugs received Emergency Use Authorization (EUA) or full approval from the FDA for the treatment of COVID-19. Here are some antiviral drugs that have been used or considered for COVID-19 treatment along with potential side effects:

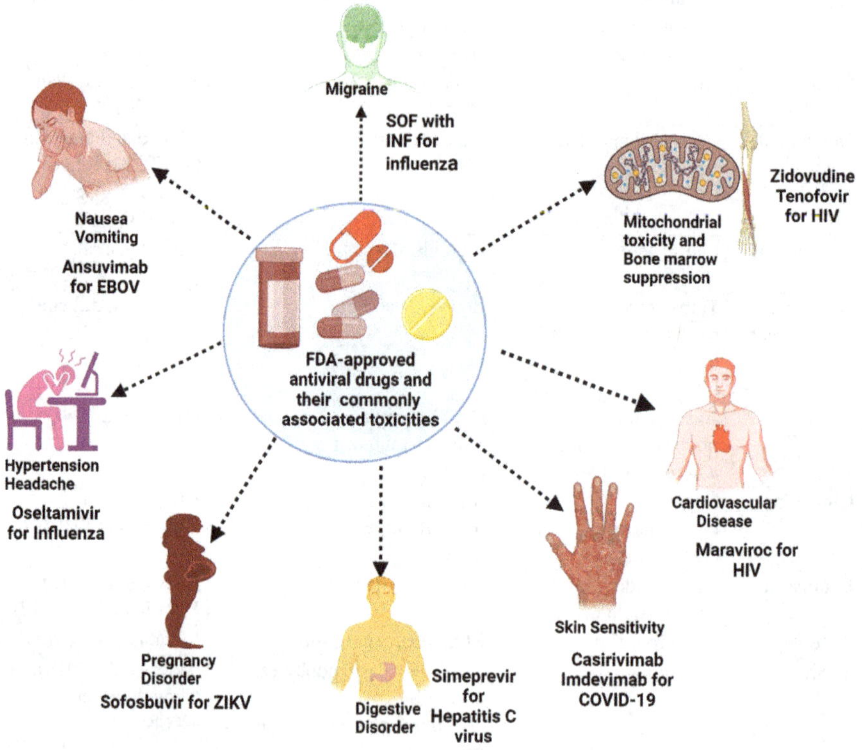

Fig. 24.2 Commonly associated toxicities of FDA-approved antiviral drugs

1. **Remdesivir:** This antiviral drug received EUA for COVID-19 treatment. Common side effects include nausea, vomiting and increased liver enzymes (Nhean et al. 2023).
2. **Monoclonal Antibodies (Casirivimab and Imdevimab, Bamlanivimab and Etesevimab):** These antibodies received EUA for the treatment of mild to moderate COVID-19. Potential side effects include allergic reactions (Nelson et al. 2000).
3. **Molnupiravir:** An antiviral drug with oral administration, it inhibits the replication of SARS-CoV-2. Side effects may include gastrointestinal symptoms and alterations in liver enzymes (Singh et al. 2021).
4. **Paxlovid (Nirmatrelvir and Ritonavir):** This combination antiviral drug received EUA for treating COVID-19. Common side effects include diarrhea and nausea (Hashemian et al. 2023).
5. **Favipiravir:** Used in some countries for COVID-19 treatment, it may cause gastrointestinal issues and increased levels of uric acid.
6. **Baricitinib and Tofacitinib:** Janus kinase inhibitors that received EUA in combination with remdesivir. Side effects may include an increased risk of infections.

24.3.1 Toxicity of FDA-Approved Drugs in the Treatment of HIV

The level of toxicity of human immunodeficiency virus (HIV) antiviral medications varies based on the particular medication or medication combination taken (Table 24.3) (Adkins and Noble 1998).

1. **Zidovudine and tenofovir** are examples of medications known as nucleoside reverse transcriptase inhibitors (NRTIs), which can have adverse effects like headache, weariness, and nausea. Use over an extended period of time may

Table 24.3 Antiviral drugs and their toxicities

S. No.	Drugs	Infection	Toxicity	Reference
1.	Ansuvimab	Ebola	Nausea, vomiting	Kaplon and Reichert (2021)
2.	Oseltamivir	Influenza	Headache, hypertension	Whitley et al. (2001)
3.	SOF with IFN	Influenza	Migraine	Whitley et al. (2001)
4.	Simeprevir	Hepatitis C	Digestive disorder	Thuppal et al. (2015)
5.	Sofosbuvir	Zika	Pregnancy disorder	Bullard-Feibelman et al. (2017)
6.	Casirivimab Imdevimab	SARS-CoV-2	Skin sensitivity, allergic reaction	Chary et al. (2023)
7.	Moravirac	HIV	Cardiovascular as well as liver disease	Emmelkamp and Rockstroh (2008)
8.	Zidovudine Tenofovir	HIV	Mitochondrial toxicity and bone marrow suppression	Thuppal et al. (2015)

cause mitochondrial toxicity and bone marrow suppression (Paton et al. 2021; Thuppal et al. 2015).
2. **Drugs like efavirenz** may cause neuropsychiatric side effects such as vivid dreams, vertigo, and difficulty in concentrating. These drugs are known as non-nucleoside reverse transcriptase inhibitors, or NNRTIs.
3. **Protease Inhibitors (PIs):** PIs, which include medications like atazanavir and ritonavir, can lead to metabolic problems like lipodystrophy, insulin resistance, and lipid-driven abnormalities.
4. **Integrase Strand Transfer Inhibitors (INSTIs):** This class of medications includes dolutegravir and raltegravir.
5. **Entry Inhibitors**: The entry inhibitor maraviroc is used as entry inhibitors has been linked to increased risk of cardiovascular events as well as liver issues (Emmelkamp and Rockstroh 2008).
6. **Fusion Inhibitors:** One fusion inhibitor that is injected and may cause injection site reactions is enfuvirtide.
7. **Products in Combination:** To make treatment easier, many antiretroviral medications are bundled into single-tablet regimens. The exact medications that are used in combination determine the side effect profile.

It is important to remember that individual responses to medications can differ, and that newer formulations and drugs may have different toxicity profiles. In addition to carefully monitoring patients for side effects, medical professionals' base treatment decisions on the patient's overall health as well as the possible risks and benefits of the selected course of action should always be taken into consideraton.

24.3.2 Toxicity of Ansuvimab for Treatment of EBOV

Ansuvimab is a single mAb initially isolated from memory B cells of two patients who survived the EBOV outbreak in Kikwit in 1995 and maintained antibodies against the EBOV surface GP for 11 years after the infection (Lee 2021). Ridgeback Biotherapeutics created ansuvimab to treat EBOV infections in both adult and pediatric patients. The US National Institute of Allergy and Infectious Diseases and the company signed a patent license agreement in December 2018 so that ansuvimab could be used to treat EVD. Ridgeback Biotherapeutics was authorized by the US Department of Health and Human Services to produce ansuvimab between September 2019 and April 2020 (Kaplon and Reichert 2021; Lee 2021). FDA approved ansuvimab on December 21, 2020, based on encouraging results from phase II/III clinical trials.

Adverse events observed in clinical trial studies for a drug differ from those observed in patients in the real world because clinical trials are carried out under different conditions. Ansuvimab at doses of 5, 25, and 50 mg/kg was administered to healthy adults in an open-label phase I study; the findings showed no fatalities or side effects. Three days following the ansuvimab injection, only four patients experienced mild side effects, such as headache, chills, nausea, and muscle or joint pain.

According to Gaudinski et al. (2019), these findings also imply that ansuvimab is safe and tolerable in healthy adults. Amidst an EBOV pandemic in the Democratic Republic of the Congo in 2018–2019, the PALM clinical trial assessed the safety and effectiveness of ansuvimab in 173 participants, comprising 119 adults and 54 pediatric patients. Fever was reported as the most frequent adverse event in at least 10% of patients who received ansuvimab, and approximately 29% of the subjects displayed hypersensitivity, including prespecified infusion-related events. Additionally, ≥40% of subjects reported experiencing diarrhea, pyrexia, abdominal pain, and vomiting, among other pre-specified symptoms that were tracked every day. Hypotension, tachycardia, rapid breathing, and chills were other frequent side effects (Mulangu et al. 2019). The FDA advises against using this medication in conjunction with a live virus vaccine to prevent EBOV. Patients should report any signs of systemic hypersensitivity reactions as soon as possible, given the side effects of ansuvimab administration that have been documented thus far. Patients should also be informed about hypersensitivity reactions, including events related to infusion during and after infusion (Infusion-Associated 2021).

24.3.3 Toxicity of Sofosbuvir, a FDA-Approved ZIKV Drug

ZIKV infection may be treated with sofosbuvir, an FDA-approved nucleotide polymerase inhibitor for the distantly related hepatitis C virus. Research using cell culture revealed that sofosbuvir effectively prevents the growth and infection of various ZIKV strains in a variety of human tumor cell lines and can be used to isolate human fetal-derived neural stem cells without causing appreciable side effects. Furthermore, oral sofosbuvir therapy shielded mice from ZIKV-induced death. Further research and development of sofosbuvir as a treatment for ZIKV infection in nonhuman primates and eventually humans may be warranted (Ojha et al. 2023). Despite the lack of evidence of fetal harm in animal studies using sofosbuvir (Bullard-Feibelman et al. 2017), there are no controlled data in human pregnancy. Sofosbuvir has been categorized as category B and is not advised for use in pregnant women. It is recommended for use in combination with ribavirin, which can result in birth defects, and/or pegylated interferon, which has abortifacient effects, to treat HCV infection. Thus, until the effects of sofosbuvir administration during pregnancy are well-understood, the immediate use of sofosbuvir as a monotherapy for preventing microcephaly and other congenital malformations seems unlikely (Levine 2019).

24.3.4 Toxicity of FDA-Approved Drugs for EBOV

As of right now, there's no proven cure for the Ebola virus (EBOV). The majority of medications and vaccines created to date are not yet authorized for use in human trials (Wilson et al. 2000). With preclinical evidence in living animals, miglustat, clomiphene, and toremifene are the most promising FDA-approved medications to treat Ebola virus infection. Significant antiviral activity against EBOV was demonstrated in vitro by three derivatives of miglustat. Considerable survival rates (50–70%) for their treatments were noted in a mouse model. After receiving treatments with Clomiphene and Toremifene, 50–90% of the mice survived. Nevertheless, clomiphene has a very poor oral absorption efficiency. Therefore, it is advised to take toremifene orally. Toremifene acts at the entry/fusion step, whereas miglustat acts at the secretion/envelopment step. Mice were given far higher effective dosages of Miglustat and Toremifene than humans are clinically prescribed. Miglustat is primarily used to treat Gaucher disease and Niemann–Pick disease type C, while toremifene is employed in breast cancer therapy. Common side effects of miglustat include digestive issues and tremors, while toremifene may cause hot flashes and changes in vision (Anemia 2013).

24.3.5 Toxicity of FDA-Approved Drugs for Hepatitis Virus C

Globally, liver diseases like hepatocellular carcinoma and liver cirrhosis are primarily caused by the chronic hepatitis C virus (HCV). Approximately 180 million people worldwide are impacted by it, or 3% of the global population (Asrani et al. 2019). Based on their target sites of action, DAAs can be divided into three main classes. The first category consists of ritonavir-boosted NS3/4A inhibitors, such as boceprevir, telaprevir, simeprevir, asunaprevir, grazoprevir, and paritaprevir. NS5A inhibitors, such as velpatasvir, ombitasvir, ledipasvir, elbasvir, and daclatasvir (DCV), comprise the second group. The third category consists of nonnucleoside polymerase inhibitors and NS5B nucleotide inhibitors, such as sofosbuvir (Geddawy et al. 2017). When SOF was used with IFN or for a longer treatment period (24 vs. 12 weeks), the most frequent side effects were documented. When it was combined with RBV in phase III trials, the side effects also included RBV. Anemia, asthenia, diarrhea, pruritus, headache, nausea, sleeplessness, and fatigue were among the side effects. The side effects included decreased appetite, influenza-like symptoms, pyrexia, chills, neutropenia, and myalgia when combined with RBV and IFN. Diarrhea, nausea, fatigue, headaches, insomnia, and elevations in both bilirubin and lipase are the most frequent side effects of LDV (>5% of patients) as reported by phase III clinical trial (Cure et al. 2013). The medication simeprevir is well tolerated. The majority of side effects that have been reported are known to be caused by concurrent medication use. When combined with RBV/IFN, the following adverse effects have been documented: myalgia, rash, nausea, photosensitivity, pruritus, and dyspnea. Headache and nausea were reported as side effects in other studies without IFN. Rash, photosensitivity, and pruritus were the documented side effects for SIM/SOF in the absence of IFN and RBV (El-Khayat et al. 2017).

24.3.6 Toxicity of FDA-Approved Drugs for Influenza Virus

Causes of influenza Seasonal outbreaks of respiratory viral infections are linked to notable rates of morbidity and death. It has been demonstrated that influenza vaccinations lower the risk of infection and protect against some of the virus's aftereffects. Similarly, for the prevention and treatment of influenza, two classes of antivirals—adamantanes, which include amantadine and rimantadine, and neuraminidase inhibitors, which include laninamivir, oseltamivir, peramivir, and zanamivir—are currently approved; numerous other classes of antivirals and immune modulators are also being researched. The emergence of resistant mutants is one of the biggest threats to our arsenal of antivirals (Ison 2011). The most frequent adverse effects Oseltamivir of that may occur are headache, nausea and vomiting. The German Society of Paediatric Infectious Diseases (DGPI) guidelines should be followed when considering antiviral therapy for newborns and young infants (less than three months) in cases of severe clinical illness (Agency 2011). Due to inadequate oral bioavailability of zanamivir, it was combined with lactose and breathed as a dry powder. Nausea, vomiting, diarrhea, and headaches are undesirable side effects. The European Medicines Agency (EMA) has approved zanamivir for the treatment and prophylaxis of influenza infections in patients as young as five. The EMA published a summary in 2011 regarding the compassionate use of intravenous (i.v.) zanamivir in a particular targeted population (ADVICE) (European Centre for Disease Prevention and Control 2017). This medication should only be used to treat critically ill adults and children whose conditions are life-threatening.

24.4 Conclusion

Antifungal drugs are essential for treating fungal infections, but they have many limitations. One significant challenge is the development of antifungal resistance. Fungi can adapt and become resistant to commonly used antifungal agents, reducing the effectiveness of treatment. This resistance is a growing concern, especially in hospitals. The narrow spectrum of many antifungal drugs is another limitation. They are often designed to target specific fungal species, so the choice of antifungal treatment relies heavily on accurate diagnosis. Misdiagnosis can result in ineffective treatment, and the development of new drugs to combat emerging fungal pathogens is a resource-intensive process. Antifungal drugs can also be associated with significant toxicities, especially when used at high doses or over extended periods. This can limit their use, particularly in patients with underlying health conditions. Moreover, some antifungal drugs can be costly, posing a barrier to access for certain patients, especially in regions with limited healthcare resources. On the other hand, alternative therapies, such as plant-based antifungals and natural remedies, offer potential advantages. They are often considered safer and may have a lower risk of resistance. However, their efficacy can vary, and they may not be suitable for severe or systemic fungal infections. Combining conventional antifungal drugs with alternative therapies may provide a more comprehensive approach to fungal infection management, potentially reducing the dosage required and minimizing the risk of resistance. Nonetheless, rigorous research is needed to establish the effectiveness of alternative therapies, and they should be used with caution, particularly in cases of serious fungal infections. In summary, even though antiviral medications have shown to be extremely helpful in the diagnosis, management, and treatment of viral infections, there is still considerable worry about their potential for toxicity. Drug development is challenging because of the delicate balance that must be struck between minimizing side effects and effectively stopping viral replication. Future developments in antiviral medication design and ongoing research should result in more specialized and less harmful therapies. In order to ensure that the advantages of antiviral treatments exceed any potential risks, creative ways to improve the safety and efficacy of these treatments will require cooperation between researchers, physicians and the pharmaceutical industry. Antiviral medication regimens could become safer and more effective in the future with continued dedication to their improvement.

Acknowledgments SKD acknowledges financial and infrastructural supports from the University of Delhi (R&D and seed grants) and Dr. B. R. Ambedkar Center for Biomedical Research [ACBR] (start-up and research grants). SKD also acknowledges the Institution of Eminence grants (grant ID: IoE/2021-23/12/FRP) from the University of Delhi, India, and ICMR extramural grant (grant ID: 5/4/1-11/CVD/2022-NCD-I) for providing financial support for the current work. BioRender was used to make the figures for the current chapter.

References

Abdullah A, Saleh I, Hasan K (2020) Molecular evaluation of fungalysins and subtlilisin gene family for clinical species of dermatophytes. Biochem Cell Arch 20(1)

Adkins JC, Noble S (1998) Efavirenz Drugs 56:1055–1064

Agency EM (2011) Summary on compassionate use for IV zanamivir. In: European medicines Agency Amsterdam, The Netherlands

Ahmed M, Hussain F (2013) Chemical composition and biochemical activity of Aloe vera (Aloe barbadensis miller) leaves. Int J Chem Biochem Sci 3(5):29–33

Ahmedi S, Pant P, Raj N, Manzoor N (2022) Limonene inhibits virulence associated traits in Candida albicans: in-vitro and in-silico studies. Phytomed Plus 2(3):100285

Akter J, Amzad Hossain M, Sano A, Takara K, Zahorul Islam M, Hou D-X (2018) Antifungal activity of various species and strains of turmeric (curcuma spp.) against fusarium Solani Sensu Lato. Pharm Chem J 52:320–325

Anemia B (2013) Toxicity & dose modification of chemotherapeutic agents. Med Diagn Treatment 1648

Asrani SK, Devarbhavi H, Eaton J, Kamath PS (2019) Burden of liver diseases in the world. J Hepatol 70(1):151–171

Basha J, Goudgaon NM (2021) A comprehensive review on pyrimidine analogs-versatile scaffold with medicinal and biological potential. J Mol Struct 1246:131168

Ben-Ami R, Kontoyiannis DP (2021) Resistance to antifungal drugs. Infect Dis Clin 35(2):279–311

Bonamigo RR, Dornelles SIT (2023) Dermatology in public health environments: a comprehensive textbook. Springer

Bullard-Feibelman KM, Govero J, Zhu Z, Salazar V, Veselinovic M, Diamond MS, Geiss BJ (2017) The FDA-approved drug sofosbuvir inhibits Zika virus infection. Antivir Res 137:134–140

Cardoso AMR, Cavalcanti YW, de Almeida LDFD, de Lima Pérez ALA, Padilha WWN (2012) Antifungal activity of plant-based tinctures on Candida. RSBO Rev Sul-Brasi Odontol 9(1):25–30

Carolus H, Pierson S, Lagrou K, Van Dijck P (2020) Amphotericin B and other polyenes—discovery, clinical use, mode of action and drug resistance. J Fungi 6(4):321

Chandrasekaran B, Cherukupalli S, Karunanidhi S, Kajee A, Aleti RR, Sayyad N et al (2019) Design and synthesis of novel heterofused pyrimidine analogues as effective antimicrobial agents. J Mol Struct 1183:246–255

Chary M, Barbuto AF, Izadmehr S, Tarsillo M, Fleischer E, Burns MM (2023) COVID-19 therapeutics: use, mechanism of action, and toxicity (vaccines, monoclonal antibodies, and immunotherapeutics). J Med Toxicol 19(2):205–218

Chen SC-A, Slavin MA, Sorrell TC (2011) Echinocandin antifungal drugs in fungal infections: a comparison. Drugs 71:11–41

Chowdhury H, Banerjee T, Walia S (2008) In vitro screening of Curcuma longa L and its derivatives sa antifungal agents against Helminthosporrum oryzae and fusarium solani. Pestic Res J 20(1):6–9

Como JA, Dismukes WE (1994) Oral azole drugs as systemic antifungal therapy. N Engl J Med 330(4):263–272

Courouble VV, Dey SK, Yadav R, Timm J, Harrison JJE, Ruiz FX et al (2021) Revealing the structural plasticity of SARS-CoV-2 nsp7 and nsp8 using structural proteomics. J Am Soc Mass Spectrom 32(7):1618–1630

Cure S, Guerra I, Dusheiko G (2013) Long-term outcomes of sofosbuvir (SOF) for the treatment of chronic hepatitis C infected (CHC) patients. Value Health 16(7):A357

De Clercq E, Li G (2016) Approved antiviral drugs over the past 50 years. Clin Microbiol Rev 29(3):695–747

Dey JK, Dey SK (2020) SARS-CoV-2 pandemic, COVID-19 case fatality rates and deaths per million population in India. J Bioinform Comput Syst Biol 2(1):110

Dey SK, Saini M, Dhembla C, Bhatt S, Rajesh AS, Anand V et al (2022) Suramin, penciclovir, and anidulafungin exhibit potential in the treatment of COVID-19 via binding to nsp12 of SARS-CoV-2. J Biomol Struct Dyn 40(24):14067–14083

Dieffenbach CW, Fauci AS (2011) Thirty years of HIV and AIDS: future challenges and opportunities. Ann Intern Med 154(11):766–771

El-Khayat HR, Fouad YM, Maher M, El-Amin H, Muhammed H (2017) Efficacy and safety of sofosbuvir plus simeprevir therapy in Egyptian patients with chronic hepatitis C: a real-world experience. Gut 66(11):2008–2012

Emmelkamp JM, Rockstroh JK (2008) Maraviroc, risks and benefits: a review of the clinical literature. Expert Opin Drug Saf 7(5):559–569

Espinel-Ingroff A (2022) Commercial methods for antifungal susceptibility testing of yeasts: strengths and limitations as predictors of resistance. J Fungi 8(3):309

European Centre for Disease Prevention and Control (2017) Expert opinion on neuraminidase inhibitors for the prevention and treatment of influenza—review of recent systematic reviews and meta-analyses. ECDC, Stockholm. https://www.ecdc.europa.eu/en/publications-data/expert-opinion-neuraminidase-inhibitors-prevention-and-treatment-influenza-review

Evans AS (2013) Viral infections of humans: epidemiology and control. Springer Science & Business Media

Farasati Far B, Bokov D, Widjaja G, Setia Budi H, Kamal Abdelbasset W, Javanshir S et al (2023) Metronidazole, acyclovir and tetrahydrobiopterin may be promising to treat COVID-19 patients, through interaction with interleukin-12. J Biomol Struct Dyn 41(10):4253–4271

Gaudinski MR, Coates EE, Novik L, Widge A, Houser KV, Burch E et al (2019) Safety, tolerability, pharmacokinetics, and immunogenicity of the therapeutic monoclonal antibody mAb114 targeting Ebola virus glycoprotein (VRC 608): an open-label phase 1 study. Lancet 393(10174):889–898

Geddawy A, Ibrahim YF, Elbahie NM, Ibrahim MA (2017) Direct acting anti-hepatitis C virus drugs: clinical pharmacology and future direction. J Transl Intern Med 5(1):8–17

Grover ND (2010) Echinocandins: a ray of hope in antifungal drug therapy. Indian J Pharmacol 42(1):9

Hammer K, Carson C, Riley T (2004) Antifungal effects of Melaleuca alternifolia (tea tree) oil and its components on Candida albicans, Candida glabrata and Saccharomyces cerevisiae. J Antimicrob Chemother 53(6):1081–1085

Hammoudi Halat D, Younes S, Mourad N, Rahal M (2022) Allylamines, benzylamines, and fungal cell permeability: a review of mechanistic effects and usefulness against fungal pathogens. Membranes 12(12):1171

Hashemian SMR, Sheida A, Taghizadieh M, Memar MY, Hamblin MR, Baghi HB et al (2023) Paxlovid (Nirmatrelvir/ritonavir): a new approach to Covid-19 therapy? Biomed Pharmacother 162:114367

Houšť J, Spížek J, Havlíček V (2020) Antifungal drugs. Metabolites 10(3):106

Infusion-Associated, Hypersensitivity Reactions Including (2021) AHFS® first release™. Am J Health Syst Pharm 78(8):649

Ismail M, Srivastava V, Marimani M, Ahmad A (2022) Carvacrol modulates the expression and activity of antioxidant enzymes in Candida auris. Res Microbiol 173(3):103916

Ison MG (2011) Antivirals and resistance: influenza virus. Curr Opin Virol 1(6):563–573. https://doi.org/10.1016/j.coviro.2011.09.002

Johnson MD, MacDougall C, Ostrosky-Zeichner L, Perfect JR, Rex JH (2004) Combination antifungal therapy. Antimicrob Agents Chemother 48(3):693–715

Kaplon H, Reichert JM (2021) Antibodies to watch in 2021. Paper presented at the MAbs

Katarina P, Snežana P, Mira S, Saša S, Zorica L, Dragana J (2012) Antifungal activity of indigenous pseudomonas isolates against Alternaria tenuissima isolated from Echinacea purpurea. Paper presented at the proceedings of the seventh conference on medicinal and aromatic plants of southeast European countries (proceedings of the 7th CMAPSEEC), Subotica, Serbia, 27–31 May, 2012

Kaur M, Shivgotra R, Bhardwaj N, Saini S, Thakur S, Jain SK (2023) Nascent Nanoformulations as an insight into the limitations of the conventional systemic antifungal therapies. Curr Drug Targets 24(2):171–190

Kaushik N, Pujalte GG, Reese ST (2015) Superficial fungal infections. Prim Care 42(4):501–516

Kovacic P, Cooksy A (2012) Novel, unifying mechanism for amphotericin B and other polyene drugs: electron affinity, radicals, electron transfer, autoxidation, toxicity, and antifungal action. MedChemComm 3(3):274–280

Kuete V (2017) Allium sativum. In: Medicinal spices and vegetables from Africa. Elsevier, pp 363–377

Lee A (2021) Ansuvimab: first approval. Drugs 81:595–598

Levine MM (2019) Monoclonal antibody therapy for Ebola virus disease. N Engl J Med 381(24):2365–2366

Manohar V, Ingram C, Gray J, Talpur NA, Echard BW, Bagchi D, Preuss HG (2001) Antifungal activities of origanum oil against Candida albicans. Mol Cell Biochem 228:111–117

Manzoor N (2019) Candida pathogenicity and alternative therapeutic strategies. In: Pathogenicity and drug resistance of human pathogens: mechanisms and novel approaches, pp 135–146

Medda S, Hajra A, Dey U, Bose P, Mondal NK (2015) Biosynthesis of silver nanoparticles from Aloe vera leaf extract and antifungal activity against Rhizopus sp. and aspergillus sp. Appl Nanosci 5:875–880

Mount HOC, Revie NM, Todd RT, Anstett K, Collins C, Costanzo M et al (2018) Global analysis of genetic circuitry and adaptive mechanisms enabling resistance to the azole antifungal drugs. PLoS Genet 14(4):e1007319

Mulangu S, Dodd LE, Davey RT, Tshiani Mbaya O, Proschan M, Mukadi D et al (2019) A randomized, controlled trial of Ebola virus disease therapeutics. N Engl J Med 381(24):2293–2303. https://doi.org/10.1056/NEJMoa1910993

Nelson P, Reynolds G, Waldron E, Ward E, Giannopoulos K, Murray P (2000) Demystified…: monoclonal antibodies. Mol Pathol 53(3):111

Nhean S, Varela ME, Nguyen Y-N, Juarez A, Huynh T, Udeh D, Tseng AL (2023) COVID-19: a review of potential treatments (corticosteroids, remdesivir, tocilizumab, bamlanivimab/etesevimab, and casirivimab/imdevimab) and pharmacological considerations. J Pharm Pract 36(2):407–417

Nicola AM, Albuquerque P, Paes HC, Fernandes L, Costa FF, Kioshima ES et al (2019) Antifungal drugs: new insights in research & development. Pharmacol Ther 195:21–38

Ojha D, Basu R, Peterson KE (2023) Therapeutic targeting of organelles for inhibition of Zika virus replication in neurons. Antivir Res 209:105464. https://doi.org/10.1016/j.antiviral.2022.105464

Ospina Salazar DI, Hoyos Sánchez RA, Orozco Sanchez F, Arango Arteaga M, Gómez Londoño LF (2015) Antifungal activity of neem (Azadirachta indica: Meliaceae) extracts against dermatophytes. Acta Biol Colombiana 20(3):181–192

Paton NI, Musaazi J, Kityo C, Walimbwa S, Hoppe A, Balyegisawa A et al (2021) Dolutegravir or darunavir in combination with zidovudine or tenofovir to treat HIV. N Engl J Med 385(4):330–341

Perlin DS (2015) Mechanisms of echinocandin antifungal drug resistance. Ann N Y Acad Sci 1354(1):1–11

Petranyi G, Meingassner JG, Mieth H (1987) Antifungal activity of the allylamine derivative terbinafine in vitro. Antimicrob Agents Chemother 31(9):1365–1368

Rice CM, Lenches EM, Eddy SR, Shin SJ, Sheets RL, Strauss JH (1985) Nucleotide sequence of yellow fever virus: implications for flavivirus gene expression and evolution. Science 229(4715):726–733

Rizwan T, Kothidar A, Meghwani H, Sharma V, Shobhawat R, Saini R et al (2022) Comparative analysis of SARS-CoV-2 envelope viroporin mutations from COVID-19 deceased and surviving patients revealed implications on its ion-channel activities and correlation with patient mortality. J Biomol Struct Dyn 40(20):10454–10469

Robbins N, Wright GD, Cowen LE (2016) Antifungal drugs: the current armamentarium and development of new agents. Microbiol Spectrum 4(5):4.5.19

Shahjahan, Dey JK, Dey SK (2023) Competitive binding of ribavirin, velpatasvir, andremdesivir to the active site of DNA polymerase canmake them repurposable drugs to combat Monkeypox. Int J Sci Eng Dev Res 8(10):545–552. Retrieved from https://www.ijsdr.org/papers/IJSDR2310119.pdf

Shreaz S, Wani WA, Behbehani JM, Raja V, Irshad M, Karched M et al (2016) Cinnamaldehyde and its derivatives, a novel class of antifungal agents. Fitoterapia 112:116–131

Singh AK, Singh A, Singh R, Misra A (2021) Molnupiravir in COVID-19: A systematic review of literature. Diabetes Metab Syndr Clin Res Rev 15(6):102329

Soderstrom K (2014) Viral replication

Soylu EM, Kurt Ş, Soylu S (2010) In vitro and in vivo antifungal activities of the essential oils of various plants against tomato grey mould disease agent Botrytis cinerea. Int J Food Microbiol 143(3):183–189

Stütz A (1987) Allylamine derivatives—a new class of active substances in antifungal chemotherapy. Angew Chem Int Ed Engl 26(4):320–328

Thomas CI, Purnell EW, Rosenthal MS (1965) Treatment of herpetic keratitis with IDU and corticosteroids. Am J Ophthalmol 60(2):204–217

Thuppal SV, Wanke CA, Noubary F, Cohen JT, Mwamburi M, Ooriapdickal AC et al (2015) Toxicity and clinical outcomes in patients with HIV on zidovudine and tenofovir based regimens: a retrospective cohort study. Trans R Soc Trop Med Hyg 109(6):379–385

Tompa DR, Immanuel A, Srikanth S, Kadhirvel S (2021) Trends and strategies to combat viral infections: a review on FDA approved antiviral drugs. Int J Biol Macromol 172:524–541

Trivedi HD, Joshi VB, Patel BY (2022) Pyrazole bearing pyrimidine analogues as the privileged scaffolds in antimicrobial drug discovery: a review. Anal Chem Lett 12(2):147–173

Tsutsumi-Arai C, Takakusaki K, Arai Y, Terada-Ito C, Takebe Y, Imamura T et al (2019) Grapefruit seed extract effectively inhibits the Candida albicans biofilms development on polymethyl methacrylate denture-base resin. PLoS One 14(5):e0217496

Tverdek FP, Kofteridis D, Kontoyiannis DP (2016) Antifungal agents and liver toxicity: a complex interaction. Expert Rev Anti-Infect Ther 14(8):765–776

Van Daele R, Spriet I, Wauters J, Maertens J, Mercier T, Van Hecke S, Brüggemann R (2019) Antifungal drugs: what brings the future? Med Mycol 57(Suppl. 3):S328–S343

Vanreppelen G, Wuyts J, Van Dijck P, Vandecruys P (2023) Sources of antifungal drugs. J Fungi 9(2):171

Wan Q, Li N, Du L, Zhao R, Yi M, Xu Q, Zhou Y (2019) Allium vegetable consumption and health: an umbrella review of meta-analyses of multiple health outcomes. Food Sci Nutr 7(8):2451–2470

Whitley RJ, Hayden FG, Reisinger KS, Young N, Dutkowski R, Ipe D et al (2001) Oral oseltamivir treatment of influenza in children. Pediatr Infect Dis J 20(2):127–133

Wilson JA, Hevey M, Bakken R, Guest S, Bray M, Schmaljohn AL, Hart MK (2000) Epitopes involved in antibody-mediated protection from Ebola virus. Science 287(5458):1664–1666

Xing Y, Li X, Xu Q, Yun J, Lu Y (2010) Antifungal activities of cinnamon oil against Rhizopus nigricans, aspergillus flavus and Penicillium expansum in vitro and in vivo fruit test. Int J Food Sci Technol 45(9):1837–1842

Yadav R, Courouble VV, Dey SK, Harrison JJE, Timm J, Hopkins JB et al (2022) Biochemical and structural insights into SARS-CoV-2 polyprotein processing by Mpro. Sci Adv 8(49):eadd2191

Natural Compound Toxicity: An Egregiously Overlooked Topic

25

Priyanka Bhardwaj, Ayesha Aiman, Faiza Iram, Israil Saifi, Seemi Farhat Basir, Imtaiyaz Hassan, Asimul Islam, and Nikhat Manzoor

25.1 Introduction

Various organisms, including bacteria, fungi, and plants, synthesize diverse organic molecules, some of which are crucial for their existence (Williams et al. 1989). While natural products play a crucial part in the process of drug development, it is important to note that certain natural compounds exhibit significant levels of toxicity. In the present context, the term "toxicity" pertains to the detrimental impact exerted on an entire organism, including animals, bacteria, and plants, as well as the influence on specific components of such organism, such as cells or organs. However, it is essential to note that the harmful impact of some chemicals can also have advantageous applications in treating certain diseases. One such example is the use of cytotoxic substances in cancer therapy. Therefore, secondary metabolites have the potential to exhibit multiple functionalities (Schoental 1965). Organisms produce a wide variety of different chemicals. Primary and secondary metabolites are the two groups these compounds fall under. Primary metabolites are compounds typically engaged in an organism's primary growth, development, and reproduction.

Ayesha Aiman and Faiza Iram contributed equally to this work.

P. Bhardwaj · A. Aiman · S. F. Basir · N. Manzoor (✉)
Department of Biosciences, Jamia Millia Islamia, New Delhi, India
e-mail: priyankabhardwaj1409@gmail.com; ayeshaaiman9654@gmail.com; sbasir@jmi.ac.in; nmanzoor@jmi.ac.in

F. Iram · I. Hassan · A. Islam
Centre for Interdisciplinary Research in Basic Sciences, Jamia Millia Islamia, New Delhi, India
e-mail: faiza2206219@st.jmi.ac.in; mihassan@jmi.ac.in; aislam@jmi.ac.in

I. Saifi
School of Engineering and Technology, Sharda University, Greater Noida, Uttar Pradesh, India
e-mail: 2022300634.israil@dr.sharda.ac.in

N. Manzoor (ed.), *Advances in Antifungal Drug Development*,
https://doi.org/10.1007/978-981-97-5165-5_25

In contrast, secondary metabolites are only indirectly involved in the metabolism. Amino acids, carbohydrates, proteins, DNA, and lipid molecules are all considered to be fundamental substances. The term secondary metabolite was coined by the A. Kossel in 1891. Although primary metabolites are in every living cell that is capable of division, secondary metabolites only occur accidentally, and they do not play an important role in the life of the organism. Primary metabolites are found in all dividing living cells. However, secondary metabolites are only present incidentally and are not crucial for the organism's survival. Several secondary metabolites are derived from primary metabolism, yet they do not constitute the fundamental molecular framework of the organism. The absence of a secondary metabolite does not immediately result in the termination of an organism's life, which contrasts with primary metabolites. Plants being the primary contributors of secondary metabolites account for around 80% of these compounds. Other significant producers include bacteria, fungi, as well as various marine organisms such as sponges, tunicates, corals, and snails (Berdy 2005). Natural products have undergone evolutionary processes spanning millions of years, leading to the development of a distinctive array of therapeutic natural compounds. The occurrence and production of secondary metabolites have been documented in the species that have undergone selection pressure within a certain evolutionary group (Tiwari and Rana 2015).

Natural products encompass a wide range of chemicals that are synthesized by living organisms. The definition of natural products provides great precision. However, a joint agreement is only sometimes incorporated. Some reports conclude all small molecules are generated during metabolic reactions, while others restrict the classification of "natural products" to secondary or nonessential metabolic products. However, in the field of life sciences, it is essential to note that the distinction between primary and secondary metabolites is sometimes ambiguous and relies on the compound's potential. As a result, natural products exhibit a wide range of biological functions and possess characteristics similar to those of pharmaceutical drugs. Hence, predating the advent of recent chemical pharmacology, natural products have been employed for many centuries as constituents of traditional medicinal practice, specifically active agents that are involved in herbal remedies. In contemporary times, numerous conventional recovery modalities, such as Indian Ayurveda, Traditional Chinese Medicines, and African Herbal Medicines, continue to serve as the predominant therapeutic approach for a significant portion of the global population. This preference can be attributed to financial constraints, personal convictions, or challenges associated with obtaining medicinal remedies. In contemporary pharmacology, natural products have emerged as a crucial asset for discovering and developing novel pharmaceuticals and scaffolds (Mirza et al. 2015; Newman and Cragg 2016; Khalifa et al. 2019). The study of natural products and their analogues in the context of food is a field of intense research (Ahmed et al. 2010; Yue et al. 2014; Naveja et al. 2018; Dagan-Wiener et al. 2019). Weekly research studies have been published in scientific journals with peer reviews on the beneficial impacts of natural products on the recovery process of many human and animal diseases. Most antibiotics as well as antifungals are developed using natural products extracted from microorganisms.

Pharmaceutical substances employed in treating several types of cancers, cardiovascular ailments, diabetes, and other medical conditions frequently consist of either novel natural products or their derivatives. For example, over the period from 1981 to 2014, more than 50% of newly created pharmaceutical drugs were derived, from novel natural products (Newman and Cragg 2016). Moreover, thc significance of natural chemicals has been investigated in the cosmetic sector (Dunkel et al. 2009; Mahesh et al. 2019) as well as in agriculture, namely in the realm of natural pesticide formulation (Sparks et al. 2019). Although natural products possess therapeutic characteristics, it is crucial to acknowledge that they can also have detrimental consequences. Unfortunately, researchers typically overlook these negative aspects due to the overshadowing good influence.

25.2 Chemistry of Natural Compounds

There is a current understanding of more than 2,140,000 secondary metabolites, which are often categorized based on their extensive variations in structure, function, and biosynthetic pathways. The classification of secondary metabolites encompasses five primary categories, namely terpenoids and steroids, fatty acid-derived compounds and polyketides, alkaloids, nonribosomal polypeptides, and enzyme cofactors (Fig. 25.1) (McMurry 2014). The major focus of the discussion here is on the fundamental classifications of natural commodities, given the existence of more than 200,000 structures some of which are given in Fig. 25.2.

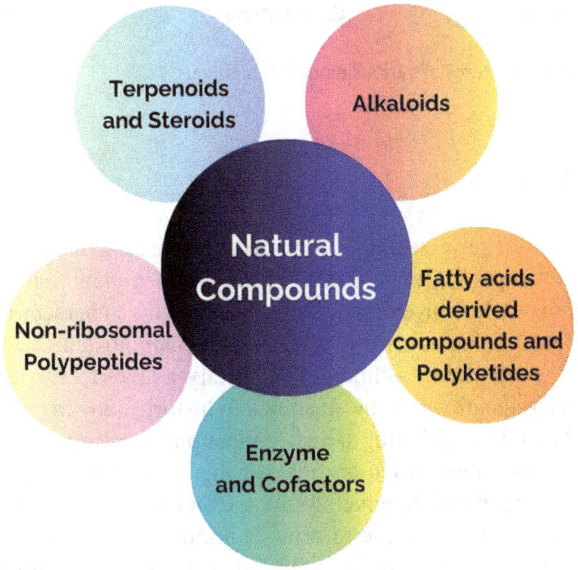

Fig. 25.1 Classification of the plant-based natural compounds

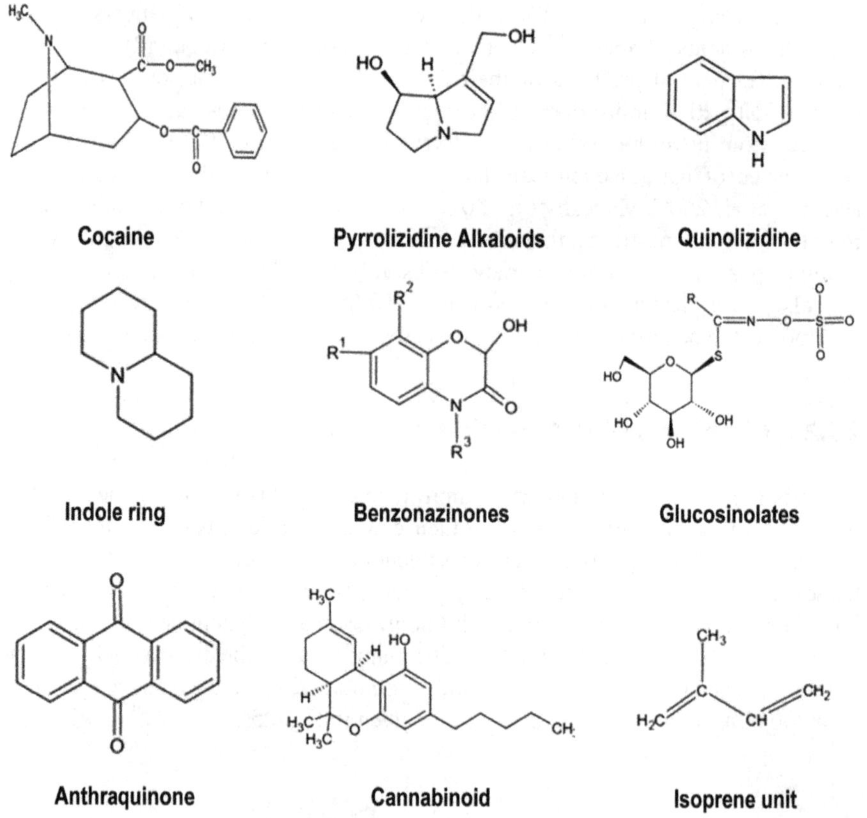

Fig. 25.2 Chemical structures of various natural compounds

25.2.1 Alkaloids

The etymology of the term "alkaloid" can be traced back to its Arabic origin, precisely the word "al-qali," which pertains to the ashes of plant material containing potassium carbonate. In conventional terms, alkaloids are characterized as heterocyclic nitrogen molecules that are synthesized from amino acids. Other chemicals are categorized as alkaloids, although not strictly adhere to the aforementioned criterion. This classification may be attributed to historical factors or their observed bioactivities. Alkaloids, including a vast collection of natural compounds, presently exceed 12,000 known structures, making them one of the largest groupings in this category (Springob and Kutchan 2009). Given the vast quantity and extensive range of alkaloids, it is unfeasible to provide a thorough overview encompassing all distinct varieties. Consequently, only selected categories are presented for introduction.

25.2.1.1 Purine Alkaloids

Purine alkaloids are a class of nitrogenous compounds that originate from the metabolic processes of nucleosides (Ashihara and Crozier 2001). They are synthesized in a diverse range of plant taxa that are taxonomically unrelated. Caffeine is the purine alkaloid that is found in the highest abundance, followed by theobromine. Additionally, there are other minor purines such as theophylline and paraxanthine. Coffee seeds, commonly referred to as "beans," typically contain around 1% caffeine, whereas young tea leaves possess a higher caffeine content ranging from 2% to 3% (Ashihara and Suzuki 2004). It potentially functions as a form of defense against herbivores and as an autotoxin by impeding the germination process of coffee seedlings (Hollingsworth et al. 2002). Caffeine, a central stimulant, is commonly ingested through various drinks such as coffee, tea, and sodas, as well as through cold medication and painkillers. The principal mode of action by which caffeine and other purine alkaloids elicit their effects is through the inhibition of adenosine receptors, resulting in the subsequent release of neurotransmitters (Benowitz 1990). At elevated doses, the activity of phosphodiesterase, the enzyme responsible for the hydrolysis of the second messenger cyclic adenosine monophosphate (cAMP), is inhibited. Nevertheless, the ingestion of caffeine-containing drinks typically does not result in attaining such blood concentrations. In recent times, there has been a heightened focus on caffeine due to its association with a decreased likelihood of developing Parkinson's disease among individuals who consume coffee (Ascherio et al. 2004). On the other hand, hypotension may arise as a result of PDE inhibition in instances of overdose or poisoning, when significantly elevated and hazardous levels are attained (Holstege et al. 2003; Kapur and Smith 2009).

25.2.1.2 Tropane Alkaloids

Tropane alkaloids are derived from the amino acid ornithine or arginine. All of them have the bicyclic tropane structure, which is composed of a seven-membered ring with a nitrogen atom bridging between carbon atoms 1 and 5. The nitrogen atom is methylated. There have also been descriptions of nortropanes that do not possess N-methylation, as well as seco-tropanes having a divided N-bridge (Griffin and Lin 2000). The highly recognized compound, cocaine, falls under this particular category. Cocaine can be described as the ester of benzoic acid and the tropane base methyl ecgonine. Cocaine is classified as a potent central stimulant with a high potential for addiction. Its mechanism of action involves the inhibition of neurotransmitter re-uptake, specifically targeting dopamine and norepinephrine at synapses. Additionally, cocaine also acts as an inhibitor of monoamine oxidase, an enzyme responsible for the degradation of dopamine, epinephrine, and norepinephrine. This phenomenon results in feelings of exhilaration, heightened levels of activity, decreased appetite, and reduced weariness. Peripheral effects encompass many physiological responses such as elevated heart rate and blood pressure, pupil dilation, increased blood glucose levels, and heightened body temperature (White and Lambe 2003). The anticholinergic poisoning occurred due to the presence of tropane alkaloid pollutants in many food items, including hamburgers, honey, and millet used for porridge, homemade wine, and Paraguay tea made from *Ilex paraguariensis*. In 1995, a widespread outbreak originated in New York City as a result of the use of heroin contaminated with scopolamine, leading to significant anticholinergic toxicity and impacting a minimum of 300 unsuspecting persons (Hamilton et al. 2000).

25.2.1.3 Pyrrolizidine Alkaloids

The primary structure of pyrrolizidine alkaloids consists of a hydroxymethylpyrrolizidine (referred to as the necine base), which is predominantly esterified with branching aliphatic mono- or dicarboxylic acids (known as necic acids). The necine base is synthesized through the biosynthesis of spermidine and putrescine, both of which are derived from arginine (Hartmann et al. 1988). The investigation of the genesis of necic acids has been limited to pyrrolizidine alkaloids of the senecionine and lycopsamine types. These necic acids are produced from the metabolism of amino acids (Hartmann and Ober 2000).

Numerous pyrrolizidine alkaloids have hepatotoxic, mutagenic, and carcinogenic properties. The administration of some substances has the potential to induce veno-occlusive disease of the liver, which can progress to the development of cirrhosis and ultimately result in liver failure. The primary causes of intoxications involving pyrrolizidines include the presence of pyrrolizidine-containing plants in cereals due to contamination, as well as the consumption of herbal medications containing these alkaloids. The genotoxicity of the compound may be attributed to certain structural characteristics, namely the presence of a double bond between carbon atoms 1 and 2 in the necine base, the existence of hydroxy groups at carbon atoms 7 and 9, and the esterification of at least one of these hydroxy groups with a branched carbon chain (Frei et al. 1992) .

25.2.1.4 Quinolizidine Alkaloids

Quinolizidine alkaloids are synthesized through the biosynthetic pathway involving lysine and its conversion to cadaverine. Except the bicyclic molecule lupinine, majority of chemicals within this category have a tricyclic or tetracyclic structure. Quinolizidine alkaloids are found in high quantities within the Fabaceae family, as well as in several other taxonomic groups such as Berberidaceae, Chenopodiacae, Ranunculaceae, Rubiaceae, and Solanaceae, despite their lack of direct relation (Wink 2002). The hazardous nature and some pharmacological characteristics of the substances can be elucidated by their ability to block sodium (Na^+) and potassium (K^+) channels, as well as their interaction with nicotinic and muscarinic acetylcholine receptors. The compound sparteine, derived from the plant species *Cytisus scoparius*, is employed as an antiarrhythmic agent. Nevertheless, the use of this substance for medical purposes is diminishing and becoming more limited because around 10% of patients exhibit an inability to metabolize this alkaloid, resulting in adverse effects of intoxication (Wink 2004).

25.2.1.5 Alkaloids Derived from Tyrosine

Tyrosine, an amino acid, serves as a precursor for a multitude of alkaloids. The benzylisoquinoline alkaloids constitute the most extensive group. Furthermore, various additional kinds of alkaloids derive from tyrosine, such as the Ipecac alkaloids and the Amaryllidaceae alkaloids. Benzylisoquinoline alkaloids are synthesized by the condensation of two molecules of tyrosine. The primary intermediary involved in the biosynthesis of benzylisoquinolines is (S)-reticuline, which can undergo several rearrangements and alterations to generate distinct structural categories. The existing assemblage of alkaloids has over 2500 identified structures, exhibiting a wide range of diversity.

Benzylisoquinolines alkaloids are mostly found in basal angiosperms, such as plants belonging to the Berberidaceae, Fumariaceae, Papaveraceae, Menispermaceae, and Ranunculaceae families, etc. Certain benzylisoquinoline alkaloids have potent pharmacological properties and have consequently been utilized in the field of medicine. Notable examples include the analgesic morphine, the antitussive and analgesic codeine, the muscle relaxant tubocurarine, and the antibacterial and anti-inflammatory agent sanguinarine. Due to their inherently heterogeneous architectures, these compounds exhibit distinct modes of action, as one would anticipate. Morphine and codeine have agonistic properties at the m-, d-, and k-opioid receptors, which are typically acted upon by endogenous ligands such as endorphins, enkephalins, and dynorphins (Schiff 2002). Coclaurine is a type of Benzylisoquinolines which exhibits convulsive toxicity, with a minimum lethal dose of 70 mg/kg for rabbits (Funayama and Cordell 2014).

25.2.1.6 Monoterpene Indole Alkaloids

This class of alkaloids is biosynthesized from tryptophan and secologanin via the central intermediate 3 α (S)-strictosidine. Over 2000 structurally diverse monoterpene indole alkaloids are known, and among them are several pharmacologically valuable compounds (O'Connor and Maresh 2006). *Catharanthus roseus*, also referred to as the Madagascar periwinkle, belongs to the Apocynaceae family and had previously been identified as *Vinca rosea* (Jacobs et al. 2004). The plant's potential as an antidiabetic agent in Jamaica prompted its evaluation for hypoglycemic action by Eli-Lilly, USA, and the Cancer Research Centre, Canada, during the late 1950s. While plant extracts were shown to be useless in treating diabetes, researchers from both institutes separately identified the anticancer properties of several bisindole alkaloids, specifically vinblastine and vincristine (Noble 1990). These chemicals can bind to tubulin and impede its process of polymerization, hence obstructing the assembly of the mitotic spindle. Consequently, this leads to the stoppage of the cell cycle at metaphase in dividing cells which may lead to various ailments in high concentrations.

25.2.2 Benzoxazinones

Benzoxazinones are generated from indole-3-glycerol phosphate, which is also the immediate precursor of tryptophan. In scholarly literature, it is common to utilize acronyms that originate from the substitution pattern of the benzoxazinone ring to differentiate and identify specific members belonging to this class. Benzoxazinones are mostly found in the monocot Poaceae family, but they are also present in certain families of eudicot plants, such as Ranunculaceae, Acanthaceae, and Plantaginaceae (Sicker et al. 2000). Benzoxazinones function as pre-existing defensive compounds, exhibiting antibacterial, antifungal, and antialgal characteristics (Bravo and Lazo 1993, 1996). Benzoxazinones possess toxicity, which enables them to serve as allelochemicals (Sicker et al. 2000).

25.2.3 Cyanogenic Glycosides

Cyanogenic glucosides consist of β-glucosides of α-hydroxynitriles, also known as cyanohydrins. These compounds are generated from five proteinogenic amino acids, namely phenylalanine, tyrosine, valine, isoleucine, and leucine, as well as the nonproteinogenic amino acid cyclopentenyl-glycine. There are two forms, namely the (R)-form and the (S)-form, of numerous cyanogenic glucosides (Hegnauer and Hegnauer 1986; Seigler 1991). Within the gastrointestinal tract of herbivorous animals, the ß-glucosidic bond has the potential to undergo hydrolysis through the activity of bacteria present in the intestines. The rationale behind the toxicity of hydrogen cyanide can be elucidated by its strong attraction to metal ions. The cyanide ions form a combination with iron (III) within the active site of cytochrome oxidase, resulting in the inhibition of the respiratory chain. Mammals can ingest limited quantities of cyanogenic glucosides and afterward eliminate their toxicity. This detoxification process mostly occurs through the action of the hepatic enzyme rhodanese, which facilitates the conversion of cyanide into thiocyanate. Prolonged use of nonlethal quantities, meanwhile, may lead to limb paralysis (known as Konzo) or neurological ailments owing to cyanide poisoning or iodine insufficiency resulting from the buildup of the iodine antagonist thiocyanate (Selmar 1999).

25.2.4 Glucosinolates

Glucosinolates are β-thioglucosides of (Z)-N-hydroximinosulfate esters. These compounds are produced from the aliphatic amino acid known as alanine. There are a total of 120 distinct structures of glucosinolates (Fahey et al. 2001). Glucosinolates disrupt liver function by affecting the expression of phase I and II enzymes (Williamson et al. 1996). The presence of glucosinolates can have a suppressive impact on the thyroid, leading to subtle symptoms such as reduced reproductive ability and growth or, in more severe instances, the development of a noticeable goiter (Taljaard 1993). Glucosinolate metabolites have the potential to cause genetic damage and may even lead to the development of cancer (Hanschen et al. 2014; Capuano et al. 2017; Abbaoui et al. 2018).

25.2.5 Polyketides

Polyketides are synthesized by making use of two-carbon units that are produced from activated acetate in the form of acetyl-CoA and malonyl-CoA. In contrast to fatty acids, which also derive from similar precursors, polyketides exhibit the retention of all or a majority of their oxygen functionalities. During the process of polyketide biosynthesis, the formation of poly-b-keto intermediates occurs, which are very reactive. These intermediates frequently undergo cyclization, resulting in the formation of either six-membered aromatic rings or 2-pyrone rings. Numerous polyketides exhibit phenolic characteristics similar to phenylpropanoids. However, these two groups may be differentiated based on the substitution pattern observed in the aromatic ring (Springob and Kutchan 2009). Patulin is a type of polyketide metabolite, similar to various significant mycotoxins such as aflatoxins, fumonisins, and ochratoxins (Birch et al. 1955).

25.2.6 Anthraquinones

Anthraquinones are derived from eight acetate units and consist of three six-membered rings. Anthrones refer to anthraquinones that have undergone reduction, together with their corresponding metabolic antecedents. The three categories of anthranoids generated from acetate in plants have hydroxy groups at positions C-1 and C-8, with the possibility of a third hydroxy group at position C-3. The compounds are conjugated with sugar, predominantly glucose, in the form of O- and C-glycosides. The chromophoric nature of anthranoids is attributed to the conjugation of a carbonyl group with an aryl moiety, resulting in yellow, orange, or red coloration (Srinivas et al. 2007). The consumption of anthraquinones and dianthrones undergo conversion into anthrones, which are biologically active metabolites, through the metabolic activity of bacteria residing in the large intestine. The regular use of anthranoid laxatives might result in the condition of dehydration, mineral deficiency, and reversible discoloration of the intestinal mucosa (*Pseudomelanosis coli*) (Srinivas et al. 2007).

The prenylated acylphloroglucinols, such as bitter acids, were found in the hop plant (*Humulus lupulus*) and hyperforin derived from St. John's wort (*Hypericum perforatum*), in addition to cannabinoids obtained from Indian hemp (*Cannabis sativa*). These chemicals exhibit lipophilic properties as a result of their terpenoid moiety and are frequently synthesized or accumulated within specialized glands or glandular trichomes (Springob and Kutchan 2009). Cannabis has been utilized as a psychoactive substance since ancient times. The current global trend is to outlaw the use of *C. sativa* because of its psychoactive properties. Cannabis preparations are commonly consumed by inhalation via smoking, while alternative methods of ingestion such as oral consumption in the form of cakes or cookies are also used. The primary psychoactive ingredient found in cannabis, known as THC, has an affinity for cannabinoid receptors present in several regions of the central nervous system, including the brain and spinal cord, as well as immune cells. The use of cannabis products induces a state of exhilaration and relaxation; nevertheless, prolonged and repetitive usage may result in the development of addictive tendencies (Dewick 2002).

25.2.7 Terpenoids and Steroids

Terpenoids, referred to as isoprenoids, are the most extensive category of naturally occurring compounds found in plants, encompassing over 40,000 distinct structures. Terpenes are organic compounds originating from isoprene units comprising five carbon atoms. These terpenes can be categorized based on the number of isoprene molecules they contain. The classification includes hemiterpenes (C5), monoterpenes (C10), sesquiterpenes (C15), diterpenes (C20), triterpenes (C30), tetraterpenes (C40), and polyterpenes (Dewick 2002). Terpenoids in plants are derived from two distinct biosynthetic pathways: the cytosolic mevalonic acid (MVA) pathway and the plastid-located deoxyxylulose phosphate (DXP) pathway, which is also known as the methylerythritol phosphate (MEP) system. Both biosynthetic pathways produce the active isoprene unit dimethylallyl diphosphate (DMAPP) and isopentenyl diphosphate (IPP). These units are capable of being connected by head-to-tail or tail-to-tail linkage and undergo cyclization and other modifications such as oxidations or rearrangements (Laule et al. 2003; Schuhr et al. 2003). Humulene is a monocyclic sesquiterpene. Humulene is a poisonous chemical that exhibits toxicity by causing an excessive synthesis of reactive oxygen species (ROS), reducing levels of glutathione (GSH), and causing oxidative damage to DNA, proteins, and lipids, ultimately leading to cell death (Agus 2021).

25.3 The Negative Facet of Natural Compounds

While the majority of natural products have beneficial effects, they can sometimes result in toxicity and adverse consequences for human health. Short-term health consequences include acute poisoning, including a spectrum of allergic responses, severe gastrointestinal discomfort, diarrhea, etc. The potential long-term health ramifications include impacts on the immunological, reproductive, and neurological systems, as well as the development of cancer. Thereby, toxicity has its influence on entire organisms, encompassing both overarching structures and specific cellular or organ components that may be deleteriously impacted. The following section focuses on the disorders associated with the toxicity of natural compounds.

25.3.1 Neurological Disorders

It is important to emphasize that many natural substances can exhibit harmful effects in humans, necessitating further research to determine the appropriate dosage range for safe consumption. Some cannabinoids and retinoids have been found to exhibit detrimental toxic effects (Kelly and Nappe 2018). Delta-9 tetrahydrocannabinol is one of the numerous cannabinoids that have been found in the species of *Cannabis Sativa*. It is also known as D9 tetrahydrocannabinol (Δ9 THC) and is the most predominant naturally occurring psychoactive molecule with antioxidative properties (Gonçalves et al. 2020). Studies on animals have shown that D9-THC causes dose-dependent toxicity as well as structural alterations in parts of the brain that have a high concentration of cannabinoid receptor type 1 regions (CB1Rs). These parts of the brain include the hippocampus, the amygdala, the cerebellum, the prefrontal cortex, and the striatum. Battistella et al. have reported persistent alterations in the hippocampus or para-hippocampal complex as well as in the amygdala regions of the brain (Battistella et al. 2014). In a similar vein, substances like epibatidine, despite being agonists of the neuronal nicotinic acetylcholine receptors, have been reported to be highly toxic and thereby, may only be used in the synthesis of analogues (Salehi et al. 2018). Stimulation of exocrine secretions (such as sialorrhea, rhinorrhea and lacrimation), hypertension, seizures, and muscle paralysis are the primary clinical indications of epibatidine toxicity (Salehi et al. 2018). Similarly, Epibatidine has been found to induce hypertension, pulmonary paralysis, and convulsions. Death can occur at doses that are only marginally higher than those required for antinociception (Traynor 1998; Wiener and Nelson 2024).

Alkaloids such as atropine, hyoscyamine and scopolamine exert their pharmacological effects by inhibiting the activity of Muscarinic acetylcholine receptors. A blockade of these receptors may lead to manifestations such as rapid heart rate, pupil dilation, impaired bowel movement, dry skin with a burning sensation, and dry mouth owing to a reduction in the amount of perspiration and saliva produced in the body (Beyer et al. 2009). In addition to its systemic effects, atropine has the potential to influence CNS (Lakstygal et al. 2018), leading to symptoms such as anxiousness, mental confusion, and delusions (Beyer et al. 2009). Cytisine has a strong affinity for nicotinic acetylcholine receptors (nAChRs), similar to nicotine. It functions as a central nervous system antagonist, leading to the manifestation of symptoms such as delirium and convulsions (Gotti and Clementi 2021). Respiratory paralysis or circulatory system failure may result in fatality. Ibogaine, a commonly used pharmacological agent, is associated with several adverse effects, including ataxia, tremors, muscular spasms, and seizures. Additionally, it has the potential to induce alterations in both blood pressure and heart rate (Koenig and Hilber 2015). Domoic acid, which has a structural resemblance to kainic acid, exhibits the ability to activate ionotropic glutamate receptors (iGluRs) and selectively affects dendritic regions within the brain. Consequently, this molecular interaction gives rise to epileptic seizures and cognitive impairments. Additionally, it induces retinal damage, lesions in the spinal cord, and clinical symptoms in the cardiovascular system as a result of the activation of N-methyl-D-aspartate receptors (NDMARs) in cardiomyocytes (Pulido 2008). Salvinorin A, a potent hallucinogenic substance, functions as a selective and highly effective agonist at kappa-opioid receptors. Salvinorin-A elicited a state of dissociation from the external environment, evoked intricate visual experiences and aural sensations, and altered the perception of internal bodily sensations (Maqueda et al. 2015; Silva et al. 2019).

25.3.2 Lathyrism

The excessive ingestion of *Lathyrus sativus* has been shown to result in the development of neurolathyrism. The ingestion of the excitatory amino acid β-N-oxalyl-L-α,β-diaminopropionic acid (BOOA) found in grass peas has been seen to result in the development of a gradually worsening condition characterized by spastic paraparesis, upper motor neuron manifestations, and instability in gait (Enneking 2011). BOOA can elicit neurolathyrism in experimental animals and functions as an antagonist at the alpha-amino-3-hydroxy-5-methyl-4-isoxazolepropionic acid (AMPA) subtype of glutamate receptors. Konzo is a neurological disorder caused by the ingestion of inadequately processed cassava (*Manihot esculenta*), that contains poisonous cyanogenic glycosides (Kashala-Abotnes et al. 2019). This condition is classified as an upper motor neuron illness. Symptoms are associated with an absence of crucial sulfur-containing amino acids, which play a protective role against oxidative damage. The repetitive ingestion of cassava roots or grass pea seeds, which are high in protein, leads to the development of neurotoxicity in females with irregular menstrual cycles. Instances of konzo have been sporadically documented in many African nations, but the occurrence of neurolathyrism persists in India as well (Artal 2015; Khandare et al. 2018, 2020; Kumar et al. 2022).

25.3.3 Embryopathic and Cytotoxic Effects

The fodder that a pregnant animal consumes is a common factor in the development of embryopathies in cattle; research on this topic is vital since it may have implications for humans. When "wild corn," or cow cabbage or *Veratrum californicum*, was given to pregnant sheep between 10 and 15 days of their pregnancies, it has been shown to cause anomalies in the development of a fetus with several unusual and potentially harmful defects (Hill et al. 2000; Welch et al. 2009). The neonate had increased weight and presented with many abnormalities, such as hydrocephalus, proboscis, cyclopean eye, and respiratory obstruction, which were attributed to an extended gestational period. Therefore, the neonate did not sustain upon delivery. Miroestrol, an estrogenic compound derived from the tuberous roots of the leguminous plant *Pueraria mirifica*, is known to possess hazardous properties and is mostly found in the Northern regions of Thailand. The medicinal use of this plant has a rich historical background, particularly in the context of rejuvenating elderly folks. Nevertheless, caution should be exercised while considering its administration in younger populations. This drug has estrogenic activity comparable to that of endogenous estrogens, even when administered through the oral route (Kakehashi et al. 2016).

25.3.4 Neoplastic Diseases

Kaposi sarcoma, as first reported by Kaposi in 1872, is a medical disorder characterized by the presence of numerous pigmented sarcomas (Cesarman et al. 2019). This phenomenon is also seen among the Italian population, although it is not prevalent among African Americans in the United States. The etiology of this phenomenon remains unidentified. Similarly, the condition of Onyalai is characterized by the presence of blood-filled vesicles on the hard palate and cheeks, as well as purpura and hemorrhaging from different bodily orifices. Cycasin is a glucoside with neurotoxic, teratogenic as well as carcinogenic and neurotoxic properties that is present in cycads, namely in species such as *Cycas revoluta* and *Zamia pumila* (Kobayashi and Matsumoto 1965; Sieber et al. 1980; Kisby et al. 2013). The clinical manifestations of poisoning include symptoms such as emesis, diarrheal episodes, debilitation, convulsions, and hepatotoxic effects. The earliest "natural" etiological basis for the high prevalence of liver disorders (including kwashiorkor, liver tumors or cirrhosis) was due to pyrrolizidine (Senecio) alkaloids. This was common among the people of Africa and in a few other countries located in the tropical and subtropical regions (Schoental 1963; Edgar et al. 2015). Similarly, a variety of mycotoxins have been identified as carcinogens. These include ochratoxin aflatoxins, zearalenone, sterigmatocystinpenicillic acid, T-2 toxin, patulin, griseofulvin, luteoskyrin, ergot, and cyclochlorotine (National Research Council 1982). A correlation has been shown between the presence of aflatoxins, which are synthesized by fungi in many agricultural products such as maize, peanuts, cottonseed, and tree nuts, and an elevated susceptibility to the development of liver cancer (Hamid et al. 2013; Kimanya et al. 2021). The primary fungi responsible for the production of aflatoxins, which are recognized as the most powerful hepatocarcinogens, are *Aspergillus flavus* and *Aspergillus parasiticus*. The presence of aflatoxin in food products has been linked to a significant prevalence of liver cancer in Africa and Asia. Aflatoxin is known to possess carcinogenic properties in several animal species, leading to the development of tumors in the kidney, liver, lung, colon and stomach. Notably, males and young individuals seem to be more susceptible to its carcinogenic effects (Diet 1982; Dhakal et al. 2020).

25.3.5 Toxic Neuropathies

The etiology of most neurological illnesses is mostly associated with prenatal and postnatal exposure to environmental toxins created by industrial activities. Several neurotoxic metals, including lead (Pb), aluminum (Al), mercury (Hg), manganese (Mn), cadmium (Cd), and arsenic (As), as well as pesticides or metal-nanoparticles, have been associated with the development of dementia and Parkinson's disease. The pollutants are recognized for their capacity to generate plaques or amyloid deposits or neurofibrillary tangles (NFTs), which serve as the primary characteristics of these neurological disorders (Nabi and Tabassum 2022). Neuropathies may arise due to several circumstances, including environmental, recreational, occupational, or iatrogenic causes. The incidence of these neuropathies may be impacted by geographical and economic variables. The leading source of drug toxicity, specifically chemotherapy treatments, is responsible for the highest occurrence rate of 68% during the first month of therapy (Seretny et al. 2014; Karam and Dyck 2015). Significant peripheral neurotoxicants include arsenic, lead, mercury, and organophosphorous chemicals, which are often encountered in impoverished nations (Trivedi et al. 2017). The relocation of industrial processes to less developed regions has resulted in the persistence of peripheral neurotoxicants such as hexane, carbon disulfide, and chlorofluorocarbon replacement agents like 1-bromopropane, which continue to pose potential hazards for peripheral neuropathy (Ichihara et al. 2004; Meyer-Baron et al. 2012; Samukawa et al. 2012). Polyneuropathy, a condition characterized by damage to several peripheral nerves, may be attributed to two prevalent factors: Type 2 diabetes and alcoholism. These causes can also be classified as toxic neuropathies since they result from the detrimental effects of excessive glucose and ethanol on the nervous system.

25.3.6 Other Diseases

Algal toxins, which are present in both marine and freshwater environments, have the potential to induce gastrointestinal disturbances, such as diarrhea and vomiting, as well as neuromuscular impairment, leading to paralysis, in people, animals, and fish (Wang 2008). Similarly, Ciguatera fish poisoning (CFP) is a condition that arises from the ingestion of fish that has been infected with dinoflagellates, which are capable of producing ciguatoxins. This toxin is responsible for inducing a range of symptoms, including nausea, vomiting, heart disease and other neurological manifestations (Rhodes et al. 2020). Cyanogenic glycosides, which are present in more than 2000 plant species, have the potential to induce cyanide poisoning in humans. Furocoumarins, which are stress-induced toxins generated as a result of physical injury, have the potential to induce gastrointestinal problems in those who are vulnerable (Melough et al. 2018). Furthermore, Mycotoxins, emanating from molds present in food, have the potential to induce both immediate and chronic health consequences, such as the development of cancer and compromised immune function (Kraft et al. 2021; Awuchi et al. 2022). In addition, plants belonging to the Solanaceae family, including tomatoes, potatoes, and eggplants, are known to possess inherent toxic compounds known as solanines and chaconine (Badowski and Urbanek-Karłowska 1999; Ostreikova et al. 2022). To mitigate solanine synthesis, it is recommended to keep potatoes in an environment characterized by darkness, low temperature, and dryness, while concurrently refraining from consuming any green or sprouting sections. Wild mushrooms have the potential to harbor toxins such as muscimol and muscarine, which may elicit adverse physiological responses including vomiting, diarrhea, cognitive impairment, visual impairments, excessive salivation, and hallucinatory experiences (Tran and Juergens 2023). The ripe fruit of *Blighia sapida*, often known as ackee, is a dietary staple in Western Africa and the Caribbean region. It includes hypoglycin-A, a strong hypoglycemic drug that has been connected to several medical conditions. *Annona muricata* and *Annona squamosa* are well-known for their neurotoxic characteristics (Höllerhage et al. 2015). The long-term use of a substance may lead to the development of Parkinsonian syndrome, characterized by symptoms such as axial stiffness, early postural instability, bradykinesia, and frontal dysfunction. The compound glycyrrhizin, which is included in the root of Licorice, has the potential to be detrimental to those who have heart or renal ailments (Omar et al. 2012). Additional botanical specimens include onions, garlic, peppermint, prunes, curcumin, and papaya. The starfruit species found in Taiwan that have been acidified are known to contain oxalic acid, which has the potential to cause renal failure (Wijayaratne et al. 2018). Meanwhile, Chinese herb nephropathy (CHN) is a recently identified form of renal fibrosis that mostly affects females. Cases of this condition have been documented in many countries including France, Spain, the United Kingdom, and Japan (Arlt et al. 2002). Table 25.1 mentions some natural products, the class to which they belong and their toxicity.

Table 25.1 Toxic effect various classes of natural components

Natural product	Class	Toxic effect	References
D9-THC	Cannabinoids	Structural alterations in brain (specifically hippocampus, amygdala, cerebellum, prefrontal cortex, and striatum)	Battistella et al. (2014)
Epibitadine	Alkaloids	Stimulates exocrine secretion, hypertension, seizures, muscle paralysis	Sestito et al. (2019)
Atropine, hyoscyamine, scopolamine	Alkaloids	Rapid heart rate, pupil dilation, impaired bowel movement	Beyer et al. (2009), Lakstygal et al. (2018)
Cytisine	Alkaloids	Delirium and convulsions, respiratory paralysis, or circulatory system failure	Gotti and Clementi (2021)
Ibogaine	Alkaloids	Ataxia, tremors, muscular spasms, seizures, alterations in both blood pressure and heart rate	Koenig and Hilber (2015)
Salvinorin A	Terpenoid	Evoked intricate visual experiences and aural sensations, and altered the perception of internal bodily sensations	Maqueda et al. (2015), Silva et al. (2019)

(continued)

Table 25.1 (continued)

Natural product	Class	Toxic effect	References
β-N-oxalyl-l-α,β-diaminopropionic acid (BOOA)	Cannabinoids	Spastic paraparesis, upper motor neuron manifestations, instability in gait, neurolathyrism	Enneking (2011)
Cassava (Manihot esculenta)	Cyanogenic glycosides	Upper motor neuronal toxicity	Kashala-Abotnes et al. (2019)
Miroestrol	Steroids	Epilepsy, diabetes, asthma, and migraine	Kakehashi et al. (2016)
Cycasin	Glucoside	Emesis, diarrheal episodes, debilitation, convulsions, and hepatotoxic effects, neurotoxic, teratogenic and carcinogenic	Kobayashi and Matsumoto (1965), Sieber et al. (1980), Kisby et al. (2013)
Pyrrolizidine (Senecio)	Alkaloid	Liver disorders (including kwashiorkor, liver tumors and cirrhosis)	Schoental, (1963), Wiedenfeld (2011), Edgar et al. (2015)

25.4 Conclusion

As previously stated, natural compounds are molecules that often have properties advantageous to human health. However, these compounds and their analogues might lead to hazardous consequences. The prevailing idea proposes that secondary metabolites are produced either as byproducts or as a mechanism for detoxification. Additionally, secondary metabolites are synthesized as a means of survival, namely as a defensive mechanism against other species; this is exemplified in situations when prey capture is involved. Nevertheless, it is important to clearly define the specific objective or purpose for which these compounds demonstrate their toxicity. The destructive effects brought on by numerous living forms, including plants, animals, and bacteria, have fascinated and alarmed humans throughout recorded history. Hence, high toxicity levels are an intrinsic attribute of the aforementioned concern. To mitigate the potential health hazards of natural products in our daily lives, which is often overlooked, it is important to take some precautions. One should not assume that the safety of anything is guaranteed only based on its inherent characteristics. It is recommended to discard food items that exhibit signs of bruising, breakage, or discoloration, specifically those that display the presence of mold. The level of interest in this particular domain is of such magnitude that a whole scientific discipline has been exclusively devoted to investigating harmful substances (toxins) generated by organisms, commonly referred to as toxicology (Wexler et al. 2015). The evaluation of the properties that contribute to the toxicological nature of medications is crucial in the realm of pharmaceutical research, particularly when investigating the potential therapeutic effects of a certain natural component for a specific ailment. Hence, it is recommended to ascertain toxicological outcomes early in the creation of novel compounds (Madariaga-Mazón et al. 2019).

References

Abbaoui B, Lucas CR, Riedl KM, Clinton SK, Mortazavi A (2018) Cruciferous vegetables, isothiocyanates, and bladder cancer prevention. Mol Nutr Food Res 62(18):1800079

Agus HH (2021) Terpene toxicity and oxidative stress. In: Toxicology. Elsevier, pp 33–42

Ahmed J, Preissner S, Dunkel M, Worth CL, Eckert A, Preissner R (2010) SuperSweet—a resource on natural and artificial sweetening agents. Nucleic Acids Res 39(Suppl. 1):D377–D382

Arlt VM, Stiborova M, Schmeiser HH (2002) Aristolochic acid as a probable human cancer hazard in herbal remedies: a review. Mutagenesis 17(4):265–277

Artal FJC (2015) Adverse neurological effects caused by the ingestion of plants, seeds, and fruits. In: Bioactive nutraceuticals and dietary supplements in neurological and brain disease. Elsevier, pp 215–219

Ascherio A, Weisskopf MG, O'Reilly EJ, McCullough ML, Calle EE, Rodriguez C, Thun MJ (2004) Coffee consumption, gender, and Parkinson's disease mortality in the cancer prevention study II cohort: the modifying effects of estrogen. Am J Epidemiol 160(10):977–984

Ashihara H, Crozier A (2001) Caffeine: a well known but little mentioned compound in plant science. Trends Plant Sci 6(9):407–413

Ashihara H, Suzuki T (2004) Distribution and biosynthesis of caffeine in plants. Front Biosci 9(2):1864–1876

Awuchi CG, Ondari EN, Nwozo S, Odongo GA, Eseoghene IJ, Twinomuhwezi H, Ogbonna CU, Upadhyay AK, Adeleye AO, Okpala COR (2022) Mycotoxins' toxicological mechanisms involving humans, livestock and their associated health concerns: a review. Toxins 14(3):167

Badowski P, Urbanek-Karłowska B (1999) Solanine and chaconine: occurrence, properties, methods for determination. Rocz Panstw Zakl Hig 50(1):69–75

Battistella G, Fornari E, Annoni J-M, Chtioui H, Dao K, Fabritius M, Favrat B, Mall J-F, Maeder P, Giroud C (2014) Long-term effects of cannabis on brain structure. Neuropsychopharmacology 39(9):2041–2048

Benowitz NL (1990) Clinical pharmacology of caffeine. Annu Rev Med 41(1):277–288

Berdy J (2005) Bioactive microbial metabolites. J Antibiot 58(1):1–26

Beyer J, Drummer OH, Maurer HH (2009) Analysis of toxic alkaloids in body samples. Forensic Sci Int 185(1–3):1–9

Birch A, Massy-Westropp R, Moye C (1955) Studies in relation to biosynthesis. VII. 2-Hydroxy-6-methylbenzoic acid in Penicillium griseofulvum Dierckx. Aust J Chem 8(4):539–544

Bravo HR, Lazo W (1993) Antimicrobial activity of cereal hydroxamic acids and related compounds. Phytochemistry 33(3):569–571

Bravo HR, Lazo W (1996) Antialgal and antifungal activity of natural hydroxamic acids and related compounds. J Agric Food Chem 44(6):1569–1571

Capuano E, Dekker M, Verkerk R, Oliviero T (2017) Food as pharma? The case of glucosinolates. Curr Pharm Des 23(19):2697–2721

Cesarman E, Damania B, Krown SE, Martin J, Bower M, Whitby D (2019) Kaposi sarcoma. Nat Rev Dis Primers 5(1):9

Dagan-Wiener A, Di Pizio A, Nissim I, Bahia MS, Dubovski N, Margulis E, Niv MY (2019) BitterDB: taste ligands and receptors database in 2019. Nucleic Acids Res 47(D1):D1179–D1185

Dewick PM (2002) Medicinal natural products: a biosynthetic approach. John Wiley & Sons

Dhakal, A., Hashmi MF, Sbar E (2020) Aflatoxin toxicity

Diet N (1982) Cancer, report of a committee on diet. In: Nutrition and cancer, assembly of life sciences. National Research Council, National Academy Press, Washington, DC

Dunkel M, Schmidt U, Struck S, Berger L, Gruening B, Hossbach J, Jaeger IS, Effmert U, Piechulla B, Eriksson R (2009) SuperScent—a database of flavors and scents. Nucleic Acids Res 37(Suppl. 1):D291–D294

Edgar JA, Molyneux RJ, Colegate SM (2015) Pyrrolizidine alkaloids: potential role in the etiology of cancers, pulmonary hypertension, congenital anomalies, and liver disease. Chem Res Toxicol 28(1):4–20

Enneking D (2011) The nutritive value of grasspea (Lathyrus sativus) and allied species, their toxicity to animals and the role of malnutrition in neurolathyrism. Food Chem Toxicol 49(3):694–709

Fahey JW, Zalcmann AT, Talalay P (2001) The chemical diversity and distribution of glucosinolates and isothiocyanates among plants. Phytochemistry 56(1):5–51

Frei H, Lüthy J, Brauchli J, Zweifel U, Würgler FE, Schlatter C (1992) Structure/activity relationships of the genotoxic potencies of sixteen pyrrolizidine alkaloids assayed for the induction of somatic mutation and recombination in wing cells of Drosophila melanogaster. Chem Biol Interact 83(1):1–22

Funayama S, Cordell GA (2014) Alkaloids: a treasury of poisons and medicines. Elsevier

Gonçalves EC, Baldasso GM, Bicca MA, Paes RS, Capasso R, Dutra RC (2020) Terpenoids, cannabimimetic ligands, beyond the cannabis plant. Molecules 25(7):1567

Gotti C, Clementi F (2021) Cytisine and cytisine derivatives. More than smoking cessation aids. Pharmacol Res 170:105700

Griffin WJ, Lin GD (2000) Chemotaxonomy and geographical distribution of tropane alkaloids. Phytochemistry 53(6):623–637

Hamid AS, Tesfamariam IG, Zhang Y, Zhang ZG (2013) Aflatoxin B1-induced hepatocellular carcinoma in developing countries: geographical distribution, mechanism of action and prevention. Oncol Lett 5(4):1087–1092

Hamilton RJ, Hamilton R, Perrone J, Hoffman R, Henretig FM, Karkevandian EH, Marcus S, Shih RD, Blok B, Nordenholz K (2000) A descriptive study of an epidemic of poisoning caused by heroin adulterated with scopolamine. J Toxicol Clin Toxicol 38(6):597–608

Hanschen FS, Lamy E, Schreiner M, Rohn S (2014) Reactivity and stability of glucosinolates and their breakdown products in foods. Angew Chem Int Ed 53(43):11430–11450

Hartmann T, Ober D (2000) Biosynthesis and metabolism of pyrrolizidine alkaloids in plants and specialized insect herbivores. In: Biosynthesis: aromatic polyketides, isoprenoids, alkaloids, pp 207–243

Hartmann T, Sander H, Adolph R, Toppel G (1988) Metabolic links between the biosynthesis of pyrrolizidine alkaloids and polyamines in root cultures of Senecio vulgaris. Planta 175:82–90

Hegnauer, R., Hegnauer R (1986) Allgemeine Literaturübersicht: (Bd. I, S. 29–40; Bd. III, S. 40–41). Chemotaxonomie der Pflanzen: Eine Übersicht über die Verbreitung und die systematische Bedeutung der Pflanzenstoffe: 10–206

Hill JR, Burghardt RC, Jones K, Long CR, Looney CR, Shin T, Spencer TE, Thompson JA, Winger QA, Westhusin ME (2000) Evidence for placental abnormality as the major cause of mortality in first-trimester somatic cell cloned bovine fetuses. Biol Reprod 63(6):1787–1794

Höllerhage M, Roesler TW, Berjas M, Luo R, Tran K, Richards KM, Sabaa-Srur AU, Maia JGS, Moraes MRD, Godoy HT (2015) Neurotoxicity of dietary supplements from Annonaceae species. Int J Toxicol 34(6):543–550

Hollingsworth RG, Armstrong JW, Campbell E (2002) Caffeine as a repellent for slugs and snails. Nature 417(6892):915–916

Holstege CP, Hunter Y, Baer AB, Savory J, Bruns DE, Boyd JC (2003) Massive caffeine overdose requiring vasopressin infusion and hemodialysis. J Toxicol Clin Toxicol 41(7):1003–1007

Ichihara G, Li W, Shibata E, Ding X, Wang H, Liang Y, Peng S, Itohara S, Kamijima M, Fan Q (2004) Neurologic abnormalities in workers of a 1-bromopropane factory. Environ Health Perspect 112(13):1319–1325

Jacobs DI, Snoeijer W, Hallard D, Verpoorte R (2004) The Catharanthus alkaloids: pharmacognosy and biotechnology. Curr Med Chem 11(5):607–628

Kakehashi A, Yoshida M, Tago Y, Ishii N, Okuno T, Gi M, Wanibuchi H (2016) Pueraria mirifica exerts estrogenic effects in the mammary gland and uterus and promotes mammary carcinogenesis in donryu rats. Toxins 8(11):275

Kapur R, Smith MD (2009) Treatment of cardiovascular collapse from caffeine overdose with lidocaine, phenylephrine, and hemodialysis. Am J Emerg Med 27(2):253. e253-253. e256

Karam C, Dyck PJB (2015) Toxic neuropathies. In: Seminars in neurology. Thieme Medical Publishers

Kashala-Abotnes E, Okitundu D, Mumba D, Boivin MJ, Tylleskär T, Tshala-Katumbay D (2019) Konzo: a distinct neurological disease associated with food (cassava) cyanogenic poisoning. Brain Res Bull 145:87–91

Kelly BF, Nappe TM (2018). Cannabinoid toxicity

Khalifa SA, Elias N, Farag MA, Chen L, Saeed A, Hegazy M-EF, Moustafa MS, Abd El-Wahed A, Al-Mousawi SM, Musharraf SG (2019) Marine natural products: a source of novel anticancer drugs. Mar Drugs 17(9):491

Khandare AL, Kumar RH, Meshram I, Arlappa N, Laxmaiah A, Venkaiah K, Rao PA, Validandi V, Toteja G (2018) Current scenario of consumption of Lathyrus sativus and lathyrism in three districts of Chhattisgarh state, India. Toxicon 150:228–234

Khandare AL, Kalakumar B, Validandi V, Qadri S, Harishankar N, Singh SS, Kodali V (2020) Neurolathyrism in goat (Capra hircus) kid: model development. Res Vet Sci 132:49–53

Kimanya ME, Routledge MN, Mpolya E, Ezekiel CN, Shirima CP, Gong YY (2021) Estimating the risk of aflatoxin-induced liver cancer in Tanzania based on biomarker data. PLoS One 16(3):e0247281

Kisby GE, Moore H, Spencer PS (2013) Animal models of brain maldevelopment induced by cycad plant genotoxins. Birth Defects Res C Embryo Today 99(4):247–255

Kobayashi A, Matsumoto H (1965) Studies on methylazoxymethanol, the aglycone of cycasin: isolation, biological, and chemical properties. Arch Biochem Biophys 110(2):373–380

Koenig X, Hilber K (2015) The anti-addiction drug ibogaine and the heart: a delicate relation. Molecules 20(2):2208–2228

Kraft S, Buchenauer L, Polte T (2021) Mold, mycotoxins and a dysregulated immune system: a combination of concern? Int J Mol Sci 22(22):12269

Kumar P, Prasad A, Giridhar F (2022) Neurolathyrism with deep vein thrombosis and bony exostosis: are they new forms of angiolathyrism and osteolathyrism? Cureus 14(8)

Lakstygal AM, Kolesnikova TO, Khatsko SL, Zabegalov KN, Volgin AD, Demin KA, Shevyrin VA, Wappler-Guzzetta EA, Kalueff AV (2018) Dark classics in chemical neuroscience: atropine, scopolamine, and other anticholinergic deliriant hallucinogens. ACS Chem Neurosci 10(5):2144–2159

Laule O, Fürholz A, Chang H-S, Zhu T, Wang X, Heifetz PB, Gruissem W, Lange M (2003) Crosstalk between cytosolic and plastidial pathways of isoprenoid biosynthesis in Arabidopsis thaliana. Proc Natl Acad Sci 100(11):6866–6871

Madariaga-Mazón A, Hernández-Alvarado RB, Noriega-Colima KO, Osnaya-Hernández A, Martinez-Mayorga K (2019) Toxicity of secondary metabolites. Phys Sci Rev 4(12):20180116

Mahesh SK, Fathima J, Veena VG (2019) Cosmetic potential of natural products: industrial applications. In: Natural bio-active compounds: volume 2: chemistry, pharmacology and health care practices, pp 215–250

Maqueda AE, Valle M, Addy PH, Antonijoan RM, Puntes M, Coimbra J, Ballester MR, Garrido M, González M, Claramunt J (2015) Salvinorin-A induces intense dissociative effects, blocking external sensory perception and modulating interoception and sense of body ownership in humans. Int J Neuropsychopharmacol 18(12):pyv065

McMurry J (2014) Organic chemistry: with biological applications. Cengage Learning Inc., CA, USA

Melough MM, Cho E, Chun OK (2018) Furocoumarins: a review of biochemical activities, dietary sources and intake, and potential health risks. Food Chem Toxicol 113:99–107

Meyer-Baron M, Kim EA, Nuwayhid I, Ichihara G, Kang S-K (2012) Occupational exposure to neurotoxic substances in Asian countries–challenges and approaches. Neurotoxicology 33(4):853–861

Mirza SB, Bokhari H, Fatmi MQ (2015) Exploring natural products from the biodiversity of Pakistan for computational drug discovery studies: collection, optimization, design and development of a chemical database (ChemDP). Curr Comput Aided Drug Des 11(2):102–109. https://doi.org/10.2174/157340991102150904101740

Nabi M, Tabassum N (2022) Role of environmental toxicants on neurodegenerative disorders. Front Toxicol 4:837579

National Research Council (1982) Diet, nutrition, and cancer. National Academies Press
Naveja JJ, Rico-Hidalgo MP, Medina-Franco JL (2018) Analysis of a large food chemical database: chemical space, diversity, and complexity. F1000Research 7
Newman DJ, Cragg GM (2016) Natural products as sources of new drugs from 1981 to 2014. J Nat Prod 79(3):629–661
Noble RL (1990) The discovery of the vinca alkaloids—chemotherapeutic agents against cancer. Biochem Cell Biol 68(12):1344–1351
O'Connor SE, Maresh JJ (2006) Chemistry and biology of monoterpene indole alkaloid biosynthesis. Nat Prod Rep 23(4):532–547
Omar HR, Komarova I, El-Ghonemi M, Fathy A, Rashad R, Abdelmalak HD, Yerramadha MR, Ali Y, Helal E, Camporesi EM (2012) Licorice abuse: time to send a warning message. Ther Adv Endocrinol Metab 3(4):125–138
Ostreikova T, Kalinkina O, Bogomolov N, Chernykh I (2022) Glycoalkaloids of plants in the family Solanaceae (nightshade) as potential drugs. Pharm Chem J 56(7):948–957
Pulido OM (2008) Domoic acid toxicologic pathology: a review. Mar Drugs 6(2):180–219
Rhodes LL, Smith KF, Murray JS, Nishimura T, Finch SC (2020) Ciguatera fish poisoning: the risk from an Aotearoa/New Zealand perspective. Toxins 12(1):50
Salehi B, Sestito S, Rapposelli S, Peron G, Calina D, Sharifi-Rad M, Sharopov F, Martins N, Sharifi-Rad J (2018) Epibatidine: a promising natural alkaloid in health. Biomol Ther 9(1):6
Samukawa M, Ichihara G, Oka N, Kusunoki S (2012) A case of severe neurotoxicity associated with exposure to 1-bromopropane, an alternative to ozone-depleting or global-warming solvents. Arch Intern Med 172(16):1257–1260
Schiff PL (2002) Opium and its alkaloids. Am J Pharm Educ 66(2):188–196
Schoental R (1963) Liver disease and "natural" hepatotoxins. Bull World Health Organ 29(6):823
Schoental R (1965) Toxicology of natural products. Food Cosmet Toxicol 3:609–620
Schuhr CA, Radykewicz T, Sagner S, Latzel C, Zenk MH, Arigoni D, Bacher A, Rohdich F, Eisenreich W (2003) Quantitative assessment of crosstalk between the two isoprenoid biosynthesis pathways in plants by NMR spectroscopy. Phytochem Rev 2:3–16
Seigler DS (1991) Cyanide and cyanogenic glycosides. GS Rosenthal & MR Berenbaum, pp. 35–77
Selmar D (1999) Biosynthesis of cyanogenic glycosides, glucosinolates and non protein amino acids. Biochem Plant Second Metab 2:79
Seretny M, Currie GL, Sena ES, Ramnarine S, Grant R, MacLeod MR, Colvin LA, Fallon M (2014) Incidence, prevalence, and predictors of chemotherapy-induced peripheral neuropathy: a systematic review and meta-analysis. Pain 155(12):2461–2470
Sestito S, Rapposelli S, Peron G, Calina D, Sharifi-Rad M, Sharopov F, Martins N, Sharifi-Rad J (2019) Epibatidine: a promising natural alkaloid in health. Biomol Ther 9
Sicker D, Frey M, Schulz M, Gierl A (2000) Role of natural benzoxazinones in the survival strategy of plants. Int Rev Cytol 198:319–346
Sieber SM, Correa P, Dalgard DW, McIntire KR, Adamson RH (1980) Carcinogenicity and hepatotoxicity of cycasin and its aglycone methylazoxymethanol acetate in nonhuman primates. J Natl Cancer Inst 65(1):177–189
Silva AR, Grosso C, Delerue-Matos C, Rocha JM (2019) Comprehensive review on the interaction between natural compounds and brain receptors: benefits and toxicity. Eur J Med Chem 174:87–115
Sparks TC, Wessels FJ, Lorsbach BA, Nugent BM, Watson GB (2019) The new age of insecticide discovery-the crop protection industry and the impact of natural products. Pestic Biochem Physiol 161:12–22
Springob K, Kutchan TM (2009) Introduction to the different classes of natural products. In: Plant-derived natural products: synthesis, function, and application, pp 3–50
Srinivas G, Babykutty S, Sathiadevan PP, Srinivas P (2007) Molecular mechanism of emodin action: transition from laxative ingredient to an antitumor agent. Med Res Rev 27(5):591–608
Taljaard T (1993) Cabbage poisoning in ruminants. J S Afr Vet Assoc 64(2):96–100
Tiwari R, Rana C (2015) Plant secondary metabolites: a review. Int J Eng Res Gen Sci 3(5):661–670
Tran HH, Juergens AL (2023) Mushroom toxicity. In: StatPearls. StatPearls Publishing

Traynor J (1998) Epibatidine and pain. Br J Anaesth 81(1):69–76
Trivedi S, Pandit A, Ganguly G, Das SK (2017) Epidemiology of peripheral neuropathy: an Indian perspective. Ann Indian Acad Neurol 20(3):173–184
Wang D-Z (2008) Neurotoxins from marine dinoflagellates: a brief review. Mar Drugs 6(2):349–371
Welch K, Panter K, Lee S, Gardner D, Stegelmeier B, Cook D (2009) Cyclopamine-induced synophthalmia in sheep: defining a critical window and toxicokinetic evaluation. J Appl Toxicol 29(5):414–421
Wexler P, Fonger GC, White J, Weinstein S (2015) Toxinology: taxonomy, interpretation, and information resources. Sci Technol Libr 34(1):67–90
White SM, Lambe CJ (2003) The pathophysiology of cocaine abuse. J Clin Forensic Med 10(1):27–39
Wiedenfeld H (2011) Plants containing pyrrolizidine alkaloids: toxicity and problems. Food Addit Contam Part A 28(3):282–292
Wiener SW, Nelson LS (2024) Cholinergic agent attack (nicotine, Epibatidine, and Anatoxin-a). In: Ciottone's disaster medicine. Elsevier, pp 725–728
Wijayaratne DR, Bavanthan V, De Silva M, Nazar A, Wijewickrama ES (2018) Star fruit nephrotoxicity: a case series and literature review. BMC Nephrol 19:1–7
Williams DH, Stone MJ, Hauck PR, Rahman SK (1989) Why are secondary metabolites (natural products) biosynthesized? J Nat Prod 52(6):1189–1208
Williamson G, Wang H, Griffiths S (1996) Glucosinolates as bioactive components of brassica vegetables: induction of cytochrome P450 1A1 in Hep G2 cells as assessed using transient transfection, vol 24. Portland Press Ltd, p 383S
Wink M (2002) Production of quinolizidine alkaloids in in vitro cultures of legumes. In: Medicinal and aromatic plants XII. Springer, pp 118–136
Wink M (2004) Allelochemical properties of quinolizidine alkaloids. In: Allelopathy: chemistry and mode of action of allelochemicals. CRC Press, Boca Raton, pp 183–200
Yue Y, Chu G-X, Liu X-S, Tang X, Wang W, Liu G-J, Yang T, Ling T-J, Wang X-G, Zhang Z-Z (2014) TMDB: a literature-curated database for small molecular compounds found from tea. BMC Plant Biol 14:1–8

Fungal Infestation and Antifungal Treatment of Organic Heritage Objects

26

Jasmine Shakir, Saiema Ahmedi, Satish Pandey, and Nikhat Manzoor

26.1 Introduction

The conservation and management of cultural, natural, and architectural heritage for future generations is referred to as heritage conservation. It includes preserving historic structures, locations, works of art, natural landscapes, and intangible heritage that are important to a community, a local area, or country. The goal of heritage conservation is to uphold the resources' worth, its integrity, and authenticity. The idea of heritage conservation was developed in response to the loss and destruction of historical cultural and natural riches. The damage brought about by conflicts, urbanization, development, abuse, and natural calamities made it clear that our shared human legacy must be safeguarded and preserved. It emphasized the duties that heritage conservators must do by integrating several sciences to extend the lifespan of the heritage asset (Mazzeo et al. 2011). The activity of scientists and other researchers who look into the study and preservation of cultural heritage is known as conservation research. As an applied field of study, conservation research focuses on the preservation of the tangible components of cultural heritage. It covers scientific research that employs historical, artistic, conservational, chemical, biological, and physical techniques (Mazzeo et al. 2011). As such, the methodology of conservation research is essentially interdisciplinary. The heritage conservation

J. Shakir · S. Pandey
Department of Conservation, Indian Institute of Heritage (Formerly National Museum Institute), Noida, Uttar Pradesh, India
e-mail: jasmineshakir64@gmail.com; conservation.nmi@gov.in; satish.pandey.nmi@gov.in

S. Ahmedi · N. Manzoor (✉)
Medical Mycology Lab, Department of Biosciences, Jamia Millia Islamia (A Central University), New Delhi, India
e-mail: saiemaahmedi1995@gmail.com; nmanzoor@jmi.ac.in

N. Manzoor (ed.), *Advances in Antifungal Drug Development*,
https://doi.org/10.1007/978-981-97-5165-5_26

in true sense is a multidisciplinary field of study where heritage conservationists combine the expertise from various disciplines, including chemistry, physics, art history, mycology, and material science to achieve their objectives.

26.2 Historical Background and Scope

Earliest conservation initiatives date back to the nineteenth century when historic structures and landscapes were destroyed indiscriminately as a consequence of the industrial revolution and fast urbanization in the middle of the nineteenth century. Intellectuals, artists, and academics who supported the preservation of historic architecture and artwork expressed alarm over this. Groups like the American Institute of Architects (AIA) in the United States, and the Society for the Protection of Ancient Buildings (SPAB) in the United Kingdom emphasized the value of conservation and preservation of the original art works of cultural significance, especially the historic architectural preservation (Stoner 2005). Significant advancements in heritage protection date back to the twentieth century. International standards and rules for heritage conservation were established by the Athens Charter for the Restoration of Historic Monuments (1931) and the Venice Charter for the Conservation and Restoration of Monuments and Sites (1964) (Stoner 2005). Global conservation efforts were further aided by the formation of international organizations including the United Nations Educational, Scientific, and Cultural Organization (UNESCO) in 1945 (Stoner 2005). The 1972 adoption of the World Heritage Convention by the UNESCO set out to identify and safeguard places with exceptional universal value, both cultural and natural, throughout the globe.

Conservation currently refers to a broader understanding of heritage, which includes intangible cultural heritage, customs, and indigenous knowledge. It acknowledges the importance of community involvement and sustainable management as well as the social, economic, and environmental components of human heritage. Conservators, architects, archaeologists, historians, urban planners, legislators, and local communities all play their specific roles in heritage conservation today. To ensure the long-term preservation and enjoyment of human cultural and natural heritage, conservationists perform a variety of activities like documentation, restoration, adaptive reuse, interpretation, and education. Heritage preservation has economic importance as well. It supports sustainable development, tourism, and cross-cultural communication in addition to strengthening the sense of cultural identity and pride of communities. It serves as a reminder of our shared history, diversity, and the need to protect and cherish human heritage for future generations. In this chapter, we summarize the process of conservation, especially the biodegradable heritage material and fungal infestation. The methods and technologies in conservation, preservation, and restoration of heritage artwork and its treatment against fungal infestation are discussed alongside new technologies such as the use of nanomaterials in conservation process (Stoner 2005).

26.3 Nature of Organic Heritage Objects

Organic materials include carbon-based compounds, derived from living organisms. They are combinations of a few of the lightest elements, such as hydrogen, carbon, nitrogen, and oxygen. When exposed to light or other radiation for an extended period, organic materials such as wood, leather, textiles, and paper become embrittled, faded, or yellowed due to the disintegration of the covalent bonding structure that many carbon-containing compounds share (Clayden et al. 2012). Artifacts or items constructed from organic materials, which come from living organisms are referred to as organic heritage objects. The qualities of the materials used to create organic historic artifacts have an impact on their nature.

(a) **Biodegradability**: Over time, organic materials are prone to deterioration and breakdown. Microorganisms, temperature, light, humidity, and other variables can all cause these items to deteriorate. It is frequently necessary to take preservation measures to stop or slow down this process.
(b) **Sensitivity to Environmental Conditions**: Variations in humidity and temperature can have an impact on organic materials. Excessive moisture content or high temperatures can cause warping, cracking, or the growth of mold and fungi, among other extreme situations.
(c) **Pest Vulnerability**: Organic materials may attract insects and other pests, which could result in infestations that do a great deal of harm. Preventive techniques, like controlled conditions and routine inspections, are frequently used by museums and collectors to shield organic heritage objects from pests.
(d) **Chemical Changes**: Over time, organic materials may experience chemical shifts such as oxidation and color fading. Certain chemicals, light, and pollution exposure can quicken these processes. Utilizing appropriate techniques to stabilize and safeguard the objects may be part of conservation efforts.

26.4 Biodegradable Heritage Material and Fungal Infestation

Heritage material which is made to naturally degrade over time is referred to as a biodegradable heritage material. This includes manuscripts, textiles, paintings, leather, wood, and bone material of historical importance. These materials are made from natural sources and degrade naturally over time especially in humid and warm climate, where the fungal deterioration has been recognized as a common problem (Savković et al. 2021; Simons 2007). The fungus can harm any biodegradable object, causing it to deteriorate, discolor, and lose its structural integrity and economic value. A sizable percentage of the world's cultural heritage artifacts including the paintings, manuscripts, textile and other material in museums and elsewhere are reportedly found in different states of deterioration especially in warm and humid climate with mold attack as the most common cause. The problem is widespread in these places as mold and fungi grow faster in humid environments; the biodegradable artwork providing a natural source of nutrients to the microorganism.

Even the oil paintings, which may take hundreds of years (up to 500 years) before the painting is completely dried, are susceptible to infestation (Menon 2021).

In fact, all types of biodegradable materials used in heritage artwork are susceptible to infestation. In paintings, both nonpigmented and pigmented materials are vulnerable to microbial attack. Artists and museums usually resort to cumbersome methods, such as wrapping paintings with silica gel bags, a technique which may be effective for small paintings, but cumbersome and difficult for large size paintings. The organic material used in paintings (canvas fabric, linseed oil) and pigments (red lac and red and yellow earths) may act as a nutrient source by the painting-associated microorganisms (Paner 2009; Caselli et al. 2018).

Canvas used in aisle paintings is particularly vulnerable to infestation. Canvas is typically made of cotton or linen, which are made up of cellulose, a complex carbohydrate and a preferred source of nutrition for fungus growth. Cotton and linen attract fungus when relative humidity (RH), temperature, and other conditions favor microbial growth. Majorly cotton is composed of complex fibers of mainly cellulose ($\approx$90%), which is a meshwork of glucose chains. Glucose is an ideal nutrient for the growth of microorganisms (Menon 2021). On fungal infestation, canvas can be digested by this eukaryotic microbe at affected sites and provide ample nutrition in the form of glucose, thereby favoring mold growth. Antifungal treatment and various strategies such as the use of "*neem*" nanoparticles can be used to effectively stop the fungal and other microbial growth on fiber materials (cellulose) such as textiles and papers. Substances like red lac and red and yellow earths in the pigments may also support and contribute to mold growth on an infested painting (Menon 2021). Even the wood and bone materials, stones and architectural heritage are not spared by fungi.

26.5 Types of Deteriorating Heritage Objects and Species of Fungi Identified

A wide range of organic materials can be colonized by the fungus. The kinds of fungi that grow on various organic items are determined by elements like the makeup of the material, the surrounding environment, and the availability of nutrients (Caselli et al. 2018). In order to reduce the possibility of fungal growth on organic artifacts, conservators and collection managers frequently take preventive measures to control ambient variables, such as temperature and humidity. For these artifacts to be preserved, timely action and routine monitoring are essential. The following are some typical fungal species that are connected to different organic objects:

(a) **Wood**: Historical wooden structures, furniture, and artifacts are frequently exposed to fungal deterioration. Fungi can digest the cellulose and lignin in the wood, causing weakening, staining, and loss of structural integrity in these structures. White Rot Fungi (*Phellinus* spp. and *Trametes* spp.) break down both cellulose and lignin in wood, resulting in a white, spongy decay (Sterflinger and Pinzari 2012). Brown Rot Fungi (*Gloeophyllum* spp. and *Fomitopsis* spp.) primarily break down cellulose, leaving behind a brown, cubical decay (Sterflinger and Pinzari 2012).

(b) **Textile**: The garments, tapestries, flags and carpets, are susceptible to fungal deterioration, resulting in discoloration, embrittlement, and loss of structural integrity. Additionally, mold growth on textiles accumulates enzymes and acids which can deteriorate the textile fibers, causing irreparable harm. Various mold species, such as *Aspergillus* spp., *Penicillium* spp., and *Cladosporium* spp., can grow on textiles, especially in damp conditions (Sterflinger and Pinzari 2012). Mildew is a common term for mold growth on textiles. It often appears as a white or gray powdery substance.

(c) **Parchment and Leather**: Historical leather goods, leather-bound books, and parchment documents are susceptible to microbial deterioration. These materials contain the animal protein collagen, which is destroyed by fungi, resulting in softening of fibers, and weakening of the material. Mold growth also produces obnoxious odors and visible stains and turns the object dark. Just like textiles, leather can be susceptible to mold and mildew growth, especially in humid environments. Various mold species, such as *Aspergillus* spp., *Penicillium* spp., and *Cladosporium* spp., can grow on leather, especially in damp conditions (Sterflinger and Pinzari 2012). Insects, like Dermestid Beetles, are also known to infest leather objects, feeding on the proteins in the material.

(d) **Paintings**: Microbial infestation can cause stains, discoloration, and pigment degradation on paintings, especially those made on organic materials like canvas or wood panels. The layers of paint can also be penetrated by fungi, which results in flaking, blistering, and loss of detail. Various mold species, such as *Aspergillus* spp., *Penicillium* spp., and *Cladosporium* spp., can grow on canvas and other surfaces used to create the painting, especially in fluctuating RH conditions (Caselli et al. 2018).

(e) **Paper and Records**: Paper-based materials (manuscripts, books, archive records) get discolored, weakened, and eventually disintegrate due to fungal infestation, additionally attracting pests like silverfish and booklice, speeding up deterioration. Mold can thrive on paper in humid conditions. *Stachybotrys, Aspergillus,* and *Penicillium* species are common culprits (Simons 2007). Foxing is a term used for reddish-brown stains on paper, often caused by fungi, although the exact mechanism is not fully understood.

(f) **Photographs**: Prints and negatives of old photographs can both suffer from fungal deterioration; for example, the gelatin emulsion can be damaged by the fungi, resulting in staining, discoloration, and loss of image quality. Photographs with mold growth on them may become unstable and suffer from surface damage.

(g) **Masonry and Stone**: The organic acids produced by the fungi harm and affect the appearance of the masonry, stone, buildings and sculpture surfaces, dissolving the minerals and causing erosion and the loss of surface details. *Aspergillus* and *Cladosporium* species have been frequently found on the moist stone surfaces.

(h) **Natural History Collections**: Collections such as taxidermy animals, dried insect, animal and plant specimens, fur, feathers and plant tissues can all deteriorate due to microbial growth, resulting in loss of structural integrity and a decline in scientific knowledge.
(i) **Ethnographic Artifacts**: Cultural artifacts and items (masks, musical instruments, baskets, and ritual objects) which are made of wood, plant fibers, or animal skins are susceptible to microbial deterioration. Microbial growth can cause discoloration, structural harm, and the loss of cultural importance of the object.
(j) **Coins and Metal Artifacts**: Indirect fungal deterioration can affect metal artifacts such as sculptures, weapons, coins, and archaeological findings. A humid atmosphere caused by fungal growth in storage spaces or display cases encourage corrosion and tarnish the metal surface. Fungi may also deteriorate organic materials that are in contact or linked to metal objects, such as wood or leather.

Table 26.1 Various fungal species and the nature of the heritage material on which they grow (Simons 2007; Sterflinger and Pinzari 2012; Caselli et al. 2018)

Genus/species	Heritage material
Alternaria spp., *Aspergillus flavus, Aspergillus sect. Niger, A. sydowii, A. versicolor, Aureobasidium pullulans, Chaetomium funicola, Cladosporium herbarum, C. cladosporioides, Eurotium chevalieri, E. rubrum, Fusarium* spp., *Mucor* spp., *Penicillium chrysogenum, P. citrinum, P. decumbens,* and many other spp.	Paintings: oil, water color, acrylic
Alternaria spp., *Aspergillus clavatus, A. flavus, A. glaucus, A. terreus, A. repens, A. ruber, A. fumigatus, A. ochraceus, A. nidulans, Aspergillus sect. Niger, Botrytis cinerea, Chaetomium globosum, C. elatum, C. indicum, Eurotium amstelodami, Fusarium* spp., *Mucor* spp., *Paecilomyces variotii, Penicillium chrysogenum, P. funinculosum, P. pupurogenum, P. rubrum, P. variabile, P. spinulosum, P. fellutatum, P. frequetans, P. citrinum, Pichia guilliermondi, Rhizopus oryzae, Stachybotrys chartarum, Toxicocladosporium irritans, Trichoderma harzianum, T. viride, Stemphilium* spp., *Ulocladium* spp.	Paper and cellulose textile: laid-paper, wood pulp paper, and cellulose textiles (cotton, linen)
Cladosporium cladosporioides, Epicoccum nigrum, Phlebiopsis gigantea, Penicillium chrysogenum, Thanatephorus cucumeris	Parchment
Absidia glauca, A. cylindrospora, A. spinosa, Acremonium spp. *Alternaria alternata, Aspergillus sydowii, A. candidus, A. clavatus, A. carneus, A. foetidus, A. flavus, A. fumigatus,* and many other species of the genus *Arthroderma, Aureobasidium pullulans, Chaetomium globosum, Chrysosporium* spp., *Coniosporium* spp., *Cladosporium cladosporioides, Cunninghamella echinulata, C. elegans, Epicoccum nigrum, Emericella* spp., *Geotrichum candidum, Mucor* spp., *Penicillium brevicompactum, Penicillium chrysogenum* and *many* other species of the genus *Phoma medicaginis, Scopulaiopsis* spp., *Stachybotrys chartarum, Trichophyton* spp., *Rhizopus* spp.	Keratinous substrates: leather, wool, feathers, fur, hair
Archaeological discoveries frequently contain a significant number of spores which reflect the diversity of the relevant soil	Archeological findings: bones, ceramics

It is crucial to keep in mind that a variety of microbial species are engaged in deterioration of cultural property, and their effects can change based on the atmosphere, moisture content, and individual material properties. A list of different heritage objects and the deteriorating microbial/fungal spp. is provided in Table 26.1.

26.6 Mechanisms of Damage and Fungal Substances Impacting Heritage Art

When microorganisms physically or chemically modify or affect the look of materials, it is referred to as biodeterioration, which can potentially colonize important cultural heritage. Recent developments in applied microbiology and biotechnology offer crucial insights into the preservation of cultural heritage. Although there have been many physical and mechanical techniques employed in the past, they are unable to completely stop organisms from growing. Fungi can seriously harm organic heritage materials using a variety of methods. For artifacts composed of materials including wood, paper, textiles, and leather, fungal damage is a prevalent worry (Gutarowska 2020). The following are the main fungal modes of damage:

(a) **Enzymatic Degradation**: One important cause of deterioration in legacy materials is enzymatic degradation by fungi, which is especially noticeable in organic materials including wood, paper, textiles, and leather (Gutarowska 2020). A range of enzymes produced by fungi degrade complex organic substances, causing historical objects to deteriorate. The main enzymes in this process are ligninases, cellulases, and hemicellulases (Allsopp et al. 2004). A variety of variables, including the type of fungal species present, the makeup of the historical material, and environmental circumstances, might affect the intricate and dynamic process of enzymatic destruction by fungus. Strategies for preventive conservation entail managing temperature and humidity, to inhibit fungal growth and enzymatic activity (Allsopp et al. 2004). Enzymes produced by fungi, such as cellulases and ligninases, decompose complex organic molecules like cellulose and lignin that are found in materials like paper and wood (Gutarowska 2020). The material's structural integrity is weakened by this enzymatic breakdown.

(b) **Hyphal Penetration**: One important way that fungi can harm historical items is through hyphal penetration. Molds, mildews, and decay fungi are examples of fungi that grow by producing hyphae and mycelial structures (Allsopp et al. 2004). The structure of organic materials can be physically broken down by these hyphae, resulting in a variety of degradation. As fungi proliferate, they produce hyphae, and pseudohyphae besides the blastoconidia. The thread-like hyphal structures pierce organic matter and form biofilms on the surfaces. This morphological form has the ability to physically disrupt the material's structure and build networks for the absorption of nutrients.

(c) **Staining and Discoloration**: Common signs of microbial activity in heritage objects are discoloration and staining, which are brought on by the fungus. The pigments and metabolic byproducts produced by fungi, such molds and mildews, can cause noticeable alterations in the color and texture of organic materials (Gutarowska 2020). During their growth and metabolism, fungi generate metabolic byproducts. The impacted material may become stained or discolored as a result of these byproducts, which will change how it looks.

(d) **Production of Secondary Metabolites**: A wide range of secondary metabolites can be produced by fungi, including those that develop on heritage materials. These substances frequently have ecological functions, such as defense mechanisms or interactions with other species, but they are not directly involved in the growth and reproduction of the fungus (Allsopp et al. 2004). The generation of secondary metabolites by fungi can affect historical items in a number of ways, including discoloration, staining, and possible health risks. Mycotoxins are very hazardous substances that are produced by certain types of fungus. These compounds may be detrimental to the surrounding environment as well as the organic material. Mycotoxins can endanger human health as well as cause the substance to deteriorate (Allsopp et al. 2004).

(e) **Fungal Communities**: Various fungus species that may colonize and interact with organic substrates form dynamic assemblages, known as fungal communities, on historic materials. These communities are essential to both the preservation and deterioration of cultural assets. The type of material, the surroundings, and the existence of other microorganisms are some of the variables that affect the makeup and traits of fungal communities (Allsopp et al. 2004). Applying successful conservation measures requires an understanding of the fungal populations on heritage items. The long-term preservation of cultural artifacts depends on routine monitoring, the identification of relevant fungal species, and the creation of focused preservation techniques.

As a result of various fungus species colonizing the same material throughout time, microbial communities may arise in succession. This may lead to cumulative and complex patterns of deterioration. A biodeteriogen selects a substrate according to its ability to degrade it, in accordance with the chemical composition and the environmental conditions. Since fungi are ubiquitous microorganisms with pronounced metabolic activities and saprophyte mode of nutrition (thriving on decaying matter), these organisms are capable of colonizing various types of microenvironments and constantly causing problems in collection and conservation of cultural heritage (Savković et al. 2019). The fungal propagules (spores and mycelial fragments) are always present in the air, with concentrations in proportion with suitable environmental factors (Kasprzyk 2008). Several fungi live in a symbiotic association with human body, making their spread easy. Due to the nature of the metabolism (saprophyte), fungal infestation can cause both aesthetic and physical damage to a variety cultural heritage, including the stone, masonry, paint, paper, wood, textile, and other materials.

Different organic and inorganic materials have been utilized historically to create works of art, and in more recent times, these materials are being currently utilized once more for an accurate restoration or conservation. The mineral pigments were combined with organic binders (egg yolk, casein, linseed, poppy, hemp, Chinese wood, or other resins) to create paint (Szczepanowska 2013). Prior to painting, linen canvas clamped to wooden frames were frequently prepared with rabbit skin, making them susceptible to mold growth. Due to the enormous variety of enzymes that fungi can create, museum exhibits are inextricably linked to mold prevention, monitoring, and treatment of mold on contaminated objects (Dicus 2000). Detachment is frequently reported on infestation if fungi move and enter cracks in the paint. They are a nuisance in paper conservation because of their capacity to secrete the enzyme cellulase which degrades the cellulose, of which the manuscript or canvas is made of. Fungi can also degrade lignin. However, the lignin-degrading fungi are infrequently found in indoor contexts. Cellulose-degrading *Serpula lacrymans* or *Conophora puteana*, on the other hand, can inflict significant harm to culturally significant objects (Bech-Andersen and Elborne 2004).

It is to be noted that even though fungi have a saprophytic mode of nutrition, they have certain preferences for food, the material on which they thrive. Their differential nutritional requirements influence the artifact directly or indirectly. Numerous enzymes secreted by the fungi to digest its food possess varied substrate specificities. Thus, the composition of the cultural/archaeological heritage and environmental circumstances influences the degree of damage by a particular fungal species. Professionals in conservation can build effective preservation and restoration techniques with the aid of accurate species identification and an understanding of their digestive mechanisms as well as the nature of the cultural heritage material. It is important to employ adequate techniques for correct species-identification and physiological characterization of autochthonous isolates to accurately assess potential threats to the heritage art (Savković et al. 2019). The fungi can produce a variety of substances including many enzymes—which include β-glucosidases, cellobiohydrolases, endoglucanases, and xylanases, to name a few. The impact of fungal deterioration on a material depends on its type, and the fungal species. Paper and textiles, for example, have abundant cellulose, which is degraded by the fungal enzyme cellulase. The cellulase-producing fungi are therefore more likely to attack these articles. Keratinous substances, on the other hand, are more likely to be attracted by other fungal groups (Table 26.1).

Fungal deterioration can damage a piece of artwork in various ways—including discoloration, bio-pitting, cracking, exfoliation, and patina formation, to name a few of the symptoms due to fungal hyphal growth and penetration into the object. Contrary to this physical damage, the chemical damage caused by fungi (chemical mechanism) involve secretion of acids, the release of extracellular enzymes, the creation of pigments, oxidation/reduction reactions, and formation of secondary mycogenic minerals, resulting in significant structural and aesthetic changes that may be irreversible and may harm artworks in the long run. Understanding complex biodeterioration processes brought on by the microorganisms, especially fungal deteriogens, requires careful isolation and identification of fungal isolates and the application of microscopic and other techniques and *in vitro* biodegradation testing.

26.7 Common Fungi of Interest to Heritage Conservators

A lengthy history of tales and mysteries shroud the relationship between mold and cultural heritage. The so-called "Curse of the Pharaoh"—the death of several archaeologists in the team of Howard Carter after discovery and opening of Tutankhamun's tomb—was later explained by the fact that the spores of the pathogenic fungi *Aspergillus niger* and *Aspergillus flavus* were present on and inside the sarcophagi, and that a lung infection or other systemic mycosis—Aspergillosis—was the possible cause of death. However, even though there is no statistical support for an increase in disease incidence in relation with the archaeological discoveries and no concrete scientific evidence to suggest Aspergillosis as a primary cause of the discoverer's death, there is some truth to this narrative. Microorganisms can deteriorate a variety of substances ranging from the building material (stone, mortar, slurries, and even concrete) to the metals, glass and several objects used in heritage art and material (Piñar and Sterflinger 2009). The historical material and artwork which is primarily made of the natural material (for example, the paper and paint) is particularly susceptible to fungal deterioration. A number of microorganisms including not only the fungi but also bacteria such as *Staphylococcus* and *Bacillus* genera are drawn to the heritage objects and architecture, where they can dwell deep or on surface and cause deterioration if not caught and treated in time (Caselli et al. 2018). The bacteria, especially cyanobacteria, algae and lichens contribute to deterioration or weathering of the stone, producing a characteristic phenomenon consisting of large green-black patches or stains.

Cyanobacteria (*Eucapsis*, *Leptolyngbya*, *Scytonema,* and *Fischerella*) can be one of the most important deterioration agents for wall paintings and inscriptions. Caselli et al. (2018) isolated the filamentous fungus *Aspergillus*, *Penicillium*, *Cladosporium*, and *Alternaria* from deteriorating pictorial layer of paintings using culture-dependent techniques. Briefly, they collected the above-mention filamentous fungi from the infested paintings and cultivated the isolates by growing them in the nutrient medium. They also analyzed the paintings by Scanning Electron Microscopy (SEM) and Energy Dispersive X-ray Spectroscopy (EDS). In another study, on some 200 oil paintings at the UST Museum of Arts and Sciences, Philippines, a total of 48 strains of fungi were isolated and found to cause a significant damage to the paintings including surface damage (81%), discoloration (14%), and structural damage (5%). *Aspergillus* was among the most common of all fungi (77%) isolated from these paintings, followed by *Penicillium* (13%), *Chaetomium* (4%), *Rhizopus* (2%), *Cunninghamella* (2%), and *Mucor* (2%) (Paner 2009).

Recent studies have demonstrated a relationship between the material stability and chemical structure of the polymeric network, in particular, a low degree of cross-linking was observed in combination with a high degree of oxidation of the polymeric network and the oil paint layer sensitivity to water (Nardelli et al. 2021). Water is a medium for the growth of microorganisms. Other fungal genera found in heritage articles include *Rhodotorula* with a high affinity to osmotic environments. *Exophiala, Aureobasidium, Coniosporium,* and *Wallemia* are few other examples of the so-called "black yeasts" and microcolonial fungi frequently discovered in conjunction with high-osmolarity materials such as the wall paints or hydrocarbons-containing materials, tensides (the agents which decrease the surface tension) silicone, or paraffin waxes.

Phylogenetically, the fungi have been divided into (1) *Chytridiomycota* (which represents many secondary aquatic fungi), (2) *Glomeromycota* (live in symbiotic relation with plant roots—has little or no significance with museum materials), (3) *Zygomycota* (biodeteriogens of grain, fruit and vegetables—sometime occurring on museum material as opportunists), and (4) *Basidiomycota* (most mushrooms and toadstools—*Serpula lacrymans*, the most important wood decaying indoor fungus that belongs to *Basidiomycota*, and damages historical buildings made of wood). In general, hyphomycetous fungi or mold, mostly *Ascomycetes,* is the most important biodeteriogen in museums, museums' storage, libraries, collections, and restoration studios (Sterflinger and Pinzari 2012). Mold is basically a term applied for the asexually reproducing form of fungi (*Anamorphs*). From the biodeterioration point of view, the fungi associated with the cultural heritage objects can be: (1) **Opportunistic** fungi—growing on practically all types of material if the environment is sufficiently humid, but are not able to degrade the material enzymatically and use it as main source of carbon, or (2) **Real "material pathogens"**—substrate-specific, able to degrade specific material of works of art such as cellulolytic fungi on paper and keratinolytic fungi on leather, hair and feathers (Meier and Petersen 2006; Blyskal 2009). The infestation spreads with the help of fungal spores and conidia which are readily carried by the wind and air flow and infest both indoor and outdoor settings, causing primarily airborne fungal infection of cultural heritage material with notable seasonal fluctuations.

26.8 Physical and Chemical Changes in Fungus-Infested Heritage Objects

The fungi primarily affect an object in two ways—mechanically and chemically, as described above. The processes often take place simultaneously and depending on the nature of the substrate (exogenic or endogenic), the effect of one process can be more prominent than the other. Notably, depending on its location, the fungal colonizers can affect a substrate either from the surface to its interior, and *vice versa* (Garg et al. 1995). Physical processes take place under the influence of the fungal hyphal apical growth or by the formation of fruiting bodies on the surface and/or the inner layers of the colonized material. If the fungal growth is superficial, it promotes the spread of mycelium which covers the substrate and changes its original appearance and aesthetic value (Garg et al. 1995). Internal fungal growth might lead to further damage and exfoliation detachment of painted layers, if the artwork happens to be a painting. The melanized micromycetes are well-known inducers of mechanical deterioration, especially of the stone substrates, since melanin provides mechanical rigidness to the fungal structure, and it enhances the turgor and facilitates hyphal penetration (Pinna and Salvadori 2008). Figure 26.1 shows scanning electron microscopic image of hyphae of *Aspergillus niger* on leather.

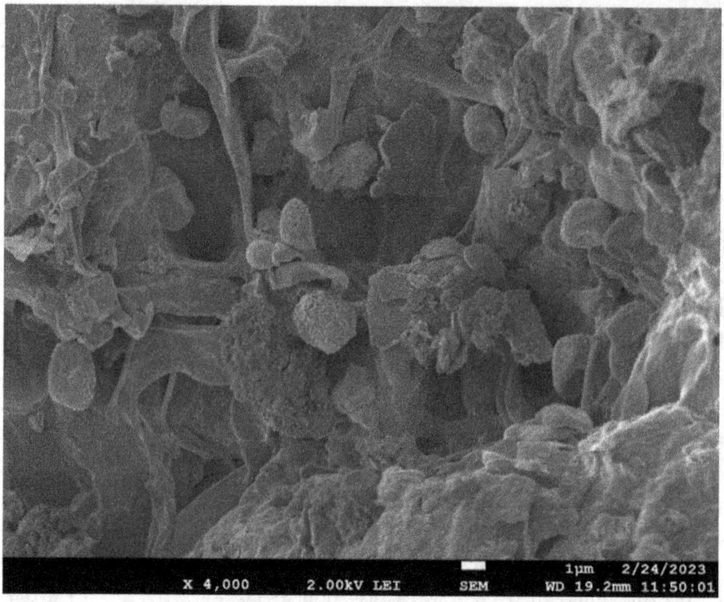

Fig. 26.1 The scanning electron microscopic image of a leather object showing the penetration of *Aspergillus niger*. The image was taken by the authors at magnification of 4000×, indicating depth of fungal infestation in the object

Chemical changes due to fungal infestation result in the breakdown of organic materials of which the art is made of. The enzymes produced by fungi to digest their food (biodegradable material) especially the cellulases and hemicellulases, hydrolyze the complex carbohydrates/polysaccharides like cellulose and hemicellulose. The cellulose fibers found in paper, textiles, and wooden artifacts deteriorate as a result of this process. Enzymes that can convert large, complex, and often water-insoluble compounds of the material (cellulose, hemicellulose, lignin, proteins, and lipids) into low-molecular-weight soluble substances such as glucose play an important role in the biodeterioration processes. Fungus attack can start oxidative processes, especially when the oxygen and moisture are present. This may cause organic compounds to oxidize, producing carbonyl groups and other reactive chemical moieties in the process, consequently causing the discoloration of the damaged material, yellowing, or turn it darker. Specific micromycetes produce pigments which vary in chemical composition and color and are species-specific (Garg et al. 1995). They are present in hyphae, conidia, or are secreted into the substrate while their production is determined by the availability of nutrients, minerals, UV radiation, pH, temperature, and other environmental factors (Florian 2002). The pigment secretion on or into the substrate material leads to appearance of different, frequently irreversible, colorations leading to the observable changes on the heritage object, thus diminishing its aesthetics and accelerating the process of deterioration (Caneva et al. 2003).

26.9 Fungal Growth Overview

Fungi thrive in different conditions, displaying growth through hyphal extension, spore formation, and mycelium development (Merad et al. 2021). This growth contributes to ecological processes, nutrient recycling, and the occurrence of fungal infections in organisms. Understanding fungal growth is crucial in various fields, including agriculture, food industry, medicine and preservation of museum artifacts and heritage items. Research studies on growth mechanisms help develop strategies to control fungal diseases and enhance our understanding of its ecology in natural ecosystems. Fungi exhibit diverse morphologies, including yeast, molds, and dimorphic forms. Yeasts are unicellular fungi, while molds are multicellular with a filamentous structure (Money 2016). Dimorphic fungi can switch between yeast and mold forms depending on environmental conditions. Fungi also play both beneficial and detrimental roles in museum environments. Understanding these roles is crucial for preserving artifacts and maintaining a healthy museum space. Fungi are known agents of biodeterioration, especially for organic materials like wood, paper, leather, and textiles (Koul and Upadhyay 2018). They can cause structural damage and deterioration, threatening the longevity of artifacts. High humidity levels in museum environments can lead to mold growth on artifacts (Savković et al. 2021). Mold not

Table 26.2 Various key features for fungal growth

S. No.	Factors	Role	References
1.	Environmental factors	Fungal growth is influenced by environmental factors such as temperature, humidity, pH, and nutrient availability. Different fungal species have specific requirements for optimal growth, and variations in these conditions can affect their development	Walker and White (2017)
2.	Nutrient absorption	Fungi are heterotrophic organisms, relying on external sources for nutrition. They secrete enzymes into their surroundings to break down complex organic molecules into simpler forms, which can then be absorbed by the fungal cells	Money (2016)
3.	Spore germination	Fungi reproduce through spores, which are often dispersed in the environment. When spores encounter suitable conditions, they germinate, giving rise to new hyphae and initiating the growth of a new fungal colony	Merad et al. (2021)
4.	Hyphal extension	Fungi typically grow as multicellular structures called hyphae, which are thin, thread-like filaments. Hyphal tips elongate through a process known as apical growth, enabling the fungus to explore and exploit its environment for nutrients	Money (2016); Walker and White (2017)
5.	Mycelium formation	Hyphae collectively form a network called mycelium, which constitutes the main body of the fungus. The mycelium functions as the primary structure for nutrient absorption and plays a crucial role in the life cycle of the fungus	Merad et al. (2021)

only damages the physical structure of objects but can also cause discoloration and create aesthetically undesirable effects. Some fungi can release spores that may pose health risks, particularly for individuals with allergies or respiratory conditions (Merad et al. 2021). This is a concern for both museum staff and visitors. Table 26.2 summarizes various key factors associated with fungal growth.

26.9.1 Media and Favorable Conditions for Mold Growth

In a museum, molds can present major challenges, impacting both the preservation of artifacts and the overall museum environment (Branysova et al. 2022). Molds, including various species of *Aspergillus*, require specific environmental conditions and media to grow. *Aspergillus* species are commonly found in indoor and outdoor environments and can grow on a variety of substrates (Walker and White 2017). Figure 26.2 shows the major factors involved in the growth of *Aspergillus* species.

26.9.2 Media and Its Components for *Aspergillus* Growth

The essential components for growing *Aspergillus* in culture media include a carbon source (e.g., glucose), a nitrogen source (e.g., ammonium sulfate), minerals (e.g., salts), vitamins, and agar as a solidifying agent (Viegas et al. 2020). These media

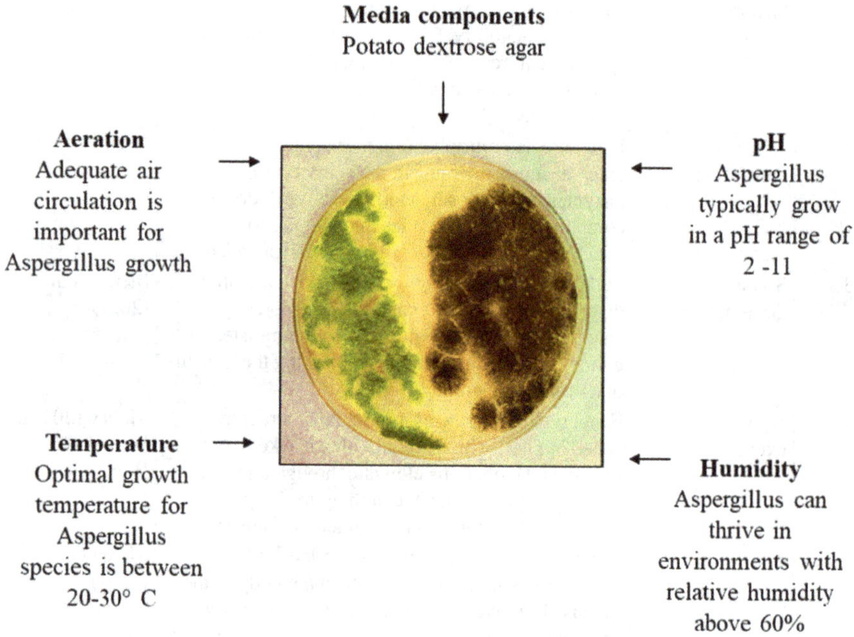

Fig. 26.2 The key factors that influence the growth of *Aspergillus*

provide the necessary nutrients to support the growth and development of *Aspergillus* species for research and laboratory purposes. In addition to the basic components mentioned, specific media formulations for *Aspergillus* growth may vary depending on the particular strain or research objectives. Some commonly used culture media for *Aspergillus* include Potato Dextrose Agar (PDA) and Sabouraud Dextrose Agar. These media are designed to provide a balanced nutritional environment with carbohydrates, proteins, and other essential nutrients, promoting robust mycelial growth and sporulation (Ali et al. 2016). Adjustments in pH, temperature, and the addition of supplements may further optimize the conditions for studying various aspects of *Aspergillus* (Viegas et al. 2020). Vitamins play a crucial role as cofactors in enzymatic reactions. Biotin, thiamine, and other vitamins may be included in the growth medium to support growth (Viegas et al. 2020). *Aspergillus* is an aerobic organism; hence, adequate aeration or shaking in liquid cultures ensures the availability of oxygen.

26.9.3 Isolation of *Aspergillus flavus* and *Aspergillus niger* from Leather Surface

Aspergillus flavus and *Aspergillus niger* are both fungal species that can be found on various surfaces, including leather (Branysova et al. 2022). When isolated from a leather surface, certain features can be observed to identify these fungi (Table 26.3, Fig. 26.3).

Table 26.3 Some characteristic properties of *Aspergillus* species

Aspergillus species	Characteristic/property	References
Aspergillus flavus	• **Colonial Morphology**: Colonies of *A. flavus* are typically fast-growing and may appear initially as greenish-yellow, yellow, or yellow-green • **Conidiophores and Conidia**: The conidiophores of *A. flavus* are vesiculate, meaning they have a swollen, balloon-like structure. Conidia (asexual spores) are typically spherical to elliptical and have a smooth surface. The color of conidia can range from yellow to green • **Sclerotia**: *A. flavus* is known to produce sclerotia, which are compact masses of hyphae. However, the presence of sclerotia may vary • **Aflatoxin Production**: *A. flavus* is known for producing aflatoxins, which are mycotoxins that can be harmful. Aflatoxins are not visually observed but may be detected through laboratory analysis	Klich (2007), Yao et al. (2015)

(continued)

Table 26.3 (continued)

Aspergillus species	Characteristic/property	References
Aspergillus niger	• **Colonial Morphology**: Colonies of *A. niger* are often fast-growing and initially appear white, turning to dark brown or black over time • **Conidiophores and Conidia**: Conidiophores of *A. niger* are typically unbranched and bear a vesicle at the apex. Conidia are produced in chains and are typically black or dark brown in color. They are often round or oval • **pH Indicator**: *A. niger* is known for its ability to produce citric acid in industrial laboratories	Scazzocchio (2009), Ellis et al. (2007), Daba (2017)

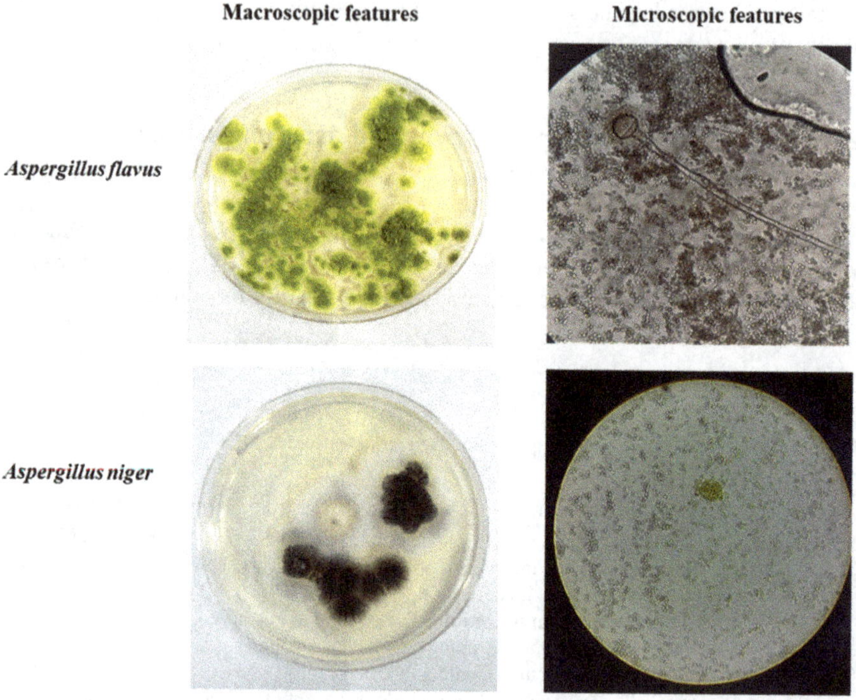

Fig. 26.3 Macroscopic and microscopic features of different *Aspergillus* species

26.9.4 Secreted Hydrolytic Enzymes

Hydrolytic enzymes are a class of enzymes that facilitate the breakdown of complex molecules into simpler compounds through the addition of water molecules (da Silva et al. 2001). These enzymes play vital roles in various biological processes, and they are categorized based on the specific type of biomolecule they target. *Aspergillus* species are known for producing a variety of hydrolytic enzymes, which play crucial roles in the degradation of complex organic substrates (Table 26.4).

Table 26.4 List of secreted hydrolytic enzymes by *Aspergillus* species

S. No.	Hydrolytic enzymes	Role	References
1.	Proteases	• Break down proteins into peptides and amino acids • Important for pathogenesis and nutrient acquisition • Employed in processes like protein hydrolysis and the production of protein-based products	Medina et al. (2005)
2.	Lipases	• Hydrolysis of lipids (fats and oils) into fatty acids and glycerol • Have applications in industries such as food processing and biodiesel production	da Silva et al. (2001)
3.	Amylases	• Hydrolyze starch into simpler sugars • Used in various industrial applications, including the food and beverage industry	Tiwari et al. (2015)
4.	Cellulases	• Breaks down cellulose into glucose units • Find applications in industries like biofuel production and textile processing	Juturu and Wu (2014)
5.	Hemicellulases	• Target hemicellulose, a complex polysaccharide present in plant cell walls • Contribute to the degradation of plant biomass	Manju and Chadha (2011)
6.	Pectinases	• Hydrolyze pectin, a complex polysaccharide found in plant cell walls • Have applications in fruit and vegetable processing, as well as in the clarification of fruit juices	Nighojkar et al. (2019)
7.	Chitinases	• Breaks down chitin, a structural component in the cell walls of fungi and insects • Involved in fungal antagonism and have potential applications in agriculture and pest control	Singh et al. (2021)
8.	Phytases	• Hydrolyze phytic acid, releasing phosphorus and making it available for plant uptake • Have applications in animal feed to enhance phosphorus utilization	Dailin et al. (2019)
9.	Oxidases	• Generate reactive oxygen species (ROS), which can cause oxidative damage to host tissues and contribute to pathogenesis	Hatinguais et al. (2021)

It is important to note that specific enzymes produced can vary among different *Aspergillus* species and strains. The study of these enzymes provides insights into the mechanisms of pathogenesis and can be valuable for developing diagnostic tools and therapeutic interventions. The pathogenesis of *Aspergillus* involves various mechanisms, and secreted enzymes play a crucial role in the invasion and colonization of host tissues. The severity of infections can vary depending on the host immune system, underlying health conditions, and other factors. Treatment typically involves antifungal medications, and early diagnosis is critical for a favorable outcome. Figure 26.4 illustrates major hydrolytic enzymes secreted by *Aspergillus* that plays crucial role in pathogenesis.

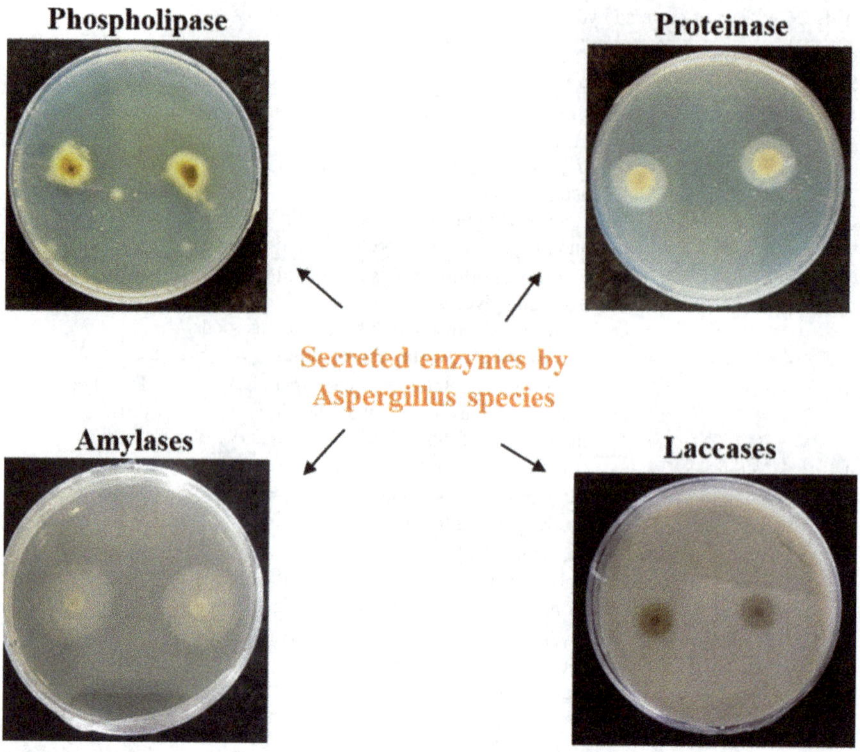

Fig. 26.4 Major secreted hydrolytic enzymes by *Aspergillus* species

26.10 Control of Microbial Growth and Precautionary Measures

Traditional methods coupled with high-end analytical tools and advancements in conservation and preservation techniques have implications in protecting cultural heritage and architecture. An integrated approach relying on climate control, material-specific cleaning and application of carefully selected biocides are all important to prevent fungal contamination or to treat and already contaminated object. The easiest strategy to handle a mold is to avoid it altogether by getting rid of it, or at least minimizing any environmental factor that encourages mold spore germination and growth. Simply closing the windows can cut the amount of outdoor mold spores by 2%. Spore levels are further diminished by 3% when an electrostatic filtering system is used, and an additional 5% reduction can be achieved by appropriate/central air-conditioning. Briefly, fungal growth and spread can be inhibited by,

(a) **Controlling the Relative Humidity or RH**. RH in museums and other display spaces should not exceed 60%. The RH can be managed by:
   - Installing dehumidifiers in spaces where art is displayed or stored
   - Tracking the RH (45–55%) at an ideal range
   - Save closed cardboard boxes for storing artwork only when doing so temporarily.

   The correct level of air moisture is crucial when utilizing large HVAC systems for heating and cooling and can be attained by employing dehumidifiers.

(b) **Controlling temperature**. Maintaining the temperature range between 64 °F and 68 °F (18 °C and 20 °C) is crucial. It is important to ensure that the art is not kept or displayed next to the exterior walls—a space of 15–18 in. is advised. Any leaky pipes, downspouts, vents, or gaps in walls, flooring, ceiling, windows, or doors should be repaired and high moisture from leaks or outside air by adequate ventilation needs to be prevented. Insulation of pipes, ventilate attics or crawl spaces, and routine maintenance of HVAC systems, changing air filters should be completed within the specified time.

(c) **Starving the mold**. Books, clothes, furniture, and draperies are frequently constructed of organic material that may be broken down into simple sugars, peptides, and amino acids, just like many works of art. Mold feeds and flourish on this nutritious material. The substrates and surfaces made of inorganic material is also vulnerable to mold infestations due to age and surface buildup from dust, filth, bug debris, and handling oils.

(d) **Conservatory treatments**. Chemical fungicides, extracts such as the essential oils, natural and bio-based compounds are potential antifungal agents. Advancements in technology such as the nanotechnology have opened up possibilities for innovative antifungal treatments. Nanoparticles with antifungal properties can be applied to the material surface to inhibit fungal growth while minimizing the alteration of the material's original structure. National Museum Institute, Delhi NCR, India is doing some cutting-edge research in this area.

(e) **High-end analytical tools**. The high-end analytical tools to study the impact of infestation and treatment need to be employed judiciously.

26.10.1 Precautionary Measures

Proper storage, frequent monitoring, and targeted treatment are crucial measures to check the microbial growth and contain fungal deterioration of museum and cultural heritage. Besides deteriorating the objects of heritage value, fungi release allergens and chemicals which are harmful to the peoples who may come into contact with the damaged material, risking their health. It is important to remember that a variety of fungal species may be to blame for the deterioration of these cultural property objects, and their impacts can vary depending on the environment, the moisture level of the material, and the specific material characteristics (Viñas 2005). Conservation practices are crucial in preventing fungal degradation of these important historical artifacts. They consist of appropriate handling, regular observation, and focused care. On historical objects, microbial decay has a significant detrimental influence. Structures may become unstable, and irreplaceable cultural objects may be destroyed, historical or artistic value may be lost, etc. Fungus development can also lead to health problems since some fungi release allergens and substances that can infect people who come into contact with the infested materials.

Fungi can survive at low water level and are well adapted to thrive in the micro-climatic niches formed due to condensation, lack of ventilation and retention of water by the hygroscopic material. Capacity to thrive at low humidity and ability to secrete a wide range of enzymes such as cellulases, glucanases, laccases, phenolases, keratinases, and monooxygenases is perhaps the most significant causes of deterioration in museums, collections, and libraries (Viñas 2005). They thrive in environments with substantially lower humidity levels; in museums, 55% relative humidity (RH) is generally regarded as the border line for fungal growth. Denying moisture, the fungus needs to germinate and proliferate is the best way to inhibit growth. The RH ought to be regularly checked and kept below 65%. At 45–55% RH, the spore germination is less likely to happen. Dehumidifiers can be used to lower the moisture in air. Similarly, temperature between 18 and 20 °C can significantly reduce microbial germination and growth. Maintaining an appropriate air circulation is also important (Viñas 2005). The climate control is adjusted below this value. Poor ventilation and dishomogeneity in surface temperatures can produce water condensation points and local micro-climates with water availability more than the rest of the indoor environment (Szczepanowska 2013). A fungal colony grows up to 4 mm/day, resulting in contamination of a whole collection within a few days.

The chemical composition of the heritage material, which contains a variety of organic substances, plays an important role in fungal growth, spore germination and colony development. The carbon source used in art material especially the paints which can be used by the fungi include the mineral pigments dispersed in organic binders (egg yolk, casein, oil, or resins) (Szczepanowska 2013). The animal glue is generally used as primer before painting on linen canvas. Ceramics have food remains that have been used as fungus-food for thousands of years. Textiles, leather, straw, clay, natural hair and feathers on sculptures and other works of art are common (Szczepanowska 2013). The books, manuscripts, and scrolls which are made of paper, papyrus, and parchment are frequently mounted with the starch paste as an adhesive. Any deteriorated brick, leaky roof, masonry pointing, leaking pipes, gutters, downspouts, and broken windows should be repaired. Silica gel and other buffers can be used in a closed environment/cabinet to regulate the RH. These buffers can release or absorb moisture into the atmosphere. The NPS Museum Handbook, Part I can be referred as a guide on the use of silica gel. Appropriate techniques should be applied to prevent the microbial growth from occurring on heritage material (Fig. 26.5).

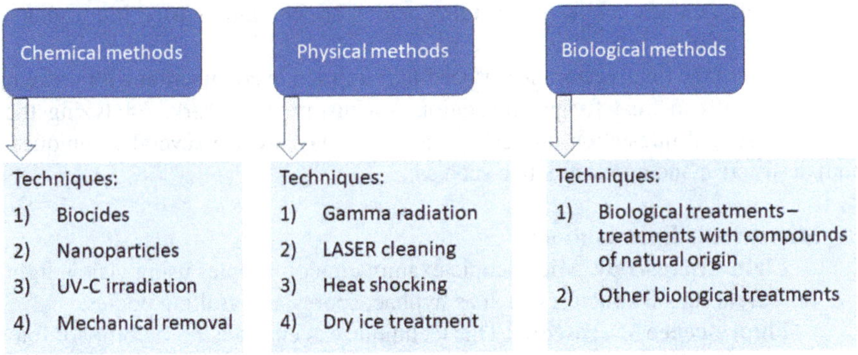

Fig. 26.5 Methods and techniques to inhibit the microbial growth on cultural heritage

Table 26.5 List of antifungal drugs used in museums to protect artifacts from fungal growth (Ali et al. 2014; Santos et al. 2019)

Antifungal agents	Mode of action	Examples	Usage
Azoles	Inhibit the activity of lanosterol 14α-demethylase and disrupts the conversion of lanosterol to ergosterol, a crucial step in the biosynthesis of fungal cell membranes	Fluconazole, itraconazole, ketoconazole, voriconazole, and posaconazole	In the form of antifungal sprays
Polyenes	Bind to ergosterol creating pores that increase fungal cell membrane permeability, leading to leakage of cellular contents and ultimately causing fungal cell death	Amphotericin B	In the form of sprays or coating with antifungal ointments/gels
Allylamines	Inhibit fungal ergosterol synthesis by targeting squalene epoxidase, disrupting cell membrane formation and leading to fungal cell death	Terbinafine	antifungal sprays
Echinocandins	Inhibit fungal cell wall synthesis by targeting β-(1,3)-D-glucan synthase, resulting in cell wall destabilization and fungal cell death	Caspofungin, Micafungin	In the form of sprays

26.11 Current Treatment and Analysis Protocols

Antifungal agents are commonly used in museums to protect artifacts from fungal growth and deterioration. Fungi can pose a significant threat to various materials, including paper, wood, textiles, and other organic substances commonly found in museums. Preservation in museums typically relies on methods to control fungal growth, such as maintaining optimal environmental conditions and employing preventive conservation strategies. Additionally, the use of chemicals, including antifungal drugs, requires careful consideration to avoid unintended side effects or damage to artifacts. Table 26.5 represents some antifungal agents mainly azoles, polyenes, allylamines, and echinocandins, which are commonly used in museums to protect fungal growth.

Majority of the time, restoration procedures are used in conjunction with preventative measures to treat fungal infestations on historical artwork. Analyzing the impact of fungal infestation on heritage artwork may involve several techniques, both destructive and nondestructive such as:

(a) **Microscopic Examination**:

Light Microscopy: Microscopic examination of samples using visible light can reveal fungal structures such as hyphae, spores, and fruiting bodies.

Fluorescence Microscopy: This technique can enhance the visibility of fungal elements by using fluorescent dyes that bind to specific structures.

(b) **Culture Techniques**:

Fungal Culture: Isolating and growing fungi from a sample on a nutrient medium allows for the identification of specific fungal species.

DNA Analysis: DNA sequencing can be used to identify fungi at the species level. Polymerase Chain Reaction (PCR) may be employed for this purpose.

(c) **Chemical Analysis**:

Infrared Spectroscopy (FTIR): FTIR can be used to analyze the chemical composition of the artwork and detect any changes caused by fungal contamination.

Gas Chromatography-Mass Spectrometry (GC-MS): This technique can identify volatile organic compounds produced by fungi, aiding in their identification.

(d) **Imaging Techniques**:

Scanning Electron Microscopy (SEM): SEM provides high-resolution images of the surface, allowing for detailed examination of fungal structures and the extent of damage.

X-ray Imaging: X-ray techniques can be used to examine the internal structure of artworks, providing information on the depth of fungal penetration.

(e) **Moisture Content Measurement**:

Gravimetric Method: Measuring the moisture content of the artwork can help assess the environmental conditions conducive to fungal growth.

(f) **Nondestructive Imaging**:

Ultraviolet (UV) Imaging: UV light can reveal fluorescence patterns associated with fungal contamination on surfaces.

Infrared Imaging: Infrared imaging can detect changes in the thermal properties of the artwork caused by fungal activity.

(g) **Analysis of Secondary Metabolites**: High-Performance Liquid Chromatography (HPLC) can be used to analyze secondary metabolites produced by fungi, aiding in their identification.

(h) **Collaboration with Mycologists**: Consulting with mycologists and microbiologists can provide additional insights into fungal identification and behavior.

Conservation professionals may combine several of these techniques to obtain a comprehensive understanding of the fungal contamination and its impact on heritage artwork. When dealing with fungal contamination on leather or any cultural heritage item, it is crucial to involve professionals in mycology and conservation to ensure proper identification and treatment. Additionally, safety precautions should be taken, especially if dealing with potentially harmful mycotoxins like those produced by *Aspergillus flavus*. It is crucial to follow ethical guidelines and ensure that analysis methods are chosen based on their suitability for the specific artwork and the preservation of its integrity. Additionally, ongoing research in conservation science may introduce new and improved techniques over time.

26.12 Need for Sustainable Conservation Interventions

The activity of conserving and restoring the artworks in order to ensure its long-term stability and aesthetic integrity entails a variety of scientific, technological and artistic methods with the goal of researching, examining, recording, and caring for the heritage works of art to lessen the damage, decay, or deterioration brought on by numerous variables including aging, environmental conditions, accidents, or prior interventions. Manuscripts, paintings, textiles, ceramics, photographs, wood and bone sculptures, and leather objects are particularly susceptible to the insects, microbial, especially fungal infestations. Recent developments in applied microbiology and biotechnology offer crucial insights into the preservation of cultural heritage. Although there have been many physical and mechanical techniques employed in the past, they are unable to completely stop organisms from growing. Commercial formulations of organic biocides are short-lived because native microflora can use them as a source of nutrients once they become resistant to the biocides (Gatenby and Townley 2003). Consequently, inorganic nanoparticles have been proven to be of success in terms of heritage conservation. However, chemicals being used for treatment may also potentially harm the object due to various factors.

Chemicals are frequently used by conservators for maintenance tasks including cleaning. When these chemicals are produced, used, or disposed of, they may have negative effects on people or the environment. Conservators can lessen the impact of conservation procedures by employing two green chemistry principles: using safer chemicals and discovering safer chemical substitutes. Like sustainable or circular chemistry, green chemistry is a branch of chemical engineering that focuses on designing goods and procedures to reduce or completely do away with the need for dangerous materials (Gatenby and Townley 2003). Green chemistry concentrates on the environmental impact of chemistry, including reducing the use of non-renewable resources and developing technical methods for pollution prevention, whereas environmental chemistry focuses on the impacts of harmful chemicals on the environment (Gatenby and Townley 2003).

Green chemistry techniques have the potential to be extremely important for conservation and restoration initiatives involving heritage objects. When using green chemistry to preserve historical materials, the following results could occur:

(a) **Minimization of Hazardous Chemicals**: Green chemistry seeks to reduce the usage of hazardous substances in order to lessen the possibility of adverse effects on the environment and cultural heritage specialists. This is especially crucial when managing priceless and delicate artifacts.

(b) **Preservation of Original Materials**: Using green chemistry helps to preserve heritage artifacts' original materials and qualities. The essential properties of the materials are less likely to be altered or damaged by conservation efforts because harsh chemicals and processes are avoided.

(c) **Sustainable Treatment Techniques**: Green chemistry promotes the creation and application of environment-friendly, sustainable treatment techniques. This covers the choice of environment-friendly consolidants, cleaning products, and solvents.

(d) **Energy Efficiency**: Conservation techniques can support overall sustainability in historical preservation efforts by utilizing energy-efficient technologies supported by green chemistry concepts.

(e) **Reduction of Waste Generation**: Green chemistry aims to produce as little waste as possible. This is important for conservation since hazardous waste disposal can have detrimental effects on the ecosystem. The main goal of sustainable practices is to recycle and reuse materials whenever possible.

(f) **Biodegradability:** Green chemistry frequently uses environment-friendly chemicals that are intended to decompose naturally. By doing this, the long-term environmental impact is decreased and the persistence of chemicals in the environment is reduced.

(g) **Encouragement of Renewable Resources**: Materials originating from sustainable sources and renewable resources are promoted by green chemistry. This is in line with the overarching objective of encouraging environmental sustainability and lowering reliance on finite resources.

(h) **Local and Traditional Knowledge**: Green chemistry concepts are in line with conservation efforts that incorporate local and traditional knowledge. This involves utilizing conventional methods, which are frequently more ecologically friendly and sustainable, along with locally sourced materials.

(i) **Research and Innovation**: Conserving cultural assets by using green chemistry encourages new ideas and research. The progress of conservation research is aided by the creation of novel, sustainable materials and techniques.

(j) **Educational Value**: Increasing public awareness of sustainable practices like green chemistry techniques in conservation cultivates ecologically responsible values in professionals, students, and the general public.

Alternative antifungals compounds and industrial solvents, that have fungicidal properties, are still being researched. Several alternatives, for preventing the growth of mold in outdoor stone heritage and indoor storage places have been tested, including tea tree oil, zosteric acid, lavender oil, and neem oil (Gatenby and Townley 2003). Traditional conservation methods have their own limitations. There are reports in literature that the current restoration methods do not completely remove the infesting microbe and their spores, and there is always a chance of recurring infestation. The use of neem and its active components as part of green chemistry treatments on heritage materials can prove to be of immense potential as neem itself is an antifungal and antifeedant treatment of cultural heritage material (Mordue and Nisbet 2000). Neem is a naturally occurring pesticide. The oil extracted from the seeds of the neem tree has strong antimicrobial and insecticidal properties, and can be used as a natural and environmentally benign defense against the microbial and insect infestations—both frequently associated with the ruining of the historical artworks, records, and structures (Mordue and Nisbet 2000). The bioactive ingredients isolated from neem can be an efficient substitute for hazardous synthetic chemicals which pose risks to both conservators and the artifacts and can have negative environmental impacts. The neem extract can be used to develop substances which can be easily applied to the surfaces of art, books, and buildings to stop microbial growth and insect infestation (Mordue and Nisbet 2000).

This approach of employing neem in heritage conservation has several benefits summarized below:

- Neem is eco-friendly, sustainable, cost effective, and safer for conservators.
- Neem has preservation prowess and shows effects against a range of infestations. It shows antibacterial, antifungal, antifeedant, and insecticidal properties, and its application can protect an art specimen against microbe and insect infestation, thus prolonging the lifespan of the heritage material (Mordue and Nisbet 2000).
- Neem-based treatments show versatility as they can be applied on a range of art materials, including paper, textiles, wood, and even the stone architecture.
- Neem-based treatments show cultural sensitivity aligning with the philosophy of preserving heritage in its authentic form, as they do not alter the appearance or composition of artifacts.
- Convenient traditional application methods like brushing, spraying, or immersion can be used, depending on the nature of the material and condition.

Thus, the current approach of conservation toward reduced energy consumption and reduced risk for people and historical artifacts can be achieved by shifting away from solvent cleaning toward alternatives that reduce dependency on chemicals. One way to eliminate stains and residues is to use gel systems constituting enzymes in place of solvents for surface cleaning. Another cleaning option for a variety of hard surfaces is using lasers.

26.13 Conclusion and Recommendations

The fungal infestations of heritage art and architecture pose a serious threat to our cultural heritage and may cause an irreparable harm. Protection of our cultural legacy has been a challenge for heritage conservationists who have been working on newer ideas for conservation and preservation such as the development of novel treatment strategies, chemical remedies, biodeteriogens, and ecologically friendly methods without endangering fragile heritage objects. Finding an ideal balance between practical preservation and the preservation of the natural qualities and aesthetics of the material continues to be difficult. Collaboration between professionals in conservation science, materials science, mycology (study of fungi), and nanotechnology is crucial for making significant advancements and create sustainable and efficient preservation methods for cultural heritage and architecture against microbial, especially the fungal infestation. The use of green chemistry in the conservation of heritage materials is crucial for promoting sustainability, reducing environmental impact, and ensuring the long-term preservation of cultural artifacts. As the field of conservation continues to evolve, incorporating green chemistry principles can contribute to a more environment-friendly and responsible approach to heritage preservation.

References

Ali DMI, Abdel-Rahman TM, El-Badawey NF, Ali EAM (2014) Control of fungal paper deterioration by antifungal drugs, essential oils, gamma and laser irradiation. Egypt J Bot 54:219–246

Ali SR, Fradi AJ, Al-Aaraji AM (2016) Comparison between different cultural medium on the growth of five Aspergillus species. World J Pharmaceut Res 5(8):9–16

Allsopp D, Seal KJ, Gaylarde CC (2004) Introduction to biodeterioration, 2nd edn. Cambridge University Press, New York, NY, p 1

Bech-Andersen J, Elborne SA (2004) The true dry rot fungus (Serpula lacrymans) from nature to houses. In: Dradacky M (ed) European research on cultural heritage state of the art studies, vol 2, pp 445–448

Blyskal B (2009) Fungi utilizing keratinous substrates. Int Biodeterior Biodegrad 63:631–653

Branysova T, Demnerova K, Durovic M, Stiborova H (2022) Microbial biodeterioration of cultural heritage and identification of the active agents over the last two decades. J Cult Herit 55:245–260

Caneva G, Maggi O, Nugari MP, Pietrini AM, Piervittori V, Ricci S, Roccardi A (2003) The biological aerosol as a factor of biodeterioration. In: Mandrioli P, Caneva G, Sabbioni C (eds) Cultural heritage and aerobiology. Methods and measurement techniques for biodeterioration monitoring. Springer Science+Business Media, Dordrecht, pp 3–29

Caselli E, Pancaldi S, Baldisserotto C, Petrucci F, Impallaria A, Volpe L, D'Accolti M, Soffritti I, Coccagna M, Sassu G, Bevilacqua F, Volta A, Bisi M, Lanzoni L, Mazzacane S (2018) Characterization of biodegradation in a 17th century easel painting and potential for a biological approach. PLoS One 13(12):e0207630. https://doi.org/10.1371/journal.pone.0207630

Clayden J, Greeves N, Warren S (2012) Organic chemistry. Oxford University Press, pp 1–15. ISBN 0-19-927029-5

da Silva MC, Bertolini MC, Ernandes JR (2001) Biomass production and secretion of hydrolytic enzymes are influenced by the structural complexity of the nitrogen source in Fusarium oxysporum and Aspergillus nidulans. J Basic Microbiol 41(5):269–280

Daba G (2017) Production of citric acid by *Aspergillus niger* using molasses as substrate. In: Proyecto de grado, Addis Ababa Institute of Technology, Ethiopia

Dailin DJ, Hanapi SZ, Elsayed EA, Sukmawati D, Azelee NIW, Eyahmalay J, El Enshasy H (2019) Fungal phytases: biotechnological applications in food and feed industries. In: Recent advancement in white biotechnology through fungi: Volume 2: Perspective for value-added products and environments, pp 65–99

Dicus DH (2000) One response to a collection wide mold outbreak: how bad can it be, how good can it get? J Am Inst Conserv 39:85–105

Ellis DH, Davis S, Alexiou H, Handke R, Bartley R (2007) Descriptions of medical fungi, vol 2. University of Adelaide, Adelaide

Florian MLE (2002) Fungal facts: solving fungal problems in heritage collections. Archetype, London, p 146

Garg KL, Kamal KJ, Mishra AK (1995) Role of fungi in the deterioration of wall paintings. Sci Total Environ 1995(167):255–271. https://doi.org/10.1016/0048-9697(95)04587-Q

Gatenby S, Townley P (2003) Preliminary research into the use of the essential oil of *Melaleuca alternifolia* (tea tree oil) in museum conservation. AICCM Bullet 28(1):67–70. https://doi.org/10.1179/bac.2003.28.1.014

Gutarowska B (2020) The use of -omics tools for assessing biodeterioration of cultural heritage: a review. J Cult Herit 45:351–361. https://doi.org/10.1016/j.culher.2020.03.006

Hatinguais R, Pradhan A, Brown GD, Brown AJ, Warris A, Shekhova E (2021) Mitochondrial reactive oxygen species regulate immune responses of macrophages to Aspergillus fumigatus. Front Immunol 12:929

Juturu V, Wu JC (2014) Microbial cellulases: engineering, production and applications. Renew Sustain Energy Rev 33:188–203

Kasprzyk I (2008) Aeromycology—main research fields of interest during the last 25 years. Ann Agric Environ Med 15:1–7

Klich MA (2007) *Aspergillus flavus*: the major producer of aflatoxin. Mol Plant Pathol 8(6):713–722

Koul B, Upadhyay H (2018) Fungi-mediated biodeterioration of household materials, libraries, cultural heritage and its control. In: Fungi and their role in sustainable development: current perspectives, pp 597–615

Manju S, Chadha BS (2011) Production of hemicellulolytic enzymes for hydrolysis of lignocellulosic biomass. In: Biofuels. Academic Press, pp 203–228

Mazzeo R, Roda A, Prati S (2011) Analytical chemistry for cultural heritage: a key discipline in conservation research. Anal Bioanal Chem 399:2885–2887. https://doi.org/10.1007/s00216-011-4672-5

Medina ML, Haynes PA, Breci L, Francisco WA (2005) Analysis of secreted proteins from *Aspergillus flavus*. Proteomics 5(12):3153–3161

Meier C, Petersen K (2006) Schimmelpilze Auf Papier—Ein Handbuch Für Restauratoren. Der Andere Verlag, Tönning, Germany, p 198

Menon A (2021) Artist gives longer life to oil paintings. https://www.thehindu.com/entertainment/art/kochi-based-artist-suresh-tr-develops-a-chemical-reagent-to-restore-oil-paintings-destroyed-by-fungus-and-mold/article34176649.ece

Merad Y, Derrar H, Belmokhtar Z, Belkacemi M (2021) Aspergillus genus and its various human superficial and cutaneous features. Pathogens 10(6):643

Money NP (2016) Fungal diversity. In: The fungi. Academic Press, pp 1–36

Mordue AJ, Nisbet AJ (2000) Azadirachtin from the neem tree *Azadirachta indica*: its action against insects. Anais Soc Entomol Brasil 29(4):615–632

Nardelli F, Martini F, Lee J, Lluvears-Tenorio A, La Nasa J, Duce C, Ormsby B, Geppi M, Bonaduce I (2021) The stability of paintings and the molecular structure of the oil paint polymeric network. Sci Rep 11(1):14202. https://doi.org/10.1038/s41598-021-93268-8

Nighojkar A, Patidar MK, Nighojkar S (2019) Pectinases: production and applications for fruit juice beverages. In: Processing and sustainability of beverages. Woodhead Publishing, pp 235–273

Paner CM (2009) Chemical control of fungi infesting easel oil paintings at the University of Santo Tomas, Museum of Arts and Sciences. University of Santo Tomas Graduate School, Manila. https://www.slideshare.net/crisenciopaner/chemical-control-of-fungi-infesting-easel-oil-paintings-at-the-university-of-santo-tomas-museum-of-arts-and-sciences

Piñar G, Sterflinger K (2009) Microbes and building materials. In: Cornejo DN, Haro JL (eds) Building materials: properties, performance and applications. Nova Science Publishers, New York, pp 163–188

Pinna D, Salvadori O (2008) Processes of biodeterioration: general mechanisms. In: Caneva G, Nugari MP, Nugari MP, Salvadori O (eds) Plant biology for cultural heritage: biodeterioration and conservation. Los Angeles, Getty Conservation Institute, pp 15–34

Santos RS, Loureiro KC, Rezende PS, Andrade LN, de Melo Barbosa R, Santini A, Severino P (2019) Innovative nanocompounds for cutaneous administration of classical antifungal drugs: a systematic review. J Dermatol Treat 30(6):617–626

Savković Ž, Stupar M, Unković N, Ivanović Ž, Blagojević J, Vukojević J (2019) In vitro biodegradation potential of airborne Aspergilli and Penicillia. Sci Nat 106(3–4):8. https://doi.org/10.1007/s00114-019-1603-3

Savković Ž, Stupar M, Unković N, Knežević A, Vukojević J, Ljaljević Grbić M (2021) Fungal deterioration of cultural heritage objects. In: Biodegradation technology of organic and inorganic pollutants, pp 267–288. https://doi.org/10.5772/intechopen.98620

Scazzocchio C (2009) Aspergillus: a multifaceted genus. Encycl Microbiol:401–421

Simons M (2007) Lascaux cave paintings threatened by fungus. https://www.nytimes.com/2007/12/09/world/europe/09iht-cave.1.8653751.html

Singh R, Upadhyay SK, Singh M, Sharma I, Sharma P, Kamboj P, Khan F (2021) Chitin, chitinases and chitin derivatives in biopharmaceutical, agricultural and environmental perspective. Biointerface Res Appl Chem 11(3):9985–10005

Sterflinger K, Pinzari F (2012) The revenge of time: fungal deterioration of cultural heritage with particular reference to books, paper and parchment. Environ Microbiol 14(3):559–566. https://doi.org/10.1111/j.1462-2920.2011.02584.x

Stoner JH (2005) Changing approaches in art conservation: 1925 to the present. In: National Academies of Sciences, Engineering, and Medicine. Scientific examination of art: modern techniques in conservation and analysis. The National Academies Press, Washington, DC. https://doi.org/10.17226/11413

Szczepanowska HM (2013) Conservation of cultural heritage: key principles and approaches. Routledge, London. ISBN 978-0415674744

Tiwari SP, Srivastava R, Singh CS, Shukla K, Singh RK, Singh P, Sharma R (2015) Amylases: an overview with special reference to alpha amylase. J Global Biosci 4(1):1886–1901

Viegas C, Dias M, Carolino E, Sabino R (2020) Culture media and sampling collection method for *Aspergillus* spp. assessment: tackling the gap between recommendations and the scientific evidence. Atmosphere 12(1):23

Viñas SM (2005) Contemporary theory of conservation. Elsevier Butterworth-Heinemann, Oxford, p 185. ISBN 978-0750662246.

Walker GM, White NA (2017) Introduction to fungal physiology. Fungi: biology and applications, pp. 1–35

Yao H, Hruska Z, Di Mavungu JD (2015) Developments in detection and determination of aflatoxins. World Mycotoxin J 8(2):181–191

Antifungal Drug Discovery Using Bioinformatics Tools

27

Rashi Verma, Disha Disha, and Luqman Ahmad Khan

27.1 Introduction

Fungal infections have emerged as a significant public health concern, ranging from superficial to life-threatening systemic infections (Bongomin et al. 2017). Immunocompromised individuals, including those with underlying conditions like COVID-19, are particularly at risk of developing severe mycoses with increased mortality (Dutta et al. 2018; Pal et al. 2021). Various fungi, such as *Candida*, *Aspergillus*, *Fusarium*, *Mucorales*, and molds, can cause opportunistic and healthcare-associated infections in vulnerable patients (Benedict 2020). Additionally, certain geographical areas experience prevalent endemic mycoses, further adding to the complexity of managing these infections (Thompson et al. 2021).

Advancements in healthcare and aggressive therapeutic technologies have contributed to the rise in life-threatening fungal infections affecting immunocompromised individuals. The at-risk population includes HIV patients, transplant recipients, and cancer patients undergoing immunosuppressive treatments (Garcia-Cuesta et al. 2014; Jayachandran et al. 2016). Moreover, urban development, changing land use, and increased international travel have led to a rise in endemic fungal

R. Verma
Medical Mycology Lab, Department of Biosciences, Jamia Millia Islamia, New Delhi, India

Department of Neuroscience, Morehouse School of Medicine, Atlanta, GA, USA
e-mail: rv.01.nip@gmail.com

D. Disha
Vocational Studies and Applied Sciences, Gautam Buddha University, Noida, Uttar Pradesh, India
e-mail: dishaasingh1991@gmail.com

L. A. Khan (✉)
Medical Mycology Lab, Department of Biosciences, Jamia Millia Islamia, New Delhi, India
e-mail: lkhan@jmi.ac.in

N. Manzoor (ed.), *Advances in Antifungal Drug Development*,
https://doi.org/10.1007/978-981-97-5165-5_27

diseases in some regions, while developing countries face a significant burden of fungal infections, particularly among HIV patients (Bongomin et al. 2017; Guinea 2014).

Early diagnosis remains a critical challenge, especially in immunocompromised individuals, to reduce high case fatality rates. Although progress has been made in nonculture-based diagnostic methods, developing simple, rapid, and cost-effective clinical tests for invasive fungal infections continues to be a challenge (Lass-Flörl 2017). Despite these challenges, there have been important strides in the field of antifungal drugs. Novel categories of drugs and enhanced administration techniques have advanced the prospects of effectively managing invasive fungal infections (Butts and Krysan 2012; Moudgal and Sobel 2010). However, resistance to existing antifungals is a concern, necessitating continued surveillance and research (Prasad et al. 2016). Overall, the scenario highlights the importance of effective management, and development of potent antifungals to mitigate their impact on vulnerable patient populations (Fig. 27.1).

Antifungal development and designing are vital areas of research aimed at combating fungal infections, which can pose significant health risks to humans and animals alike. To develop effective and safe antifungal treatments, researchers employ a diverse range of methods and techniques (Mueller et al. 2021). In this context, three key approaches stand out: in silico, in vivo, and in vitro methods. In silico methods utilize computational tools to predict drug interactions with fungal targets, providing valuable insights into potential antifungal agents (Cairns et al. 2016). In vivo experiments involve studying drug efficacy and safety in living organisms, while in vitro techniques allow researchers to explore drug-fungus interactions in controlled laboratory environments (Espinel-Ingroff and Shadomy 1989). This integration of approaches enables a comprehensive understanding of antifungal compounds, paving the way for the development of more potent and targeted therapies to combat fungal infections effectively.

In recent decades, bioinformatics plays a significant role in drug design by providing valuable insights into new chemical structures, functions, and potential drug targets with greater precision and accuracy (Wishart 2005). The concept of rational drug design, which involves computational technologies related to structural biology to design target proteins, has been actively explored since the 1980s (Peláez 2011). One of the major advantages of computational methods is their cost-effectiveness and rapid acquisition of information, making them invaluable in complementing experimental research to optimize scientific endeavors. Bioinformatics has found applications in various scientific fields, including biochemistry, biophysics, genetics, pharmacology, toxicology, immunology, and agriculture. It has played a pivotal role in the development of new drugs, as is evident from the methods frequently utilized to search and design drugs against bacteria, viruses, and parasites (Tamay-Cach et al. 2016).

Given the current surge in antimicrobial resistance, particularly in antifungal agents, the discovery of antifungals and their biological activities has become paramount for ensuring safe and effective treatment of infections. Researchers have complemented experimental and computational studies on antifungal agents and

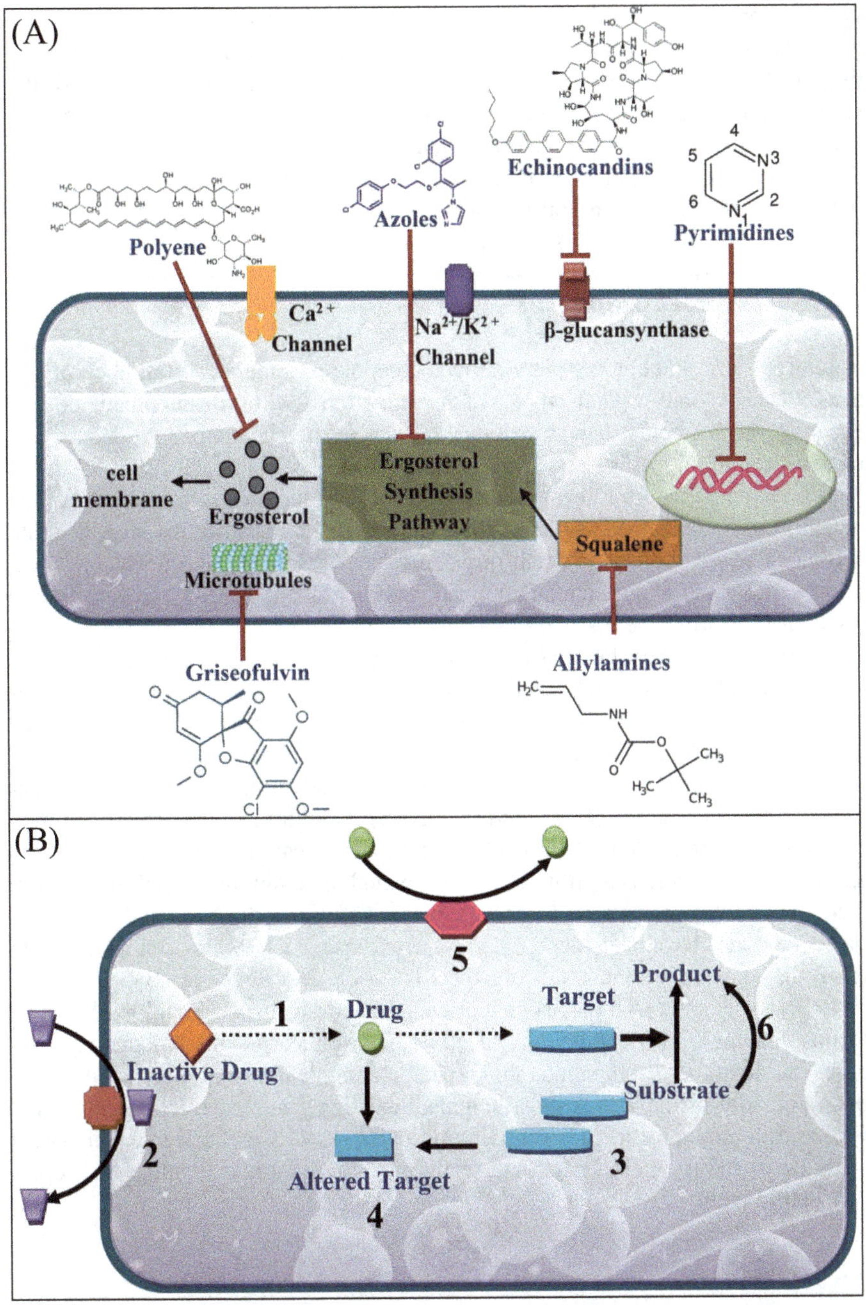

Fig. 27.1 Antifungal drug discovery and resistance. (**a**) Various antifungal drugs like polyenes, pyrimidine analogs, azoles, echinocandins, allylamines, morpholines, and thiocarbamates that have been approved to treat *Candida* infections. Nystatin and 5-flucytosine disrupt cell membranes, while azoles inhibit ergosterol synthesis. Echinocandins target glucan synthesis, and allyl-

updated contributions of bioinformatics to pharmacology of natural and synthetic products (Barceló et al. 2017a, b; Peralta et al. 2012; Perez et al. 2003). These examines are being conducted in accordance with World Health Organization (WHO) recommendations, which encourage the discovery of antifungals from both natural and synthetic sources. In the light of increasing prevalence of multidrug resistance, the purpose of this chapter is to update our knowledge about these compounds which were in existence as antifungals with application of computational methods.

27.2 Natural Compounds

Natural resources, including plant-derived compounds and those from microorganisms, have become crucial targets of investigation for the development of new agents in various fields such as medicine, pharmacy, biochemistry, agriculture, cosmetology, and more (Fig. 27.2) (Atanasov et al. 2021; Gunatilaka 2006; Sun et al. 2021). The vital need for new therapeutic agents and the growing awareness of the potential of natural products have driven increased research in this area. Overall, natural resources hold significant promise for the discovery and development of novel compounds with potential applications in various industries and for improving human health.

Since ancient times, humans have utilized natural resources for both nutritional and medicinal purposes (Petrovska 2012). Plants have been consumed for their nutritive substances like proteins, carbohydrates, lipids, and crude fibers. Numerous studies have explored the nutritional quality of diverse plants. In recent years, there has been increasing interest in functional foods and nutraceuticals due to their potential health benefits (Daliri and Lee 2015). Scientific studies have identified compounds responsible for therapeutic effects, and research has been conducted on antimicrobial as well as antifungal activity related to certain medicinal plants. Folk medicine employs numerous plants to counteract various diseases, and antimicrobial compounds have been identified in some of these plants, which can be extracted easily in aqueous medium, making their isolation cost-effective (Bandoni et al. 1976; Penna et al. 2001). Distinct natural compounds such as flavonoids, alkaloids, lectins, saponins, and peptides, have been studied for their potential medicinal properties. Some of these compounds show specific biological activities that could make them potential medicaments to complement existing pharmaceutical drugs. Some well-known examples include digoxin, morphine, vincristine, and eugenol. These are widely used for their cardiotonic, analgesic, antitumoral, and antiseptic properties, respectively.

Fig. 27.1 (continued) amines/thiocarbamates block squalene oxidation, all causing cell death. (**b**) Resistance mechanisms that include: (1) drug inactivation, (2) overactive drug efflux pumps, (3) increased target production, (4) altered target enzymes, (5) strengthened cell walls, and (6) adaptive stress responses or metabolic changes. Addressing these mechanisms is crucial in managing *Candida* infections (Verma et al. 2021)

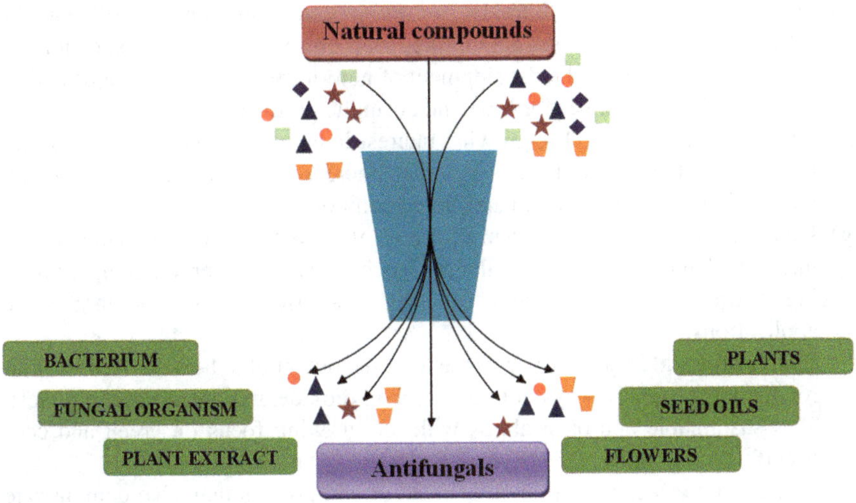

Fig. 27.2 Plant-derived compounds (plant extract, fruits, leaves, seed oils, and flowers) and microorganisms (bacterial and fungal organism) as antifungal agents

Microalgae have emerged as a promising source of bioactive compounds, including those with antifungal properties. Some microalgae species have been found to possess antioxidant, antitumoral, and acetyl cholinesterase inhibition properties, along with a highly unsaturated fatty acids profile, making them potential sources of biomolecules for pharmaceutical and food industries (Custódio et al. 2014). These microscopic photosynthetic organisms produce a wide array of secondary metabolites, some of which exhibit significant antifungal activity against various fungal pathogens. Some key points about microalgae as potential antifungal agents (Khavari et al. 2021) are:

(a) **Bioactive compounds**: Microalgae synthesize a diverse range of bioactive compounds, including fatty acids, pigments, peptides, terpenoids, polyphenols, and polysaccharides. These compounds are responsible for various biological activities, including antifungal effects.
(b) **Mechanisms of action**: The antifungal mechanisms of microalgal compounds may involve disrupting fungal cell membranes, inhibiting fungal enzymes, interfering with cellular signaling pathways, or inducing oxidative stress within fungal cells.
(c) **Selective activity**: Some microalgal compounds have demonstrated selective antifungal activity, targeting fungal pathogens while sparing beneficial microorganisms. This selective action is desirable for reducing side effects and minimizing disruptions to the natural microbial community.
(d) **Broad-spectrum activity**: Certain microalgal extracts or isolated compounds have shown broad-spectrum antifungal activity, effective against multiple fungal species, including *Candida*, *Aspergillus*, and *Cryptococcus*.

(e) **Potential applications**: Microalgal antifungal compounds have potential applications in pharmaceuticals, agriculture, food preservation, and biotechnology. They could be used in the development of new antifungal drugs, biopesticides, and natural preservatives for food and cosmetic products.
(f) **Commercial interest**: The growing interest in natural and sustainable alternatives to synthetic antifungals has led to increased research and commercial interest in microalgal-derived antifungal compounds.
(g) **Challenges**: Despite their promise, the development and commercialization of microalgal antifungals face challenges, such as scaling up production, optimizing extraction methods, and ensuring cost-effectiveness for large-scale applications.
(h) **Environmental impact**: Microalgae have the potential to be cultivated in environmentally friendly ways, utilizing carbon dioxide and other waste products. This sustainable approach aligns with the growing focus on green and eco-friendly technologies.
(i) **Synergistic effects**: Microalgal extracts or compounds may also demonstrate synergistic effects when combined with conventional antifungal drugs, enhancing overall antifungal efficacy, and potentially reducing the development of drug resistance.

Overall, microalgae represent a promising and relatively untapped resource for antifungal discovery. Ongoing research into the identification and characterization of specific microalgal compounds with potent antifungal properties will likely lead to their integration into various applications for human health, agriculture, and environmental protection (Mutanda et al. 2020). However, further studies are needed to validate the efficacy and safety of microalgal-derived antifungals and to overcome technical challenges related to large-scale production and extraction.

Seed oils from various plants have been studied for their potential antifungal properties. These oils contain a wide array of bioactive compounds, including fatty acids, phenolic compounds, terpenoids, and other secondary metabolites, which contribute to their antifungal activity (Dhifi et al. 2016). Tea tree oil derived from the seeds of *Melaleuca alternifolia* contains terpinen-4-ol and other terpenoids that exhibit potent antifungal effects against various fungal pathogens, including *Candida* species (Mielczarek et al. 2023). *Nigella sativa* contains thymoquinone and other bioactive compounds with antifungal properties, which have shown activity against *Candida albicans* and other fungi (Shokri 2016). *Azadirachta indica* is known for its broad-spectrum antifungal properties. It contains azadirachtin and other limonoids that inhibit the growth of several pathogenic fungi (Wylie and Merrell 2022). Grapefruit seed extract restrains polyphenolic compounds, such as naringenin and hesperidin, which exhibit antifungal effects against various fungal species (Takeoka et al. 2001). *Ricinus communis* has ricinoleic acid, which demonstrated antifungal properties against dermatophytes and *Candida* species (Franke et al. 2019). *Cucurbita pepo* includes phytosterols and tocopherols, which have shown antifungal activity against *Candida* species and other fungi (Gedi 2022). Pomegranate seed oil contains bioactive compounds like punicic acid and ellagic acid, which have demonstrated antifungal activity against *C. albicans* and other fungal pathogens (Badr et al. 2020). *Helianthus annuus* contains fatty acids like linoleic acid, which has been shown to have antifungal properties against certain fungi (Guo et al. 2017). *Sesamum indicum* and its constituent lignans have been studied for their potential use in antifungal formulations (Andargie et al. 2021). It is important to note that, while seed oils show promising antifungal activity, their effectiveness can vary depending on the specific fungal species and strain, as well as the concentration and formulation used. Additionally, some seed oils may exhibit antifungal activity against certain fungi but not others.

27.3 FDA-Approved Natural Antifungals

Some of the most used natural products as antifungals include (Butts and Krysan 2012):

(a) **Polyenes**: Natural polyene antifungals, such as amphotericin B and nystatin, are produced by the bacterium *Streptomyces nodosus* and are effective against a wide range of fungal infections. They bind to the fungal cell membrane, causing pore formation and leakage of cellular contents, leading to fungal cell death.
(b) **Azoles**: Many azole antifungals, like fluconazole, itraconazole, and voriconazole, have natural origins or are derived from natural compounds. They inhibit the enzyme lanosterol 14α-demethylase, essential for fungal ergosterol synthesis, disrupting cell membrane integrity and fungal growth.
(c) **Echinocandins**: Echinocandin antifungals, including caspofungin, micafungin, and anidulafungin, are semisynthetic derivatives of naturally occurring lipopeptides produced by fungi. They target the synthesis of β-glucan, a critical component of the fungal cell wall, leading to cell wall damage and eventual cell death.
(d) **Herbal Antifungals**: Various herbal products, such as garlic (allicin), tea tree oil, oregano oil (carvacrol and thymol), and pau d'Arco, have been used traditionally for their antifungal properties. They may interfere with fungal cell membranes, enzymes, or other vital processes, inhibiting fungal growth.
(e) **Flucytosine (5-FC)**: Flucytosine is a synthetic fluorinated pyrimidine analogue that, while not entirely natural, is often grouped with natural products due to its fungal origin. It is converted into its active form, 5-fluorouracil, inside the fungal cell, disrupting RNA and DNA synthesis and inhibiting fungal growth.
(f) **Black walnut extract**: Black walnut extract contains juglone, a compound with antifungal properties. It has been traditionally used for its efficacy against various fungal infections.

27.4 Contribution of Bioinformatics

Numerous research studies employ informatics techniques for the analysis of both natural and synthetic compounds. This section encompasses significant searches spanning from 2011 to the present day, showcasing the substantial role of bioinformatics in the development of antifungal strategies. Some examples and applications are stated below.

2011–2015: To explore the potential loss-of-function mechanism, Jin et al. (2012) employed mutant variants of the ABC transporter. They conducted a comprehensive assessment of the transporter's effectiveness by juxtaposing outcomes obtained through both computational and empirical approaches. As a result, researchers disseminated three-dimensional model of the transporter mutations that served as the basis for conducting a rigorous investigation utilizing molecular docking techniques with diverse compounds. Barceló et al. (2014) analyzed the possible molecular interactions between CDR1 and CDR2 with ligands 6-prenylpinocembrine (6PP), fluconazole and adenosine triphosphate which revealed the distinct ways of competence among 6PP and fluconazole for specific site on CDR pump in *C. albicans*. Irfan and Abid (2015) conducted docking studies to study the interaction between CYP51 with 18 newly synthesized triazoles. The study showed compounds containing quinoline ring binds to CYP51 with higher affinity. Guerrero-Perilla et al. (2015) docked 32 natural compounds to N-myristoyl transferase (NMT) among which flavonoids, alkaloids, quinones and xanthones exhibited a stronger interaction with NMT. Singh et al. (2015) showed a broad spectrum of Wortmannin on antifungal activity and irreversible inhibition of phosphoinositide 3-kinase (PI3K) of *Candida* sp., *Cunninghamella terreus*, *Trichophyton rubrum*, *Rhizopus oryzae*, *Rhizopus versicolor*, *Aspergillus fumigatus*, *Aspergillus pullulans*, *Fusarium moniliforme*, and *Saccharomyces cerevisiae*. The researchers disclosed that Wortmannin produced by *Penicillium radicum* could be used as efficient antifungal after structural modifications to improve stability. Abadio et al. (2015) build a library of 3000 small molecules out of which 12 compounds were selected against *Paracoccidiodes* spp. On experimental evaluation, three of the compounds tested showed inhibitory activity against the fungal enzymes and were patented.

2016–2020: Tetrazolium chloride and trimethyl tin chloride were identified by Nim et al. (2016) as new substrates. Tamay-Cach et al. (2016) reviewed various in silico techniques and their application in the field of biological systems and antimicrobial agents. Bencurova et al. (2018) explained distinct in silico methods like virtual screening, docking simulation, molecular dynamics (MD) simulations, sequence alignment, quantitative structure activity relationship (QSAR, mtQSAR, mtk-QSBER). Abu-Izneid et al. (2018) synthesized new bioactive compounds and evaluated their therapeutic relevance using computational approaches which established its antifungal activity against fungal strains. Rogozhin et al. (2018) characterized antimicrobial hairpin-like peptide EcAMP1 isolated with *Echinochloa crus-galli* seeds. On modification, they reported a significantly weaker activity of EcAMP1 toward pathogenic fungus *Fusarium solani*. Zhao et al. (2019) explored the inhibitory effect of IAMU80070 on *Aspergillus flavus* at the transcriptional level and showed that IAMU80070 exhibited the highest antifungal activity. Tyagi et al. (2019) developed a dedicated platform which systematically catalogues 2585 peptide entries among which 510 are unique plant-origin peptides along with their antifungal properties. Yang et al. (2019) indicated that *Sarocladium brachiariae* could potentially synthesize a variety of unknown-function secondary metabolites, which may play an important role in the adaptation to its endophytic lifestyle and antifungal activity. Lopes et al. (2019) investigated sequences of Osmotin and thaumatin-like proteins. They indicated that peptide "CAADIVGQCPAKLK" can be a valuable target for the development of a desired antimicrobial agent against *C. albicans*. Magoulas et al. (2021) synthesized series of new thieno [2,3-d] pyrimidin-4(3H)-one derivatives and evaluated their activity against eight fungal species. Researchers reported excellent antifungal activity by most of the compounds as compared to control. Verma et al. (2020) performed comparative and subtractive genomics approach to identify novel target. They included genome scale metabolic model reconstruction and in silico gene knockout which revealed FAS2 as a potent target to inhibit the growth of *C. albicans*. Ji et al. (2020) carried out bioinformatics analysis of the W10 protein from *Bacillus licheniformis* which disclosed its antifungal activity against the fungal plant pathogens.

2021–2023: Marbán-González et al. (2021) explored the octahydroisoindolone scaffold as a potential new antifungal series by docking and simulation analysis and proposed compound 5 as the best antifungal with pharmacodynamic and pharmacokinetic character. R et al. (2021) assessed the antifungal action of *Bacillus amyloliquefaciens* isolated VB7 against *Fusarium oxysporum*. Renganathan et al. (2021) synthesized silver nanoparticles from aqueous flower extract of *Bauhinia tomentosa Linn* and proven its effectivity as an antifungal plant. Prajapati et al. (2021) uncovered the antifungal mode of action of curcumin by potential inhibition of CYP51B of *Rhizopus oryzae*.

Sharma et al. (2022) demonstrated increased interaction of phytocompounds from *Curcuma longa* leaves against fungal pathogens. The compounds 3,7-Cyclodecadien-1-one, and 3,7-dimethyl-10-(1-methylethylidene) were notably identified as safer antifungal agents. Madanchi et al. (2022) designed and assessed antifungal peptides, highlighting Aurein N3 with low cytotoxicity and strong ergosterol binding. Feng et al. (2022) unveiled ten secondary metabolite biosynthetic gene clusters in *Bacillus halotolerans* LDFZ001, including a novel kijanimicin biosynthetic cluster. *Talaromyces purpureogenus* CX11 genome analysis indicated moderate antifungal activity of compounds 1 and 2 against *Fusarium oxysporum*. The studies of Baptista et al. (2022) revealed that *Bacillus velezensis* CMRP 4489 has antifungal potential with several biosynthetic gene clusters capable of producing antifungal molecules. Hussen et al. (2022) screened 158 antifungal phytochemicals and compared them to isomaltotriose as a positive control. Akhtar et al. (2022) evaluated a multivalent epitope-based vaccine targeting secreted aspartic protease 2 (SAP2) protein of *C. tropicalis*. Kisa et al. (2022) extracted active compounds from *Turanecio hypochionaeus*, and found that chlorogenic acid had significant binding energy against HMG_CoA R, α-amylase, α-glucosidase, and lipase. Yousafi et al. (2022) docked *Camellia sinensis* essential oils with *Fusarium oxysporum* polyketide synthase β-ketoacyl synthase domain, leading to the selection of CMNPD24498 (FW054-1) from Verrucosispora as a potential lead compound against *Fusarium oxysporum*. Verma et al. (2022) repurposed FDA-approved drugs and proposed arbutamin, hydroxychloroquine, and carindacilin as potent inhibitor against TRR1, TOM40, and YHB1 of *C. albicans*. Kamboj et al. (2022) explored possible targets of α-pinene, eugenol, berberine, and curcumin against *Curvularia lunata* using computational approach. Cheema et al. (2022) showed that ZnO nanoparticles significantly reduced the fungal spore germination and caused disintegration of hyphae for both fungal strains. Mandal et al. (2022) performed a knowledge-based search of known metallo-β-lactamase inhibitors to identify starting points for early engagement of medicinal chemistry.

Faisal et al. (2023) identified Luteolin and Rosmarinic acid as the top contenders for potential bioactivity with acceptable drug-like properties. Singh et al. (2023) showed six compounds were potentially active, among which compound 6 was the most potent antifungal as shown by in vitro analysis. Jaiswal and Kumar (2023) developed a pipeline to unveil the underlying hidden smORFs and domain analysis that revealed antimicrobial peptide as antifungal agents. Prasad et al. (2023) carried out in silico screening of phytochemicals isolated from arabica coffee and found naringin (a flavonoid glycoside) as an ideal compound against *Aspergillus fumigatus*. Zhao et al. (2023) modified peptides and synthesized Ap920-WI as an effective antifungal. Nakada et al. (2023) designed tetrazole derivatives to determine structure activity relationship which divulged that tetrazole backbone-containing compounds could be developed as novel antifungal drugs against *Cryptococcus spp.* Sathiyamoorthy et al. (2023) modeled the ergosterol biosynthetic enzymes and docked them with furanone derivatives which showed potential interaction between ligand 6 and lanosterol 14 α-demethylase. Metzner et al. (2023) repurposed FDA-approved drugs and displayed the blockage of hyphal transition by 33 drugs among which NSC 697923 targeted eIF3 with highest activity against resistant strains of *C. albicans*. Mahalapbutr et al. (2023) showed that FDA-approved amphotericin B exhibited the highest binding efficiency against human tyrosinase. Wadhwa et al. (2023) developed an integrated computational and biological screening platform that selected sesamin as the most promising phytochemical endowed with a potential antifungal profile. Kushveer et al. (2023) showed the potential of binding 2, 4-di-tertbutylphenol to β-tubulin in *F. verticillioides*. Khichi et al. (2023) generated and evaluated analogues of fluconazole using molecular docking. Melavanki et al. (2023) docked a pyridine derivative, 3DPP, to show the potential information regarding antifungal activity.

Akash et al. (2023) displayed coptisine derivatives as safe and potentially effective against Black fungus, Monkeypox virus, and Marburg virus. Viana et al. (2023) revealed that chitinase enzymes can be potential targets for these compounds. Khan et al. (2023) suggested camptothecin and GKK1032A2 as potential antifungal natural compounds. Ciociola et al. (2023) tested two arginine- and proline-rich peptides against planktonic cells and biofilms of *C. albicans* and *C. glabrata* strains. Consequently, these peptides showed similar potent activity. Senra et al. (2023) identified two CS-αβ defensins with probable antifungal activity. Yang et al. (2023)

presented potential of biocontrol agents isolated from bacterium KJ-34 as antifungals through comparative and bioinformatics analyses. Zhan et al. (2023) showed honokiol possesses antifungal and anti-inflammatory effects in *Aspergillus fumigatus* keratitis. Krishnan et al. (2023) used *P. hysterophorus* biomass and demonstrated 2,4-Di-tertbutylphenol and 1H-Pyrazole, 4-ethyl-3,5-dimethyl could be potent antifungal agents against Trr1 gene of *Cryptococcus neoformans*, which encodes for thioredoxin reductase enzyme. Sama-ae et al. (2023) identified 46 compounds derived from different sources of which 15 molecules had strong binding affinity for CYP51. Molecular docking simulation studies suggested that, among these didymellamides could be a promising inhibitor.

27.5 Bioinformatics Tools and Antifungal Development

Bioinformatic tools have revolutionized the study of herbal antifungals, providing researchers with a powerful means to analyze large datasets and gain a deeper understanding of the molecular mechanisms underlying their antifungal activities (Fig. 27.3). These computational approaches have accelerated the identification of potential compounds, improved our knowledge, and paved the way for the development of new and effective antifungal therapies. As technology continues to advance, bioinformatics will remain a crucial asset in the exploration of herbal antifungals and their potential applications in clinical settings.

27.5.1 In Silico Screening of Herbal Compounds

One of the primary challenges in drug discovery is the vast number of compounds to test. In silico screening that enables researchers to filter and prioritize potential compounds is based on their structural features and predicted activities. It is a

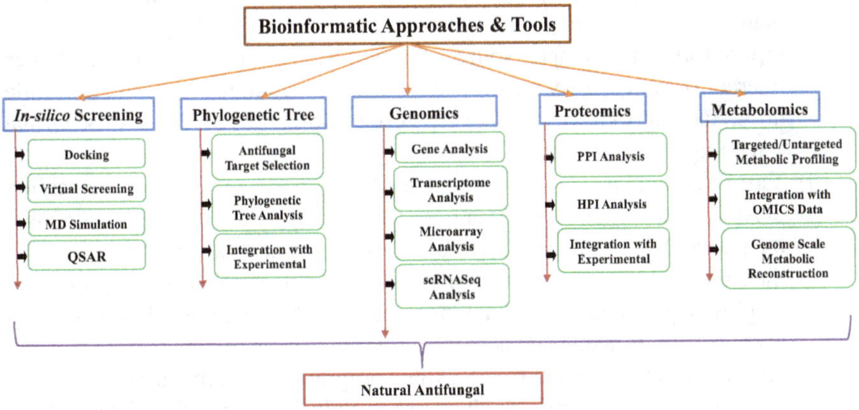

Fig. 27.3 Contribution of bioinformatics in antifungal drug discovery

computational approach widely used in drug discovery to identify potential lead compounds from large chemical databases (Miozza et al. 2020; Sharma et al. 2017). This technique has also been applied to study herbal compounds and their potential antifungal activities. It offers a cost-effective and time-efficient way to narrow down the vast number of compounds before experimental testing.

1. **Molecular docking**

   Molecular docking is a key component of in silico screening that predicts the binding orientation and affinity of a small molecule (herbal compound) to a target protein (e.g., a fungal enzyme or receptor). The 3D structure of the target protein is obtained from experimental methods like X-ray crystallography or NMR spectroscopy. The 3D structure of the herbal compound is generated or retrieved from databases.

   During experimental determination, when the 3D structure of target proteins, such as fungal enzymes or receptors, is unavailable, homology modeling becomes an indispensable technique. This computational method leverages the structural information of homologous proteins with known 3D structures to construct plausible models for the target proteins of interest. The ensuing homology models, obtained through sequence alignment and refinement, seamlessly integrate into molecular docking studies. Docking softwares, such as AutoDock, AutoDock Vina, or GOLD, are then employed to computationally dock the herbal compound into the binding site of the target protein (Joshi et al. 2021; Shahinozzaman et al. 2020). Besides these, several other tools are commonly used for virtual screening. Some of these are mentioned below:

   (a) **DOCK** (University of California, San Francisco) is a widely used molecular docking program that focuses on the prediction of ligand-protein interactions. It employs an incremental construction algorithm for ligand binding site matching.
   (b) **Schrödinger Suite** (Schrödinger, LLC) includes various tools for molecular modeling and drug discovery, such as Glide for high throughput docking, LigPrep for ligand preparation, and Virtual Screening Workflow for automated virtual screening.
   (c) **Open Babel** is an open-source cheminformatics toolbox that facilitates interconversion between different molecular structure file formats. While not a docking tool itself, it is often used in preprocessing steps for virtual screening.
   (d) **VinaLC** (AutoDock Vina for Ligand Chemistry) is a ligand-centric version of AutoDock Vina, designed for high-throughput virtual screening. It focuses on accelerating the docking process for large compound libraries.
   (e) **Surflex-Dock** (Tripos International) is a high-performance docking tool that employs a patented scoring function and a fast heuristic search algorithm. It is known for its speed and accuracy in virtual screening.
   (f) **rDock** is an open-source docking program that emphasizes a high-throughput virtual screening. It supports multiple scoring functions and can be utilized for the screening of large compound libraries.

(g) **MGLTools** (The Scripps Research Institute) is a software suite for the analysis and visualization of molecular structures. It includes tools such as AutoGrid for grid-based docking and AutoDockTools for preparing ligands and proteins for docking.
(h) **Pharmer** is a ligand pharmacophore search tool that helps in the identification of compounds with specific pharmacophoric features. It can be used as a complementary tool in virtual screening workflows.

The docking software evaluates the binding energy and interaction patterns between the compound and the protein, which allows the identification of potential herbal compounds with strong binding affinity to the target. This amalgamation of molecular docking and homology modeling in the bioinformatics-driven exploration of antifungal drug candidates exemplifies a synergistic strategy that transcends the limitations imposed by the absence of experimental structures for certain target proteins, ultimately enhancing the efficiency and scope of antifungal drug discovery efforts.

2. **Molecular Dynamics (MD) Simulations**

MD simulations are used to study the dynamic behavior of molecular systems over time. After molecular docking analysis identifies potential herbal compounds, MD simulations can be employed to study stability of the protein–compound complex and flexibility of the binding site. MD simulations use physics-based force fields to calculate the movement of atoms and molecules, providing insights into conformational changes and binding stability of the herbal compound within the protein active site (Chmiela et al. 2018). In addition to force fields several key parameters and considerations come into play for assessment of complex stability:

(a) **Force fields** are paramount in MD simulations as they dictate the mathematical representation of atomic interactions. Choosing an appropriate force field is critical for accurately capturing the dynamics and stability of the protein–compound complex. Common force fields include AMBER, CHARMM, and GROMOS.
(b) The **integration time step** is a numerical parameter that determines the time interval over which the equations of motion are solved. A suitable time step is essential to balance computational efficiency and accuracy in capturing the dynamics of the molecular system.
(c) Maintaining a **constant temperature and pressure** during MD simulations is crucial for mimicking realistic physiological conditions. Thermostats and barostats are employed to regulate temperature and pressure, respectively.
(d) Inclusion of explicit **solvent molecules** in MD simulations is vital to mimic the effects of the surrounding environment. Water is a common solvent, and different solvent models, such as TIP3P or SPC, can be employed depending on the simulation requirements.
(e) Accurate representation of **electrostatic and van der Waals interactions** is crucial for capturing the forces between atoms and molecules. Long-range electrostatic interactions are often treated using methods like Particle Mesh Ewald (PME).

(f) **Achieving convergence** in MD simulations involves running simulations for a sufficiently long time to sample a representative ensemble of conformations. **Adequate sampling** is critical for obtaining meaningful insights into the stability and dynamics of the protein–compound complex.
(g) Postsimulation **analysis tools** are employed to extract relevant information from MD trajectories. These tools help in assessing parameters such as root-mean-square deviation (RMSD), root-mean-square fluctuation (RMSF), and hydrogen bonding patterns, providing a comprehensive understanding of complex stability.

3. **Quantitative Structure–activity relationship analysis**

Quantitative structure–activity relationship (QSAR) is a statistical modeling technique used to correlate the chemical structure of compounds with their biological activities (Chatterjee and Roy 2023). In the context of herbal antifungals, QSAR analysis can identify specific molecular features associated with antifungal activity. QSAR models are built based on a dataset of herbal compounds with known antifungal activity and their corresponding chemical descriptors (e.g., molecular weight, hydrogen bond donors/acceptors, hydrophobicity). The model is then used to predict the activity of new herbal compounds with similar structural features, prioritizing those with the most favorable predicted activity for further experimental testing.

27.5.2 Phylogenetic Analysis of Antifungal Targets

To understand the mechanisms of action of herbal antifungals, it is essential to investigate the molecular targets they interact with. Bioinformatics enables the construction of phylogenetic trees that investigating the evolutionary relationships of antifungal targets across different fungal species, shedding light on the mechanisms of action of herbal antifungals. Several bioinformatics tools are commonly used for phylogenetic analysis, aiding in the construction of phylogenetic trees (Challa and Neelapu 2019; Kapli et al. 2020). Some notable tools are:

(a) **MEGA** (Molecular Evolutionary Genetics Analysis) is a user-friendly software with a wide range of tools for phylogenetic analysis. It allows the construction of phylogenetic trees based on various methods, including neighbor-joining, maximum likelihood, and Bayesian analysis.
(b) **RAxML** (Randomized Axelerated Maximum Likelihood) is a fast and efficient tool for maximum likelihood-based phylogenetic analysis. It is particularly useful for large datasets and can handle a variety of substitution models.
(c) **PhyML** (Phylogenetic estimation using Maximum Likelihood) is a software for phylogenetic tree construction using maximum likelihood. It provides options for selecting substitution models and branch support estimation.
(d) **IQ-TREE** (Iterative Quadratic Tree Estimation) is a versatile tool that combines maximum likelihood and Bayesian methods for phylogenetic tree con-

struction. It offers a range of models for substitution rate estimation and branch support.

(e) **MrBayes** is a Bayesian inference tool for phylogenetic analysis. It uses a Markov Chain Monte Carlo (MCMC) approach to estimate phylogenetic trees and their parameters.

By constructing phylogenetic trees, researchers can **identify homologous proteins** and **discern conserved regions** critical for the biological functions of these targets. This information is particularly relevant in the context of antifungal drug discovery, as conserved regions often represent essential domains or active sites susceptible to inhibition. Furthermore, the analysis facilitates an **understanding of the evolutionary divergence** between pathogenic and nonpathogenic fungi, enabling the identification of target proteins that may be specific to pathogens, minimizing potential impact on beneficial fungi or nonpathogenic counterparts. Thus, phylogenetic analysis serves as a powerful bioinformatics tool, guiding the rational selection of antifungal targets and informing the design of therapeutics with enhanced specificity and efficacy.

The phylogenetic analysis includes several steps:

1. **Selection of antifungal targets**

   Before conducting phylogenetic analysis, researchers first identify potential antifungal targets. These targets could be essential enzymes, receptors, or proteins involved in vital cellular processes within the fungal pathogen. For example, targets may include ergosterol biosynthesis enzymes, cell wall synthesis proteins, chitin synthases, or fungal-specific kinases.
2. **Sequence retrieval and alignment**

   The next step involves retrieving protein sequences of the identified targets from various fungal species. These sequences are obtained from public databases (UniProt—https://www.uniprot.org/, NCBI—https://www.ncbi.nlm.nih.gov/protein/, Ensembl Fungi—https://fungi.ensembl.org/index.html, MycoCosm—https://mycocosm.jgi.doe.gov/, FungiDB—https://fungidb.org/fungidb/) or through experimental techniques such as RNA-seq or proteomics. Multiple sequence alignment tools, like ClustalW or MUSCLE, are then used to align the sequences, ensuring that equivalent amino acids are aligned in the same columns (https://www.ebi.ac.uk/Tools/msa/).
3. **Construction of phylogenetic trees**

   With the aligned sequences in hand, researchers employ various phylogenetic tree construction methods, such as Neighbor-Joining, Maximum Likelihood, or Bayesian methods (Munjal et al. 2018). These algorithms use the sequence information to build trees that represent the evolutionary relationships among the target proteins from different fungi.
4. **Analyzing phylogenetic trees**

   Once the phylogenetic trees are constructed, researchers can analyze and interpret the results (Brown 2002). Several important observations can be made:

(a) **Conservation**: If a target protein clusters together in the tree with homologs from diverse fungal species, it indicates a high degree of conservation among different fungi. This suggests that interfering with the function of such targets might have broad-spectrum antifungal effects.
(b) **Divergence**: On the other hand, if a target protein forms a distinct branch with homologs from specific pathogenic fungi, it may suggest a potential drug target specific to those pathogens. Targeting such proteins could lead to selective antifungal therapies, minimizing harm to nonpathogenic fungi.
(c) **Gene duplication and specialization**: Phylogenetic analysis can also reveal instances of gene duplication events, where certain fungi may have multiple copies of a target gene. These duplicated genes may have undergone functional specialization, leading to differences in drug susceptibility among fungi.
(d) **Horizontal gene transfer**: In some cases, phylogenetic analysis can identify horizontal gene transfer events, where genetic material is transferred between different species. Such events can influence the emergence of antifungal resistance and highlight potential challenges in drug development.

5. **Integration with experimental data**

   Phylogenetic analysis is often complemented with experimental data, such as gene knockout studies, RNA interference experiments, or gene expression profiling. By combining computational insights with experimental evidence, researchers can validate the significance of identified antifungal targets and better understand their roles in fungal biology and drug response (Cárdenas et al. 2020).

27.5.3 Transcriptomics and Differential Gene Expression

Herbal antifungals may exert their effects by modulating gene expression in the fungal cells. High-throughput techniques such as RNA sequencing (RNA-seq) can be used to profile the gene expression changes in fungal cells treated with herbal compounds. Bioinformatics tools are then employed to analyze the vast amount of transcriptomic data, identifying differentially expressed genes and pathways impacted by the treatment (D'Agostino et al. 2022).

1. **RNA sequencing (RNA-seq)**

   RNA-seq is the primary experimental technique used in transcriptomics. It involves the high-throughput sequencing of cDNA (complementary DNA) molecules derived from RNA samples. The RNA-seq data generated provides a snapshot of the transcriptome, revealing the abundance of various RNA molecules, including mRNAs, noncoding RNAs, and alternatively spliced variants (Hrdlickova et al. 2017; Wang et al. 2009). To study the effects of designed antifungals on gene expression, researchers typically design experiments with two or more groups: a control group (untreated fungi) and one or more treatment groups (fungi treated with herbal antifungals). The fungal cells are harvested at specific time points after treatment, and RNA is extracted and prepared for sequencing. The RNA-seq includes several steps:

(a) **Data preprocessing**

Raw RNA-seq data contains sequence reads that must be processed before analysis. This preprocessing involves quality control, removing adapter sequences, and aligning the reads to the reference genome or transcriptome. The aligned reads are then counted to determine the expression levels of genes and other RNA molecules.

(b) **Differential gene expression analysis**

Differential gene expression analysis compares the gene expression levels between different experimental groups. Statistical tests, such as the edgeR or DESeq2 algorithms, are commonly used to identify genes that show significant differences in expression between the control and treatment groups (Ji and Sadreyev 2018). Genes with a fold change above a certain threshold and a p-value below a specified significance level are considered differentially expressed.

(c) **Functional enrichment analysis**

After identifying the differentially expressed genes, functional enrichment analysis is performed to gain insight into the biological processes, molecular functions, and cellular components that are overrepresented among the differentially expressed genes (Reimand et al. 2019). This analysis involves comparing the gene list to biological databases and gene ontology resources to identify enriched pathways and functions.

(d) **Validation of differential gene expression**

Differential gene expression results are often validated through other experimental techniques, such as quantitative real-time PCR (qRT-PCR) or Western blotting. These validation methods help confirm the RNA-seq results and strengthen the conclusions drawn from the transcriptomics analysis.

27.5.4 Proteomics and Protein–Protein Interactions

Proteomics is another powerful tool used in the study of herbal antifungals. By employing mass spectrometry, researchers can identify and quantify the proteins present in fungal cells before and after treatment with herbal compounds. Bioinformatic analysis of this data can provide valuable insights into the protein–protein interactions and cellular pathways affected by the herbal antifungals.

(a) **Experimental techniques, protein identification, and quantification**

Mass spectrometry (MS) is the primary experimental technique used in proteomics. There are different variations of MS, including liquid chromatography-mass spectrometry (LC-MS) and tandem mass spectrometry (MS/MS). In shotgun proteomics, protein mixtures are digested into peptides, which are then analyzed by MS. The resulting mass spectra are used to identify and quantify the proteins present in the sample. Proteomics data analysis involves matching the acquired mass spectra against protein databases using search algorithms such as SEQUEST, Mascot, or MaxQuant (Chen et al. 2020; Schiebenhoefer et al. 2020). This process identifies the proteins present in the sample based on their unique peptide sequences. Protein quantification is achieved by comparing the abundance of peptides corresponding to each protein between different experimental conditions.

(b) **Protein–protein interaction analysis**

Protein–protein interaction (PPI) analysis is a critical aspect of proteomics that explores the physical interactions between proteins within a cell (De Las Rivas and Fontanillo 2010; Ding and Kihara 2019). Understanding PPIs is crucial for elucidating cellular pathways, protein complexes, and signaling networks. Analyzing PPI data involves the construction of protein interaction networks, where nodes represent proteins, and edges represent their interactions. Network analysis can reveal key proteins (hubs) that have many connections, potentially indicating important regulatory or signaling roles. Clustering algorithms can be applied to identify functional modules or protein complexes within the network.

In the context of PPI networks, nodes represent individual proteins. Each protein in the network is assigned a node, and these nodes are interconnected based on the physical interactions observed between the corresponding proteins. The properties and characteristics of nodes, such as their degree (number of connections), centrality, and functional annotations, are crucial for understanding the overall structure and dynamics of the protein interaction network. Edges in a PPI network denote the physical interactions between proteins. An edge connects two nodes, indicating that the corresponding proteins have been experimentally observed or predicted to interact within a cellular context. The edges can be weighted to represent the strength or confidence of the interaction. The nature of the edges provides insight into the specificity and strength of the interactions, contributing to the overall architecture of the PPI network. Hubs are proteins within a PPI network that exhibit a high degree of connectivity, meaning they have many interacting partners compared to other proteins in the network. Hubs play a crucial role in the network structure and function, as they often correspond to key regulatory proteins or proteins with essential cellular functions. The identification of hubs through network analysis is valuable for understanding the hierarchical organization of the cellular machinery. Hubs may act as central points in signaling pathways, facilitating communication between different cellular components.

In brief, analyzing these elements collectively provides a systems-level understanding of cellular processes, including the identification of important regulatory proteins, the delineation of functional modules or complexes, and insights into the overall organization of cellular signaling networks.

27.5.5 Metabolomics for Profiling Metabolic Changes

Metabolomics is a field of bioinformatics that focuses on the comprehensive analysis of small molecules, known as metabolites, present in cells, tissues, or organisms (Azad and Shulaev 2018). It provides a snapshot of the cellular metabolism and can be used to study the metabolic changes induced by herbal antifungals. Metabolomics is a valuable tool for understanding the global effects of herbal compounds on fungal metabolism and identifying key metabolites and metabolic pathways influenced by the treatment.

1. **Untargeted and targeted metabolomics.**

   Metabolomics can be performed by two main approaches: untargeted and targeted metabolomics (Shen et al. 2022). Untargeted metabolomics aims to profile as many metabolites as possible, including known and unknown compounds, providing a global view of metabolic changes (Souza and Patti 2021). On the other hand, targeted metabolomics focuses on quantifying a specific set of metabolites, often those involved in specific pathways or processes, allowing for more precise and quantitative analysis (Roberts et al. 2012). The metabolic profiling includes (Chen et al. 2022):

   (a) **Experimental design**

   In the context of herbal antifungals, metabolomics experiments are designed with control (untreated) and treatment groups (fungi treated with herbal compounds). Fungal cells are harvested at specific time points after treatment, and metabolites are extracted from the samples. The extracted metabolites are then subjected to MS or NMR analysis.

   (b) **Data analysis**

   Metabolomics data analysis involves the identification and quantification of metabolites and the statistical comparison of metabolite levels between control and treatment groups. Multivariate statistical methods, such as principal component analysis (PCA) and partial least squares-discriminant analysis (PLS-DA), are commonly used to visualize and analyze the complex metabolomics datasets.

   (c) **Identifying altered metabolic pathways**

   By comparing metabolite profiles between control and treated samples, researchers can identify specific metabolites that change significantly in response to herbal antifungals. These metabolites can then be mapped to metabolic pathways using databases such as Kyoto Encyclopedia of Genes and Genomes (KEGG) (https://www.genome.jp/kegg/), BioCyc (https://www.biocyc.org/) or MetaCyc (https://metacyc.org/). Such analysis can highlight the affected metabolic pathways and reveal potential points of intervention for herbal antifungals.

2. **Integration with other omics data**

   The integration of metabolomics data with other omics data, such as genomics, proteomics, and transcriptomics, allows for a comprehensive understanding of the molecular mechanisms underlying the effects of herbal antifungals (Cao and Gao 2022; Subramanian et al. 2004). Integrative analysis can reveal the connections between altered gene expression, protein abundance, and metabolite levels, providing a holistic view of the cellular response to treatment.

3. **Genome scale metabolic model reconstruction**

   Genome-scale metabolic models (GSMMs) are comprehensive representations of all metabolic reactions occurring in an organism based on its genomic information (Gu et al. 2019). GSMM reconstruction involves the systematic collection of metabolic reactions and associated enzymes, along with the corresponding genes, from genome annotations and other data sources. These models serve as powerful tools to study the metabolism of organisms, including fungi, and can be utilized in various areas of research, including antifungal drug development (Mirhakkak et al. 2023; Verma et al. 2021). GSMMs have been successfully constructed for several fungal species, including pathogenic fungi responsible for human and plant infections. These models provide a detailed overview of the metabolic capabilities of fungi, offering insights into the utilization of carbon sources, production of biomass, and the production of secondary metabolites like toxins and antifungal compounds (Mendoza et al. 2019). The GSMM can be used for several purposes:

   (a) **Predicting essential metabolic genes**

   By simulating the growth of fungi using GSMMs, researchers can identify essential metabolic genes that are crucial for the survival and proliferation of the pathogen. Disruption of these essential genes could be a potential target for the development of novel antifungal drugs. By focusing on essential metabolic pathways, GSMMs aid in identifying vulnerabilities in the fungal metabolism that can be exploited for therapeutic purposes.

   (b) **Drug target identification**

   GSMMs can be used to predict potential drug targets in the fungal metabolism. By analyzing the impact of inhibiting specific reactions in the model, researchers can identify reactions whose inhibition would significantly hinder fungal growth without affecting host cells. These reactions can be evaluated as potential drug targets for antifungal development.

   (c) **Antifungal mechanism of action**

   GSMMs can provide valuable insights into the mechanisms of action of existing antifungal drugs. By simulating the effects of antifungal agents on the metabolic pathways of fungi, researchers can understand how these drugs disrupt the fungal metabolism and hinder growth or survival.

   (d) **Rational drug design**

   GSMMs can aid in the rational design of new antifungal drugs. By predicting metabolic pathways specific to fungi or distinguishing between fungal and host metabolism, researchers can develop compounds that selectively target fungal pathogens while minimizing adverse effects on the host.

   (e) **Combination therapies**

   GSMMs can also be used to optimize combination therapies involving multiple antifungal drugs. By analyzing the synergistic effects of different drug combinations on the fungal metabolism, researchers can identify optimal drug combinations that enhance efficacy and minimize the development of drug resistance.

27.6 Conclusion

In conclusion, the realm of bioinformatics has emerged as a pivotal force in the field of antifungal development, encompassing both natural and synthetic sources. While historical methods of discovering antifungals have proven laborious and time-consuming, the integration of high-throughput technologies and bioinformatics strategies offers a transformative pathway. By harnessing the power of extensive data generation in biosciences, researchers can expedite the identification of antifungals from both natural and synthetic origins, resulting in more efficient and cost-effective antifungal development. The dynamic synergy between bioinformatics and experimental approaches not only enhances the understanding of compound pharmacological activities but also addresses critical challenges like antifungal resistance. As we navigate the ever-evolving landscape of antifungal discovery, the fusion of bioinformatics and traditional methods promises a more streamlined and effective route to novel antifungal solutions.

References

Abadio AKR, Kioshima ES, Leroux V, Martins NF, Maigret B, Felipe MSS (2015) Identification of new antifungal compounds targeting thioredoxin reductase of paracoccidioides genus. PLoS One 10:e0142926. https://doi.org/10.1371/journal.pone.0142926

Abu-Izneid T, Rauf A, Bawazeer S, Wadood A, Patel S (2018) Anti-dengue, cytotoxicity, antifungal, and in silico study of the newly synthesized 3-o-phospo-α-D-glucopyranuronic acid compound. Biomed Res Int 2018:8648956. https://doi.org/10.1155/2018/8648956

Akash S, Hossain A, Mukerjee N, Sarker MMR, Khan MF, Hossain MJ, Rashid MA, Kumer A, Ghosh A, León-Figueroa DA, Barboza JJ, Padhi BK, Sah R (2023) Modified coptisine derivatives as an inhibitor against pathogenic Rhizomucor miehei, Mycolicibacterium smegmatis (Black Fungus), Monkeypox, and Marburg virus by molecular docking and molecular dynamics simulation-based drug design approach. Front Pharmacol 14:1140494. https://doi.org/10.3389/fphar.2023.1140494

Akhtar N, Singh A, Upadhyay AK, Mannan MA-U (2022) Design of a multi-epitope vaccine against the pathogenic fungi Candida tropicalis using an in silico approach. J Genet Eng Biotechnol 20:140. https://doi.org/10.1186/s43141-022-00415-3

Andargie M, Vinas M, Rathgeb A, Möller E, Karlovsky P (2021) Lignans of Sesame (Sesamum indicum L.): a comprehensive review. Molecules 26:883. https://doi.org/10.3390/molecules26040883

Atanasov AG, Zotchev SB, Dirsch VM, Supuran CT (2021) Natural products in drug discovery: advances and opportunities. Nat Rev Drug Discov 20:200–216. https://doi.org/10.1038/s41573-020-00114-z

Azad RK, Shulaev V (2018) Metabolomics technology and bioinformatics for precision medicine. Brief Bioinform 20:1957–1971. https://doi.org/10.1093/bib/bbx170

Badr AN, Ali HS, Abdel-Razek AG, Shehata MG, Albaridi NA (2020) Bioactive components of pomegranate oil and their influence on mycotoxin secretion. Toxins (Basel) 12:748. https://doi.org/10.3390/toxins12120748

Bandoni AL, Mendiondo ME, Rondina RVD, Coussio JD (1976) Survey of argentine medicinal plants: Folklore and phytochemical—screening. II Econ Bot 30:161–185

Baptista JP, Teixeira GM, de Jesus MLA, Bertê R, Higashi A, Mosela M, da Silva DV, de Oliveira JP, Sanches DS, Brancher JD, Balbi-Peña MI, de Padua Pereira U, de Oliveira AG (2022) Antifungal activity and genomic characterization of the biocontrol agent Bacillus velezensis CMRP 4489. Sci Rep 12:17401. https://doi.org/10.1038/s41598-022-22380-0

Barceló S, Peralta MA, Ortega MG, Cabrera JL, del Perez C (2014) Interacciones moleculares de un flavonoide prenilado con transportadores de antimicóticos dependientes de ATP

Barceló S, Miozza V, Passero P, Farah E, Pérez C (2017a) Aplicaciones de la informática en el estudio de productos naturales. Rev Fac Odontol (B. Aires):22–31

Barceló S, Peralta M, Calise M, Finck S, Ortega G, Diez RA, Cabrera JL, Pérez C (2017b) Interactions of a prenylated flavonoid from *Dalea elegans* with fluconazole against azole-resistant *Candida albicans*. Phytomedicine 32:24–29. https://doi.org/10.1016/j.phymed.2017.05.001

Bencurova E, Gupta SK, Sarukhanyan E, Dandekar T (2018) Identification of antifungal targets based on computer modeling. J Fungi 4:81. https://doi.org/10.3390/jof4030081

Benedict K (2020) Public awareness of invasive fungal diseases—United States, 2019. MMWR Morb Mortal Wkly Rep 69. https://doi.org/10.15585/mmwr.mm6938a2

Bongomin F, Gago S, Oladele RO, Denning DW (2017) Global and multi-national prevalence of fungal diseases-estimate precision. J Fungi (Basel) 3:57. https://doi.org/10.3390/jof3040057

Brown TA (2002) Molecular phylogenetics. In: Genomes, 2nd edn. Wiley-Liss

Butts A, Krysan DJ (2012) Antifungal drug discovery: something old and something new. PLoS Pathog 8:e1002870. https://doi.org/10.1371/journal.ppat.1002870

Cairns TC, Studholme DJ, Talbot NJ, Haynes K (2016) New and improved techniques for the study of pathogenic fungi. Trends Microbiol 24:35–50. https://doi.org/10.1016/j.tim.2015.09.008

Cao Z-J, Gao G (2022) Multi-omics single-cell data integration and regulatory inference with graph-linked embedding. Nat Biotechnol 40:1458–1466. https://doi.org/10.1038/s41587-022-01284-4

Cárdenas R, Martínez-Seoane J, Amero C (2020) Combining experimental data and computational methods for the non-computer specialist. Molecules 25:4783. https://doi.org/10.3390/molecules25204783

Challa S, Neelapu NRR (2019) Phylogenetic trees: applications, construction, and assessment. In: Hakeem KR, Shaik NA, Banaganapalli B, Elango R (eds) Essentials of bioinformatics, volume III: in silico life sciences: agriculture. Springer International Publishing, Cham, pp 167–192. https://doi.org/10.1007/978-3-030-19318-8_10

Chatterjee M, Roy K (2023) Chapter 1: Quantitative structure-activity relationships (QSARs) in medicinal chemistry. In: Roy K (ed) Cheminformatics, QSAR and machine learning applications for novel drug development. Academic Press, pp 3–38. https://doi.org/10.1016/B978-0-443-18638-7.00029-3

Cheema AI, Ahmed T, Abbas A, Noman M, Zubair M, Shahid M (2022) Antimicrobial activity of the biologically synthesized zinc oxide nanoparticles against important rice pathogens. Physiol Mol Biol Plants 28:1955–1967. https://doi.org/10.1007/s12298-022-01251-y

Chen C, Hou J, Tanner JJ, Cheng J (2020) Bioinformatics methods for mass spectrometry-based proteomics data analysis. Int J Mol Sci 21:2873. https://doi.org/10.3390/ijms21082873

Chen Y, Li E-M, Xu L-Y (2022) Guide to metabolomics analysis: a bioinformatics workflow. Metabolites 12:357. https://doi.org/10.3390/metabo12040357

Chmiela S, Sauceda HE, Müller K-R, Tkatchenko A (2018) Towards exact molecular dynamics simulations with machine-learned force fields. Nat Commun 9:3887. https://doi.org/10.1038/s41467-018-06169-2

Ciociola T, Giovati L, De Simone T, Bergamaschi G, Gori A, Consalvi V, Conti S, Vitali A (2023) Novel arginine- and proline-rich candidacidal peptides obtained through a bioinformatic approach. Antibiotics (Basel) 12:472. https://doi.org/10.3390/antibiotics12030472

Custódio L, Soares F, Pereira H, Barreira L, Vizetto-Duarte C, Rodrigues MJ, Rauter AP, Alberício F, Varela J (2014) Fatty acid composition and biological activities of Isochrysis galbana T-ISO,

Tetraselmis sp. and Scenedesmus sp.: possible application in the pharmaceutical and functional food industries. J Appl Phycol 26:151–161. https://doi.org/10.1007/s10811-013-0098-0
D'Agostino N, Li W, Wang D (2022) High-throughput transcriptomics. Sci Rep 12:20313. https://doi.org/10.1038/s41598-022-23985-1
Daliri EB-M, Lee BH (2015) Current trends and future perspectives on functional foods and nutraceuticals. In: Liong M-T (ed) Beneficial microorganisms in food and nutraceuticals, microbiology monographs. Springer International Publishing, Cham, pp 221–244. https://doi.org/10.1007/978-3-319-23177-8_10
De Las Rivas J, Fontanillo C (2010) Protein–protein interactions essentials: key concepts to building and analyzing interactome networks. PLoS Comput Biol 6:e1000807. https://doi.org/10.1371/journal.pcbi.1000807
de Oliveira Viana J, Silva E Souza E, Sbaraini N, Vainstein MH, Gomes JNS, de Moura RO, Barbosa EG (2023) Scaffold repositioning of spiro-acridine derivatives as fungi chitinase inhibitor by target fishing and in vitro studies. Sci Rep 13:7320. https://doi.org/10.1038/s41598-023-33279-9
Dhifi W, Bellili S, Jazi S, Bahloul N, Mnif W (2016) Essential oils' chemical characterization and investigation of some biological activities: a critical review. Medicines (Basel) 3:25. https://doi.org/10.3390/medicines3040025
Ding Z, Kihara D (2019) Computational identification of protein-protein interactions in model plant proteomes. Sci Rep 9:8740. https://doi.org/10.1038/s41598-019-45072-8
Dutta BS, Sharma A, Hazarika NK, Barua P, Begum S (2018) Distribution of Candida species amongst various clinical samples from immunocompromised patients attending Tertiary Care Hospital, Assam. Int J Curr Microbiol App Sci 7:1004–1019
Espinel-Ingroff A, Shadomy S (1989) In vitro and in vivo evaluation of antifungal agents. Eur J Clin Microbiol Infect Dis 8:352–361. https://doi.org/10.1007/BF01963469
Faisal S, Tariq MH, Ullah R, Zafar S, Rizwan M, Bibi N, Khattak A, Noora A, Abdullah (2023) Exploring the antibacterial, antidiabetic, and anticancer potential of Mentha arvensis extract through in-silico and in-vitro analysis. BMC Complement Med Ther 2023(23):267. https://doi.org/10.1186/s12906-023-04072-y
Feng Z, Xu M, Yang J, Zhang R, Geng Z, Mao T, Sheng Y, Wang L, Zhang J, Zhang H (2022) Molecular characterization of a novel strain of Bacillus halotolerans protecting wheat from sheath blight disease caused by Rhizoctonia solani Kühn. Front Plant Sci 13:1019512. https://doi.org/10.3389/fpls.2022.1019512
Franke H, Scholl R, Aigner A (2019) Ricin and Ricinus communis in pharmacology and toxicology-from ancient use and "Papyrus Ebers" to modern perspectives and "poisonous plant of the year 2018". Naunyn Schmiedeberg's Arch Pharmacol 392:1181–1208. https://doi.org/10.1007/s00210-019-01691-6
Garcia-Cuesta C, Sarrion-Pérez M-G, Bagán JV (2014) Current treatment of oral candidiasis: a literature review. J Clin Exp Dent 6:e576–e582. https://doi.org/10.4317/jced.51798
Gedi MA (2022) Chapter 14: Pumpkin seed oil components and biological activities. In: Mariod AA (ed) Multiple biological activities of unconventional seed oils. Academic Press, pp 171–184. https://doi.org/10.1016/B978-0-12-824135-6.00030-1
Gu C, Kim GB, Kim WJ, Kim HU, Lee SY (2019) Current status and applications of genome-scale metabolic models. Genome Biol 20:121. https://doi.org/10.1186/s13059-019-1730-3
Guerrero-Perilla C, Bernal F, Coy-Barrera E (2015) Molecular docking study of naturally occurring compounds as inhibitors of N-myristoyl transferase towards antifungal agents discovery. Revista Colombiana de Ciencias Químico Farmacéuticas 44:162–178. https://doi.org/10.15446/rcciquifa.v44n2.56291
Guinea J (2014) Global trends in the distribution of Candida species causing candidemia. Clin Microbiol Infect 20:5–10. https://doi.org/10.1111/1469-0691.12539
Gunatilaka AAL (2006) Natural products from plant-associated microorganisms: distribution, structural diversity, bioactivity, and implications of their occurrence. J Nat Prod 69:509–526. https://doi.org/10.1021/np058128n

Guo S, Ge Y, Na Jom K (2017) A review of phytochemistry, metabolite changes, and medicinal uses of the common sunflower seed and sprouts (Helianthus annuus L.). Chem Cent J 11:95. https://doi.org/10.1186/s13065-017-0328-7

Hamaamin Hussen N, Hameed Hasan A, Jamalis J, Shakya S, Chander S, Kharkwal H, Murugesan S, Ajit Bastikar V, Pyarelal Gupta P (2022) Potential inhibitory activity of phytoconstituents against black fungus: in silico ADMET, molecular docking and MD simulation studies. Comput Toxicol 24:100247. https://doi.org/10.1016/j.comtox.2022.100247

Hrdlickova R, Toloue M, Tian B (2017) RNA-Seq methods for transcriptome analysis. Wiley Interdiscip Rev RNA 8. https://doi.org/10.1002/wrna.1364

Irfan M, Abid M (2015) Three dimensional structure modeling of lanosterol 14-α demethylase of Candida albicans and docking studies with new triazole derivatives. Chem Inform 1:4

Jaiswal M, Kumar S (2023) smAMPsTK: a toolkit to unravel the smORFome encoding AMPs of plant species. J Biomol Struct Dyn 18:1–13. https://doi.org/10.1080/07391102.2023.2235605

Jayachandran AL, Katragadda R, Thyagarajan R, Vajravelu L, Manikesi S, Kaliappan S, Jayachandran B (2016) Oral candidiasis among cancer patients attending a Tertiary Care Hospital in Chennai, South India: an evaluation of clinicomycological association and antifungal susceptibility pattern. Can J Infect Dis Med Microbiol. https://doi.org/10.1155/2016/8758461

Ji F, Sadreyev RI (2018) RNA-seq: basic bioinformatics analysis. Curr Protoc Mol Biol 124:e68. https://doi.org/10.1002/cpmb.68

Ji Z-L, Peng S, Chen L-L, Liu Y, Yan C, Zhu F (2020) Identification and characterization of a serine protease from Bacillus licheniformis W10: a potential antifungal agent. Int J Biol Macromol 145:594–603. https://doi.org/10.1016/j.ijbiomac.2019.12.216

Jin MS, Oldham ML, Zhang Q, Chen J (2012) Crystal structure of the multidrug transporter P-glycoprotein from Caenorhabditis elegans. Nature 490(7421):566–569. https://doi.org/10.1038/nature11448

Joshi C, Chaudhari A, Joshi C, Joshi M, Bagatharia S (2021) Repurposing of the herbal formulations: molecular docking and molecular dynamics simulation studies to validate the efficacy of phytocompounds against SARS-CoV-2 proteins. J Biomol Struct Dyn 40(18):8405–8419. https://doi.org/10.1080/07391102.2021.1922095

Kamboj H, Gupta L, Kumar P, Sen P, Sengupta A, Vijayaraghavan P (2022) Gene expression, molecular docking, and molecular dynamics studies to identify potential antifungal compounds targeting virulence proteins/genes VelB and THR as possible drug targets against Curvularia lunata. Front Mol Biosci 9:1055945. https://doi.org/10.3389/fmolb.2022.1055945

Kapli P, Yang Z, Telford MJ (2020) Phylogenetic tree building in the genomic age. Nat Rev Genet 21:428–444. https://doi.org/10.1038/s41576-020-0233-0

Khan MA, Ahsan A, Khan MA, Sanjana JM, Biswas S, Saleh MA, Gupta DR, Hoque MN, Sakif TI, Rahman MM, Islam T (2023) In-silico prediction of highly promising natural fungicides against the destructive blast fungus Magnaporthe oryzae. Heliyon 9:e15113. https://doi.org/10.1016/j.heliyon.2023.e15113

Khavari F, Saidijam M, Taheri M, Nouri F (2021) Microalgae: therapeutic potentials and applications. Mol Biol Rep 48:4757–4765. https://doi.org/10.1007/s11033-021-06422-w

Khichi A, Jakhar R, Dahiya S, Arya J, Dangi M, Chhillar AK (2023) In silico and in vitro evaluation of designed fluconazole analogues as lanosterol 14α-demethylase inhibitors. J Biomol Struct Dyn:1–14. https://doi.org/10.1080/07391102.2023.2220808

Kısa D, Imamoglu R, Kaya Z, Taskin-Tok T, Taslimi P (2022) Turanecio hypochionaeus: determination of its polyphenol contents, and bioactivities potential assisted with in silico studies. Chem Biodivers 19:e202200109. https://doi.org/10.1002/cbdv.202200109

Krishnan S, Gupta K, Sivaraman S, Venkatachalam P, Yennamalli RM, Shanmugam SR (2023) Waste to drugs: identification of pyrolysis by-products as antifungal agents against Cryptococcus neoformans. J Biomol Struct Dyn 41(24):15386–15399. https://doi.org/10.1080/07391102.2023.2188960

Kushveer JS, Sharma R, Samantaray M, Amutha R, Sarma VV (2023) Purification and evaluation of 2, 4-di-tert butylphenol (DTBP) as a biocontrol agent against phyto-pathogenic fungi. Fungal Biol 127:1067–1074. https://doi.org/10.1016/j.funbio.2023.05.002

Lass-Flörl C (2017) Current challenges in the diagnosis of fungal infections. Methods Mol Biol 1508:3–15. https://doi.org/10.1007/978-1-4939-6515-1_1

Lopes FES, da Costa HPS, Souza PFN, Oliveira JPB, Ramos MV, Freire JEC, Jucá TL, Freitas CDT (2019) Peptide from thaumatin plant protein exhibits selective anticandidal activity by inducing apoptosis via membrane receptor. Phytochemistry 159:46–55. https://doi.org/10.1016/j.phytochem.2018.12.006

Madanchi H, Rahmati S, Doaei Y, Sardari S, Mousavi Maleki MS, Rostamian M, Ebrahimi Kiasari R, Seyed Mousavi SJ, Ghods E, Ardekanian M (2022) Determination of antifungal activity and action mechanism of the modified Aurein 1.2 peptide derivatives. Microb Pathog 173:105866. https://doi.org/10.1016/j.micpath.2022.105866

Magoulas GE, Kalopetridou L, Ćirić A, Kritsi E, Kouka P, Zoumpoulakis P, Chondrogianni N, Soković M, Prousis KC, Calogeropoulou T (2021) Synthesis, biological evaluation and QSAR studies of new thieno[2,3-d]pyrimidin-4(3H)-one derivatives as antimicrobial and antifungal agents. Bioorg Chem 106:104509. https://doi.org/10.1016/j.bioorg.2020.104509

Mahalapbutr P, Sabuakham S, Nasoontorn S, Rungrotmongkol T, Silsirivanit A, Suriya U (2023) Discovery of amphotericin B, an antifungal drug as tyrosinase inhibitor with potent anti-melanogenic activity. Int J Biol Macromol 246:125587. https://doi.org/10.1016/j.ijbiomac.2023.125587

Mandal M, Xiao L, Pan W, Scapin G, Li G, Tang H, Yang S-W, Pan J, Root Y, de Jesus RK, Yang C, Prosise W, Dayananth P, Mirza A, Therien AG, Young K, Flattery A, Garlisi C, Zhang R, Chu D, Sheth P, Chu I, Wu J, Markgraf C, Kim H-Y, Painter R, Mayhood TW, DiNunzio E, Wyss DF, Buevich AV, Fischmann T, Pasternak A, Dong S, Hicks JD, Villafania A, Liang L, Murgolo N, Black T, Hagmann WK, Tata J, Parmee ER, Weber AE, Su J, Tang H (2022) Rapid evolution of a fragment-like molecule to pan-metallo-beta-lactamase inhibitors: initial leads toward clinical candidates. J Med Chem 65:16234–16251. https://doi.org/10.1021/acs.jmedchem.2c00766

Marbán-González A, Hernández-Mendoza A, Ordóñez M, Razo-Hernández RS, Viveros-Ceballos JL (2021) Discovery of octahydroisoindolone as a scaffold for the selective inhibition of Chitinase B1 from Aspergillus fumigatus. In Silico Drug Des Studies Mol 26:7606. https://doi.org/10.3390/molecules26247606

Melavanki R, Kusanur R, Sharma K, Sadasivuni KK, Koppal VV, Patil NR (2023) Exploration of spectroscopic, computational, fluorescence turn-off mechanism, molecular docking and in silico studies of pyridine derivative. Photochem Photobiol Sci 22:1991–2003. https://doi.org/10.1007/s43630-023-00427-z

Mendoza SN, Olivier BG, Molenaar D, Teusink B (2019) A systematic assessment of current genome-scale metabolic reconstruction tools. Genome Biol 20:158. https://doi.org/10.1186/s13059-019-1769-1

Metzner K, O'Meara MJ, Halligan B, Wotring JW, Sexton JZ, O'Meara TR (2023) Imaging-based screening identifies modulators of the eIF3 translation initiation factor complex in Candida albicans. Antimicrob Agents Chemother 67:e0050323. https://doi.org/10.1128/aac.00503-23

Mielczarek M, Marchewka J, Kowalski K, Cieniek Ł, Sitarz M, Moskalewicz T (2023) Effect of tea tree oil addition on the microstructure, structure and selected properties of chitosan-based coatings electrophoretically deposited on zirconium alloy substrates. Appl Surf Sci 609:155266. https://doi.org/10.1016/j.apsusc.2022.155266

Miozza V, Barcelo S, Passero P, Farah E, Perez C (2020) Contributions of bioinformatics to study natural antifungals: review in a pharmacological context

Mirhakkak MH, Chen X, Ni Y, Heinekamp T, Sae-Ong T, Xu L-L, Kurzai O, Barber AE, Brakhage AA, Boutin S, Schäuble S, Panagiotou G (2023) Genome-scale metabolic modeling of Aspergillus fumigatus strains reveals growth dependencies on the lung microbiome. Nat Commun 14:4369. https://doi.org/10.1038/s41467-023-39982-5

Moudgal V, Sobel J (2010) Antifungals to treat Candida albicans. Expert Opin Pharmacother 11:2037–2048. https://doi.org/10.1517/14656566.2010.493875

Mueller SW, Kedzior SK, Miller MA, Reynolds PM, Kiser TH, Krsak M, Molina KC (2021) An overview of current and emerging antifungal pharmacotherapy for invasive fungal infections. Expert Opin Pharmacother 22:1355–1371. https://doi.org/10.1080/14656566.2021.1892075

Munjal G, Hanmandlu M, Srivastava S (2018) Phylogenetics algorithms and applications. Ambient Commun Comput Syst 904:187–194. https://doi.org/10.1007/978-981-13-5934-7_17

Mutanda T, Naidoo D, Bwapwa JK, Anandraj A (2020) Biotechnological applications of microalgal oleaginous compounds: current trends on microalgal bioprocessing of products. Front Energy Res 8

Nakada N, Miyazaki T, Mizuta S, Hirayama T, Nakamichi S, Takeda K, Mukae H, Kohno S, Tanaka Y (2023) Screening and synthesis of tetrazole derivatives that inhibit the growth of cryptococcus species. ChemMedChem 18(18):e202300157. https://doi.org/10.1002/cmdc.202300157

R S, Nakkeeran S, Saranya N, Senthilraja C, Renukadevi P, Krishnamoorthy AS, El Enshasy HA, El-Adawi H, Malathi VG, Salmen SH, Ansari MJ, Khan N, Sayyed RZ (2021) Mining the Genome of Bacillus velezensis VB7 (CP047587) for MAMP Genes and Non-Ribosomal Peptide Synthetase Gene Clusters Conferring Antiviral and Antifungal Activity. Microorganisms 9:2511. https://doi.org/10.3390/microorganisms9122511

Nim S, Lobato LG, Moreno A, Chaptal V, Rawal MK, Falson P, Prasad R (2016) Atomic modelling and systematic mutagenesis identify residues in multiple drug binding sites that are essential for drug resistance in the major Candida transporter Cdr1. Biochim Biophys Acta 1858(11):2858–2870. https://doi.org/10.1016/j.bbamem.2016.08.011

Pal R, Singh B, Bhadada SK, Banerjee M, Bhogal RS, Hage N, Kumar A (2021) COVID-19-associated mucormycosis: an updated systematic review of literature. Mycoses 64:1452–1459. https://doi.org/10.1111/myc.13338

Peláez F (2011) Paradigmas actuales en las etapas tempranas del proceso de descubrimiento y desarrollo de nuevos fármacos. Anales de Química de la RSEQ:36–45

Penna C, Marino S, Vivot E, Cruañes MC, Muñoz JD, Cruañes J, Ferraro G, Gutkind G, Martino V (2001) Antimicrobial activity of Argentine plants used in the treatment of infectious diseases. Isolation of active compounds from Sebastiania brasiliensis. J Ethnopharmacol 77:37–40. https://doi.org/10.1016/S0378-8741(01)00266-5

Peralta MA, Calise M, Fornari MC, Ortega MG, Diez RA, Cabrera JL, Pérez C (2012) A prenylated flavanone from Dalea elegans inhibits rhodamine 6 G efflux and reverses fluconazole-resistance in Candida albicans. Planta Med 78:981–987. https://doi.org/10.1055/s-0031-1298627

Perez C, Tiraboschi IN, Ortega MG, Agnese AM, Cabrera JL (2003) Further antimicrobial studies of 2′4′-dihidroxy-5′-(1?-dimethylallyl)-6-prenylpinocembrin from Dalea elegans. Pharm Biol 41:171–174. https://doi.org/10.1076/phbi.41.3.171.15090

Petrovska BB (2012) Historical review of medicinal plants' usage. Pharmacogn Rev 6:1–5. https://doi.org/10.4103/0973-7847.95849

Prajapati J, Rao P, Poojara L, Goswami D, Acharya D, Patel SK, Rawal RM (2021) Unravelling the antifungal mode of action of curcumin by potential inhibition of CYP51B: A computational study validated in vitro on mucormycosis agent, Rhizopus oryzae. Arch Biochem Biophys 712:109048. https://doi.org/10.1016/j.abb.2021.109048

Prasad R, Shah AH, Rawal MK (2016) Antifungals: mechanism of action and drug resistance. Adv Exp Med Biol 892:327–349. https://doi.org/10.1007/978-3-319-25304-6_14

Prasad SK, Bhat SS, Koskowska O, Sangta J, Ahmad SF, Nadeem A, Sommano SR (2023) Naringin from coffee inhibits foodborne Aspergillus fumigatus via the NDK pathway: evidence from an In Silico study. Molecules 28(13):5189. https://doi.org/10.3390/molecules28135189

Reimand J, Isserlin R, Voisin V, Kucera M, Tannus-Lopes C, Rostamianfar A, Wadi L, Meyer M, Wong J, Xu C, Merico D, Bader GD (2019) Pathway enrichment analysis and visualization of omics data using g: profiler, GSEA, Cytoscape and EnrichmentMap. Nat Protoc 14:482–517. https://doi.org/10.1038/s41596-018-0103-9

Renganathan S, Subramaniyan S, Karunanithi N, Vasanthakumar P, Kutzner A, Kim P-S, Heese K (2021) Antibacterial, antifungal, and antioxidant activities of silver nanoparticles biosynthesized from Bauhinia tomentosa Linn. Antioxidants (Basel) 10:1959. https://doi.org/10.3390/antiox10121959

Roberts LD, Souza AL, Gerszten RE, Clish CB (2012) Targeted metabolomics. Curr Protoc Mol Biol. https://doi.org/10.1002/0471142727.mb3002s98. Chapter 30, Unitas 30.2.1-24

Rogozhin E, Zalevsky A, Mikov A, Smirnov A, Egorov T (2018) Characterization of hydroxyproline-containing hairpin-like antimicrobial peptide EcAMP1-Hyp from Barnyard Grass (Echinochloa crusgalli L.) seeds: structural identification and comparative analysis of antifungal activity. Int J Mol Sci 19:3449. https://doi.org/10.3390/ijms19113449

Sama-ae I, Pattaranggoon NC, Tedasen A (2023) *In silico* prediction of Antifungal compounds from natural sources towards Lanosterol 14-alpha demethylase (CYP51) using Molecular docking and Molecular dynamic simulation. J Mol Graph Model 121:108435. https://doi.org/10.1016/j.jmgm.2023.108435

Sathiyamoorthy J, Rathore SS, Mohan S, Uma Maheshwari C, Ramakrishnan J (2023) Elucidation of furanone as ergosterol pathway inhibitor in Cryptococcus neoformans. J Biomol Struct Dyn 5:1–14. https://doi.org/10.1080/07391102.2023.2230301

Schiebenhoefer H, Schallert K, Renard BY, Trappe K, Schmid E, Benndorf D, Riedel K, Muth T, Fuchs S (2020) A complete and flexible workflow for metaproteomics data analysis based on MetaProteomeAnalyzer and prophane. Nat Protoc 15:3212–3239. https://doi.org/10.1038/s41596-020-0368-7

Senra MVX (2023) In silico characterization of cysteine-stabilized αβ defensins from neglected unicellular microeukaryotes. BMC Microbiol 23:82. https://doi.org/10.1186/s12866-023-02817-w

Shahinozzaman M, Ishii T, Ahmed S, Halim MA, Tawata S (2020) A computational approach to explore and identify potential herbal inhibitors for the p21-activated kinase 1 (PAK1). J Biomol Struct Dyn 38:3514–3526. https://doi.org/10.1080/07391102.2019.1659855

Sharma Y, Chumber SK, Kaur M (2017) Studying the prevalence, species distribution, and detection of in vitro production of phospholipase from Candida isolated from cases of invasive candidiasis. J Glob Infect Dis 9:8–11. https://doi.org/10.4103/0974-777X.199995

Sharma N, Gupta N, Orfali R, Kumar V, Patel CN, Peng J, Perveen S (2022) Evaluation of the antifungal, antioxidant, and anti-diabetic potential of the essential oil of Curcuma longa leaves from the North-Western Himalayas by in vitro and in silico analysis. Molecules 27:7664. https://doi.org/10.3390/molecules27227664

Shen S, Huang J, Li T, Wei Y, Xu S, Wang Y, Ning J (2022) Untargeted and targeted metabolomics reveals potential marker compounds of an tea during storage. LWT 154:112791. https://doi.org/10.1016/j.lwt.2021.112791

Shokri H (2016) A review on the inhibitory potential of Nigella sativa against pathogenic and toxigenic fungi. Avicenna J Phytomed 6:21–33

Singh V, Praveen V, Tripathi D, Haque S, Somvanshi P, Katti SB, Tripathi CKM (2015) Isolation, characterization and antifungal docking studies of Wortmannin isolated from Penicillium radicum. Sci Rep 5:11948. https://doi.org/10.1038/srep11948

Singh N, Islam MU, Sharma H (2023) Schiff base: a review of pharmacological activities. IJPSR 2023, 14(2)

Souza AL, Patti GJ (2021) A protocol for untargeted metabolomic analysis: from sample preparation to data processing. Methods Mol Biol 2276:357–382. https://doi.org/10.1007/978-1-0716-1266-8_27

Subramanian SV, Nandy S, Kelly M, Gordon D, Davey Smith G (2004) Patterns and distribution of tobacco consumption in India: cross sectional multilevel evidence from the 1998-9 national family health survey. BMJ 328:801–806. https://doi.org/10.1136/bmj.328.7443.801

Sun F-J, Li M, Gu L, Wang M-L, Yang M-H (2021) Recent progress on anti-Candida natural products. Chin J Nat Med 19:561–579. https://doi.org/10.1016/S1875-5364(21)60057-2

Takeoka G, Dao L, Wong RY, Lundin R, Mahoney N (2001) Identification of benzethonium chloride in commercial grapefruit seed extracts. J Agric Food Chem 49:3316–3320. https://doi.org/10.1021/jf010222w

Tamay-Cach F, Villa-Tanaca ML, Trujillo-Ferrara JG, Alemán-González-Duhart D, Quintana-Pérez JC, González-Ramírez IA, Correa-Basurto J (2016) In Silico studies most employed in the discovery of new antimicrobial agents. Curr Med Chem 23:3360–3373

Thompson GR, Le T, Chindamporn A, Kauffman CA, Alastruey-Izquierdo A, Ampel NM, Andes DR, Armstrong-James D, Ayanlowo O, Baddley JW, Barker BM, Bezerra LL, Buitrago MJ, Chamani-Tabriz L, Chan JFW, Chayakulkeeree M, Cornely OA, Cunwei C, Gangneux J-P, Govender NP, Hagen F, Hedayati MT, Hohl TM, Jouvion G, Kenyon C, Kibbler CC, Klimko N, Kong DCM, Krause R, Lee LL, Meintjes G, Miceli MH, Rath P-M, Spec A, Queiroz-Telles F, Variava E, Verweij PE, Schwartz IS, Pasqualotto AC (2021) Global guideline for the diagnosis and management of the endemic mycoses: an initiative of the European Confederation of Medical Mycology in cooperation with the International Society for Human and Animal Mycology. Lancet Infect Dis 21:e364–e374. https://doi.org/10.1016/S1473-3099(21)00191-2

Tyagi A, Pankaj V, Singh S, Roy S, Semwal M, Shasany AK, Sharma A (2019) PlantAFP: a curated database of plant-origin antifungal peptides. Amino Acids 51:1561–1568. https://doi.org/10.1007/s00726-019-02792-5

Verma R, Pradhan D, Maseet M, Singh H, Jain AK, Khan LA (2020) Genome-wide screening and in silico gene knockout to predict potential candidates for drug designing against Candida albicans. Infect Genet Evol 80:104196. https://doi.org/10.1016/j.meegid.2020.104196

Verma R, Pradhan D, Singh H, Jain AK, Khan LA (2021) Metabolic network modeling for rational drug design against Candida albicans, Candida albicans. IntechOpen. https://doi.org/10.5772/intechopen.96749

Verma R, Pradhan D, Nayek A, Singh H, Jain AK, Khan LA (2022) Target-based drug repurposing against Candida albicans—a computational modeling, docking, and molecular dynamic simulations study. J Cell Biochem 123:289–305. https://doi.org/10.1002/jcb.30163

Wadhwa K, Kaur H, Kapoor N, Brogi S (2023) Identification of Sesamin from Sesamum indicum as a potent antifungal agent using an integrated in silico and biological screening platform. Molecules 28:4658. https://doi.org/10.3390/molecules28124658

Wang Z, Gerstein M, Snyder M (2009) RNA-Seq: a revolutionary tool for transcriptomics. Nat Rev Genet 10:57–63. https://doi.org/10.1038/nrg2484

Wishart DS (2005) Bioinformatics in drug development and assessment. Drug Metab Rev 37:279–310. https://doi.org/10.1081/dmr-55225

Wylie MR, Merrell DS (2022) The antimicrobial potential of the Neem Tree Azadirachta indica. Front Pharmacol 13:891535. https://doi.org/10.3389/fphar.2022.891535

Yang Y, Liu X, Cai J, Chen Y, Li B, Guo Z, Huang G (2019) Genomic characteristics and comparative genomics analysis of the endophytic fungus Sarocladium brachiariae. BMC Genomics 20:782. https://doi.org/10.1186/s12864-019-6095-1

Yang C, Wang Z, Wan J, Qi T, Zou L (2023) Burkholderia gladioli strain KJ-34 exhibits broad-spectrum antifungal activity. Front Plant Sci 14:1097044. https://doi.org/10.3389/fpls.2023.1097044

Yousafi Q, Bibi S, Saleem S, Hussain A, Hasan MM, Tufail M, Qandeel A, Khan MS, Mazhar S, Yousaf M, Moustafa M, Al-Shehri M, Khalid M, Kabra A (2022) Identification of novel and safe fungicidal molecules against Fusarium oxysporum from plant essential oils: in vitro and computational approaches. Biomed Res Int 2022:5347224. https://doi.org/10.1155/2022/5347224

Zhan L, Tian X, Lin J, Zhang Y, Zheng H, Peng X, Zhao G (2023) Honokiol reduces fungal burden and ameliorate inflammation lesions of Aspergillus fumigatus keratitis via Dectin-2 downregulation. Int Immunopharmacol 118:109849. https://doi.org/10.1016/j.intimp.2023.109849

Zhao Y, Zhang C, Folly YME, Chang J, Wang Y, Zhou L, Zhang H, Liu Y (2019) Morphological and transcriptomic analysis of the inhibitory effects of Lactobacillus plantarum on Aspergillus flavus growth and aflatoxin production. Toxins (Basel) 11:636. https://doi.org/10.3390/toxins11110636

Zhao L, Islam MS, Song P, Zhu L, Dong W (2023) Isolation and optimization of a broad-spectrum synthetic antimicrobial peptide, Ap920-WI, from Arthrobacter sp. H5 for the biological control of plant diseases. Int J Mol Sci 24(13):10598. https://doi.org/10.3390/ijms241310598